€98
'18
Math

Lecture Notes in Computational Science and Engineering 119

Editors:

More information about this series at http://www.springer.com/series/3527

Marco L. Bittencourt • Ney A. Dumont •
Jan S. Hesthaven
Editors

Spectral and High Order Methods for Partial Differential Equations ICOSAHOM 2016

Selected Papers from the ICOSAHOM Conference, June 27-July 1, 2016, Rio de Janeiro, Brazil

Springer

Editors
Marco L. Bittencourt
School of Mechanical Engineering
University of Campinas
Campinas, Brazil

Ney A. Dumont
Department of Civil Engineering
Pontifical Catholic University of Rio de Janeiro
Rio de Janeiro, Brazil

Jan S. Hesthaven
EPFL-SB-MATHICSE-MCSS
Ecole Polytechnique Fédérale de Lausanne
Lausanne, Switzerland

ISSN 1439-7358 ISSN 2197-7100 (electronic)
Lecture Notes in Computational Science and Engineering
ISBN 978-3-319-65869-8 ISBN 978-3-319-65870-4 (eBook)
DOI 10.1007/978-3-319-65870-4

Library of Congress Control Number: 2017955943

Mathematics Subject Classification (2010): Primary: 65M70, 65N35, 65N30, 74S25, 76M10, 76M22, 78M10, 78M22; Secondary: 33Cxx, 41Axx, 65Cxx, 65Dxx, 65Lxx, 65Mxx, 65Nxx, 74Jxx, 76Mxx, 78Mxx, 80Mxx

Cover illustration: Multi-level vorticity particle method with penalization around complex geometries: 3D vorticity passed a wind turbine (Courtesy of Professor Georges-Henri Cottet, University Grenoble Alpes and Institut Universitaire de France, and Professor Petros Koumoutsakos, ETH Zurich).

Printed on acid-free paper

This Springer imprint is published by Springer Nature
The registered company is Springer International Publishing AG
The registered company address is: Gewerbestrasse 11, 6330 Cham, Switzerland

Preface

This volume presents selected papers from the eleventh International Conference on Spectral and High-Order Methods (ICOSAHOM'16) that was held in Rio de Janeiro, RJ, Brazil, during the week of June 27th to July 1st, 2016. These selected papers were refereed by a member of scientific committee of ICOSAHOM as well as by other leading scientists.

The first ICOSAHOM conference was held in Como, Italy, in 1989 and marked the beginning of an international conference series in Montpellier, France (1992); Houston, USA (1995); Tel Aviv, Israel (1998); Uppsala, Sweden (2001); Providence, USA (2004); Beijing, China (2007); Trondheim, Norway (2009); Gammarth, Tunisia (2012); and Salt Lake City, USA (2014).

ICOSAHOM has established itself as the main meeting place for researchers with interests in the theoretical, applied, and computational aspects of high-order methods for the numerical solution of partial differential equations.

With about 200 participants, ICOSAHOM'16 took place in the Othon Palace Hotel at the Copacabana Beach. The program consisted of nine invited lectures spread out through the week, 16 mini-symposia, hosting approximately 155 talks, and 22 contributed talks.

The content of these proceedings is organized as follows. First, contributions from the invited speakers are included. The remainder of the volume consists of refereed selected papers highlighting the broad spectrum of topics presented at ICOSAHOM'16.

The success of the meeting was ensured through the financial support given by the University of Campinas (UNICAMP), Pontifical Catholic University of Rio de Janeiro (PUC-RJ), the Brazilian Research Council (CNPq), and the Coordination for the Improvement of Higher Education Personnel (CAPES).

Special thanks go to members of our local organizing committee Philippe Devloo, Alvaro Coutinho, and Saulo Pomponet de Oliveira. We would like also to thank Creacteve Eventos, in special Alessandra Leitão and Michele Christinni, and Sócrates Duarte from SWGE for their invaluable help in the organization of the conference. Thanks also to PhD students from UNICAMP and PUC-RJ who helped during the event.

Campinas, SP, Brazil — Marco L. Bittencourt
Rio de Janeiro, RJ, Brazil — Ney A. Dumont
Lausanne, Switzerland — Jan S. Hesthaven

Contents

Part I
Invited Papers

hp-Version Discontinuous Galerkin Approximations of the Elastodynamics Equation

Paola F. Antonietti, Alberto Ferroni, Ilario Mazzieri, and Alfio Quarteroni

Abstract In this paper we extend the results contained in Antonietti et al. (J Sci Comput 68(1):143-170, 2016) and consider the problem of approximating the elastodynamics equation by means of *hp*-version discontinuous Galerkin methods. For the resulting semi-discretized schemes we derive stability bounds as well as *hp* error estimates in the energy and L^2-norms. Our theoretical estimates are verified through three dimensional numerical experiments.

1 Introduction

The present paper deals with the numerical modeling through the (linear) elastodynamics equation of seismic wave propagation phenomena in complex, three-dimensional media. Currently, the numerical methods mostly employed to tackle seismic wave propagation include finite differences, pseudo-spectral, spectral element, and high–order/spectral element discontinuous (DG) Galerkin techniques. In particular Spectral Element methods, firstly introduced for fluid dynamics problems in the seminal paper [28], have become one of the most effective and powerful approaches for solving three-dimensional seismic wave propagation problems in strongly heterogeneous media thanks to their geometrical flexibility and high order accuracy, which made them well suited to correctly approximate the wave field. We refer to [14, 20, 22, 25, 35] for the first development of Spectral Element methods for the elastodynamics equation, and, for example, to [21, 23, 36, 39] for its application in computational seismology. In recent years, *displacement-based* high–order/spectral element discontinuous Galerkin methods have also been

P.F. Antonietti (✉) • A. Ferroni • I. Mazzieri
MOX, Dipartimento di Matematica, Politecnico di Milano, Piazza Leonardo da Vinci 32, I-20133 Milano, Italy
e-mail: paola.antonietti@polimi.it; alberto.ferroni@polimi.it; ilario.mazzieri@polimi.it

A. Quarteroni
CMCS, Ecole Polytechnique Federale de Lausanne (EPFL), Station 8, 1015 Lausanne, Switzerland
e-mail: alfio.quarteroni@epfl.ch

M.L. Bittencourt et al. (eds.), *Spectral and High Order Methods for Partial Differential Equations ICOSAHOM 2016*, Lecture Notes in Computational Science and Engineering 119, DOI 10.1007/978-3-319-65870-4_1

developed for linear and nonlinear (visco) elastic wave propagation problems, mainly because the discretization parameters, i.e. the mesh-size and/or the polynomial approximation degree, can be naturally tailored to the region of interests; see e.g. [3–6, 12, 24, 26, 31–33]. Additionally, high–order/spectral element DG methods feature very low dispersion and dissipation errors. A dispersion/dissipation analysis based on the approach of [1] for DG approximations of elastic wave propagation problems has been carried out in [3, 11] and [4], respectively, on two-dimensional quadrilateral/triangular meshes: the extension to three-dimensions has been addressed recently in [15]. Another interesting feature is their being embarrassingly parallel and therefore naturally oriented towards high performance parallel computing. DG methods are thus very well suited to deal with *i)* the intrinsic multi-scale nature of seismic wave propagation problems, involving a relative broad range of wavelengths; *ii)* the complexity of the geometrical constraints. The aim of this paper is to extend to the *hp*-version the theoretical analysis developed in [5] as well as to prove approximation bounds in the L^2 norm. For the sake of brevity, here we focus only on *displacement* DG formulation, but the present analysis can be extended also to *displacement-stress* formulations. We show that, also in the *hp*-version setting, stability and approximation properties hold without the need to introduce an extra term that penalizes the time derivative of the displacement besides the displacement itself, as considered in previous works [31–33]. Our semidiscrete analysis represents an intermediate but essential step towards the analysis of stability of the fully discrete scheme resulting after time integration.

The remaining part of manuscript is organized as follows. In Sect. 2 we introduce the model problem and its *hp*-version discontinuous Galerkin approximation. The stability analysis is presented in Sect. 3, whereas in Sect. 4 we present the *hp*−version *a priori* error estimates in both the energy and L^2 norms. Three-dimensional numerical experiments verifying the theory are presented in Sect. 5.

2 Problem Statement and its *hp*-Version Discontinuous Galerkin Approximation

Let $\Omega \subset \mathbf{R}^d$, $d = 2, 3$, be an open, bounded convex region with Lipschitz boundary $\partial\Omega$. Throughout the paper, $[H^m(\Omega)]^d$ and $[H^m(\Omega)]^{d\times d}_{\text{sym}}$ denote the standard Sobolev spaces of vector–valued and symmetric tensor-valued functions defined over Ω, respectively, and $(\cdot,\cdot)_\Omega$ denote the standard inner product in any of the spaces $[L^2(\Omega)]^d$ or $[L^2(\Omega)]^{d\times d}_{\text{sym}}$. For given $T > 0$ and $\mathbf{f} = \mathbf{f}(x,t) \in L^2((0,T];[L^2(\Omega)]^d)$, we consider the problem of approximating the variational formulation of the linear elastodynamics equation with homogeneous Dirichlet boundary conditions: for all $t \in (0,T]$ find $\mathbf{u} = \mathbf{u}(t) \in \mathbf{V} \equiv [H^1_0(\Omega)]^d$ such that:

$$(\rho\mathbf{u}_{tt},\mathbf{v})_\Omega + (\mathscr{D}\boldsymbol{\varepsilon}(\mathbf{u}),\boldsymbol{\varepsilon}(\mathbf{v}))_\Omega = (\mathbf{f},\mathbf{v})_\Omega \qquad \forall\,\mathbf{v} \in \mathbf{V}, \tag{1}$$

subjected to the (regular enough) initial conditions $\mathbf{u}_0$ and $\mathbf{u}_1$. Here, $\mathbf{u} : \Omega \times [0,T] \longrightarrow \mathbb{R}^d$ is the displacement vector field and $\boldsymbol{\varepsilon}(\mathbf{u}) : \Omega \longrightarrow \mathbb{R}^{d\times d}_{\text{sym}}$ is the symmetric gradient. Moreover, ρ is the mass density, which is supposed to be a strictly positive and uniformly bounded function, and $\mathscr{D} = \mathscr{D}(x) : \mathbb{R}^{d\times d}_{\text{sym}} \longrightarrow \mathbb{R}^{d\times d}_{\text{sym}}$ is the inverse of the *compliance* tensor defined as $\mathscr{D}\boldsymbol{\tau} = 2\mu\boldsymbol{\tau} + \lambda \text{tr}(\boldsymbol{\tau})\mathbb{I} \quad \forall\, \boldsymbol{\tau} \in \mathbb{R}^{d\times d}_{\text{sym}}$. $\mathbb{I} \in \mathbb{R}^{d\times d}$ and $\text{tr}(\cdot)$ are the identity and trace operators, respectively, and $\lambda, \mu \in L^\infty(\Omega)$, $\lambda, \mu > 0$, being the Lamé parameters.

Henceforth, C denotes a generic positive constant independent of the discretization parameters, but that can depend on the physical quantities ρ, $\mathscr{D}$ as well as on the final observation time T. Moreover, $x \lesssim y$ and $x \gtrsim y$ will signify $x \leq Cy$ and $x \geq Cy$, respectively, with C as before.

2.1 Mesh, Trace Operators, and Discrete Spaces

We consider a sequence $\{\mathscr{T}_h\}_h$ of shape-regular (not-necessarily matching) partitions of Ω into disjoint open elements K such that $\overline{\Omega} = \cup_{K\in\mathscr{T}_h}\overline{K}$, where each $K \in \mathscr{T}_h$ is the affine image of a fixed master element $\widehat{K}$, *i.e.*, $K = F_K(\widehat{K})$, $\widehat{K}$ being either the open unit d-simplex or the open unit hypercube in $\mathbb{R}^d$, $d = 2, 3$. An interior face (for $d = 2$, "face" means "edge") of $\mathscr{T}_h$ is defined as the (non–empty) interior of $\partial\overline{K}^+ \cap \partial\overline{K}^-$, where $K^\pm$ are two adjacent elements of $\mathscr{T}_h$. Similarly, a boundary face of $\mathscr{T}_h$ is defined as the (non-empty) interior of $\partial\overline{K} \cap \overline{\Omega}$, where K is a boundary element of $\mathscr{T}_h$. We collect the interior and boundary faces in the sets $\mathscr{F}_h^I$ and $\mathscr{F}_h^B$, respectively, and define $\mathscr{F}_h = \mathscr{F}_h^I \cup \mathscr{F}_h^B$. We also assume the following mesh-regularity: *i)* for any $K \in \mathscr{T}_h$ and for all $F \in \mathscr{F}_h$, $F \subset \partial K$, $h_K \lesssim h_F$; *ii)* for any pair of elements $K^\pm \in \mathscr{T}_h$ sharing a $(d-1)$–dimensional face $h_{K^-} \lesssim h_{K^+} \lesssim h_{K^-}$: cf. [16, 29] for example.

Next, we introduce suitable trace operators, cf. [8]. Let F be an interior face shared by two elements $K^\pm$ of $\mathscr{T}_h$, and let $\mathbf{n}^\pm$ denote the normal unit vectors on F pointing outward $K^\pm$, respectively. For (regular enough) vector-valued and symmetric tensor-valued functions $\mathbf{v}$ and $\boldsymbol{\tau}$, respectively, we define the *weighted average* and *jump* operators as

$$\begin{aligned}
&\{\mathbf{v}\}_\delta = \delta\mathbf{v}^+ + (1-\delta)\mathbf{v}^-, \qquad &&\{\boldsymbol{\tau}\}_\delta = \delta\boldsymbol{\tau}^+ + (1-\delta)\boldsymbol{\tau}^-, \quad \delta \in [0,1],\\
&[\![\mathbf{v}]\!] = \mathbf{v}^+ \odot \mathbf{n}^+ + \mathbf{v}^- \odot \mathbf{n}^-, \qquad &&[\![\boldsymbol{\tau}]\!] = \boldsymbol{\tau}^+\mathbf{n}^+ + \boldsymbol{\tau}^-\mathbf{n}^-,
\end{aligned} \tag{2}$$

where $\mathbf{v}^\pm$ and $\boldsymbol{\tau}^\pm$ denote the traces of $\mathbf{v}$ and $\boldsymbol{\tau}$ on F taken within the interior of $K^\pm$, respectively, and where $\mathbf{v} \odot \mathbf{n} = (\mathbf{v}\mathbf{n}^T + \mathbf{n}\mathbf{v}^T)/2$. Notice that $[\![\mathbf{v}]\!]$ is a symmetric tensor-valued function. On a boundary face $F \in \mathscr{F}_h^B$, we set analogously

$$\{\mathbf{v}\}_\delta = \mathbf{v}, \quad \{\boldsymbol{\tau}\}_\delta = \boldsymbol{\tau}, \quad [\![\mathbf{v}]\!] = \mathbf{v} \odot \mathbf{n}, \quad [\![\boldsymbol{\tau}]\!] = \boldsymbol{\tau}\mathbf{n}. \tag{3}$$

When $\delta = 1/2$, we drop the subindex and simply write $\{\cdot\}$.

Finally, to any element $K \in \mathscr{T}_h$ we assign a polynomial approximation degree $p_K \geq 1$, and define the *hp*-discontinuous finite element space

$$\mathbf{V}_{hp} = \{\mathbf{u} \in [L^2(\Omega)]^d \ : \ \mathbf{u} \circ F_K \in [\mathbb{M}^{p_K}(\widehat{K})]^d \quad \forall\, K \in \mathscr{T}_h\},$$

where $\mathbb{M}^{p_K}(\widehat{K})$ is either the space $\mathbb{P}^{p_K}(\widehat{K})$ of polynomials of degree at most p_K on $\widehat{K}$, if $\widehat{K}$ is the reference d-simplex, or the space $\mathbb{Q}^{p_K}(\widehat{K})$ of tensor–product polynomials on $\widehat{K}$ of degree p_K in each coordinate direction, if $\widehat{K}$ is the unit reference hypercube in $\mathbb{R}^d$. In the following we also assume that the following *local bounded* variation holds: $p_{K^-} \lesssim p_{K^+} \lesssim p_{K^-}$ for any pair of elements $K^\pm \in \mathscr{T}_h$ sharing a $(d-1)$–dimensional face, cf. [29] for example.

Given a face $F \in \mathscr{F}_h$ of an element $K \in \mathscr{T}_h$, i.e., $F \subset \partial K$ the following inverse inequality holds:

$$\|v\|^2_{L^2(F)} \lesssim \frac{p_K^2}{h_K} \|v\|^2_{L^2(K)} \quad \forall v \in \mathbb{M}^{p_K}(K),$$

cf. [34]. Finally, we recall the following interpolation estimates, cf. [9].

Lemma 1 *For any real number $s_K \geq 0$ and for any function $\mathbf{v} \in [H^{s_K}(K)]^d$, $K \in \mathscr{T}_h$, there exists $\Pi_h \mathbf{v} \in \mathbf{V}_{hp}$ such that*

$$\sum_{K \in \mathscr{T}_h} \|\mathbf{v} - \Pi_{hp}\mathbf{v}\|_{H^r(K)} \lesssim \sum_{K \in \mathscr{T}_h} \frac{h_K^{\min(s_K, p_K+1)-r}}{p_K^{s_K - r}} \|\mathbf{v}\|_{H^{s_K}(K)} \qquad \forall r,\ 0 \leq r \leq s,$$

$$\sum_{K \in \mathscr{T}_h} \|D^\xi(\mathbf{v} - \Pi_{hp}\mathbf{v})\|_{L^2(\partial K)} \lesssim \sum_{K \in \mathscr{T}_h} \frac{h_K^{\min(s_K, p_K+1)-|\xi|-1/2}}{p_K^{s_K-|\xi|-1/2}} \|\mathbf{v}\|_{H^{s_K}(K)} \quad \forall \xi,\ 0 \leq |\xi| \leq k, \tag{4}$$

where $\xi \in \mathbb{N}_0^d$ is a multi-index of length $|\xi|$. Here, the second inequality holds provided $s_K > 1/2$ and k is the largest non-negative integer strictly less than $s - 1/2$.

2.2 Semi-Discrete and Fully-Discrete Formulations

We are now ready to state the semi-discrete weak formulation: For any time $t \in (0, T]$, find $\mathbf{u}^h = \mathbf{u}^h(t) \in \mathbf{V}_{hp}$ such that

$$(\rho \mathbf{u}^h_{tt}, \mathbf{v})_{\mathscr{T}_h} + \mathscr{A}(\mathbf{u}^h, \mathbf{v}) = (\mathbf{f}, \mathbf{v})_{\mathscr{T}_h} \quad \forall\, \mathbf{v} \in \mathbf{V}_{hp}, \tag{5}$$

subjected to the initial conditions $\mathbf{u}^h_0$ and $\mathbf{u}^h_1$, being $\mathbf{u}^h_0, \mathbf{u}^h_1 \in \mathbf{V}_{hp}$ suitable approximations in $\mathbf{V}_{hp}$ of the initial data $\mathbf{u}_0, \mathbf{u}_1$, respectively. The bilinear form $\mathscr{A}(\cdot, \cdot)$:

$\mathbf{V}_{hp} \times \mathbf{V}_{hp} \longrightarrow \mathbb{R}$ in (16) is given by

$$\mathscr{A}(\mathbf{w}, \mathbf{v}) = (\boldsymbol{\varepsilon}(\mathbf{w}), \mathscr{D}\boldsymbol{\varepsilon}(\mathbf{v}))_{\mathscr{T}_h} - \langle \{\mathscr{D}\boldsymbol{\varepsilon}(\mathbf{w})\}_\delta, [\![\mathbf{v}]\!]\rangle_{\mathscr{F}_h} \\ - \langle [\![\mathbf{w}]\!], \{\mathscr{D}\boldsymbol{\varepsilon}(\mathbf{v})\}_\delta\rangle_{\mathscr{F}_h} + \langle \sigma [\![\mathbf{w}]\!], [\![\mathbf{v}]\!]\rangle_{\mathscr{F}_h}, \quad (6)$$

where we have used the shorthand notation $(\mathbf{w}, \mathbf{v})_{\mathscr{T}_h} = \sum_{K \in \mathscr{T}_h} (\mathbf{w}, \mathbf{v})_K$ and $(\mathbf{w}, \mathbf{v})_{\mathscr{F}_h} = \sum_{F \in \mathscr{F}_h} (\mathbf{w}, \mathbf{v})_F$. The above method corresponds to the family of Interior Penalty (IP) methods: for $\delta = 1/2$, we get the Symmetric Interior Penalty (SIP) method [7, 41], whereas for $\delta \neq 1/2$ we obtain the *weighted* SIP method of Stenberg, [38]. In (6) the stabilization function $\sigma \in L^\infty(\mathscr{F}_h)$ is defined facewise as

$$\sigma = \sigma(x) = \begin{cases} \alpha\{\mathscr{D}\} \dfrac{\max(p_{K^+}^2, p_{K^-}^2)}{\min(h_{K^+}, h_{K^-})} & \text{if } x \in \partial\overline{K^+} \cap \partial\overline{K^-}, \\ \alpha\{\mathscr{D}\} \dfrac{p_K^2}{h_K} & \text{if } x \in \partial\overline{K} \cap \partial\Omega, \end{cases} \quad (7)$$

By defining $[H^s(\mathscr{T}_h)]^d$ as the space of elementwise $[H^s(K)]^d$ functions, $s \geq 0$, and endowing the $\mathbf{V}_{hp}$ and $\mathbf{V}_{hp} + [H^2(\mathscr{T}_h)]^d$ spaces with the (mesh-dependent) norms

$$\begin{aligned} \|\mathbf{v}\|_{\mathrm{DG}}^2 &= \|\mathscr{D}^{1/2}\boldsymbol{\varepsilon}(\mathbf{v})\|_{L^2(\mathscr{T}_h)}^2 + \|\sigma^{1/2}[\![\mathbf{v}]\!]\|_{L^2(\mathscr{F}_h)}^2 && \forall \mathbf{v} \in \mathbf{V}_{hp}, \\ |\!|\!|\mathbf{v}|\!|\!|_{\mathrm{DG}}^2 &= \|\mathbf{v}\|_{\mathrm{DG}}^2 + \|\sigma^{-1/2}\{\boldsymbol{\varepsilon}(\mathbf{v})\}\|_{L^2(\mathscr{F}_h)}^2 && \forall \mathbf{v} \in \mathbf{V}_{hp} + [H^2(\mathscr{T}_h)]^d, \end{aligned} \quad (8)$$

respectively, with $\|\mathbf{w}\|_{L^2(\mathscr{T}_h)} = \sqrt{(\mathbf{v}, \mathbf{v})_{\mathscr{T}_h}}$ and $\|\mathbf{w}\|_{L^2(\mathscr{F}_h)} = \sqrt{(\mathbf{v}, \mathbf{v})_{\mathscr{F}_h}}$, with standard arguments it is easy to prove the following result.

Lemma 2 *The bilinear form* $\mathscr{A}(\cdot,\cdot) : \mathbf{V}_{hp} \times \mathbf{V}_{hp} \longrightarrow \mathbb{R}$ *defined as in* (6) *satisfies*

$$|\mathscr{A}(\mathbf{w}, \mathbf{v})| \lesssim \|\mathbf{v}\|_{\mathrm{DG}} \|\mathbf{w}\|_{\mathrm{DG}}, \quad \mathscr{A}(\mathbf{v}, \mathbf{v}) \gtrsim \|\mathbf{v}\|_{\mathrm{DG}}^2 \quad \forall \mathbf{w}, \mathbf{v} \in \mathbf{V}_{hp}, \quad (9)$$

where the second estimates holds provided that the penalty parameter α *is chosen large enough, cf.* (7). *Moreover,*

$$|\mathscr{A}(\mathbf{w}, \mathbf{v})| \lesssim |\!|\!|\mathbf{v}|\!|\!|_{\mathrm{DG}} |\!|\!|\mathbf{w}|\!|\!|_{\mathrm{DG}}, \quad \forall \mathbf{w}, \mathbf{v} \in \mathbf{V}_{hp} + [H^2(\mathscr{T}_h)]^d.$$

We remark that a sharp estimate on the minimum value α so that the second estimate in (9) holds can be obtained based on employing the results of [2].

The semi-discrete algebraic formulation of problem (16) reads as

$$\mathbf{M}\ddot{\mathbf{U}}(t) + \mathbf{A}\mathbf{U}(t) = \mathbf{F}(t), \quad (10)$$

supplemented with initial conditions $\mathbf{U}(0) = \mathbf{U}^0$ and $\dot{\mathbf{U}}(0) = \mathbf{V}^0$. The vector $\mathbf{U} = \mathbf{U}(t)$ contains, for any time t, the expansion coefficients of $\mathbf{u}_h(t) \in V_{hp}$ in a chosen basis. Analogously, $\mathbf{M}$ and $\mathbf{A}$ are the matrix representations of the mass and stiffness bilinear forms, respectively. By fixing a time-step $\Delta t > 0$ and denoting by $\mathbf{U}^i \approx \mathbf{U}(t_i)$ the approximation of $\mathbf{U}$ at time $t_i = i\Delta t$, we discretize (10) by the leap-frog method

$$\mathbf{M}\mathbf{U}^{n+1} = (2\mathbf{M} - \Delta t^2\mathbf{A})\mathbf{U}^n - \mathbf{M}\mathbf{U}^{n-1} + \Delta t^2\mathbf{F}^n, \quad n = 1, \ldots.$$

with $\mathbf{M}\mathbf{U}^1 = (\mathbf{M} - \frac{\Delta t^2}{2}\mathbf{A})\mathbf{U}^0 + \Delta t\mathbf{M}\mathbf{V}^0 + \frac{\Delta t^2}{2}\mathbf{F}^0$.

3 Stability of the Semi-Discrete Formulation

We now prove stability in the following natural energy norm induced by the DG methods described in the previous section:

$$\|\mathbf{v}\|_{\mathrm{E}}^2 = \|\rho^{1/2}\mathbf{v}_t\|_{0,\mathscr{T}_h}^2 + \|\mathbf{v}\|_{\mathrm{DG}}^2 \quad \forall \mathbf{v} = \mathbf{v}(t) \in C^2([0,T];\mathbf{V}_{hp}) \quad \forall \in [0,T]. \tag{11}$$

First, we recall the following classical result, cf. [30, pag. 28].

Lemma 3 *Let $\xi \in L^2(0,T)$ a positive function and $\eta \in C^0(0,T)$ a non-negative function such that*

$$\eta^2(t) \le C + \int_0^t \xi(\tau)\eta(\tau)\,d\tau \quad \forall t \in (0,T)$$

with C a non-negative constant. Then,

$$\eta(t) \le \sqrt{C} + \frac{1}{2}\int_0^t \xi(\tau)\,d\tau \quad \forall t \in (0,T)$$

For the forthcoming analysis we will assume that the (possible) discontinuities of the piecewise constant stiffness tensor $\mathscr{D}$ are aligned with the mesh partition $\mathscr{T}_h$.

Proposition 1 *Let $\mathbf{u}^h \in C^2((0,T];\mathbf{V}_{hp})$ be the approximate solution obtained with the SIP(δ) method* (16), *for a sufficiently large penalty parameter α, cf.* (7). *Then,*

$$\|\mathbf{u}^h(t)\|_{\mathrm{E}}^2 \lesssim \|\mathbf{u}^h(0)\|_{\mathrm{E}} + \int_0^t \|\mathbf{f}(\tau)\|_{L^2(\Omega)}\,d\tau \quad 0 < t \le T.$$

Proof We take $\mathbf{v} = \mathbf{u}_t^h \in \mathbf{V}_{hp}$ in (16) to obtain

$$\frac{1}{2}\frac{d}{dt}\left(\|\mathbf{u}^h\|_{\mathrm{E}}^2 - 2\langle\{\mathscr{D}\boldsymbol{\varepsilon}(\mathbf{u}^h)\}_\delta, [\![\mathbf{u}^h]\!]\rangle_{\mathscr{F}_h}\right) = (\mathbf{f}, \mathbf{u}_t^h)_{\mathscr{T}_h}. \tag{12}$$

Integrating in time between 0 and t leads to

$$\|\mathbf{u}^h\|_{\mathrm{E}}^2 - 2\langle\{\mathscr{D}\boldsymbol{\varepsilon}(\mathbf{u}^h)\}_\delta, [\![\mathbf{u}^h]\!]\rangle_{\mathscr{F}_h} = \|\mathbf{u}_0^h\|_{\mathrm{E}}^2 - 2\langle\{\mathscr{D}\boldsymbol{\varepsilon}(\mathbf{u}^h(0))\}_\delta, [\![\mathbf{u}^h(0)]\!]\rangle_{\mathscr{F}_h} + 2\int_0^t (\mathbf{f}, \mathbf{u}_\tau^h)_{\mathscr{T}_h}\, d\tau. \quad (13)$$

We first observe that, for any $F \in \mathscr{F}_h$, and any $\mathbf{w}, \mathbf{v} \in \mathbf{V}_{hp}$, the Cauchy-Schwarz inequality gives

$$\begin{aligned}
\sum_{F\in\mathscr{F}_h} \left|\langle\{\mathscr{D}\varepsilon(\mathbf{w})\}_\delta, [\![\mathbf{v}]\!]\rangle_F\right| &\le \|\sigma^{-1/2}\{\mathscr{D}\boldsymbol{\varepsilon}(\mathbf{w})\}\|_{0,\mathscr{F}_h}\|\sigma^{1/2}[\![\mathbf{v}]\!]\|_{0,\mathscr{F}_h} \\
&\lesssim \frac{1}{\sqrt{\alpha}}\|\mathscr{D}^{1/2}\boldsymbol{\varepsilon}(\mathbf{w})\|_{0,\mathscr{T}_h}\|\sigma^{1/2}[\![\mathbf{v}]\!]\|_{0,\mathscr{F}_h} \\
&\le \frac{1}{\sqrt{\alpha}}\|\mathbf{w}\|_{\mathrm{DG}}\|\mathbf{v}\|_{\mathrm{DG}} \le \frac{1}{\sqrt{\alpha}}\|\mathbf{w}\|_{\mathrm{E}}\|\mathbf{v}\|_{\mathrm{E}},
\end{aligned}$$

where in the second step we have employed the definition (7) of the penalty function σ, the local bounded variation property of the discretization parameters, together with the trace-inverse inequality (2.1). From the Young inequality, we obtain

$$\|\mathbf{u}^h\|_{\mathrm{E}}^2 - 2\langle\{\mathscr{D}\boldsymbol{\varepsilon}(\mathbf{u}^h)\}_\delta, [\![\mathbf{u}^h]\!]\rangle_{\mathscr{F}_h} \gtrsim \|\mathbf{u}^h\|_{\mathrm{E}}^2,$$

provided that the penalty parameter α is chosen sufficiently large. This leads to

$$\|\mathbf{u}^h\|_{\mathrm{E}}^2 \lesssim \|\mathbf{u}^h(0)\|_{\mathrm{E}}^2 + \int_0^t (\mathbf{f}, \mathbf{u}_\tau^h)_{\mathscr{T}_h}\, d\tau.$$

Next, we observe that, from the Cauchy-Schwarz inequality we have

$$\int_0^t (\mathbf{f}, \mathbf{u}_\tau^h)_{\mathscr{T}_h}\, d\tau \le \int_0^t \|\mathbf{f}\|_{\mathscr{T}_h,0}\|\rho^{1/2}\mathbf{u}_\tau^h\|_{\mathscr{T}_h,0} \le \int_0^t \|\mathbf{f}\|_{\mathscr{T}_h,0}\|\mathbf{u}^h\|_{\mathrm{E}}\, d\tau,$$

which leads to

$$\|\mathbf{u}^h(t)\|_{\mathrm{E}}^2 \lesssim \|\mathbf{u}^h(0)\|_{\mathrm{E}}^2 + \int_0^t \|\mathbf{f}\|_{\mathscr{T}_h,0}\|\mathbf{u}^h\|_{\mathrm{E}}\, d\tau.$$

The theorem follows by Lemma 3.

4 Error Analysis of the Semi-Discrete Formulation

Before stating the main result of this section, we recall some preliminary results that will be needed for the forthcoming analysis.

Lemma 4 *For any* $\mathbf{v} \in [H^{s_K}(K)]^d$, $s_K \geq 0$, $K \in \mathscr{T}_h$, *there exists* $\Pi_{hp}\mathbf{v} \in \mathbf{V}_{hp}$ *s.t.*

$$|||\mathbf{v} - \Pi_{hp}\mathbf{v}|||_{\mathrm{DG}}^2 \lesssim \sum_{K\in\mathscr{T}_h} \frac{h_K^{2\min(s_K,p_K+1)-2}}{p_K^{2s_K-3}} \|\mathbf{v}\|_{H^{s_K}(K)}^2. \tag{14}$$

Moreover, if $\mathbf{v}, \mathbf{v}_t \in [H^{s_K}(K)]^d$, *for any* $K \in \mathscr{T}_h$, *then*

$$\|\mathbf{v} - \Pi_{hp}\mathbf{v}\|_{\mathrm{E}}^2 \lesssim \sum_{K\in\mathscr{T}_h} \frac{h_K^{2\min(s_K,p_K+1)-2}}{p_K^{2s_K-3}} \left(\|\mathbf{v}_t\|_{H^{s_K}(K)}^2 + \|\mathbf{v}\|_{H^{s_K}(K)}^2\right). \tag{15}$$

Proof We only show (15), as (14) is a particular case. Recalling the definition of the energy norm $\|\cdot\|_{\mathrm{E}}$ and employing the estimates of Lemma 1 we obtain

$$\|\rho^{1/2}(\mathbf{v}_t - \Pi_{hp}\mathbf{v}_t)\|_{0,\mathscr{T}_h}^2 \lesssim \sum_{K\in\mathscr{T}_h} \frac{h_K^{2\min(s_K,p_K+1)}}{p_K^{2s_K}} \|\mathbf{v}_t\|_{H^{s_K}(K)}^2,$$

$$\|\mathscr{D}^{1/2}\boldsymbol{\varepsilon}(\mathbf{v} - \Pi_{hp}\mathbf{v})\|_{L^2(\mathscr{T}_h)}^2 \lesssim \sum_{K\in\mathscr{T}_h} \frac{h_K^{2\min(s_K,p_K+1)-2}}{p_K^{2s_K-2}} \|\mathbf{v}\|_{H^{s_K}(K)}^2,$$

$$\|\sigma^{1/2}[\![\mathbf{v} - \Pi_{hp}\mathbf{v}]\!]\|_{L^2(\mathscr{F}_h)}^2 \lesssim \sum_{K\in\mathscr{T}_h} \frac{h_K^{2\min(s_K,p_K+1)-2}}{p_K^{2s_K-3}} \|\mathbf{v}\|_{H^{s_K}(K)}^2,$$

that is

$$\|\mathbf{v} - \Pi_{hp}\mathbf{v}\|_{\mathrm{E}}^2 \lesssim \sum_{K\in\mathscr{T}_h} \frac{h_K^{2\min(s_K,p_K+1)-2}}{p_K^{2s_K-3}} \left(\frac{h_K^2}{p_K^3}\|\mathbf{v}_t\|_{H^{s_K}(K)}^2 + \frac{1}{p_K}\|\mathbf{v}\|_{H^{s_K}(K)}^2 + \|\mathbf{v}\|_{H^{s_K}(K)}^2\right)$$

$$\lesssim \sum_{K\in\mathscr{T}_h} \frac{h_K^{2\min(s_K,p_K+1)-2}}{p_K^{2s_K-3}} (\|\mathbf{v}_t\|_{H^s(K)}^2 + \|\mathbf{v}\|_{H^{s_K}(K)}^2),$$

where the last step follows by observing that $\frac{h_K^2}{p_K^3} < 1$ and $\frac{1}{p_K} < 1$ for any $K \in \mathscr{T}_h$.

4.1 Error Estimates in the Energy Norm

In this section we present *a priori* error estimates in the natural energy norm. Assuming that the exact solution $\mathbf{u}$ is regular enough, i.e., $\mathbf{u}|_K \in [H^{s_K}(K)]^d$ for any $K \in \mathscr{T}_h$, with $s_K \geq 2$, with standard arguments it is also possible to show that formulation (16) is *strongly consistent*, i.e.,

$$(\rho \mathbf{u}_{tt}, \mathbf{v})_{\mathscr{T}_h} + \mathscr{A}(\mathbf{u}, \mathbf{v}) = (\mathbf{f}, \mathbf{v})_{\mathscr{T}_h} \quad \forall \, \mathbf{v} \in \mathbf{V}_{hp}. \tag{16}$$

From the above identity, we can obtain the following relation for the error $\mathbf{e} = \mathbf{u} - \mathbf{u}^h$

$$(\rho \mathbf{e}_{tt}, \mathbf{v})_{\mathscr{T}_h} + \mathscr{A}(\mathbf{e}, \mathbf{v}) = \mathbf{0} \quad \forall \, \mathbf{v} \in \mathbf{V}_{hp}, \tag{17}$$

which serves as the basis for the forthcoming error estimates.

Theorem 1 (A-Priori Error Estimate in the Energy Norm) *Let* $\mathbf{u}$ *be the exact solution of problem* (1) *and let* $\mathbf{u}^h$ *be its approximation based on employing the semidiscrete DG formulation given in* (16), *with a penalty parameter* α *chosen large enough, cf.* (7). *If, for any time* $t \in [0, T]$, *the exact solution* $\mathbf{u}(t)$ *and its two first temporal derivatives belong* $[H^{s_K}(K)]^d$, $K \in \mathscr{T}_h$, $s_K \geq 2$, *then*

$$\sup_{t \in (0,T]} \|\mathbf{e}(t)\|_{\mathrm{E}}^2 \lesssim \sup_{t \in (0,T]} \left\{ \sum_{K \in \mathscr{T}_h} \frac{h_K^{2\min(s_K, p_K+1)-2}}{p_K^{2s_K-3}} \left(\|\mathbf{u}_t(t)\|_{H^{s_K}(K)}^2 + \|\mathbf{u}(t)\|_{H^{s_K}(K)}^2 \right) \right\}$$
$$+ \int_0^T \sum_{K \in \mathscr{T}_h} \left\{ \frac{h_K^{2\min(s_K, p_K+1)-2}}{p_K^{2s_K-3}} \left(\|\mathbf{u}_{tt}(\tau)\|_{H^{s_K}(K)}^2 + \|\mathbf{u}_t(\tau)\|_{H^{s_K}(K)}^2 \right) \right\} d\tau \, .$$

Before reporting the proof of Theorem 1 we recall the integration by parts formula

$$\int_0^t (\mathbf{w}, \mathbf{v}_\tau)_* d\tau = (\mathbf{w}(t), \mathbf{v}(t))_* - (\mathbf{w}(0), \mathbf{v}(0))_* - \int_0^t (\mathbf{w}_\tau, \mathbf{v})_* d\tau, \tag{18}$$

that holds for $\mathbf{w}, \mathbf{v}$ regular enough and for any scalar product $(\cdot, \cdot)_*$

Proof Let $\Pi_{hp}\mathbf{u} \in \mathbf{V}_{hp}$ be the interpolant defined as in Lemma 4. By decomposing the error as $\mathbf{e} = \mathbf{e}^\pi - \mathbf{e}^h$, with $\mathbf{e}^\pi = \mathbf{u} - \Pi_{hp}\mathbf{u}$ and $\mathbf{e}^h = \mathbf{u}^h - \Pi_{hp}\mathbf{u}$, (17) becomes:

$$(\rho \mathbf{e}_{tt}^h, \mathbf{v})_{\mathscr{T}_h} + \mathscr{A}(\mathbf{e}^h, \mathbf{v}) = (\rho \mathbf{e}_{tt}^\pi, \mathbf{v})_{\mathscr{T}_h} + \mathscr{A}(\mathbf{e}^\pi, \mathbf{v}) \quad \forall \, \mathbf{v} \in \mathbf{V}_{hp}.$$

By taking $\mathbf{v} = \mathbf{e}^h$ in the above identity, we have

$$\frac{1}{2} \frac{d}{dt} \left(\|\mathbf{e}^h\|_{\mathrm{E}}^2 - 2 \langle [\![\mathbf{e}^h]\!], \{ \mathscr{D}\boldsymbol{\varepsilon}(\mathbf{e}^h) \}_\delta \rangle_{\mathscr{F}_h} \right) = (\rho \mathbf{e}_{tt}^\pi, \mathbf{e}_t^h)_{\mathscr{T}_h} + \mathscr{A}(\mathbf{e}^\pi, \mathbf{e}_t^h) \, . \tag{19}$$

Reasoning as in the proof of Theorem 1, we have

$$\|\mathbf{e}^h\|_{\mathrm{E}}^2 - 2 \langle [\![\mathbf{e}^h]\!], \{ \mathscr{D}\boldsymbol{\varepsilon}(\mathbf{e}^h) \}_\delta \rangle_{\mathscr{F}_h} \gtrsim \|\mathbf{e}^h\|_{\mathrm{E}}^2,$$

provided that the penalty parameter α is chosen large enough; cf. (7). Integrating (19) in time between 0 and t and using that $\mathbf{e}^h(0) = \mathbf{u}^h(0) - \mathbf{u}^\pi(0) = \mathbf{0}$, we get

$$\begin{aligned}
\|\mathbf{e}^h\|_{\mathrm{E}}^2 &\lesssim \int_0^t (\rho\mathbf{e}_{tt}^\pi, \mathbf{e}_t^h)_{\mathscr{T}_h}\, d\tau + \int_0^t \mathscr{A}(\mathbf{e}^\pi, \mathbf{e}_t^h)\, d\tau \\
&\lesssim \int_0^t \|\mathbf{e}_t^\pi\|_{\mathrm{E}}\|\mathbf{e}^h\|_{\mathrm{E}}\, d\tau + \mathscr{A}(\mathbf{e}^\pi, \mathbf{e}^h) - \int_0^t \mathscr{A}(\mathbf{e}_t^\pi, \mathbf{e}^h)\, d\tau \\
&\lesssim \int_0^t \|\mathbf{e}_t^\pi\|_{\mathrm{E}}\|\mathbf{e}^h\|_{\mathrm{E}}\, d\tau + |||\mathbf{e}^\pi|||_{\mathrm{DG}}\|\mathbf{e}^h\|_{\mathrm{DG}} + \int_0^t |||\mathbf{e}_t^\pi|||_{\mathrm{DG}}\|\mathbf{e}^h\|_{\mathrm{DG}}\, d\tau,
\end{aligned} \tag{20}$$

where the second step follows based on employing the Cauchy-Schwarz inequality together with integration by parts formula (18) with $\mathbf{w} = \mathbf{e}^\pi$, $\mathbf{v} = \mathbf{e}^h$ and $(\cdot,\cdot)_* = \mathscr{A}(\cdot,\cdot)$, whereas the third one follows from Lemma (8). From the Young inequality

$$|||\mathbf{e}^\pi|||_{\mathrm{DG}}\|\mathbf{e}^h\|_{\mathrm{DG}} \le \frac{1}{\epsilon}|||\mathbf{e}^\pi|||_{\mathrm{DG}}^2 + \epsilon\|\mathbf{e}^h\|_{\mathrm{DG}}^2 \le \frac{1}{\epsilon}|||\mathbf{e}^\pi|||_{\mathrm{DG}}^2 + \epsilon\|\mathbf{e}^h\|_{\mathrm{E}}^2,$$

we can suitably choose ϵ and rewrite (25) as

$$\|\mathbf{e}^h\|_{\mathrm{E}}^2 \lesssim +|||\mathbf{e}^\pi|||_{\mathrm{DG}}^2 + \int_0^t (|||\mathbf{e}_t^\pi|||_{\mathrm{DG}} + \|\mathbf{e}_t^\pi\|_{\mathrm{E}})\|\mathbf{e}^h\|_{\mathrm{DG}}\, d\tau. \tag{21}$$

Applying Gronwall's Lemma 3 we get

$$\|\mathbf{e}^h(t)\|_{\mathrm{E}} \lesssim \sup_{t\in[0,T]} |||\mathbf{e}^\pi(t)|||_{\mathrm{DG}}^2 + \int_0^t \|\mathbf{e}_t^\pi(\tau)\|_{\mathrm{E}}\, d\tau \qquad \forall t \in (0,T].$$

Finally, from the Cauchy-Schwarz inequality and the above bound, and taking the supremum over $t \in (0,T]$

$$\sup_{t\in(0,T]} \|\mathbf{e}(t)\|_{\mathrm{E}}^2 \lesssim \sup_{t\in(0,T]} \left\{\|\mathbf{e}^\pi(t)\|_{\mathrm{E}}^2 + |||\mathbf{e}^\pi(t)|||_{\mathrm{DG}}^2\right\} + \int_0^T \|\mathbf{e}_t^\pi(\tau)\|_{\mathrm{E}}^2\, d\tau\ .$$

The proof is completed by applying Lemma 4.

Remark 1 If the mesh size is quasi uniform, i.e. $h = \max_{K\in\mathscr{T}_K} h_K \approx h_K$ for any $K \in \mathscr{T}_h$, the polynomial approximation degree is uniform, i.e. $p_K = p$ for any $K \in \mathscr{T}_h$, and the exact solution satisfies $\mathbf{u}|_K, \mathbf{u}_t|_K, \mathbf{u}_{tt}|_K \in [H^s(K)]^d$ for any $K \in \mathscr{T}_h$

and for any $t \in [0, T]$, with $s \geq p + 1$, the error estimate of Theorem 1 becomes

$$\sup_{t\in(0,T]} \|\mathbf{e}(t)\|_{\mathrm{E}}^2 \lesssim \frac{h^{2p}}{p^{2s-3}} \sup_{t\in(0,T]} \left\{\|\mathbf{u}_t(t)\|_{H^s(\Omega)}^2 + \|\mathbf{u}(t)\|_{H^s(\Omega)}^2\right\} + \frac{h^{2p}}{p^{2s-3}} \int_0^T \left\{\|\mathbf{u}_{tt}(\tau)\|_{H^s(\Omega)}^2 + \|\mathbf{u}_t(\tau)\|_{H^s(\Omega)}^2\right\} d\tau \,.$$

The above bounds are optimal in h and suboptimal in p by a factor $p^{1/2}$; see, e.g., [19, 29] for analogous bounds for stationary (scalar) second order elliptic problems. Optimal error estimates with respect to the polynomial approximation degree can be shown either using the projector of [17] provided the solution belongs to a suitable augmented space, or whenever a continuous interpolant can be built; cf. [37].

4.2 *Error Estimates in the L^2 Norm*

In this section we present *a priori* error estimates in the L^2 norm. We follow the approach of [13] for second order hyperbolic equations, and introduce, for a regular enough vector-valued function $\mathbf{w}$, the elliptic-projection operator $\Pi\mathbf{w}$ defined as

$$\mathscr{A}(\Pi\mathbf{w}, \mathbf{v}) = \mathscr{A}(\mathbf{w}, \mathbf{v}) \quad \forall \mathbf{v} \in \mathbf{V}_{hp}. \tag{22}$$

We immediately have

$$\|\mathbf{u} - \Pi\mathbf{u}\|_{\mathrm{DG}} \leq \|\mathbf{u} - \Pi_{hp}\mathbf{u}\|_{\mathrm{DG}} + \|\Pi_{hp}\mathbf{u} - \Pi\mathbf{u}\|_{\mathrm{DG}} \lesssim |||\Pi_{hp}\mathbf{u} - \mathbf{u}|||_{\mathrm{DG}}, \tag{23}$$

where Π_{hp} is the interpolant of Lemma 4, and where the second step follows from Lemma 2 and the definition (22)

$$\begin{aligned}\|\Pi_{hp}\mathbf{u} - \Pi\mathbf{u}\|_{\mathrm{DG}}^2 &\lesssim \mathscr{A}(\Pi_{hp}\mathbf{u} - \Pi\mathbf{u}, \Pi_{hp}\mathbf{u} - \Pi\mathbf{u}) = \mathscr{A}(\Pi_{hp}\mathbf{u} - \mathbf{u}, \Pi_{hp}\mathbf{u} - \Pi\mathbf{u}) \\ &\lesssim |||\Pi_{hp}\mathbf{u} - \mathbf{u}|||_{\mathrm{DG}} \|\Pi_{hp}\mathbf{u} - \Pi\mathbf{u}\|_{\mathrm{DG}}.\end{aligned}$$

We also recall the following Poincaré–Friedrichs inequality valid for piecewise vector–valued H^1 functions

$$\|\mathbf{v}\|_{L^2(\mathscr{T}_h)}^2 \lesssim \sum_{K\in\mathscr{T}_h} \|\nabla\mathbf{v}\|_{L^2(K)}^2 + \sum_{F\in\mathscr{F}_h} \frac{1}{h_F}\|[\![\mathbf{v}]\!]\|_{L^2(F)}^2 \quad \forall \mathbf{v} \in [H^1(\mathscr{T}_h)]^d,$$

cf. [10]. Using that $\sum_{K\in\mathscr{T}_h} \|\nabla\mathbf{v}\|_{L^2(K)} \leq \|\varepsilon(\mathbf{v})\|_{L^2(\mathscr{T}_h)}$, and from the definition of the DG norm and of the stabilization function (12), it immediately follows

$$\|\mathbf{v}\|_{L^2(\mathscr{T}_h)}^2 \lesssim \|\mathbf{v}\|_{\mathrm{DG}} \quad \forall \mathbf{v} \in [H^1(\mathscr{T}_h)]^d. \tag{24}$$

Theorem 2 (A-Priori Error Estimate in the L^2 Norm) *Under the Assumptions of Theorem 1, it holds*

$$\sup_{t\in(0,T]} \|\mathbf{e}(t)\|^2_{L^2(\Omega)} \lesssim \sup_{t\in(0,T]} \left\{ \frac{h^{2\min(s,p+1)}}{p^{2s-2}} \left(\|\mathbf{u}_t(t)\|^2_{H^s(\Omega)} + \|\mathbf{u}(t)\|^2_{H^s(\Omega)} \right) \right\}$$
$$+ \int_0^T \left\{ \frac{h^{2\min(s,p+1)}}{p^{2s-2}} \left(\|\mathbf{u}_{tt}(\tau)\|^2_{H^s(\Omega)} + \|\mathbf{u}_t(\tau)\|^2_{H^s(\Omega)} \right) \right\} d\tau \, .$$

with $h = \max_{K\in\mathcal{T}_h} h_K$, $p = \min_{K\in\mathcal{T}_h} p_K$ *and* $s = \min_{K\in\mathcal{T}_h} s_K$.

Proof As in the proof of Theorem 1 we decompose the error as $\mathbf{e} = \mathbf{e}^\pi - \mathbf{e}^h$, where now $\mathbf{e}^\pi = \mathbf{u}^h - \Pi\mathbf{u}$ and $\mathbf{e}^h = \mathbf{u} - \Pi\mathbf{u}$, $\Pi\mathbf{u}$ being the elliptic projector defined in (22). With the above decomposition, the error equation (17) becomes:

$$(\rho\mathbf{e}^h_{tt}, \mathbf{v})_{\mathcal{T}_h} + \mathcal{A}(\mathbf{e}^h, \mathbf{v}) = (\rho\mathbf{e}^\pi_{tt}, \mathbf{v})_{\mathcal{T}_h} \quad \forall\, \mathbf{v} \in \mathbf{V}_{hp}.$$

By taking $\mathbf{v} = \mathbf{e}^h$ in the above identity and reasoning as in the proof of Theorem 1, we have

$$\|\mathbf{e}^h\|^2_{\mathrm{E}} - 2\langle [\![\mathbf{e}^h]\!], \{\mathcal{D}\boldsymbol{\varepsilon}(\mathbf{e}^h)\}_\delta\rangle_{\mathcal{F}_h} \gtrsim \|\mathbf{e}^h\|^2_{\mathrm{E}},$$

provided that the penalty parameter α is chosen large enough; cf. (7). Therefore, integrating in time between 0 and t and using that $\mathbf{e}^h(0) = \mathbf{u}^h(0) - \mathbf{u}^\pi(0) = \mathbf{0}$, we get

$$\|\mathbf{e}^h\|^2_{\mathrm{E}} \lesssim \int_0^t (\rho\mathbf{e}^\pi_{tt}, \mathbf{e}^h_t)_{\mathcal{T}_h}\, d\tau \quad \lesssim \int_0^t \|\mathbf{e}^\pi_{tt}\|_{L^2(\Omega)} \|\mathbf{e}^h\|_{\mathrm{E}}\, d\tau \tag{25}$$

where the second step follows based on employing the Cauchy-Schwarz inequality. Applying Gronwall's Lemma 3 we get

$$\|\mathbf{e}^h(t)\|_{\mathrm{E}} \lesssim \int_0^t \|\mathbf{e}^\pi_{tt}(\tau)\|_{L^2(\Omega)}\, d\tau \qquad \forall t \in (0,T].$$

Next, from the Cauchy-Schwarz inequality, the above bound and the Poincaré–Friedrichs inequality (24), we immediately get

$$\|\mathbf{u} - \mathbf{u}^h\|_{L^2(\mathcal{T}_h)} \lesssim \|\mathbf{e}^\pi\|_{L^2(\mathcal{T}_h)} + \|\mathbf{e}^h\|_{L^2(\mathcal{T}_h)} \lesssim \|\mathbf{e}^\pi\|_{L^2(\mathcal{T}_h)} + \|\mathbf{e}^h\|_{\mathrm{DG}}$$
$$\leq \|\mathbf{e}^\pi\|_{L^2(\mathcal{T}_h)} + \|\mathbf{e}^h\|_{\mathrm{E}} \lesssim \|\mathbf{e}^\pi\|_{L^2(\mathcal{T}_h)} + \int_0^t \|\mathbf{e}^\pi_{tt}(\tau)\|_{L^2(\Omega)}\, d\tau.$$

The estimate of the terms on the right hand sides is based on employing a duality argument; cf [13]. Let $\boldsymbol{\xi}$ be the solution of the problem

$$\nabla \cdot \sigma(\boldsymbol{\xi}) = \mathbf{e}^{\pi} \quad \text{in } \Omega, \quad \boldsymbol{\xi} = \mathbf{0} \quad \text{on } \partial\Omega.$$

As Ω is convex, the above problem is well-posed and its unique solution $\boldsymbol{\xi} \in [H^2(\Omega)]^d$ and satisfies $\|\boldsymbol{\xi}\|_{H^2(\Omega)} \lesssim \|\mathbf{e}^{\pi}\|_{L^2(\Omega)}$. Moreover, it holds

$$\|\mathbf{e}^{\pi}\|^2_{L^2(\Omega)} = (\mathbf{e}^{\pi}, \mathbf{e}^{\pi})_{L^2(\Omega)} = \mathscr{A}(\boldsymbol{\xi}, \mathbf{e}^{\pi}) = \mathscr{A}(\boldsymbol{\xi} - \boldsymbol{\xi}^{\pi}, \mathbf{e}^{\pi}) \lesssim |||\boldsymbol{\xi} - \boldsymbol{\xi}^{\pi}|||_{\mathrm{DG}} |||\mathbf{e}^{\pi}|||_{\mathrm{DG}}$$

where $\boldsymbol{\xi}^{\pi} \in \mathbf{V}_{hp}$ is the interpolant of Lemma 4, and where the last steps follows from Lemma 2. Employing the interpolation estimates of Lemma 4 we have

$$\|\mathbf{e}^{\pi}\|^2_{L^2(\Omega)} \lesssim \frac{h}{p^{1/2}} \|\boldsymbol{\xi}\|_{H^2(\Omega)} |||\mathbf{e}^{\pi}|||_{\mathrm{DG}} \lesssim \frac{h}{p^{1/2}} \|\mathbf{e}^{\pi}\|_{L^2(\Omega)} |||\mathbf{e}^{\pi}|||_{\mathrm{DG}},$$

where $h = \max_{K \in \mathscr{T}_h} h_K$ and $p = \min_{K \in \mathscr{T}_h} p_K$. The proof is completed by employing the error bounds of Theorem 1.

5 Numerical Results

The results of this section have been obtained with *SPEED* (http://speed.mox.polimi.it/), an open source Fortran code developed at Politecnico di Milano by the Laboratory for Modeling and Scientific Computing MOX of the Department of Mathematics and the Department of Civil and Environmental Engineering. SPEED is specifically designed for the simulation of seismic waves propagation problems, including both the ground motion induced by large scale earthquakes and soil-structure interaction in urban areas; see, e.g., [27]. Throughout the section we have set $\Delta t = 10^{-5}$ so that the temporal component of the error is negligible compared to the spatial one.

In the first example we consider an elastic wave propagation problem in $\Omega = (0,1)^3$, with $\rho = \lambda = \mu = 1$. The source term $\mathbf{f}$ and the initial data are chosen so that the exact solution of problem (1) is given by

$$\mathbf{u}_{\mathrm{ex}}(x, y, z, t) = \sin(3\pi t) \begin{bmatrix} -\sin^2(\pi x)\sin(2\pi y)\sin(2\pi z) \\ \sin^2(\pi y)\sin(2\pi x)\sin(2\pi z) \\ \sin^2(\pi z)\sin(2\pi x)\sin(2\pi y) \end{bmatrix}.$$

We first we consider both a tetrahedral and a hexahedral grid with mesh size $h = 0.5$ and let p varies from 2 to 8. In Fig. 1a we report the error computed in the energy norm $\|\cdot\|_{\mathrm{E}}$ at $t = T = 0.05$ and as a function of the polynomial degree. As expected, an exponential convergence is observed. For the sake of comparison Fig. 1a also

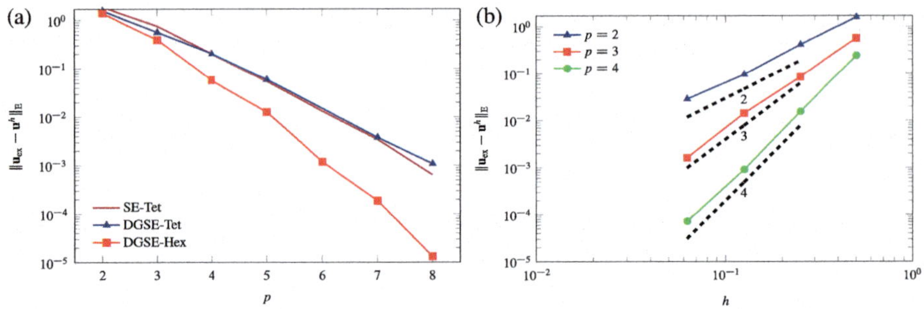

Fig. 1 Example 1. (**a**) Computed errors versus p: computed errors measured in the energy norm $\|\cdot\|_{\mathrm{E}}$ at $t = T = 0.05$ versus the polynomial degree p for a tetrahedral mesh (DG-Tet) and a hexahedral grid (DG-Hex). The results are also compared with the corresponding one based on employing conforming Spectral Element method on the same tetahedral grid (SE-Tet). (**b**) Computed errors versus h: computed errors measured in the energy norm $\|\cdot\|_{\mathrm{E}}$ at $t = T$ versus the mesh size for $p = 2, 3, 4$. The *dashed lines* denote the expected slopes of the error curves

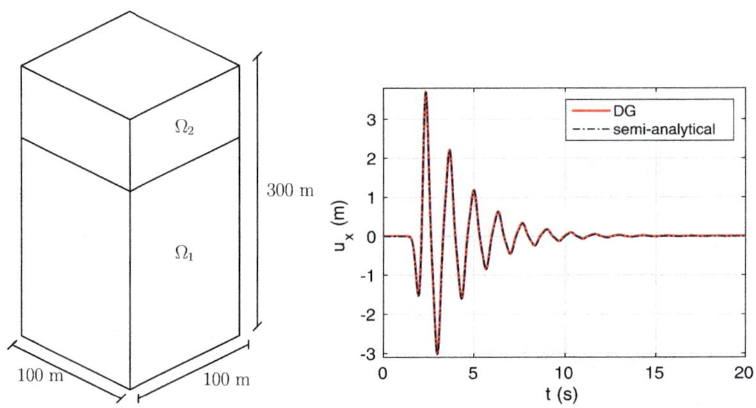

Fig. 2 Example 2. *Left*: Computational domain $\Omega = \Omega_1 \cup \Omega_2$. The elastic wave propagates from the bottom of Ω_1 to the top surface of Ω_2. *Right*: Computed time history of the x component of the displacement $\mathbf{u}_x$ recorded at $R = (50, 50, 0)$ m. The results are compared with a reference semi-analytical solution $\mathbf{u}_{TH}$ obtained with the Thomson-Haskell propagation matrix method

reports the corresponding computed errors obtained with a conforming Spectral Element method on the same tetrahedral grid. Next, we investigate the behavior of the error as a function of the grid size h for different polynomial degrees. We consider a sequence of uniformly refined tetrahedral grids starting from an initial decomposition of size $h_0 = 0.5$. In Fig. 1b we report the computed errors measured in energy norm $\|\cdot\|_{\mathrm{E}}$ at the final observation time $t = T$ versus the grid size for $p = 2, 3, 4$. As expected, the results confirm a convergence rate of order p.

In the second test we consider a plane wave propagating along the vertical direction in a layered elastic half-space $\Omega = (0, 100) \times (0, 100) \times (-300, 0)\, m$, see Fig. 2 (left). In Table 1 we report the depth and the material properties of the

Table 1 Example 2. Material properties

Layer	Depth $[m]$	$\rho\ [Kg/m^3]$	$c_P\ [m/s]$	$c_S\ [m/s]$	Dumping $\xi\ [1/s]$
Ω_1	200	2200	4000	2000	$\pi \times 10^{-3}$
Ω_2	100	1800	600	300	$\pi \times 10^{-2}$

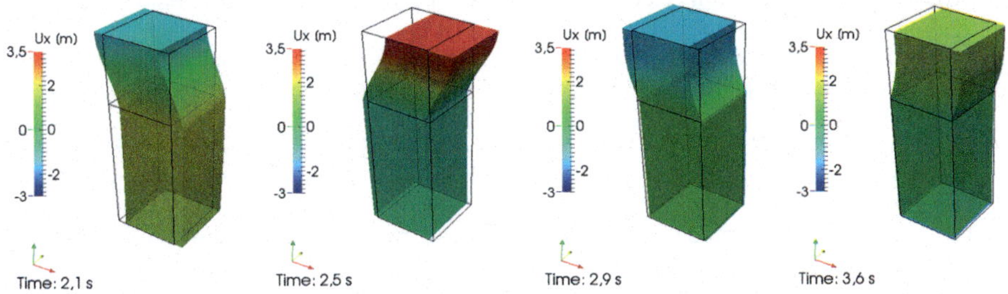

Fig. 3 Example 2. Snapshots of the x-component of the displacement $\mathbf{u}_x$.The deformed domain (colored) is compared with the non distorted one (*black line*)

half-space Ω_1 and the layer Ω_2. The source plane wave is polarized in the x direction and its time dependency is given by a unit amplitude Ricker wave with peak frequency at 1 Hz. A dumping term proportional to $2\rho\xi\mathbf{u}_t + 2\xi^2\mathbf{u}$, with ξ as in Table 1, is also added to the equation to take into account viscoelastic effects. The subdomains Ω_1 and Ω_2 are discretized with a hexahedral and a tetrahedral mesh, respectively, and the computational grids are built in order to have at least five grid points per wavelength, with $p = 4$ in both Ω_1 and Ω_2. Finally, we impose absorbing boundary conditions on the bottom surface, a free surface condition on the top surface, and homogeneous Dirichlet conditions for the y and z component of the displacement on the remaining boundaries. In Fig. 2 (right) we report the computed solution which is also compared with a reference semi-analytical solution $\mathbf{u}_{TH}$ based on the Thomson-Haskell propagation matrix method, cf. [18, 40]. More precisely, Fig. 2 (right) shows the time history of the x component of the displacement $\mathbf{u}_x$ recorded at the point $R = (50, 50, 0)\ m$. Finally, in Fig. 3 we report four snapshots of the deformed computational domain when invested by the plane wave. Two relevant physical effects can be observed: *i)* the wave field is amplified at the top of the domain due to the free surface condition; *ii)* reflections of the wave field take place inside the layer Ω_2 characterized by a softer material with respect to the half space Ω_1.

Acknowledgements Paola F. Antonietti and Ilario Mazzieri have been partially supported by the research grant no. 2015-0182 funded by Fondazione Cariplo and Regione Lombardia. Part of this work has been completed while Paola F. Antonietti was visiting the Institut Henri Poincaré (IHP), Paris. She thanks the Institute for the kind hospitality.

References

1. M. Ainsworth, Dispersive and dissipative behaviour of high order discontinuous Galerkin finite element methods. J. Comput. Phys. **198**(1), 106–130 (2004)
2. M. Ainsworth, R. Rankin, Technical note: a note on the selection of the penalty parameter for discontinuous Galerkin finite element schemes. Numer. Methods Partial Differential Equations **28**(3), 1099–1104 (2012)
3. P.F. Antonietti, I. Mazzieri, A. Quarteroni, F. Rapetti, Non-conforming high order approximations of the elastodynamics equation. Comput. Methods Appl. Mech. Eng. **209–212**, 212–238 (2012)
4. P.F. Antonietti, C. Marcati, I. Mazzieri, A. Quarteroni, High order discontinuous Galerkin methods on simplicial elements for the elastodynamics equation. Numer. Algorithms **71**(1), 181–206 (2016)
5. P.F. Antonietti, B. Ayuso de Dios, I. Mazzieri, A. Quarteroni, Stability analysis of discontinuous Galerkin approximations to the elastodynamics problem. J. Sci. Comput. **68**(1), 143–170 (2016)
6. P.F. Antonietti, A. Ferroni, I. Mazzieri, R. Paolucci, A. Quarteroni, C. Smerzini, M. Stupazzini, Numerical modeling of seismic waves by discontinuous Spectral Element methods. MOX Report 9/2017 (Submitted, 2017)
7. D.N. Arnold, An interior penalty finite element method with discontinuous elements. SIAM J. Numer. Anal. **19**(4), 742–760 (1982)
8. D.N. Arnold, F. Brezzi, B. Cockburn, L.D. Marini, Unified analysis of discontinuous Galerkin methods for elliptic problems. SIAM J. Numer. Anal. **39**(5), 1749–1779 (2001)
9. I. Babuŝka, M. Suri, The *hp* version of the finite element method with quasiuniform meshes. Math. Model. Numer. Anal. **21**(2), 199–238 (1987)
10. S.C. Brenner, Korn's inequalities for piecewise H^1 vector fields. Math. Comput. **73**(247), 1067–1087 (2004)
11. J.D. De Basabe, M.K. Sen, M.F. Wheeler, The interior penalty discontinuous Galerkin method for elastic wave propagation: grid dispersion. Geophys. J. Int. **175**(1), 83–93 (2008)
12. M. Dumbser, M. Käser, An arbitrary high-order discontinuous galerkin method for elastic waves on unstructured meshes - ii. the three-dimensional isotropic case. Geophys. J. Int. **167**(1), 319–336 (2006)
13. T. Dupont, L^2-estimates for Galerkin methods for second order hyperbolic equations. SIAM J. Numer. Anal. **10**, 880–889 (1973)
14. E. Faccioli, F. Maggio, R. Paolucci, A. Quarteroni, 2d and 3d elastic wave propagation by a pseudo-spectral domain decomposition method. J. Seimol. **1**(3), 237–251 (1997)
15. A. Ferroni, P.F. Antonietti, I. Mazzieri, A. Quarteroni, Dispersion-dissipation analysis of 3D continuous and discontinuous Spectral Element methods for the elastodynamics equation. MOX Report 18/2016 (Submitted)
16. E.H. Georgoulis, E. Hall, P. Houston, Discontinuous Galerkin methods for advection-diffusion-reaction problems on anisotropically refined meshes. SIAM J. Sci. Comput. **30**(1), 246–271 (2007/2008)
17. E.H. Georgoulis, E. Süli, Optimal error estimates for the *hp*-version interior penalty discontinuous Galerkin finite element method. IMA J. Numer. Anal. **25**(1), 205–220 (2005)
18. N.A. Haskell, The dispersion of surface waves on multi-layered media. Bull. Seismol. Soc. Am. **43**, 17–34 (1953)
19. P. Houston, C. Schwab, E. Süli, Discontinuous *hp*-finite element methods for advection-diffusion-reaction problems. SIAM J. Numer. Anal. **39**(6), 2133–2163 (2002)
20. D. Komatitsch, J. Tromp, Introduction to the spectral element method for three-dimensional seismic wave propagation. Geophys. J. Int. **139**(3), 806–822 (1999)
21. D. Komatitsch, J. Tromp, Spectral-element simulations of global seismic wave propagation - i. validation. Geophys. J. Int. **149**(2), 390–412 (2002)

22. D. Komatitsch, J.-P. Vilotte, R. Vai, J. Castillo-Covarrubias, F. Snchez-Sesma, The spectral element method for elastic wave equations - application to 2-d and 3-d seismic problems. Int. J. Numer. Meth. Eng. **45**(9), 1139–1164 (1999)
23. D. Komatitsch, J. Ritsema, J. Tromp, Geophysics: the spectral-element method, beowulf computing, and global seismology. Science **298**(5599), 1737–1742 (2002)
24. M. Kser, M. Dumbser, An arbitrary high-order discontinuous Galerkin method for elastic waves on unstructured meshes - i. the two-dimensional isotropic case with external source terms. Geophys. J. Int. **166**(2), 855–877 (2006)
25. D.J.P. Lahaye, F. Maggio, A. Quarteroni, Hybrid finite element–spectral element approximation of wave propagation problems. East-West J. Numer. Math. **5**(4), 265–289 (1997)
26. I. Mazzieri, M. Stupazzini, R. Guidotti, C. Smerzini, Speed: spectral elements in elastodynamics with discontinuous Galerkin: a non-conforming approach for 3d multi-scale problems. Int. J. Numer. Meth. Eng. **95**(12), 991–1010 (2013)
27. I. Mazzieri, M. Stupazzini, R. Guidotti, C. Smerzini, Speed: spectral elements in elastodynamics with discontinuous Galerkin: a non-conforming approach for 3D multi-scale problems. Int. J. Numer. Methods Eng. **95**(12), 991–1010 (2013)
28. A.T. Patera, Spectral methods for spatially evolving hydrodynamic flows, in *Spectral Methods for Partial Differential Equations (Hampton, VA, 1982)* (SIAM, Philadelphia, 1984), pp. 239–256
29. I. Perugia, D. Schötzau, An *hp*-analysis of the local discontinuous Galerkin method for diffusion problems. J. Sci. Comput. **17**(1–4), 561–571 (2002)
30. A. Quarteroni, *Numerical models for differential problems*, vol. 8, 2nd edn. MS&A. Modeling, Simulation and Applications. (Springer, Milan, 2014). Translated from the fifth (2012) Italian edition by Silvia Quarteroni
31. B. Rivière, M.F. Wheeler, Discontinuous finite element methods for acoustic and elastic wave problems, in *Current Trends in Scientific Computing (Xi'an, 2002)*, vol. 329, Contemporary Mathematics (American Mathematical Society, Providence, 2003), pp. 271–282
32. B. Rivière, S. Shaw, M.F. Wheeler, J.R. Whiteman, Discontinuous Galerkin finite element methods for linear elasticity and quasistatic linear viscoelasticity. Numer. Math. **95**(2), 347–376 (2003)
33. B. Rivière, S. Shaw, J.R. Whiteman, Discontinuous Galerkin finite element methods for dynamic linear solid viscoelasticity problems. Numer. Methods Partial Differential Equations **23**(5), 1149–1166 (2007)
34. C. Schwab, *p- and hp-Finite Element Methods*. Numerical Mathematics and Scientific Computation (The Clarendon Press, Oxford University Press, New York, 1998). Theory and applications in solid and fluid mechanics
35. G. Seriani, E. Priolo, A. Pregarz, Modelling waves in anisotropic media by a spectral element method, in *Mathematical and Numerical Aspects of Wave Propagation (Mandelieu-La Napoule, 1995)* (SIAM, Philadelphia, 1995), pp. 289–298
36. C. Smerzini, R. Paolucci, M. Stupazzini, Experimental and numerical results on earthquake-induced rotational ground motions. J. Earthq. Eng. **13**(Suppl. 1), 66–82 (2009)
37. B. Stamm, T.P. Wihler, *hp*-optimal discontinuous Galerkin methods for linear elliptic problems. Math. Comput. **79**(272), 2117–2133 (2010)
38. R. Stenberg, Mortaring by a method of J. A. Nitsche, in *Computational Mechanics (Buenos Aires, 1998)* (Centro Internac. Métodos Numér. Ing., Barcelona, 1998)
39. M. Stupazzini, R. Paolucci, H. Igel, Near-fault earthquake ground-motion simulation in the grenoble valley by a high-performance spectral element code. Bull. Seismol. Soc. Am. **99**(1), 286–301 (2009)
40. W.T. Thomson, Transmission of elastic waves through a stratified solid medium. J. Appl. Phys. **21**, 89–93 (1950)
41. M.F. Wheeler, An elliptic collocation-finite element method with interior penalties. SIAM J. Numer. Anal. **15**(1), 152–161 (1978)

A Polynomial Spectral Calculus for Analysis of DG Spectral Element Methods

David A. Kopriva

Abstract We introduce a polynomial spectral calculus that follows from the summation by parts property of the Legendre-Gauss-Lobatto quadrature. We use the calculus to simplify the analysis of two multidimensional discontinuous Galerkin spectral element approximations.

1 Introduction

The discontinuous Galerkin Spectral Element Method (DGSEM) introduced by Black [4, 5] has the desired properties of spectral accuracy, geometric flexibility, and excellent phase and dissipation properties [10, 21]. Spectral accuracy comes from the use of high order polynomial approximations to the solutions and fluxes, and high order Gauss quadratures for the inner products, e.g. [20]. Geometric flexibility comes from the multi-element subdivision of the domain. The DGSEM is now developed to the point of being efficient for large scale engineering level computations, e.g. [1, 3, 8], among others.

Robustness, however, has been an issue with the DGSEM at high order. It usually works, but it can go unstable even when the solutions are smooth. For nonlinear problems, this is probably not surprising. Examples are demonstrated in the computation of the Taylor-Green vortex problem, where instability at high orders is seen [11]. But instability arises even in linear problems when the coefficients are variable, which can come from inherent variability [2] or from variability introduced by curved elements [16]. The instability, we will show, comes from aliasing errors associated with the products of polynomials and insufficient Gauss quadrature precision.

Robust (provably stable) versions of the DGSEM that start from a split form of the partial differential equation (PDE) have recently been developed for linear hyperbolic systems for static [16] and moving domains [19]. In addition to stability, the approximations match the additional conservative and constant state preserving

D.A. Kopriva (✉)
Department of Mathematics, The Florida State University, Tallahassee, FL 32306, USA
e-mail: kopriva@math.fsu.edu

M.L. Bittencourt et al. (eds.), *Spectral and High Order Methods for Partial Differential Equations ICOSAHOM 2016*, Lecture Notes in Computational Science and Engineering 119, DOI 10.1007/978-3-319-65870-4_2

properties of the PDE [17]. The approach is applicable to nonlinear problems, where, depending on the equations and split form, the methods are energy or entropy stable [9, 12, 13].

In this paper, we introduce a polynomial spectral calculus that allows us to mirror the continuous PDE analysis to show stability of Black's and the split-form approximations with a simple, compact notation applicable to any number of space dimensions. For the split form method, we also show how to use the calculus to demonstrate conservation and constant state preservation. The key starting point of the calculus is the summation by parts property satisfied by the Gauss-Lobatto quadrature [15], which allows us to write discrete versions of the Gauss law and its variants. Those discrete Gauss laws, in turn, allow us to write algebraically equivalent forms of the approximations, with which we can easily analyze their properties.

2 Linear Hyperbolic Problems on Bounded Domains

As examples of the use of the discrete calculus, we will analyze two discontinuous Galerkin spectral element approximations to the linear system of conservation laws

$$\mathbf{u}_t + \nabla \cdot \overrightarrow{\mathbf{f}} = 0, \tag{1}$$

where $\mathbf{u}\left(\overrightarrow{x}, t\right) = \mathbf{u}\left(x_1, x_2, x_3, t\right) = [u_1\ u_2\ \ldots\ u_p]^T$ is the state vector and

$$\overrightarrow{\mathbf{f}}\left(\mathbf{u}\right) = \sum_{m=1}^{3} \mathcal{A}^{(m)}\left(\overrightarrow{x}\right) \mathbf{u}\hat{x}_m \equiv \overrightarrow{\mathcal{A}}\mathbf{u} \tag{2}$$

is the linear flux space-state vector, where $\hat{x}_m$ is the unit vector in the mth coordinate direction. For simplicity we will assume that the system has been symmetrized and is hyperbolic so that

$$\mathcal{A}^{(m)} = \left(\mathcal{A}^{(m)}\right)^T \quad \text{and} \quad \sum_{m=1}^{3} \alpha_m \mathcal{A}^{(m)} = \mathcal{R}\left(\overrightarrow{\alpha}\right) \Lambda\left(\overrightarrow{\alpha}\right) \mathcal{R}^{-1}\left(\overrightarrow{\alpha}\right) \tag{3}$$

for any $\left\|\overrightarrow{\alpha}\right\|_2^2 = \sum_{m=1}^{3} \alpha_m^2 \neq 0$ and some real diagonal matrix Λ. We will also assume that the matrices $\mathcal{A}^{(m)}$ have bounded derivatives in the sense that

$$\left\|\nabla \cdot \overrightarrow{\mathcal{A}}\right\|_2 < \infty, \tag{4}$$

where $\|\cdot\|_2$ is the matrix 2-norm. Additional constraints on the coefficient matrices need to be added later to ensure that the derivatives of their interpolants converge in the maximum norm. The product rule applied to (1) leads to the nonconservative form of the system

$$\mathbf{u}_t + \left(\nabla \cdot \overrightarrow{\mathcal{A}}\right)\mathbf{u} + \overrightarrow{\mathcal{A}} \cdot \nabla \mathbf{u} = 0. \tag{5}$$

With appropriate initial and characteristic boundary conditions on a bounded domain $\Omega \in \mathbb{R}^3$ the problem is (i) well posed, (ii) conservative, and, under conditions on $\overrightarrow{\mathcal{A}}$, (iii) preserves a constant state. These properties are demonstrated from a weak form of the average of the conservative, (1), and nonconservative, (5), forms of the equation, the so-called "split-form". To write the weak form, we define the $\mathbb{L}^2$ inner product and norm

$$(\mathbf{u}, \mathbf{v}) = \int_\Omega \mathbf{u}^T \mathbf{v}\, dxdydz, \qquad \|\mathbf{u}\| = \sqrt{(\mathbf{u}, \mathbf{u})}. \tag{6}$$

Then for any state vector $\boldsymbol{\phi} \in \mathbb{L}^2(\Omega)$,

$$(\mathbf{u}_t, \boldsymbol{\phi}) + \frac{1}{2}\left(\nabla \cdot \overrightarrow{\mathbf{f}}, \boldsymbol{\phi}\right) + \frac{1}{2}\left\{\left(\left(\nabla \cdot \overrightarrow{\mathcal{A}}\right)\mathbf{u}, \boldsymbol{\phi}\right) + \left(\overrightarrow{\mathcal{A}} \cdot \nabla \mathbf{u}, \boldsymbol{\phi}\right)\right\} = 0. \tag{7}$$

From vector calculus, we have the extended Gauss law,

$$\int_\Omega \mathbf{u}^T \nabla \cdot \overrightarrow{\mathbf{f}}\, dxdydz = \int_{\partial\Omega} \mathbf{u}^T \overrightarrow{\mathbf{f}} \cdot \hat{n} dS - \int_\Omega (\nabla \mathbf{u})^T \cdot \overrightarrow{\mathbf{f}}\, dxdydz, \tag{8}$$

where $\hat{n}$ is the outward unit normal. We write (8) in inner product form as

$$\left(\mathbf{u}, \nabla \cdot \overrightarrow{\mathbf{f}}\right) = \int_{\partial\Omega} \mathbf{u}^T \overrightarrow{\mathbf{f}} \cdot \hat{n} dS - \left(\nabla \mathbf{u}, \overrightarrow{\mathbf{f}}\right). \tag{9}$$

We can apply the extended Gauss law to the inner products in the braces in (7) and use the fact that $\overrightarrow{\mathcal{A}}$ is symmetric to get an equivalent form that separates the boundary and volume contributions

$$(\mathbf{u}_t, \boldsymbol{\phi}) + \int_{\partial\Omega} \overrightarrow{\mathbf{f}} \cdot \hat{n}\boldsymbol{\phi} dS - \frac{1}{2}\left(\overrightarrow{\mathbf{f}}, \nabla \boldsymbol{\phi}\right) + \frac{1}{2}\left\{\left(\left(\nabla \cdot \overrightarrow{\mathcal{A}}\right)\mathbf{u}, \phi\right) - \left(\mathbf{u}, \nabla \cdot \left(\overrightarrow{\mathcal{A}}\boldsymbol{\phi}\right)\right)\right\} = 0. \tag{10}$$

Constant state preservation, conservation and well-posedness are shown with judicious choices of $\mathbf{u}$ and $\boldsymbol{\phi}$. To find under what conditions a constant state is preserved, set $\mathbf{u} = \mathbf{c} =$ constant in (7) to see that

$$\left(\mathbf{u}_t, \phi\right) + \left(\left(\nabla \cdot \overrightarrow{\mathcal{A}}\right) \mathbf{c}, \phi\right) = 0, \tag{11}$$

from which it follows that $\mathbf{u}_t = 0$ if $\nabla \cdot \overrightarrow{\mathcal{A}} = 0$.

Global conservation is shown by selectively choosing each component of the state vector $\boldsymbol{\phi}$ in (10) to be unity and again noting that the coefficient matrices are symmetric to see that the terms in the braces cancel to leave

$$\frac{d}{dt} \int_{\Omega} \mathbf{u} dx dy dz = - \int_{\partial\Omega} \overrightarrow{\mathbf{f}} \cdot \hat{n} dS. \tag{12}$$

To find conditions under which the initial boundary value problem is well-posed, we choose $\boldsymbol{\phi} = \mathbf{u}$ in (7) and note that

$$\left(\nabla \cdot \overrightarrow{\mathbf{f}} + \overrightarrow{\mathcal{A}} \cdot \nabla \mathbf{u}, \mathbf{u}\right) = \int_{\Omega} \nabla \cdot \left(\mathbf{u}^T \overrightarrow{\mathcal{A}} \mathbf{u}\right) d\overrightarrow{x}. \tag{13}$$

Replacing those terms in (7) and multiplying the equation by two gives

$$\frac{d}{dt} \|\mathbf{u}\|^2 + \int_{\Omega} \nabla \cdot \left(\mathbf{u}^T \overrightarrow{\mathcal{A}} \mathbf{u}\right) d\overrightarrow{x} + \left(\left(\nabla \cdot \overrightarrow{\mathcal{A}}\right) \mathbf{u}, \mathbf{u}\right) = 0. \tag{14}$$

Gauss' theorem allows us to replace the second term by a surface integral so

$$\frac{d}{dt} \|\mathbf{u}\|^2 + \int_{\partial\Omega} \mathbf{u}^T \overrightarrow{\mathcal{A}} \cdot \hat{n} \mathbf{u} dS = -\left(\left(\nabla \cdot \overrightarrow{\mathcal{A}}\right) \mathbf{u}, \mathbf{u}\right). \tag{15}$$

We bound the right hand side by

$$-\left(\left(\nabla \cdot \overrightarrow{\mathcal{A}}\right) \mathbf{u}, \mathbf{u}\right) \leqslant \max_{\Omega} \left\|\nabla \cdot \overrightarrow{\mathcal{A}}\right\|_2 \|\mathbf{u}\|^2 \equiv 2\gamma \|\mathbf{u}\|^2 \tag{16}$$

so

$$\frac{d}{dt} \left(e^{-2\gamma t} \|\mathbf{u}\|^2\right) \leq e^{-2\gamma t} \int_{\partial\Omega} \mathbf{u}^T \overrightarrow{\mathcal{A}} \cdot \hat{n} \mathbf{u} dS. \tag{17}$$

Integrating over the time interval $[0, T]$ we write the energy in terms of the initial value and a boundary integral

$$\|\mathbf{u}(T)\|^2 \le e^{2\gamma T} \|\mathbf{u}(0)\|^2 + \int_0^T \int_{\partial\Omega} e^{2\gamma(T-t)} \mathbf{u}^T \overrightarrow{\mathcal{A}} \cdot \hat{n} \mathbf{u} dS dt. \tag{18}$$

To properly pose the problem we must impose appropriate boundary conditions. From (3), we separate the waves traveling to the left and right of the boundary relative to $\hat{n}$ as

$$\overrightarrow{\mathcal{A}} \cdot \hat{n} = \sum_{m=1}^{3} \mathcal{A}^{(m)} \hat{n}_m = \mathcal{R}\Lambda\mathcal{R}^{-1} = \mathcal{P}\Lambda^+\mathcal{R}^{-1} + \mathcal{R}\Lambda^-\mathcal{R}^{-1} \equiv \mathcal{A}^+ + \mathcal{A}^-, \tag{19}$$

where $\Lambda^{\pm} = \Lambda \pm |\Lambda|$ and we have left off the explicit dependence on $\hat{n}$. When we replace the values of $\mathbf{u}$ along the boundary associated with the incoming Λ^- waves with a boundary state, $\mathbf{g}$, the solution can be bounded in terms of the initial and boundary data,

$$\begin{aligned} \|\mathbf{u}(T)\|^2 + \int_0^T \int_{\partial\Omega} \mathbf{u}^T \mathcal{A}^+ \mathbf{u} dS dt &\le e^{2\gamma T} \|\mathbf{u}(0)\|^2 + \int_0^T \int_{\partial\Omega} e^{2\gamma(T-t)} \mathbf{g}^T |\mathcal{A}^-| \mathbf{g} dS dt \\ &\le e^{2\gamma T} \left\{ \|\mathbf{u}(0)\|^2 + \int_0^T \int_{\partial\Omega} \mathbf{g}^T |\mathcal{A}^-| \mathbf{g} dS dt \right\}. \end{aligned} \tag{20}$$

Furthermore, if $\nabla \cdot \overrightarrow{\mathcal{A}} = 0$, $\gamma = 0$ and the energy does not grow in time except for energy introduced at the boundaries,

$$\|\mathbf{u}(T)\|^2 + \int_0^T \int_{\partial\Omega} \mathbf{u}^T \mathcal{A}^+ \mathbf{u} dS dt \le \|\mathbf{u}(0)\|^2 + \int_0^T \int_{\partial\Omega} \mathbf{g}^T |\mathcal{A}^-| \mathbf{g} dS dt. \tag{21}$$

3 A Polynomial Spectral Calculus

To follow the continuous problem analysis as closely as possible, we introduce a discrete calculus that looks and behaves like the continuous one as much as possible. We define the calculus for the reference domain $E = [-1, 1]^3$ with coordinates $\overrightarrow{\xi} = (\xi, \eta, \zeta) = \xi\hat{\xi} + \eta\hat{\eta} + \zeta\hat{\zeta} = \sum_{m=1}^{3} \xi^{(m)}\hat{\xi}^m$. Corresponding forms hold for two dimensional problems.

We represent functions of the reference domain coordinates by polynomials of degree N or less, i.e. as elements of $\mathbb{P}^N(E) \subset \mathbb{L}^2(E)$. A basis for the polynomials on E is the tensor product of the one dimensional Lagrange basis. Using that basis, we write a polynomial, U, in terms of nodal values $U_{ijk} = U\left(\xi_i, \eta_j, \zeta_k\right)$ as an upper case letter, which for three space dimensions is

$$U = \sum_{i,j,k=0}^{N} U_{ijk}\ell_i(\xi)\ell_j(\eta)\ell_k(\zeta), \tag{22}$$

where

$$\ell_l\left(s\right) = \prod_{i=0;i\neq l}^{N} \frac{s - s_i}{s_l - s_i} \tag{23}$$

is the one-dimensional Lagrange interpolating polynomial with the property $\ell_l\left(s_m\right) = \delta_{lm},\ l,m = 0,1,2,\ldots,N$. The points $s_i,\ \ i = 0,1,2,\ldots,N$ are the interpolation points, whose locations are chosen below. We also write the interpolation operator, $\mathbb{I}^N : \mathbb{L}^2 \rightarrow \mathbb{P}^N$, which projects square integrable functions on E onto polynomials, as

$$\mathbb{I}^N\left(u\right) = \sum_{i,j,k=0}^{N} u_{ijk}\ell_i(\xi)\ell_j(\eta)\ell_k(\zeta). \tag{24}$$

The use of the tensor product means that one and two dimensions are special cases of three dimensions, which is why we concentrate on three dimensional geometries here.

Derivatives of polynomials on E evaluated at the nodes can be represented by matrix-vector multiplication. For instance,

$$\left.\frac{\partial U}{\partial \xi}\right|_{nml} = \sum_{i,j,k=0}^{N} U_{ijk}\ell_i'(\xi_n)\ell_j(\eta_m)\ell_k(\zeta_l) = \sum_{i=0}^{N} U_{iml}\ell'_i(\xi_n) \equiv \sum_{i=0}^{N} U_{iml}\mathcal{D}_{ni}, \tag{25}$$

where $\mathcal{D}$ is the derivative matrix. The gradient and divergence of a polynomial in three space dimensions evaluated at a point $\left(\xi_n, \eta_m, \zeta_l\right)$ are therefore

$$\begin{aligned}
\nabla U|_{nml} &= \sum_{i=0}^{N} U_{iml}\mathcal{D}_{ni}\hat{\xi} + \sum_{j=0}^{N} U_{njl}\mathcal{D}_{mj}\hat{\eta} + \sum_{k=0}^{N} U_{nmk}\mathcal{D}_{lk}\hat{\zeta}, \\
\nabla \cdot \vec{F}\Big|_{nml} &= \sum_{i=0}^{N} F^{(1)}_{iml}\mathcal{D}_{ni} + \sum_{j=0}^{N} F^{(2)}_{njl}\mathcal{D}_{mj} + \sum_{k=0}^{N} F^{(3)}_{nmk}\mathcal{D}_{lk}.
\end{aligned} \tag{26}$$

The use of the calculus that we develop depends on the choice that the interpolation nodes, s_i, are the nodes of the Legendre-Gauss-Lobatto (LGL) quadrature. We represent the one dimensional LGL quadrature of a function $g(s)$ using the notation

$$\int_{-1}^{1} gds \approx \sum_{i=0}^{N} g\left(s_i\right)\omega_i \equiv \int_N gds, \tag{27}$$

where the ω_i are the LGL quadrature weights. The quadrature is exact if $g \in \mathbb{P}^{2N-1}$. By tensor product extension, we write three dimensional volume integral approximations as

$$\int_{E,N} gd\xi d\eta d\zeta \equiv \sum_{i,j,k=0}^{N} g_{ijk}\omega_{ijk}, \tag{28}$$

where $\omega_{ijk} = \omega_i\omega_j\omega_k$. Two-dimensional surface integral approximations are

$$\begin{aligned}\int_{\partial E,N} \vec{g}\cdot\hat{n}dS &= \sum_{i,j=0}^{N} \omega_{ij}g^{(1)}\left(\xi,\eta_i,\zeta_j\right)\Big|_{\xi=-1}^{1} + \sum_{i,j=0}^{N} \omega_{ij}g^{(2)}\left(\xi_i,\eta,\zeta_j\right)\Big|_{\eta=-1}^{1}\\ &\quad + \sum_{i,j=0}^{N} \omega_{ij}g^{(3)}\left(\xi_i,\eta_j,\zeta\right)\Big|_{\zeta=-1}^{1}\\ &\equiv \int_N g^{(1)}d\eta d\zeta\Big|_{\xi=-1}^{1} + \int_N g^{(2)}d\xi d\zeta\Big|_{\eta=-1}^{1} + \int_N g^{(3)}d\xi d\eta\Big|_{\zeta=-1}^{1}.\end{aligned} \tag{29}$$

Two space dimensional areas and edge integrals are defined similarly.

We define the discrete inner product of two functions f and g and the discrete norm of f from the quadrature

$$(f,g)_{E,N} = \int_{E,N} fgd\xi d\eta d\zeta \equiv \sum_{i,j,k=0}^{N} f_{ijk}g_{ijk}\omega_{ijk}, \quad \|f\|_{E,N} = \sqrt{(f,f)_{E,N}}. \tag{30}$$

The definition is extended for vector arguments like

$$\vec{\mathbf{f}} = \sum_{m=1}^{3} \mathbf{f}^{(m)}\hat{\xi}^m, \tag{31}$$

for a state vector $\mathbf{f}^{(m)} = [f_1^{(m)}\ f_2^{(m)}\ \ldots\ f_p^{(m)}]^T$ as

$$\left(\vec{\mathbf{f}},\vec{\mathbf{g}}\right)_N = \int_{E,N} \sum_{m=1}^{3} \left(\mathbf{f}^{(m)}\right)^T \mathbf{g}^{(m)}d\xi d\eta d\zeta = \sum_{i,j,k=0}^{N} \omega_{ijk} \sum_{m=1}^{3} \left(\mathbf{f}_{ijk}^{(m)}\right)^T \mathbf{g}_{ijk}^{(m)}, \tag{32}$$

and similarly for other arguments.

The Lagrange basis functions are orthogonal with respect to the discrete inner product defined in (30) [7]. In one space dimension, for instance, $\left(\ell_i, \ell_j\right)_{E,N} = \omega_j \delta_{ij}$. Also, from the definitions of the interpolation operator and the discrete inner product,

$$(f, g)_{E,N} = \left(\mathbb{I}^N(f), \mathbb{I}^N(g)\right)_{E,N}. \tag{33}$$

Finally, the discrete norm is equivalent to the continuous norm [6] in that for $U \in \mathbb{P}^N$,

$$\|U\|_E \leqslant \|U\|_{E,N} \leqslant C\|U\|_E, \tag{34}$$

where C is a constant.

The crucial property for the analysis of the discrete approximation is the *summation by parts* (SBP) property satisfied by the LGL quadrature. Let $U, V \in \mathbb{P}^N$. Then the LGL quadrature, which is exact for polynomials of degree $2N-1$, satisfies

$$\int_N UV'dx = UV|_{-1}^{1} - \int_N U'Vdx \quad (\textit{Summation By Parts}). \tag{35}$$

The result extends to all space dimensions [15] with

$$\begin{aligned}
\left(U_\xi, V\right)_N &= \int_N UVd\eta d\zeta|_{\xi=-1}^{1} - \left(U, V_\xi\right)_N \\
\left(U_\eta, V\right)_N &= \int_N UVd\xi d\zeta|_{\eta=-1}^{1} - \left(U, V_\eta\right)_N \\
\left(U_\zeta, V\right)_N &= \int_N UVd\xi d\eta|_{\zeta=-1}^{1} - \left(U, V_\zeta\right)_N.
\end{aligned} \tag{36}$$

We can use (35) and (36) to formulate a discrete integral calculus. If we replace U in (36) by the components of a vector $\vec{F}$, and sum, we get the *Discrete Extended Gauss Law (DXGL)* originally derived in [15]: For any vector of polynomials $\vec{F} \in \mathbb{P}^N$ and any polynomial $V \in \mathbb{P}^N$,

$$\left(\nabla \cdot \vec{F}, V\right)_N = \int_{\partial E,N} \vec{F} \cdot \hat{n} V dS - \left(\vec{F}, \nabla V\right)_N \quad (\textit{Discrete Extended Gauss Law}), \tag{37}$$

where $\hat{n}$ is the unit outward normal at the faces of E. Carrying this further, if we set $V = 1$ we get the *Discrete Gauss Law* (DGL)

$$\left(\nabla \cdot \vec{F}, 1\right)_N = \int_{E,N} \nabla \cdot \vec{F} d\xi d\eta d\zeta = \int_{\partial E,N} \vec{F} \cdot \hat{n} dS \quad (\textit{Discrete Gauss Law}). \tag{38}$$

The DGL is exact for polynomial arguments of degree $2N - 1$. By using the appropriate definitions for the inner products, both discrete Gauss laws extend to hold for state vectors $\overleftrightarrow{\mathbf{F}}$ and $\mathbf{V}$.

Next, we see that if we replace the vector flux $\vec{F}$ in (37) with $\nabla\Phi \in \mathbb{P}^N$, then we get the discrete version of Green's first identity,

$$\left(\nabla^2\Phi, V\right)_N + (\nabla\Phi, \nabla V)_N = \int_{\partial E,N} \nabla\Phi \cdot \hat{n} V dS \quad (\textit{Discrete Green's First Identity}). \tag{39}$$

Swapping the variables Φ and V and subtracting from the original gives Green's second identity

$$\left(\nabla^2\Phi, V\right)_N - \left(\nabla^2 V, \Phi\right)_N$$
$$= \int_{\partial E,N} \left(\nabla\Phi \cdot \hat{n} V - \nabla V \cdot \hat{n}\Phi\right) dS \quad (\textit{Discrete Green's Second Identity}) \tag{40}$$

The discrete Green's identities would be useful to prove stability of continuous Galerkin spectral element methods of second order problems.

Other identities that do not involve quadratic products of polynomial arguments hold discretely through exactness of the LGL quadrature. For instance,

$$\int_{E,N} \nabla V d\xi d\eta d\zeta = \int_{\partial E,N} V \hat{n} dS \tag{41}$$

and

$$\int_{E,N} \nabla \times \vec{F} d\xi d\eta d\zeta = \int_{\partial E,N} \hat{n} \times \vec{F} dS. \tag{42}$$

What we see, then, is that the well-known integral identities hold due to either integration or summation by parts.

Whereas integration rules hold discretely, product differentiation rules do not usually hold because differentiation and interpolation do not always commute. For instance, the product rule does not generally hold. That is, for polynomials U, V,

$$\nabla\left(\mathbb{I}^N(UV)\right) \neq \mathbb{I}^N(U\nabla V) + \mathbb{I}^N(V\nabla U) \tag{43}$$

unless the product $UV \in \mathbb{P}^N$. [7].

4 Discontinuous Galerkin Spectral Element Approximations

We now use the polynomial calculus introduced in section 3 to formulate and analyze discontinuous Galerkin spectral element approximations in three space dimensions. The steps to derive two dimensional approximations are identical. The domain Ω is subdivided into N_{el} nonoverlapping hexahedral elements, $e^r, r = 1, 2, \ldots, N_{el}$. We assume here that the subdivision is conforming. Each element is mapped from the reference element E by a transformation $\vec{x} = \vec{X}\left(\vec{\xi}\right)$. From the transformation, we define the three covariant basis vectors

$$\vec{a}_i = \frac{\partial \vec{X}}{\partial \xi^i} \quad i = 1, 2, 3, \tag{44}$$

and (volume weighted) contravariant vectors, formally written as

$$\mathcal{J}\vec{a}^i = \vec{a}_j \times \vec{a}_j, \quad (i, j, k) \text{ cyclic}, \tag{45}$$

where

$$\mathcal{J} = \vec{a}_1 \cdot \left(\vec{a}_2 \times \vec{a}_3\right) \tag{46}$$

is the Jacobian of the transformation.

Under the mapping, the divergence of a spatial vector flux can be written compactly in terms of the reference space variables as

$$\nabla \cdot \vec{\mathbf{f}} = \frac{1}{\mathcal{J}} \sum_{i=1}^{3} \frac{\partial}{\partial \xi^i} \left(\mathcal{J}\vec{a}^i \cdot \vec{\mathbf{f}}\right) = \frac{1}{\mathcal{J}} \sum_{i=1}^{3} \frac{\partial \tilde{\mathbf{f}}^i}{\partial \xi^i} = \frac{1}{\mathcal{J}} \nabla_\xi \cdot \tilde{\mathbf{f}}. \tag{47}$$

The vector $\tilde{\mathbf{f}}$ is the volume weighted contravariant flux whose components are $\tilde{\mathbf{f}}^i = \mathcal{J}\vec{a}^i \cdot \vec{\mathbf{f}}$.

The conservation law is then represented on the reference domain by another conservation law

$$\mathcal{J}\mathbf{u}_t + \nabla_\xi \cdot \left(\tilde{\mathcal{A}}\mathbf{u}\right) = 0, \tag{48}$$

where we have defined the (volume weighted) contravariant coefficient matrices

$$\mathcal{A}^i = \mathcal{J}\vec{a}^i \cdot \vec{\mathcal{A}} \tag{49}$$

and

$$\tilde{\mathcal{A}} = \sum_{i=1}^{3} \mathcal{A}^i \hat{\xi}^i. \tag{50}$$

We can also construct the nonconservative form of the system on the reference domain using the chain rule,

$$\mathcal{J}\mathbf{u}_t + \left(\nabla_\xi \cdot \tilde{\mathcal{A}}\right)\mathbf{u} + \tilde{\mathcal{A}} \cdot \nabla_\xi \mathbf{u} = 0. \tag{51}$$

We construct weak forms of the conservative and nonconservative equations by taking the inner product of the equations with a test function $\boldsymbol{\phi} \in \mathbb{L}^2(E)$ and applying extended Gauss Law to the space derivative terms,

$$(\mathcal{J}\mathbf{u}_t, \boldsymbol{\phi})_E + \int_E \tilde{\mathbf{f}} \cdot \hat{n}^T \boldsymbol{\phi} dS - \left(\tilde{\mathbf{f}}, \nabla_\xi \boldsymbol{\phi}\right)_E = 0 \tag{52}$$

and

$$(\mathcal{J}\mathbf{u}_t, \boldsymbol{\phi}) + \int_E \tilde{\mathbf{f}} \cdot \hat{n}^T \boldsymbol{\phi} dS - \left(\mathbf{u}, \nabla_\xi \cdot \tilde{\mathbf{f}}(\boldsymbol{\phi})\right)_E + \left(\left(\nabla_\xi \cdot \tilde{\mathcal{A}}\right)\mathbf{u}, \boldsymbol{\phi}\right)_E = 0. \tag{53}$$

When we average the two equations (52) and (53) we get the split weak form

$$\begin{aligned}(\mathcal{J}\mathbf{u}_t, \boldsymbol{\phi}) \; &-\frac{1}{2}\left\{\left(\tilde{\mathbf{f}}(\mathbf{u}), \nabla_\xi \boldsymbol{\phi}\right)_E + \left(\mathbf{u}, \nabla_\xi \cdot \tilde{\mathbf{f}}(\boldsymbol{\phi})\right)_E - \left(\left(\nabla_\xi \cdot \tilde{\mathcal{A}}\right)\mathbf{u}, \boldsymbol{\phi}\right)_E\right\} \\ &+ \int_{\partial E} \left(\tilde{\mathbf{f}} \cdot \hat{n}\right)^T \boldsymbol{\phi} dS = 0.\end{aligned} \tag{54}$$

4.1 The DGSEM

The original DG spectral element method introduced by Black [4] starts from the conservative weak form (52). We use the calculus now to show that it is stable if the coefficient matrices $\tilde{\mathcal{A}}$ are constant. If, in addition, characteristic boundary conditions are used at physical boundaries, the approximation is optimally stable in the sense that the global energy discretely matches (21).

To construct the approximation, one approximates the solutions, fluxes, coefficient matrices and Jacobian with polynomial interpolants on element $e^r \to E$ by

$$\begin{aligned} &\mathbf{u} \approx \mathbf{U}^r \in \mathbb{P}^N \\ &\tilde{\mathbf{f}} \approx \tilde{\mathbf{F}}^r(\mathbf{U}) = \mathbb{I}^N\left(\mathbb{I}^N\left(\tilde{\mathcal{A}}\right)\mathbf{U}\right) = \sum_{i,j,k=0}^{N} \tilde{\mathcal{A}}_{ijk}\mathbf{U}_{ijk}\ell_i(\xi)\,\ell_j(\eta)\,\ell_k(\zeta) \\ &\tilde{\mathcal{A}} \approx \mathbb{I}^N\left(\tilde{\mathcal{A}}\right) \\ &\mathcal{J}^r \approx J^r = \mathbb{I}^N\left(\mathcal{J}^r\right). \end{aligned} \tag{55}$$

From this point, we leave off the superscripts r and subscripts ξ on ∇_ξ unless necessary.

To continue the construction, one replaces the continuous inner products by the discrete inner products, here being Gauss-Lobatto quadratures. The normal boundary flux is replaced by a consistent numerical flux, $\tilde{\mathbf{f}} \leftarrow \tilde{\mathbf{F}}^*\left(\mathbf{U}^L, \mathbf{U}^R; \hat{n}\right)$ where $\mathbf{U}^{L,R}$ are the left and right states at the element boundary, measured with respect to the outward normal, $\hat{n}$. The numerical flux ensures continuity of the normal flux at element faces. Finally, $\boldsymbol{\phi}$ is restricted to elements of $\mathbb{P}^N$. The result of the approximations is the formal statement of the method

$$[DGSEM] \quad (J\mathbf{U}_t, \boldsymbol{\phi})_N + \int_{\partial E,N} \tilde{\mathbf{F}}^{*,T}\boldsymbol{\phi}\, dS - \left(\tilde{\mathbf{F}}(\mathbf{U}), \nabla\boldsymbol{\phi}\right)_N = 0. \tag{56}$$

Details for going from the formal statement to the form to implement can be found in [14].

Alternate, yet algebraically equivalent forms of the DGSEM can be derived by applying the DXGL. For instance, if we apply the DXGL to the last inner product in (56) we get the algebraically equivalent form

$$(J\mathbf{U}_t, \boldsymbol{\phi})_N + \int_{\partial E,N} \left\{\tilde{\mathbf{F}}^* - \tilde{\mathbf{F}}\cdot\hat{n}\right\}^T \boldsymbol{\phi}\, dS + \left(\nabla\cdot\tilde{\mathbf{F}}(\mathbf{U}), \boldsymbol{\phi}\right)_N = 0. \tag{57}$$

If the contravariant coefficient matrices are constant, implying that the original problem is constant coefficient and the elements are rectangular in shape, then the DGSEM approximation is strongly stable. To show stability, we set $\boldsymbol{\phi} = \mathbf{U}$ in (57) and define the volume weighted norm

$$\|\mathbf{U}\|_{J,N}^2 \equiv (J\mathbf{U}, \mathbf{U})_N. \tag{58}$$

Then

$$\frac{1}{2}\frac{d}{dt}\|\mathbf{U}\|_{J,N}^2 + \int_{\partial E,N} \left\{\tilde{\mathbf{F}}^* - \tilde{\mathbf{F}}\cdot\hat{n}\right\}^T \mathbf{U}\, dS + \left(\nabla\cdot\tilde{\mathbf{F}}(\mathbf{U}), \mathbf{U}\right)_N = 0. \tag{59}$$

With constant coefficients, the volume term in (59) can be converted to a surface quadrature. The coefficient matrices being constant and symmetric implies that

$$\begin{aligned}\left(\nabla\cdot\tilde{\mathbf{F}}(\mathbf{U}),\mathbf{U}\right)_N &= \underbrace{\left(\nabla\cdot\mathbb{I}^N\left(\tilde{\mathcal{A}}\mathbf{U}\right),\mathbf{U}\right)_N = \left(\tilde{\mathcal{A}}\cdot\nabla\mathbf{U},\mathbf{U}\right)_N}_{*} = \left(\nabla\mathbf{U},\tilde{\mathcal{A}}\mathbf{U}\right)_N \\ &= \left(\nabla\mathbf{U},\tilde{\mathbf{F}}(\mathbf{U})\right)_N.\end{aligned} \tag{60}$$

The key step is the second marked with the "*", where the product rule applies because $\tilde{\mathcal{A}}\mathbf{U}\in\mathbb{P}^N$ when $\tilde{\mathcal{A}}$ is constant. We then substitute the equivalence (60) into the DXGL to see that

$$\left(\nabla\cdot\tilde{\mathbf{F}}(\mathbf{U}),\mathbf{U}\right)_N = \frac{1}{2}\int_{\partial E,N}\left(\tilde{\mathbf{F}}\cdot\hat{n}\right)^T\mathbf{U}dS. \tag{61}$$

Therefore, the local energy changes according to

$$\frac{1}{2}\frac{d}{dt}\|\mathbf{U}\|^2_{J,N} + \int_{\partial E,N}\left\{\tilde{\mathbf{F}}^* - \frac{1}{2}\tilde{\mathbf{F}}\cdot\hat{n}\right\}^T\mathbf{U}dS = 0, \tag{62}$$

and stability depends solely on what happens on the element faces.

The change in the total energy is found by summing over all the elements. Although the numerical flux is continuous at element interfaces, the solution and flux are discontinuous. If we define the jump in a quantity with the usual notation $[\![V]\!] = V^R - V^L$, then

$$\begin{aligned}\frac{d}{dt}\left(\sum_{r=1}^{N_{el}}\|\mathbf{U}^r\|^2_{J,N}\right) \leqslant -2\Bigg\{&\sum_{\substack{Boundary\\Faces}}\int_{\partial E,N}\left(\mathbf{F}^* - \frac{1}{2}\mathbf{F}\cdot\hat{n}\right)^T\mathbf{U}dS \\ &- \sum_{\substack{Interior\\Faces}}\int_{\partial E,N}\left(\mathbf{F}^{*,T}[\![\mathbf{U}]\!] - \frac{1}{2}\left[\!\left[(\mathbf{F}\cdot\hat{n})^T\mathbf{U}\right]\!\right]\right)dS\Bigg\}.\end{aligned} \tag{63}$$

Stability is determined, therefore, only by the influence of the jumps at the element boundaries and the physical boundary approximations through the numerical flux. For linear problems, it is natural to choose an upwinded or central flux,

$$\tilde{\mathbf{F}}^*\left(\mathbf{U}^L,\mathbf{U}^R;\hat{n}\right) = \frac{1}{2}\left\{\tilde{\mathbf{F}}\left(\mathbf{U}^L\right)\cdot\hat{n} + \tilde{\mathbf{F}}\left(\mathbf{U}^R\right)\cdot\hat{n}\right\} - \sigma\frac{\left|\tilde{\mathcal{A}}\cdot\hat{n}\right|}{2}\left\{\mathbf{U}^R - \mathbf{U}^L\right\}, \tag{64}$$

where $\sigma = 0$ is the central flux and $\sigma = 1$ is the fully upwind flux. With this flux [16],

$$\tilde{\mathbf{F}}^{*,T}[\![\mathbf{U}]\!] - \frac{1}{2}\left[\!\!\left[\left(\tilde{\mathbf{F}}\cdot\hat{n}\right)^T\mathbf{U}\right]\!\!\right] = -\frac{\sigma}{2}[\![\mathbf{U}]\!]^T\left|\tilde{\mathcal{A}}\cdot\hat{n}\right|[\![\mathbf{U}]\!] \leq 0, \tag{65}$$

so that the interior face terms in (63) are dissipative. To match the PDE energy bound, (21), the fully upwind flux must be used at the physical boundaries. With exterior values $\mathbf{g}$ set along incoming characteristics [18] and when $\sigma = 1$,

$$\left(\mathbf{F}^* - \frac{1}{2}\tilde{\mathbf{F}}\cdot\hat{n}\right)^T\mathbf{U} = \frac{1}{2}\mathbf{U}^T\mathcal{A}^+\mathbf{U} + \frac{1}{2}\left\|\sqrt{|\mathcal{A}^-|}\mathbf{U} - \sqrt{|\mathcal{A}^-|}\mathbf{g}\right\|_2^2 - \frac{1}{2}\mathbf{g}^T\left|\mathcal{A}^-\right|\mathbf{g}. \tag{66}$$

If we define the total energy by

$$\|\mathbf{U}\|_{J,N}^2 = \sum_{r=1}^{K}\|\mathbf{U}^r\|_{J,N}^2, \tag{67}$$

and integrate (63) in time, the total energy satisfies (c.f. (20))

$$\begin{aligned}\|\mathbf{U}(T)\|_{J,N}^2 + \sum_{\substack{Boundary\\Faces}}\int_0^T\int_{\partial E,N}\mathbf{U}^T\mathcal{A}^+\mathbf{U}dSdt \leqslant \|\mathbf{U}(0)\|_{J,N}^2 \\ + \sum_{\substack{Boundary\\Faces}}\int_0^T\int_{\partial E,N}\mathbf{g}^T\left|\mathcal{A}^-\right|\mathbf{g}dSdt.\end{aligned} \tag{68}$$

Finally, if the interpolant of the Jacobian is bounded from below, $J > 0$, then for some positive constants c and C [16],

$$c\,\|\mathbf{U}\|_{L^2(\Omega)}^2 \leqslant \|\mathbf{U}\|_{J,N}^2 \leqslant C\,\|\mathbf{U}\|_{L^2(\Omega)}^2, \tag{69}$$

which says that, like the continuous solution, the energy approximate solution is bounded by the data in the continuous norm over the entire domain.

4.2 Stabilization by Split Form

If the contravariant coefficient matrices are not constant, then the key step in (60) does not hold because interpolation and differentiation do not commute. We show now that stability hangs on whether or not the dissipation introduced by the numerical flux at the element interfaces and by the characteristic boundary conditions is sufficient to counterbalance the aliasing errors associated with the volume term that remains. That balance shows why the approximation [DGSEM] can be, but does not have to be, stable for variable coefficient problems or curved elements.

The use of the polynomial calculus allows us to quickly and compactly construct four algebraically equivalent representations of a split form approximation [16] that is strongly stable, constant state preserving and globally conservative for non-constant coefficient problems where the coefficient variation is due to inherent variability in the PDE and/or due to variability introduced by the coefficient mappings from curved elements to the reference element. It also allows us to simplify the analysis done, for example, in [17].

The result of applying the approximations (55) and LGL quadrature to (54) is the first split form of the DGSEM used in [16]. In accordance to common terminology, this is the "weak" form

$$[W] \quad (J\mathbf{U}_t,\boldsymbol{\phi})_N - \frac{1}{2}\left\{\left(\tilde{\mathbf{F}}(\mathbf{U}),\nabla\boldsymbol{\phi}\right)_N + \left(\mathbf{U},\nabla\cdot\tilde{\mathbf{F}}(\boldsymbol{\phi})\right)_N - \left(\nabla\cdot\left(\mathbb{I}^N\left(\tilde{A}\right)\right)\mathbf{U},\boldsymbol{\phi}\right)_N\right\} + \int_{\partial E,N}\tilde{\mathbf{F}}^{*,T}\boldsymbol{\phi}\,dS = 0. \tag{70}$$

We get alternative, yet algebraically equivalent forms by applying the DXGL (37) to selected terms in (70). When we apply the DXGL to the first two inner products in the braces and use the fact that the coefficient matrices are symmetric we get the "strong" form

$$[S] \quad (J\mathbf{U}_t,\boldsymbol{\phi})_N + \frac{1}{2}\left\{\left(\nabla\cdot\tilde{\mathbf{F}}(\mathbf{U}),\boldsymbol{\phi}\right)_N + \left(\mathbb{I}^N\left(\tilde{A}\right)\cdot\nabla\mathbf{U},\boldsymbol{\phi}\right)_N + \left(\nabla\cdot\left(\mathbb{I}^N\left(\tilde{A}\right)\right)\mathbf{U},\boldsymbol{\phi}\right)_N\right\} + \int_{\partial E,N}\left\{\tilde{\mathbf{F}}^* - \tilde{\mathbf{F}}\cdot\hat{n}\right\}^T\boldsymbol{\phi}\,dS = 0. \tag{71}$$

If we rearrange the terms in [S] to "strong+correction" form

$$[SC] \quad (J\mathbf{U}_t,\boldsymbol{\phi})_N + \int_{\partial E,N}\left\{\tilde{\mathbf{F}}^* - \tilde{\mathbf{F}}\cdot\hat{n}\right\}^T\boldsymbol{\phi}\,dS + \left(\nabla\cdot\tilde{\mathbf{F}}(\mathbf{U}),\boldsymbol{\phi}\right)_N + \frac{1}{2}\left(\left\{\mathbb{I}^N\left(\tilde{A}\right)\cdot\nabla\mathbf{U} + \nabla\cdot\left(\mathbb{I}^N\left(\tilde{A}\right)\right)\mathbf{U} - \nabla\cdot\tilde{\mathbf{F}}(\mathbf{U})\right\},\boldsymbol{\phi}\right)_N = 0, \tag{72}$$

we see that the split form approximation is the strong form of the original DGSEM (57) plus a correction term in the amount by which the product rule (43) does not hold. When the product rule does hold, such as when the contravariant coefficient matrices are constant, the correction term vanishes and we are back to the original scheme of Black, [DGSEM].

We get a fourth algebraically equivalent "directly stable" form by applying the DXGL to only the first inner product in the braces of the weak form [*W*],

$$[DS]\quad (J\mathbf{U}_t,\boldsymbol{\phi})_N+\frac{1}{2}\left\{\left(\nabla\cdot\tilde{\mathbf{F}}(\mathbf{U}),\boldsymbol{\phi}\right)_N-\left(\mathbf{U},\nabla\cdot\tilde{\mathbf{F}}(\boldsymbol{\phi})\right)_N+\left(\nabla\cdot\mathbb{I}^N\left(\tilde{A}\right)\mathbf{U},\boldsymbol{\phi}\right)_N\right\}$$
$$+\int_{\partial E,N}\left\{\tilde{\mathbf{F}}^*-\frac{1}{2}\tilde{\mathbf{F}}\cdot\hat{n}\right\}^T\boldsymbol{\phi}dS=0. \tag{73}$$

Any of the four equivalent forms [*W*] $\Leftrightarrow$ [*S*] $\Leftrightarrow$ [*SC*] $\Leftrightarrow$ [*DS*] can be used as is convenient for computation or theory. For instance, to show conservation, choose the form [*W*] and selectively set each component of $\boldsymbol{\phi}$ to one. Then the first inner product in the braces vanishes and the second and third cancel leaving

$$\int_{E,N} J\mathbf{U}_t d\boldsymbol{\xi}=-\int_{\partial E,N}\tilde{\mathbf{F}}^* dS. \tag{74}$$

Summing over all elements, the interior face contributions cancel leaving the global conservation statement

$$\frac{d}{dt}\sum_{r=1}^{N_{el}}\int_{E,N} J^r\mathbf{U}^r d\boldsymbol{\xi}=-\sum_{\substack{Boundary\\Faces}}\int_{\partial E,N}\tilde{\mathbf{F}}^{*,r} dS. \tag{75}$$

To find conditions under which the approximation is constant state preserving, use the form [*S*] with $\mathbf{U}=\mathbf{c}=$ const in all elements. The first and third inner products in the braces vanish provided that $\nabla\cdot\left(\mathbb{I}^N\left(\tilde{A}\right)\right)=0$, and the second is explicitly zero. Consistency of the numerical flux implies that $\tilde{\mathbf{F}}^*\left(\mathbf{c},\mathbf{c};\hat{n}\right)=\tilde{\mathbf{F}}\cdot\hat{n}$. Therefore, $(J\mathbf{U}_t,\boldsymbol{\phi})_N=0$ for all $\boldsymbol{\phi}\in\mathbb{P}^N$, which implies that at each node *nml* in each element *r*, $d\mathbf{U}^r_{nml}/dt=0$.

Finally, the split form approximation is optimally stable in the sense that with the numerical flux (64), the norm of the approximate solution satisfies an energy statement like (20). We show stability using [*DS*] and $\boldsymbol{\phi}=\mathbf{U}$. With the substitution, the volume terms represented by the first two inner products in the braces immediately cancel. The third inner product in the braces can be bounded

$$\left(\nabla\cdot\left(\mathbb{I}^N\left(\tilde{\mathcal{A}}\right)\right)\mathbf{U},\mathbf{U}\right)_N \leqslant \max_E \left\| \frac{\nabla\cdot\left(\mathbb{I}^N\left(\tilde{\mathcal{A}}\right)\right)}{J} \right\|_2 (J\mathbf{U},\mathbf{U})_N \equiv 2\hat{\gamma}(J\mathbf{U},\mathbf{U})_N, \tag{76}$$

and under assumptions on the smoothness of $\overrightarrow{\mathcal{A}}$ [22] and positivity of the Jacobian [16] the coefficient $\hat{\gamma}$ will converge spectrally to γ. (If the divergence of the interpolant vanishes, then $\hat{\gamma} = 0$.) With the bound on the divergence of the coefficient matrices,

$$\frac{1}{2}\frac{d}{dt}\left\|\mathbf{U}^r\right\|_{J,N}^2 \leqslant -\int_{\partial E,N}\left\{\tilde{\mathbf{F}}^* - \frac{1}{2}\tilde{\mathbf{F}}^r\cdot\hat{n}\right\}^T \mathbf{U}^r dS + \frac{1}{2}2\hat{\gamma}^r\left\|\mathbf{U}^r\right\|_{J,N}^2. \tag{77}$$

The change in the total energy is again found by summing over all the elements. If we introduce the integrating factor $\hat{\gamma} = \max_r \hat{\gamma}^r$,

$$\begin{aligned}\frac{d}{dt}\left(e^{-2\hat{\gamma}t}\sum_{r=1}^{N_{el}}\left\|\mathbf{U}^r\right\|_{J,N}^2\right) \leqslant -2e^{-2\hat{\gamma}t}\Bigg\{&\sum_{\substack{Boundary\\Faces}}\int_{\partial E,N}\left(\tilde{\mathbf{F}}^* - \frac{1}{2}\mathbf{F}\cdot\hat{n}\right)^T\mathbf{U}dS\\ &-\sum_{\substack{Interior\\Faces}}\int_{\partial E,N}\left(\tilde{\mathbf{F}}^{*,T}[\![\mathbf{U}]\!] - \frac{1}{2}\left[\!\left[(\mathbf{F}\cdot\hat{n})^T\mathbf{U}\right]\!\right]\right)dS\Bigg\}.\end{aligned} \tag{78}$$

The interface and boundary terms on the right hand side of (78) are identical to what appeared in the original DGSEM, (63). Therefore, the total energy satisfies

$$\begin{aligned}&\left\|\mathbf{U}(T)\right\|_{J,N}^2 + \sum_{\substack{Boundary\\Faces}}\int_0^T\int_{\partial E,N}\mathbf{U}^T\mathcal{A}^+\mathbf{U}dSdt\\ &\qquad\leqslant e^{2\hat{\gamma}T}\left\|\mathbf{U}(0)\right\|_{J,N}^2 + \sum_{\substack{Boundary\\Faces}}\int_0^T\int_{\partial E,N}e^{-2\hat{\gamma}(T-t)}\mathbf{g}^T\left|\mathcal{A}^-\right|\mathbf{g}dSdt.\end{aligned} \tag{79}$$

As with the continuous solution, (21), if $\nabla\cdot\mathbb{I}^N\left(\tilde{\mathcal{A}}\right) = 0$, $\hat{\gamma} = 0$, and

$$\left\|\mathbf{U}(T)\right\|_{J,N}^2 + \sum_{\substack{Boundary\\Faces}}\int_0^T\int_{\partial E,N}\mathbf{U}^T\mathcal{A}^+\mathbf{U}dSdt \leqslant \left\|\mathbf{U}(0)\right\|_{J,N}^2 + \sum_{\substack{Boundary\\Faces}}\int_0^T\int_{\partial E,N}\mathbf{g}^T\left|\mathcal{A}^-\right|\mathbf{g}dSdt. \tag{80}$$

Applying the norm equivalence (69), we see that the split form approximation is strongly stable for variable coefficient problems and/or curved elements.

Finally, we can use the stability analysis of the split form approximation to write the conditions needed for the original DGSEM to be stable when the coefficients are variable and/or the elements are curved. For simplicity, let us suppose that $\nabla \cdot \overrightarrow{\mathcal{A}} = 0$ and external boundary states $\mathbf{g} = 0$ so that the global energy should not increase and instability is not masked by natural growth. Let us also assume that $\nabla \cdot \mathbb{I}^N \left(\tilde{\mathcal{A}}\right) = 0$ so that $\hat{\gamma}$ also vanishes. Then by (79), the energy of the split form approximation does not grow. With $[DS] \Leftrightarrow [SC]$ and $[DGSEM] \Leftrightarrow [SC] - [C]$, where $[C]$ is the correction term

$$\frac{1}{2}\left(\left\{\mathbb{I}^N\left(\tilde{\mathcal{A}}\right)\cdot\nabla\mathbf{U} + \nabla\cdot\left(\mathbb{I}^N\left(\tilde{\mathcal{A}}\right)\right)\mathbf{U} - \nabla\cdot\tilde{\mathbf{F}}(\mathbf{U})\right\}, \boldsymbol{\phi}\right)_N, \tag{81}$$

the elemental energy for the DGSEM satisfies

$$\frac{1}{2}\frac{d}{dt}\left\|\mathbf{U}\right\|_{J,N}^2 \leq -\int_{\partial E,N}\left\{\tilde{\mathbf{F}}^* - \frac{1}{2}\tilde{\mathbf{F}}\cdot\hat{n}\right\}^T \mathbf{U}dS + \frac{1}{2}\left|\left(\left\{\mathbb{I}^N\left(\tilde{\mathcal{A}}\right)\cdot\nabla\mathbf{U} - \nabla\cdot\tilde{\mathbf{F}}(\mathbf{U})\right\}, \mathbf{U}\right)_N\right|. \tag{82}$$

The term in the braces of the volume term on the right of (82) is non-zero unless the product rule holds. Therefore, for the DGSEM to be stable when the coefficients are variable, the surface terms (including the dissipation arising from the physical boundaries seen in (66)) must be sufficiently large to counteract any destabilizing influence of (growth from) the volume term, which might require trying more dissipative numerical fluxes than the characteristic upwind flux. Practice has shown that at least at low order one can often find numerical fluxes for which the influence of the surface terms is sufficiently dissipative. But (82) shows that the approximation can be unstable if the aliasing growth contribution is larger than the dissipation contribution from the element faces.

5 Summary

In this paper, we described a discrete integral spectral calculus for polynomial spectral methods using Legendre-Gauss-Lobatto quadrature. This calculus allowed us to write and analyze discontinuous Galerkin spectral element approximations in a compact notation consistent with the continuous version. In particular, it is possible to easily derive four algebraically equivalent forms of a split form approximation for linear hyperbolic systems. These four equivalent forms can then be used to show global conservation, constant state preservation (when applicable) and, most importantly, strong stability of the split form approximation for variable coefficient problems on curved elements.

References

1. C. Altmann, A. Beck, A. Birkefeld, F. Hindenlang, M. Staudenmaier, G. Gassner, C.-D. Munz, *Discontinuous Galerkin for High Performance Computational Fluid Dynamics (HPCDG)*. (Springer, Berlin, 2012), pp. 277–288
2. A. Beck, G. Gassner, C.-D. Munz, High order and underresolution, in *Recent Developments in the Numerics of Nonlinear Hyperbolic Conservation Laws*, ed. by R. Ansorge, H. Bijl, A. Meister, T. Sonar. Notes on Numerical Fluid Mechanics and Multidisciplinary Design, vol. 120 (Springer, Berlin, 2013), pp. 41–55
3. A.D. Beck, T. Bolemann, D. Flad, H. Frank, G.J. Gassner, F. Hindenlang, C.-D. Munz, High-order discontinuous galerkin spectral element methods for transitional and turbulent flow simulations. Int. J. Numer. Methods Fluids **76**(8), 522–548 (2014)
4. K. Black, A conservative spectral element method for the approximation of compressible fluid flow. Kybernetika **35**(1), 133–146 (1999)
5. K. Black, Spectral element approximation of convection-diffusion type problems. Appl. Numer. Math. **33**(1–4), 373–379 (2000)
6. C. Canuto, A. Quarteroni, Approximation results for orthogonal polynomials in Sobolev spaces. Math. Comput. **38**(157), 67–86 (1982)
7. C. Canuto, M.Y. Hussaini, A. Quarteroni, T.A. Zang, *Spectral Methods: Fundamentals in Single Domains* (Springer, Berlin, 2006)
8. H.M. Frank, C.-D. Munz, Direct aeroacoustic simulation of acoustic feedback phenomena on a side-view mirror. J. Sound Vib. **371**, 132–149 (2016)
9. G. Gassner, A skew-symmetric discontinuous Galerkin spectral element discretization and its relation to SBP-SAT finite difference methods. SIAM J. Sci. Comput. **35**(3), A1233–A1253 (2013)
10. G. Gassner, D.A. Kopriva, A comparison of the dispersion and dissipation errors of Gauss and Gauss–Lobatto discontinuous Galerkin spectral element methods. SIAM J. Sci. Comput. **33**, 2560–2579 (2011)
11. G.J. Gassner, A.D. Beck, On the accuracy of high-order discretizations for underresolved turbulence simulations. Theor. Comput. Fluid Dyn. **27**(3–4), 221–237 (2013)
12. G.J. Gassner, A.R. Winters, D.A. Kopriva, Split form nodal discontinuous Galerkin schemes with summation-by-parts property for the compressible Euler equations. J. Comput. Phys. **327**, 39–66 (2016)
13. G.J. Gassner, A.R. Winters, D.A. Kopriva, A well balanced and entropy conservative discontinuous Galerkin spectral element method for the shallow water equations. Appl. Math. Comput. **272**(Part 2), 291–308 (2016)
14. D.A. Kopriva, *Implementing Spectral Methods for Partial Differential Equations*, Scientific Computation (Springer, Berlin, 2009)
15. D.A. Kopriva, G. Gassner, On the quadrature and weak form choices in collocation type discontinuous Galerkin spectral element methods. J. Sci. Comput. **44**(2), 136–155 (2010)
16. D.A. Kopriva, G. Gassner, An energy stable discontinuous Galerkin spectral element discretization for variable coefficient advection problems. SIAM J. Sci. Comput. **36**(4), A2076–A2099 (2014)
17. D.A. Kopriva, G.J. Gassner, Geometry effects in nodal discontinuous Galerkin methods on curved elements that are provably stable. Appl. Math. Comput. **272**(Part 2), 274–290 (2016)
18. D.A. Kopriva, A.R. Winters, M. Bohm, G.J. Gassner, A provably stable discontinuous Galerkin spectral element approximation for moving hexahedral meshes. Comput. Fluids. **139**, 148–160 (2016)

19. D.A. Kopriva, A.R. Winters, M. Bohm, G.J. Gassner, A provably stable discontinuous Galerkin spectral element approximation for moving hexahedral meshes. Comput. Fluids **139**, 148–160 (2016). doi:10.1016/j.compfluid.2016.05.023
20. D.A. Kopriva, J. Nordström, G.J. Gassner, Error boundedness of discontinuous Galerkin spectral element approximations of hyperbolic problems. J. Sci. Comput. **72**(1), 1–17 (2017)
21. D. Stanescu, D.A. Kopriva, M.Y. Hussaini, Dispersion analysis for discontinuous spectral element methods. J. Sci. Comput. **15**(2), 149–171 (2001)
22. Z. Xie, L.-L. Wang, X. Zhao, On exponential convergence of Gegenbauer interpolation and spectral differentiation. Math. Comput. **82**, 1017–1036 (2013)

Certified Reduced Basis Method for Affinely Parametric Isogeometric Analysis NURBS Approximation

Denis Devaud and Gianluigi Rozza

Abstract In this work we apply reduced basis methods for parametric PDEs to an isogeometric formulation based on NURBS. We propose an integrated and complete work pipeline from CAD to parametrization of domain geometry, then from full order to certified reduced basis solution. IsoGeometric Analysis (IGA), as well as reduced basis methods for parametric PDEs growing research themes in scientific computing and computational mechanics. Their combination enhances the solution of some class of problems, especially the ones characterized by parametrized geometries. This work shows that it is also possible for some class of problems to deal with affine geometrical parametrization combined with a NURBS IGA formulation. In this work we show a certification of accuracy and a complete integration between IGA formulation and parametric certified greedy RB formulation by introducing two numerical examples in heat transfer with different parametrization.

1 Introduction and Motivation

In this work we apply reduced basis methods for parametric PDEs to an isogeometric formulation based on NURBS. The motivation for this work is an integrated and complete work pipeline from CAD to parametrization of domain geometry, then from full order to certified reduced basis solution. IsoGeometric Analysis (IGA) is a growing research theme in scientific computing and computational mechanics, as well as reduced basis methods for parametric PDEs. Their combination enhances the solution of some class of problems, especially the ones characterized by parametrized geometries we introduced in this work. For a general overview on Reduced Basis (RB) methods we recall [7, 14] and on IGA [3]. This work wants to

D. Devaud (✉)
ETHZ SAM, Seminar for Applied Mathematics, CH-8092, Zurich, Switzerland
e-mail: denis.devaud@sam.math.ethz.ch

G. Rozza
SISSA, International School for Advanced Studies, Mathematics Area, mathLab, Via Bonomea 265, 34136 Trieste, Italy
e-mail: grozza@sissa.it

M.L. Bittencourt et al. (eds.), *Spectral and High Order Methods for Partial Differential Equations ICOSAHOM 2016*, Lecture Notes in Computational Science and Engineering 119, DOI 10.1007/978-3-319-65870-4_3

demonstrate that it is also possible for some class of problems to deal with affine geometrical parametrization combined with a NURBS IGA formulation. This is what this work brings as original ingredients with respect to other works dealing with reduced order methods and IGA (set in a non-affine formulation, and using a POD [2] sampling without certification: see for example for potential flows [11] and for Stokes flows [16]). In this work we show a certification of accuracy and a complete integration between IGA formulation and parametric certified greedy RB formulation. Section 2 recalls the abstract setting for parametrized PDEs, Sect. 3 recalls IGA setting, Sect. 4 deals with RB formulation, and Sect. 5 illustrates two numerical examples in heat transfer with different parametrization.

2 Elliptic Coercive Parametrized Partial Differential Equations

In what follows, elliptic coercive parametrized partial differential equations are introduced [12, 13, 15]. We consider the following problem: given a parameter $\boldsymbol{\mu} \in \mathcal{D}$, evaluate

$$s(\boldsymbol{\mu}) = l(u(\boldsymbol{\mu})), \tag{2.1}$$

where $u(\boldsymbol{\mu}) \in X$ is the solution of

$$a(u(\boldsymbol{\mu}), v; \boldsymbol{\mu}) = f(v), \qquad \forall v \in X. \tag{2.2}$$

Here $a(\cdot, \cdot; \boldsymbol{\mu}) : X \times X \to \mathbb{R}$ is a bilinear, continuous and coercive form associated to a parametrized partial differential equation for every $\boldsymbol{\mu} \in \mathcal{D}$. The space $X := X(\Omega)$ is a Hilbert space on the computational domain $\Omega \subset \mathbb{R}^d$ endowed with the scalar product $(\cdot, \cdot)_X$ for $d = 2, 3$. Since second-order partial differential equations for scalar problems are considered, we have $H_0^1(\Omega) \subset X \subset H^1(\Omega)$, where $H^1(\Omega) := \{v : \Omega \to X \big| \ v \in L^2(\Omega), \nabla v \in L^2(\Omega)^d\}$ and $H_0^1(\Omega)$ is the space of functions in $H^1(\Omega)$ whose traces vanish on the boundary. The space $L^2(\Omega)$ denotes the set of square integrable functions. We require moreover that Ω admits a (multipatches) NURBS representation. This is explained in more details in the next section. The functions $f : X \to \mathbb{R}$ and $l : X \to \mathbb{R}$ are linear and continuous functionals. Finally, the set $\mathcal{D}$ denotes the parameter domain and is assumed to be finite-dimensional. More precisely, we write $\mathcal{D} := [a_1, b_1] \times \cdots \times [a_P, b_P] \subset \mathbb{R}^P$ for $a_i, b_i \in \mathbb{R}$, $i = 1, \ldots, P$. We consider here both physical and geometrical parameters. The geometrical case is further investigated in Sect. 3.3. For a sake of simplicity, the so-called compliant case is considered, that is (i) a is symmetric and (ii) $l = f$.

One of the crucial assumptions to apply the reduced basis method is that a admits an affine decomposition with respect to the parameter $\boldsymbol{\mu}$, that is

$$a(u, v; \boldsymbol{\mu}) = \sum_{q=1}^{Q} \Theta^q(\boldsymbol{\mu}) a^q(u, v). \tag{2.3}$$

Here $\Theta^q : \mathcal{D} \to \mathbb{R}$ denotes a (smooth) $\boldsymbol{\mu}$-dependent function and $a^q : X \times X \to \mathbb{R}$ is a $\boldsymbol{\mu}$-independent bilinear continuous form for $q = 1, \ldots, Q$. Since the compliant case is considered, we require moreover that a^q is symmetric. We do not make any further assumption on the coercivity of a^q. Note that we have assumed that the right-hand side of equation (2.2) is parameter-independent but in practice f may depend on the parameter $\boldsymbol{\mu}$. In that case, we express $f(v; \boldsymbol{\mu})$ as a sum of Q_f products of $\boldsymbol{\mu}$-dependent functions and $\boldsymbol{\mu}$-independent linear continuous forms on X.

For the bilinear form $a(\cdot, \cdot; \boldsymbol{\mu})$, we define its continuity and coercivity constants for every $\boldsymbol{\mu} \in \mathcal{D}$ as

$$\gamma(\boldsymbol{\mu}) := \sup_{v \in X} \sup_{w \in X} \frac{a(v, w; \boldsymbol{\mu})}{\|v\|_X \|w\|_X},$$

and

$$\alpha(\boldsymbol{\mu}) := \inf_{v \in X} \frac{a(v, v; \boldsymbol{\mu})}{\|v\|_X^2},$$

where $\|\cdot\|_X$ is the norm on X induced by the scalar product $(\cdot, \cdot)_X$. Since a is continuous and coercive, there exists $0 < \alpha_0 \leq \gamma_0 < \infty$ such that $\alpha_0 \leq \alpha(\boldsymbol{\mu}) \leq \gamma(\boldsymbol{\mu}) \leq \gamma_0$ for all $\boldsymbol{\mu} \in \mathcal{D}$.

In the following section, we introduce a NURBS approximation of the problem (2.1)–(2.2). Since it is computationally unaffordable to compute such solution for every input parameter, we then consider a RB approximation of it.

3 Isogeometric Analysis NURBS Approximation

In this section, we introduce non-uniform rational B-splines (NURBS) approximation for the problem (2.1)–(2.2). First, a brief survey of B-splines and NURBS functions is conducted and the proper approximation is introduced [1, 3, 8]. We then present in Sect. 3.3 the affine preconditioning for parameter-dependent domains. This is a necessary assumption to obtain the affine decomposition (2.3) which in turn is crucial to perform RB approximation.

3.1 B-Splines

The B-splines functions are the basis to define NURBS. We give a brief introduction to B-splines in what follows. In the context of isogeometric analysis, the notion of *patches* is very important. They play the role of subdomains and material properties are assumed to be uniform in each patch. Unlike standard finite element (FE) analysis, the B-splines and NURBS basis functions are local to patches and not elements.

The FE basis functions map the reference element in the parametric domain to each element in the physical space. B-splines functions take a patch (a set of elements) in the parameter space and map it to multiple elements in the physical domain.

Let us define a *knot vector* in one dimension as a set of non-decreasing coordinates in the parameter domain denoted $\Xi = \{\xi_1, \ldots, \xi_{n+p+1}\}$, where $\xi_i \in \mathbb{R}$ is called the *i*th *knot*, $i = 1, \ldots, n + p + 1$. Here, p denotes the polynomial order of the B-splines and n the number of basis functions. The B-splines are completely defined by the knot vector Ξ, the number of basis functions n and their order p. Since this does not affect the construction of B-splines we set by convention $\xi_1 = 0$ and $\xi_{n+p+1} = 1$. Note that repetitions are allowed in the knot vector and are used to control the local regularity across each knot. A knot vector in which ξ_1 and ξ_{n+p+1} are repeated $p + 1$ times is called *open* knot vector. In what follows, we consider only open knot vectors but the construction is the same for general knot vectors. Moreover, we may refer a patch as a subdomain and an element as a knot span, i.e. an interval of the form $[\xi_i, \xi_{i+1}]$.

The B-spline functions are constructed recursively with respect to the polynomial order. For $p = 0$ and an open knot vector Ξ, we define

$$N_{i,0}(x) := \begin{cases} 1 \text{ if } \xi_i \le x \le \xi_{i+1}, \\ 0 \text{ otherwise.} \end{cases}$$

For $p = 1, 2, \ldots,$ we define recursively the B-spline basis functions as

$$N_{i,p}(x) := \frac{x - \xi_i}{\xi_{i+p} - \xi_i} N_{i,p-1}(x) + \frac{\xi_{i+p+1} - x}{\xi_{i+p+1} - \xi_{i+1}} N_{i+1,p-1}(x). \tag{3.1}$$

We present in Fig. 1 an example of B-spline basis functions for $n = 10$, $p = 3$ and the knot vector $\xi = \{0, 0, 0, 0, 0.25, 0.25, 0.25, 0.5, 0.75, 0.75, 1, 1, 1, 1\}$. Equation (3.1) is called the Cox-de-Boor recursion formula [4, 5]. Note that for $p = 0, 1$, the B-spline basis functions coincide with the FE ones. The B-splines constitute a partition of the unity, that is

$$\sum_{i=1}^{n} N_{i,p}(x) = 1, \qquad \forall x \in [0, 1].$$

A second feature is that they are pointwise non-negative, i.e. $N_{i,p}(x) \ge 0\ \forall x \in [0, 1]$. This implies that the coefficients of the mass matrix are greater or equal than zero. The support of $N_{i,p}$ is $[\xi_i, \xi_{i+p+1}]$. The basis function $N_{i,p}$ has $p - m_i$ continuous derivatives, where m_i is the multiplicity of ξ_i, i.e. the number of repetitions of ξ_i. An important remark is that the B-spline basis functions are not interpolatory at the location of knot values ξ_i unless the multiplicity of ξ_i is exactly p.

We are now in position to define B-spline curves, surfaces and solids in $\mathbb{R}^d$. Let us assume that we are given three sets of B-spline basis functions $\{N_{i,p}\}$, $\{M_{j,q}\}$ and $\{L_{k,r}\}$ constructed on the knot vectors $\{\xi_1, \ldots, \xi_{n+p+1}\}$, $\{\eta_1, \ldots, \eta_{m+q+1}\}$ and $\{\zeta_1, \ldots, \zeta_{l+r+1}\}$ for $i = 1, \ldots, n, j = 1, \ldots, m$ and $k = 1, \ldots, l$, respectively.

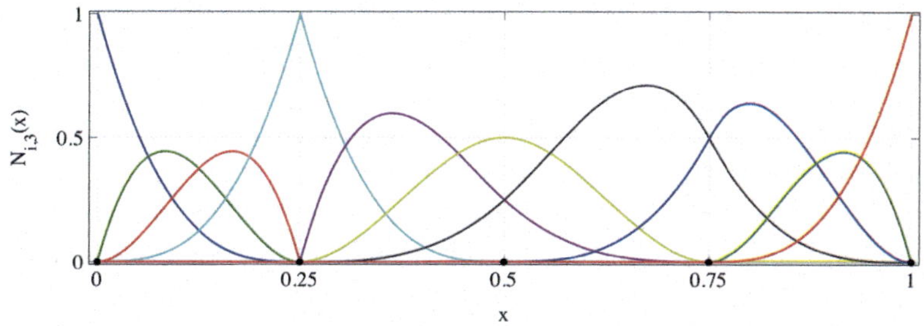

Fig. 1 Example of B-spline basis functions for $\xi = \{0, 0, 0, 0, 0.25, 0.25, 0.25, 0.5, 0.75, 0.75, 1, 1, 1, 1\}$, $n = 10$ and $p = 3$. We see that the regularity is related to the multiplicity of each ξ_i. Moreover, for $\xi_i = 0.25$ we have $m_i = p$ and we see that the basis function is interpolatory at this knot

The B-spline curves are obtained by considering linear combinations of B-spline basis functions. Let $C_i \in \mathbb{R}^d$ be the coefficients referred as *control points*, for $i = 1, \ldots, n$. We then define a B-spline curve as

$$S(x) := \sum_{i=1}^{n} N_{i,p}(x) C_i.$$

Such curves have at least as many continuous derivatives across an element boundary than its underlying B-spline basis function has across the corresponding knot value. A crucial property of the B-spline curves is that an affine transformation of the curve is obtained by applying the transformation to the control points. It is the so-called *affine covariance* and play an important role in the affine decomposition (2.3) when considering parameter-dependent domains. Now that the univariate B-splines have been introduced, we generalize the definition to higher dimensions by considering a tensor product structure.

Given a so-called *control net* $\{C_{i,j}\} \subset \mathbb{R}^d$ for $i = 1, \ldots, n$ and $j = 1, \ldots, m$, we define a B-spline surface as

$$S(x, y) := \sum_{i=1}^{n} \sum_{j=1}^{m} N_{i,p}(x) M_{j,q}(y) C_{i,j}.$$

Several properties of the B-spline surfaces result from their tensor product structures. For instance, the basis also forms a partition of the unity and the number of continuous partial derivatives are determined from the underlying one-dimensional knot vectors and polynomial orders. The local support is also deducted from the one-dimensional basis, that is the support of $N_{i,p}(x)M_{j,q}(y)$ is $[\xi_i, \xi_{i+p+1}] \times [\eta_j, \eta_{j+q+1}]$.

Finally, we introduce the definition of a B-spline solid. Considering a control lattice $\{C_{i,j,k}\}$ for $i = 1, \ldots, n$, $j = 1, \ldots, m$ and $k = 1, \ldots, l$, it is defined as

$$S(x, y, z) := \sum_{i=1}^{n} \sum_{j=1}^{m} \sum_{k=1}^{l} N_{i,p}(x) M_{j,q}(y) L_{k,r}(z) C_{i,j,k}.$$

The properties of the B-spline solids are a direct extension of those presented in the case of surfaces. In particular, the affine covariance property still holds for B-spline surfaces and solids. Note that what has been presented here is valid for a single patch. The case of multipatches geometries is introduced in the context of NURBS basis functions in the next section.

For the purpose of the analysis presented in Sect. 3.4, we briefly discuss the notions of h-refinement, p-refinement and k-refinement. A complete discussion can be found in [3, 8]. The notion of h-refinement in FE analysis is similar to the *knot insertion* in IGA. Let us consider a knot vector $\Xi = \{\xi_1, \dots, \xi_{n+p+1}\}$ and associated control points $\{B_1, \dots, B_n\}$. Considering a knot $\bar{\xi} \in [\xi_k, \xi_{k+1}[$, we then build the new knot vector as $\bar{\Xi} = \{\xi_1, \dots, \xi_k, \bar{\xi}, \xi_{k+1}, \dots, \xi_{n+p+1}\}$ and the associated control points $\{\bar{B}_1, \dots, \bar{B}_{n+1}\}$ as

$$\bar{B}_i := \alpha_i B_i + (1 - \alpha_i) B_{i-1}, \tag{3.2}$$

where

$$\alpha_i := \begin{cases} 1, & 1 \le i \le k - p, \\ \frac{\bar{\xi} - \xi_i}{\xi_{i+p} - \xi_i}, & k - p + 1 \le i \le k, \\ 0, & k + 1 \le i \le n + p + 2. \end{cases} \tag{3.3}$$

By choosing the new control points as (3.2) and (3.3), it is possible to maintain the continuity of the original basis functions. Note that it is possible to insert repetition of already existing knot values. This will decrease the regularity of the basis functions at this knot. An important remark is that the solution spanned by the increased basis functions based contains the one spanned by the original B-splines. This allows to keep the geometry unchanged by inserting new knots.

The second concept introduced here is the FE p-refinement, which analogous is *order elevation*. It is possible to increase the polynomial order of the basis functions. To keep the regularity of the previous B-splines, it is necessary to repeat each knot value of the knot vector. As in the case of knot insertion, the new span contains the one from the original basis functions.

The last notion is the one of k-refinement which does not have an analogous in FE analysis. It is based on the principle that order elevation and knot insertion do not commute. If we insert a new knot value $\bar{\xi}$, the continuity of the basis functions at this knot will be C^{p-1}. If then we further increase the order of the basis, the multiplicity of $\bar{\xi}$ increase to keep this continuity. Instead, if we first increase the order of the basis to q and then insert a new knot value, the continuity will be C^{q-1} at this knot. This second process is called k-refinement. It allows to control the number of new basis functions. Hence the number of degrees of freedom associated to the B-splines will also be kept under control, which in turn allows to keep low computational costs.

3.2 Non-Uniform Rational B-Splines

The introduction of NURBS allows us to exactly represent domains that it is not possible to describe considering polynomials. The construction of such geometries in $\mathbb{R}^d$ are obtained by projective transformations in $\mathbb{R}^{d+1}$. It is then possible to construct for instance conic sections. Such projective transformation yields rational polynomial functions.

The process to construct NURBS basis functions is presented here and follows mainly [3, 8]. Let us consider a knot vector Ξ, a number of basis functions n, a polynomial order p and a set of control points $\{B_i^w\}$ in $\mathbb{R}^{d+1}$ defining a B-spline curve. Such points are called *projective* control points for the associated NURBS curve. We then define the control points of the NURBS curve as follows

$$\begin{aligned} w_i &:= (B_i^w)_{d+1}, && i = 1, \ldots, d, \\ (B_i)_j &:= \left(B_i^w\right)_j / w_i, && i, j = 1, \ldots, d, \end{aligned}$$

where $(B_i)_j$ is the jth component of the vector B_i. The scalars w_i are called weights. Let $\{N_{i,p}\}$ be the B-spline basis functions associated to Ξ, n and p. Based on the definition of the control points, we can introduce the NURBS basis functions defined as

$$R_i^p(x) := \frac{N_{i,p}(x) w_i}{\sum_{i'=1}^{n} N_{i',p}(x) w_{i'}}. \tag{3.4}$$

The associated NURBS curve is then defined as

$$C(x) := \sum_{i=1}^{n} R_i^p(x) B_i.$$

Considering the basis functions defined by (3.4), we define NURBS surfaces and solids in the same manner. To do this, we define rational basis functions for surfaces and solids. Let $\{M_{j,q}\}$ and $\{L_{k,r}\}$ be B-spline basis functions for $1 \leq j \leq m$ and $1 \leq k \leq l$. Moreover, consider projective control nets and lattices $\{B_{i,j}^w\}$ and $\{B_{i,j,k}^w\}$ in $\mathbb{R}^{d+1}$, respectively. The weights to construct the NURBS basis functions are given by

$$\begin{aligned} w_{i,j} &:= \left(B_{i,j}^w\right)_{d+1}, && i, j = 1, \ldots, d, \\ w_{i,j,k} &:= \left(B_{i,j,k}^w\right)_{d+1}, && i, j, k = 1, \ldots, d. \end{aligned}$$

Fig. 2 Examples of NURBS solids obtained considering multiple patches. The number of patches used for each example are 3 (**a**), 4 (**b**), 3 (**c**), 4 (**d**) and 3 (**e**), respectively

We then define NURBS basis functions for surfaces and solids as

$$R_{i,j}^{p,q}(x,y) := \frac{N_{i,p}(x)M_{j,q}(y)w_{i,j}}{\sum_{i'=1}^{n}\sum_{j'=1}^{m} N_{i',p}(x)M_{j',q}(y)w_{i',j'}}, \qquad i,j=1\ldots,d,$$

$$R_{i,j,k}^{p,q,r}(x,y,z) := \frac{N_{i,p}(x)M_{j,q}(y)L_{k,r}(z)w_{i,j,k}}{\sum_{i'=1}^{n}\sum_{j'=1}^{m}\sum_{k'=1}^{l} N_{i',p}(x)M_{j',q}(y)L_{k',r}(z)w_{i',j',k'}}, \qquad i,j,k=1\ldots,d.$$

The properties stated for the B-spline basis functions also hold for the NURBS. In particular, they form a partition of the unity and their continuity and support are the same as the underlying B-splines. The affine covariance property also holds for NURBS functions. Moreover, the basis functions are interpolatory at knot values where the multiplicity is equal to the order. The notions of h-, p- and k-refinement generalize to NURBS functions. Note that if all the weights are equal, the NURBS coincide with the underlying B-splines due to the partition of the unity property. In nearly all the practical applications, it is necessary to have multiple patches to describe the domain with NURBS functions. This also allows to have different material properties, each associated to a different patch. The only feature to pay attention to is the regularity of the basis across the patches interfaces. Usually, C^0 is the only regularity guaranteed, but techniques can be used to increase it [3]. In Fig. 2, we present several examples of NURBS solids obtained considering multipatches representations.

To simplify the notations, we denote by $R_{i,p}$, $1 \leq i \leq n$ the NURBS basis functions and $\{B_i\}$ the associated control points for curves, surfaces and solids. We also use the notation Ξ for the associated knot vectors. Note that it is a slight abuse of notation because in the case of surfaces and solids, p and Ξ are vectors and matrices, respectively.

For the purpose of our analysis, we require that the computational domain Ω can be obtained through a NURBS parametrization. To introduce the notations, we impose that Ω is parameter-independent. The parameter-dependent case is treated in the next section. Let us consider the following decomposition of the domain

$$\overline{\Omega} = \bigcup_{k=1}^{P_{\text{dom}}} \overline{\Omega}^k, \tag{3.5}$$

where $\Omega^j \cap \Omega^k = \emptyset$ for $1 \leq j < k \leq P_{\mathrm{dom}}$. We require that for every subdomain Ω^k, there exist p_k, n_k, Ξ_k, NURBS basis functions $\{R^k_{i,p}\}$ and associated control points $\mathcal{B}^k := \{B^k_i\}$ such that for every $y \in \Omega^k$, there exists $x \in \mathcal{H}^d$ satisfying

$$y = F^k(x) := \sum_{i=1}^{n_k} R^k_{i,p_k}(x) B^k_i. \tag{3.6}$$

Here $\mathcal{H}^d = [0,1]^d$ denotes the unit hypercube in d-dimension and $F^k : (0,1)^d \to \Omega^k$. Considering for every $1 \leq k \leq P_{\mathrm{dom}}$ the function F^k defined above, we construct a global mapping $F : (0,1)^d \to \Omega$ which describes the whole computational domain. We assume that F is smooth and invertible. In that case, we say that Ω admits a NURBS representation through F. So far, we have only considered parameter-independent geometries. In the next section, we introduce the affine preconditioning conditions for parameter-dependent domains.

3.3 *Affine Preconditioning for Parameter-Dependent Domains*

In many applications, it is of great interest to consider parameter-dependent geometries. We introduce here the conditions that need to be fulfilled in that case to be able to perform the RB method presented in this paper. In particular, it is important that the affine decomposition (2.3) of the bilinear form a still holds. Let us consider the domain splitting introduced in (3.5). The computational domain for an input parameter $\boldsymbol{\mu} \in \mathcal{D}$ is denoted $\Omega_o(\boldsymbol{\mu})$. Here, the subscript o stands for the original domain.

The domain $\Omega_o(\boldsymbol{\mu})$ needs to be represented as the image of a reference domain through an affine mapping. Let us choose $\boldsymbol{\mu}_{\mathrm{ref}} \in \mathcal{D}$ as a parameter that represents our reference domain, i.e. $\Omega = \Omega_o(\boldsymbol{\mu}_{\mathrm{ref}})$. Moreover, we denote $\Omega^k = \Omega^k_o(\boldsymbol{\mu}_{\mathrm{ref}})$ while considering the decomposition (3.5). We need that for every $1 \leq k \leq P_{\mathrm{dom}}$, there exists an affine mapping $\mathcal{T}^k(\cdot;\boldsymbol{\mu}) : \Omega^k \to \Omega^k_o(\boldsymbol{\mu})$ such that

$$\overline{\Omega}^k_o(\boldsymbol{\mu}) = \mathcal{T}^k(\overline{\Omega}^k;\boldsymbol{\mu}).$$

The mappings $\mathcal{T}^k(\cdot;\boldsymbol{\mu})$ have to be bijective and collectively continuous, that is

$$\mathcal{T}^k(x;\boldsymbol{\mu}) = \mathcal{T}^l(x;\boldsymbol{\mu}), \qquad \forall x \in \overline{\Omega}^k \cap \overline{\Omega}^l,\ 1 \leq k < l \leq P_{\mathrm{dom}}. \tag{3.7}$$

Due to the affine covariance property of the NURBS functions, we only need to require that the control points can be obtained as the image of reference control points through an affine mapping. More formally, let us denote by $\{B^k_i(\boldsymbol{\mu})\}$ the control points associated with the subdomains $\Omega^k_o(\boldsymbol{\mu})$. We then require that for every $1 \leq k \leq P_{\mathrm{dom}}$, there exists an affine mapping $\mathcal{T}^k(\cdot;\boldsymbol{\mu}) : \Omega^k \to \Omega^k_o(\boldsymbol{\mu})$ such that

$$B^k_i(\boldsymbol{\mu}) = \mathcal{T}^k(B^k_i;\boldsymbol{\mu}),$$

where $\{B_i^k\}$ are the control points associated to the reference subdomains Ω^k. Turning to the condition (3.7), we require that

$$\mathcal{T}^k(B_i^k; \boldsymbol{\mu}) = \mathcal{T}^l(B_i^k; \boldsymbol{\mu}), \qquad \forall B_i^k \in \mathcal{B}^k \cap \mathcal{B}^l, \ 1 \leq k < l \leq P_{\text{dom}}.$$

In other words, we only need to ensure continuity of the mappings through the control points defining the interfaces of patches to obtain the continuity on the whole interface. More explicitly, we define the affine mappings $\mathcal{T}^k$ for every $x \in \overline{\Omega}^k$ and $\boldsymbol{\mu} \in \mathcal{D}$ as

$$\mathcal{T}^k(x; \boldsymbol{\mu}) := C^k(\boldsymbol{\mu}) + G^k(\boldsymbol{\mu})x,$$

where $C^k : \mathcal{D} \to \mathbb{R}^d$ and $G^k : \mathcal{D} \to \mathbb{R}^{d \times d}$ for every $1 \leq k \leq P_{\text{dom}}$. To define the affine decomposition of the bilinear form a, we need to define the Jacobians and inverse of the transformations as

$$J^k(\boldsymbol{\mu}) := |\det(G^k(\boldsymbol{\mu}))|, \tag{3.8}$$

$$D^k(\boldsymbol{\mu}) := \left(G^k(\boldsymbol{\mu})\right)^{-1}, \tag{3.9}$$

for $1 \leq k \leq P_{\text{dom}}$. Based on the $\mathcal{T}^k$ transformations, we can define a global affine mapping $\mathcal{T} : \Omega \to \Omega_o(\boldsymbol{\mu})$ as

$$\mathcal{T}(x; \boldsymbol{\mu}) := \mathcal{T}^k(x; \boldsymbol{\mu}), \qquad k = \min\left\{1 \leq l \leq P_{\text{dom}} \middle| \, x \in \overline{\Omega}^l\right\}.$$

The mapping $\mathcal{T}$ is globally bijective and piecewise affine. The choice of the minimum is arbitrary and could be chosen differently.

In what follows, we give an example of an affine transformation applied to a 3-dimensional toroidal solid. It is built on four patches, which yields $P_{\text{dom}} = 4$, and to every patch are associated 27 control points. Based on that, it is possible to uniquely determine C^k and G^k for $1 \leq k \leq P_{\text{dom}}$. In that case, the transformations are given by

$$C^k = \begin{pmatrix} 0 \\ 0 \\ 0 \end{pmatrix}, \qquad G^k = \begin{pmatrix} \mu & 0 & 0 \\ 0 & 1 & 0 \\ 0 & 0 & 1 \end{pmatrix}, \qquad 1 \leq k \leq 4, \tag{3.10}$$

where we have considered the single parameter μ that controls the semi-axis x. In Fig. 3, the original domain and the transformed one for $\mu = 1.5$ are depicted together with their lattices of control points. We see that the transformation is exactly applied to the control points.

Now that our affine preconditioning assumption has been stated, we need to express our bilinear form on the reference domain. The problem (2.1)–(2.2) is defined on the original domain $\Omega_o(\boldsymbol{\mu})$. To be able to obtain the affine expansion (2.3)

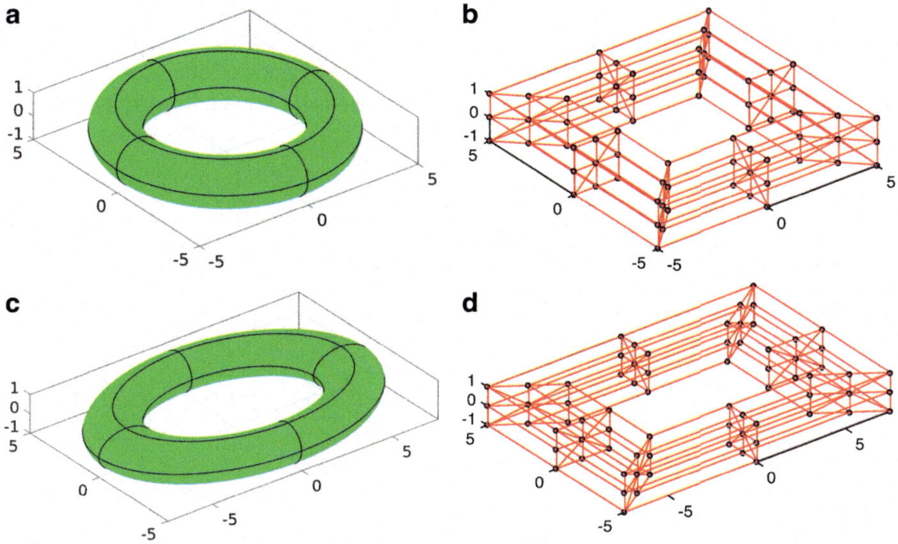

Fig. 3 Example of the affine transformation (3.10) applied to a torus. The original domain and its lattice of control points are presented in (**a**) and (**b**), respectively. The torus after transformation is depicted in (**c**) while its lattice is presented in (**d**)

for the bilinear form a arising from the weak formulation of a second-order PDE, we need that the underlying integrals are defined on the reference domain. This is presented in details in what follows for two-dimensional problems but the case $d = 3$ is treated analogously. On the original domain, the problem is the following one: given a parameter $\boldsymbol{\mu} \in \mathcal{D}$, evaluate

$$s_o(\boldsymbol{\mu}) = l(u_o(\boldsymbol{\mu})),$$

where $u_o(\boldsymbol{\mu}) \in X_o$ is the solution of

$$a_o(u_o(\boldsymbol{\mu}), v; \boldsymbol{\mu}) = f_o(v), \qquad \forall v \in X.$$

Since we are considering second-order partial differential, we require that a_o can be written as

$$a_o(v, w; \boldsymbol{\mu}) = \sum_{k=1}^{P_{\text{dom}}} \int_{\Omega_o^k(\boldsymbol{\mu})} \left[\frac{\partial v}{\partial x_1} \; \frac{\partial v}{\partial x_2} \; v \right] A_o^k(\boldsymbol{\mu}) \begin{bmatrix} \frac{\partial w}{\partial x_1} \\ \frac{\partial w}{\partial x_2} \\ v \end{bmatrix},$$

where $A_o^k : \mathcal{D} \to \mathbb{R}^{3\times 3}$ is a symmetric positive semi-definite matrix. We express the right-hand side f_o in the same way, that is

$$f_o(v) = \sum_{k=1}^{P_{\text{dom}}} \int_{\Omega_0^k(\boldsymbol{\mu})} f_o^k v, \tag{3.11}$$

where $f_o^k \in \mathbb{R}$. Note that it is possible to have a parameter-dependent right-hand side by simply replacing f_o^k by $f_o^k(\boldsymbol{\mu})$. As already discussed, we need $\boldsymbol{\mu}$-independent integrals to be able to fulfill the affine assumption (2.3) for the bilinear form a. We then consider the problem (2.1)–(2.2) with the bilinear form a expressed as

$$a(v, w; \boldsymbol{\mu}) = \sum_{k=1}^{P_{\text{dom}}} \int_{\Omega^k} \begin{bmatrix} \frac{\partial v}{\partial x_1} & \frac{\partial v}{\partial x_2} & v \end{bmatrix} A^k(\boldsymbol{\mu}) \begin{bmatrix} \frac{\partial w}{\partial x_1} \\ \frac{\partial w}{\partial x_2} \\ v \end{bmatrix}, \tag{3.12}$$

where the $A^k : \mathcal{D} \to \mathbb{R}^{3\times3}$ are defined for $\boldsymbol{\mu} \in \mathcal{D}$ as

$$A^k(\boldsymbol{\mu}) = J^k(\boldsymbol{\mu})\mathcal{G}^k(\boldsymbol{\mu})A_o^k(\boldsymbol{\mu}) \left(\mathcal{G}^k(\boldsymbol{\mu})\right)^T, \qquad 1 \leq k \leq P_{\text{dom}}.$$

Here the matrices $\mathcal{G}^k(\boldsymbol{\mu})$ are defined as

$$\mathcal{G}^k(\boldsymbol{\mu}) := \begin{pmatrix} D^k(\boldsymbol{\mu}) & \begin{matrix} 0 \\ 0 \end{matrix} \\ \begin{matrix} 0 & 0 \end{matrix} & 1 \end{pmatrix}, \qquad 1 \leq k \leq P_{\text{dom}},$$

where $J^k(\boldsymbol{\mu})$ and $D^k(\boldsymbol{\mu})$ are defined by (3.8) and (3.9), respectively. Note that this holds under the assumptions presented at the begin of this section. In the same manner, the right-hand side is expressed as

$$f(v) = \sum_{k=1}^{P_{\text{dom}}} \int_{\Omega^k} f^k(\boldsymbol{\mu})v,$$

where $f^k : \mathcal{D} \to \mathbb{R}$ is defined by

$$f^k(\boldsymbol{\mu}) = J^k(\boldsymbol{\mu})f_o^k, \qquad 1 \leq k \leq P_{\text{dom}}.$$

We can then explicitly expand (3.12) to obtain the affine decomposition (2.3) for the bilinear form a. In the development presented here, the $A^k(\boldsymbol{\mu})$ and $f^k(\boldsymbol{\mu})$ are local to patches and may represent different material properties and geometry variations.

3.4 Isogeometric Analysis NURBS Approximation of Elliptic Coercive Parametrized PDEs

We present in this section the isogeometrical analysis NURBS approximation of the problem (2.1)–(2.2). In this context, the isoparametric concept is considered, that is the solution is represented in the same space as the geometry. In that case, the mesh of the NURBS is defined as the product of the knot vectors and the elements are the knot spans. The degrees of freedom associated with the basis functions are called *control variables*.

Let us assume that Ω admits a NURBS parametrization through F as defined in (3.6). To simplify the notations, we consider a single set of indices $\{1, \dots, \mathcal{N}\}$ for the degrees of freedom and we write

$$F(x) = \sum_{i=1}^{\mathcal{N}} \tilde{R}_{i,p}(x) B_i, \qquad x \in (0,1)^d, \tag{3.13}$$

for NURBS basis functions $\{\tilde{R}_{i,p}\}$ and associated control points $\{B_i\}$. To represent our solution in a finite dimensional space, we need to define the basis functions

$$R_{i,p} := \tilde{R}_{i,p} \circ F^{-1}, \tag{3.14}$$

where F is the invertible mapping defined by (3.13). Based on that representation, we construct the NURBS approximation space

$$X^{\mathcal{N}} := \operatorname{span}\left\{R_{i,p}\right\}_{1 \leq i \leq \mathcal{N}} \subset X. \tag{3.15}$$

As already discussed in Sect. 3.1, the process of knot insertion does not change the underlying geometry. In that setting, increasing $\mathcal{N}$ does not change the shape of the parametrized domain and so we keep the exact parametrization while refining the mesh. For approximation properties of NURBS approximation spaces, we refer the reader to [1]. We approximate the solution of (2.1)–(2.2) by an element of $X^{\mathcal{N}}$. The approximate problem is the following one: given a parameter $\boldsymbol{\mu} \in \mathcal{D}$, evaluate

$$s^{\mathcal{N}}(\boldsymbol{\mu}) = l(u^{\mathcal{N}}(\boldsymbol{\mu})), \tag{3.16}$$

where $u^{\mathcal{N}}(\boldsymbol{\mu}) \in X^{\mathcal{N}}$ is the solution of

$$a(u^{\mathcal{N}}(\boldsymbol{\mu}), v; \boldsymbol{\mu}) = f(v), \qquad \forall v \in X^{\mathcal{N}}. \tag{3.17}$$

Considering the basis $\{R_{i,p}\}$ for $X^{\mathcal{N}}$ defined by (3.14), we extend the NURBS solution $u^{\mathcal{N}}(\boldsymbol{\mu})$ for $\boldsymbol{\mu} \in \mathcal{D}$ as

$$u^{\mathcal{N}}(x, \boldsymbol{\mu}) = \sum_{i=1}^{\mathcal{N}} u_i^{\mathcal{N}}(\boldsymbol{\mu}) R_{i,p}(x), \qquad x \in \Omega,$$

where the coefficients $u_i^{\mathcal{N}}(\boldsymbol{\mu})$ are called control variables. The regularity of $u^{\mathcal{N}}(\boldsymbol{\mu})$ follows from that of the NURBS basis. For instance, the continuity of the solution across element boundaries depends on the continuity of the underlying basis functions across the associated knot span.

Our goal then becomes to solve the problem (3.16)–(3.17) with high precision. However, for real-time context and many query problems, it would be computationally unaffordable to approximate the solution for each input parameter. For that reason, we introduce in the next section a method to approximate such solution with reduced computation costs.

4 Reduced Basis Method for Isogeometric Analysis NURBS Approximation

As it has already been pointed out, it is computationally unaffordable to compute a new NURBS solution for every input parameter $\boldsymbol{\mu}$. The goal of the RB method is then to approximate the NURBS solution $u^{\mathcal{N}}(\boldsymbol{\mu})$ with reduced computational costs. Considering $\mathcal{N}$ sufficiently large, we have that $u^{\mathcal{N}}(\boldsymbol{\mu})$ is close enough to $u(\boldsymbol{\mu})$ in a certain norm so that the NURBS approximation can be viewed as the "truth" solution.

Given a positive integer $N_{\max} \ll \mathcal{N}$, we construct a sequence of approximation spaces

$$X_1^{\mathcal{N}} \subset X_2^{\mathcal{N}} \subset \cdots \subset X_{N_{\max}}^{\mathcal{N}} \subset X^{\mathcal{N}}. \tag{4.1}$$

Those spaces are obtained considering a Greedy algorithm presented more in details in Sect. 4.1. The hierarchical hypothesis (4.1) is important to ensure the efficiency of the method. Several spaces can be considered to construct such sequence, but they all focus on the smooth parametric manifold $\mathcal{M}^{\mathcal{N}} := \left\{u^{\mathcal{N}}(\boldsymbol{\mu}) \middle| \, \boldsymbol{\mu} \in \mathcal{D}\right\}$. If it is smooth enough, we can expect it to be well approximated by low-dimensional spaces. In what follows, we consider the special case of Lagrange reduced basis spaces built using a master set of parameter points $\boldsymbol{\mu}^n \in \mathcal{D}$, $1 \leq n \leq N_{\max}$. Other examples such as the POD spaces [15] could be considered. For $1 \leq N \leq N_{\max}$, we define $S^N := \left\{\boldsymbol{\mu}^1, \dots, \boldsymbol{\mu}^N\right\}$ and the associated Lagrange RB spaces

$$X_N^{\mathcal{N}} := \operatorname{span}\left\{u^{\mathcal{N}}(\boldsymbol{\mu}^n) \middle| \, 1 \leq n \leq N\right\}.$$

The selection of the snapshots $u^{\mathcal{N}}(\boldsymbol{\mu}^n)$ is one of the crucial points of the RB method and is further investigated in the next section. We apply the Gram-Schmidt process in the $(\cdot,\cdot)_X$ inner product to the snapshots $u^{\mathcal{N}}(\boldsymbol{\mu}^n)$ in order to obtain mutually orthonormal functions $\zeta_n^{\mathcal{N}}$. In that case, we have $X_N^{\mathcal{N}} = \operatorname{span}\left\{\zeta_n^{\mathcal{N}} \middle| \, 1 \leq n \leq N\right\}$. Since colinearities are avoided using the Gram-Schmidt process, we are ensured that the N obtained is minimal. The RB approximation of the problem (3.16)–(3.17) is obtained considering Galerkin projection: given $\boldsymbol{\mu} \in \mathcal{D}$, evaluate

$$s_N^{\mathcal{N}}(\boldsymbol{\mu}) = f(u_N^{\mathcal{N}}(\boldsymbol{\mu})), \tag{4.2}$$

where $u_N^{\mathcal{N}}(\boldsymbol{\mu}) \in X_N^{\mathcal{N}}$ is the solution of

$$a(u_N^{\mathcal{N}}(\boldsymbol{\mu}), v; \boldsymbol{\mu}) = f(v), \qquad \forall v \in X_N^{\mathcal{N}}. \tag{4.3}$$

Since the particular compliant case is considered, we obtain

$$s^{\mathcal{N}}(\boldsymbol{\mu}) - s_N^{\mathcal{N}}(\boldsymbol{\mu}) = \|u^{\mathcal{N}}(\boldsymbol{\mu}) - u_N^{\mathcal{N}}(\boldsymbol{\mu})\|_{\boldsymbol{\mu}}^2, \tag{4.4}$$

where $\|\cdot\|_{\boldsymbol{\mu}}$ is the energy norm induced by the inner product $a(\cdot,\cdot;\boldsymbol{\mu})$. In Sect. 4.2, we present an example of an inexpensive and efficient a posteriori error estimator $\Delta_N(\boldsymbol{\mu})$ for $\|u^{\mathcal{N}}(\boldsymbol{\mu}) - u_N^{\mathcal{N}}(\boldsymbol{\mu})\|_{\boldsymbol{\mu}}$ on which the Greedy algorithm is based. Due to the relation (4.4), it is possible to ensure that the error arising from the RB approximation on our output of interest is bounded by a prescribed tolerance.

Since $u_N^{\mathcal{N}}(\boldsymbol{\mu}) \in X_N^{\mathcal{N}} = \text{span}\left\{\zeta_n^{\mathcal{N}} \mid 1 \le n \le N\right\}$, we expand it as

$$u_N^{\mathcal{N}}(\boldsymbol{\mu}) = \sum_{m=1}^{N} u_{N,m}^{\mathcal{N}}(\boldsymbol{\mu})\zeta_m^{\mathcal{N}}. \tag{4.5}$$

The unknowns then become the coefficients $u_{N,m}^{\mathcal{N}}(\boldsymbol{\mu})$. Inserting (4.5) in (4.2) and (4.3) and using the hypothesis that f is linear and a bilinear, we obtain

$$s_N^{\mathcal{N}}(\boldsymbol{\mu}) = \sum_{m=1}^{N} u_{N,m}^{\mathcal{N}}(\boldsymbol{\mu})f(\zeta_m^{\mathcal{N}}),$$

and

$$\sum_{m=1}^{N} u_{N,m}^{\mathcal{N}}(\boldsymbol{\mu})a(\zeta_m^{\mathcal{N}}, \zeta_n^{\mathcal{N}}; \boldsymbol{\mu}) = f(\zeta_n^{\mathcal{N}}), \qquad 1 \le n \le N. \tag{4.6}$$

The stiffness matrix associated to the system (4.6) is of size $N \times N$ with $N \le N_{\max} \ll \mathcal{N}$. It yields a considerably smaller computational effort than to solve the system associated to (3.17), which matrix is of size $\mathcal{N} \times \mathcal{N}$. However, the formation of the stiffness matrix involves the computation of the $\zeta_m^{\mathcal{N}}$ associated with the $\mathcal{N}$-dimensional NURBS space.

This drawback is avoided by constructing an Offline-Online procedure taking advantage of the affine decomposition (2.3). In the Offline stage, the Greedy algorithm is used to construct the set of parameters S^N. Then, the $u^{\mathcal{N}}(\boldsymbol{\mu}^n)$ and the $\zeta_n^{\mathcal{N}}$ are built for $1 \le n \le N$. The $f(\zeta_n^{\mathcal{N}})$ and $a^q(\zeta_m^{\mathcal{N}}, \zeta_n^{\mathcal{N}})$ are also formed and stored. Note the importance here of the affine decomposition. It implies that the vector and matrices stored are independent of the input parameter $\boldsymbol{\mu}$.

In the Online part, the stiffness matrix associated to (4.6) is assembled considering the affine decomposition (2.3). This yields

$$a(\zeta_m^{\mathcal{N}}, \zeta_n^{\mathcal{N}}; \boldsymbol{\mu}) = \sum_{q=1}^{Q} \Theta^q(\boldsymbol{\mu})a^q(\zeta_m^{\mathcal{N}}, \zeta_n^{\mathcal{N}}), \qquad 1 \le m, n \le N.$$

The same process is applied to the right-hand side f. The $N \times N$ system (4.3) is then solved to obtain $u^{\mathcal{N}}_{N,m}(\boldsymbol{\mu})$, $1 \leq m \leq N$. Finally, the output of interest (4.2) is computed considering the coefficients obtained.

As already discussed, one of the main feature of the RB method is that we have a posteriori error estimators $\Delta_N(\boldsymbol{\mu})$ for $\|u^{\mathcal{N}}(\boldsymbol{\mu}) - u^{\mathcal{N}}_N(\boldsymbol{\mu})\|^2_{\boldsymbol{\mu}} = s^{\mathcal{N}}(\boldsymbol{\mu}) - s^{\mathcal{N}}_N(\boldsymbol{\mu})$ whose computation costs are independent of $\mathcal{N}$. It allows us to certify our method and make it reliable. A discussion on such estimators is presented in Sect. 4.2.

4.1 Greedy Algorithm for the Snapshots Selection

One of the most important step taking place in the Offline stage is the selection of the parameters $\boldsymbol{\mu}^n$, $1 \leq n \leq N$. Several algorithms are available in the literature [15] but we introduce here a greedy procedure for completeness. The general idea of this procedure is to retain at iteration N the snapshot $u^{\mathcal{N}}(\boldsymbol{\mu}^N)$ which approximation by $X^{\mathcal{N}}_{N-1}$ is the worst. Let us assume that we are given a finite sample of points $\Xi \subset \mathcal{D}$ and pick randomly a first parameter $\boldsymbol{\mu}^1 \in \Xi$. Then for $N = 2, \ldots, N_{\max}$, compute

$$\boldsymbol{\mu}^N := \arg\max_{\boldsymbol{\mu} \in \Xi} \Delta_{N-1}(\boldsymbol{\mu}),$$

where $\Delta_N(\boldsymbol{\mu})$ is a sharp and inexpensive a posteriori error estimator for $\|u^{\mathcal{N}}(\boldsymbol{\mu}) - u^{\mathcal{N}}_N(\boldsymbol{\mu})\|_{H^1_0(\Omega)}$ or $\|u^{\mathcal{N}}(\boldsymbol{\mu}) - u^{\mathcal{N}}_N(\boldsymbol{\mu})\|_{\boldsymbol{\mu}}$. The algorithm is typically stopped when $\Delta_N(\boldsymbol{\mu})$ is smaller than a prescribed tolerance for every $\boldsymbol{\mu} \in \Xi$. It is clear that the precision of the approximation spaces obtain increase with the size of the sample considered.

Since $X^{\mathcal{N}}_{N-1} \subset X^{\mathcal{N}}_N$, we expect to have $\Delta_N(\boldsymbol{\mu}) \leq \Delta_{N-1}(\boldsymbol{\mu})$, which ensures that $N_{\max} < \infty$. Even if this procedure has not been proven to convergence, it is widely used and many examples have been presented to illustrate its convergence. The derivation of $\Delta_N(\boldsymbol{\mu})$ is crucial for the Greedy and we introduce an example of such estimator in the next section.

4.2 A Posteriori Error Estimators for Elliptic Coercive Partial Differential Equations

The main ingredient of the Greedy algorithm procedure is the computation of the error estimator, which has to be independent of $\mathcal{N}$. In fact, it is used online to certify that the error of our RB approximation with respect to the truth solution is under control. For completeness, the derivation of such estimator is presented here when the so-called compliant case is considered, i.e. a is symmetric and $f = l$. See e.g. [15] for the non-compliant case. Let us introduce the error $e^{\mathcal{N}}(\boldsymbol{\mu}) = u^{\mathcal{N}}(\boldsymbol{\mu}) - u^{\mathcal{N}}_N$

$(\boldsymbol{\mu}) \in X^{\mathcal{N}}$ which satisfies the following equation

$$a(e^{\mathcal{N}}(\boldsymbol{\mu}), v; \boldsymbol{\mu}) = f(v; \boldsymbol{\mu}) - a(u_N^{\mathcal{N}}(\boldsymbol{\mu}), v; \boldsymbol{\mu}) =: r(v; \boldsymbol{\mu}), \ \forall v \in X^{\mathcal{N}} \tag{4.7}$$

where $r(\cdot; \boldsymbol{\mu}) \in \left(X^{\mathcal{N}}\right)'$ is the residual and $\left(X^{\mathcal{N}}\right)'$ denotes the dual space of $X^{\mathcal{N}}$. To define our a posteriori error estimator, we need to have a lower bound $\alpha_{\mathrm{LB}}^{\mathcal{N}}(\boldsymbol{\mu})$ of $\alpha^{\mathcal{N}}(\boldsymbol{\mu})$ such that $0 < \alpha_{\mathrm{LB}}^{\mathcal{N}}(\boldsymbol{\mu}) \leq \alpha^{\mathcal{N}}(\boldsymbol{\mu}) \ \forall \boldsymbol{\mu} \in \mathcal{D}$ and the online costs to compute $\alpha_{\mathrm{LB}}^{\mathcal{N}}(\boldsymbol{\mu})$ are independent of $\mathcal{N}$. We then define the following a posteriori error estimator

$$\Delta_N(\boldsymbol{\mu}) := \frac{\|r(\cdot; \boldsymbol{\mu})\|_{(X^{\mathcal{N}})'}}{\alpha_{\mathrm{LB}}^{\mathcal{N}}(\boldsymbol{\mu})}.$$

To compute $\|r(\cdot; \boldsymbol{\mu})\|_{(X^{\mathcal{N}})'}$, the main ingredients are to use the affine assumption (2.3) on a and the expansion (4.5) of $u_N^{\mathcal{N}}(\boldsymbol{\mu})$ in the space $X_N^{\mathcal{N}}$. Then, using the definition (4.7) of the residual, this leads to a system depending only on N for every $\boldsymbol{\mu}$, which makes the computation independent of $\mathcal{N}$.

The procedure used to compute the coercivity lower bound $\alpha_{\mathrm{LB}}^{\mathcal{N}}(\boldsymbol{\mu})$ is the so-called successive constraint method (SCM) [10]. Considering sets based on parameter samples and the terms Θ^q of the affine decomposition (2.3), it is possible to reduce this problem to a linear optimization problem. This method works by taking into account neighbour informations for the parameters and its precision increases with the size of the neighbourhood considered. The SCM also creates a coercivity upper bound $\alpha_{\mathrm{UB}}^{\mathcal{N}}(\boldsymbol{\mu})$ of $\alpha^{\mathcal{N}}(\boldsymbol{\mu})$ in the same manner. The algorithm is stopped when $\max_{\boldsymbol{\mu} \in \Xi} \left(\alpha_{\mathrm{UB}}^{\mathcal{N}}(\boldsymbol{\mu}) - \alpha_{\mathrm{LB}}^{\mathcal{N}}(\boldsymbol{\mu})\right) / \alpha_{\mathrm{UB}}^{\mathcal{N}}(\boldsymbol{\mu})$ is smaller than a prescribed tolerance ε.

We emphasize on the fact that it is very important that the costs associated to the computation of Δ_N are independent of $\mathcal{N}$. That allows us to develop the Offline-Online procedure discussed in Sect. 4, which is a crucial ingredient for the reduced basis method.

5 Numerical Illustrations

In this section, we present several numerical illustrations of the method introduced in this paper. The first example considered is a case of heat conduction involving only physical parameters, i.e. we use different conductivity coefficients in regions of the domain. Then a case containing geometrical parameters is introduced. The aim of the first two illustrations is to present the possibilities that are allowed while using NURBS basis functions. For this reason, both are computed over curvy three dimensional domains. In particular, the second example illustrates the theory developed in Sect. 3.3 to treat parameter dependent geometries. All computations have been performed using the `Matlab` [17] packages `GeoPDEs` [6] and `rbMIT` [9] for the NURBS and RB approximations, respectively.

Our goal in this section is to present standard examples to show that the method under consideration yields indeed good results. For this reason, all cases involve simple elliptic equations of the form

$$\begin{aligned} -\nabla\left(\sigma(\boldsymbol{\mu},x)\nabla u\right) &= f, \text{ in } \Omega(\boldsymbol{\mu}), \\ u &= g, \text{ on } \Gamma_D(\boldsymbol{\mu}), \\ \sigma(\boldsymbol{\mu},x)\tfrac{\partial u}{\partial n} &= h, \text{ on } \Gamma_N(\boldsymbol{\mu}), \end{aligned} \tag{5.1}$$

with $\Gamma_N \cap \Gamma_D = \emptyset$ and $\partial\Omega = \Gamma_N \cup \Gamma_D$. In particular, note that f, g and h are parameter independent. Moreover, we only deal with piecewise constant conductivity coefficients $\sigma(\boldsymbol{\mu}, x)$. For all examples, the prescribed tolerance for the greedy algorithm is 10^{-6}.

5.1 *Physical Parameters for Heat Conduction in a Pipeline*

In this first example, we consider heat conduction in a pipeline. The domain under consideration is depicted in Fig. 4. It is built on 5 different patches, one for every straight part and one for each of the curvy ones. The domain is parameter independent and we consider three parameters $\boldsymbol{\mu} = (\mu_1, \mu_2, \mu_3) \in [1,5]^3$, each one being the conductivity coefficient in one of the straight portion. More precisely,

$$a(\boldsymbol{\mu}, x) := \mu_1\chi_{\Omega_1} + \chi_{\Omega_2} + \mu_2\chi_{\Omega_3} + \chi_{\Omega_4} + \mu_3\chi_{\Omega_5},$$

where χ_{Ω_i} is the characteristic function over the ith patch, $1 \le i \le 5$. Let us denote the input boundary by Γ_{in}, the output by Γ_{out} and the inner and outer circular ones by Γ_{curve}. The functions are given by $f = 0$, $g = 0$ and $h = \chi_{\Gamma_{\text{out}}}$ and the associated boundaries are given by $\Gamma_D := \Gamma_{\text{in}}$ and $\Gamma_N := \Gamma_{\text{curve}} \cup \Gamma_{\text{out}}$. This simulation can be interpreted as heat conduction in a metal pipe where different metals constitute the structure and an imposed temperature is considered on one of the flat faces.

Fig. 4 Computation domain for the pipeline test case. Five patches were necessary to build the structure, one for each straight part and one for each angle

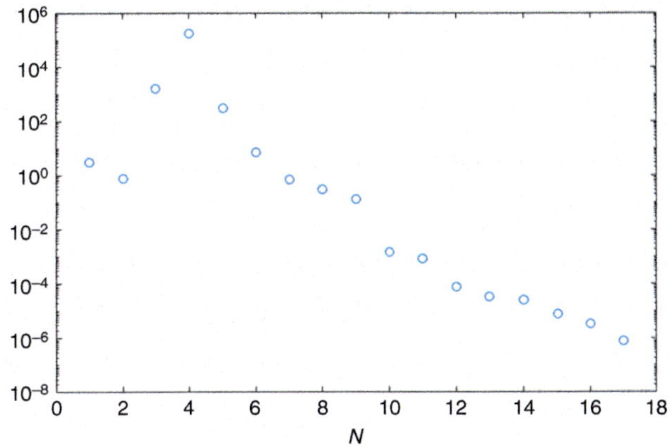

Fig. 5 Convergence of the greedy algorithm (see Sect. 4.1) for the pipeline test case of Sect. 5.1

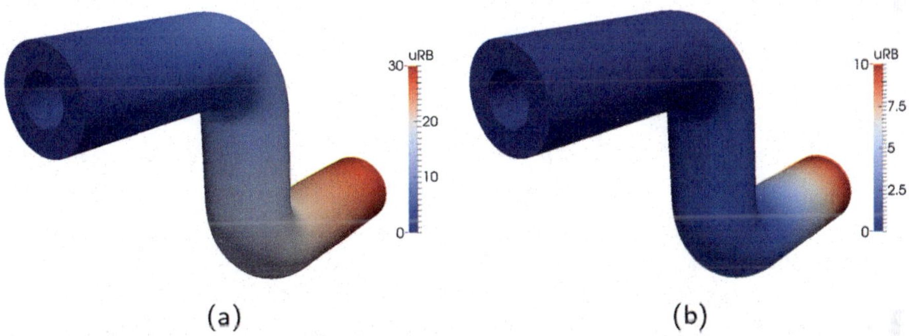

Fig. 6 Reduced basis approximation of the pipeline test case for the physical parameters (**a**) $\mu = (1, 1, 1)$ and (**b**) $\mu = (3, 2, 5)$. Note that the scale is not the same in both cases. The first case gives rise to a perfect linear approximation, which is the expected behavior. The second one displays a lack of smoothness at the interfaces of the patches. This is due to the fact that we have C^1 continuity inside each patch while only continuity is guaranteed at the interfaces

The number of degrees of freedom for the NURBS approximation is $\mathcal{N} = 16650$ while the size of the RB space is 17, which yields a big reduction of the computational costs. The computation time to perform the offline step is 27 min and the average evaluation time for the RB approximation is $5 \cdot 10^{-4}$ s. We present in Fig. 5 the convergence of the greedy algorithm.

Finally, in Fig. 6, we show the solution on the whole domain for different values of the parameters. Note that the solution is not completely smooth at the interfaces of the patches. This comes from the fact that we have C^1 continuity in each of the patch while we only have C^0 continuity at the interfaces. Methods exist to obtain more regularity at the interfaces (see e.g. [3]).

5.2 Geometrical Parameters for Heat Conduction in a Cylinder

We present here the case of a parameter dependent geometry. The reference domain is a cylinder of radius 2 and height 1 oriented in the z direction, i.e. $\Omega(\boldsymbol{\mu}_{\text{ref}}) := \{(r, \theta, z) \mid r \in [0, 2], \theta \in [0, 2\pi], z \in [0, 1]\}$ where (r, θ, z) denote the cylindrical coordinates in $\mathbb{R}^3$. The reference cylinder is depicted in Fig. 7a. To build it, four patches were necessary.

The transformations under consideration are scaling with respect to the y and z axis. More precisely, three parameters $\boldsymbol{\mu} = (\mu_1, \mu_2, \mu_3) \in [1, 5]^3$ are considered, where μ_1 scales the portion of the domain for which $y > 0$, μ_2 the one where $y < 0$, and μ_3 scales in the z direction. In other words, the transformation in the part of the domain where $y > 0$ is given by

$$C^k = \begin{pmatrix} 0 \\ 0 \\ 0 \end{pmatrix}, \qquad G^k = \begin{pmatrix} 1 & 0 & 0 \\ 0 & \mu_1 & 0 \\ 0 & 0 & \mu_3 \end{pmatrix},$$

while in the region $y < 0$ it reads

$$C^k = \begin{pmatrix} 0 \\ 0 \\ 0 \end{pmatrix}, \qquad G^k = \begin{pmatrix} 1 & 0 & 0 \\ 0 & \mu_2 & 0 \\ 0 & 0 & \mu_3 \end{pmatrix}.$$

In Fig. 7, we present the domain after application of the affine transformation for different values of the parameters.

Considering the conductivity coefficient in (5.1), we have $\sigma(\boldsymbol{\mu}, x) = 1$. In order to describe the boundary conditions, let us denote by Γ_{bot}, Γ_{top} and Γ_{curve} the bottom, top and curvy boundaries, respectively. We impose homogeneous Dirichlet boundary conditions on $\Gamma_D := \Gamma_{\text{curve}}$ and unitary Neumann conditions on $\Gamma_N := \Gamma_{\text{top}} \cup \Gamma_{\text{bot}}$. Finally the right-hand side function is $f = 10\chi_B$, where B is the ball of radius 0.2 centered at $(0, 0, 0.5)$.

Turning to the computational costs, the number of degrees of freedom for the NURBS approximation is $\mathcal{N} = 3240$ and the one of the RB is $N = 57$. The whole

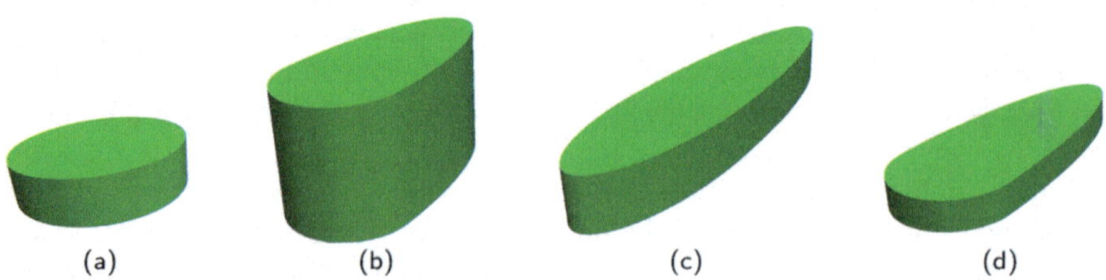

Fig. 7 Computational domain for the cylinder test case for different values of the parameters. The original domain is presented in (**a**). Four patches were necessary to build the structure, one for each quarter of the cylinder. The affine transformation from Sect. 5.2 were considered for (**b**) $\boldsymbol{\mu} = (1, 3, 4)$, (**c**) $\boldsymbol{\mu} = (3, 5, 2)$ and (**d**) $\boldsymbol{\mu} = (1, 4, 1)$

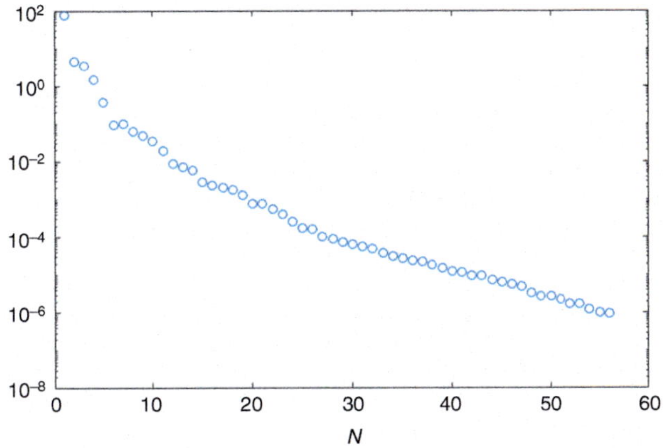

Fig. 8 Convergence of the greedy algorithm (see Sect. 4.1) for the cylinder test case of Sect. 5.2

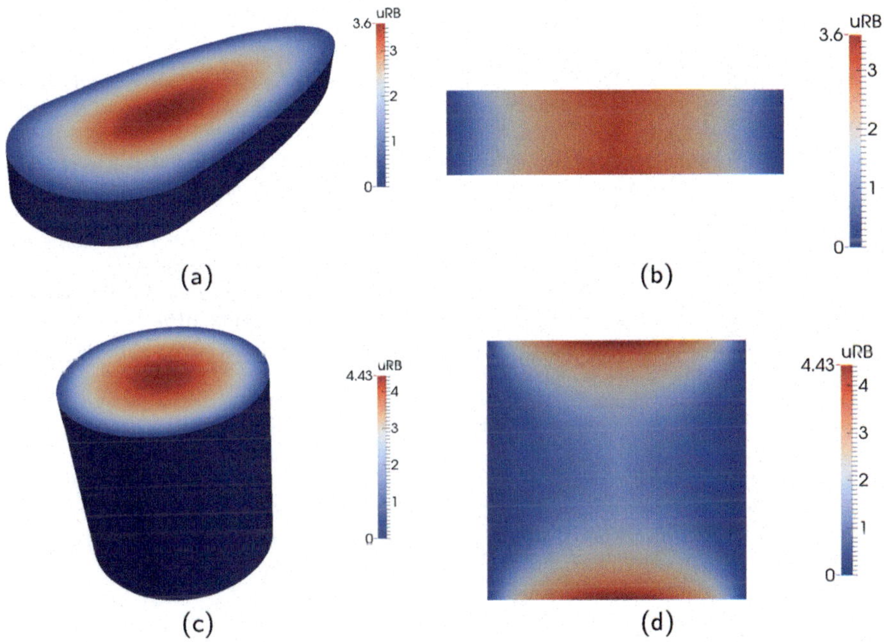

Fig. 9 Reduced basis approximation of the cylinder test case for (**a**) $\boldsymbol{\mu} = (1, 4, 1)$ and (**c**) $\boldsymbol{\mu} = (1, 1, 4)$. The pictures (**b**) and (**d**) represent the value of the field on the plane $\{(x, y, z) \in \Omega \mid y = 0\}$ for the values considered in (**a**) and (**c**), respectively. Note that different scales have been used for the different values of the parameters

offline procedure took 30 min while the average RB evaluation takes $5 \cdot 10^{-4}$ s. In Fig. 8, we show the convergence of the greedy algorithm for this case.

The solution for several values is depicted in Fig. 9. We present the RB approximation on the whole domain as well as its evaluation on the plane $\{(x, y, z) \in \Omega \mid y = 0\}$.

References

1. Y. Bazilevs, L. Beirao da Veiga, J.A. Cottrell, T.J.R. Hughes, G. Sangalli, Isogeometric analysis: approximation, stability and error estimates for h-refined meshes. Math. Models Methods Appl. Sci. **16**(7), 1031–1090 (2006)
2. F. Chinesta, A. Huerta, G. Rozza, K. Willcox, Model Order Reduction. Encyclopedia of Computational Mechanics (Elsevier, Amsterdam, 2016)
3. J.A. Cottrell, T.J.R. Hughes, Y. Bazilevs, Isogeometric analysis: toward integration of CADand FEA (John Wiley & Sons, Chichester, 2009)
4. M.G. Cox, The numerical evaluation of b-splines. IMA J. Appl. Math. **10**(2), 134–149 (1972)
5. C. De Boor, On calculating with B-splines. J. Approx. Theory **6**(1), 50–62 (1972)
6. C. De Falco, A. Reali, R. Vázquez, GeoPDEs: a research tool for isogeometric analysis of PDEs. Adv. Eng. Softw. **42**(12), 1020–1034 (2011)
7. J.S. Hesthaven, G. Rozza, B. Stamm, *Certified Reduced Basis Methods for Parametrized Partial Differential Equations*. Springer Briefs in Mathematics (Springer, Berlin, 2015)
8. T.J.R. Hughes, J.A. Cottrell, Y. Bazilevs, Isogeometric analysis: cad, finite elements, nurbs, exact geometry and mesh refinement. Comput. Methods Appl. Mech. Eng. **194**(39), 4135–4195 (2005)
9. D.B.P. Huynh, N.C. Nguyen, G. Rozza, A.T. Patera, rbMIT software: copyright MIT. Technology Licensing Office (2006/2007), http://augustine.mit.edu/
10. D.B.P. Huynh, G. Rozza, S. Sen, A.T. Patera, A successive constraint linear optimization method for lower bounds of parametric coercivity and inf-sup stability constants. C.R. Math. **345**(8), 473–478 (2007)
11. A. Manzoni, F. Salmoiraghi, L. Heltai, Reduced basis isogeometric methods (RB-IGA) for the real-time simulation of potential flows about parametrized NACA airfoils. Comput. Methods Appl. Mech. Eng. **284**, 1147–1180 (2015)
12. A.T. Patera, G. Rozza, *Reduced Basis Approximation and a Posteriori Error Estimation for Parametrized Partial Differential Equations*. MIT (2007), http://augustine.mit.edu/
13. A. Quarteroni, G. Rozza, A. Manzoni, Certified reduced basis approximation for parametrized partial differential equations and applications. J. Math. Ind. **1**(1), 1–49 (2011)
14. G. Rozza, Reduced basis approximation and error bounds for potential flows in parametrized geometries. Commun. Comput. Phys. **9**, 1–48 (2011)
15. G. Rozza, D.B.P. Huynh, A.T. Patera, Reduced basis approximation and a posteriori error estimation for affinely parametrized elliptic coercive partial differential equations. Arch. Comput. Meth. Eng. **15**(3), 229–275 (2008)
16. F. Salmoiraghi, F. Ballarin, L. Heltai, G. Rozza, Isogeometric analysis-based reduced order modelling for incompressible linear viscous flows in parametrized shapes. Adv. Model. Simul. Eng. Sci. **3**, 21 (2016)
17. The MathWorks Inc. Matlab. version 8.1.0.604 (R2013a) (2013)

Towards *p*-Adaptive Spectral/*hp* Element Methods for Modelling Industrial Flows

D. Moxey, C.D. Cantwell, G. Mengaldo, D. Serson, D. Ekelschot, J. Peiró, S.J. Sherwin, and R.M. Kirby

Abstract There is an increasing requirement from both academia and industry for high-fidelity flow simulations that are able to accurately capture complicated and transient flow dynamics in complex geometries. Coupled with the growing availability of high-performance, highly parallel computing resources, there is therefore a demand for scalable numerical methods and corresponding software frameworks which can deliver the next-generation of complex and detailed fluid simulations to scientists and engineers in an efficient way. In this article we discuss recent and upcoming advances in the use of the *spectral/hp element method* for addressing these modelling challenges. To use these methods efficiently for such applications, is critical that computational resolution is placed in the regions of the flow where it is needed most, which is often not known *a priori*. We propose the use of spatially and temporally varying polynomial order, coupled with appropriate error estimators, as key requirements in permitting these methods to achieve computationally efficient high-fidelity solutions to complex flow problems in the fluid dynamics community.

D. Moxey (✉)
College of Engineering, Mathematics and Physical Sciences, University of Exeter, Exeter EX4 4QF, UK
e-mail: d.moxey@exeter.ac.uk

C.D. Cantwell • D. Serson • D. Ekelschot • J. Peiró • S.J. Sherwin
Department of Aeronautics, Imperial College London, London SW7 2AZ, UK
e-mail: c.cantwell@imperial.ac.uk; d.serson14@imperial.ac.uk; d.ekelschot12@imperial.ac.uk; j.peiro@imperial.ac.uk; s.sherwin@imperial.ac.uk

G. Mengaldo
Division of Engineering and Applied Sciences, California Institute of Technology, Pasadena, CA 91125, USA
e-mail: mengaldo@caltech.edu

R.M. Kirby
Scientific Computing and Imaging Institute, University of Utah, Salt Lake City, UT 84112, USA
e-mail: kirby@cs.utah.edu

M.L. Bittencourt et al. (eds.), *Spectral and High Order Methods for Partial Differential Equations ICOSAHOM 2016*, Lecture Notes in Computational Science and Engineering 119, DOI 10.1007/978-3-319-65870-4_4

1 Introduction

Computational modelling is now regularly used in the fluid dynamics community, giving insight into flow problems where experimentation is too difficult, impractical or costly to realise. The complex geometries and time constraints involved in modern industrial studies imply that, to date, most numerical simulations are restricted to being steady in time. This limits their capabilities, particularly when the problem of interest involves fundamentally unsteady flow dynamics, such as vortex shedding. However, with the wider availability and reducing cost of large-scale computing power, academic and industrial fluid dynamicists are increasingly looking to perform finely-detailed unsteady simulations. These high-fidelity simulations will allow us to obtain deeper insight into many challenging engineering problems, where steady-state solvers struggle to capture the relevant unsteady flow structures.

One of the main challenges in conducting such simulations is that the complex geometries that are a natural consequence of studying industrial problems will inherently generate flow structures across a large range of time and length scales. From a practical perspective, it becomes difficult or impossible to predict where numerical resolution is required in the computational domain before the simulation is run in order to accurately resolve the flow. Since uniform refinement across very large domains is computationally prohibitive, the community is turning to adaptive methods, where resolution is dynamically adjusted within the domain as a function of time, in order to overcome this issue.

The spectral/*hp* element method [11] – in which an unstructured elemental decomposition capable of resolving complex geometries is equipped with high-order polynomial bases are used to give routes to convergence in terms of element size h and polynomial order p – has been used in academic applications for several years. However, it is now emerging as one of the enabling technologies for the simulation of high-fidelity industrial simulations. From a numerical perspective, these methods offer attractive properties such as low diffusion and dispersion errors, meaning that for smooth solutions fewer degrees of freedom are required to attain the same accuracy as compared to traditional low-order methods [19]. From a computational perspective, the use of a higher polynomial order leads to compact data structures and enables a balance between the computational and memory intensiveness of the method. This is increasingly becoming a key factor in the efficient use of modern many-core hardware.

On the whole, the development of adaptive methods has been mostly focused around h-adaption, where the elements are refined or coarsened in order to adjust the numerical resolution. The use of p-adaption, on the other hand, has received far less attention. Most of the work in this area has focused on *hp*-adaption, which has been an area of significant attention with various works investigating these techniques for elliptic problems [4, 8, 24] that are not necessarily immediately applicable for fluid-based problems. However, p-adaption has been shown to be a viable technique in a study by Li and Jameson [14], where adaption in p was shown to provide the highest accuracy with respect to the numerical resolution and computing time.

However high-order methods have presented inherent difficulties that have only started to be overcome in the last few years. These challenges are both mathematical and practical. On the theoretical side, there has been a need to overcome stability issues arising due to aliasing of the solution [17] and timestep size [6]; investigate the generation of curved meshes which conform to the boundary of complex three-dimensional domains [20, 22]; and investigate parallel scaling of these methods [27]. On the practical side, the mathematical complexity of the methods has necessitated the development of software frameworks [3] to improve accessibility to academia and industry. These developments now mean that these high-order methods are being applied in very high Reynolds number flows that are of significant interest to, for example, the aerodynamics and aeronautics community [15].

In this article, we will discuss some practicalities of implementing spectral/*hp* element solvers which use a spatially variable polynomial order across computational domain. We do this both in the context of incompressible and compressible flow. For the former we use a continuous Galerkin approach to solving a semi-implicit form of the incompressible Navier-Stokes equations; for the latter we use a discontinuous Galerkin projection with an explicit time-stepping method. Section 2 discusses the formulation of these methods and how variable polynomial orders are handled in each case. Section 3 illustrates the capabilities of adaptivity in *p*, before concluding with a brief outlook in Sect. 4.

2 Formulation

This section begins with a brief discussion of the formulation of the spectral/*hp* element method, the basis being used to represent elemental expansions and how this relates to discontinuous and continuous formulations. We then describe how this formulation can be adapted to allow variable polynomial order across the computational domain, provide implementation details and give an overview of the techniques required to make this approach computationally tractable.

2.1 Domain Discretisation

The domain Ω is subdivided into N_{el} non-overlapping elements Ω^e, such that $\Omega = \bigcup_{e=1}^{N_{\mathrm{el}}} \Omega^e$. In two dimensions, these elements are a mixture of quadrilaterals and triangles; in three dimensions, a mixture of hexahedra, triangular prisms, square-based pyramids and tetrahedra are considered. We define a standard element Ω_{st} for each shape. For example, a standard quadrilateral is defined by $\Omega_{\mathrm{st}} = \{(\xi_1, \xi_2) | \xi_1, \xi_2 \in [-1, 1]\}$. We equip each standard region with a set of polynomial basis functions ϕ_n with which to approximate functions. A scalar function u on an element Ω^e is represented by an expansion

$$u(\mathbf{x}) = \sum_{n=1}^{M(e,P)} \hat{u}_n^e \phi_n(\boldsymbol{\xi}), \tag{1}$$

where points $\boldsymbol{\xi} \in \Omega_{\mathrm{st}}$, $\mathbf{x} \in \Omega^e$, and the two are related through an invertible mapping $\boldsymbol{\chi}^e : \Omega_{\mathrm{st}} \to \Omega^e$ such that $\mathbf{x} = \boldsymbol{\chi}^e(\boldsymbol{\xi})$. The upper bound of the summation, $M(e, P)$, defines the number of modes that represent the solution in the element Ω^e and is a function of both the polynomial order and the element type. We let $\mathbb{P}_k(\Omega^e)$ denote the polynomial space spanned by the $M(e, P)$ basis functions, with k the maximum polynomial order, on the e-th elemental region.

In order to represent a function across the entire domain Ω, we must select an appropriate function space to represent our approximation. In this work we will consider two classic discretisations: the continuous (CG) and discontinuous Galerkin (DG) methods, which require the spaces

$$D^{\mathrm{CG}}(\Omega) = \{v \in C^0(\Omega) \mid v|_{\Omega^e} \in \mathbb{P}_k(\Omega^e)\}, \tag{2}$$

$$D^{\mathrm{DG}}(\Omega) = \{v \in L^2(\Omega) \mid v|_{\Omega^e} \in \mathbb{P}_k(\Omega^e)\} \tag{3}$$

with C^0 and L^2 being the usual spaces of continuous and square-integrable functions respectively and k initially considered spatially constant across elements. We note that in the context of discontinuous spectral element methods, significant effort has recently been spent in the development of high-order flux reconstruction schemes [10, 25]. While they are in principle different, these schemes can be cast within the same framework as the discontinuous Galerkin method [9, 18]. Therefore, the adaption technique described hereafter can be directly extended to the flux reconstruction method.

2.1.1 Choice of Basis

The choice of the basis ϕ is particularly important when variable polynomial order across elements is required. We opt to use a set of functions that augment the usual linear finite element modes with higher-order polynomials, defined as

$$\psi_p(\xi) = \begin{cases} \frac{1-\xi}{2}, & p = 0, \\ \frac{1+\xi}{2}, & p = 1, \\ \frac{1-\xi}{2}\frac{1+\xi}{2}P_{p-2}^{(1,1)}(\xi), & p \geq 2, \end{cases} \tag{4}$$

where $P_p^{(\alpha,\beta)}(\xi)$ is the p-th order Jacobi polynomial with coefficients α and β. In one dimension on the segment $[-1, 1]$, we have that $\phi_n = \psi_n$ in (1). In higher dimensions, quadrilaterals and hexahedral expansion bases are defined using a

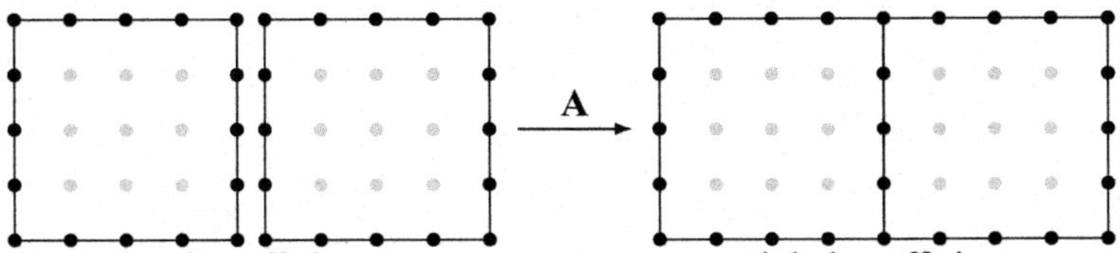

Fig. 1 Diagram describing assembly operation between two $\mathbb{P}_4$ quadrilaterals

tensor product of these one-dimensional functions. Other element types use a similar choice of basis that still permits a tensorial expansion (for more details, see [11]).

There are several advantages to this choice of basis in the context of a mesh of variable polynomial order. The first is that it results in a topological decomposition of the basis, so that the modes of an element can be classified into vertex, edge-interior, face-interior and volume-interior modes. Only vertex, edge and face modes have support which extends to the boundary of the element; interior modes are zero on the boundary. This is depicted for an order 4 quadrilateral in Fig. 1, where black circles represent the boundary modes and grey the interior. When we discuss the modification of any contributions along an edge of the element, this only therefore requires the modification of coefficients along that edge, as opposed to across the entire element. Additionally, this set of modes is *hierarchical*; that is, the degree of each basis polynomial $\phi_p(\xi)$ increases as a function of p. This is in contrast to, for example, a classical spectral element method in which Lagrange interpolants define a nodal basis depending on a choice of nodes ξ_j. At order P these are defined as

$$\phi_p(\xi) = \ell_p(\xi) = \prod_{q\neq p}^{q=P} \frac{\xi - \xi_q}{\xi_p - \xi_q}$$

so that every basis function is of the same polynomial order P, whilst still yielding a boundary-interior decomposition.

2.2 *Implementation Details*

2.2.1 Continuous Galerkin Formulation

The key operation of the CG formulation is assembly, wherein local elemental contributions are gathered to impose the C^0-continuity of the underlying function space, as depicted visually in Fig. 1. The assembly operation associates a vector of concatenated local elemental coefficients $\hat{\mathbf{u}}_l = (\hat{\mathbf{u}}^1, \ldots, \hat{\mathbf{u}}^{N_{\rm el}})$ to their global counterparts $\hat{\mathbf{u}}_g$ through an injective map. Here, we note that each $\hat{\mathbf{u}}^e$ corresponds to the vector of local coefficients in Eq. 1. The coefficients in $\hat{\mathbf{u}}_g$ describe the

Algorithm 1 Continuous C^0 assembly operation

for $e = 1 \to N_{\text{el}}$ **do**
  for $i = 1 \to M(e)$ **do**
    $\hat{\mathbf{u}}_g[\text{map}[e][i]] \mathrel{+}= \text{sign}[e][i]\ \hat{\mathbf{u}}_l^e[\text{i}]$
  end for
end for

contribution to the solution of the modes which span $D^{\text{CG}}(\Omega)$. Mathematically, this operation is expressed through a sparse matrix-vector operation $\hat{\mathbf{u}}_g = \mathbf{A}\hat{\mathbf{u}}_l$. For a uniform polynomial order mesh, the columns of $\mathbf{A}$ are non-zero where local degrees of freedom meet to form global degrees of freedom, and zero otherwise, so that the valency of a global degree of freedom i is defined as the number of non-zero columns in the i-th row of $\mathbf{A}$. In practice, the high sparsity of $\mathbf{A}$ means that we use array indirection to implement the action of $\mathbf{A}$ without explicitly constructing it, as defined in Algorithm 1.
We note that two arrays are required:

- map[e][i] stores the index of the global degree of freedom corresponding to mode i of element Ω^e;
- sign[e][i] stores either 1 or -1 to align modes that are of odd polynomial orders such that the basis remains continuous (see [11] for more details).

Throughout the rest of this section we will consider a Helmholtz problem

$$\nabla^2 u + \lambda u = f \tag{5}$$

which is later used for incompressible simulations through the use of an operator splitting scheme [13]. This is put into a weak form by defining appropriate finite-dimensional test and trial spaces, multiplying each term by a test function and integrating over the domain. After applying integration by parts we obtain the equation

$$(\mathbf{L} + \lambda\mathbf{M})\hat{\mathbf{u}}_g = \hat{\mathbf{f}}$$

where $\mathbf{L}$ and $\mathbf{M}$ are the global Laplacian and mass matrices, respectively, and $\hat{\mathbf{f}}$ is the Galerkin projection of f onto $D^{\text{CG}}(\Omega)$. The assembly map is used not only to calculate $\hat{\mathbf{f}}$, but also to construct the matrices $\mathbf{L}$ and $\mathbf{M}$ from their constituent elemental matrices, through the relationship

$$\mathbf{L} = \mathbf{A}\left[\bigoplus_{e=1}^{N_{\text{el}}} \mathbf{L}^e\right]\mathbf{A}^\top. \tag{6}$$

We note that in practice, even at moderately low polynomial orders, $\mathbf{L} + \lambda\mathbf{M}$ is rarely explicitly constructed. The use of the mapping above allows us to apply the

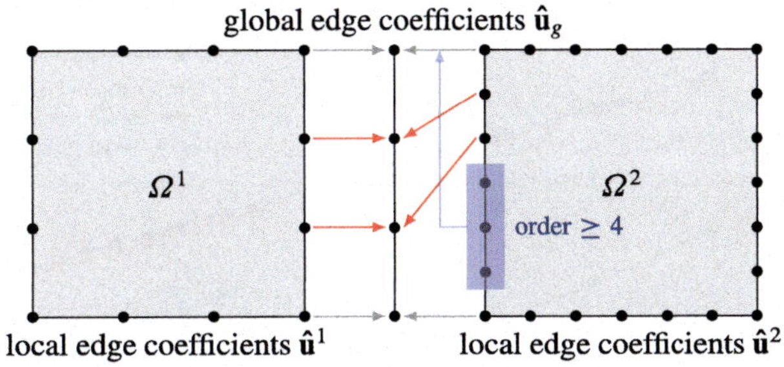

Fig. 2 Diagram describing assembly operation between a $\mathbb{P}_3$ and $\mathbb{P}_6$ quadrilateral. The nodes here correspond to vertex and edge modes of the hierarchical basis. *Red arrows* indicate the usual connectivity; *blue arrows* indicate modes that are zeroed using the sign array

action of this operator and leverage the computational optimisations possible due to the rich structure of the elemental matrices.

To modify this procedure for spatially varying polynomial orders, we must address the situation depicted in Fig. 2, where two elements meet that differ in polynomial orders; in this case a $\mathbb{P}_3$ and $\mathbb{P}_6$ quadrilateral. In the global space, the edge connecting these elements (depicted in the middle of the figure) should be at most an order 3 polynomial and so some additional logic is required to discard the higher degrees of freedom contributed by the $\mathbb{P}_6$ quadrilateral in the assembly process. To this end, we note that since we are using a hierarchical basis, Algorithm 1 can remain unchanged by altering the sign and mapping arrays to easily filter out the higher-order contributions. We impose that on the common edge, the coefficients of the sign array on the $\mathbb{P}_6$ element are set to zero for the highlighted modes corresponding to a polynomial degrees between 4 and 6. This ensures that in the assembly operation, no contribution from these high-frequency modes is included. The corresponding coefficients in the mapping array are set to point to one of the known vertex coefficients to avoid memory overflow errors. We note that if the basis were not to be of a hierarchical construction, then in general, all of the modes along an edge can be of equal polynomial order. In this case, the above procedure needs to be modified to perform a polynomial interpolation onto the correct space, rather than simply zeroing elements of the sign array.

As a test of the validity of this approach, we consider the Helmholtz problem in the a square $[-1, 1]^2$, in which $\mathbf{f}$ is defined to obtain a prescribed solution $u(x, y) = \sin(\pi x)\sin(\pi y)$. We consider a series of meshes with h elements in each direction. We then solve Eq. (5) using the continuous Galerkin formulation for four cases. uniform polynomial orders of $P = 6$ and $P = 9$, and then a mixed order where half of the elements are set to $P = 6$, and half to $P = 9$. Figure 3 shows the L^2 error of these simulations, where we clearly observe the same convergence rate for all simulations, and the mixed order case has a slightly lower error than the $P = 6$ case as expected. Increasing the mixed order case to $P = 7$ lowers the error so that it lies between the two uniform simulations.

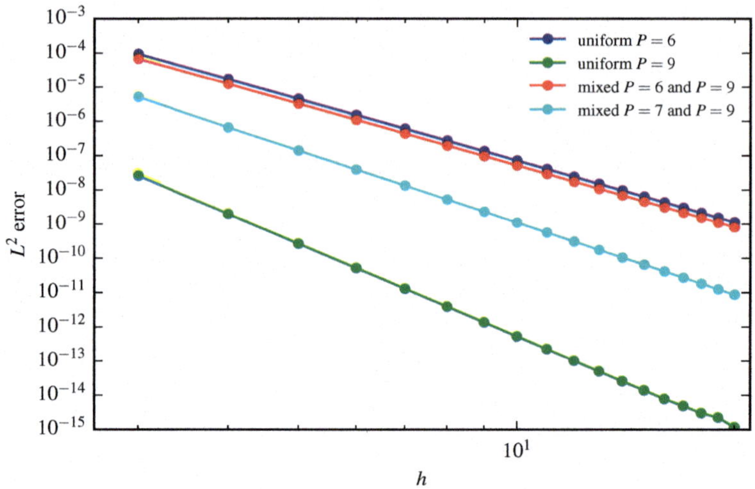

Fig. 3 Convergence of Helmholtz problem for simple square case

2.2.2 Discontinuous Galerkin Formulation

We now briefly discuss the implementation of variable polynomial order in the discontinuous Galerkin (DG) formulation, which is described in greater detail in [5]. The use of DG is widely increasing in modern fluid dynamics codes and is especially popular for discretising hyperbolic or mixed hyperbolic-parabolic systems, such as the compressible Euler and Navier-Stokes equations, which form the cornerstone of modern aerodynamics problems. To illustrate the discretisation we consider a simple scalar conservation law

$$\frac{\partial u}{\partial t} + \nabla \cdot \mathbf{F}(u) = 0.$$

Using the variational form of the problem together with the function space $D^{DG}(\Omega)$ defined in Eq. (3) leads to the discontinuous Galerkin method, wherein we consider for each element the ODE system

$$\frac{\mathrm{d}}{\mathrm{d}t}\int_{\Omega^e} u\phi \,\mathrm{d}\mathbf{x} + \int_{\partial\Omega^e} \phi\mathbf{F}(u)\cdot\mathbf{n}\,\mathrm{d}s = \int_{\Omega^e} \mathbf{F}(u)\cdot\nabla\phi\,\mathrm{d}x$$

where ϕ is a test function lying in $\mathbb{P}_k(\Omega^e)$ and $\mathbf{n}$ denotes the normal vector to the element boundary $\partial\Omega^e$. We also assume that these ODEs are discretised explicitly in time, so that at each timestep we must calculate the volume term on the right hand side, calculate the flux term on the left hand side, and then incorporate the flux term into the volume term. The remaining first term on the left hand side, in an explicit timestepping setting, corresponds to the action of the elemental mass matrix. The

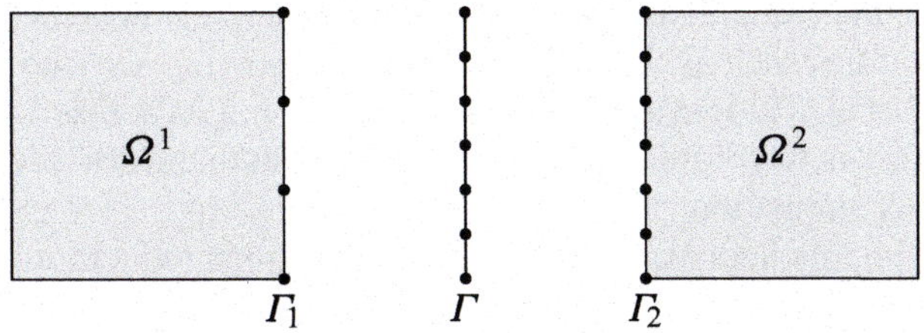

Fig. 4 Diagram describing treatment of variable polynomial order in DG for quadrilateral elements

only place in which we need to consider the application of variable polynomial order is therefore the second part of this process.

We again consider the problem of two quadrilaterals of different orders in Fig. 4. We first note that since the discretisation in time considers all elemental degrees of freedom, we must consider the boundary terms at the higher polynomial order to avoid stability issues, otherwise there are degrees of freedom within the higher-order element that become undetermined. Additionally, we wish to preserve the locally conservative nature of the DG method, implying that in the notation of the figure, we require

$$\int_{\Gamma_1} \mathbf{F}(u)\,\mathrm{d}s = \int_{\Gamma_2} \mathbf{F}(u)\,\mathrm{d}s$$

where Γ_1 and Γ_2 are the edges of the two elements that intersect to make the trace element Γ.

To project the trace contributions back into the volume consistently, on the higher side we may simply copy the coefficients directly from Γ to Γ_2. For Γ_1, we have a higher-degree polynomial that must be incorporated into a lower degree edge. To do this in a conservative fashion, we perform a change of basis of the elemental coefficients from the hierarchical basis of Eq. (4) onto an orthogonal space of Legendre polynomials. We then apply a low-pass filter, by zeroing the unwanted high-frequency polynomials. This is necessary since the basis functions given in Eq. (4) are not orthogonal and performing a filtering in this space will alter the mean flux, leading to a loss of conservation. Finally, we perform a change of basis back to the lower-order hierarchical basis.

2.3 *Efficiency Across a Range of Polynomial Orders*

As a final note on implementation considerations, a clear observation that can be made when using a variable polynomial order is that the sizes of elemental matrices can vary drastically, particularly when considering three-dimensional elements. This is important since operator evaluations, such as the Laplacian matrix of Eq. (6), form the bulk of the computational cost of the spectral/*hp* element method, either in computing quantities such as the inner product or in solving a system of equations in an iterative fashion. Efficient evaluation of these operators across a wide range of polynomial orders is therefore an important component to the efficacy of a variable polynomial order simulation.

The underlying mathematical formulation and tensor-product form of the basis admits a number of different implementation choices for the evaluation of these operators, each of which admits differing performance across polynomial orders and choice of hardware [1, 2, 16, 26]. Furthermore, the results of [21] suggest that for modern hardware, where memory bandwidth is a valuable commodity, elemental operations should be amalgamated wherever possible to minimise data transfer and efficiently utilise the memory hierarchy. In the context of variable polynomial orders, the amalgamation of elements that are of the same type and polynomial order, combined with an appropriate implementation strategy as described in [21], should be performed to maximise the computational performance of the method.

3 Results

This section gives a brief overview of results achieved to date using adaption in the polynomial order p with the compressible and incompressible formulations, focusing on error indicators and how this affects the ability to capture the underlying flow physics, and on the computational cost of these approaches.

3.1 Incompressible Flow

In this section, we present an example of a simulation employing adaptive polynomial order for solving the incompressible Navier-Stokes equations, which can be represented as

$$\frac{\partial \mathbf{u}}{\partial t} = -(\mathbf{u}\cdot\nabla)\mathbf{u} - \nabla p + \nu\nabla^2\mathbf{u}, \quad \nabla\cdot\mathbf{u} = 0 \tag{7}$$

where $\mathbf{u}$ is the velocity, p is the pressure, and ν is the kinematic viscosity and, without loss of generality, we set the density to be unity. Given a reference length L and a reference velocity U, the Reynolds number is defined as $Re = \frac{LU}{\nu}$. We solve these equations using a CG-approach and a semi-implicit velocity-correction scheme [13], whereby (7) is separated into an explicit convective term, an implicit Poisson equation for pressure and three further implicit Helmholtz equations for the velocity components.

In the procedure we employed, the polynomial order is adjusted during the solution based on an estimate of the discretisation error in each element. This estimate for the error (sometimes called *sensor*) was based on the one used for shock capture in [23]. In the present work, this is defined as

$$S_e = \frac{\|u_P - u_{P-1}\|_{2,e}^2}{\|u_P\|_{2,e}^2}, \tag{8}$$

where u_P is the solution obtained for the u velocity using the current polynomial order P, u_{P-1} is the projection of this solution to a polynomial of order $P-1$, $\|\cdot\|_2$ is the L_2 norm and the subscript e indicates that this refers to a single element.

Considering this estimate for the discretisation error, the adaptive procedure can be summarized as:

1. Advance the equations for n_{steps} time steps.
2. Calculate S_e for each element.
3. Modify the polynomial order in each element:
   - if $S_e \geq \epsilon_u$ and $P < P_{\text{max}}$, increase P by 1;
   - if $S_e \leq \epsilon_l$ and $P > P_{\text{min}}$, decrease P by 1;
   - maintain same P if none of the above is true.
4. Project the solution to the new polynomial space.
5. Repeat for n_{runs}.

In the above, ϵ_u is the tolerance above which the polynomial order is increased, $\epsilon_l \leq \epsilon_u$ is the tolerance below which the polynomial order is decreased and P_{min} and P_{max} are the minimum and maximum polynomial orders imposed on the procedure.

It is important to note that changing the polynomial order during the solution is costly, due to the need to assemble and decompose the linear systems for the implicit part of the method. Therefore, the choice of n_{steps} plays a key role in obtaining an efficient solution. A lower value of n_{steps} will lead to the refinement step being performed more frequently, at the expense of a higher average computational cost per timestep.

To illustrate this method, we consider quasi-3D simulations of the incompressible flow around a NACA0012 profile, shown in Fig. 5b, with Reynolds number $Re = 50{,}000$ and angle of attack $\alpha = 15°$. A spectral/hp discretisation is applied in the xy plane, with the span direction discretised by a Fourier series, as proposed in [12]. The adaptive procedure was employed only in the spectral/hp plane, with a fixed number of modes used in the Fourier direction.

Figure 5 shows the distribution of polynomial order obtained using $n_{\text{steps}} = 4{,}000$, $P_{\text{min}} = 2$ and $P_{\text{max}} = 9$. It is clear that the boundary layers and the regions of turbulent separated flow are represented by high order polynomials, while lower orders are used in regions of laminar flow far from the wing. In this case, the average number of degrees of freedom per element is approximately 49, which is equivalent to the value for a constant $P = 6$ simulation.

Table 1 compares the cost of this simulation using the adaptive procedure with the cost for several different values of constant polynomial order, and with using

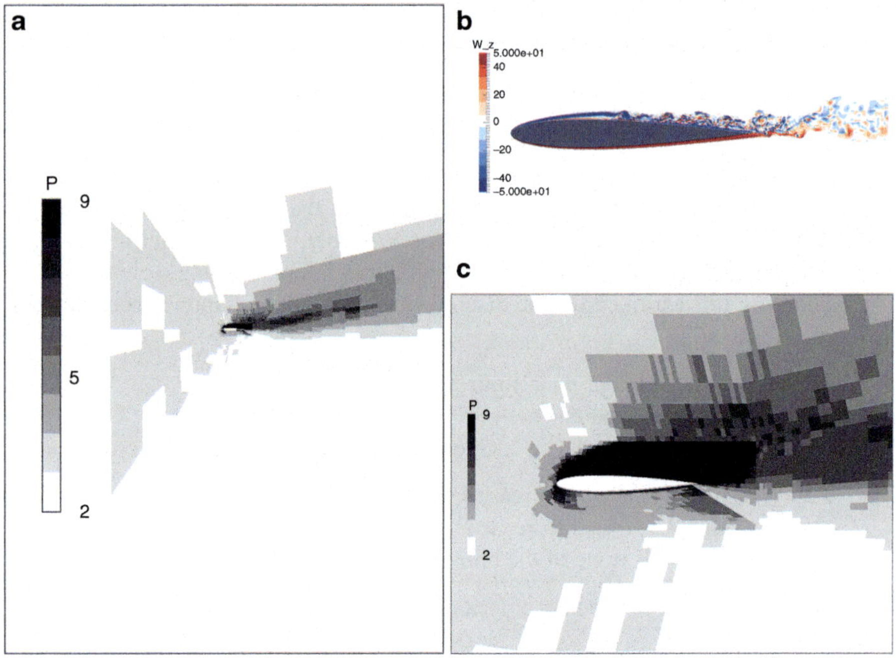

Fig. 5 Polynomial order distribution obtained for incompressible flow around a NACA0012 profile with $Re = 50{,}000$ and $\alpha = 15°$. (**a**) Macro. (**b**) Representative flow solution. (**c**) Detail

Table 1 Comparison of the computational cost of adaptive order case of Fig. 5 with constant uniform polynomial order and with variable order without adaptive procedure

Case	*Cost*	$\frac{1}{Cost}$
$P = 5$	0.60	1.66
$P = 6$	0.72	1.39
$P = 7$	1.08	0.93
$P = 8$	1.19	0.84
$P = 9$	1.53	0.65
Variable order (fixed)	0.95	1.05
Adaptive order	1.00	1.00

The computational costs are normalized with respect to the adaptive order case

the same variable polynomial order distribution without performing the adaptive procedure. We note that for this value of n_{steps}, the refinement procedure corresponds to 5% of the computational cost. This is more than offset by the gains obtained from using a more efficient distribution of degrees of freedom, with the adaptive case presenting roughly the same cost as the $P = 7$ case, and being 35% faster than the $P = 9$ case.

3.2 Compressible Flow Using Explicit Timestepping

The accurate solution of compressible flow is an important topic in a number of application areas. For instance, the aeronautical community is concerned with accurately predicting the lift and drag coefficients of different wing configurations whilst keeping the computational cost low. This allows considering a wide range of geometries during the design lifecycle and provides the basis for aerodynamic shape optimization. In these applications, the key to accurately predicting lift and drag lies in determining the regions of the domain which influence these coefficients the most. Adaptive methods, combined with appropriate error estimators, are one route to producing fast, accurate and reliable results. This section describes progress made in [5], where a goal-based error estimator based on an adjoint problem derived from the underlying equations has been applied together with the p-adaptive techniques described in the previous section. The error estimator derives an adjoint problem from a coarsely-resolved base flow, the solution of which determines the areas of the domain which have the greatest sensitivity to the lift and drag coefficients. Although this technique has been explored previously, a review of which can be found in [7], this has mostly focused around h-adaptivity where the element size is refined or coarsened and p-adaptivity at low values of p. The purpose of this work has been to consider a wider range of polynomial orders for this problem.

3.2.1 Governing Equations

We consider the compressible Navier-Stokes equations written in conservative form

$$\frac{\partial \mathbf{U}}{\partial t} + \nabla \cdot \mathbf{F}(\mathbf{U}) = \nabla \cdot \mathbf{F}_v(\mathbf{U}),$$

where $\mathbf{U} = [\rho, \rho u_1, \rho u_2, \rho u_3, E]^\top$ is the vector of conserved variables, ρ is the density, (u_1, u_2, u_3) the velocity components and E is the specific total energy. $\mathbf{F}(\mathbf{U})$ and $\mathbf{F}_v(\mathbf{U})$ denote the usual inviscid and viscous flux terms respectively, where the ideal gas law is used to close the system. For a more detailed outline, see [5].

3.2.2 Adaptive Procedure

Summarising the process at a very high level, the adaptivity procedure runs as follows for this problem:

- Run a low-order simulation to obtain a steady flow field.
- Use this flow field to solve a goal-based adjoint problem by considering an infinitesimal perturbation to the flow field.
- Compute a distribution of the polynomial order according to a goal-based error estimator based on the adjoint solution.
- Using the techniques of Sect. 2, perform the simulation again to compute a solution with a lower error of the lift or drag.

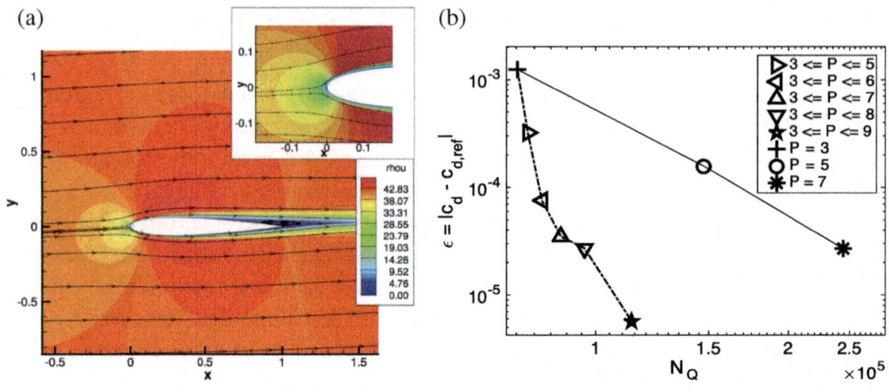

Fig. 6 Variable polynomial order simulations of a compressible laminar NACA0012 wing, taken from [5]. (**a**) x-momentum. (**b**) Convergence for different polynomial orders

For an in-depth overview of all of the techniques used in the computation of the adjoint and error estimator, the interested reader should consult [5]. To highlight the resolution capability of this adaptive method, a series of simulations have been performed to compare the use of variable p with an appropriate error estimator against a uniform refinement in p. We consider the simulation presented in [5], where the laminar subsonic flow over a classical NACA0012 wing geometry is studied at an angle of attack $\alpha = 2°$, Mach number of 0.1 and Reynolds number 5,000. A number of simulations are considered:

- a high resolution case at $P = 9$ is used as a reference solution, the obtained solution for the x-momentum for which can be seen in Fig. 6a;
- uniform polynomial order simulations are performed at $P = 3, 5$ and 7;
- variable polynomial orders are performed with $3 \leq P \leq 5 \to 9$.

To compare these simulations we calculate the error as $\varepsilon = \|c_d - c_{d,\mathrm{ref}}\|$ where

$$c_d = \frac{2}{\rho_\infty u_\infty^2 A} \oint_\Gamma \mathbf{u} \cdot [\cos\alpha, \sin\alpha]\, \mathrm{d}s$$

is the drag coefficient, ρ_∞ and u_∞ are the farfield density and velocity, A the frontal area of the wing and $c_{d,\mathrm{ref}}$ denoting the drag coefficient of the reference $P = 9$ case.

The error obtained using these cases can be seen in Fig. 6b, where it is viewed against the number of degrees of freedom N_Q of the resulting mesh, where two distinct trends can be observed. We see that increasing the polynomial order uniformly does reduce the error obtaining in the drag coefficient at a reasonably constant rate. However, the use of the goal-based error estimator, coupled with the use of a variable polynomial order, allows us to greatly reduce the resolution (and therefore the cost) required for these simulations. For example, the simulations at $3 \leq P \leq 8$ and $P = 7$ have very comparable values of ε. The main difference is

Table 2 Summary of normalised CPU cost and error in drag c_d for various constant and spatially variable polynomial orders, compared to a uniform simulation at $P = 9$

Case	Cost	ε
$P = 3$	0.28	1.2×10^{-3}
$P = 5$	0.29	1.57×10^{-4}
$P = 7$	0.64	2.69×10^{-5}
$P = 9$	1.0	–
$3 \leq P \leq 5$	0.31	3.19×10^{-4}
$3 \leq P \leq 6$	0.32	7.44×10^{-5}
$3 \leq P \leq 7$	0.34	3.47×10^{-5}
$3 \leq P \leq 8$	0.36	2.71×10^{-5}
$3 \leq P \leq 9$	0.45	5.63×10^{-6}

that whereas the uniform case has around 2.5×10^5 degrees of freedom, the variable case needs only 1×10^5 to produce a comparable error, which represents a significant saving in the cost of the simulation. This can be observed in Table 2, where the CPU time for each simulation is reported as a proportion of the reference $P = 9$ case.

4 Conclusions

In this article we have discussed the use and implementation of adaptive polynomial order in the spectral/*hp* element method. The canonical flows considered here show the clear benefits of this adaptive process, bringing a reduction in both the computational cost and the number of degrees of freedom required to resolve a given problem. However, there are still a number of challenges that need to be addressed before these methods can be brought to bear on extremely large-scale problems. Numerically, future work should focus around the development of more robust error estimators, particularly in the context of unsteady simulations, perhaps based around an unsteady formulation of the adjoint approach used for compressible simulations in Sect. 3. We note that this is inherently more expensive than the sub-cell estimator, however it will give a better indication of error throughout the domain. More sophisticated techniques also need to be developed for parallel simulations. In particular, the efficient preconditioning of these systems remains an open problem, and very large-scale simulations require the development of adaptive load-balancing techniques that can be used to re-distribute the workload evenly across processors as the polynomial order changes. Finally, when dealing with complex geometries, techniques need to be developed to couple the change in polynomial order to the treatment of curvilinear surfaces and the elements that connect to them, in order to preserve the accurate representation of the underlying geometry.

Acknowledgements D.M. acknowledges support from the EU Horizon 2020 project ExaFLOW (grant 671571) and the PRISM project under EPSRC grant EP/L000407/1. D.S. is grateful for the support received from CNPq (grant 231787/2013–8) and FAPESP (grant 2012/23493-0). D.E. acknowledges support from the EU ITN project ANADE (grant PITN-GA-289428). S.J.S.

acknowledges Royal Academy of Engineering support under their research chair scheme. R.M.K. acknowledges support from the US Army Research Office under W911NF1510222 (overseen by Dr. M. Coyle). Computing resources supported by the UK Turbulence Consortium (EPSRC grant EP/L000261/1) and the Imperial College HPC service.

References

1. C. Cantwell, S. Sherwin, R. Kirby, P. Kelly, From h to p efficiently: strategy selection for operator evaluation on hexahedral and tetrahedral elements. Comput. Fluids **43**(1), 23–28 (2011)
2. C.D. Cantwell, S.J. Sherwin, R.M. Kirby, P.H.J. Kelly, From *h* to *p* efficiently: selecting the optimal spectral/*hp* discretisation in three dimensions. Math. Mod. Nat. Phenom. **6**, 84–96 (2011)
3. C.D. Cantwell, D. Moxey, A. Comerford, A. Bolis, G. Rocco, G. Mengaldo, D. de Grazia, S. Yakovlev, J.E. Lombard, D. Ekelschot, B. Jordi, H. Xu, Y. Mohamied, C. Eskilsson, B. Nelson, P. Vos, C. Biotto, R.M. Kirby, S.J. Sherwin, Nektar++: an open-source spectral/*hp* element framework. Comput. Phys. Commun. **192**, 205–219 (2015)
4. L. Demkowicz, W. Rachowicz, P. Devloo, A fully automatic *hp*-adaptivity. J. Sci. Comput. **17**(1), 117–142 (2002)
5. D. Ekelschot, D. Moxey, S.J. Sherwin, J. Peiró, A *p*-adaptation method for compressible flow problems using a goal-based error estimator. Comput. Struct. **181**, 55–69 (2017)
6. E. Ferrer, D. Moxey, S.J. Sherwin, R.H.J. Willden, Stability of projection methods for incompressible flows using high order pressure-velocity pairs of same degree: continuous and discontinuous Galerkin formulations. Commun. Comput. Phys. **16**(3), 817–840 (2014)
7. K.J. Fidkowski, D.L. Darmofal, Review of output-based error estimation and mesh adaptation in computational fluid dynamics. AIAA J. **49**(4), 673–694 (2011)
8. G. Giorgiani, S. Fernández-Méndez, A. Huerta, Goal-oriented *hp*-adaptivity for elliptic problems. Int. J. Numer. Methods Fluids **72**(1), 1244–1262 (2013)
9. D. de Grazia, G. Mengaldo, D. Moxey, P. Vincent, S.J. Sherwin, Connections between the discontinuous Galerkin method and high-order flux reconstruction schemes. Int. J. Numer. Methods Fluids **75**(12), 860–877 (2014)
10. H.T. Huynh, A flux reconstruction approach to high-order schemes including discontinuous Galerkin methods, in: 18th AIAA Computational Fluid Dynamics Conference, p. 4079 (2007)
11. G. Karniadakis, S. Sherwin, *Spectral/hp Element Methods for Computational Fluid Dynamics*, 2nd edn. (Oxford University Press, Oxford, 2005)
12. G.E. Karniadakis, Spectral element-Fourier methods for incompressible turbulent flows. Comput. Methods Appl. Mech. Eng. **80**(1–3), 367–380 (1990)
13. G.E. Karniadakis, M. Israeli, S.A. Orszag, High-order splitting methods for the incompressible Navier-Stokes equations. J. Comput. Phys. **97**(2), 414–443 (1991)
14. L.Y. Li, Y. Allaneau, A. Jameson, Comparison of *h*- and *p*-adaptations for spectral difference methods, in: 40th AIAA Fluid Dynamics Conference and Exhibit (2010)
15. J.E.W. Lombard, D. Moxey, S.J. Sherwin, J.F.A. Hoessler, S. Dhandapani, M.J. Taylor, Implicit large-eddy simulation of a wingtip vortex. AIAA J. **54**(2), 506–518 (2016)
16. G. Markall, A. Slemmer, D. Ham, P. Kelly, C. Cantwell, S. Sherwin, Finite element assembly strategies on multi-core and many-core architectures. Int. J. Numer. Methods Fluids **71**(1), 80–97 (2013)
17. G. Mengaldo, D. de Grazia, D. Moxey, P.E. Vincent, S.J. Sherwin, Dealiasing techniques for high-order spectral element methods on regular and irregular grids. J. Comput. Phys. **299**, 56–81 (2015)
18. G. Mengaldo, D. de Grazia, P.E. Vincent, S.J. Sherwin, On the connections between discontinuous Galerkin and flux reconstruction schemes: extension to curvilinear meshes. J. Sci. Comput. **67**(3), 1272–1292 (2016)

19. R.C. Moura, G. Mengaldo, J. Peiró, S.J. Sherwin, On the eddy-resolving capability of high-order discontinuous Galerkin approaches to implicit LES/under-resolved DNS of Euler turbulence. J. Comput. Phys. **330**, 615–623 (2017)
20. D. Moxey, M.D. Green, S.J. Sherwin, J. Peiró, An isoparametric approach to high-order curvilinear boundary-layer meshing. Comput. Methods Appl. Mech. Eng. **283**, 636–650 (2015)
21. D. Moxey, C.D. Cantwell, R.M. Kirby, S.J. Sherwin, Optimizing the performance of the spectral/*hp* element method with collective linear algebra operations. Comput. Methods Appl. Mech. Eng. **310**, 628–645 (2016)
22. D. Moxey, D. Ekelschot, Ü. Keskin, S.J. Sherwin, J. Peiró, High-order curvilinear meshing using a thermo-elastic analogy. Comput. Aided Des. **72**, 130–139 (2016)
23. P.O. Persson, J. Peraire, Sub-cell shock capturing for discontinuous Galerkin methods. AIAA paper **112** (2006)
24. P. Solín, L. Demkowicz, Goal-oriented *hp*-adaptivity for elliptic problems. Comput. Methods Appl. Mech. Eng. **193**(1), 449–468 (2004)
25. P.E. Vincent, P. Castonguay, A. Jameson, A new class of high-order energy stable flux reconstruction schemes. J. Sci. Comput. **47**(1), 50–72 (2011)
26. P.E. Vos, S.J. Sherwin, R.M. Kirby, From *h* to *p* efficiently: implementing finite and spectral/*hp* element methods to achieve optimal performance for low- and high-order discretisations. J. Comput. Phys. **229**(13), 5161–5181 (2010)
27. S. Yakovlev, D. Moxey, S.J. Sherwin, R.M. Kirby, To CG or to HDG: a comparative study in 3D. J. Sci. Comput. **67**(1), 192–220 (2016)

A Perfect Absorbing Layer for High-Order Simulation of Wave Scattering Problems

Li-Lian Wang and Zhiguo Yang

Abstract We report a novel approach to design artificial absorbing layers for spectral-element discretisation of wave scattering problems with bounded scatterers. It is essentially built upon two techniques: (i) a complex compression coordinate transformation that compresses all outgoing waves in the open space into the artificial layer, and then forces them to be attenuated and decay exponentially; (ii) a substitution (for the unknown) that removes the singularity induced by the transformation, and diminishes the oscillations near the inner boundary of the layer. As a result, the solution in the absorbing layer has no oscillation and is well-behaved for arbitrary high wavenumber and very thin layer. It is therefore well-suited and perfect for high-order simulations of scattering problems.

1 Introduction

Many partial differential equations (PDEs) are naturally set in unbounded domains. In order to solve them numerically, one has to truncate or reduce the infinite physical domains in some way. A critical issue is how to carry out this without inducing significant artificial errors to the solutions. A direct domain truncation with a hard-wall or periodic boundary condition is a viable option for problems with rapidly decaying solutions in space. For problems with decaying but slowly varying solutions (e.g., elliptic and diffusion equations), a reliable approach is to compress the solution at infinity to a finite domain by using a suitable coordinate transformation, and then solve the transformed PDE in a finite domain with a

L.-L. Wang (✉)
Division of Mathematical Sciences, School of Physical and Mathematical Sciences,
Nanyang Technological University, Singapore 637371
e-mail: lilian@ntu.edu.sg

Z. Yang
Division of Mathematical Sciences, School of Physical and Mathematical Sciences,
Nanyang Technological University, Singapore 637371

Present address: Department of Mathematics, Purdue University, West Lafayette, IN 47907, USA
e-mail: yang0347@e.ntu.edu.sg, yang1508@purdue.edu

M.L. Bittencourt et al. (eds.), *Spectral and High Order Methods for Partial Differential Equations ICOSAHOM 2016*, Lecture Notes in Computational Science and Engineering 119, DOI 10.1007/978-3-319-65870-4_5

hard-wall boundary condition. However, these techniques fail to work for wave problems as the underlying solutions are typically oscillating and decay slowly. Indeed, Johnson [13] remarked that "any real coordinate mapping from an infinite to a finite domain will result in solutions that oscillate infinitely fast as the boundary is approached – such fast oscillations cannot be represented by any finite-resolution grid, and will instead effectively form a reflecting hard wall." In practice, the reduction of an unbounded domain by artificial boundary conditions [12] and perfectly matched layers (PMLs) [3, 9] has been intensively studied for the scattering problems.

In this report, we offer a new absorbing layer that is well-suited for high-order discretisation of wave scattering problems. The idea stems from the concept of an inside-out (or inverse) invisibility cloak for electromagnetic waves, first proposed by Zharova et al. [18], which was based on a coordinate transformation that compresses an open space to a finite cloaking layer with physically meaningful medium. Such a layer was expected to prevent waves inside the enclosed region from propagating outside of the layer. Ideally, the cloaking layer could be a perfect absorbing layer for scattering problems. However, it was far from perfect, as the material parameters therein were highly singular and the approximation of the solution suffered from *the curse of infinite oscillation* [13]. We introduce two techniques to surmount these obstacles: (i) complex compression coordinate transformation; and (ii) variable substitution. This leads to a transformed problem in the absorbing layer with the remarkable features: (i) its solution has no oscillation; and (ii) it is nearly definite for arbitrary high wavenumber, as opposite to the strong indefiniteness of the Helmholtz and Maxwell's equations. To fix the idea, we focus on the two-dimensional Helmholtz problem with a circular absorbing layer, and outline the extension to the rectangular layer. We demonstrate that the proposed absorbing layer is completely non-reflective and perfect for very thin layer, arbitrary high wavenumber and incident angle.

It is noteworthy that (i) the idea of using complex transformations to damp the waves is similar to complex stretching of PMLs [3, 8–10], but the transformation herein compresses all outgoing waves into the layer, and also maps the far-field boundary condition to the outer boundary naturally; and (ii) the use of substitution $u = ve^{\mathrm{i}k\rho}/\sqrt{\rho}$ is found in the context of infinite element methods for scattering problems to capture the decay rate of outgoing wave, see e.g., [11], but the substitution in (20) is adopted for different purpose with a different power in ρ.

2 Time-Harmonic Acoustic Scattering Problem

Consider the time-harmonic wave scattering governed by the Helmholtz equation:

$$\Delta u + k^2 u = 0 \quad \text{in } \Omega_\infty := \mathbb{R}^2 \setminus \bar{D}; \tag{1a}$$

$$u = g \quad \text{on } \partial D; \quad \partial_r u - \mathrm{i}ku = o(r^{-1/2}) \ \text{ as } r = |\mathbf{x}| \to \infty, \tag{1b}$$

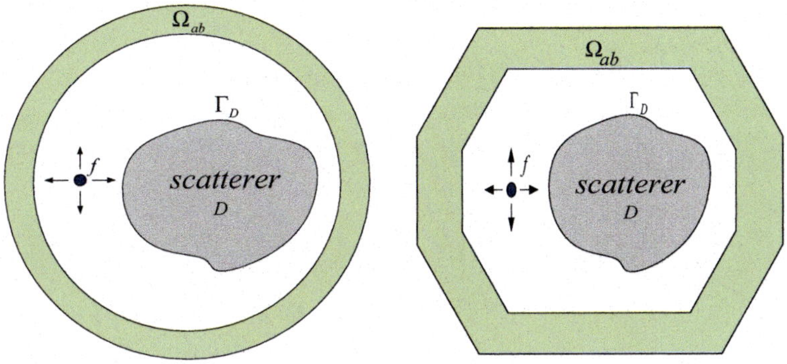

Fig. 1 Schematic illustration of an absorbing layer Ω_{ab}. *Left*: annular layer. *Right*: polygonal layer

where the wavenumber $k > 0$, $D \subset \mathbb{R}^2$ is a bounded scatterer with Lipschitz boundary $\Gamma_D = \partial D$, and the data $g \in H^{1/2}(\Gamma_D)$ is generated by the incident wave. In fact, the technique can be applied to solve the Helmholtz-type problems in inhomogeneous, anisotropic media or with an external source, which are confined in a bounded domain Ω_a enclosing $\bar{D}$, that is,

$$\nabla \cdot \big(\mathbf{C}\, \nabla u\big) + k^2 n\, u = f \quad \text{in } \Omega_\infty, \tag{2}$$

in place of (1a). Here, $\mathbf{C} \in \mathbb{C}^{2\times 2}$ is a symmetric matrix, and $n > 0$ the reflective index. Assume that exterior to Ω_a, $\mathbf{C} = \mathbf{I}_2, n = 1$ and $f = 0$.

To numerically solve the exterior problem (1) or (2) with (1b), we reduce the infinite domain by surrounding the computational domain $\Omega_f := \Omega_a \setminus \bar{D}$ via an artificial layer Ω_{ab} with a finite thickness. Without loss of generality, we consider two types of layers: (i) $\Omega_{\text{ab}} = \{a < r < b\}$ is a circular annulus (cf. Fig. 1 (left)); and (ii) Ω_{ab} is a polygonal annulus (cf. Fig. 1 (right)). The former is more convenient to illustrate the idea and to compare with the PML techniques in [3, 7, 10], while the latter is more practical and flexible to the geometry of the scatterer. In what follows, we focus on the derivation of the PDE in Ω_{ab} that couples with the Helmholtz problem in Ω_f to achieve the aforementioned goals.

The form of the transformed Helmholtz operator under a generic coordinate transformation finds useful later on, which can be verified by knowledge of calculus.

Lemma 1 *Define the Helmholtz operator:*

$$\tilde{\mathscr{H}}[\tilde{u}] = \Delta \tilde{u} + k^2 \tilde{u}. \tag{3}$$

Given a coordinate transformation between $\tilde{\mathbf{x}} = (\tilde{x}, \tilde{y})$ *and* $\mathbf{x} = (x, y)$ *with the Jacobian matrix*

$$x = x(\tilde{\mathbf{x}}), \quad y = y(\tilde{\mathbf{x}}); \quad \mathbf{J} := \frac{\partial(x, y)}{\partial(\tilde{x}, \tilde{y})} = \begin{bmatrix} \partial_{\tilde{x}} x & \partial_{\tilde{y}} x \\ \partial_{\tilde{x}} y & \partial_{\tilde{y}} y \end{bmatrix}, \tag{4}$$

we have the transformed Helmholtz operator

$$\mathscr{H}[u] = \frac{1}{n}\{\nabla \cdot (\mathbf{C}\nabla u) + k^2\, n\, u\}, \tag{5}$$

where $u(\mathbf{x}) = \tilde{u}(\tilde{\mathbf{x}})$ *and*

$$\mathbf{C} = \begin{bmatrix} C_{11} & C_{12} \\ C_{12} & C_{22} \end{bmatrix} = \frac{\mathbf{J}\,\mathbf{J}^t}{\det(\mathbf{J})}, \quad n = \frac{1}{\det(\mathbf{J})}. \tag{6}$$

2.1 Real Compression Coordinate Transformation

We start with the "compression" coordinate transformation for the inside-out invisibility cloak in [18]:

$$r = b - \frac{(b-a)^2}{\rho + b - 2a} \quad \text{or} \quad \rho = \frac{s(r)}{b-r}, \quad s(r) := a^2 + r(b-2a), \tag{7}$$

for $\rho \in [a, \infty)$ and $r \in [a, b)$. This one-to-one mapping compresses the open space exterior to a disk of radius $\rho = a$ into the annulus $a \le r < b$, where the inner circle $\rho = a (= r)$ remains unchanged, while $\rho = \infty$ corresponds to $r = b$.

We now derive the equation in the compressed layer $\Omega_{\rm ab}$ by using Lemma 1. By the chain rule involving the original Cartesian coordinates-$(\tilde{x}, \tilde{y})$ with the polar coordinates-(ρ, θ); and the physical Cartesian coordinates-(x, y) with the polar coordinates-(r, θ), we have

$$\mathbf{J} = \frac{\partial(x, y)}{\partial(\tilde{x}, \tilde{y})} = \frac{\partial(x, y)}{\partial(r, \theta)} \frac{\partial(r, \theta)}{\partial(\rho, \theta)} \frac{\partial(\rho, \theta)}{\partial(\tilde{x}, \tilde{y})}. \tag{8}$$

A direct calculation leads to

$$\mathbf{J} = \mathbf{R}\,\mathbf{J}_0\,\mathbf{R}^t \quad \text{with} \quad \mathbf{J}_0 = \begin{bmatrix} dr/d\rho & 0 \\ 0 & r/\rho \end{bmatrix}, \quad \mathbf{R} = \begin{bmatrix} \cos\theta & -\sin\theta \\ \sin\theta & \cos\theta \end{bmatrix}. \tag{9}$$

Then by (6),

$$\mathbf{C}_0 = \mathbf{R} \begin{bmatrix} c_0 & 0 \\ 0 & 1/c_0 \end{bmatrix} \mathbf{R}^t, \quad n_0 = \frac{\rho}{r}\frac{d\rho}{dr}, \quad c_0 := \frac{\rho}{r}\frac{dr}{d\rho}. \tag{10}$$

As a consequence of Lemma 1, we obtain the modified Helmholtz equation:

$$\nabla \cdot (\mathbf{C}_0 \nabla u) + k^2\, n_0\, u = 0 \quad \text{in } \Omega_{ab}. \tag{11}$$

Noting from (7) that

$$\frac{d\rho}{dr} = \Big(\frac{b-a}{b-r}\Big)^2 = \Big(\frac{dr}{d\rho}\Big)^{-1}, \quad r \in (a,b), \tag{12}$$

we have

$$n_0 = \frac{s(r)}{r}\frac{(b-a)^2}{(b-r)^3}, \qquad c_0 = \frac{s(r)}{r}\frac{b-r}{(b-a)^2}. \tag{13}$$

It is evident that $1/c_0, n \to \infty$, when $r \to b^-$. This implies the wavenumber becomes infinitely large near the outer boundary of layer $\Omega_{\rm ab}$. In other words, the solution u has infinite oscillation. It is no wonder that all the outgoing waves in the open space are compressed into the finite layer $\Omega_{\rm ab}$, so this induces the so-called *curse of infinite oscillation*. Thus, it is advisable to use a *complex compression coordinate transformation* to attenuate the waves.

2.2 Complex Compression Coordinate Transformation

Different from (7), we introduce the complex compression mapping

$$\tilde\rho(r) = \rho(r) + {\rm i}\sigma_0(\rho(r) - a), \quad \rho(r) = \frac{s(r)}{b-r}, \quad r \in [a,b), \tag{14}$$

where $s(r) = a^2 + r(b-2a)$ as before, and $\sigma_0 > 0$ is a tuning parameter. For notational convenience, we denote

$$\alpha := 1 + {\rm i}\sigma_0 = \frac{d\tilde\rho}{d\rho}, \quad \beta := 1 + {\rm i}\sigma_0\Big(1 - \frac{a}{\rho}\Big) = \frac{\tilde\rho}{\rho}. \tag{15}$$

Using Lemma 1, we can derive following PDE in $\Omega_{\rm ab}$.

Theorem 1 *Using the transformation* (14), *we derive the Helmholtz-type problem:*

$$\nabla\cdot\big(\mathbf{C}\nabla u\big) + k^2\, n\, u = 0 \quad \text{in } \Omega_{ab}, \tag{16}$$

$$u = \Psi \ \text{ at } r = a; \quad \frac{1}{\alpha}\frac{dr}{d\rho}\partial_r u - {\rm i}ku = o(|\tilde\rho|^{-1/2}) \ \text{ as } r \to b^-, \tag{17}$$

where Ψ is from the solution of the interior Helmholtz equation at the inner boundary $r = a$, and

$$\mathbf{C} = \mathbf{R}\begin{bmatrix} c & 0 \\ 0 & 1/c \end{bmatrix}\mathbf{R}^t, \quad n = \alpha\beta\frac{s(r)}{r}\frac{(b-a)^2}{(b-r)^3}, \quad c = \frac{\beta}{\alpha}\frac{s(r)}{r}\frac{b-r}{(b-a)^2}. \tag{18}$$

Moreover, if $\psi \in L^2(0, 2\pi)$, *we have the following point-wise bounds for all* $r \in (a, b)$,

$$\|u(r, \cdot)\|_{L^2(0,2\pi)} \leq \exp\Big\{-k\sigma_0(\rho - a)\Big(1 - \frac{a^2}{k^2\rho^2 + k^2\sigma_0^2(\rho - a)^2}\Big)^{1/2}\Big\}\|\Psi\|_{L^2(0,2\pi)}. \tag{19}$$

It is important to point out that the solution in the point-wise sense (19) decays exponentially like $O(e^{-k\sigma_0/(b-r)})$ as $r \to b^-$. We also observe from (18) that the coefficients $1/c, n \to \infty$ as $r \to b^-$. Though the product nu is well-behaved, the problem (16)–(17) is still challenging for numerical solution due to the involved singular coefficients.

2.3 Variable Substitution

To handle the singularity and remove essential oscillations of u, we introduce the following substitution in $\Omega_{\rm ab}$:

$$u = vw, \quad w = \Big(\frac{a}{\rho}\Big)^{3/2} e^{{\rm i}k(\rho - a)}, \tag{20}$$

where $\rho = \rho(r)$ is as in (7). It is important to remark that

(i) We incorporate the complex exponential to capture the oscillation of u, so that v essentially has no oscillation for arbitrary high wavenumber and very thin layer (see Fig. 2 below).

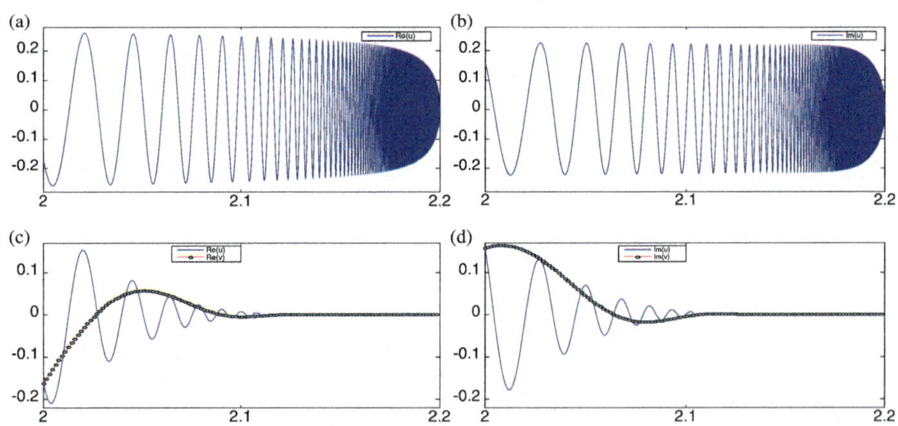

Fig. 2 Profiles of the solution (33) with $k = 200$, $\theta_0 = \pi/4$ and $a_0 = 1$ under the real compression mapping (7) and complex compression mapping (14), and the substitution (20) with $r \in (2, 2.2)$ and along $\theta = 0$. (**a**) $\mathrm{Re}\{u(\rho(r), 0)\}$ under (7). (**b**) $\mathrm{Im}\{u(\rho(r), 0)\}$ under (7). (**c**) $\mathrm{Re}\{u(\tilde\rho(r), 0)\}$ under (14) vs. $\mathrm{Re}\{v\}$ in (20). (**d**) $\mathrm{Im}\{u(\tilde\rho(r), 0)\}$ under (14) vs. $\mathrm{Im}\{v\}$ in (20)

(ii) In real implementation, we can build in the substitution into the basis functions, and formally approximate u by non-conventional basis:

$$u_N \in \operatorname{span}\{\psi_j = w\phi_j \,:\, 0 \le j \le N\}, \tag{21}$$

where v is essentially approximated by the usual polynomial or piecewise polynomial basis $\{\phi_j\}$ in spectral/spectral-element methods.

Remark 1 Recall that for fixed m and large $|z|$ (cf. [1]),

$$H_m^{(1)}(z) \sim \sqrt{\frac{2}{\pi z}}\, e^{\mathrm{i}(z-\frac{1}{2}m\pi-\frac{1}{4}\pi)}, \quad -\pi < \arg(z) < 2\pi. \tag{22}$$

By (60), we have the asymptotic estimates for fixed m :

$$|\hat{u}_m(r)| = |\hat{\psi}_m| \left|\frac{H_m^{(1)}(k\tilde{\rho}(r))}{H_m^{(1)}(ka)}\right| \sim \sqrt{\frac{a}{|\tilde{\rho}|}}\, e^{-k\sigma_0(\rho-a)} |\hat{\psi}_m|\, e^{\mathrm{i}k(\rho-a)}. \tag{23}$$

In view of this, the complex exponential in (20) captures the oscillations of u near the inner boundary $r = a$, so we expect v has no oscillation and decays exponentially in the layer $\Omega_{\rm ab}$. □

We find it is more convenient to carry out the substitution through the variational form. Let $L^2_\omega(\Omega)$ be a weighted space of square integrable functions with the inner product and norm denoted by $(\cdot,\cdot)_{\omega,\Omega}$ and $\|\cdot\|_{\omega,\Omega}$ as usual. Define the trace integral $\langle u, v\rangle_{\Gamma_b} := \oint_{\Gamma_b} u\, \bar{v}\, d\gamma$. Let $\Omega = \Omega_f \cup \Omega_{\rm ab}$, and assume $g = 0$. Formally, we define the bilinear form associated with (1) in Ω_f coupled with (16):

$$\mathbb{B}_\Omega(u,\phi) := \mathbb{B}_{\Omega_f}(u,\phi) + \mathbb{B}_{\Omega_{\rm ab}}(u,\phi) \quad \text{with} \quad \mathbb{B}_{\Omega_f}(u,\phi) = (\nabla u, \nabla\phi)_{\Omega_f} - k^2(u,\phi)_{\Omega_f},$$

$$\mathbb{B}_{\Omega_{\rm ab}}(u,\phi) = (\mathbf{C}\nabla u, \nabla\phi)_{\Omega_{\rm ab}} - k^2(n\,u,\phi)_{\Omega_{\rm ab}} - \langle \mathbf{C}\nabla u\cdot\mathbf{n}, \phi\rangle_{\Gamma_b}, \tag{24}$$

where $\mathbf{n} = (\cos\theta, \sin\theta)^t$ is the unit outer normal to Γ_b.

Theorem 2 *With the substitution $u = vw$ and $\phi = \psi w$ in (20), we have*

$$\mathbb{B}_{\Omega_{\rm ab}}(u,\phi) = (\varpi_1 \mathbf{C}\nabla v, \nabla\psi)_{\Omega_{\rm ab}} + \frac{1}{\alpha}(\beta\partial_{\mathbf{n}} v, \psi\varpi_2)_{\Omega_{\rm ab}} + \frac{1}{\alpha}(\beta\varpi_2 v, \partial_{\mathbf{n}}\psi)_{\Omega_{\rm ab}} + (\varpi_3 v, \psi)_{\Omega_{\rm ab}}, \tag{25}$$

where $\partial_{\mathbf{n}} = \mathbf{n}\cdot\nabla$ is the directional derivative along the normal direction, and

$$\varpi_1 = \frac{a^3}{\rho^3}, \quad \varpi_2 = \frac{a^3}{r}\frac{1}{\rho^2}\Big(-\frac{3}{2\rho} + \mathrm{i}k\Big), \quad \varpi_3 = \frac{a^3(b-a)^2}{rs^2(r)}\Big(\Big(\frac{\beta}{\alpha} - \alpha\beta\Big)k^2 + \frac{\beta}{\alpha}\frac{9}{4\rho^2}\Big),$$

$$\alpha = 1 + \mathrm{i}\sigma_0, \quad \beta = 1 + \mathrm{i}\sigma_0\Big(1 - \frac{a}{\rho}\Big), \quad \frac{1}{\rho} = \frac{b-r}{s(r)}, \quad s(r) = a^2 + r(b-2a), \quad r \in (a,b). \tag{26}$$

Remark 2 Some remarks are in order.

(i) Compared with the singular coefficients in (18), we observe from (26) that the involved coefficients become regular. In particular, ϖ_3 is uniformly bounded above and below away from zero.
(ii) The DtN boundary condition is transformed to the outer boundary $r = b$, this naturally eliminate the boundary term in (24).
(iii) When u is approximated by a non-conventional basis (21), we can use (25)–(26) to compute the matrices of the linear system. In fact, $\Omega_{\rm ab}$ can be replaced by any element of a non-overlapping partition of $\Omega_{\rm ab}$. □

Remarkably, the transformed problem in v is nearly definite for any wavenumber $k > 0$, as opposite to the indefiniteness of the original problem.

Theorem 3 *With the substitution $u = vw$ in* (20)*, we have*

$$\begin{aligned}\mathrm{Re}\{\mathbb{B}_{\Omega_{\rm ab}}(u,u)\} \ge\, & c_1(1-\epsilon^{-1})\|\partial_r v\|^2_{\omega^2} + c_2\|\partial_\theta v\|^2_{\omega} + c_3\|v(a,\cdot)\|^2_{L^2(0,2\pi)} \\ & + a^3|I|^2k^2\int_0^{2\pi}\!\!\int_a^b \frac{\Theta(r)}{s^2(r)}|v|^2 dr d\theta,\end{aligned} \tag{27}$$

where $\omega = b - r$, $|I| = b - a$, $\varepsilon > 1$ and

$$c_1 = \frac{a^3}{b\bar c^2|I|^4}\frac{1}{1+\sigma_0^2},\quad c_2 = \frac{a^3}{b\bar c^4|I|^2},\quad c_3 = \frac{3}{2}\frac{1}{1+\sigma_0^2},\quad \bar c = \max\{a,|I|\},$$

$$\Theta(r) = \frac{15}{4a^2k^2}\frac{\sigma_0^2}{1+\sigma_0^2}t^3 - \frac{9}{4a^2k^2}t^2 - \frac{\sigma_0^2}{1+\sigma_0^2}(\sigma_0^2-\epsilon+2)t + (\sigma_0^2-\epsilon),\quad t = \frac{a}{\rho}. \tag{28}$$

For simplicity, we denote the coefficients (up to a sign) of the cubic polynomial in t by $\{\gamma_i\}_{i=0}^3$, and define

$$\widetilde\Theta(t) := \Theta(r) = \gamma_3 t^3 - \gamma_2 t^2 - \gamma_1 t + \gamma_0,\quad t\in(0,1]. \tag{29}$$

One verifies readily that

$$\widetilde\Theta'(t) = 3\gamma_3\Big\{\Big(t-\frac{\gamma_2}{3\gamma_3}\Big)^2 - \frac{\gamma_2^2}{9\gamma_3^2} - \frac{\gamma_1}{3\gamma_3}\Big\},\quad \frac{\gamma_2}{3\gamma_3} = \frac{1+\sigma_0^2}{5\sigma_0},$$

$$\widetilde\Theta'(0) = -\gamma_1 < 0,\quad \widetilde\Theta'(1) < 3\gamma_3\Big(\frac{3}{5} - \frac{4a^2k^2}{45}(\sigma_0^2-\epsilon+2)\Big).$$

If $k^2 \ge k_0^2 := 27/(4a^2(\sigma_0^2-\epsilon+2))$ and $1<\epsilon<\sigma_0^2$, then $\widetilde\Theta'(t) < 0$, and

$$\widetilde\Theta(1) < \widetilde\Theta(t) < \widetilde\Theta(0) = \sigma_0^2 - \epsilon,\quad t\in(0,1);\quad \widetilde\Theta(t) \ge \widetilde\Theta(t_*) > 0,\quad t\in(0,t_*], \tag{30}$$

where

$$t_* = \frac{\gamma_0}{\gamma_1} = 1 - \frac{1+\epsilon\sigma_0^{-2}}{1+(2-\epsilon)\sigma_0^{-2}}\frac{1}{\sigma_0^2}, \quad t_* = 1 - \frac{1}{\sigma_0^2} + 2(\epsilon-1)\frac{1}{\sigma_0^4} + O(\sigma_0^{-6}). \tag{31}$$

This implies

$$\Theta(r) > 0 \quad \text{if} \quad \rho \geq \frac{a}{t_*} \quad \text{or} \quad a < \frac{b-a+b(t_*^{-1}-1)}{b-a+a(t_*^{-1}-1)}a \leq r < b. \tag{32}$$

In particular, if $\sigma_0 \gg 1$, we have $\Theta(r) > 0$ for all $r \in (a,b)$.

2.4 *Numerical Results*

2.4.1 Illustration of the Solution in Ω_{ab} Under Different Transformations

Consider the exterior problem

$$\Delta u + k^2 u = 0, \quad \rho > a_0; \quad u|_{\rho=a_0} = g; \quad \partial_\rho u - iku = o(\rho^{-1/2}), \tag{33}$$

where we take $g = -\exp\big(ika_0\cos(\theta-\theta_0)\big)$ with the incident angle θ_0. It is known that it admits a unique series solution $u(\rho,\theta)$. As before, we reduce the unbounded domain by an artificial annular layer Ω_{ab} with radius $a > a_0$.

We plot in Fig. 2 the profiles of the solution under different transformations. We see that the infinite oscillation of the solution in the layer Ω_{ab} by the real compression transformation (7). The solution decays exponentially with the complex compression transformation, but it oscillates near $r = a$. However, with the substitution (20), v becomes well-behaved in the layer, which actually we approximate.

2.4.2 Spectral-Element Methods for Scattering Problems

We demonstrate that the proposed absorbing layer is totally non-reflective, and robust for high wavenumber and very thin layer. To show the high accuracy, we solve (1) with the scatterer D being a disk of radius a_0, which is reduced to two annuluses: $\Omega = \Omega_f \cup \Omega_{ab}$. Here, we use Fourier approximation in θ direction, and spectral-element method in radial direction [14]. Note that for $r \in [a,b]$, we use the non-standard basis $\psi_j = w\phi_j$ with ϕ_j being the usual polynomial nodal or modal basis as in (21).

We also intend to compare our approach with the PML technique using the complex coordinate stretching

$$\tilde{r} = r + i\int_a^r \sigma(t)dt, \quad r \in (a,b), \quad \sigma(t) > 0. \tag{34}$$

Typically, there are two choices of the absorbing function $\sigma(t)$.

(i) Regular function (see, e.g., [7, 10]):

$$\sigma(t) = \sigma_1\Big(\frac{t-a}{b-a}\Big)^n, \quad \text{so} \quad \tilde{r} = r + \mathrm{i}\sigma_1 \frac{b-a}{n+1}\Big(\frac{r-a}{b-a}\Big)^{n+1}, \quad r \in (a,b), \tag{35}$$

where n is a positive integer and $\sigma_1 > 0$ is a tuning parameter.

(ii) Singular function (or unbounded absorbing function (see, e.g., [4, 5]):

$$\sigma(t) = \frac{\sigma_2}{b-t}, \quad \text{so} \quad \tilde{r} = r + \mathrm{i}\sigma_2 \ln\Big(\frac{b-a}{b-r}\Big), \quad r \in (a,b), \tag{36}$$

where $\sigma_2 > 0$ plays the same role as σ_1.

Observe from (14) and (36) that the imaginary parts of both transformations involve two different one-to-one mappings between $(0,\infty)$ and (a,b), i.e.,

$$z = \frac{\rho(r)-a}{b-a} = \frac{r-a}{b-r}, \quad r = \frac{a+bz}{1+z}, \quad r \in (a,b), \quad z \in (0,\infty), \tag{37}$$

and

$$z = \ln\Big(\frac{b-a}{b-r}\Big), \quad r = b-(b-a)\frac{1}{e^z}, \quad r \in (a,b), \quad z \in (0,\infty). \tag{38}$$

It is noteworthy that the algebraic mapping (37) has been used for mapped spectral methods in unbounded domains see, e.g., [6, 14], where at times one employs the following logarithmic mapping similar to (38):

$$z = \ln\Big(\frac{b-2a+r}{b-r}\Big), \quad r = b-(b-a)\frac{2}{1+e^z}, \quad r \in (a,b), \quad z \in (0,\infty). \tag{39}$$

Indeed, one can choose any of these singular mappings in (35) for the PML, but the singularity of the coefficients in the PML equation is very different between (38)–(39) and (37). In fact, the authors [4, 5] suggested the use of e.g., Gauss-quadrature rules to avoid sampling the singular values at $r = b$, but it should be pointed out the logarithmic singularity induced by (38) is more challenging to deal with than the algebraic mapping (37).

We reiterate the significant differences of our approach from the PML: (i) we use the compression transformation for both the real and imaginary parts in (36) so we can directly transform the far-field radiation conditions to $r = b$; and (ii) more importantly, the substitution allows us to remove the singularity and oscillation in the layer leading to well-behaved functions which can be accurately approximated by standard approximation tools.

In the test, we take g to be the same as in (33) with $a_0 = 1, \theta_0 = 0$, and use Theorem 2 to compute the matrices related to the artificial layer. Let M be the cut-off number of the Fourier modes, and $\mathbf{N} = (N_1, N)$ be the number of points in r-direction of two layers, respectively. We measure the maximum pointwise error in Ω_f. We take $N_1 = 200, M = ka$ with $a = 2$, $b = 2.2$ and vary N so that the waves in the interior layer can be well-resolved, and the error should be dominated by the approximation in the outer annulus. In Fig. 3a–b, we compare the accuracy of the solver with PAL ($\sigma_0 = 1.5$), PML ($n = 1, \sigma_1 = 1.89, 1.43$ for $k = 150, 200$, respectively: optimal value based on the rule in [7]) and UPML using unbounded absorbing function (36) ($\sigma_2 = 1/k$: optimal value suggested by [5]). Observe that our approach outperforms the PML with two choices of the absorbing functions, and the advantage is even significant for high wavenumber. In addition, the effect of the singularity related to UPML is observable for slightly large N.

We also study the influence of the thickness of the absorbing layer. In Fig. 3c, we vary the thickness of the layer $b - a = 0.02, 0.05, 0.1, 0.5$ and plot the error against $N = 5, 10, \cdots, 40$ with $k = 100$. For a fixed N, we observe the thinner the layer the smaller the error, which shows the result is insensitive to the thickness. In Fig. 3d, we plot $\mathrm{Re}(u_{\mathbf{N}})|_{\Omega_f}$ and $\mathrm{Re}(v_{\mathbf{N}})|_{\Omega_{\rm ab}}$ with $b - a = 0.02$ and $N = 40$. Notice that the approximation of v has no oscillation and is well-behaved in the layer.

In Fig. 3e–f, we further test PAL with a perfect conducting ellipse D with $\partial D := \{(x, y) = \zeta(\cosh\xi \cos\theta, \sinh\xi \sin\theta), \theta \in [0, 2\pi)\}$ and fix $(\zeta, \xi) = (0.8, 0.5)$ with $k = 50$, $(a, b) = (2, 2.2)$. We partition $\Omega = \{\Omega_f^{(i)}\}_{i=1}^8 \cup \{\Omega_{\rm ab}^{(i)}\}_{i=1}^8$ into 16 non-overlapping (curved) quadrilateral elements as shown in Fig. 3e. Using the Gordon-Hall elemental transformation $\{T_f^i, T_{\rm ab}^i\} : [-1, 1]^2 \mapsto \{\Omega_f^{(i)}, \Omega_{\rm ab}^{(i)}\}$, we define the

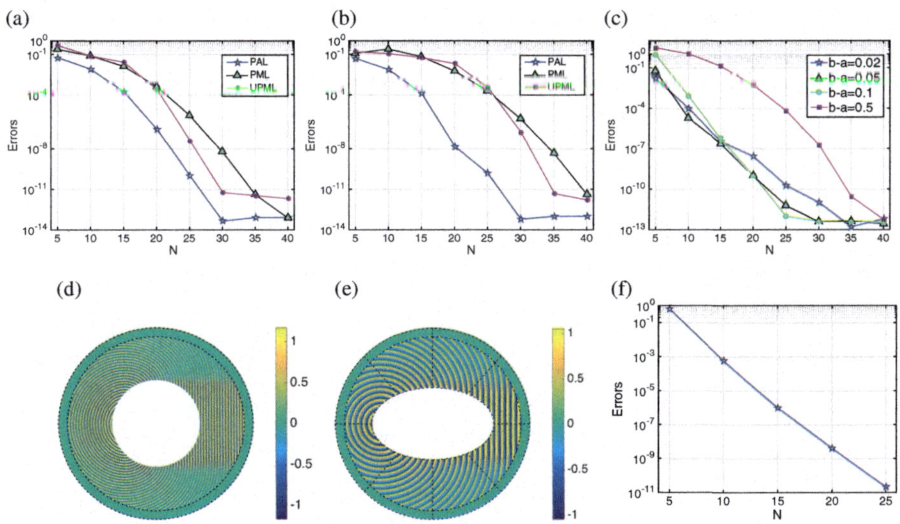

Fig. 3 In (**a**)–(**b**): $b = 2.2$. In (**c**): $k = 100$. In (**d**): $k = 100, b = 2.2, N = 40$. In (**e**)–(**f**): $k = 50, N_1 = 60$, $(a, b) = (2, 2.2)$, $\partial D := \{(x, y) = \zeta(\cosh\xi \cos\theta, \sinh\xi \sin\theta), \theta \in [0, 2\pi)\}$ with $(\zeta, \xi) = (0.8, 0.5)$, $\sigma_0 = 1.5$, and for (**e**): $N = 25$. (**a**) PAL vs PML ($k = 150$). (**b**) PAL vs PML ($k = 200$). (**c**) Errors vs thickness of $\Omega_{\rm ab}$. (**d**) $\mathrm{Re}(u_N)$ and $\mathrm{Re}(v_{\mathbf{N}})$. (**e**) $\mathrm{Re}(u_N)$ and $\mathrm{Re}(v_{\mathbf{N}})$. (**f**) Error of (**e**) against N

approximation space

$$u_{\mathbf{N}} \in V_{\mathbf{N}} = \Big\{u \in H^1(\Omega) : u|_{\Omega_f^{(i)}} \circ T_f^i \in \mathbb{P}_{N_1} \times \mathbb{P}_{N_1},\ u|_{\Omega_{\rm ab}^{(i)}} = v_{\mathbf{N}} w,\ v_{\mathbf{N}}|_{\Omega_{\rm ab}^{(i)}} \circ T_{\rm ab}^i \in \mathbb{P}_{N_1} \times \mathbb{P}_N\Big\}. \tag{40}$$

In Fig. 3e, we plot $\mathrm{Re}(u_{\mathbf{N}})|_{\Omega_f}$ and $\mathrm{Re}(v_{\mathbf{N}})|_{\Omega_{\rm ab}}$ with $(N_1, N) = (60, 25)$. In Fig. 3f, we take $N_1 = 60$ (the interior layer can be well-resolved) and vary $N = 5, 10, \cdots, 25$ so that the maximum point-wise error in Ω_f should be dominated by N (the number of points along the radial direction in $\Omega_{\rm ab}$). We see the errors decay exponentially for the spectral-element approximation, and a high accuracy can be achieved with a small N.

2.4.3 Simulation of Cylindrical Inside-Out Cloak

We illustrate that with the lossy and dispersive materials in the cloaking layer $\Omega_{\rm ab}$, we can achieve the perfectness of the aforementioned inside-out cloak. Assume that the scatterer D in (1) is penetrable, and place an active "point" source centred at (x_0, y_0) in the disk $r < a$:

$$f(x,y) = A \exp\Big(-\frac{(x-x_0)^2 + (y-y_0)^2}{2\sigma^2}\Big), \tag{41}$$

with $(A, x_0, y_0, \sigma) = (10^5, -0.3, -0.3, 0.01)$. We take $k = 50$, $(a, b) = (1, 1.5)$ and $\sigma_0 = 0.1$. In Fig. 4a, we plot $\mathrm{Re}(u)|_{\Omega_f}$ and $\mathrm{Re}(v)|_{\Omega_{\rm ab}}$ with $(N_1, N) = (50, 30)$. We depict in Fig. 4b–c the extracted profiles along x-axis. We see that the waves radiated by the active source are completely absorbed by the cloaking layer $\Omega_{\rm ab}$. Indeed, the unknown v in the layer is very well-behaved.

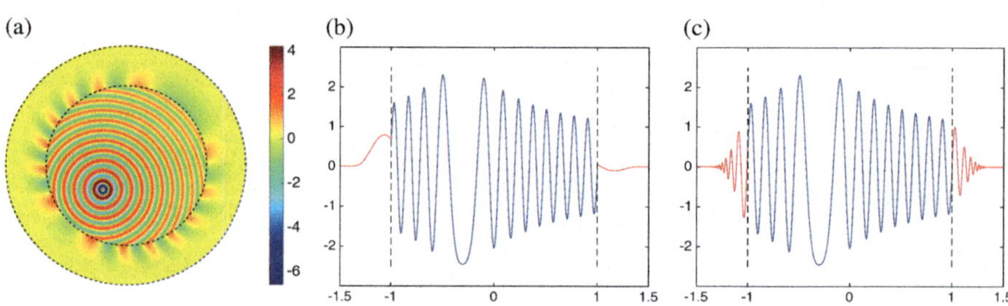

Fig. 4 Inside-out cloaking phenomenon generated by a point source defined in (41) with $k = 50$, $(a, b) = (1, 1.5)$, $\sigma_0 = 0.1$, $M = ka$ and $\mathbf{N} = (50, 30)$. (**a**) Cloaking of a point source. (**b**) Profile of u & v along x-axis. (**c**) Profile of u along x-axis

3 Rectangular/Polygonal Absorbing Layer

In practice, the rectangular/polygonal layer is more desirable and flexible for e.g., elongated scatterers and for element methods. In fact, the two techniques for designing the perfect annular absorbing layer can be extended to this setting. To fix the idea, we set

$$\Omega_f = \{\mathbf{x} \in \mathbb{R}^2 : |x_i| < L_i,\ i = 1, 2\}, \quad \Omega = \{\mathbf{x} \in \mathbb{R}^2 : |x_i| < L_i + d_i,\ i = 1, 2\},$$

with $L_1/L_2 = d_1/d_2$. Then, the absorbing layer consists of four trapezoidal pieces: $\Omega_{\rm ab} = \Omega \setminus \bar{\Omega}_f = \Omega^r \cup \Omega^l \cup \Omega^t \cup \Omega^b$, whose non-parallel sides are rays from the origin O, as illustrated in Fig. 5a.

Like (14), the complex compression coordinate transformation for the right and top pieces Ω^r and Ω^t, respectively, takes the form:

$$\tilde{x}_1 = \rho_1(x_1) + {\rm i}\sigma_0(\rho_1(x_1) - L_1), \quad \tilde{x}_2 = \tilde{x}_1 x_2/x_1, \quad \mathbf{x} \in \Omega^r, \tag{42}$$

$$\tilde{x}_2 = \rho_2(x_2) + {\rm i}\sigma_0(\rho_2(x_2) - L_2), \quad \tilde{x}_1 = \tilde{x}_2 x_1/x_2, \quad \mathbf{x} \in \Omega^t, \tag{43}$$

and for the left and bottom pieces Ω^r and Ω^t, we transform by symmetry:

$$\big(\tilde{x}_1(x_1), \tilde{x}_2(x_1, x_2)\big)|_{\Omega^l} = \big(-\tilde{x}_1(-x_1), \tilde{x}_2(-x_1, x_2)\big)|_{\Omega^r}, \tag{44}$$

$$\big(\tilde{x}_1(x_1, x_2), \tilde{x}_2(x_2)\big)|_{\Omega^b} = \big(\tilde{x}_1(x_1, -x_2), -\tilde{x}_2(-x_2)\big)|_{\Omega^t}. \tag{45}$$

In the above, we have

$$\rho_1(x_1) = \frac{L_1^2 + (d_1 - L_1)x_1}{L_1 + d_1 - x_1}, \quad \rho_2(x_2) = \frac{L_2^2 + (d_2 - L_2)x_2}{L_2 + d_2 - x_2}. \tag{46}$$

Like (7), the real transformation $\breve{x}_1 = \rho_1(x_1)$ maps $\breve{x}_1 \in [L_1, \infty)$ to $x_1 \in [L_1, L_1 + d_1)$. As a result, the trapezoid Ω^r on the right is compressed along radial direction

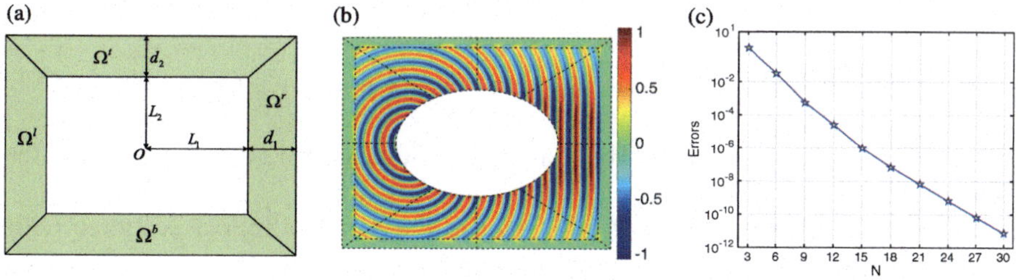

Fig. 5 In (b)–(c), $\partial D := \{(x, y) = \zeta(\cosh\xi \cos\theta, \sinh\xi \sin\theta), \theta \in [0, 2\pi)\}$ with $(\zeta, \xi) = (0.8, 0.5)$, $\sigma_0 = 1.5$. **(a)** Schematic illustration of $\Omega_{\rm ab}$. **(b)** Re($u_{\rm N}$) and Re($v_{\rm N}$). **(c)** Error of (a) against N

from an open "trapezoid" with $L_1 \le \breve{x}_1 < \infty$ and two infinitely-long, non-parallel sides on the same rays as Ω^r. Likewise for three other trapezoidal pieces, they are compressed from open "trapezoids".

Using Lemma 1, we can derive the Helmholtz-type PDE as with that in Theorem 1. Thanks to the symmetry of the layer, one only needs to calculate the material parameters in Ω^r and Ω^t.

Theorem 4 *By the transformation* (42)–(43)*, we have* **C** *and* **n** *take the form*

$$C_{11} = \frac{\beta_1}{\alpha}\frac{\rho_1}{x_1\rho_1'}, \quad C_{22} = \frac{\alpha}{\beta_1}\frac{x_1\rho_1'}{\rho_1} + \frac{\beta_1}{\alpha}\frac{\rho_1\rho_1'}{x_1}\Big(\frac{x_2}{x_1}\Big)^2\Big(\frac{1}{\rho_1'} - \frac{\alpha}{\beta_1}\frac{x_1}{\rho_1}\Big)^2, \tag{47a}$$

$$C_{12} = \frac{x_2}{x_1}\Big(\frac{\beta_1}{\alpha}\frac{\rho_1}{x_1\rho_1'} - 1\Big), \quad n = \alpha\beta_1\frac{\rho_1\rho_1'}{x_1}, \quad \text{in } \Omega^r, \tag{47b}$$

and

$$C_{11} = \frac{\alpha}{\beta_2}\frac{x_2\rho_2'}{\rho_2} + \frac{\beta_2}{\alpha}\frac{\rho_2\rho_2'}{x_2}\Big(\frac{x_1}{x_2}\Big)^2\Big(\frac{1}{\rho_2'} - \frac{\alpha}{\beta_2}\frac{x_2}{\rho_2}\Big)^2, \quad C_{22} = \frac{\beta_2}{\alpha}\frac{\rho_2}{x_2\rho_2'}, \tag{48a}$$

$$C_{12} = \frac{x_1}{x_2}\Big(\frac{\beta_2}{\alpha}\frac{\rho_2}{x_2\rho_2'} - 1\Big), \quad n = \alpha\beta_2\frac{\rho_2\rho_2'}{x_2}, \quad \text{in } \Omega^t, \tag{48b}$$

where $\alpha = 1 + \sigma_0 \mathrm{i}$, *and* $\beta_i = \tilde{x}_i/\rho_i\ (i = 1, 2)$. *With the symmetric relations* (44)–(45)*, we have*

$$\{C_{11}, C_{22}, n\}(x_1, x_2)|_{\Omega^l} = \{C_{11}, C_{22}, n\}(-x_1, x_2)|_{\Omega^r}, \quad C_{12}(x_1, x_2)|_{\Omega^l} = -C_{12}(-x_1, x_2)|_{\Omega^r}, \tag{49}$$

$$\{C_{11}, C_{22}, n\}(x_1, x_2)|_{\Omega^b} = \{C_{11}, C_{22}, n\}(x_1, -x_2)|_{\Omega^t}, \quad C_{12}(x_1, x_2)|_{\Omega^t} = -C_{12}(x_1, -x_2)|_{\Omega^t}. \tag{50}$$

We shall provide the derivations in a forthcoming work. Like (20), we use the following substitution to diminish the singularity and essential oscillations:

$$u = vw, \quad w = \big(L_1/\rho_1\big)^{3/2} e^{\mathrm{i}k\frac{r}{x_1}(\rho_1 - L_1)} \text{ in } \Omega^r, \quad w = \big(L_2/\rho_2\big)^{3/2} e^{\mathrm{i}k\frac{r}{x_2}(\rho_2 - L_2)} \text{ in } \Omega^t, \tag{51}$$

$$w(x_1, x_2)|_{\Omega^l} = w(-x_1, x_2)|_{\Omega^r}, \quad w(x_1, x_2)|_{\Omega^b} = w(x_1, -x_2)|_{\Omega^t}, \quad \text{with } r = \sqrt{x_1^2 + x_2^2}. \tag{52}$$

This can be implemented as in Theorem 2. The details shall be reported in a later work.

To test our proposed method, we enclose the same elliptical scatterer with the same setting as in Fig. 3e by a rectangular layer with $(L_1, L_2) = (1, 0.8)$ and $(d_1, d_2) = (0.1, 0.08)$. We partition $\Omega = \{\Omega_f^{(i)}\}_{i=1}^8 \cup \{\Omega_{\rm ab}^{(i)}\}_{i=1}^8$ into 16 non-overlapping quadrilateral elements as shown in Fig. 5b. Once again, the spectral-element scheme can be implemented by the unconventional basis in (21) and $u_{\mathrm{N}} \in V_{\mathrm{N}}$ in (40) with w defined in (51)–(52). Let $\theta_0 = 0$, $k = 50$, and $\sigma_0 = 1.5$, we plot $\mathrm{Re}(u_{\mathrm{N}})|_{\Omega_f}$ and $\mathrm{Re}(v_{\mathrm{N}})|_{\Omega_{\mathrm{ab}}}$ with $(N_1, N) = (60, 30)$ in Fig. 5b. We plot the maximum error in Ω_f with fixed $N_1 = 60$ and $N = 3, 6, \cdots, 30$ in Fig. 5c. Observe that the error decays exponentially as N increases, and the approximation in the layer has no oscillation and is well-behaved.

4 Extensions and Discussions

We discuss various extensions and relevant futures works to conclude this report.

- The complex compression coordinate transformation (14) can be directly applied to construct three-dimensional spherical absorbing layer. However, the substitution (20) should be replaced by

$$u = vw, \quad w = \left(\frac{a}{\rho}\right)^2 e^{\mathrm{i}k(\rho - a)}. \tag{53}$$

- For 3D polyhedral layers, we can compress the outgoing waves of the open space in radial direction as with the polygonal layer outlined previously. The related real compression transformation can also be viewed as an inside-out polyhedral cloak version of that for the polyhedral cloak in [16].
- It is of interest and necessity to theoretically analyse the well-posedness of the reduced problem, and conduct the related error estimates, which the analysis in [7, 8, 15] can shed light on, and we shall report in future works.
- Time-dependent formulations of the equation in the absorbing layer can be obtained by taking the inverse Fourier transform in time of the time-harmonic counterparts as with the PML technique, see e.g., [9]. Remarkably, Daniel et al. in [2] proposed a high-order super-grid-scale absorbing layer, whose limiting case can be viewed as the real compression mapping discussed in Sect. 2.1, together with an artificial viscosity term to damp the waves. Different from the PAL technique and the above idea, which only involve spatial coordinate transformations, Zenginoğlu constructed a hyperboloidal layer in [17] by using a space-time coordinate transformation along characteristic lines. The comparison of the accuracy and efficiency between these methods is worthy of deep investigation.

Acknowledgements The authors thank Dr. Bo Wang from Hunan Normal University, China, for discussions at the early stage of this topic. L. Wang would like to thank the Scientific Committee and local organizers of ICOSAHOM 2016 for the conference invitation to Rio de Janeiro, Brazil. The work of two authors was partially supported by Singapore MOE AcRF Tier 1 Grant (RG 27/15) and MOE AcRF Tier 2 Grant (MOE 2013-T2-1-095, ARC 44/13).

Appendix 1. Proof of Theorem 1

Proof Given the transformation (14), (8) becomes

$$\mathbf{J} = \frac{\partial(x,y)}{\partial(\tilde{x},\tilde{y})} = \frac{\partial(x,y)}{\partial(r,\theta)} \frac{\partial(r,\theta)}{\partial(\tilde{\rho},\theta)} \frac{\partial(\tilde{\rho},\theta)}{\partial(\tilde{x},\tilde{y})}. \tag{54}$$

With $\tilde{\rho}$ in place of ρ in (9)–(10), we have

$$\mathbf{J} = \mathbf{R}\,\mathbf{J}_1\mathbf{R}^t \quad \text{with} \quad \mathbf{J}_1 = \begin{bmatrix} dr/d\tilde{\rho} & 0 \\ 0 & r/\tilde{\rho} \end{bmatrix}, \tag{55}$$

and

$$\mathbf{C} = \mathbf{R}\begin{bmatrix} c & 0 \\ 0 & 1/c \end{bmatrix}\mathbf{R}^t, \quad n = \frac{\tilde{\rho}}{r}\frac{d\tilde{\rho}}{dr} = \alpha\beta\frac{\rho}{r}\frac{d\rho}{dr}, \quad c := \frac{\tilde{\rho}}{r}\frac{dr}{d\tilde{\rho}} = \frac{\beta}{\alpha}\frac{\rho}{r}\frac{dr}{d\rho}. \tag{56}$$

Then we can work out the explicit expressions of n, c in (18) as (13).

Note that the asymptotic boundary condition at $r = b$ is transformed from the Sommerfeld radiation condition in (1b).

We now derive the estimate (19). For this purpose, we expand the solution and data in Fourier series:

$$\{u, \Psi\} = \sum_{|m|=0}^{\infty} \{\hat{u}_m(r), \hat{\psi}_m(r)\} e^{\mathrm{i}m\theta}, \tag{57}$$

where $\{\hat{u}_m(r), \hat{\psi}_m(r)\}$ are the Fourier coefficients. Then we can reduce the problem (16)–(17) to

$$\frac{1}{r}\big(rc\,\hat{u}_m'\big)' - \frac{m^2}{r^2c}\hat{u}_m + k^2 n\,\hat{u}_m = 0, \quad r \in [a,b), \ \ |m| = 0,1,2,\cdots, \tag{58}$$

$$\hat{u}_m = \hat{\psi}_m \ \text{ at } \ r = a; \quad \frac{1}{\alpha}\frac{dr}{d\rho}\hat{u}_m' - \mathrm{i}ku = o(|\tilde{\rho}|^{-1/2}) \ \text{ as } \ r \to b^-. \tag{59}$$

One can verify by using the Bessel equation of Hankel function (cf. [1]):

$$r^2y'' + ry' + (r^2 - m^2)y = 0, \quad y = H_m^{(1)}(r),$$

that the unique solution of (16)–(17) is

$$u = \sum_{|m|=0}^{\infty} \hat{u}_m(r)e^{\mathrm{i}m\theta} \ \text{ with } \ \hat{u}_m(r) = \hat{\psi}_m \frac{H_m^{(1)}(k\tilde{\rho})}{H_m^{(1)}(ka)}. \tag{60}$$

We next resort to a uniform estimate of Hankel functions first derived in [7, Lemma 2.2]: *For any complex* z *with* $\mathrm{Re}(z), \mathrm{Im}(z) \geq 0$, *and for any real* Θ *such that* $0 < \Theta \leq |z|$, *we have for any real order* ν,

$$|H_\nu^{(1)}(z)| \leq e^{-\mathrm{Im}(z)\left(1-\frac{\Theta^2}{|z|^2}\right)^{1/2}} |H_\nu^{(1)}(\Theta)|, \tag{61}$$

which implies

$$\max_{|m|\geq 0} \left| \frac{H_m^{(1)}(k\tilde{\rho})}{H_m^{(1)}(ka)} \right| \leq \exp\Big\{-k\sigma_0(\rho-a)\Big(1-\frac{a^2}{k^2\rho^2+k^2\sigma_0^2(\rho-a)^2}\Big)^{1/2}\Big\}, \quad \rho > a. \tag{62}$$

Therefore, we can derive (19) by using the Parseval's identity of Fourier series and (62). □

Appendix 2. Proof of Theorem 2

Proof We first deal with the boundary term $\langle \mathbf{C}\nabla u \cdot \mathbf{n}, \phi\rangle_{\Gamma_b}$ in (24). By a direct calculation and (56), we have

$$(\partial_r u, r^{-1}\partial_\theta u)^t = \mathbf{R}^t \nabla u, \quad \mathbf{C}\nabla u \cdot \mathbf{n} = \mathbf{R}\,\mathrm{diag}(c, c^{-1})\,\mathbf{R}^t\,\nabla u \cdot \mathbf{n} = c\,\partial_r u. \tag{63}$$

Thus, using (56) and the substitutions: $\phi = w\psi$ and $u = wv$, we can write

$$\langle \mathbf{C}\nabla u \cdot \mathbf{n}, \phi\rangle_{\Gamma_b} = \langle c u_r, \phi\rangle_{\Gamma_b} = \langle c\bar{w} u_r, \psi\rangle_{\Gamma_b} = a^{3/2}\Big\langle \frac{\beta}{r}\sqrt{\frac{b-r}{s(r)}}e^{-\mathrm{i}k(\rho-a)}\frac{1}{\alpha}\frac{dr}{d\rho}u_r, \psi\Big\rangle_{\Gamma_b}$$

$$= a^{3/2}\Big\langle \frac{\beta}{r}\sqrt{\frac{b-r}{s(r)}}e^{-\mathrm{i}k(\rho-a)}\Big(\frac{1}{\alpha}\frac{dr}{d\rho}u_r - \mathrm{i}ku\Big), \psi\Big\rangle_{\Gamma_b} + \mathrm{i}ka^{3/2}\Big\langle \frac{\beta}{r}\sqrt{\frac{b-r}{s(r)}}e^{-\mathrm{i}k(\rho-a)}u, \psi\Big\rangle_{\Gamma_b}$$

$$= a^{3/2}\Big\langle \frac{\beta}{r}\sqrt{\frac{b-r}{s(r)}}e^{-\mathrm{i}k(\rho-a)}\Big(\frac{1}{\alpha}\frac{dr}{d\rho}u_r - \mathrm{i}ku\Big), \psi\Big\rangle_{\Gamma_b} + \mathrm{i}ka^{3}\Big\langle \frac{\beta}{r}\frac{(b-r)^2}{s^2(r)}v, \psi\Big\rangle_{\Gamma_b}.$$

Noting that the integral along Γ_b is in θ, we obtain from the transformed Sommerfeld radiation condition (17) that $\langle \mathbf{C}\nabla u \cdot \mathbf{n}, \phi\rangle_{\Gamma_b} \to 0$ as $r \to b^-$.

We next deal with the other two terms in (24). Using the basic differentiation rules

$$\nabla u = w\,\nabla v + v\,\nabla w, \quad \nabla\bar{\phi} = \bar{w}\,\nabla\bar{\psi} + \bar{\psi}\,\nabla\bar{w},$$

we derive from (24) and a direct calculation that

$$\mathbb{B}_{\Omega_{\rm ab}}(u,\phi) = \big(|w|^2\mathbf{C}\nabla v, \nabla\psi\big)_{\Omega_{\rm ab}} + \big(w\,\mathbf{C}\nabla v\cdot\nabla\bar{w}, \psi\big)_{\Omega_{\rm ab}} + \big(v\,\bar{w}\mathbf{C}\,\nabla w, \nabla\psi\big)_{\Omega_{\rm ab}} + \big(\mathbf{C}\,\nabla w\cdot\nabla\bar{w}\,v, \psi\big)_{\Omega_{\rm ab}} - k^2\big(|w|^2\,n\,v, \psi\big)_{\Omega_{\rm ab}}. \tag{64}$$

As $\mathbf{C}$ is symmetric, one verifies readily that for any vectors $\mathbf{a}$ and $\mathbf{b}$ with two components, we have $(\mathbf{Ca})\cdot\mathbf{b} = (\mathbf{Cb})\cdot\mathbf{a}$. Thus, we can rewrite

$$\big(w\,\mathbf{C}\nabla v\cdot\nabla\bar{w}, \psi\big)_{\Omega_{\rm ab}} = \big(w\,\mathbf{C}\nabla\bar{w}\cdot\nabla v, \psi\big)_{\Omega_{\rm ab}}. \tag{65}$$

As w is independent of θ, we immediately get $\nabla w = \frac{dw}{dr}\mathbf{n}$. Then by (56),

$$\mathbf{C}\nabla w = \frac{dw}{dr}\mathbf{R}\,\mathrm{diag}(c, c^{-1})\,\mathbf{R}^t\,\mathbf{n} = c\frac{dw}{dr}\mathbf{n}. \tag{66}$$

Thus, we have

$$w\mathbf{C}\nabla\bar{w} = c\,w\frac{d\bar{w}}{dr}\mathbf{n}, \quad \bar{w}\mathbf{C}\nabla w = c\,\bar{w}\frac{dw}{dr}\mathbf{n}, \quad \mathbf{C}\,\nabla w\cdot\nabla\bar{w} = c\Big|\frac{dw}{dr}\Big|^2. \tag{67}$$

Introducing

$$\varpi_1 = |w|^2, \quad \varpi_2 = c\,\bar{w}\frac{\alpha}{\beta}\frac{dw}{dr}, \quad \varpi_3 = c\Big|\frac{dw}{dr}\Big|^2 - k^2|w|^2 n, \quad \partial_{\mathbf{n}} = \mathbf{n}\cdot\nabla, \tag{68}$$

we can derive (25) from (64)–(65) and (67)–(68). By (20),

$$\frac{dw}{dr} = w\frac{d\rho}{dr}\Big(-\frac{3}{2\rho} + \mathrm{i}k\Big). \tag{69}$$

We can work out $\{\varpi_j\}_{j=1}^3$ by using (12), (56) and (69). □

Appendix 3. Proof of Theorem 3

Proof We take $v = \psi$ in (25). By (56) and (63), we have

$$\begin{aligned}\mathrm{Re}\big(\varpi_1\mathbf{C}\nabla v, \nabla v\big)_{\Omega_{\rm ab}} &= \mathrm{Re}\int_0^{2\pi}\!\!\int_a^b \Big\{c|v_r|^2 + \frac{1}{cr^2}|v_\theta|^2\Big\}\varpi_1\, rdrd\theta \\ &= \int_0^{2\pi}\!\!\int_a^b \Big\{\mathrm{Re}\Big(\frac{\beta}{\alpha}\Big)\frac{a^3}{r\rho^2}\frac{dr}{d\rho}\Big\}|v_r|^2 rdrd\theta + \int_0^{2\pi}\!\!\int_a^b \Big\{\mathrm{Re}\Big(\frac{\alpha}{\beta}\Big)\frac{a^3}{r\rho^4}\frac{d\rho}{dr}\Big\}|v_\theta|^2 rdrd\theta.\end{aligned} \tag{70}$$

Using (26) and integration by parts leads to

$$
\begin{aligned}
&\mathrm{Re}\Big\{\frac{1}{\alpha}(\beta D_{\mathbf{n}}v, v\varpi_2)_{\Omega_{\mathrm{ab}}} + \frac{1}{\alpha}(\beta v\varpi_2, D_{\mathbf{n}}v)_{\Omega_{\mathrm{ab}}}\Big\} = 2\int_0^{2\pi}\!\!\int_a^b \mathrm{Re}\Big(\frac{\beta}{\alpha}\Big)\,\mathrm{Re}(\varpi_2 v\bar{v}_r)rdrd\theta \\
&= \int_0^{2\pi}\!\!\int_a^b \mathrm{Re}\Big(\frac{\beta}{\alpha}\Big)\,\mathrm{Re}(\varpi_2)(\partial_r|v|^2)rdrd\theta - 2\int_0^{2\pi}\!\!\int_a^b \mathrm{Re}\Big(\frac{\beta}{\alpha}\Big)\,\mathrm{Im}(\varpi_2)\mathrm{Im}(v\bar{v}_r)rdrd\theta \\
&= \frac{3}{2}\frac{1}{1+\sigma_0^2}\|v(a,\cdot)\|^2_{L^2(0,2\pi)} + \frac{3}{2}\int_0^{2\pi}\!\!\int_a^b \Big\{\frac{a^3}{r\rho^4}\Big(\frac{4\sigma_0^2}{1+\sigma_0^2}\frac{a}{\rho} - 3\Big)\frac{d\rho}{dr}\Big\}|v|^2 rdrd\theta \\
&\quad - 2k\int_0^{2\pi}\!\!\int_a^b \mathrm{Re}\Big(\frac{\beta}{\alpha}\Big)\frac{a^3}{\rho^2}\mathrm{Im}(v\bar{v}_r)drd\theta.
\end{aligned}
\tag{71}
$$

It is evident that

$$
\mathrm{Re}\big(\varpi_3 v, v\big)_{\Omega_{\mathrm{ab}}} = \int_0^{2\pi}\!\!\int_a^b \mathrm{Re}(\varpi_3)|v|^2 rdrd\theta. \tag{72}
$$

Note from (15) that

$$
\mathrm{Re}\Big(\frac{\beta}{\alpha}\Big) = 1 - \frac{\sigma_0^2}{1+\sigma_0^2}\frac{a}{\rho} > \frac{1}{1+\sigma_0^2}, \quad \mathrm{Re}\Big(\frac{\alpha}{\beta}\Big) = 1 + \frac{\sigma_0(1-a/\rho)}{1+\sigma_0^2(1-a/\rho)^2}\frac{a}{\rho} > 1. \tag{73}
$$

Using the Cauchy-Schwarz inequality, we obtain

$$
\begin{aligned}
2k\int_0^{2\pi}\!\!\int_a^b \mathrm{Re}\Big(\frac{\beta}{\alpha}\Big)\frac{a^3}{\rho^2}\mathrm{Im}(v\bar{v}_r)drd\theta &\le \frac{1}{\epsilon}\int_0^{2\pi}\!\!\int_a^b \Big\{\mathrm{Re}\Big(\frac{\beta}{\alpha}\Big)\frac{a^3}{r\rho^2}\frac{dr}{d\rho}\Big\}|v_r|^2 rdrd\theta \\
&\quad + \epsilon k^2\int_0^{2\pi}\!\!\int_a^b \Big\{\mathrm{Re}\Big(\frac{\beta}{\alpha}\Big)\frac{a^3}{r\rho^2}\frac{d\rho}{dr}\Big\}|v|^2 rdrd\theta,
\end{aligned}
\tag{74}
$$

where ϵ is a positive constant independent of k. Thus, by (25), (70)–(74) and collecting the terms, we obtain

$$
\begin{aligned}
\mathrm{Re}\big\{\mathbb{B}_{\Omega_{\mathrm{ab}}}(u,u)\big\} &\ge \Big(1-\frac{1}{\epsilon}\Big)\int_0^{2\pi}\!\!\int_a^b \Big\{\mathrm{Re}\Big(\frac{\beta}{\alpha}\Big)\frac{a^3}{r\rho^2}\frac{dr}{d\rho}\Big\}|v_r|^2 rdrd\theta \\
&+ \int_0^{2\pi}\!\!\int_a^b \Big\{\mathrm{Re}\Big(\frac{\alpha}{\beta}\Big)\frac{a^3}{r\rho^4}\frac{d\rho}{dr}\Big\}|v_\theta|^2 rdrd\theta + \frac{3}{2}\frac{1}{1+\sigma_0^2}\|v(a,\cdot)\|^2_{L^2(0,2\pi)} \\
&+ \int_0^{2\pi}\!\!\int_a^b \Big\{\mathrm{Re}(\varpi_3) + \Big(\frac{3}{2}\frac{1}{\rho^2}\Big(\frac{4\sigma_0^2}{1+\sigma_0^2}\frac{a}{\rho} - 3\Big) - \epsilon k^2\mathrm{Re}\Big(\frac{\beta}{\alpha}\Big)\Big)\frac{a^3}{r\rho^2}\frac{d\rho}{dr}\Big\}|v|^2 rdrd\theta.
\end{aligned}
\tag{75}
$$

We next work out and estimate the functions in the brackets. We have

$$s(r) = a^2 + r(b-2a) \le |I|\bar{c}, \quad \bar{c} := \max\{a, |I|\}, \quad |I| := b-a. \tag{76}$$

By (12), (26) and (76),

$$\frac{a^3}{r\rho^2}\frac{dr}{d\rho} = \frac{a^3}{|I|^2}\frac{(b-r)^4}{rs^2(r)} \ge \frac{a^3}{b\,\bar{c}^2|I|^4}(b-r)^4; \quad \frac{a^3}{r\rho^4}\frac{d\rho}{dr} = \frac{a^3|I|^2}{r}\frac{(b-r)^2}{s^4(r)} \ge \frac{a^3}{b\,\bar{c}^4|I|^2}(b-r)^2, \tag{77}$$

so we can obtain the lower bounds of the first two terms.

With a careful calculation, we can work out the summation in the curly brackets of the last term in (75) by using (12), (15) and (26). □

References

1. M. Abramowitz, I.A. Stegun, (eds.) *Handbook of Mathematical Functions with Formulas, Graphs, and Mathematical Tables*. A Wiley-Interscience Publication (John Wiley & Sons Inc., New York, 1984). Reprint of the 1972 edition, Selected Government Publications.
2. D. Appelö, T. Colonius, A high-order super-grid-scale absorbing layer and its application to linear hyperbolic systems. J. Comput. Phys. **228**(11), 4200–4217 (2009)
3. J.P. Berenger, A perfectly matched layer for the absorption of electromagnetic waves. J. Comput. Phys. **114**(2), 185–200 (1994)
4. A. Bermúdez, L. Hervella-Nieto, A. Prieto, R. Rodríguez, An exact bounded perfectly matched layer for time-harmonic scattering problems. SIAM J. Sci. Comput. **30**(1), 312–338 (2007)
5. A. Bermúdez, L. Hervella-Nieto, A. Prieto, R. Rodríguez, An optimal perfectly matched layer with unbounded absorbing function for time-harmonic acoustic scattering problems. J. Comput. Phys. **223**(2), 469–488 (2007)
6. J.P. Boyd, *Chebyshev and Fourier Spectral Methods*, 2nd en. (Dover Publications, Mineola, 2001)
7. Z.M. Chen, X.Z. Liu, An adaptive perfectly matched layer technique for time-harmonic scattering problems. SIAM J. Numer. Anal. **43**(2), 645–671 (2005)
8. Z.M. Chen, X.M. Wu, Long-time stability and convergence of the uniaxial perfectly matched layer method for time-domain acoustic scattering problems. SIAM J. Numer. Anal. **50**(5), 2632–2655 (2012)
9. W.C. Chew, W.H. Weedon, A 3D perfectly matched medium from modified Maxwell's equations with stretched coordinates. Microw. Opt. Technol. Lett. **7**(13), 599–604 (1994)
10. F. Collino, P. Monk, The perfectly matched layer in curvilinear coordinates. SIAM J. Sci. Comput. **19**(6), 2061–2090 (1998)
11. L. Demkowicz, J. Shen, A few new (?) facts about infinite elements. Comput. Methods Appl. Mech. Eng. **195**(29–32), 3572–3590 (2006)
12. T. Hagstrom, Radiation boundary conditions for the numerical simulation of waves, in *Acta Numerica, 1999*, vol. 8 (Cambridge University Press, Cambridge, 1999), pp. 47–106
13. S. Johnson, Notes on perfectly matched layers. Technical Report, Massachusetts Institute of Technology, Cambridge, MA (2010)
14. J. Shen, T. Tang, L.L. Wang, *Spectral Methods: Algorithms, Analysis and Applications*, Springer Series in Computational Mathematics, vol. 41 (Springer, Berlin, 2011)

15. L.L. Wang, B. Wang, X.D. Zhao, Fast and accurate computation of time-domain acoustic scattering problems with exact nonreflecting boundary conditions. SIAM J. Appl. Math. **72**(6), 1869–1898 (2012)
16. Z.G. Yang, L.L. Wang, Z.J. Rong, B. Wang, B.L. Zhang, Seamless integration of global Dirichlet-to-Neumann boundary condition and spectral elements for transformation electromagnetics. Comput. Methods Appl. Mech. Eng. **301**, 137–163 (2016)
17. A. Zenginoğlu, Hyperboloidal layers for hyperbolic equations on unbounded domains. J. Comput. Phys. **230**(6), 2286–2302 (2011)
18. N.A. Zharova, L.V. Shadrivov, Y.S. Kivshar, Inside-out electromagnetic cloaking. Opt. Express **16**(7), 4615–4620 (2008)

High Order Semi-Lagrangian Particle Methods

Georges-Henri Cottet and Petros Koumoutsakos

Abstract Semi-lagrangian (or remeshed) particle methods are conservative particle methods where the particles are remeshed at each time-step. The numerical analysis of these methods show that their accuracy is governed by the regularity and moment properties of the remeshing kernel and that their stability is guaranteed by a lagrangian condition which does not rely on the grid size. Turbulent transport and more generally advection dominated flows are applications where these features make them appealing tools. The adaptivity of the method and its ability to capture fine scales at minimal cost can be further reinforced by remeshing particles on adapted grids, in particular through wavelet-based multi-resolution analysis.

1 Accuracy Issues in Particle Methods

Particle methods are not in general associated with the concept of high accuracy. They are instead viewed as numerical models, able to reproduce qualitative features of advection dominated phenomena even with few particles, in particular in situations with strongly unsteady dynamics. Examples of early applications of particle methods which illustrate these capabilities in flow simulations are transition to turbulence in wall bounded flows [23] or the study of vortex reconnection [25]. Free surface or compressible flows are other examples where particle methods can give an intuitive qualitative understanding of the flow dynamics in situations where Direct Numerical Simulations with classical discretization methods would require prohibitive computational resources. This is in particular true for applications in computer graphics [17] or in astrophysics [12, 20].

The numerical analysis of particle methods allows to understand the accuracy issues that these methods face. Particle methods are based on the concept that Dirac

G.-H. Cottet (✉)
University Grenoble Alpes and Institut Universitaire de France, Grenoble, France
e-mail: georges-henri.cottet@univ-grenoble-alpes.fr

P. Koumoutsakos
ETH Zurich, Zurich, Switzerland
e-mail: petros@ethz.ch

M.L. Bittencourt et al. (eds.), *Spectral and High Order Methods for Partial Differential Equations ICOSAHOM 2016*, Lecture Notes in Computational Science and Engineering 119, DOI 10.1007/978-3-319-65870-4_6

masses give exact weak solutions to advection equations written in conservation form. For simple linear advection equations, the approximations of exact solutions by particles, measured on distribution spaces, therefore only relies on quadrature estimates for initial conditions and right hand sides. A typical error estimate for the solution U of an advection equation reads

$$\|(\mathbf{U} - \mathbf{U}_h)(\cdot, t)\|_{W^{-m,p}} \leq C_1 \exp{(C_2 T)}\, h^m$$

for $t \leq T$, where h is the initial inter-particle spacing, m is the order of the quadrature rule using particle initial locations as quadrature points, C_1 and C_2 are positive constants depending on the flow regularity, and $W^{-m,p}$ is the dual of $W^{m,q}$, with $1/p + 1/q = 1$. When one wishes to recover smooth quantities $\mathbf{U}_h^\epsilon$ from the Dirac masses carried by the particles, one needs to pay for the regularization involved in the process and a typical error estimate becomes :

$$\|(\mathbf{U} - \mathbf{U}_h^\epsilon)(\cdot, t)\|_{L^p} \leq C(\epsilon^r + h^m/\epsilon^m),$$

where r is the approximation order of the regularization used to mollify the particles (the reference [6] provide detailed proofs of the above estimates in the context of vortex methods).

This estimate immediately shows the dilemma of particle methods : the regularization size must be small, and controls the overall accuracy of the method, but it must contain enough particles so that the term h^m/ϵ^m does not compromise the convergence of the method. This constraint is even more stringent if particles are involved in additional terms of the model, in particular pressure gradient terms (in compressible flows) or diffusion terms. In the later case error terms of the form h^m/ϵ^{m+2} arise. In any case, proper convergence would require, on top of $\epsilon \to 0$, that $h/\epsilon \to 0$, the so-called overlapping condition. This condition is in practice difficult to satisfy, in particular for 3D flows. Instead, a constant ratio h/ϵ is most often chosen in refinement studies.

On the other hand, even if the overlapping condition is not satisfied, particle methods still enjoy conservation properties and some kind of adaptivity which goes with the belief that "particles go where they are needed". This belief is however often more a hand-waving argument than a reasonable assumption based on solid grounds. Figure 1 shows very simple examples which illustrate the shortcomings of particle methods in the simulation of 2D inviscid vortex flows. In this case the 2D Euler incompressible Euler equations in vorticity form

$$\frac{\partial \omega}{\partial t} + \mathrm{div}\,(\mathbf{u}\omega) = 0$$

are discretized by particles of vorticity. In the left picture, a typical particle distribution is shown for an initial vorticity field with support in an ellipse. This figure shows that particles tend to align along directions related to the flow strain, creating gaps in the vorticity support. The right picture corresponds to an axisymmetric initial vorticity field, leading to a stationary solution. In this example $\omega_0(x) = (1 - |x|^2)^3$. Error curves, in the energy norm for the particle velocities,

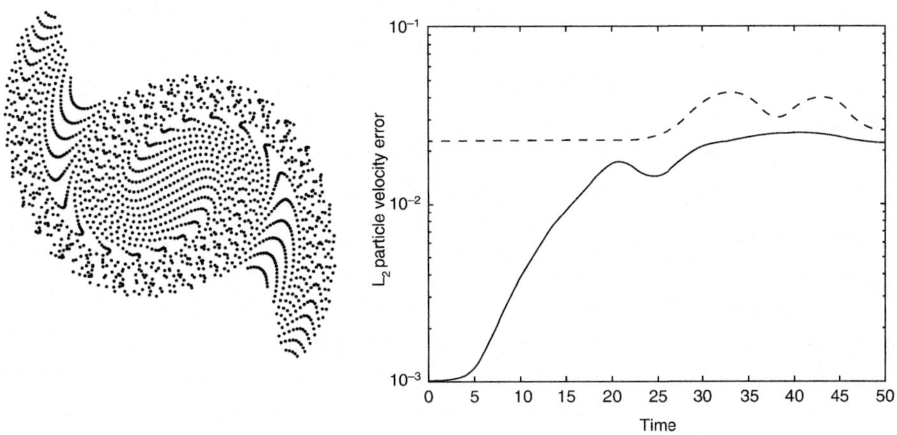

Fig. 1 Simulation with a grid-free vortex particle methods [7]. *Left picture*: particle distribution in the simulation of an elliptical vortex patch. *Right picture*: error curves for a stationary axisymmetric vortex. *Solid line* $h = 0.005$, *dashed line* $h = 0.01$. $\epsilon/h = 1.5$

corresponding to $h = h_0 = 0.1$ and $h = h_0/2$ are plotted, with a constant ratio h/ϵ. This figure shows that, despite the smoothness of the solution, the expected initial gain in accuracy is almost completely lost after a short time due to the distortion of the particle distribution.

Whenever point-wise values are required with some accuracy (for instance to recover satisfactory spectra in turbulent flows or local pressure or vorticity values on an obstacle) the overlapping condition cannot be ignored.

2 Remeshing and Semi-Lagrangian Particles

Although several methods have been considered to overcome the lack of overlapping of particles while keeping their grid-free nature, particle remeshing is to our knowledge the only tool which so far allowed to deliver in a clear-cut way accurate results for complex two and three-dimensional dynamics, in particular in incompressible flows. Remeshing was already used in early simulations for some specific flow topology, like vortex sheets [15] or filaments [19], but its first systematic use goes probably back to [13, 14], where pioneering results where obtained for flow past a cylinder at challenging Reynolds numbers and for the problem of axisymmetrization of elliptical vortices. The first numerical comparisons of these methods with spectral methods in 3D turbulent flows were performed in [9].

Remeshing consists of redistributing particle masses on nearby grid points, in a way that conserves as many moments of the particles as possible. The number of moments, and hence the accuracy, dictates the size of the remeshing kernel. Conserving the 3 first moments (including mass) in some sense guarantees that remeshing has not a diffusive net effect. This has been considered as a minimal requirement in the references already cited and in all following works.

The remeshing frequency can be a point of debate. However there are two aspects to consider. On the one hand remeshing a particle distribution which has already been highly distorted is likely to produce numerical noise. On the other hand, the time scale on which particles are distorted is the same as the one on which the particle advection should be discretized, namely $1/|\nabla \mathbf{a}|_\infty$, where $\mathbf{a}$ is the advecting velocity field. It is therefore natural to remesh particles every few time-steps and numerical truncation errors coming from remeshing must be accounted for in the numerical analysis. Recently [10] particle methods have been analyzed from this point of view - in other words as semi-lagrangian methods.

To describe the method and discuss its accuracy, let us consider the 1D model linear advection problem - which is somehow the engine of particle methods in all applications :

$$\theta_t + (a\theta)_x = 0, x \in \mathbf{R}, t > 0, \tag{1}$$

where a is a given smooth velocity field. A particle method where particles are remeshed at each time step can be recast as

$$\theta_i^{n+1} = \sum \theta_j^n \Gamma\left(\frac{x_j^{n+1} - x_i}{\Delta x}\right), i \in \mathbf{Z}^d, n \geq 0. \tag{2}$$

In the above equation Δx is the grid size on which particles are remeshed (assuming a regular grid), x_j are the grid points and Γ is the remeshing interpolating kernel. x_j^{n+1} is the result of the advection at time t_{n+1} of the particle located at x_j at time t_n.

Note that to generalize the method to several dimensions one may use similar formulas with remeshing kernels obtained by tensor products of 1D kernels (this is the traditional way) or, following [18], one can alternate the advection steps in successive directions, with classical recipes to increase the accuracy of the splitting involved in this process. This later method is economical when one uses high order kernels (with large supports) as, in 3D, its computational cost scales like $O(3M)$ instead of $O(M^3)$ for a kernel involving M points in each direction.

The moment conservation properties mentioned earlier to be satisfied by the remeshing kernel Γ can be expressed as

$$\sum_{k\in\mathbf{Z}} (x-k)^\alpha \Gamma(x-k) = \begin{cases} 1 & \text{if } \alpha = 0 \\ 0 & \text{if } 1 \leq \alpha \leq p \end{cases}, x \in \mathbf{R}, \tag{3}$$

for a given value of $p \geq 1$. An additional requirement is that Γ is globally in $W^{r+1,\infty}$ and of class C^∞ in each integer interval (in practice Γ is a polynomial in these intervals), and satisfies the interpolation property : $\Gamma(i-j) = \delta_{ij}$. In the simple case of an Euler explicit scheme to advect particles, $x_j^{n+1} = x_j + a(x_j, t_n)\Delta t$ and when the time step satisfies the condition

$$\Delta t < |a'|_{L^\infty}^{-1}, \tag{4}$$

on can prove [10] that the consistency error of the semi-lagrangian method is bounded by $O(\Delta t + \Delta x^\beta)$ where $\beta = \min(p, r)$. Using higher order Runge-Kutta schemes increase the time accuracy, as expected. Moreover, at least for kernels of order up to 4, under appropriate decay properties for the kernel Γ one can prove for a large class of kernels the stability of the method under the sole assumption (4).

Let us give a sketch of the consistency proof for the case $r = p = 1$ if a is only a function of x and the Euler scheme is used to advance particles. We start from (2) and assume that $\theta_j^n = \theta(x_j, t_n)$ where θ is the exact smooth solution to the advection equation and we want to prove that $\theta_i^{n+1} = \theta(x_i, t_{n+1}) + O(\Delta t^2 + \Delta x^2)$.

We write $j = i + k$ and, in (2),

$$\theta_j^n = \theta_i^n + k\Delta x\, \theta_x(x_i, t_n) + O(\Delta x^2).$$

Particle advection with the Euler scheme gives

$$x_j^{n+1} = x_j + a_j\Delta t = x_i + k\Delta x + a_i\Delta t + [a(x_i + k\Delta x) - a(x_i)]$$

and thus

$$\begin{aligned}\Gamma\left(\frac{x_j^{n+1} - x_i}{\Delta x}\right) &= \Gamma\left(k + \lambda_i + \nu[a(x_i + k\Delta x) - a(x_i)]\right)\\ &= \Gamma(k + \lambda_i) + k\Delta t\, a'(x_i)\, \Gamma'(k + \lambda_i) + O(\Delta x^2),\end{aligned}$$

where we have used the notations

$$\nu = \Delta t/\Delta x\ , \quad \lambda_i = a_i\nu.$$

We thus obtain

$$\theta_i^{n+1} = \sum_k \left[\theta(x_i, t_n) + k\Delta x\, \theta_x(x_i, t_n) + O(\Delta x^2)\right]$$
$$\left[\Gamma(k + \lambda_i) + k\Delta t\, a'(x_i)\, \Gamma'(k + \lambda_i) + O(\Delta x^2)\right].$$

The moment properties of order 0 and 1 yield

$$\sum_k k\,\Gamma(k + \lambda_i) = -\lambda_i\ , \quad \sum_k k\,\Gamma'(k + \lambda_i) = -1.$$

We therefore have

$$\begin{aligned}\theta_i^{n+1} &= \theta(x_i,t_n) - \Delta t\, a(x_i)\, \theta_x(x_i,t_n) - \Delta t\, a'(x_i)\, \theta(x_i,t_n) + O(\Delta t^2 + \Delta x^2) \\ &= \theta(x_i,t_n) - \Delta t (a\,\theta)_x(x_i,t_n) + O(\Delta t^2 + \Delta x^2)\end{aligned}$$

and, by (1),

$$\theta_i^{n+1} = \theta(x_i,t_{n+1}) + O(\Delta t^2 + \Delta x^2)$$

which proves our claim.

In the general consistency result mentioned above, one can also check that the order of spatial accuracy is p whenever one can ensure that after the advection step each grid cell contains exactly one particle. [10] contains explicit expressions of kernels of order up to 6, denoted by $\Lambda_{p,r}$ where p and r measure the moment and regularity properties of the kernel. It also contains a number of refinement studies which suggests that in practice for a kernel $\Lambda_{p,r}$ the observed order of accuracy is between min (p,r) and p. Figure 2 shows the results of a typical refinement study on a 2D level set benchmark. This case consists of a level set function corresponding to a disk of radius 0.15 centered at (0.5, 0.15) in the periodic box $[0,1]^2$. The velocity field is given by

$$\mathbf{a}(x_1,x_2,t) = f(t)\left(-\sin^2(\pi x_1)\sin(2\pi x_2), \sin(2\pi x_1)\sin^2(\pi x_2)\right), \tag{5}$$

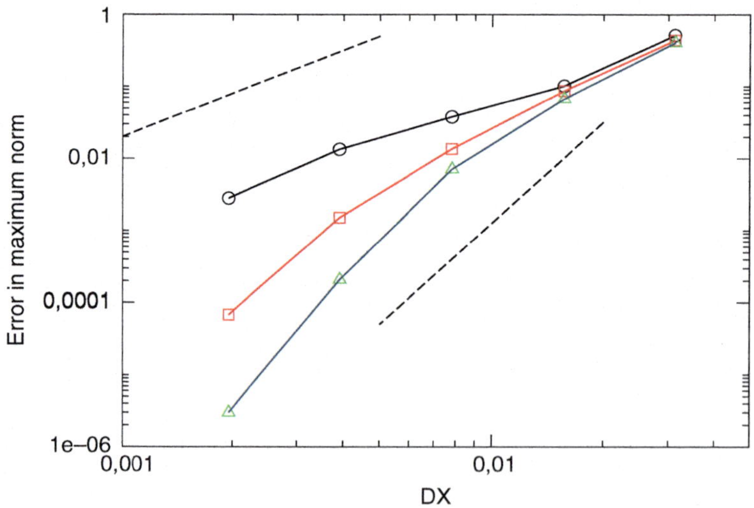

Fig. 2 Refinement study for the flow (5). CFL value is equal to 12. *Black-circle curve* : kernel $\Lambda_{2,1}$; *red-square* : kernel $\Lambda_{4,2}$; *blue-triangle* : kernel $\Lambda_{6,4}$; *dashed lines* indicate slopes corresponding to second and fourth order convergence

with $f(t) = \cos(\pi t/12)$. This field produces a strong filamentation of the solution culminating at $t = 6$ then drives the solution back to its initial state at $t = 12$, where numerical errors can be recorded. The actual convergence rate observed in this example for the kernels $\Lambda_{2,1}$, $\Lambda_{4,2}$ and $\Lambda_{6,4}$ are respectively 1.87, 3.17 and 5.92.

It is interesting to note that several authors have recently advocated the use of particle methods to correct dissipative effects of finite-difference of finite-volume level set methods. Roughly speaking the idea is to seed particles at sub-grid levels near the interface and use these particles to rectify the location of the interface (see [17] for instance). However [10] shows that, both in 2D and 3D, plain semi-lagrangian particle methods, with appropriate remeshing kernels (second order is actually enough) deliver better results with fewer points and larger time-steps.

The possibility of combining high accuracy with stability non constrained by the grid size makes semi-lagrangian particle methods appealing tools for turbulent transport. In [16] this was exploited to investigate universal scaling laws for passive scalars advected in turbulent flows. In this study the accuracy of particle methods was first compared to classical spectral methods. Figure 3 shows a typical comparison of scalar spectra and of the pdf of the scalar dissipation for a turbulent flow. In this experiment a second order kernel was used for the particle method. These results, and several other diagnostics, indicate that except for the very tail of the spectra most of the scales are well captured by the particle method with the same resolution as for the spectral method. A factor 1.2 between the grids was found sufficient to resolve satisfactorily the scalar also in the dissipative range. In the case of high Schmidt numbers (ratio between flow viscosity and scalar diffusivity) even with this requirement for slightly increased resolution, the gain in CPU time over the spectral method resulting from the use of large time steps in the particle method reached a factor 80.

3 Adaptive Semi-Lagrangian Particles

Remeshing particles somehow detracts particle methods from self adaptivity (however illusive that notion might be, as we have seen). To restore some kind of adaptivity in the method it is natural to rely on the grid on which particles are remeshed. Like for grid-based methods, one may envision several ways of doing so. One way is to assume a priori that grid refinement is desirable in some parts of the flow, typically zones which are close to fixed boundaries. Another way is to adapt the grid to the smoothness of the solution itself.

3.1 Semi-Lagrangian Particle Methods on Non-Uniform Grids

In that case we assume that a non-uniform grid (the physical space) is obtained by a predefined mapping from a cartesian uniform grid (the reference space). The

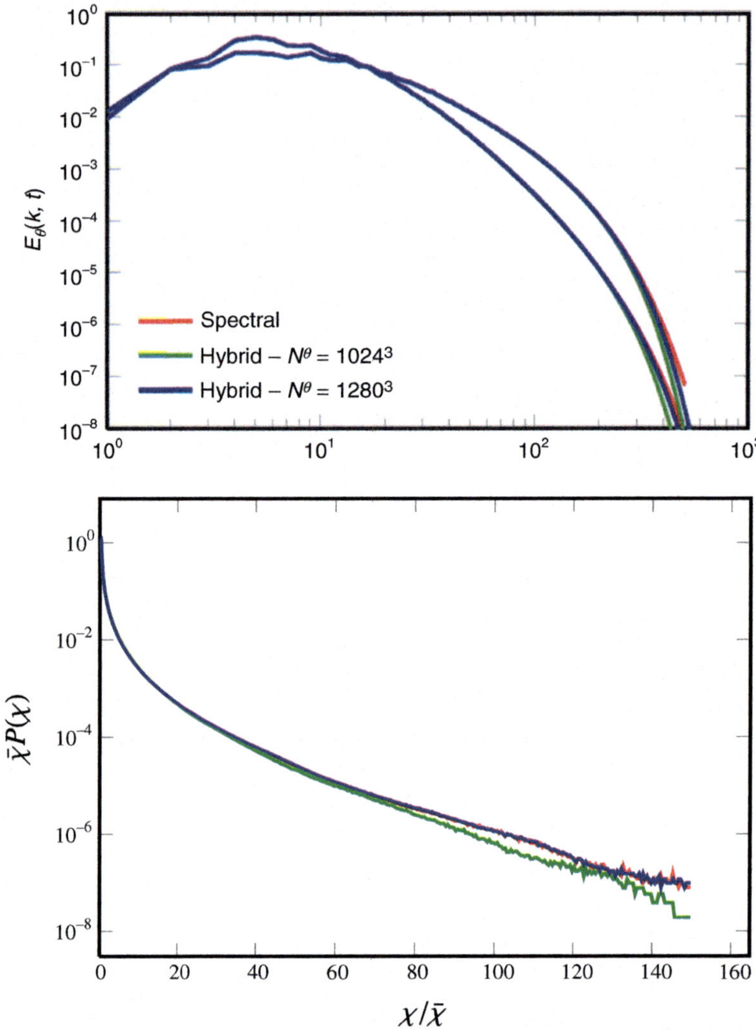

Fig. 3 Comparison of spectral and particle methods in a turbulent flow with a Schmidt number equal to 50 [16]. *Top picture*: spectra of the scalar variance. *Bottom picture*: pdf of the scalar dissipation. *Red curves*: spectral method with 1024^3 points; *green curves*: particle method with 1024^3 points ; *blue curves*: particle method with 1280^3 points

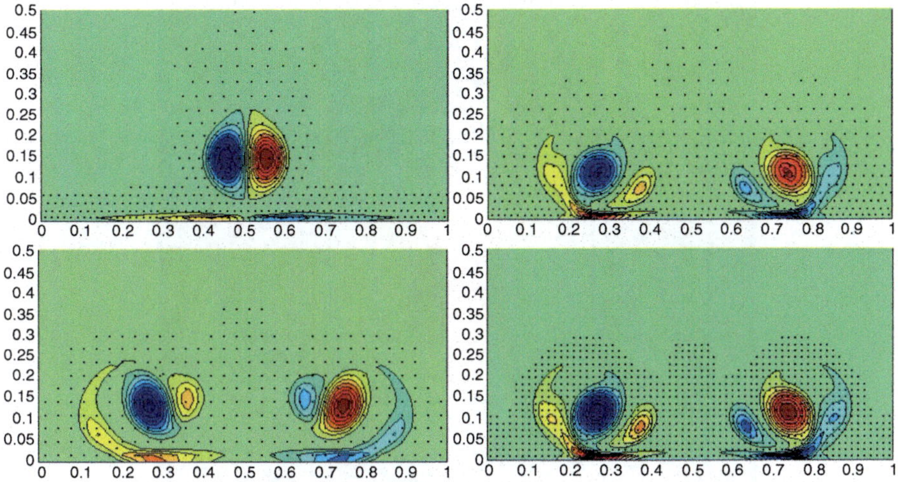

Fig. 4 Vortex dipole impinging on a wall [8]. *Top pictures*: results at two successive times with remeshing on an exponentially stretched grid. *Bottom pictures*: results with uniform grids at the later time for the coarsest (*left*) and finest (*right*) grids. Only a small percentage of the active particles are shown by dots

method works as follows [8]. At each time step, particles are pushed in the physical space, then mapped in the reference space. In this space regular remeshing formulas are used on cartesian grids, and particle locations are meshed back to the physical space. Figure 4 is an illustration of this method in the context of 2D vortex particle methods for the simulation of the rebound of a vortex dipole impinging on a wall. In this example the grid is stretched in both directions in an exponential manner with respect to the distance to the wall. Applications in 3D flows of similar technics can for instance be found in [21].

3.2 *Particle Methods with Adaptive Mesh Refinement*

Particle remeshing also enables to incorporate Adaptive Mesh Refinement (AMR) finite difference techniques [2] to adapt the particle discretization to the solution itself. Bergdorf et al. [3] describes how to define, move and remesh patches of particles at different resolution in a consistent way. Figure 5 illustrates how the method allows to capture filaments ejected by an elliptical vortex in a 2D inviscid flow. In this example the refinement was based on the vorticity gradients. We do not enter more into the details of this method as we believe that it is outperformed by the more recent wavelet-based method described below.

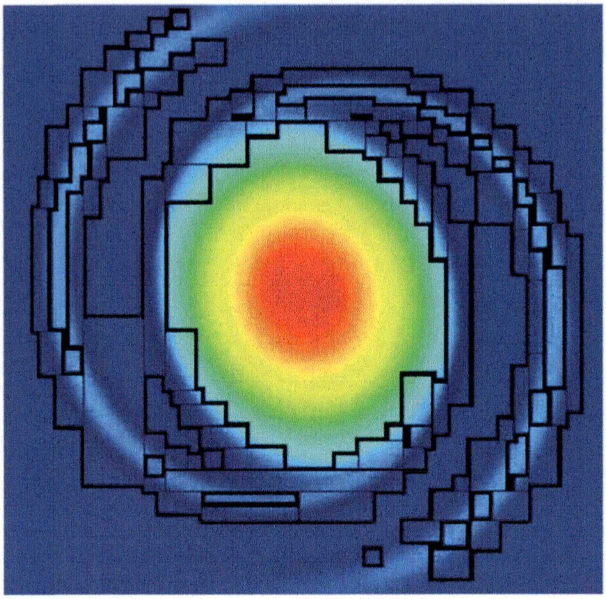

Fig. 5 Blocks of refined particles in an AMR implementation of particle methods for the simulation of an elliptical vortex [3]

3.3 *Wavelet-Based Multi-Level Particle Methods*

The concept of multi-resolution semi-lagrangian particle methods was recently pushed further in [4], using wavelet tools. The idea of combining particle and wavelet methods can actually already been found in [1]. In this reference wavelet served as particle shapes in a grid-free method. The method however did not find practical ways to address the issue of interacting and recombining wavelets. In [4] instead, because particles are remeshed at every time-step on regular grids, particle methods can inherit concepts and techniques used in the context of finite-difference methods (see for instance [24]). Nonetheless the semi-lagrangian character of the method introduces original and interesting features. To describe the method and understand how it works, let us consider again the 1D advection equation on the real line, first with constant velocity then in the general case. The multi-dimensional case and the application to incompressible flows will be next outlined.

We consider Eq. (1). The method is based on the following classical wavelet decomposition, in the framework of interpolating bi-orthogonal wavelets [5]:

$$\theta(x) = \sum_j c_k^0 \phi_j^0(x) + \sum_j \sum_{l=l_0}^{L-1} d_j^l \psi_k^l(x) \tag{6}$$

where l_0 (resp. $l = L$) corresponds to the coarsest (resp. finest) level. In the above equation ϕ_j^l and ψ_j^l are respectively the scaling and wavelet (or detail) basis functions, centered around the grid point $x_j^l = j\Delta x_l$, where $\Delta x_l = 2^{L_0-l}\Delta x_0$ is the grid size at level l. This decomposition allows to define the solution at the different scales :

$$\theta_l(x) = \sum_j c_j^l \phi_j^l(x). \tag{7}$$

Scale and detail coefficients are related between successive levels through classical filtering operations :

$$c_i^l = \sum_j \tilde{h}(2i-j)c_j^{l+1}, d_i^l = \sum_j \tilde{g}(2i-j)c_j^{l+1}, c_i^{l+1} = \sum_j [h(2j-i)c_j^l + g(2j-i)d_j^l] \tag{8}$$

where the filter functions g, h, $\tilde{g}$, $\tilde{h}$ depend on the particular wavelet system chosen.

Let us now describe one time-step of the algorithm, first for a constant velocity value. The methods advances the solution scale by scale in the following manner. Assume at time $t_n = n\Delta t$ the solution is known on grid points belonging to the nested grids $(x_j^l)_{j,l}$. For each level $l \in [L_0, L]$

- a wavelet analysis selects grid points which correspond to detail coefficients d_j^l above a given threshold
- particles are initialized on these grid points with, for grid values, the corresponding scale parameters c_j^l
- particles are pushed and remeshed on the grid $(x_k^l)_k$.

For the remeshing step above to be consistent one needs to ensure that active particles are always surrounded by "enough" particles carrying consistent values of θ_l. This is done in a way similar to what would be done in a finite-difference method by creating ghost particles in the neighborhood of the active particles. The values of the solution assigned at time t_n to these grid points is obtained by interpolation from θ_{l-1}. In order to make sure that only relevant values at level l are retained at the end of the iteration, a tag value equal to 0 is assigned to the ghost particles while the value 1 is assigned to the active particles. These tag values are pushed and remeshed along with the particles. After remeshing, only particles with tag values different from 0 are retained.

Finally, when all active particles have been moved and remeshed at all scales, values of the function at a given scale are updated by values at the next scale on even points when available.

Figure 6 illustrates the method in the case of a translating top-hat profile. The remeshing kernel is $\Lambda_{2,1}$ which for a constant velocity corresponds to a second order method. The scaling function is a piecewise linear function and the detail coefficients correspond to central finite-differences of the second derivatives. In the left picture we compare, after the time needed to travel 15 times the width of the

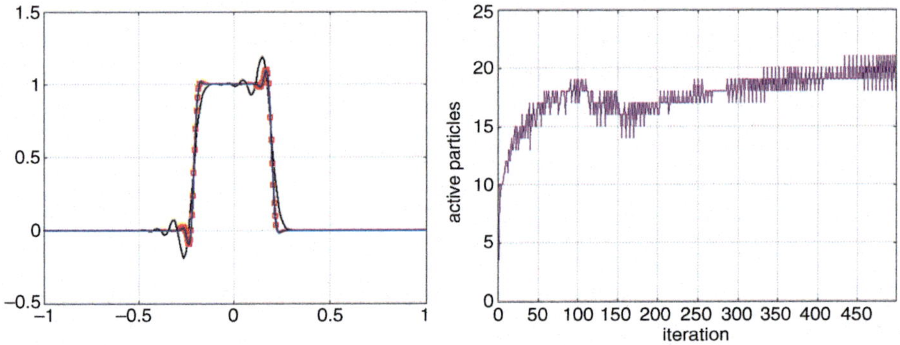

Fig. 6 Advection of top-hat profile in a uniform field by a 2-level particle method. *Left picture*: *black curve* : result for uniform coarse resolution; *blue curve*: result at the coarsest level for a 2-level method; *red squares* active particle at the fine level. *Right picture*: number of active particles as a function of time

top-hat, the results obtained with uniform one-level resolution at $\Delta x_0 = 0.01$ and $\Delta x_1 = 0.005$ and the wavelet-based method using these two levels. For the two-level case, only the coarse resolution is showed together with active particles at the higher resolution. One can see that all solutions exhibit overshoots, as expected from a second order method, but the two-level solution limits the overshoot even at the coarse level and is very close to the higher resolution solution. The number of active particles, shown on the right picture of the figure, increases to respond to the oscillations created by the remeshing, then stabilizes.

Let us now consider the case of a non uniform velocity. In this case, scales are not advected in an independent fashion but interact as a result of compression and dilatation. To allow fine scales to appear from coarse scales, there are two options. The first one is similar to what would be done in a finite difference method [24]. It consists of considering for each scale at the beginning of each step additional ghost particles at the level immediately above on which values of solutions are interpolated from the coarse scale. Another option, simpler and more specific to particle methods, is to remesh each scale on a scale twice smaller. At the end of the time-step, the value of the solution on this smaller scale is chosen to be either the result of the remeshing either of the coarse grid or the fine grid, depending on the value of the flag described earlier.

It is important to note that, in both options, the time scale on which the scale $l + 1$ appears from scale l is governed by $1/|a'|_\infty$ which is consistent with the maximum time step allowed for the particle method. In other words for time steps corresponding to large CFL numbers, only scale $l + 1$ can appear form scale l if the condition (4) is satisfied. Figure 7 illustrates the method for two scales for an initial condition consisting on a sine wave advected in a velocity given by $a(x) = 1 + cos(2\pi x)$. As the wave travels to the right it is subject to successive compressions/amplification and dilatation/damping. As a result, the number of

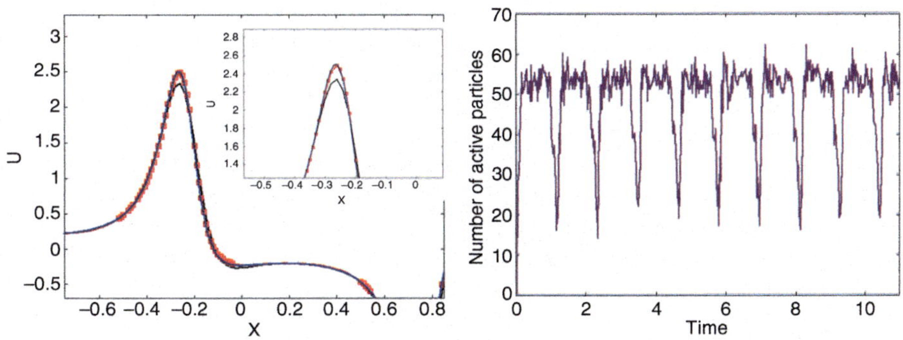

Fig. 7 Multilevel particle method in a compression/dilatation flow. See Fig. 6 for captions

active particles oscillates (right picture). The left picture shows that the adaptive method allows to recover the high resolution results.

To go beyond the 1d toy problem just considered, there are again two options. One is to use the same ideas but to rely on multidimensional tensor product wavelets, with the added complexity that one grid point is associated to several wavelets. This is the option followed in [4]. The other one, presumably simpler and which only uses 1D wavelets, would be to split advection into three successive directional advections, along the lines of [18].

In vortex flows, where these methods have been primarily applied, an additional work is to compute velocities created by multi-level vortices. The solution chosen in [4] is to use a fast multipole method, where each particle associated to the value of the solution and the associated grid size is accounted for in the Biot Savart law. The left picture of Fig. 8 illustrates the method for the flow around swimming fishes [11]. The power of the method is here fully apparent. The complex geometries and the associated boundary conditions are dealt with by a penalization method an the nested multi-level cartesian grids allow to capture at a minimal cost the fine vortices created by the swimmers, despite the low accuracy of the penalization method. This method has been implemented for both 2D and 3D geometries [22]. It has been in particular applied in combination with optimization technics to determine efficient swimming strategies [11].

4 Conclusion

Particle methods with particle remeshing at each time-step can be analyzed as semi-lagrangian conservative methods. The accuracy of the method is governed by moment and regularity properties of the remeshing kernel and high order kernels can be derived in a systematic fashion. Adaptivity in the method can be reinforced in particular by using wavelet-based multi-resolution analysis. In any case, the semi-

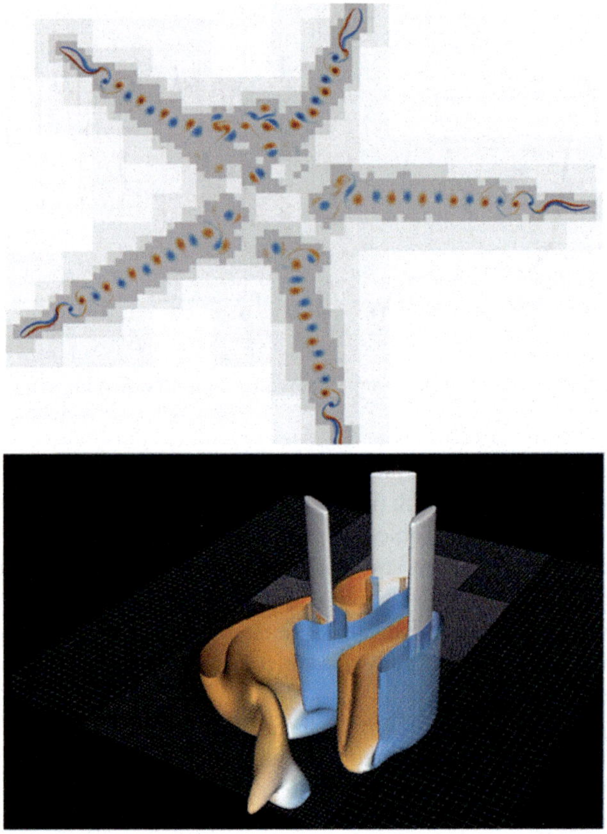

Fig. 8 Multi-level vorticity particle method with penalization around complex geometries. *Top figure* : 2D calculation around multiple fishes. *Bottom picture*: 3D vorticity passed a wind turbine (from [22])

lagrangian nature of the method allows to use time-steps which are not constrained by the grid size. In many applications this can lead to substantial computational savings.

References

1. C. Basdevant, M. Holschneider, V. Perrier, Methode des ondelettes mobiles. C. R. Acad. Sci. Paris I **310**, 647–652 (1990)
2. M. Berger, J. Oliger, Adaptive mesh refinement for hyperbolic partial differential equations. J. Comput. Phys. **3**, 484–512 (1984)
3. M. Bergdorf, G.-H. Cottet, P. Koumoutsakos, Multilevel adaptive particle methods for convection-diffusion equations. SIAM Multiscale Model. Simul. **4**, 328–357 (2005)

4. M. Bergdorf, P. Koumoutsakos, A Lagrangian particle-wavelet method. SIAM Multiscale Model. Simul. **5**(3), 980–995 (2006)
5. A. Cohen, I. Daubechies, J.C. Feauveau, Biorthogonal bases of compactly supported wavelets. Commun. Pure Appl. Math. **45**, 485–560 (1992)
6. G.-H. Cottet, A new approach for the analysis of vortex methods in 2 and 3 dimensions. Ann. Inst. Henri Poincaré **5**, 227–285 (1988)
7. G.-H. Cottet, P. Koumoutsakos, *Vortex Methods* (Cambridge University Press, Cambridge, 2000)
8. G.-H. Cottet, P. Koumoutsakos, M. Ould-Salihi, Vortex methods with spatially varying cores. J. Comput. Phys. **162**, 164–185 (2000)
9. G.-H. Cottet, B. Michaux, S. Ossia, G. Vanderlinden, A comparison of spectral and vortex methods in three-dimensional incompressible flows. J. Comput. Phys. **175**, 702–712 (2002)
10. G.-H. Cottet, J.-M. Etancelin, F. Perignon, C. Picard, High order Semi-Lagrangian particles for transport equations: numerical analysis and implementation issues. ESAIM: Math. Model. Numer. Anal. **48**, 1029–1060 (2014)
11. M. Gazzola, B. Hejazialhosseini, P. Koumoutsakos, Reinforcement learning and wavelet adapted vortex methods for simulations of self-propelled swimmers. SIAM J. Sci. Comput. **36**, 622–639 (2014)
12. R.A. Kerr, Planetary origins : a quickie birth of jupiters and saturns. Science **298**, 1698–1689 (2002)
13. P. Koumoutsakos, Inviscid axisymmetrization of an elliptical vortex. J. Comput. Phys. **138**, 821–857 (1997)
14. P. Koumoutsakos, A. Leonard, High resolution simulations of the flow around an impulsively started cylinder using vortex methods. J. Fluid Mech. **296**, 1–38 (1995)
15. R. Krasny, Desingularization of periodic vortex sheet roll-up. J. Comput. Phys. **65**, 292–313 (1986)
16. J.-B. Lagaert, G. Balarac, G.-H. Cottet, Hybrid spectral particle method for the turbulent transport of a passive scalar. J. Comput. Phys. **260**, 127–142 (2014)
17. F. Lossaso, J.O. Talton, N. Kwatra, R. Fedkiw, Two-way coupled SPH and particle level set fluid dynamics. IEEE Trans. Vis. Comput. Graph. **14**, 797–804 (2008)
18. A. Magni, G.-H. Cottet, Accurate, non-oscillatory remeshing schemes for particle methods. J. Comput. Phys. **231**(1), 152–172 (2012)
19. J.E. Martin, E. Meiburg, Numerical investigation of three-dimensional evolving jets subject to axisymmetric and azimuthal perturbation. J. Fluid Mech. **230**, 271 (1991)
20. J.J. Monaghan, Particle methods for hydrodynamics. Comput. Phys. Rep. **3**, 71–124 (1985)
21. P. Ploumhans, G.S. Winckelmans, J.K. Salmon, A. Leonard , M.S. Warren, Vortex methods for direct numerical simulation of three-dimensional bluff body flows: application to the sphere at Re = 300, 500, and 1000. J. Comput. Phys. **165**, 354–406 (2000)
22. D. Rossinelli, B. Hejazialhosseini, W. van Rees, M. Gazzola, M. Bergdorf, P. Koumoutsakos, MRAG-I2D: multi-resolution adapted grids for remeshed vortex methods on multicore architectures. J. Comput. Phys. **288**, 1–18 (2015)
23. J. Sethian, A. Ghoniem, Validation study of vortex methods. J. Comput. Phys. **54**, 425–456 (1984)
24. O. Vasilyev, Solving multi-dimensional evolution problems with localized structures using second generation wavelets. Int. J. Comput. Fluid Dyn. **17**(2), 151–168, 17, 151–168 (2003)
25. G. Winckelmans, A. Leonard, Contributions to vortex methods for the computation of three dimensional incompressible unsteady flows. J. Comput. Phys. **109**, 247–273 (1993)

Part II
Contributed Papers

Energy-Minimized High-Order Surface Meshes

Karsten Bock and Jörg Stiller

Abstract The construction of suitable curvilinear meshes for high-order methods in computational fluid dynamics still remains a challenge. This paper investigates a strictly local construction and optimization method for high-order surface meshes. The optimization procedure combines fitting and minimization of energy functionals related to bending and stretching. The weight of the energy functionals in this combination is gradually reduced during the process by application of blending functions. We apply the method to analytically defined smooth surfaces as well as triangulated scanning data. For both classes of test cases the method improves the mesh quality notably and preserves the accuracy of least-squares fitting. Three different blending functions for the energy weighting have been investigated. Furthermore, we incorporated and tested methods to reduce the additional computational costs of performing the optimization.

1 Introduction

High order methods like spectral element or discontinuous Galerkin methods are popular in computational fluid dynamics for their superior convergence properties compared with lower order methods. Yet, problems of interest in engineering involve geometrically complex domains, which have to be represented accurately with coarse meshes. This has established an increasing research interest in curved high-order meshes. So far, commercial mesh generators do not feature high-order mesh generation. Typically, high-order mesh generation starts from linear meshes obtained by these generators and curves them subsequently. During this process the meshes can get distorted, so that undulations and artifacts may even render the mesh invalid. Different techniques intending to ensure the quality and validity of curved meshes have been proposed, e.g. using elastic analogies [1, 2] or optimization algorithms [3, 4]. It is worth noting, that these methods are motivated globally, although mesh deformation is frequently carried out in a localized manner.

K. Bock (✉) • J. Stiller
Institute of Fluid Mechanics (ISM), Technische Universität Dresden, 01062 Dresden, Germany
e-mail: karsten.bock@tu-dresden.de; joerg.stiller@tu-dresden.de

M.L. Bittencourt et al. (eds.), *Spectral and High Order Methods for Partial Differential Equations ICOSAHOM 2016*, Lecture Notes in Computational Science and Engineering 119, DOI 10.1007/978-3-319-65870-4_7

The present paper describes our strictly local algorithm to construct quality optimized, high-accuracy surface meshes. The approach involves quadratic energy functionals in the fitting process. Our optimization method is applied to analytically defined as well as scattered data test cases. In contrast to other optimization methods, e.g. put forward in [5], our approach does not need the target surface parametrization, which renders it particularly useful for scattered data surfaces. This class of surface definition, often provided by scanning methods, is relevant in fields like bio-medicine or engineering science.

In this paper we show that our method is capable of quality optimization while simultaneously preserving accuracy. Furthermore, we present strategies to reduce the additional computational costs invested in optimization.

The paper is organized as follows: In Sect. 2 we extend least-squares fitting to an incremental mesh optimization algorithm based on surface entity energies. Section 3 presents the results for analytically defined smooth surface as well as scattered data examples. Section 4 concludes the paper.

2 Surface Mesh Construction

Typically, curved mesh construction methods work in a bottom-up sequence starting from an initial straight-sided mesh: (1) curving the boundary edges, (2) generating the curved boundary patches and, (3) building the curved volume elements. Our approach to build optimized high-order surface meshes follows this route too, naturally, up to step (2). First, it is ensured that the vertices of the initial, linear mesh are located on the surface. Sequentially, any entities built in a previous step are utilized as boundary conditions in the following steps.

2.1 Incremental Curve Construction

As common in Computer Aided Geometric Design (CAGD) Literature [6, 7], we use the Bernstein Bézier form to define curved mesh edges. A Bézier curve of order n is written as

$$\mathbf{c}(t) = \sum_{i=0}^{n} \mathbf{b}_i B_i^n(t) \ , \tag{1}$$

where B_i^n are the Bernstein polynomials and $\mathbf{b}_i$ the control points defining the form of the curves. Since the incremental curve construction method was described in detail in [8], we will only revisit it briefly here. Curved mesh edges are constructed by fitting them to surface-bound sampling points. The main concept in the computation of optimized curves relies on the adjustment of these sampling points in an incremental procedure minimizing curve energy. We achieve fitting by minimizing the average squared distance

$$J_x(\mathbf{b}_I; t_J, \mathbf{x}_J) = \frac{1}{m} \sum_{j=1}^{m} \left(\mathbf{c}(t_j) - \mathbf{x}_j\right)^2 \tag{2}$$

of the curve to a set of m surface points $\mathbf{x}_J$. In (2) t_J denotes the according set of samples in parameter space and $\mathbf{b}_I$ the set of the interior control points of the curve. The surface-bound sampling points $\mathbf{x}_J$ are obtained by projection $\mathcal{P}$ to the surface. This projection is carried out iteratively in an approximated surface normal direction. At the moment we employ two projections: One for exact surface definitions, and another one for application with discrete surface data in a facet representation, typically originating from scanning methods. For a more detailed description of the projection we refer to [8, 9].

To achieve energy-minimization we incorporate the energy functionals

$$E_1 = \int_0^1 \dot{\mathbf{c}}^2(t)\,\mathrm{d}t \tag{3}$$

and

$$E_2 = \int_0^1 \ddot{\mathbf{c}}^2(t)\mathrm{d}t \tag{4}$$

into the fitting in addition to the averaged squared distance. These functionals are frequently used in CAGD and especially in surface fairing [10, 11] and relate to physical stretching and bending energies, respectively.

Combining the energy functionals with (2) we obtain the curves control points $\mathbf{b}_I$ by minimizing

$$J(\mathbf{b}_I; t_J, \mathbf{x}_J, w_{\mathrm{E}}) = (1 - w_{\mathrm{E}})\bar{J}_x + w_{\mathrm{E}}\left((1 - \alpha_{\mathrm{c}})\bar{E}_1 + \alpha_{\mathrm{c}}\bar{E}_2\right), \tag{5}$$

where the overbars indicate normalizations of the aforementioned functionals. In this equation w_{E} weights the energies and the distance to each other and the parameter α_{c} balances E_1 and E_2. Using equal weighting of energies, i.e. $\alpha_{\mathrm{c}} = 0.5$ has proven a useful choice [8].

The core ingredient of the curve optimization strategy is creating an incremental fitting process during which (5) is minimized in every step. A blending function reduces the weight w_{E} continuously with the step count k of the process. We start out using $w_{\mathrm{E}} = 1$, meaning pure energy-minimization, and then gradually decrease w_{E}, finally reaching least-squares distance fitting with $w_{\mathrm{E}} = 0$ in the last step. We remark, that in every step k of the process the sampling point set $\mathbf{x}_J$ is reconstructed from the parameter set t_J and the previous steps curve by projection to the target surface. Therefore, this method results in an optimized set of sampling points, leading to energy-optimized curves and, furthermore, preserves the polynomial convergence rate, as previously shown by the authors in [8]. Generally, the blending function $w(k)$ for w_{E} can be chosen arbitrarily. In contrast to previous studies, we compare three different blending functions here. Using the substitution $b = \frac{k-1}{k_{\max}-1}$, which includes the step count k and the maximum number of steps performed $k_{\max}$, we define: A linear blending function (labeled: lin)

$$w_{\mathrm{lin}}(k) = 1 - b \ , \tag{6}$$

a smooth rational (rat)

$$w_{\mathrm{rat}}(k) = \frac{(1-b)^2}{b^2 + (1-b)^2} \tag{7}$$

and an exponentially decaying blending (exp)

$$w_{\exp}(k) = \begin{cases} p^{qb} & b \neq 1 \\ 0 & b = 1 \ . \end{cases} \tag{8}$$

The blending functions are shown in comparison in Fig. 1, where the exponential blending is shown with $p = \frac{1}{2}$ and $q = 10$, which represents the version used in this paper.

2.2 *Incremental Fitting of Triangular Patches*

Triangular patches are built in Bernstein Bézier form

$$\mathbf{s}(\boldsymbol{\tau}) = \sum_{i+j+k=n} \mathbf{b}_{ijk} B^n_{ijk}(\boldsymbol{\tau}) \ , \tag{9}$$

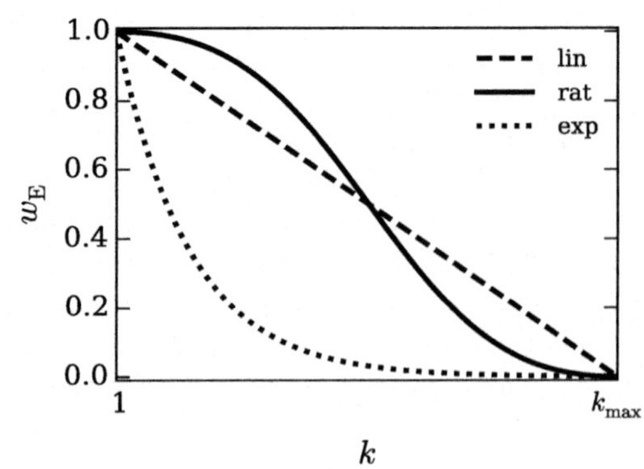

Fig. 1 Energy blending functions used in incremental curve and patch construction methods. A linear, a smooth rational and an exponential formulation are used and shown in comparison

which uses barycentric coordinates $\boldsymbol{\tau} = (\tau_1, \tau_2, \tau_3)^T$, the triangle control points $\mathbf{b}_{ijk}$ and the Bernstein polynomials $B^n_{ijk}(\boldsymbol{\tau})$. The latter are polynomials of order n and defined as

$$B^n_{ijk}(\boldsymbol{\tau}) = \frac{n!}{i!j!k!} \tau_1^i \, \tau_2^j \, \tau_3^k \tag{10}$$

and commonly applied to triangular patches. Since the methods presented here require the formulation of patches with a set of two independent parametric coordinates, the transformation

$$\boldsymbol{\tau} = \Big(1 - u - v, \, u, \, v\Big) \tag{11}$$

is used, wherein $u = \tau_2$ and $v = \tau_3$ were chosen. With this the directional derivatives

$$\partial_u \mathbf{s} = \partial_{\tau_2} \mathbf{s} - \partial_{\tau_1} \mathbf{s} \, , \tag{12}$$

$$\partial_v \mathbf{s} = \partial_{\tau_3} \mathbf{s} - \partial_{\tau_1} \mathbf{s} \tag{13}$$

follow and are applied during high-order patch construction.

The patch construction method is basically an extension of our curve construction. Equivalently to curve fitting, a squared distance functional

$$J_x\Big(\mathbf{b}_{\mathrm{I}}; \boldsymbol{\tau}_{\mathrm{J}}, \mathbf{x}_{\mathrm{J}}\Big) = \frac{1}{m} \sum_{j=1}^{m} \Big(\mathbf{s}(\boldsymbol{\tau}_j) - \mathbf{x}_j\Big)^2 \tag{14}$$

is used for triangles. The computation of a set of triangle control points b_{I} with minimization of (14) requires a set of m sampling points in parameter space $\boldsymbol{\tau}_{\mathrm{J}}$ and a corresponding set of surface-bound points $\mathbf{x}_{\mathrm{J}}$. The latter are obtained by projection operator $\mathscr{P}$.

Furthermore, we include patch energy functionals into the fitting process. Since physical membrane stretching and plate bending energies are complicated to minimize, we substitute them by the simplified quadratic approximations

$$E_1 = \int_0^1 \int_0^{1-v} \Big[(\partial_u \mathbf{s})^2 + (\partial_v \mathbf{s})^2 \Big] \, \mathrm{d}u \, \mathrm{d}v \tag{15}$$

and

$$E_2 = \int_0^1 \int_0^{1-v} \Big[(\partial_{uu} \mathbf{s})^2 + 2\, (\partial_{uv} \mathbf{s})^2 + (\partial_{vv} \mathbf{s})^2 \Big] \, \mathrm{d}u \, \mathrm{d}v \, , \tag{16}$$

respectively. Combining the functionals (14)–(16), where, again, the overbar indicates normalization by a reference value, yields

$$J\Big(\mathbf{b}_\mathrm{I}; \boldsymbol{\tau}_\mathrm{J}, \mathbf{x}_\mathrm{J}, \alpha_\mathrm{p}, w_\mathrm{E},\Big) = (1 - w_\mathrm{E})\bar{J}_x + w_\mathrm{E}\Big[(1-\alpha_\mathrm{p})\bar{E}_1 + \alpha_\mathrm{p}\bar{E}_2\Big]\,. \tag{17}$$

Minimization of functional (17) in each step of an incremental process results in energy-optimized surface-fitted triangles. Specifically, the patch energies are balanced by the factor α_p and the weight w_E between energies and squared distance is reduced from 1 to 0 during the optimization by application of an blending function $w(k)$.

3 Results

In the following we examine two classes of examples to study the curve and patch curving methods described earlier. The first test case is an explicitly defined screw surface, two scattered data surfaces follow—triangulations of a human left atrium and a statue head.

We assess the geometric accuracy by the L^2 error, which for curves is defined as

$$\varepsilon_x = \sqrt{\sum_e \int_0^1 \Big(\mathbf{c}_e - \mathscr{P}[\mathbf{c}_e]\Big)^2 \mathrm{d}t} \tag{18}$$

and evaluated over all curved edges $\mathbf{c}_e$. Equivalently, we utilize the L^2 error of all triangular patches $\mathbf{s}_f$

$$\varepsilon_x = \sqrt{\sum_f \int_0^1 \int_0^{1-v} \Big(\mathbf{s}_f - \mathscr{P}\big[\mathbf{s}_f\big]\Big)^2 \mathrm{d}u\,\mathrm{d}v}\,. \tag{19}$$

Both error formulations include the projection operator $\mathscr{P}$, briefly described in Sect. 2.1. Moreover, curve fairness is closely linked to curve energies and, therefore, evaluated using the energy norms

$$\varepsilon_{1,2} = \sqrt{\sum_e E_{1,2}(\mathbf{c}_e)} \tag{20}$$

incorporating the curve energy functionals (3) and (4), respectively. Lastly, triangular patch quality is evaluated by the quality measure

$$q = \min_{\Omega_f}\left(\frac{|\partial_u \mathbf{s} \times \partial_v \mathbf{s}|}{\max\limits_{\Omega_f} |\partial_u \mathbf{s} \times \partial_v \mathbf{s}|}\right) \tag{21}$$

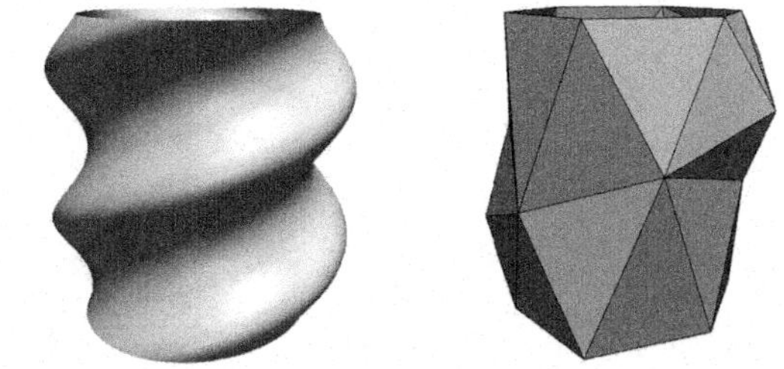

Fig. 2 Screw surface example and the coarse mesh used here. The mesh includes 32 triangles with an average mesh spacing of ca. 11 times the minimum curvature radius r_c of the surface

suggested in [12], which is based on the surface Jacobian. As a global quality measure for the surface mesh we use $q_{\min} = \min_f (q)$ the minimal element quality of all faces.

3.1 Smooth Surfaces with Analytical Definition

In this section we use the analytically defined screw surface

$$F(\mathbf{x}) = \left(x\cos(\pi z) + y\sin(\pi z)\right)^2 + \frac{16}{9}\left(x\sin(\pi z) - y\cos(\pi z)\right)^2 - 1 = 0 \quad (22)$$

as a test case to study the curve and patch construction methods and especially their dependency on the choice of blending functions mentioned in Sect. 2.1. Figure 2 depicts the screw surface for the interval $0 \leq z \leq 2$ as well as the mesh utilized as an example here. This extremely coarse mesh includes 32 triangles, featuring an average mesh spacing of approximately 11 times the minimum curvature radius $r_c \approx 0.117$ of the screw surface. During the following tests, the coarse mesh was curved using an order of $n = 12$ and the energies were balanced equally, i.e. $\alpha_c = \alpha_p = 0.5$.

In a first test scenario, we focus on curve construction. Therefore, the edges of the mesh were curved using the incremental procedure described earlier. Throughout this study the curve construction was carried out with different numbers of incremental steps $k_{c,\max}$ using the linear, rational and exponentially decaying blending functions $w(k)$. Figure 3 contains the results of this study. Accuracy is increased when allowing for a higher number of incremental steps during the construction, as can be seen in the error plot in Fig. 3a. Evidently, the blending function influences the error reduction rate. Using linear blending decreases the error slower than applying an exponential blending. Curve energy norms ε_1 and ε_2 are shown in Fig. 3b, c. Since ε_1 is linked to stretching and curve length, it can only be decreased slightly. However, considerable improvements are shown in ε_2 related to curve bending. Again, linear blending exhibits the lowest rate of energy decrease, while exponentially decaying blending is able to decrease the energy faster and to lower absolute values. From these results we conclude that using the exponentially

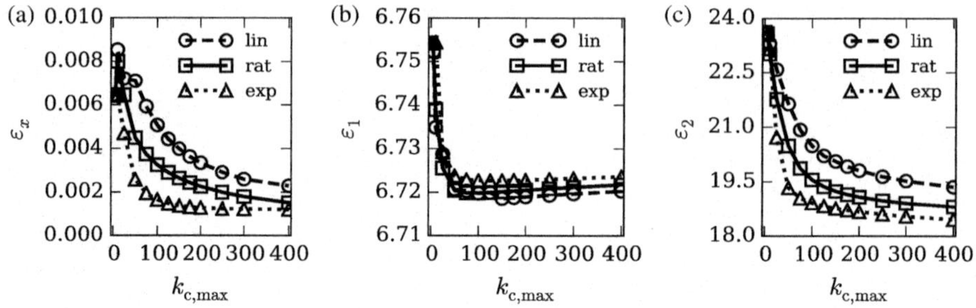

Fig. 3 Error and energy norm plots for different blendings in incremental curve optimization for to the screw example ($n = 12$, $\alpha_c = 0.5$). (**a**) Edge L^2 error. (**b**) Energy norm related to stretching. (**c**) Energy norm related to bending

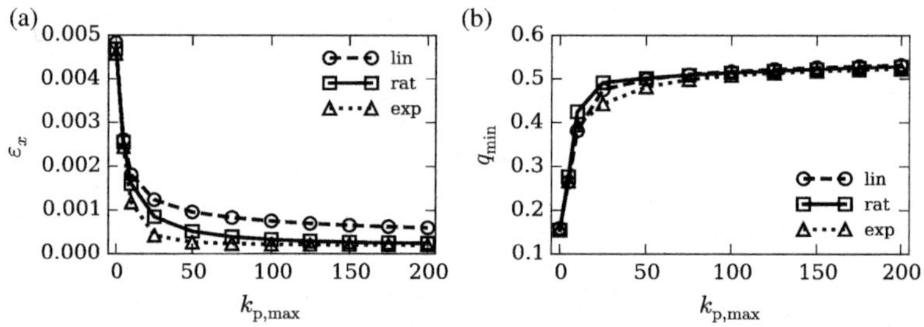

Fig. 4 Error and quality plots for different blendings in incremental patch optimization applied to the screw surface example ($n = 12$, $k_{c,max} = 150$, $\alpha_c = \alpha_p = 0.5$). (**a**) Triangle L^2 error. (**b**) Minimum element quality

decaying blending achieves the desired results, curves of high accuracy with low energies, with less steps, i.e. computational work, in comparison to utilizing the rational or linear blending. Further tests not shown here confirmed these trendings using different polynomial degrees and mesh spacings.

Figure 4 shows the results for the curved triangles computed with the three blending functions when allowing for increasing number $k_{p,max}$ of incremental steps during patch construction. The number of steps employed during the preceding curve construction was kept at $k_{c,max} = 150$ for this test. The L^2 error plotted in Fig. 4a decreases around an order of magnitude with as little as 25–50 patch construction steps, which is reached slightly faster with the exponentially decaying than with the rational blending. Linear blending is not as effective as the other two. Minimum patch qualities q_{min} exhibit lesser dependency to the choice of blending, as displayed in Fig. 4b. We remark, that the quality is improved considerably from $q_{min} \approx 0.15$ to 0.5 with the application of the incremental optimization process using 50 or more steps.

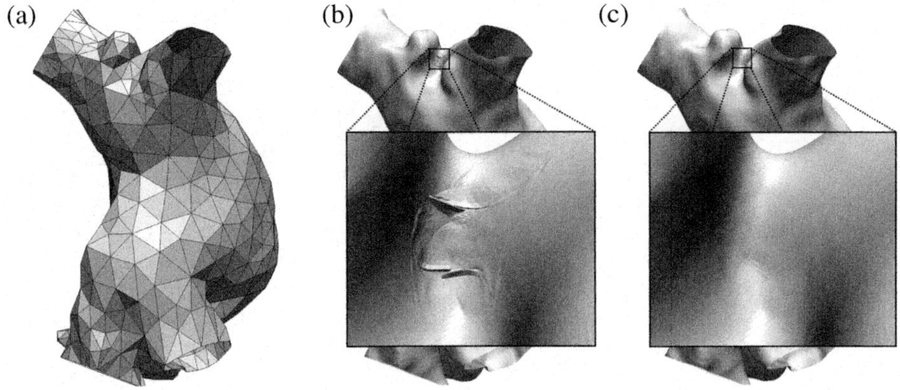

Fig. 5 Initial mesh of a human atrium with 1040 triangles and the resulting meshes of order $n = 15$ with and without application of incremental optimization ($k_{c,\max} = 150$, $k_{p,\max} = 100$, $\alpha_c = \alpha_p = 0.5$). Underlying fine mesh courtesy of Spencer Sherwin and Chris Cantwell, Imperial College London. (**a**) Initial coarse mesh. (**b**) No optimization. (**c**) Optimization applied

3.2 Scattered Surface Data

This section focuses on results obtained with scattered data surface definitions. These are frequently encountered in engineering and medicine in conjunction with scanning methods like computer tomography (CT) or magnetic resonance imaging (MRI). We use triangulated fine meshes as an exact surface representation. These were enhanced with an cubic interpolation, the PN Triangles proposed by Vlachos et al. [13], wherein we compute the required vertex normals using the method of Max [14]. The projection $\mathscr{P}$ is performed onto this interpolated surface employing Phong normals [15] based on the linear fine mesh. For details we refer to [9].

The first example is a human atrium as pictured in Fig. 5. The fine mesh is derived from CT scans and contains approximately 60,000 elements. From the fine grid we derived a coarse one consisting of 1040 elements (Fig. 5a), from which the high-order mesh $n = 15$ is build. Pure distance fitting results in the surface shown in Fig. 5b. Zooming to the detail reveals artifacts and undulations, practically rendering this mesh useless for computational purposes. After application of the incremental optimization for curves and patches ($\alpha_c = \alpha_p = 0.5$, $k_c = 100$ and $k_p = 50$ with exponentially decaying blending), the mesh presented in Fig. 5c follows. It is of visually higher quality and without artifacts. Especially for high polynomial orders and coarse meshes the optimization can push the limits of order and mesh coarseness for which suitable meshes can be obtained. Nevertheless, this comes along with an increase in computational cost, since the curve construction was performed 100 times and the patch construction 50 times in the process.

To reduce computational costs, we propose restricting the optimizations to the curves and triangles that actually need to be faired. Pursuing this idea, curves are only optimized, if their length change exceeds a certain threshold during the initial

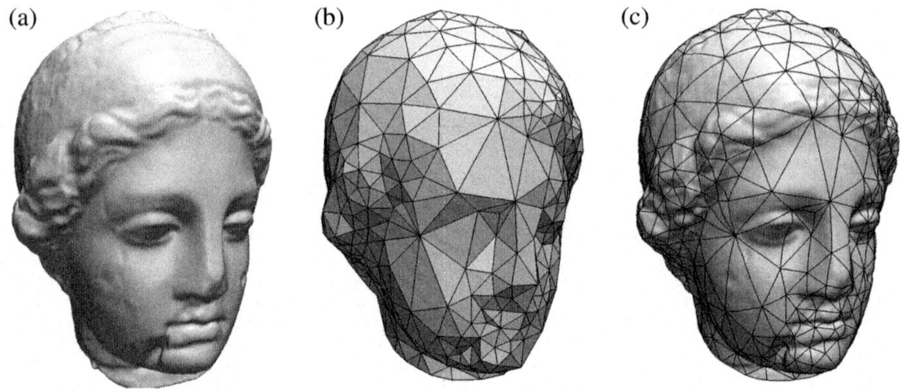

Fig. 6 High-order mesh generation for a statue head example. Fine and coarse mesh are shown as well as the resulting optimized mesh of order $n = 18$. $\alpha_c = \alpha_p = 0.5$. Underlying fine mesh courtesy of Stefan Gumhold, TU Dresden. (**a**) Fine mesh. (**b**) Coarse mesh. (**c**) High-order mesh

curving of the linear edge by distance fitting. Subsequent to the initial construction of triangular patches without optimization, the patch quality q is accessible. The number of incremental steps to perform during the construction of a patch can then be linked to its quality. With this strategy patches of high quality are not optimized, applying the more optimization steps to other patches, the poorer their quality is.

This concept was tested using the scattered data example pictured in Fig. 6, a scanned statue head. The fine mesh contains 381,236 elements, the coarse only 1043 triangles. Applying the optimization leads to an artifact-free high-order mesh of order $n = 12$, pictured in Fig. 6c. When allowing full optimization for every curve and patch, surface construction takes 3202 s on an *Intel Core i7-860* CPU. For each of the 1573 edges 150 increments were passed and equivalently 100 for each of the 1043 triangles. Employing partial optimization, 150 increments were used for 732 edges and 754 triangles were optimized using an average of 35 steps. The computational time could be reduced to 993 s.

The limitation of optimization to elements of poorer quality and highly curved edges is able to reduce computational costs. Furthermore, this yields a high-order mesh of comparable quality to using full optimization, as the histogram in Fig. 7 exhibits. Optimizing the high-order mesh in full or partially improves element qualities significantly, especially preventing any poor quality elements below $q = 0.35$.

We remark, that the strict locality of our method enables every curve and element to be processed individually in parallel. Revisiting the atrium example briefly, Fig. 8 shows the speedup and efficiency of an *OpenMP*-parallelized implementation tested on a HPC node with two *Intel Xeon X5660* CPUs. Two different examples have been tested: The 1040 triangle mesh shown in Fig. 5a and a finer one consisting of approximately 5000 elements. For the former a high-order mesh using $n = 15$ and partial optimization has been computed (labeled: 1k, partial). The latter was curved to order $n = 5$ applying the optimization for every mesh entity (label: 5k, full).

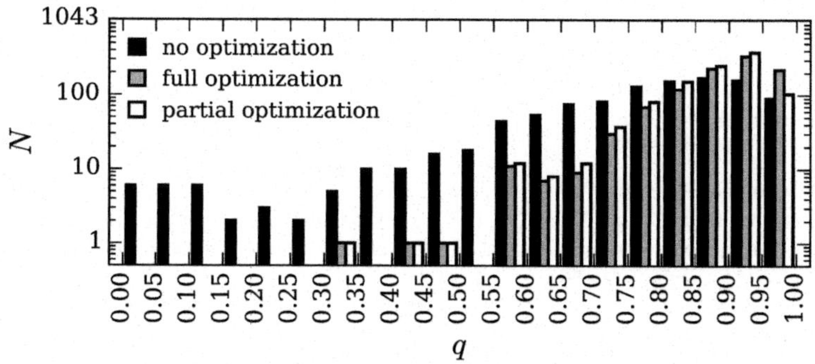

Fig. 7 Statue example quality histogram showing the number of triangles N in classes of quality q. Comparison of high-order mesh construction with no optimization performed, a full optimization and a partial optimization

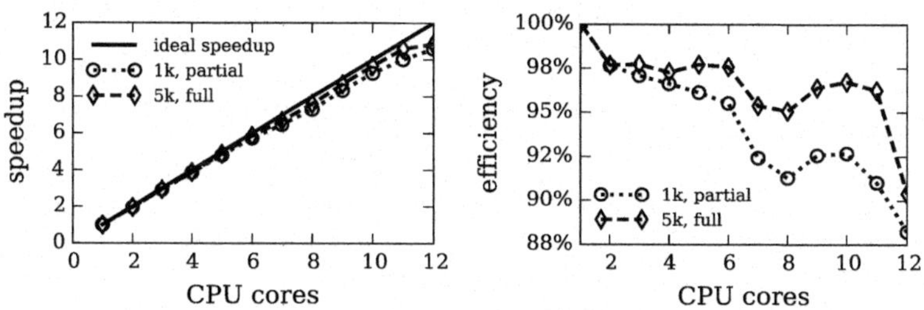

Fig. 8 Parallel speedup and efficiency of the mesh optimization method

All tests were repeated 50 times and the averaged computing times were used. Both cases show reasonable scaling and efficiency.

4 Conclusions

Following the common route in high-order mesh generation we curve initially linear starting meshes. In order to prevent artificial undulations, artifacts and to ensure mesh quality we employ an optimization strategy based on energy-minimization. The method combines least-squares fitting and energy-minimization yielding an incremental optimization method. A key ingredient of this method is the sequential reduction of the energy weighting in the mixed functional that is minimized. Different blending functions have been analysed for this purpose: a linear, a rational and an exponential one. Two different types of surface examples were addressed: an analytically defined surface and a triangulations of surface scans. The results illustrate that the optimization leads to improvement of visual appearance and mesh quality. Moreover, the optimization preserves the fitting accuracy, sometimes improving it. The mesh generation method is strictly local and requires only the projection to the target surface. This renders it particularly useful for surface definitions without available parametrizations. An important class of such definitions are triangulations obtained by scanning, as frequently encountered in bio-medical and engineering applications.

Acknowledgements The authors gratefully acknowledge the funding of this project by the German Research Foundation (DFG, STI 157/4-1). We thank the Center for Information Services and High Performance Computing (ZIH) at TU Dresden for generous allocations of computer time.

References

1. P.-O. Persson, J. Peraire, Curved mesh generation and mesh refinement using Lagrangian solid mechanics, in *Proceedings of the 47th AIAA Aerospace Sciences Meeting and Exhibit* (2009)
2. D. Moxey, D. Ekelschot, Ü. Keskin, S.J. Sherwin, J. Peiró, High-order curvilinear meshing using a thermo-elastic analogy. Comput. Aided Des. **72**, 130–139 (2015)
3. A. Gargallo-Peiró, X. Roca, J. Peraire, J. Sarrate, Optimization of a regularized distortion measure to generate curved high-order unstructured tetrahedral meshes. Int. J. Numer. Methods Eng. **103**(5), 342–363 (2015)
4. J.-F. Remacle, J. Lambrechts, C. Geuzaine, T. Toulorge, Optimizing the geometrical accuracy of 2D curvilinear meshes. Procedia Eng. **82**, 228–239 (2014). 23rd International Meshing Roundtable (IMR23)
5. E. Ruiz-Gironés, J. Sarrate, X. Roca, Generation of curved high-order meshes with optimal quality and geometric accuracy. Procedia Eng. **163**, 315–327 (2016)
6. G. Farin, *Curves and Surfaces for CAGD - A Practical Guide*, 5th edn. (Academic, New York, 2002)
7. J. Hoschek, D. Lasser, *Fundamentals of Computer Aided Geometric Design* (A.K. Peters, Wellesley, 1996)
8. K. Bock, J. Stiller, Energy-minimizing curve fitting for high-order surface mesh generation. Appl. Math. **5**, 3318–3327 (2014)
9. K. Bock, J. Stiller, Generation of high-order polynomial patches from scattered data, in *Spectral and High Order Methods for Partial Differential Equations - ICOSAHOM 2012*. Lecture Notes in Computational Science and Engineering, vol. 95 (Springer International Publishing, Berlin, 2014)
10. G. Celniker, D. Gossard, Deformable curve and surface finite-elements for free-form shape design. SIGGRAPH Comput. Graph. **25**(4), 257–266 (1991)
11. G. Greiner, Surface construction based on variational principles, in *Wavelets, Images, and Surface Fitting* (CRC Press, Boca Raton, 1994), pp. 277–286
12. S. Dey, R.M. O'Bara, M.S. Shephard, Curvilinear mesh generation in 3D, in *Proceedings of the Eighth International Meshing Roundtable* (Wiley, New York, 1999), pp. 407–417
13. A. Vlachos, J. Peters, C. Boyd, J.L. Mitchell, Curved PN triangles, in *Proceedings of the 2001 Symposium on Interactive 3D Graphics*, I3D '01 (ACM, New York, NY, 2001), pp. 159–166
14. N. Max, Weights for computing vertex normals from facet normals. J. Graph. GPU Game Tools **4**(2), 1–6 (1999)
15. B.T. Phong, Illumination for computer generated pictures. Commun. ACM **18**(6), 311–317 (1975)

Stabilization of (G)EIM in Presence of Measurement Noise: Application to Nuclear Reactor Physics

J.P. Argaud, B. Bouriquet, H. Gong, Y. Maday, and O. Mula

Abstract The Empirical Interpolation Method (EIM) and its generalized version (GEIM) can be used to approximate a physical system by combining data measured from the system itself and a reduced model representing the underlying physics. In presence of noise, the good properties of the approach are blurred in the sense that the approximation error no longer converges but even diverges. We propose to address this issue by a least-squares projection with constrains involving some a priori knowledge of the geometry of the manifold formed by all the possible physical states of the system. The efficiency of the approach, which we will call Constrained Stabilized GEIM (CS-GEIM), is illustrated by numerical experiments dealing with the reconstruction of the neutron flux in nuclear reactors. A theoretical justification of the procedure will be presented in future works.

1 General Overview and Motivation of the Paper

For the sake of clarity, we shall start by formulating the goal of the paper in general terms containing statements that will be clarified in the forthcoming sections.

Let $\mathcal{X}$ be a Banach space over a domain $\Omega \subset \mathbb{R}^d$ ($d \geq 1$) being equipped with the norm $\|.\|_{\mathcal{X}}$. Our goal is to approximate functions f from a given compact set $\mathcal{S} \subset \mathcal{X}$ which represent the states of a physical system taking place in Ω. For this,

J.P. Argaud (✉) • B. Bouriquet • H. Gong
R&D, Électricité de France, 7 boulevard Gaspard Monge, 91120 Palaiseau, France
e-mail: jean-philippe.argaud@edf.fr; bertrand.bouriquet@edf.fr; helin.gong@edf.fr

Y. Maday
Labo. J.-L. Lions, Sorbonne Université, UPMC Univ Paris 06, UMR 7598, 75005 Paris, France
Institut Universitaire de France, Paris, France

Division of Applied Mathematics, Brown University, Providence, RI, USA
e-mail: maday@ann.jussieu.fr

O. Mula
CEREMADE, PSL Research University, CNRS, UMR 7534, Université Paris-Dauphine, 75016 Paris, France
e-mail: mula@ceremade.dauphine.fr

M.L. Bittencourt et al. (eds.), *Spectral and High Order Methods for Partial Differential Equations ICOSAHOM 2016*, Lecture Notes in Computational Science and Engineering 119, DOI 10.1007/978-3-319-65870-4_8

any $f \in \mathcal{S}$ will be approximated by combining two ingredients. The first is the use of a certain amount m of measurements of f collected directly from the system itself. We represent them as linear functionals of $\mathcal{X}'$ (the dual of $\mathcal{X}$) evaluated on f. The second ingredient is the use of a (family of) subspace(s) V_n of finite dimension n which is assumed to approximate well the set $\mathcal{S}$. To limit the complexity in the approximation and also economize in the amount of sensors to place in the system, a desirable feature is to find appropriate sensors and the appropriate spaces V_n for which m and n are moderate. A necessary hypothesis to allow this is to assume some properties on the geometry of $\mathcal{S}$ expressed in terms of a rapid decay of the Kolmogorov n-width of $\mathcal{S}$ in $\mathcal{X}$,

$$d_n(\mathcal{S}, \mathcal{X}) := \inf_{\substack{X \subset \mathcal{X} \\ \dim(X) \leq n}} \max_{u \in \mathcal{S}} \min_{v \in X} \|u - v\|_{\mathcal{X}}.$$

Under this hypothesis on the decay of $d_n(\mathcal{S}, \mathcal{X})$, one can in principle build a sequence $\{X_n\}_n$ s.t. $\mathrm{dist}(\mathcal{S}, X_n) := \max_{u \in \mathcal{S}} \min_{v \in X} \|u - v\|_{\mathcal{X}} \leq \varepsilon$, where $\dim(X_n) = n \equiv n(\varepsilon)$ is moderate.

Algorithms to build $\{X_n\}_n$ (or at least the first spaces in the sequence allowing to approximate beyond a given accuracy) and find appropriate linear functionals have been proposed in the community of reduced modeling (see [1–3]). Note that, even if this is not required in the previous statements, the construction of the spaces X_n is then recursive, i.e. $X_{n-1} \subset X_n$. There, the approximation of $f \in \mathcal{S}$ is done by interpolation or related approximations. The methods (in practice mainly based on a greedy procedure) are however not robust with respect to noise in the measurements and this paper introduces a constrained least squares approximation for which numerical experiments indicate its potential to address this obstruction.

2 Mathematical Setting

Let us assume that $\mathcal{M} = \overline{\mathrm{span}(\mathcal{S})}$ (where the $\overline{\mathcal{B}}$ denotes the closure in $\mathcal{X}$ of the set $\mathcal{B}$) admits a Schauder basis $\{q_i\}_i$, i.e., for every $f \in \mathcal{X}$ there exists a unique sequence $\{c_i(f)\}$ of scalars such that $\lim_{n \to \infty} \|f - \sum_{i=1}^{n} c_i(f) q_i\|_{\mathcal{X}} = 0$. For every $n \geq 1$, we define the n-dimensional subspace $X_n := \mathrm{span}\{q_1, \ldots, q_n\}$. Let us formulate in a different manner the hypothesis made involving the Kolmogorov n-width of $\mathcal{S}$ in $\mathcal{X}$: let us assume that the error in approximating the functions of $\mathcal{S}$ in X_n is $\max_{f \in \mathcal{S}} \mathrm{dist}(f, X_n) \leq \varepsilon_n$, where the sequence $(\varepsilon_n)_n$ decays at a nice rate with n.

Let now $\{\lambda_i\}$ be the set of linear functionals of $\mathcal{X}'$ (of unity norm in $\mathcal{X}'$) such that for every $n \geq 1$, $\{\lambda_1, \ldots, \lambda_n\}$ and $\{q_1, \ldots, q_n\}$ are such that, for every $n \geq 1$ and every $1 \leq j \leq n$, $\forall i,\ 1 \leq i \leq j \quad \lambda_i(q_j) = \delta_{i,j}$. For any $n \geq 1$, we can now define a (generalized) interpolation operator $\mathcal{J}_n : \mathcal{X} \to X_n$ such that for all $f \in \mathcal{X}$ $\lambda_i(f) = \lambda_i\left(\mathcal{J}_n(f)\right), \quad i \in \{1, \ldots, n\}$. By construction, for any $n \geq 1$ and any $f \in \mathcal{X}$, $\mathcal{J}_n(f) = \mathcal{J}_{n-1}(f) + c_n(f) q_n$. where $c_n(f) = \lambda_n\left(f - \mathcal{J}_{n-1}(f)\right)$ and, for notational coherence, we set $\mathcal{J}_0 = 0$.

Using $\mathcal{J}_n$ to approximate the functions of $\mathcal{S}$ yields the error bound

$$\max_{f\in\mathcal{S}} \|f - \mathcal{J}_n(f)\|_{\mathcal{X}} \leq (1 + \Lambda_n)\,\varepsilon_n, \tag{1}$$

where $\Lambda_n := \sup_{f\in\mathcal{X}} \|\mathcal{J}_n(f)\|_{\mathcal{X}}/\|f\|_{\mathcal{X}}$ is the Lebesgue constant. The value of Λ_n diverges at a certain rate so the behavior of $\max_{f\in\mathcal{S}} \|f - \mathcal{J}_n[f]\|$ with the dimension is dictated by the trade-off between the rate of divergence of (Λ_n) (that is generally slow) and the convergence of (ε_n) (that is generally very fast). Also, for any $f \in \mathcal{S}$

$$|c_n(f)| \leq (1 + \Lambda_{n-1})\varepsilon_{n-1}, \quad n \geq 1 \tag{2}$$

where $\Lambda_0 = 0$ and $\varepsilon_0 = \max_{f\in\mathcal{S}} \|f\|$.

For any $\alpha > 0$, let us define the cone $\mathcal{K}_n(\alpha) := \{v \in V_n \;:\; v = \sum_{i=1}^n c_i q_i \;|c_i| \leq \alpha(1+\Lambda_{i-1})\varepsilon_{i-1}\}$. We have for any $n \geq 1$ and any $f \in \mathcal{S}$ $\mathcal{J}_n(f) \in \mathcal{K}_n(1)$. In presence of noise in the measurements, we assume that we receive values $\eta_1(f), \ldots, \eta_n(f)$ such that $\eta_i(f) \sim \lambda_i(f) + \mathcal{N}(0, \sigma^2)$ for $i \in \{1, \ldots, n\}$. Interpolating from these values yields an element in X_n denoted as $\mathcal{J}_n(f; \mathcal{N})$ that satisfies blurred error bound with respect to (2) that, depending on the precise definition of σ and the norm of $\mathcal{X}$ can be

$$\mathbb{E}\left(\max_{f\in\mathcal{S}} \|f - \mathcal{J}_n(f;\mathcal{N})\|\right) \leq (1 + \Lambda_n)\,\varepsilon_n + (1 + \Lambda_n)\,\sqrt{n}\sigma. \tag{3}$$

The second term of the bound diverges as n increases and shows that the method is not asymptotically robust in presence of noise. An illustration of this can be found in the numerical results below. This motivates the search for other methods which would ideally yield a bound of the form $(1 + \Lambda_n)\,\varepsilon_n + \sigma$ and for which the error is asymptotically at the level of the noise σ.

We propose to correct the interpolation operator by using more the structure of the manifold $\mathcal{S}$ that, at the discrete level, is expressed in the fact that the approximation should belong to $\mathcal{K}_n$. Indeed, the belonging of $\mathcal{J}_n(f;\mathcal{N})$ to $\mathcal{K}_n$ is not satisfied any more except if there exists $\tilde{f}$ in $\mathcal{S}$ such that $\lambda_i(\tilde{f}) = \eta_i(f)$ for any i, $1 \leq i \leq n$ (which is rarely the case). In addition, in order to minimize the effect of the noise, we can increase the number of measurements and use m larger than n linear functional evaluations at a given dimension n. This leads to propose a least-squares projection on $\mathcal{K}_n$. For a given n, we now collect the values $\eta_{n,1}(f), \ldots, \eta_{n,m(n)}(f)$ with $m(n) \geq n$ and such that $\eta_{n,i}(f) \sim \lambda_i(f) + \mathcal{N}(0, \sigma^2)$, $\quad 1 \leq i \leq m(n)$. Any $f \in \mathcal{S}$ is now approximated by

$$A_n(f) = \arg\min_{v\in\mathcal{K}_n(\alpha)} \sum_{i=1}^{m(n)} \left(\lambda_{n,i}(v) - \eta_{n,i}(f)\right)^2 \tag{4}$$

where $\alpha > 1$ is suitably chosen.

In this paper, we are running the greedy algorithm of the so-called Generalized Empirical Interpolation Method (GEIM, [1]), to generate a basis $\{q_i\}_i$ and the linear functionals $\{\lambda_i\}_i$. This method is reported to have a nice behavior for the Lebesgue constant, at least in case it is trained on a set $\mathcal{S}$ with small Kolmogorov dimension (see [3]). This approach allows an empirical optimal selections of the positions of the sensors that provide (in case where no noise pollutes the measures) a stable representation of the physical system. The precise algorithm is documented elsewhere (see [1] and [4]). Then, we approximate any $f \in \mathcal{S}$ with the function $A_n(f)$ defined in (4) with $\alpha = 2$. We call this scheme Constrained Stabilized GEIM (CS-GEIM).

Note that the above approach could also be used with a POD approach to provide the imbedded spaces $\{X_n\}_n$ (that are more expensive to provide than the greedy GEIM approach but are more accurate) and well chosen linear functionals $\lambda_{n,i}$ the choice of which infer on the behavior of the Lebesgue constant Λ_n.

3 Numerical Results

3.1 Modelling the Physical Problem

For the physical problem that we consider in this paper, the model is the two group neutron diffusion equation: the flux ϕ has two energy groups $\phi = (\phi_1, \phi_2)$. Index 1 denotes the high energy group and 2 the thermal energy 1. These are modeled by the following parameter dependent PDE model:

$$\begin{cases} -\nabla\left(D_1\nabla\phi_1\right) + (\Sigma_{a,1} + \Sigma_{s,1\to2})\phi_1 = \frac{1}{k_{\text{eff}}}\left(\chi_1\nu\Sigma_{f,1}\phi_1 + \chi_1\nu\Sigma_{f,2}\phi_2\right) \\ -\nabla\left(D_2\nabla\phi_2\right) + \Sigma_{a,2}\phi_2 - \Sigma_{s,1\to2}\phi_1 = \frac{1}{k_{\text{eff}}}\left(\chi_2\nu\Sigma_{f,1}\phi_1 + \chi_2\nu\Sigma_{f,2}\phi_2\right), \end{cases} \tag{5}$$

here k_{eff} is the so-called multiplication factor and is not a data but an unknown of the problem,[1] and the given parameters are

- D_i is the diffusion coefficient of group i with $i \in \{1, 2\}$.
- $\Sigma_{a,i}$ is the macroscopic absorption cross section of group i.
- $\Sigma_{s,1\to2}$ is the macroscopic scattering cross section from group 1 to 2.
- $\Sigma_{f,i}$ is the macroscopic fission cross section of group i.
- ν is the average number of neutrons emitted per fission.
- χ_i is the fission spectrum of group i.

they are condensed in $\mu = \{D_1, D_2, \Sigma_{a,1}, \Sigma_{a,2}, \Sigma_{s,1\to2}, \nu\Sigma_{f,1}, \nu\Sigma_{f,2}, \chi_1, \chi_2\}$.

[1]We omit here the technical details on the meaning of k_{eff} and refer to general references like [5].

We assume that the parameters of our diffusion model range in, say,

$$D_1 \in [D_{1,\min}, D_{1,\max}], \; D_2 \in [D_{2,\min}, D_{2,\max}], \ldots, \chi_2 \in [\chi_{2,\min}, \chi_{2,\max}],$$

then $\mathcal{D} := [D_{1,\min}, D_{1,\max}] \times \cdots \times [\chi_{2,\min}, \chi_{2,\max}]$ is the set of all parameters and the set of all possible states of the flux is given by

$$\mathcal{S} := \{(\phi_1, \phi_2, P)(\mu) \; : \; \mu \in \mathcal{D}\}, \tag{6}$$

where the power $P(\mu)$ is defined from $(\phi_1, \phi_2)(\mu)$ as $P(\mu)(x) := \nu\Sigma_{f,1}\phi_1(\mu)(x) + \nu\Sigma_{f,2}\phi_2(\mu)(x), \quad \forall x \in \Omega$ We assume (see [6] for elements sustaining this hypothesis) that the Kolmogorov-width decays rapidly, hence, it is possible to approximate all the states of the flux (given by $\mathcal{S}$) with an accuracy ε in well-chosen subspaces $X_n \subset \mathcal{X}$ of relatively small dimension $n(\varepsilon)$.

To ensure enough stability in the reconstruction and minimize the approximation error, it is necessary to find the optimal placement of the sensors in the core. The selection is done with GEIM. If we denote $\sigma(\phi_i, x), \quad i \in \{1, 2\}$, the measurement of ϕ_i at a position $x \in \Omega$ by a certain sensor, this measurement can be modeled by a local average over ϕ_i centered at $x \in \Omega$. Another possibility is to directly assume that the value $\phi_i(x)$ at point x is $\sigma(\phi_i, x)$.[2] Note that, in principle, the measurement could depend on other parameters apart from the position. We could imagine for instance that we have sensors with different types of accuracy or different physical properties. This flexibility is not included in the current notation but the reader will be able to extrapolate from the current explanations.

A specificity of the approach here is that $\mathcal{S}$ is composed of vectorial functions $(\phi_1, \phi_2, P)(\mu)$. We deliberately choose to take measurements only on one of the components (say ϕ_2) and thus reconstruct the whole field ϕ_1, ϕ_2 and P with the only knowledge of thermal flux measurements.

Another specificity of our approach is on the spatial location of the measurements. We consider two cases:

- Case I: the sensors can be placed at any point in the domain of definition of ϕ_2.
- Case II: the admissible sensor locations are restricted to be deployed in a restricted part of that domain;

We have already reported in [7] that these two specificities are well supported by the (G)EIM approach, as long as the greedy method is taught to achieve the goal of reconstructing the whole field $(\phi_1, \phi_2, P)(\mu)$.

Our aim here is to show that the noise can be controlled through our Constrained Stabilized (G)EIM approach.

[2]In the following part of this work, we directly assume that the value $\phi_i(x)$ at point x is $\sigma(\phi_i, x)$ as measurement.

3.2 Description of the 2D IAEA Benchmark

We consider the classical 2D IAEA Benchmark Problem [8], the core geometry which can be seen in Fig. 1. The problem conditions and the requested results are stated in page 437 of reference [8]. It is identified with the code 11-A2, and its descriptive title is Two-dimensional LWR Problem, also known as 2D IAEA Benchmark Problem. This problem represents the mid-plane $z = 190$ cm of the 3D IAEA Benchmark Problem, that is used by references [9] and show in application within [10].

The reactor domain is $\Omega = \text{region}(1, 2, 3, 4)$. The core and the reflector are $\Omega_{\text{core}} = \text{region}(1, 2, 3)$ and $\Omega_{\text{refl}} = \text{region}(4)$ respectively. We consider only the value of $D_1|_{\Omega_{\text{refl}}}$ in the reflector Ω_{refl} as a parameter (so $p = 1$ and $\mu = D_1|_{\Omega_{\text{refl}}}$). We assume that $D_1|_{\Omega_{\text{refl}}} \in [1.0, 3.0]$. The rest of the coefficients of the diffusion model (5) (including $D_1|_{\Omega_{\text{core}}}$) are fixed to the values indicated in Table 1. In principle, one could also consider these coefficients as parameters but we have decided to focus only on $D_1|_{\Omega_{\text{refl}}}$ because of its crucial role in the physical estate of the core: its variation can be understood as a change in the boundary conditions in Ω_{core} which, up to a certain extent, allows to compensate the bias of the diffusion

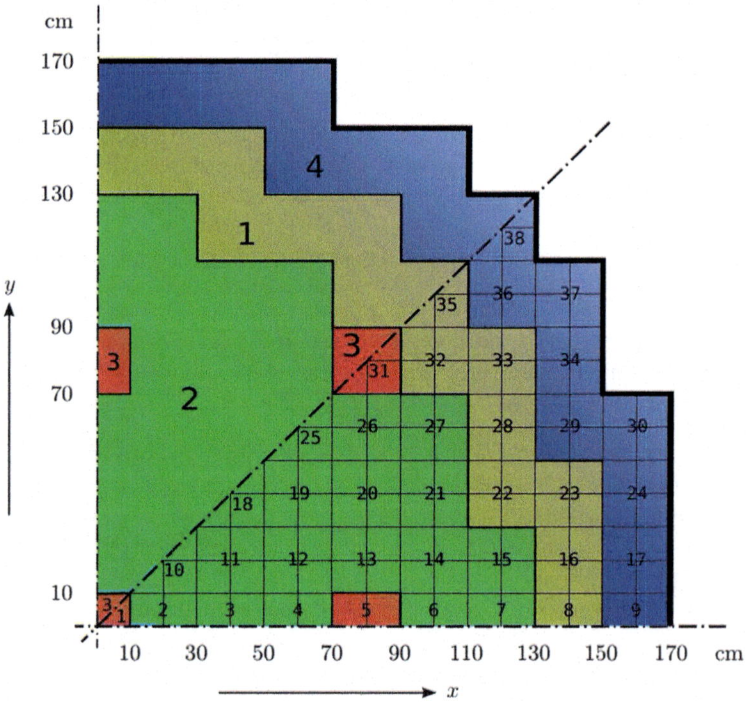

Fig. 1 Geometry of 2D IAEA benchmark, upper octant: region assignments, lower octant: fuel assembly identification (from reference [10])

Table 1 Coefficient values: diffusion coefficients D_i (in cm) and macroscopic cross sections (in cm^{-1})

Region	D_1	D_2	$\Sigma_{1\to 2}$	Σ_{a1}	Σ_{a2}	$\nu\Sigma_{f2}$	Material[a]
1	1.5	0.4	0.02	0.01	0.080	0.135	Fuel 1
2	1.5	0.4	0.02	0.01	0.085	0.135	Fuel 2
3	1.5	0.4	0.02	0.01	0.130	0.135	Fuel 2 + rod
4	[1.0, 3.0] or 2.0[b]	0.3	0.04	0	0.010	0	Reflector

[a] Axial buckling $B_{zg}^2 = 0.8 \cdot 10^{-4}$ for all regions and energy groups
[b] Here 2.0 is the exact value from reference [10]

model with respect to reality. We shall report in a future paper more extended variations of the parameters and more involved problems, like the one addressed in [11, 12] where a reduced basis is built to approximate the flux distribution when the position of the control rods varies. In comparison to their approach, the current methodology brings the additional ingredient of incorporating measurement information to the reconstruction. Note that these papers support the idea that the solution manifold in this frame of physics has a small Kolmogorov n-width.

3.3 Hypothesis of (CS-)GEIM Application

We propose to reconstruct (ϕ_1, ϕ_2, P) as explained above (see [7]) i.e., ϕ_2 will be approximated with its direct interpolant $\mathcal{J}_n[\phi_2]$ while ϕ_1 and P will be reconstructed from the measurements of ϕ_2, using the same coefficients in a coherent basis set. These are denoted as $\widetilde{\mathcal{J}_n}[\phi_1]$ and $\widetilde{\mathcal{J}_n}[P]$. Fig. 2 shows the sensor locations given by the GEIM greedy algorithm in cases I and II.

3.4 Numerical Results

Let us now turn to the analysis of the results. We study the performance of the reconstruction strategy by considering first of all the decay of the errors

$$e_n^{(\text{training})}(\phi_2) := \max_{\mu \in \mathcal{D}^{(\text{training})}} \|\phi_2(\mu) - \mathcal{J}_n[\phi_2](\mu)\|_{L^2(\Omega)} \tag{7}$$

in the greedy algorithm. Since both Case I and Case II yield very similar results, we only present plots of Case II for the sake of concision. In Fig. 3a, the decay is compared to an indicator of the optimal performance in $L^2(\Omega)$ which is obtained by a singular value decomposition of the snapshots $\phi_2(\mu)$, $\forall \mu \in \mathcal{D}^{(\text{training})}$. Note that $e_n^{(\text{training})}(\phi_2)$ decays at a similar rate as the SVD which suggests that GEIM behaves in a quasi-optimal way (see [3]). We now estimate the accuracy to reconstruct $(\phi_1, \phi_2, P)(D_1|_{\Omega_{\text{refl}}})$ for any $D_1|_{\Omega_{\text{refl}}} \in [0.5, 2.0]$ which does not necessary belong

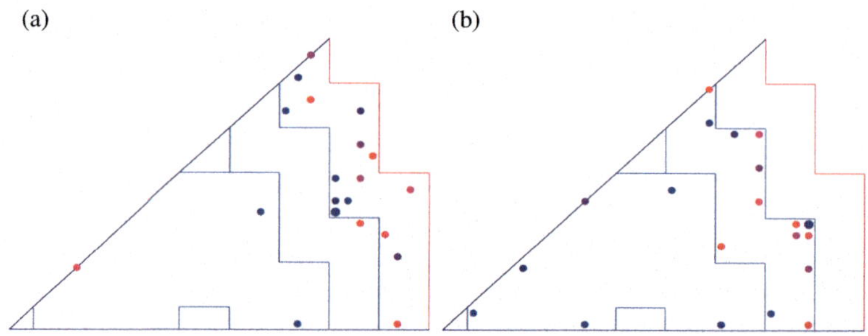

Fig. 2 Locations of the sensors chosen by the greedy EIM algorithm. (**a**) Case I (selection in Ω). (**b**) Case II (selection in Ω_{core})

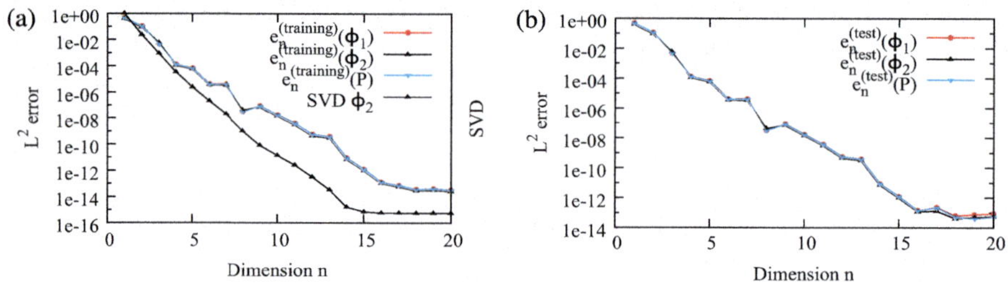

Fig. 3 Case II, $L^2(\Omega)$ norm: Reconstruction of $(\phi_1, \phi_2, P)(\mu)$ with $\left(\widetilde{\mathcal{J}}_n[\phi_1], \mathcal{J}_n[\phi_2], \widetilde{\mathcal{J}}_n[P]\right)(\mu)$. (**a**) Decay of SVD modes and of $e_n^{(\text{training})}(\phi_2)$. (**b**) Decay of $e_n^{(\text{test})}(\phi_1)$, $e_n^{(\text{test})}(\phi_2)$ and $e_n^{(\text{test})}(P)$

to the training set of snapshots. For this, we consider a test set of 300 parameters $\mathcal{D}^{(\text{test})}$ different from $\mathcal{D}^{(\text{training})}$ and compute the errors

$$\begin{cases} e_n^{(\text{test})}(\phi_1) & := \max_{\mu \in \mathcal{D}^{(\text{test})}} \|\phi_1(\mu) - \widetilde{\mathcal{J}}_n[\phi_1](\mu)\|_{L^2(\Omega)} \\ e_n^{(\text{test})}(\phi_2) & := \max_{\mu \in \mathcal{D}^{(\text{test})}} \|\phi_1(\mu) - \mathcal{J}_n[\phi_2](\mu)\|_{L^2(\Omega)} \\ e_n^{(\text{test})}(P) & := \max_{\mu \in \mathcal{D}^{(\text{test})}} \|P(\mu) - \widetilde{\mathcal{J}}_n[P](\mu)\|_{L^2(\Omega)} \end{cases} \tag{8}$$

The decay of the errors (8) is plotted in Fig. 3b. The fact that $e_n^{(\text{test})}(\phi_2)$ decays very similarly to $e_n^{(\text{training})}(\phi_2)$ confirms that the set of 300 training snapshots was representative enough of the whole manifold $\mathcal{S}$. Also, the fast decay of $e_n^{(\text{test})}(\phi_1)$ and $e_n^{(\text{test})}(P)$ shows that the use of the operator $\widetilde{\mathcal{J}}_n$ to approximate ϕ_1 and P is accurate enough.

Instead of working with $L^2(\Omega)$, it is also possible to work with other norms (provided some spacial regularity in the manifold). A particularly relevant case in neutronics is $L^\infty(\Omega)$. Figure 4 shows the results of the reconstruction procedure when working in this norm and Fig. 5 shows the behavior when considering the $H^1(\Omega)$ and working with its classical semi-norm $|u|_{H^1(\Omega)} = \int_\Omega |\nabla u|^2$.

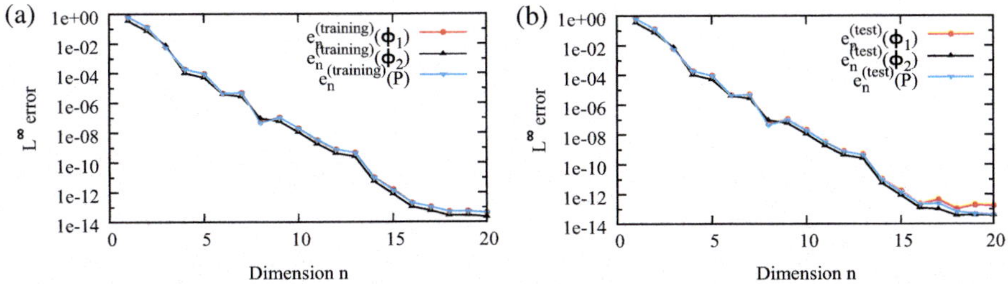

Fig. 4 Case II, $L^\infty(\Omega)$ norm: Reconstruction of $(\phi_1, \phi_2, P)(\mu)$ with $\left(\widetilde{\mathcal{J}}_n[\phi_1], \mathcal{J}_n[\phi_2], \widetilde{\mathcal{J}}_n[P]\right)(\mu)$. (**a**) Decay of the greedy errors $e_n^{\text{(training)}}(\phi_2)$. (**b**) Decay of $e_n^{\text{(test)}}(\phi_1)$, $e_n^{\text{(test)}}(\phi_2)$ and $e_n^{\text{(test)}}(P)$

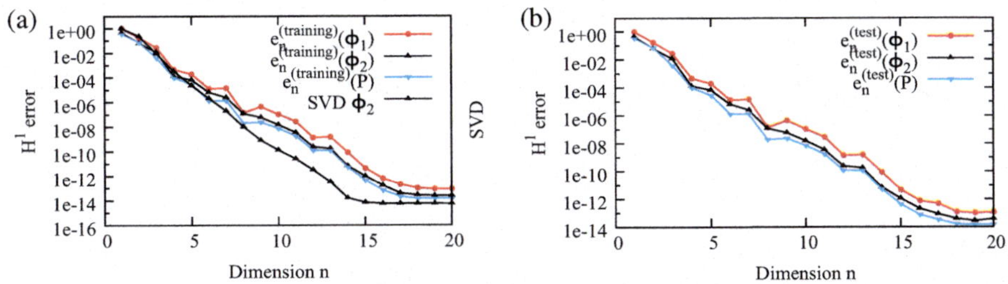

Fig. 5 Case II, $H^1(\Omega)$ norm: reconstruction of $(\phi_1, \phi_2, P)(\mu)$ with $\left(\widetilde{\mathcal{J}}_n[\phi_1], \mathcal{J}_n[\phi_2], \widetilde{\mathcal{J}}_n[P]\right)(\mu)$. (**a**) Decay of SVD modes and of $e_n^{\text{(training)}}(\phi_2)$. (**b**) Decay of $e_n^{\text{(test)}}(\phi_1)$, $e_n^{\text{(test)}}(\phi_2)$ and $e_n^{\text{(test)}}(P)$

Figure 6a shows the behavior of the Lebesgue constants in both cases. We can find that (1) the Lebesgue constant increases with GEIM interpolation function dimension, (2) if detectors are limited in a domain part (Case II), the Lebesgue constant gets worse, as an effect of the extrapolation that is required here, nevertheless the increase is still moderate.

Figure 6b shows the coefficients upper limits described as $r_n(x_n, \mu_n) = (1 + \Lambda_n)\varepsilon_n$ (see (2)) (so $|c_n| \leq r_n(x_n, \mu_n)$) for Case I and Case II, which decreases quickly with n.

We still focus on the 300 parameters $\mathcal{D}^{(test)}$, and compute the errors with Eq. (8), for each test case, we perform the interpolation process with CS-GEIM a number of times. Figure 7 shows the averaged $L^2(\Omega)$ norm for the decay of $e_n^{\text{(test)}}(\phi_1)$, $e_n^{\text{(test)}}(\phi_2)$ and $e_n^{\text{(test)}}(P)$, with noise amplitude 10^{-2} for Case I and Case II. For different measurement noise amplitude, the averaged errors in $L^2(\Omega)$ norm, $L^\infty(\Omega)$ norm and $H^1(\Omega)$ norm are shown in Figs. 8, 9 and 10 respectively, for Case I and Case II. The main conclusions are: in the noisy case, (1) CS-GEIM improves the interpolation, with the error comparable to the noise input level, (2) in extrapolation case, CS-GEIM reduces the interpolation error dramatically, which extends GEIM practical use.

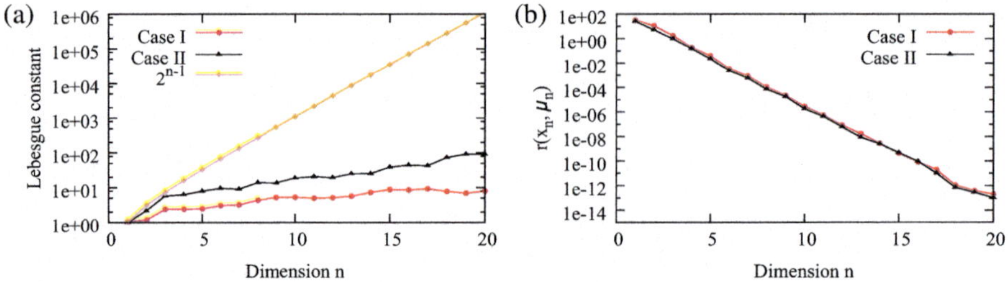

Fig. 6 The Lebesgue constant and $r_n(x_n, \mu_n)$ from GEIM. (**a**) The Lebesgue constant for Case I and Case II. (**b**) The coefficients upper limits $r_n(x_n, \mu_n)$, for Case I and Case II

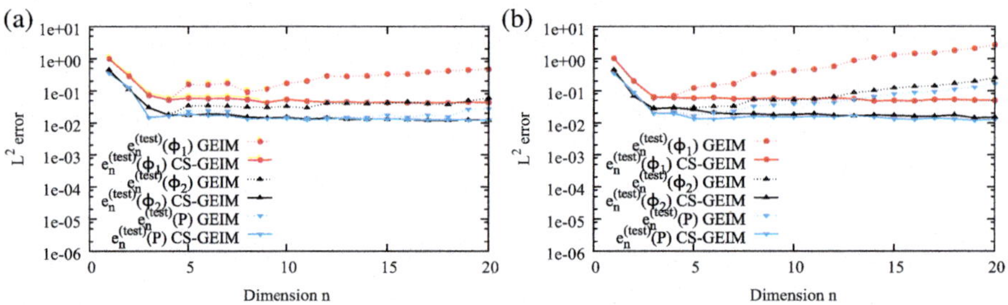

Fig. 7 $L^2(\Omega)$ norm: decay of $e_n(\phi_2)$ and $e_n^{(\mathrm{test})}(P)$. Reconstruction of $(\phi_1, \phi_2, P)(\mu)$ with $\left(\widetilde{\mathcal{J}}_n[\phi_1], \mathcal{J}_n[\phi_2], \widetilde{\mathcal{J}}_n[P]\right)(\mu)$, with noise amplitude 10^{-2}. (**a**) Case I. (**b**) Case II

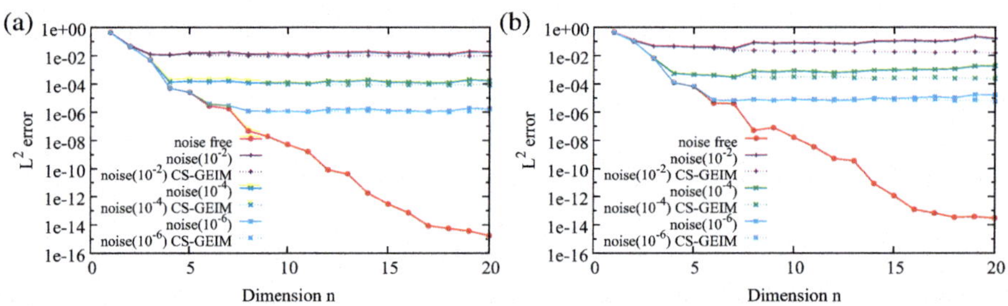

Fig. 8 $L^2(\Omega)$ norm: decay of $e_n^{(\mathrm{training})}(\phi_2)$, with noise amplitude $10^{-2}, 10^{-4}, 10^{-6}$. (**a**) Case I. (**b**) Case II

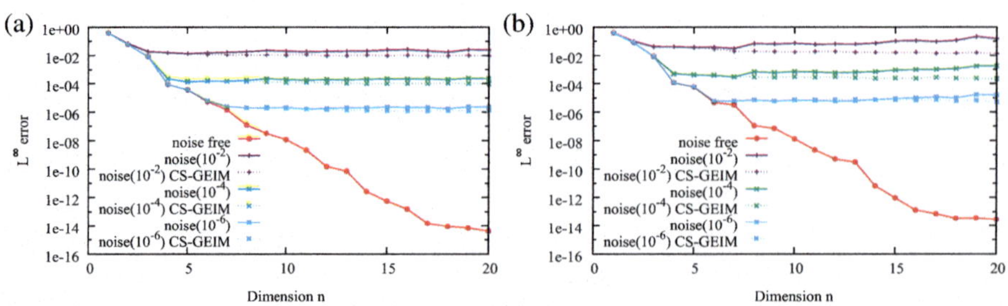

Fig. 9 $L^{\infty}(\Omega)$ norm: decay of $e_n^{(\mathrm{training})}(\phi_2)$, with noise amplitude $10^{-2}, 10^{-4}, 10^{-6}$. (**a**) Case I. (**b**) Case II

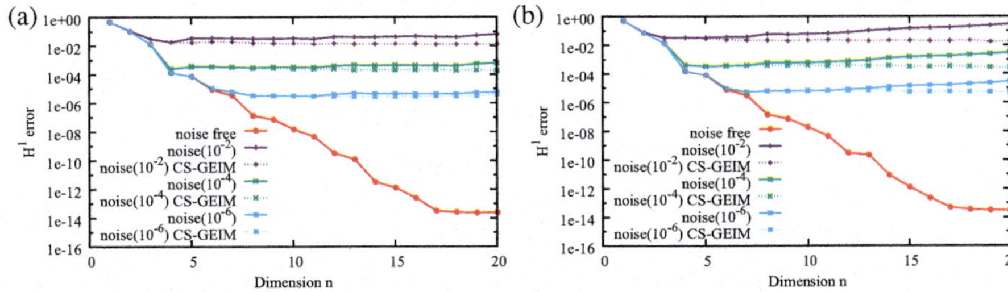

Fig. 10 $H^1(\Omega)$ norm: decay of $e_n^{(\text{training})}(\phi_2)$, with noise amplitude $10^{-2}, 10^{-4}, 10^{-6}$. **(a)** Case I. **(b)** Case II

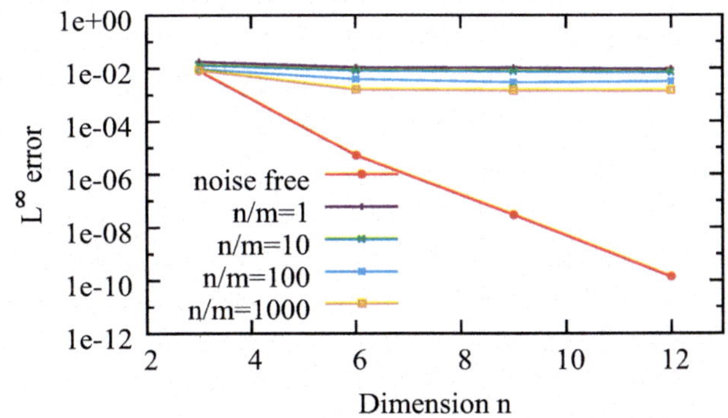

Fig. 11 CS-GEIM with different n/m ratio, the input noise level is 10^{-2}, for Case I

If we take more measurements with fixed number of interpolation functions, the ratio n/m of the number of measurements n to the number of interpolation functions m increases, so it is expected to have the same effect than to repeat independent measure at the same point in order to measure the evaluation of the measure. We consider the analytical function $g(x,\mu) \equiv \mathcal{V}((x_1,x_2);(\mu_1,\mu_2)) \equiv ((x_1-\mu_1)^2 + (x_2-\mu_2)^2)^{-1/2}$ for $x \in \Omega \equiv]0,1[^2$ and $\mu \in \mathcal{D} \equiv [-1,-0.01]^2$; we choose for $\mathcal{D}^{(training)}$ a uniform discretization sample of 400 points[3]. Then we change the ratio n/m of the number of measurements n to the number of interpolation functions m with CS-GEIM process, see Figs.11 and 12 also show the error converges with $\sim n^{-\frac{1}{2}}$.

[3]We replace the synthetic neutron problem here by the above analytical function so as to be able to have a more thorough and extensive numerical analysis

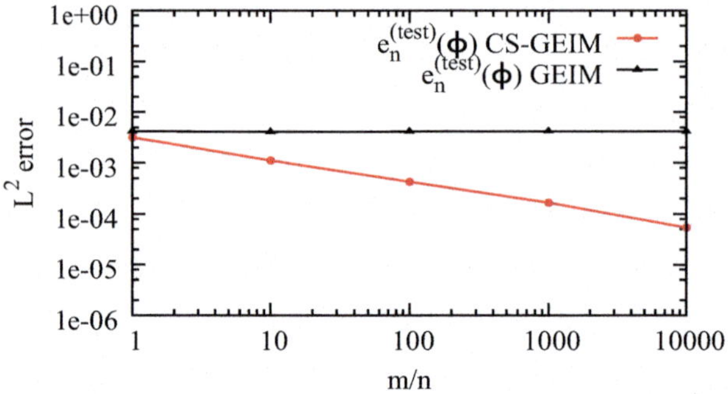

Fig. 12 CS-GEIM with different n/m ratio, the input noise level is 10^{-2}, with the function $g(x, \mu) \equiv ((x_1 - \mu_1)^2 + (x_2 - \mu_2)^2)^{-1/2}$, the error converges with $\sim n^{-\frac{1}{2}}$

4 Conclusions and Future Works

We have presented some results obtained with the Empirical Interpolation Method (EIM) for the reconstruction of the whole field, solution to a simple but representative problem in nuclear reactor physics as an example of a set of parameterized functions. With EIM, a high accuracy can be achieved in reconstructing the physical fields, and also a better sensors deployment is proposed with which most information can be extracted in a given precision even if only part of the field (either in space or in component) is included in the measurement process. Then an improved Empirical Interpolation Method (CS-(G)EIM) is proposed. With CS-(G)EIM, (1) the behavior of the interpolant is improved when measurements suffer from noise, (2) the error is dramatically improved in noisy extrapolation case, (3) it is possible to decrease the error by increasing the number of measurements.

Further works and perspective are ongoing: (1) mathematical analysis of the stable and accurate behavior of this stabilized approach, (2) in this work, our first assumption is the model is perfect (i.e. we work on in silico solutions; a broader class of methods which couple reduced models with measured data named PBDW [13] are able to correct the bias of the model and use real data.

References

1. Y. Maday, O. Mula, A generalized empirical interpolation method: application of reduced basis techniques to data assimilation, in *Analysis and Numerics of Partial Differential Equations*, ed. by F. Brezzi, P.C. Franzone, U. Gianazza, G. Gilardi. Springer INdAM Series, vol. 4 (Springer Milan, Heidelberg, 2013), pp. 221–235
2. Y. Maday, A.T. Patera, J.D. Penn, M. Yano, A parameterized-background data-weak approach to variational data assimilation: formulation, analysis, and application to acoustics. Int. J. Numer. Methods Eng. **102**(5), 933–965 (2015)

3. Y. Maday, O. Mula, G. Turinici, Convergence analysis of the generalized empirical interpolation method. SIAM J. Numer. Anal. **54**(3), 1713–1731 (2016)
4. H. Gong, J.P. Argaud, B. Bouriquet, Y. Maday, The empirical interpolation method applied to the neutron diffusion equations with parameter dependence, in *Proceedings of Physor* (2016)
5. A. Hebert, *Applied Reactor Physics* (Presses inter Polytechnique, Montreal, 2009)
6. A. Cohen, R. DeVore, Kolmogorov widths under holomorphic mappings. IMA J. Numer. Anal. **36**, 1–12 (2015). dru066
7. J.P. Argaud, B. Bouriquet, H. Gong, Y. Maday, O. Mula, Sensor placement in nuclear reactors. Submitted paper (under review)
8. Computational Benchmark Problem Comitee for the Mathematics and Computation Division of the American Nuclear Society. Argonne Code Center: Benchmark problem book (1977)
9. PARCS. IAEA 3D PWR problem. https://engineering.purdue.edu/PARCS/Code/TestSuite/CalculationMode/StandAloneMode/Eigenvalue/IAEA3DPWR
10. G. Theler, F.J. Bonetto, A. Clausse, Solution of the 2D IAEA PWR Benchmark with the neutronic code Milonga, in *Actas de la Reunión Anual de la Asociación Argentina de Tecnología Nuclear*, vol. XXXVIII (2011)
11. A. Sartori, D. Baroli, A. Cammi, D. Chiesa, L. Luzzi, R. Ponciroli, E. Previtali, M.E. Ricotti, G. Rozza, M. Sisti, Comparison of a modal method and a proper orthogonal decomposition approach for multi-group time-dependent reactor spatial kinetics. Ann. Nucl. Energy **71**, 217–229 (2014)
12. A. Sartori, D. Baroli, A. Cammi, L. Luzzi, G. Rozza, A reduced order model for multi-group time-dependent parametrized reactor spatial kinetics, in *2014 22nd International Conference on Nuclear Engineering* (American Society of Mechanical Engineers, New York, 2014), pp. V005T17A048–V005T17A048
13. Y. Maday, A.T. Patera, J.D. Penn, M. Yano, PBDW state estimation: noisy observations; configuration-adaptive background spaces, physical interpretations. ESAIM: Proc. Surv. **50**, 144–168 (2015)

Coupling DG-FEM and BEM for a Time Harmonic Eddy Current Problem

Ana Alonso Rodríguez, Salim Meddahi, and Alberto Valli

Abstract We introduce and analyze a discontinuous Galerkin FEM/BEM method for a time-harmonic eddy current problem written in terms of the magnetic field. We use standard finite elements on a partition of the conductor domain coupled with continuous boundary elements on the transmission interface. We prove quasi-optimal error estimates in the energy norm.

1 Introduction

The usual setting of an eddy current problem distinguishes between a bounded conductive region and the surrounding unbounded air region. When using the finite element method for the numerical approximation of an eddy current problem it is necessary to introduce a bounded computational domain and to approximate the decay of the solution at infinity by imposing homogeneous condition on its boundary. A more accurate strategy is to reduce the computational domain to the conductor by considering non-local boundaries conditions provided by an integral formulation of the exterior problem. The numerical approximation of this formulation couples finite elements (FEM) and boundary elements (BEM). This idea has been introduced in [4] by Bossavit and more recently in [1, 10, 11]. Our aim here is to revisit the FEM/BEM formulation given in [11] in order to provide an interior penalty discontinuous Galerkin (IPDG) approximation of the magnetic field in the conducting domain.

Discontinuous Galerkin (DG) methods can provide efficient solvers for electromagnetic problems in domains with complex geometry, see [5]. However, we only found few works applying DG methods to eddy current problems (see [14] for the

A. Alonso Rodríguez (✉) • A. Valli
Department of Mathematics, University of Trento, I-38123 Trento, Italy
e-mail: ana.alonso@unitn.it; alberto.valli@unitn.it

S. Meddahi
Departamento de Matemáticas, Facultad de Ciencias, Universidad de Oviedo, Calvo Sotelo s/n, Oviedo, Spain
e-mail: salim@uniovi.es

M.L. Bittencourt et al. (eds.), *Spectral and High Order Methods for Partial Differential Equations ICOSAHOM 2016*, Lecture Notes in Computational Science and Engineering 119, DOI 10.1007/978-3-319-65870-4_9

time-harmonic regime and [2] for a time-domain problem) and we are not aware about any DG-FEM/BEM formulation for this problem.

Due to the nonlocal character of the boundary integral operators, continuous Galerkin approximations are usually used on the boundary. As a consequence, the major difficulty that is encountered in the design of a DG-FEM/BEM method (cf. [6, 8, 9] and [16, Section 4]) is the mismatch that occurs between the interior and the boundary unknowns on the transmission interface. In our case, this difficulty appears in the transmission condition (5) where we have two variables of different nature. From one side (as the discrete variable representing ψ is $\mathrm{H}^1(\Gamma)$-conforming) we have a globally surface-divergence free function, from the other side the tangential trace of the DG approximation of the magnetic field is not $\mathrm{H}(\mathrm{div}_\Gamma)$-conforming. This impedes one to merge the two variables at the discrete level as in [11]. To address this problem, we exploit the ability of DG-methods to incorporate essential boundary conditions into the variational formulation and impose (5) weakly. As a result, in comparison with [11], we have one further independent unknown on the boundary. We show that the resulting IPDG-FEM/BEM scheme is uniformly stable with respect to the mesh parameter in an adequate DG-norm. Moreover, under suitable regularity assumptions, we provide quasi-optimal asymptotic error estimates.

We end this section with some notations that will be useful in the sequel. Given a real number $r \geq 0$ and a polyhedron $\mathcal{O} \subset \mathbb{R}^d$, $(d = 2, 3)$, we denote the norms and seminorms of the usual Sobolev space $\mathrm{H}^r(\mathcal{O})$ by $\|\cdot\|_{r,\mathcal{O}}$ and $|\cdot|_{r,\mathcal{O}}$, respectively. We use the convention $\mathrm{L}^2(\mathcal{O}) := \mathrm{H}^0(\mathcal{O})$. We recall that, for any $t \in [-1,\ 1]$, the spaces $\mathrm{H}^t(\partial\mathcal{O})$ have an intrinsic definition by localization on the Lipschitz surface $\partial\mathcal{O}$ (this is due to their invariance under Lipschitz coordinate transformations). Moreover, for all $0 < t \leq 1$, $\mathrm{H}^{-t}(\partial\mathcal{O})$ is the dual of $\mathrm{H}^t(\partial\mathcal{O})$ with respect to the pivot space $\mathrm{L}^2(\partial\mathcal{O})$. Finally we consider $\mathbf{H}(\mathbf{curl}, \mathcal{O}) := \{\mathbf{v} \in \mathrm{L}^2(\mathcal{O})^3 \ : \ \mathbf{curl}\mathbf{v} \in \mathrm{L}^2(\mathcal{O})^3\}$ and endow it with its usual Hilbertian norm $\|\mathbf{v}\|^2_{\mathbf{H}(\mathbf{curl},\mathcal{O})} := \|\mathbf{v}\|^2_{0,\mathcal{O}} + \|\mathbf{curl}\mathbf{v}\|^2_{0,\mathcal{O}}$.

2 The Model Problem

Let $\Omega \subset \mathbb{R}^3$ be a bounded polyhedral domain with a Lipschitz boundary Γ. We denote by $\mathbf{n}$ the unit normal vector on Γ that points towards $\Omega^e := \mathbb{R}^3 \setminus \overline{\Omega}$. For the sake of simplicity, we assume that Ω is simply connected and that Γ is connected. We consider the eddy current problem

$$\begin{aligned}
\imath\omega\mu\mathbf{h} + \mathbf{curl}\,\mathbf{e} &= \mathbf{0} && \text{in } \Omega\\
\mathbf{e} &= \sigma^{-1}(\mathbf{curl}\,\mathbf{h} - \mathbf{j}_e) && \text{in } \Omega\\
\mathbf{h}\times\mathbf{n} &= \nabla p \times \mathbf{n} && \text{on } \Gamma\\
\mu\mathbf{h}\cdot\mathbf{n} &= \mu_0\frac{\partial p}{\partial \mathbf{n}} && \text{on } \Gamma\\
-\Delta p &= 0 && \text{in } \Omega^e\\
p &= O\big(1/|\mathbf{x}|\big) && \text{as } |\mathbf{x}| \to \infty,
\end{aligned} \tag{1}$$

where $\omega > 0$ is the angular frequency, μ_0 is the magnetic permeability of the free space, and the conductivity σ and the magnetic permeability μ in the conductor Ω are positive and piecewise constant functions with respect to a partition of the domain Ω into Lipschitz polyhedra. Here $\mathbf{j}_e$ denotes the (complex valued) applied current density, $\mathbf{e}$ and $\mathbf{h}$ are the electric field and the magnetic field, respectively, and p is the scalar magnetic potential in the exterior region Ω_e, namely, $\mathbf{h} = \nabla p$ in Ω_e.

A finite element formulation of problem (1) requires the approximation of the decay of p at infinity by imposing a homogeneous Dirichlet boundary condition on an artificial boundary Σ located sufficiently far from the conductor Ω. A more accurate strategy for solving problem (1) consists in reducing the computational domain to the conductor Ω. This can be achieved by considering non-local boundary conditions provided by the following integral equations relating the Cauchy data $\lambda := \dfrac{\partial p}{\partial \mathbf{n}}$ and $\psi := p|_\Gamma$ on Γ (see, e.g., [15, Chap. 3]):

$$\psi = \left(\tfrac{1}{2}I + K\right)\psi - V\lambda \tag{2}$$

$$\lambda = -W\psi + \left(\tfrac{1}{2}I - K^{\mathrm{t}}\right)\lambda \tag{3}$$

where V, K, K^{t} are the boundary integral operators representing the single layer, double layer and adjoint of the double layer operators, respectively, and W is the hypersingular operator. This yields to an exact formulation of problem (1) that is adequate for a coupled FEM-BEM discretization strategy as:

$$\imath\omega\mu\mathbf{h} + \mathbf{curl}\left(\sigma^{-1}(\mathbf{curl}\,\mathbf{h} - \mathbf{j}_e)\right) = \mathbf{0} \qquad \text{in } \Omega \tag{4}$$

$$\mathbf{h}\times\mathbf{n} = \mathbf{curl}_\Gamma\psi \qquad \text{on } \Gamma \tag{5}$$

$$\frac{\mu}{\mu_0}\mathbf{h}\cdot\mathbf{n} = -W\psi + \left(\tfrac{1}{2}I - K^{\mathrm{t}}\right)\lambda \qquad \text{on } \Gamma \tag{6}$$

$$V\lambda + \left(\tfrac{1}{2}I - K\right)\psi = 0 \qquad \text{on } \Gamma, \tag{7}$$

where $\mathbf{curl}_\Gamma$ is the curl operator on the surface Γ, namely, $\mathbf{curl}_\Gamma\psi = \nabla_\Gamma\psi \times \mathbf{n}$..

In [11] it is shown that, using (5), the unknown ψ can be eliminated from (6) and (7), and that the weak formulation of the reduced problem admits a unique solution $(\mathbf{h}, \lambda) \in \mathbf{H}(\mathbf{curl}, \Omega) \times \mathrm{H}_0^{-1/2}(\Gamma)$, where $\mathrm{H}_0^{-1/2}(\Gamma) := \left\{\eta \in \mathrm{H}^{-1/2}(\Gamma);\ \langle\eta, 1\rangle_\Gamma = 0\right\}$. Here, $\langle\cdot,\cdot\rangle_\Gamma$ stands for the duality pairing between $\mathrm{H}^{-1/2}(\Gamma)$ and $\mathrm{H}^{1/2}(\Gamma)$. Then $\psi \in \mathrm{H}^{1/2}(\Gamma)$ is uniquely determined, up to an additive constant, from (5), so it is unique in $\mathrm{H}_0^{1/2}(\Gamma) := \left\{\varphi \in \mathrm{H}^{1/2}(\Gamma); \int_\Gamma \varphi = 0\right\}$.

Once the Cauchy data λ and ψ are known, the solution is computed in the exterior domain Ω^e by using the integral representation formula

$$p(\mathbf{x}) = \int_\Gamma \frac{\partial E(|\mathbf{x}-\mathbf{y}|)}{\partial \mathbf{n}_\mathbf{y}}\psi(\mathbf{y})\,ds_\mathbf{y} - \int_\Gamma E(|\mathbf{x}-\mathbf{y}|)\lambda(\mathbf{y})\,ds_\mathbf{y} \quad \text{in } \Omega^e,$$

where $E(|\mathbf{x}|) := \frac{1}{4\pi}\frac{1}{|\mathbf{x}|}$ is the fundamental solution of the Laplace operator. Let us recall some important properties of the boundary integral operators, see [15] for details. They are formally defined at almost every point $\mathbf{x} \in \Gamma$ by

$$V\xi(\mathbf{x}) := \int_\Gamma E(|\mathbf{x}-\mathbf{y}|)\xi(\mathbf{y})\,ds_{\mathbf{y}}, \qquad K\varphi(\mathbf{x}) := \int_\Gamma \frac{\partial E(|\mathbf{x}-\mathbf{y}|)}{\partial \mathbf{n}_{\mathbf{y}}}\varphi(\mathbf{y})\,ds_{\mathbf{y}},$$

$$K^{\mathrm{t}}\xi(\mathbf{x}) := \int_\Gamma \frac{\partial E(|\mathbf{x}-\mathbf{y}|)}{\partial \mathbf{n}_{\mathbf{x}}}\xi(\mathbf{y})\,ds_{\mathbf{y}}, \quad W\varphi(\mathbf{x}) := -\frac{\partial}{\partial \mathbf{n}_{\mathbf{x}}}\int_\Gamma \frac{\partial E(|\mathbf{x}-\mathbf{y}|)}{\partial \mathbf{n}_{\mathbf{y}}}\varphi(\mathbf{y})\,ds_{\mathbf{y}}.$$

They are bounded as mappings $V : \mathrm{H}^{-1/2}(\Gamma) \to \mathrm{H}^{1/2}(\Gamma)$, $K : \mathrm{H}^{1/2}(\Gamma) \to \mathrm{H}^{1/2}(\Gamma)$ and $W : \mathrm{H}^{1/2} \to \mathrm{H}^{-1/2}(\Gamma)$. Moreover, there exist constants $C_V > 0$ and $C_W > 0$ such that

$$\langle \bar{\chi}, V\chi\rangle_\Gamma \geq C_V\,\|\chi\|^2_{-1/2,\Gamma} \quad \forall\,\chi \in \mathrm{H}^{-1/2}(\Gamma) \tag{8}$$

and

$$\langle W\varphi, \bar{\varphi}\rangle_\Gamma + \left|\int_\Gamma \varphi\right|^2 \geq C_W\,\|\varphi\|^2_{1/2,\Gamma} \quad \forall\,\varphi \in \mathrm{H}^{1/2}(\Gamma). \tag{9}$$

3 The DG-FEM/BEM Formulation

We consider a sequence $\{\mathscr{T}_h\}_h$ of conforming and shape-regular triangulations of $\overline{\Omega}$. We assume that each partition $\mathscr{T}_h$ consists of tetrahedra K of diameter h_K; the unit outward normal vector to ∂K is denoted by $\mathbf{n}_K$. We also assume that the meshes $\{\mathscr{T}_h\}_h$ are aligned with the discontinuities of the piecewise constant coefficients σ and μ. The parameter $h := \max_{K\in\mathscr{T}_h}\{h_K\}$ represents the mesh size.

We denote by $\mathscr{F}_h$ the set of faces of the tetrahedra of the mesh, by $\mathscr{F}_h^0$ the sets of interior faces and by $\mathscr{F}_h^\Gamma$ the set of boundary faces. Clearly $\mathscr{F}_h := \mathscr{F}_h^0 \cup \mathscr{F}_h^\Gamma$. We notice that $\left\{\mathscr{F}_h^\Gamma\right\}_h$ is a shape-regular family of triangulations of Γ composed by triangles T of diameter h_T; therefore from now on we will denote by T the faces on Γ.

Let $\mathscr{O}_h$ be either $\mathscr{T}_h$ or $\mathscr{F}_h^\Gamma$ and E be a generic element of $\mathscr{O}_h$. We introduce for any $s \geq 0$ the broken Sobolev spaces

$$\mathrm{H}^s(\mathscr{O}_h) := \prod_{E\in\mathscr{O}_h}\mathrm{H}^s(E) \quad \text{and} \quad \mathbf{H}^s(\mathscr{O}_h) := \prod_{E\in\mathscr{O}_h}\mathrm{H}^s(E)^3.$$

For each $w := \{w_E\} \in \mathrm{H}^s(\mathscr{O}_h)$, the components w_E represents the restriction $w|_E$. When no confusion arises, the restrictions will be written without any subscript. The space $\mathrm{H}^s(\mathscr{O}_h)$ is endowed with the Hilbertian norm

$$\|w\|^2_{s,\mathscr{O}_h} := \sum_{E\in\mathscr{O}_h}\|w_E\|^2_{s,E}.$$

We use the same notation for the norm of the vectorial version $\mathbf{H}^s(\mathcal{O}_h)$. We use the standard conventions $\mathrm{L}^2(\mathcal{O}_h) := \mathrm{H}^0(\mathcal{O}_h)$ and $\mathbf{L}^2(\mathcal{O}_h) := \mathbf{H}^0(\mathcal{O}_h)$ and introduce the bilinear forms

$$(w,z)_{\mathcal{O}_h} = \sum_{E\in\mathcal{O}_h} \int_E w_E z_E, \quad \text{and} \quad (\mathbf{w},\mathbf{z})_{\mathcal{O}_h} = \sum_{E\in\mathcal{O}_h} \int_E \mathbf{w}_E \cdot \mathbf{z}_E.$$

Hereafter, given an integer $k \geq 0$ and a domain $D \subset \mathbb{R}^3$, $\mathcal{P}_k(D)$ denotes the space of polynomials of degree at most k on D. Let $h_{\mathcal{F}} \in \prod_{F\in\mathcal{F}_h} \mathcal{P}_0(F)$ be defined by $h_{\mathcal{F}}|_F := h_F$, $\forall F \in \mathcal{F}_h$, where h_F represents the diameter of the face F. We also introduce $\mathtt{s}_{\mathcal{F}} \in \prod_{F\in\mathcal{F}_h} \mathcal{P}_0(F)$ defined by $\mathtt{s}_F := \min(\sigma|_K, \sigma|_{K'})$, if $F = \partial K \cap \partial K' \in \mathcal{F}_h^0$ and $\mathtt{s}_F := \sigma|_K$, if $F = \partial K \cap \Gamma \in \mathcal{F}_h^\Gamma$.

We introduce, for $m \geq 1$, the finite element spaces

$$\mathbf{X}_h := \prod_{K\in\mathcal{T}_h} \mathcal{P}_m(K)^3, \quad \Lambda_h := \left\{ \lambda \in \prod_{T\in\mathcal{F}_h^\Gamma} \mathcal{P}_{m-1}(T); \quad \int_\Gamma \lambda = 0 \right\}$$

and

$$\Psi_h := \left\{ \phi \in \mathcal{C}^0(\Gamma);\ \phi|_T \in \mathcal{P}_{m+1}(T)\ \forall T \in \mathcal{F}_h^\Gamma,\ \int_\Gamma \phi = 0 \right\}.$$

Given $\mathbf{v} \in \mathbf{H}^{1+s}(\mathcal{T}_h)$, with $s > 1/2$, we consider $\mathbf{curl}_h \mathbf{v} \in \mathbf{H}^s(\mathcal{T}_h)$ given by $(\mathbf{curl}_h \mathbf{v})|_K = \mathbf{curl}\, \mathbf{v}_K$, for all $K \in \mathcal{T}_h$ and introduce

$$\mathbf{H}^s(\mathbf{curl}, \mathcal{T}_h) := \{ \mathbf{v} \in \mathbf{H}^s(\mathcal{T}_h); \quad \mathbf{curl}_h \mathbf{v} \in \mathbf{H}^s(\mathcal{T}_h) \}.$$

For $(\mathbf{v}, \varphi) \in \mathbf{H}^s(\mathcal{T}_h) \times \mathrm{H}^1(\mathcal{F}_h^\Gamma)$, $s > 1/2$, we introduce the jumps $[\![(\mathbf{v}, \varphi)]\!]$ by

$$[\![(\mathbf{v},\varphi)]\!] := \begin{cases} [\![\mathbf{v}\times\mathbf{n}]\!]_F := \mathbf{v}_K \times \mathbf{n}_K + \mathbf{v}_{K'} \times \mathbf{n}_{K'} & \text{if } F = K \cap K' \in \mathcal{F}_h^0(\Omega) \\ \mathbf{v}|_T \times \mathbf{n} - \mathbf{curl}_T \varphi & \text{if } T \in \mathcal{F}_h^\Gamma \end{cases}$$

and the averages $\{\mathbf{v}\} \in \mathbf{L}^2(\mathcal{F}_h)$ by

$$\{\mathbf{v}\} = \begin{cases} (\mathbf{v}_K + \mathbf{v}_{K'})/2 & \text{if } F = K \cap K' \in \mathcal{F}_h^0 \\ \mathbf{v}_K & \text{if } T \subset \partial K \in \mathcal{F}_h^\Gamma \end{cases}.$$

In order to derive the DG-FEM/BEM discretization of (4)–(7) we assume that $\mathbf{h} \in \mathbf{H}(\mathbf{curl}, \Omega) \cap \mathbf{H}^s(\mathbf{curl}, \mathcal{T}_h)$ and $\mathbf{j}_e \in \mathbf{H}^s(\mathcal{T}_h)$ with $s > 1/2$. We test (6) with $\varphi \in \mathrm{H}^{1/2}(\Gamma) \cap \mathrm{H}^1(\mathcal{F}_h^\Gamma)$, obtaining

$$\left\langle -W\psi + \left(\tfrac{1}{2}I - K^{\mathrm{t}}\right)\lambda, \varphi \right\rangle_\Gamma = \left\langle \frac{\mu}{\mu_0}\mathbf{h}\cdot\mathbf{n}, \varphi \right\rangle_\Gamma,$$

and use (4) together with an integration by parts on Γ to find

$$\left\langle -W\psi + \left(\tfrac{1}{2}I - K^{\mathrm{t}}\right)\lambda, \varphi \right\rangle_\Gamma = \frac{-1}{\imath\omega\mu_0}\int_\Gamma \mathbf{curl}\,\mathbf{e}\cdot\mathbf{n}\,\varphi = \frac{-1}{\imath\omega\mu_0}\int_\Gamma \mathbf{e}\cdot\mathbf{curl}_\Gamma\varphi, \tag{10}$$

where, for economy of notations, we reintroduced here the electric field $\mathbf{e} := \sigma^{-1}(\mathbf{curl}\,\mathbf{h} - \mathbf{j}_e)$. Moreover, we deduce from (4) that, for all $\mathbf{v} \in \mathbf{H}^s(\mathbf{curl}, \mathcal{T}_h)$,

$$\sum_{K\in\mathcal{T}_h}\left(\int_K (\imath\omega\mu\mathbf{h}\cdot\mathbf{v} + \mathbf{e}\cdot\mathbf{curl}\,\mathbf{v}) + \int_{\partial K}\mathbf{e}\cdot\mathbf{v}\times\mathbf{n}_K\right) = 0, \tag{11}$$

We also obtain from (4) that $\mathbf{curl}\,\mathbf{e} \in \mathrm{L}^2(\Omega)^3$. Consequently, the jumps of the tangential components of $\mathbf{e} \in \mathbf{H}^s(\mathcal{T}_h) \cap \mathbf{H}(\mathbf{curl}, \Omega)$ vanish across the internal faces $F \in \mathcal{F}_h^0$ and

$$\sum_{K\in\mathcal{T}_h}\int_{\partial K}\mathbf{e}\cdot\mathbf{v}\times\mathbf{n}_K = \sum_{F\in\mathcal{F}_h^0}\int_F \{\mathbf{e}\}\cdot[\![\mathbf{v}\times\mathbf{n}]\!] + \sum_{T\in\mathcal{F}_h^\Gamma}\int_T \mathbf{e}\cdot\mathbf{v}\times\mathbf{n}.$$

Inserting this identity in (11) and adding the resulting equation to (10), due to the fact that for $T \in \mathcal{F}_h^\Gamma$ one has $[\![(\mathbf{v},\varphi)]\!]_{|T} = \mathbf{v}|_T \times \mathbf{n} - \mathbf{curl}_T\varphi$ we easily get

$$\begin{aligned}(\imath\omega\mu\mathbf{h}, \mathbf{v})_{\mathcal{T}_h} + (\mathbf{e}, \mathbf{curl}_h\mathbf{v})_{\mathcal{T}_h} + \langle\{\mathbf{e}\}, [\![(\mathbf{v},\varphi)]\!]\rangle_{\mathcal{F}_h}&\\ + \imath\omega\mu_0\left\langle W\psi - \left(\tfrac{1}{2}I - K^{\mathrm{t}}\right)\lambda, \varphi\right\rangle_\Gamma &= 0.\end{aligned} \tag{12}$$

Finally, testing (7) with $\eta \in \mathrm{H}^{-1/2}(\Gamma)$ gives

$$\imath\omega\mu_0\langle\eta, \left(\tfrac{1}{2}I - K\right)\psi\rangle_\Gamma + \imath\omega\mu_0\langle\eta, V\lambda\rangle_\Gamma = 0. \tag{13}$$

Inspired from (12) and (13) we propose the following DG-FEM/BEM formulation for problem (1): Find $(\mathbf{u}_h, \psi_h) \in \mathbf{X}_h \times \Psi_h$ and $\lambda_h \in \Lambda_h$ such that

$$\begin{aligned} A_h((\mathbf{u}_h, \psi_h), (\mathbf{v}, \varphi)) - \imath\omega\mu_0\,\langle\lambda_h, (\tfrac{1}{2}I - K)\varphi\rangle_\Gamma &= L_h((\mathbf{v}, \varphi)) \\ \imath\omega\mu_0\,\langle\eta, (\tfrac{1}{2}I - K)\psi_h\rangle_\Gamma + \imath\omega\,\mu_0\langle\eta, V\lambda_h\rangle_\Gamma &= 0, \end{aligned} \tag{14}$$

for all $(\mathbf{v}, \varphi) \in \mathbf{X}_h \times \Psi_h$ and $\eta \in \Lambda_h$, where

$$
\begin{aligned}
A_h((\mathbf{u}_h, \psi_h), (\mathbf{v}, \varphi)) := {} & \imath\omega(\mu\mathbf{u}_h,\ \mathbf{v})_{\mathscr{T}_h} + (\sigma^{-1}\mathbf{curl}_h\mathbf{u}_h, \mathbf{curl}_h\mathbf{v})_{\mathscr{T}_h} + \imath\omega\mu_0\langle W\psi_h, \varphi\rangle_\Gamma \\
& + \langle\{\sigma^{-1}\mathbf{curl}_h\mathbf{u}_h\}, [\![(\mathbf{v}, \varphi)]\!]\rangle_{\mathscr{F}_h} - \overline{\langle\{\sigma^{-1}\mathbf{curl}_h\mathbf{v}\}, [\![(\mathbf{u}_h, \psi_h)]\!]\rangle_{\mathscr{F}_h}} \\
& + \alpha\langle \mathrm{s}_{\mathscr{F}}^{-1} h_{\mathscr{F}}^{-1} [\![(\mathbf{u}_h, \psi_h)]\!], [\![(\mathbf{v}, \varphi)]\!]\rangle_{\mathscr{F}_h},
\end{aligned}
$$

with a parameter $\alpha \geq 0$ and

$$
L_h((\mathbf{v}, \varphi)) := (\sigma^{-1}\mathbf{j}_e, \mathbf{curl}_h\mathbf{v})_{\mathscr{T}_h} + \Big\langle\{\sigma^{-1}\mathbf{j}_e\}, [\![(\mathbf{v}, \varphi)]\!]\Big\rangle_{\mathscr{F}_h}.
$$

The following proposition shows that the DG-FEM/BEM scheme (14) is consistent.

Proposition 1 *Let* $((\mathbf{h}, \psi), \lambda) \in [\mathbf{H}(\mathbf{curl}, \Omega) \times \mathrm{H}_0^{1/2}(\Gamma)] \times \mathrm{H}_0^{-1/2}(\Gamma)$ *be the solution of* (4)–(7). *Assume that* $\sigma^{-1}\mathbf{j}_e \in \mathbf{H}^s(\mathscr{T}_h)$ *and that* $(\mathbf{h}, \psi) \in \mathbf{H}^s(\mathbf{curl}, \mathscr{T}_h) \times \mathrm{H}^1(\mathscr{F}_h^\Gamma)$, *with* $s > 1/2$. *Then*

$$
\begin{aligned}
A_h((\mathbf{h}, \psi), (\mathbf{v}, \varphi)) \ - \ \imath\omega\mu_0\langle\lambda, (\tfrac{1}{2}I - K)\varphi\rangle_\Gamma &= L_h((\mathbf{v}, \varphi)), \\
\imath\omega\mu_0\langle\eta, (\tfrac{1}{2}I - K)\psi\rangle_\Gamma \ + \ \imath\omega\mu_0\langle\eta, V\lambda\rangle_\Gamma &= 0,
\end{aligned}
$$

for all $(\mathbf{v}, \varphi) \in \mathbf{X}_h \times \Psi_h$ *and* $\eta \in \Lambda_h$.

Proof The result is a direct consequence of identities (12) and (13), having used the fact that $\mathbf{h} \in \mathbf{H}(\mathbf{curl}, \Omega)$, so that $[\![\mathbf{h} \times \mathbf{n}]\!]_F = 0$ for each $F \in \mathscr{F}_h^0$, and Eq. (5), which furnishes $\mathbf{h}_{|T} \times \mathbf{n} = \mathbf{curl}_T\psi$ for each $T \in \mathscr{F}_h^\Gamma$.

4 Convergence Analysis

We introduce the bilinear form

$$
\begin{aligned}
\mathbb{A}_h\big(((\mathbf{u}, \phi), \zeta), ((\mathbf{v}, \varphi), \eta)\big) := {} & A_h((\mathbf{u}, \phi), (\mathbf{v}, \varphi)) - \imath\omega\mu_0\langle\zeta, (\tfrac{1}{2}I - K)\varphi\rangle_\Gamma \\
& + \imath\omega\mu_0\langle\bar{\eta}, (\tfrac{1}{2}I - K)\bar{\phi}\rangle_\Gamma + \imath\omega\mu_0\langle\bar{\eta}, V\bar{\zeta}\rangle_\Gamma
\end{aligned}
$$

and define in $\big(\mathbf{H}^s(\mathbf{curl}, \mathscr{T}_h) \times [\mathrm{H}^{1/2}(\Gamma) \cap \mathrm{H}^1(\mathscr{F}_h^\Gamma)]\big) \times \mathrm{H}^{-1/2}(\Gamma)$ the norms

$$
\begin{aligned}
|||((\mathbf{v}, \varphi), \eta)|||^2 := {} & \|(\omega\mu)^{1/2}\mathbf{v}\|_{0,\Omega}^2 + \|\sigma^{-1/2}\mathbf{curl}_h\mathbf{v}\|_{0,\Omega}^2 + \|\mathrm{s}_{\mathscr{F}}^{-1/2} h_{\mathscr{F}}^{-1/2}[\![(\mathbf{v}, \varphi)]\!]\|_{0,\mathscr{F}_h}^2 \\
& + \omega\mu_0\|\varphi\|_{1/2,\Gamma}^2 + \omega\mu_0\|\eta\|_{-1/2,\Gamma}^2
\end{aligned}
$$

and

$$|||((\mathbf{v},\varphi),\eta)|||_*^2 := |||((\mathbf{v},\varphi),\eta)|||^2 + \|\mathrm{s}_{\mathscr{F}}^{1/2} h_{\mathscr{F}}^{1/2}\{\sigma^{-1}\mathbf{curl}_h\mathbf{v}\}\|_{0,\mathscr{F}_h}^2.$$

The following discrete trace inequality is standard, (see, e.g., [7, Lemma 1.46]).

Lemma 1 *For all integer $k \geq 0$ there exists a constant $C^* > 0$, independent of h, such that,*

$$h_K\|v\|_{0,\partial K}^2 \leq C^*\|v\|_{0,K}^2 \quad \forall v \in \mathscr{P}_k(K), \quad \forall K \in \mathscr{T}_h. \tag{15}$$

It allow us to prove the following result.

Lemma 2 *For all $k \geq 0$, there exists a constant $C_\Omega > 0$, independent of the mesh size and the coefficients, such that*

$$\|\mathrm{s}_{\mathscr{F}}^{1/2} h_{\mathscr{F}}^{1/2}\{\sigma^{-1}\mathbf{w}\}\|_{0,\mathscr{F}_h} \leq C_\Omega\|\sigma^{-1/2}\mathbf{w}\|_{0,\Omega}, \quad \mathbf{w} \in \prod_{K\in\mathscr{T}_h} \mathscr{P}_k(K)^3 \tag{16}$$

Proof By definition of $\mathrm{s}_{\mathscr{F}}$, for any $\mathbf{w} \in \prod_{K\in\mathscr{T}_h} \mathscr{P}_k(K)^3$ we obtain

$$\begin{aligned}\|\mathrm{s}_{\mathscr{F}}^{1/2} h_{\mathscr{F}}^{1/2}\{\sigma^{-1}\mathbf{w}\}\|_{0,\mathscr{F}_h}^2 &= \sum_{F\in\mathscr{F}_h} h_F\|\mathrm{s}_F^{1/2}\{\sigma^{-1}\mathbf{w}\}_F\|_{0,F}^2 \\ &\leq \sum_{K\in\mathscr{T}_h}\sum_{F\in\mathscr{F}(K)} h_F\|\mathrm{s}_F^{1/2}\sigma_K^{-1}\mathbf{w}_K\|_{0,F}^2 \leq \sum_{K\in\mathscr{T}_h} h_K\|\sigma_K^{-1/2}\mathbf{w}_K\|_{0,\partial K}^2,\end{aligned}$$

where $\mathscr{F}(K)$ denotes the set of faces composing the boundary of K, namely, $\mathscr{F}(K) := \{F \in \mathscr{F}_h; \quad F \subset \partial K\}$. Then the result follows from (15).

Proposition 2 *There exists a constant $M^* > 0$, independent of h, such that*

$$|\mathbb{A}_h\left(((\mathbf{u},\phi),\zeta),((\mathbf{v},\varphi),\eta)\right)| \leq M^*|||((\mathbf{u},\phi),\zeta)|||_*|||((\mathbf{v},\varphi),\eta)|||$$

for all $((\mathbf{v},\varphi),\eta) \in (\mathbf{X}_h \times \Psi_h) \times \Lambda_h$ and for all $((\mathbf{u},\phi),\zeta) \in \left(\mathbf{H}^s(\mathbf{curl},\mathscr{T}_h) \times [\mathrm{H}^{1/2}(\Gamma) \cap \mathrm{H}^1(\mathscr{F}_h^\Gamma)]\right) \times H^{-1/2}(\Gamma)$, with $s > 1/2$.

Proof The result follows immediately from the Cauchy-Schwarz inequality, the boundedness of the maps $V : \mathrm{H}^{-1/2}(\Gamma) \to \mathrm{H}^{1/2}(\Gamma)$, $K : \mathrm{H}^{1/2}(\Gamma) \to \mathrm{H}^{1/2}(\Gamma)$ and $W : \mathrm{H}^{1/2}(\Gamma) \to \mathrm{H}^{-1/2}(\Gamma)$ and from the fact that the norms $|||\cdot|||$ and $|||\cdot|||_*$ are equivalent in $(\mathbf{X}_h \times \Psi_h) \times \Lambda_h$ (as a consequence of Lemma 2).

Proposition 3 *There exists a constant $\beta^* > 0$, independent of h, such that*

$$\mathrm{Re}\left[(1-\imath)\mathbb{A}_h\left(((\mathbf{v},\varphi),\eta),((\bar{\mathbf{v}},\bar{\varphi}),\bar{\eta})\right)\right] \geq \beta^*|||((\mathbf{v},\varphi)\eta)|||^2$$

for all $((\mathbf{v},\varphi),\eta) \in (\mathbf{X}_h \times \Psi_h) \times \Lambda_h$.

Proof By the definition of $\mathbb{A}_h(\cdot,\cdot)$ it follows

$$\mathrm{Re}\Big[(1-\imath)\mathbb{A}_h\big(((\mathbf{v},\varphi),\eta),((\bar{\mathbf{v}},\bar{\varphi}),\bar{\eta})\big)\Big] = \omega\|\mu^{1/2}\mathbf{v}\|_{0,\Omega}^2 + \|\sigma^{-1/2}\mathbf{curl}_h\mathbf{v}\|_{0,\Omega}^2$$
$$+\alpha\|\mathtt{s}_{\mathscr{F}}^{-1/2}h_{\mathscr{F}}^{-1/2}[\![(\mathbf{v},\varphi)_h]\!]\|_{0,\mathscr{F}_h}^2 + \omega\mu_0\langle\eta,V\bar{\eta}\rangle_\Gamma + \omega\mu_0\langle W\bar{\varphi},\varphi\rangle_\Gamma$$

and using (8) and (9) we deduce the result with $\beta^* = \min(1,\alpha,C_V,C_W)$.

The following Céa estimate is readily deduced from the consistency of the DG-FEM/BEM scheme, Propositions 2 and 3.

Theorem 1 *Assume that* $\sigma^{-1}\mathbf{j}_e \in \mathbf{H}^s(\mathscr{T}_h)$, *with* $s > 1/2$. *Then, the DG-FEM/BEM formulation* (14) *has a unique solution for any parameter* $\alpha \geq 0$. *Moreover if* $((\mathbf{h},\psi),\lambda) \in [\mathbf{H}(\mathbf{curl},\Omega)\times H_0^{1/2}(\Gamma)]\times H_0^{-1/2}(\Gamma)$ *and* $((\mathbf{u}_h,\psi_h),\lambda_h) \in ((\mathbf{X}_h\times\Psi_h)\times\Lambda_h)$ *are the solutions of* (4)–(7) *and* (14), *respectively, and* $(\mathbf{h},\psi) \in \mathbf{H}^s(\mathbf{curl},\mathscr{T}_h)\times \mathrm{H}^1(\mathscr{F}_h^\Gamma)$ *with* $s > 1/2$, *then,*

$$\big|\!\big|\!\big|((\mathbf{h}-\mathbf{u}_h,\psi-\psi_h),\lambda-\lambda_h)\big|\!\big|\!\big| \leq (1+\frac{M^*}{\beta^*})\big|\!\big|\!\big|((\mathbf{h}-\mathbf{v},\psi-\varphi),\lambda-\eta)\big|\!\big|\!\big|_*,$$

for all $((\mathbf{v},\varphi),\eta) \in (\mathbf{X}_h\times\Psi_h)\times\Lambda_h$.

5 Asymptotic Error Estimates

We denote by $\boldsymbol{\Pi}_h^{\mathrm{curl}}$ the m-order $\mathbf{H}(\mathbf{curl},\Omega)$-conforming Nédélec interpolation operator of the second kind, see for example [3, 13] or [12, Section 8.2]. It is well known that $\boldsymbol{\Pi}_h^{\mathrm{curl}}$ is bounded on $\mathbf{H}(\mathbf{curl},\Omega)\cap\mathbf{H}^s(\mathbf{curl},\mathscr{T}_h)$ for $s > 1/2$. Moreover, there exists a constant $C > 0$, independent of h, such that (cf. [12])

$$\|\mathbf{u}-\boldsymbol{\Pi}_h^{\mathrm{curl}}\mathbf{u}\|_{\mathbf{H}(\mathbf{curl},\Omega)} \leq Ch^{\min(s,m)}\big(\|\mathbf{u}\|_{s,\mathscr{T}_h} + \|\mathbf{curl}_h\mathbf{u}\|_{s,\mathscr{T}_h}\big). \tag{17}$$

For each triangle $T \in \mathscr{F}_h^\Gamma$ we define the interpolation operator π_T^Γ : $\mathrm{H}^{1/2+s}(T) \to \mathscr{P}_{m+1}(T)$, $s > 1/2$, uniquely determined by the conditions

$$\pi_T^\Gamma\varphi(\mathbf{a}_T) = \varphi(\mathbf{a}_T), \quad \text{for all vertices } \mathbf{a}_T \text{ of } T, \tag{18}$$

$$\int_e \pi_T^\Gamma\varphi q = \int_e \varphi q \qquad \forall q \in \mathscr{P}_{m-1}(e), \quad \text{for all edges e of T,} \tag{19}$$

$$\int_T \pi_T^\Gamma\varphi q = \int_T \varphi q \qquad \forall q \in \mathscr{P}_{m-2}(T). \tag{20}$$

The corresponding global interpolation operator π_h^Γ is $\mathrm{H}^1(\Gamma)$-conforming and satisfies the following interpolation error estimate.

Lemma 3 *Assume that* $\varphi \in H^{1/2+s}(\mathscr{F}_h^\Gamma) \cap H^{1/2}(\Gamma)$ *with* $s > 1/2$, *then*

$$\|\varphi - \pi_h^\Gamma \varphi\|_{t,\Gamma} \leq Ch^{\min\{1/2+s,m+2\}-t} \|\varphi\|_{1/2+s,\mathscr{F}_h^\Gamma}, \quad t \in \{0, 1, 1/2\} \tag{21}$$

with a constant $C > 0$ *independent of* h.

Proof We notice that, as $s > 1/2$, $\mathrm{H}^{1/2+s}(\mathscr{F}_h^\Gamma) \cap \mathrm{H}^{1/2}(\Gamma) \subset \mathscr{C}^0(\Gamma)$. Hence, π_h^Γ is bounded on $\mathrm{H}^{s+1/2}(\mathscr{F}_h^\Gamma) \cap \mathrm{H}^{1/2}(\Gamma)$. The interpolation error estimates for $t = 0$ and $t = 1$ are standard. The case $t = 1/2$ is obtained from the interpolation inequality

$$\|\phi\|_{1/2,\Gamma}^2 \leq \|\phi\|_{0,\Gamma} \|\phi\|_{1,\Gamma} \quad \forall \phi \in \mathrm{H}^1(\Gamma).$$

We introduce $\mathbf{L}_t^2(\Gamma) = \{\boldsymbol{\varphi} \in \mathrm{L}^2(\Gamma)^3;\ \boldsymbol{\varphi} \cdot \mathbf{n} = 0\}$ and consider the m-order Brezzi-Douglas-Marini (BDM) (see [3, 12]) finite element approximation of

$$\mathbf{H}(\mathrm{div}_\Gamma, \Gamma) := \left\{\boldsymbol{\varphi} \in \mathbf{L}_t^2(\Gamma); \quad \mathrm{div}_\Gamma \boldsymbol{\varphi} \in L^2(\Gamma)\right\}$$

relatively to the mesh $\mathscr{F}_h^\Gamma$, where div_Γ is the divergence operator on the surface Γ. This approximation space is given by

$$\mathrm{BDM}(\mathscr{F}_h^\Gamma) = \left\{\boldsymbol{\varphi} \in \mathbf{H}(\mathrm{div}_\Gamma, \Gamma); \quad \boldsymbol{\varphi}|_T \in \mathscr{P}_m(T)^2, \quad \forall T \in \mathscr{F}_h^\Gamma\right\}.$$

The corresponding interpolation operator Π_h^{BDM} is bounded on $\mathbf{H}(\mathrm{div}_\Gamma, \Gamma) \cap \prod_{T\in\mathscr{F}_h^\Gamma} \mathrm{H}^\delta(T)^2$ for all $\delta > 0$, and it is not difficult to check that is related to Π_h^{curl} through the following commuting diagram property:

$$(\boldsymbol{\Pi}_h^{\mathrm{curl}} \mathbf{v}) \times \mathbf{n} = \Pi_h^{\mathrm{BDM}}(\mathbf{v} \times \mathbf{n}) \quad \forall \mathbf{v} \in \mathbf{H}(\mathbf{curl}, \Omega) \cap \mathbf{H}^s(\mathbf{curl}, \mathscr{T}_h), \quad s > 1/2. \tag{22}$$

Moreover the following result holds true.

Proposition 4 *Let* $((\mathbf{h}, \psi), \lambda) \in [\mathbf{H}(\mathbf{curl}, \Omega) \times \mathrm{H}_0^{1/2}(\Gamma)] \times \mathrm{H}_0^{-1/2}(\Gamma)$ *be the solution of* (4)–(7). *Assume that* $(\mathbf{h}, \psi) \in \mathbf{H}^s(\mathbf{curl}, \mathscr{T}_h) \times H^{1/2+s}(\mathscr{F}_h^\Gamma)$ *with* $s > 1/2$. *Then,*

$$(\boldsymbol{\Pi}_h^{\mathrm{curl}} \mathbf{h}) \times \mathbf{n} = \mathbf{curl}_\Gamma(\pi_h^\Gamma \psi) \quad \text{on } \Gamma. \tag{23}$$

Proof Let us first prove that

$$\Pi_h^{\mathrm{BDM}}(\mathbf{curl}_\Gamma \psi) = \mathbf{curl}_\Gamma(\pi_h^\Gamma \psi). \tag{24}$$

It is clear that $\mathbf{curl}_\Gamma \pi_h^\Gamma \psi \in \mathrm{BDM}(\mathscr{F}_h^\Gamma)$ and it can be shown that the tangential fields $\Pi_h^{\mathrm{BDM}}(\mathbf{curl}_\Gamma \psi)$ and $\mathbf{curl}_\Gamma \pi_h^\Gamma \psi$ have the same BDM-degrees of freedom in each $F \in \mathscr{F}_h^\Gamma$, which gives (24). We deduce now (23) from (22), (24) and the transmission condition (5).

We will also use the best $L^2(\Gamma)$-approximation in Λ_h of a function $\eta \in H^r(\mathscr{F}_h^\Gamma)$, with $r > 0$, the $\mathbf{L}^2(\mathscr{T}_h)$-orthogonal projection onto $\prod_{K\in\mathscr{T}_h} \mathscr{P}_{m-1}(K)^3$ of a function $\mathbf{w} \in \mathbf{H}^r(\mathscr{T}_h)$, with $r > 1/2$, and the following estimates:

Lemma 4 *Assume that $\eta \in H^r(\mathscr{F}_h^\Gamma)$ for some $r \geq 0$. Then,*

$$\|\eta - \pi_{\Lambda_h}\eta\|_{-1/2,\Gamma} \leq Ch^{\min\{r,m\}+1/2}\|\eta\|_{r,\mathscr{F}_h^\Gamma}, \tag{25}$$

where $\pi_{\Lambda_h}\eta$ the best $L^2(\Gamma)$-approximation of η in Λ_h.

Proof See [15, Theorem 4.3.20].

Lemma 5 *Let $\mathbf{P}_h^{m-1}$ be the $\mathbf{L}^2(\mathscr{T}_h)$-orthogonal projection onto $\prod_{K\in\mathscr{T}_h} \mathscr{P}_{m-1}(K)^3$. For all $K \in \mathscr{T}_h$ and $\mathbf{w} \in \mathbf{H}^r(K)$, $r > 1/2$, we have*

$$h_K^{1/2}\|\mathbf{w} - \mathbf{P}_h^{m-1}\mathbf{w}\|_{0,\partial K} + \|\mathbf{w} - \mathbf{P}_h^{m-1}\mathbf{w}\|_{0,K} \leq Ch_K^{\min\{r,m\}}\|\mathbf{w}\|_{r,K}, \tag{26}$$

with a constant $C > 0$ independent of h.

Proof See [7, Lemmas 1.58 and 1.59].

We are now ready to prove the main theorem of this section.

Theorem 2 *Let $((\mathbf{h}, \psi), \lambda) \in [\mathbf{H}(\mathbf{curl}, \Omega) \times H_0^{1/2}(\Gamma)] \times H_0^{-1/2}(\Gamma)$ be the solution of (4)–(7) and let $((\mathbf{h}_h, \psi_h), \lambda_h) \in ((\mathbf{X}_h \times \Psi_h) \times \Lambda_h)$ be the solution of (14). Assume that $\sigma^{-1}\mathbf{j}_e \in \mathbf{H}^s(\mathscr{T}_h)$ and that $(\mathbf{h}, \psi) \in \mathbf{H}^s(\mathbf{curl}, \mathscr{T}_h) \times \mathrm{H}^{s+1/2}(\mathscr{F}_h^\Gamma)$, $\lambda \in \mathrm{H}^{s-1/2}(\mathscr{F}_h^\Gamma)$ with $s > 1/2$. Then, there exists $C > 0$, independent of h, such that*

$$\begin{aligned}|||((\mathbf{h} - \mathbf{h}_h, \psi - \psi_h), \lambda - \lambda_h)||| \leq Ch^{\min(s,m)}\Big(&\|\mathbf{h}\|_{s,\mathscr{T}_h} + \|\mathbf{curl}\,\mathbf{h}\|_{s,\mathscr{T}_h}\\ &+ \|\psi\|_{s+1/2,\mathscr{F}_h^\Gamma} + \|\lambda\|_{s-1/2,\mathscr{F}_h^\Gamma}\Big).\end{aligned}$$

Proof We deduce from Theorem 1 that

$$\begin{aligned}&|||((\mathbf{h} - \mathbf{h}_h, \psi - \psi_h), \lambda - \lambda_h)|||\\ &\leq (1 + \frac{M^*}{\beta^*})|||((\mathbf{h} - \boldsymbol{\Pi}_h^{\mathrm{curl}}\mathbf{h}, \psi - \pi_h^\Gamma\psi), \lambda - \pi_{\Lambda_h}\lambda)|||_*.\end{aligned}$$

By virtue of (23), we have

$$
\begin{aligned}
&|||((\mathbf{h}-\boldsymbol{\Pi}_h^{\mathrm{curl}}\mathbf{h},\psi-\pi_h^{\Gamma}\psi),\lambda-\pi_{\Lambda_h}\lambda)|||_*^2 = \|(\omega\mu)^{1/2}(\mathbf{h}-\boldsymbol{\Pi}_h^{\mathrm{curl}}\mathbf{h})\|_{0,\Omega}^2 \\
&+\|\sigma^{-1/2}\mathbf{curl}_h(\mathbf{h}-\boldsymbol{\Pi}_h^{\mathrm{curl}}\mathbf{h})\|_{0,\Omega}^2+\omega\mu_0\|\psi-\pi_h^{\Gamma}\|_{1/2,\Gamma}^2+\omega\mu_0\|\lambda-\pi_{\Lambda_h}\lambda\|_{-1/2,\Gamma}^2 \\
&\qquad +\|\mathrm{s}_{\mathscr{F}}^{1/2}h_{\mathscr{F}}^{1/2}\{\sigma^{-1}\mathbf{curl}_h(\mathbf{h}-\boldsymbol{\Pi}_h^{\mathrm{curl}}\mathbf{h})\}\|_{0,\mathscr{F}_h}^2. \qquad (27)
\end{aligned}
$$

We deduce from the triangle inequality that

$$
\begin{aligned}
&\|\mathrm{s}_{\mathscr{F}}^{1/2}h_{\mathscr{F}}^{1/2}\{\sigma^{-1}\mathbf{curl}(\mathbf{h}-\boldsymbol{\Pi}_h^{\mathrm{curl}}\mathbf{h})\}\|_{0,\mathscr{F}_h}=\|\mathrm{s}_{\mathscr{F}}^{1/2}h_{\mathscr{F}}^{1/2}\{\sigma^{-1}(I-\mathbf{P}_h^{m-1})\mathbf{curl}\,\mathbf{h})\}\|_{0,\mathscr{F}_h} \\
&\quad +\|\mathrm{s}_{\mathscr{F}}^{1/2}h_{\mathscr{F}}^{1/2}\{\sigma^{-1}(\mathbf{P}_h^{m-1}\mathbf{curl}\,\mathbf{h}-\mathbf{curl}\,\boldsymbol{\Pi}_h^{\mathrm{curl}}\mathbf{h})\}\|_{0,\mathscr{F}_h}=A_\Omega+B_\Omega\,. \qquad (28)
\end{aligned}
$$

Using (16) yields

$$
\begin{aligned}
&B_\Omega\le C_\Omega\|\sigma^{-1/2}(\mathbf{P}_h^{m-1}\mathbf{curl}\,\mathbf{h}-\mathbf{curl}\,\boldsymbol{\Pi}_h^{\mathrm{curl}}\mathbf{h})\|_{0,\Omega} \\
&=C_\Omega\|\sigma^{-1/2}\mathbf{P}_h^{m-1}\mathbf{curl}(\mathbf{h}-\boldsymbol{\Pi}_h^{\mathrm{curl}}\mathbf{h})\|_{0,\Omega}\le C_\Omega\|\sigma^{-1/2}\mathbf{curl}(\mathbf{h}-\boldsymbol{\Pi}_h^{\mathrm{curl}}\mathbf{h})\|_{0,\Omega}, \qquad (29)
\end{aligned}
$$

and it is straightforward to see that

$$
A_\Omega^2\le\sum_{K\in\mathscr{T}_h}h_K\|\sigma_K^{-1/2}(\mathbf{curl}\,\mathbf{h}-\mathbf{P}_h^{m-1}\mathbf{curl}\,\mathbf{h})\|_{0,\partial K}^2. \qquad (30)
$$

Combining (27)–(30) we deduce that

$$
\begin{aligned}
&|||((\mathbf{h}-\boldsymbol{\Pi}_h^{\mathrm{curl}}\mathbf{h},\psi-\pi_h^{\Gamma}\psi),\lambda-\pi_{\Lambda_h}\lambda)|||_*^2\le\|(\omega\mu)^{1/2}(\mathbf{h}-\boldsymbol{\Pi}_h^{\mathrm{curl}}\mathbf{h})\|_{0,\Omega}^2 \\
&\quad+(1+C_\Omega^2)\|\sigma^{-1/2}\mathbf{curl}(\mathbf{h}-\boldsymbol{\Pi}_h^{\mathrm{curl}}\mathbf{h})\|_{0,\Omega}^2+\omega\mu_0\|\psi-\pi_h^{\Gamma}\psi\|_{1/2,\Gamma}^2 \\
&\quad+\omega\mu_0\|\lambda-\pi_{\Lambda_h}\lambda\|_{-1/2,\Gamma}^2+\sum_{K\in\mathscr{T}_h}h_K\|\sigma_K^{-1/2}(\mathbf{curl}\,\mathbf{h}-\mathbf{P}_h^{m-1}\mathbf{curl}\,\mathbf{h})\|_{0,\partial K}^2.
\end{aligned}
$$

Finally, applying the interpolation error estimates (17), (26), (21) and (25) we obtain

$$
\begin{aligned}
&|||((\mathbf{h}-\boldsymbol{\Pi}_h^{\mathrm{curl}}\mathbf{h},\psi-\pi_h^{\Gamma}\psi),\lambda-\pi_{\Lambda_h}\lambda)|||_*\le C\Big(h^{\min(s,m)}(\|\mathbf{h}\|_{s,\mathscr{T}_h}+\|\mathbf{curl}\,\mathbf{h}\|_{s,\mathscr{T}_h}) \\
&\quad+h^{\min\{s+1/2,m+2\}-1/2}\|\psi\|_{s+1/2,\mathscr{F}_h^{\Gamma}}+h^{\min\{s-1/2,m\}+1/2}\|\lambda\|_{s-1/2,\mathscr{F}_h^{\Gamma}}\Big),
\end{aligned}
$$

and the result follows.

Remark 1 It is well-known that different choices of finite elements could be chosen for the approximation of $\mathbf{H}(\mathbf{curl}, \Omega)$ and $H^{1/2}(\Gamma)$. For instance, let us consider, for $m \geq 1$, $\mathbf{X}_h^{(0)} := \prod_{K \in \mathscr{T}_h} \mathbf{ND}_m(K)$ and

$$\Psi_h^{(0)} := \left\{ \phi \in \mathscr{C}^0(\Gamma);\ \phi|_T \in \mathscr{P}_m(T)\ \forall T \in \mathscr{F}_h^\Gamma,\ \int_\Gamma \phi = 0 \right\},$$

where $\mathbf{ND}_m(K) \subset \mathscr{P}_m(K)^3$ is the mth-order (local) Nédélec finite element space of the first kind, see for example [3, 13]. The DG-FEM/BEM scheme (14) formulated in terms of the discrete spaces $(\mathbf{X}_h^{(0)} \times \Psi_h^{(0)}) \times \Lambda_h$ provides, under the regularity assumption of Theorem 2, the same order of convergence with less degrees of freedom. However, in this case, the non-standard basis functions of $\mathbf{ND}_m(K)$ are required for the implementation of the scheme.

Acknowledgements Partial support by the University of Trento, and Spain's Ministry of Economy through Project MTM2013-43671-P.

References

1. A. Alonso Rodríguez, A. Valli, A FEM-BEM approach for electro-magnetostatics and time-harmonic eddy-current problems. Appl. Numer. Math. **59**, 2036–2049 (2009)
2. S. Außerhofer, O. Bíró, K. Preis, Discontinuous Galerkin formulation for eddy-current problems. COMPEL **28**, 1081–1090 (2009)
3. D. Boffi, F. Brezzi, M. Fortin, *Mixed Finite Element Methods and Applications*. Springer Series in Computational Mathematics, vol. 44 (Springer, Heidelberg, 2013)
4. A. Bossavit, The computation of eddy-currents, in dimension 3, by using mixed finite elements and boundary elements in association. Math. Comput. Model. **15**, 33–42 (1991)
5. C. Carstensen, R.H.W. Hoppe, N. Sharma, T. Warburton, Adaptive hybridized interior penalty discontinuous Galerkin methods for H(curl)-elliptic problems. Numer. Math. Theory Methods Appl. **4**, 13–37 (2011)
6. B. Cockburn, F.-J. Sayas, The devising of symmetric couplings of boundary element and discontinuous Galerkin methods. IMA J. Numer. Anal. **32**, 765–794 (2012)
7. D.A. Di Pietro, A. Ern, *Mathematical Aspects of Discontinuous Galerkin Methods* of Mathématiques & Applications (Berlin) [Mathematics & Applications], vol. 69 (Springer, Heidelberg, 2012)
8. G.N. Gatica, N. Heuer, F.-J. Sayas, A direct coupling of local discontinuous Galerkin and boundary element methods. Math. Comput. **79**, 1369–1394 (2010)
9. N. Heuer, S. Meddahi, F.-J. Sayas, Symmetric coupling of LDG-FEM and DG-BEM. J. Sci. Comput. **68**, 303–325 (2016)
10. R. Hiptmair, Symmetric coupling for eddy current problems. SIAM J. Numer. Anal. **40**, 41–65 (2002)
11. S. Meddahi, V. Selgas, A mixed-FEM and BEM coupling for a three-dimensional eddy current problem. M2AN Math. Model. Numer. Anal. **37**, 291–318 (2003)
12. P. Monk, *Finite Element Methods for Maxwell's Equations* (Oxford University Press, Oxford, 2003)
13. J.-C. Nédélec, A new family of mixed finite elements in $\mathbf{R}^3$. Numer. Math. **50**, 57–81 (1986)

14. I. Perugia, D. Schötzau, The hp-local discontinuous Galerkin method for low-frequency time-harmonic Maxwell equations. Math. Comput. **72**, 1179–1214 (2003)
15. S.A. Sauter, C. Schwab, *Boundary Element Methods*. Springer Series in Computational Mathematics, vol. 39 (Springer Berlin, 2011)
16. A. Zaghdani, C. Daveau, On the coupling of LDG-FEM and BEM methods for the three dimensional magnetostatic problem. Appl. Math. Comput. **217**, 1791–1810 (2010)

An LES Setting for DG-Based Implicit LES with Insights on Dissipation and Robustness

Rodrigo C. Moura, Gianmarco Mengaldo, Joaquim Peiró, and Spencer J. Sherwin

Abstract We suggest a new interpretation of implicit large eddy simulation (iLES) approaches based on discontinuous Galerkin (DG) methods by analogy with the LES-PLB framework (Pope, Fluid mechanics and the environment: dynamical approaches. Springer, Berlin, 2001), where PLB stands for 'projection onto local basis functions'. Within this framework, the DG discretization of the unfiltered compressible Navier-Stokes equations can be recognized as a Galerkin solution of a PLB-based (and hence filtered) version of the equations with extra terms originating from DG's implicit subgrid-scale modelling. It is shown that for under-resolved simulations of isotropic turbulence at very high Reynolds numbers, energy dissipation is primarily determined by the property-jump term of the Riemann flux employed. Additionally, in order to assess how this dissipation is distributed in Fourier space, we compare energy spectra obtained from inviscid simulations of the Taylor-Green vortex with different Riemann solvers and polynomial orders. An explanation is proposed for the spectral 'energy bump' observed when the Lax-Friedrichs flux is employed.

1 Introduction

Despite the rapid dissemination of DG-based implicit LES in recent years [1–3], there is still a lack of fundamental research on DG's suitability for under-resolved turbulence simulations in general. More traditional (low-order) implicit LES approaches advocate that suitable methods should have some sort of built-in subgrid-scale model in their formulation [4]. For example, in [5], modified equation analysis is applied to a particular finite volume scheme to reveal that truncation terms of dissipative character implicitly play the role of a turbulence model. A preliminary assessment on the potential of modified equations analysis for DG-based iLES proved discouraging due to Taylor series convergence issues

R.C. Moura (✉) • G. Mengaldo • J. Peiró • S.J. Sherwin
Department of Aeronautics, Imperial College London, London SW7 2BY, UK
e-mail: r.moura13@imperial.ac.uk

M.L. Bittencourt et al. (eds.), *Spectral and High Order Methods for Partial Differential Equations ICOSAHOM 2016*, Lecture Notes in Computational Science and Engineering 119, DOI 10.1007/978-3-319-65870-4_10

observed in the linear advection case [6]. However, recent studies on eigensolution (dispersion-diffusion) analysis [7, 8] showed significant potential in clarifying why and how to use DG-iLES. One of the relevant results is that DG's upwind dissipation seems to mimic the behaviour of hyper-viscosity, especially at higher polynomial orders, where dissipation is only relevant at large wavenumbers.

This work presents an alternative way to analyse DG-iLES by analogy with the LES-PLB framework proposed by Pope in [9], where PLB stands for 'projection onto local basis functions'. From this analogy, a DG discretization of the unfiltered Navier-Stokes equations is shown to be equivalent to a standard Galerkin discretization of a filtered (projected) version of the equations with additional terms related to DG's implicit subgrid-scale model. Subsequently, we consider isotropic turbulence at very high Reynolds numbers and show that these extra terms stem from the dissipative part of the Riemann flux employed. In order to gain additional insight on DG's dissipation in limit of vanishing viscosity, a numerical assessment of the inviscid Taylor-Green vortex [10, 11] problem is performed. Different polynomial orders and Riemann solvers are considered. The assessment is complemented with results from eigensolution analysis which provides insight into the distribution of dissipation in spectral space and on robustness issues.

2 LES-PLB Fundamentals

In the LES-PLB framework, the resolved field $\overline{q}$ is defined as a Galerkin projection of the actual field q and has a basis-function representation (with, say, N modes) in the form

$$\overline{q}(\mathbf{x}, t) = \mathscr{P}\{q(\mathbf{x}, t)\} = \sum_{n=1}^{N} \widehat{q}_n(t)\phi_n(\mathbf{x}) \, , \tag{1}$$

where $\mathscr{P}\{\cdot\}$ stands for the Galerkin projection. The basis functions $\phi_n(\mathbf{x})$ are chosen to be local, i.e. non-zero only at bounded regions in space, and the coefficients $\widehat{q}_n(t)$ are obtained from the projection procedure, namely

$$\widehat{q}_n(t) = \mathscr{P}_n\{q(\mathbf{x}, t)\} \, , \tag{2}$$

in which $\mathscr{P}_n\{\cdot\}$ denotes the local projection operator associated to basis function ϕ_n. The residual field is naturally given by $q'(\mathbf{x}, t) = q(\mathbf{x}, t) - \overline{q}(\mathbf{x}, t)$ and has a vanishing projection, since $\mathscr{P}\{q'\} = \mathscr{P}\{q\} - \mathscr{P}\{\overline{q}\} = \overline{q} - \overline{q} = 0$.

Let the governing equations for $q(\mathbf{x}, t)$ be compactly written as

$$\frac{\partial q}{\partial t} = \mathscr{G}(q) \, , \tag{3}$$

whereby the evolution equation for $\overline{q}(\mathbf{x}, t)$ is given by

$$\frac{\partial \overline{q}}{\partial t} = \mathscr{P}\{\mathscr{G}(q)\} = \mathscr{P}\{\mathscr{G}(\overline{q} + q')\} , \tag{4}$$

which, since q' is not known in the LES framework, can be rewritten as

$$\frac{\partial \overline{q}}{\partial t} = \mathscr{P}\{\mathscr{G}(\overline{q})\} + \mathscr{R}(q) , \quad \mathscr{R}(q) = \mathscr{P}\{\mathscr{G}(q)\} - \mathscr{P}\{\mathscr{G}(\overline{q})\} , \tag{5}$$

where $\mathscr{R}(q)$ embodies the residual motions. For the solution coefficients, one has

$$\frac{d\widehat{q}_n}{dt} = \mathscr{P}_n\{\mathscr{G}(\overline{q})\} + \mathscr{R}_n(q) , \quad \mathscr{R}_n(q) = \mathscr{P}_n\{\mathscr{G}(q)\} - \mathscr{P}_n\{\mathscr{G}(\overline{q})\} . \tag{6}$$

When the number of basis functions employed is sufficient to resolve q accurately (DNS limit), both q' and $\mathscr{R}(q)$ become negligible and Eq. (5) reduces to

$$\frac{\partial \overline{q}}{\partial t} = \mathscr{P}\{\mathscr{G}(\overline{q})\} , \tag{7}$$

which, as emphasized by Pope in [9], is precisely the standard Galerkin method's statement for the solution of Eq. (3). Prior to the DNS limit, therefore, LES-PLB amounts formally to a Galerkin method with an added source term associated to the residual motions, $\mathscr{R}(q)$, which requires modelling.

3 DG-Based iLES as LES-PLB

The compressible Navier-Stokes equations are given by Eq. (3) with

$$\mathscr{G}(q) = \nabla \cdot \mathbf{F}_v(q, \nabla q) - \nabla \cdot \mathbf{F}_i(q) , \tag{8}$$

where $\mathbf{F}_v$ and $\mathbf{F}_i$ are the viscous and inviscid flux vectors, respectively, and q stands for the conserved variables array. In DG, the numerical solution is approximated through an *hp* discretization, namely

$$q(\mathbf{x}, t) \approx \overline{q}(\mathbf{x}, t) = \sum_e \sum_{m=1}^{N} \widehat{q}_m^e(t) \phi_m^e(\mathbf{x}) , \tag{9}$$

where index e runs through all mesh elements Ω_e composing the physical domain Ω. Inside each Ω_e, the solution is represented through the element-wise coefficients $\widehat{q}_m^e$ and polynomial basis functions ϕ_m^e, which are zero outside Ω_e.

Since DG requires specific treatment for the viscous terms to ensure numerical stability [12], typical DG formulations solve the original problem through the form

$$\frac{\partial q}{\partial t} = \nabla \cdot \mathbf{F}_v(q, \mathbf{g}) - \nabla \cdot \mathbf{F}_i(q) \equiv \nabla \cdot \mathbf{F}(q, \mathbf{g}) \, , \tag{10}$$

where the gradient variable $\mathbf{g}$ is introduced as an approximation to ∇q. The gradient variable is obtained through the solution of the auxiliary equation according to the numerics of the chosen viscous scheme. Usually, $\overline{\mathbf{g}} = \mathbf{g}\{\overline{q}\}$ differs from $\nabla \overline{q}$.

The semi-discrete DG formulation at element Ω_e can be written as

$$\int_{\Omega_e} \phi_n \frac{\partial \overline{q}}{\partial t} \, d\Omega = \int_{\Omega_e} \phi_n \nabla \cdot \mathbf{F}(\overline{q}, \overline{\mathbf{g}}) \, d\Omega + \oint_{\partial \Omega_e} \phi_n [\widetilde{\mathbf{F}} - \mathbf{F}] \cdot \mathbf{n} \, d\ell \, , \tag{11}$$

where $\widetilde{\mathbf{F}}$ is a numerical flux based on information from both sides of the considered interfaces and has viscous and inviscid contributions, see Eq. (10). Note that the above is equivalent to either of the so-called "weak-weak" or "weak-strong" DG forms, see e.g. [13], as long as nearly exact integrations are employed. This might take however a very large number of quadrature points for under-resolved flows.

At this point, by comparing Eqs. (6) and (11), one is tempted to recognize the boundary integral in Eq. (11) as the residual motions term $\mathscr{R}_n(q)$ of Eq. (6). However, since the remainder of Eq. (11) does not hold as a valid Galerkin method due to the lack of inter-element communication (especially at very high Reynolds when $\mathbf{F}(q, \mathbf{g})$ looses its dependence on $\mathbf{g}$), an interface contribution is still missing. This is a subtle caveat since the exact form of interface contribution required to recover a standard Galerkin method in a discontinuous setting is not known in general. We will therefore assume that this contribution can be represented at least approximately by a *symmetrical* interface flux, hereafter denoted by $\check{\mathbf{F}}$. We remark that a symmetrical contribution is consistent with the continuous Galerkin discretization, a well-known standard Galerkin method.

The second step towards conformity with Eq. (6) is the replacement of $\overline{\mathbf{g}}$ with $\nabla \overline{q}$ in the argument of the volume integral on the right-hand side of Eq. (11), since a standard Galerkin method should only rely explicitly $\overline{q}$, see Eq. (7). By taking into account those two steps, one can rewrite Eq. (11) as

$$\begin{aligned} \int_{\Omega_e} \phi_n \frac{\partial \overline{q}}{\partial t} \, d\Omega = & \int_{\Omega_e} \phi_n \nabla \cdot \mathbf{F}(\overline{q}, \nabla \overline{q}) \, d\Omega \; + \oint_{\partial \Omega_e} \phi_n \check{\mathbf{F}} \cdot \mathbf{n} \, dS \; + \\ + \int_{\Omega_e} \phi_n \nabla \cdot [\mathbf{F}_v(\overline{q}, \overline{\mathbf{g}}) - \mathbf{F}_v(\overline{q}, \nabla \overline{q})] \, d\Omega \; & + \oint_{\partial \Omega_e} \phi_n [\widetilde{\mathbf{F}} - \check{\mathbf{F}} - \mathbf{F}] \cdot \mathbf{n} \, dS \, . \end{aligned} \tag{12}$$

Comparing Eqs. (12) and (6) clearly shows that the DG approximation of the unfiltered Navier-Stokes equations corresponds to an LES-PLB solution of the large-eddy fields, whereas the residual motions are accounted for by the last two integrals in Eq. (12). In fact, the first term in Eq. (12) corresponds to $d\widehat{q}_n/dt$, the following two terms represent $\mathscr{P}_n\{\mathscr{G}(\overline{q})\}$ and the last two integrals correspond to $\mathscr{R}_n(q)$, being therefore identified as DG's implicit subgrid-scale modelling terms.

Unfortunately, Eq. (12) gives no obvious clue about how well the terms corresponding to $\mathscr{R}_n(q)$ can perform regarding subgrid-scale modelling. Also, Eq. (12) is not in PDE form, which further complicates the physical interpretation of these terms. However, these terms are designed upon numerical, but also physical considerations, especially the inviscid flux term which relies on the solution of a Riemann problem. The role of the inviscid fluxes will be discussed in the context of homogeneous isotropic turbulence (HIT) in the next section. It is hoped that subsequent works may help to clarify the role of different terms on physical grounds.

Interpreting DG-iLES through the LES-PLB setting is not only a way to assess DG's implicit subgrid-scale modelling, but also might provide insights on how to adapt or design, for instance, numerical fluxes with improved turbulence-capturing physics. Moreover, simply knowing how to interpret the resolved fields of DG-iLES solutions as local Galerkin projections of the exact (DNS) solution might prove useful in the construction of explicit LES models to be used with DG. At last, the interpretation proposed for the resolved fields might also be relied upon when post-processing DG-iLES results and analysing turbulence data.

4 Dissipation for HIT at Very High Reynolds Numbers

We now consider Eq. (12) in the limit of vanishing viscosity, which corresponds to the DG discretization (in LES-PLB form) of the compressible Euler equations,

$$\int_{\Omega_e} \phi_n \frac{\partial \overline{q}}{\partial t}\, d\Omega + \int_{\Omega_e} \phi_n \nabla \cdot \mathbf{F}_i(\overline{q})\, d\Omega + \oint_{\partial\Omega_e} \phi_n \widetilde{\mathbf{F}}_i \cdot \mathbf{n}\, dS = \oint_{\partial\Omega_e} \phi_n [\mathbf{F}_i + \breve{\mathbf{F}}_i - \widetilde{\mathbf{F}}_i] \cdot \mathbf{n}\, dS \tag{13}$$

whose right-hand side terms constitute $\mathscr{R}_n(q)$.

Pre-multiplying Eq. (13) by $\widehat{q}_n^e$ and adding up conveniently the resulting equations for $n = 1, \ldots, N$, yields, upon summation over all elements Ω_e,

$$\frac{\partial}{\partial t} \int_{\Omega} \frac{||q||^2}{2}\, d\Omega + \sum_e \int_{\Omega_e} q^T \nabla \cdot \mathbf{F}_i\, d\Omega + \sum_e \oint_{\partial\Omega_e} q^T \breve{\mathbf{F}}_i \cdot \mathbf{n}\, dS =$$

$$= \sum_f \oint_{S_f} \left\{ q_1^T [\mathbf{F}_1 + \breve{\mathbf{F}}_{12} - \widetilde{\mathbf{F}}_{12}] \cdot \mathbf{n}_{12} + q_2^T [\mathbf{F}_2 + \breve{\mathbf{F}}_{21} - \widetilde{\mathbf{F}}_{21}] \cdot \mathbf{n}_{21} \right\} dS\,, \tag{14}$$

where the overbar on $\overline{q}$ and the index i of the (inviscid) interface fluxes have been omitted for simplicity. Summation on the right-hand side is performed over all interfaces f of the domain, where fluxes from either sides of the corresponding surfaces S_f are taken into account (as denoted by indices 1 and 2). Term $\partial_t ||q||^2/2 = q\, \partial_t q$ can be regarded as the rate of change of a 'solution energy', as it embodies energies (in the L^2 sense) of all the variables of state vector q.

We assume Riemann solvers whose numerical fluxes can be written in the form

$$\widetilde{\mathbf{F}}_{12} = \overline{\mathbf{F}}_{12} - \frac{1}{2}\,|\mathbf{J}|\,(q_2 - q_1)\,\mathbf{n}_{12}\,, \quad \widetilde{\mathbf{F}}_{21} = \overline{\mathbf{F}}_{21} - \frac{1}{2}\,|\mathbf{J}|\,(q_1 - q_2)\,\mathbf{n}_{21}\,, \tag{15}$$

where $\overline{\mathbf{F}}_{12} = \overline{\mathbf{F}}_{21} = \frac{1}{2}(\mathbf{F}_1 + \mathbf{F}_2)$ and $|\mathbf{J}|$ is a solver-specific matrix possibly related to the Jacobian $|\partial\mathbf{F}/\partial q|$. While many Riemann fluxes are compatible with the above form, see e.g. [14], it obviously does not account for all existing solvers. In particular, the form above is compatible with Roe and (local) Lax-Friedrichs fluxes [15], which are arguably the most common Riemann solvers used with DG methods.

Now by rewriting the right-hand side (RHS) of Eq. (14), which relates to DG's implicit dissipation of 'solution energy' at very high Reynolds numbers, one has

$$\mathrm{RHS} = \sum_f \oint_{S_f} \Big\{ \big[\, \mathscr{A}(q_1, q_2) + \mathscr{B}(q_1, q_2) - \mathscr{C}(q_1, q_2)\,\big] \cdot \mathbf{n}_{12} + \mathscr{D}(q_1, q_2) \Big\}\, dS\,, \tag{16}$$

where $\mathscr{A}(q_1, q_2)$, $\mathscr{B}(q_1, q_2)$ and $\mathscr{C}(q_1, q_2)$ are anti-symmetrical and given by

$$\mathscr{A} = q_1^T\mathbf{F}_1 - q_2^T\mathbf{F}_2\,, \quad \mathscr{B} = q_1^T\check{\mathbf{F}}_{12} - q_2^T\check{\mathbf{F}}_{21}\,, \quad \mathscr{C} = q_1^T\overline{\mathbf{F}}_{12} - q_2^T\overline{\mathbf{F}}_{21}\,, \tag{17}$$

whereas $\mathscr{D}(q_1, q_2)$ is symmetrical and defined as

$$\mathscr{D} = q_1^T\frac{|\mathbf{J}|}{2}(q_2 - q_1) + q_2^T\frac{|\mathbf{J}|}{2}(q_1 - q_2) = -\,\delta q^T\frac{|\mathbf{J}|}{2}\,\delta q\,, \tag{18}$$

in which $\delta q = \pm(q_1 - q_2)$, as the above expression holds with either sign for δq.

We note that the anti-symmetry of $\mathscr{A}$, $\mathscr{B}$ and $\mathscr{C}$ is preserved upon statistical averaging within a turbulent flow solution, i.e. $\mathscr{A},\!\mathscr{B},\!\mathscr{C}\,(q_1, q_2) = -\mathscr{A},\!\mathscr{B},\!\mathscr{C}\,(q_2, q_1)$ implies $\langle\, \mathscr{A},\!\mathscr{B},\!\mathscr{C}\,(q_1, q_2)\,\rangle = -\langle\, \mathscr{A},\!\mathscr{B},\!\mathscr{C}\,(q_2, q_1)\,\rangle$. On the other hand, by the rotational invariance property of isotropic turbulence, axis rotation or reflection can not alter statistics. Since axis reflection at interfaces amounts to the swapping of internal and external states, one has $\langle\, \mathscr{A},\!\mathscr{B},\!\mathscr{C}\,(q_1, q_2)\,\rangle = \langle\, \mathscr{A},\!\mathscr{B},\!\mathscr{C}\,(q_2, q_1)\,\rangle$. The only possibility is therefore $\langle\, \mathscr{A},\!\mathscr{B},\!\mathscr{C}\,\rangle = 0$. As a result, averaging Eq. (16) leads to

$$\langle\, \mathrm{RHS}\,\rangle = -\sum_f \oint_{S_f} \langle \delta q^T \frac{|\mathbf{J}|}{2}\,\delta q\rangle\, dS\,. \tag{19}$$

The conclusion obtained upon statistical averaging is that, at very high Reynolds number, the effect of DG's implicit LES model on the variation of solution energy—cf. RHS in Eq. (14)—is primarily determined by the property-jump term of the Riemann flux employed. This suggests that DG's built-in model is mainly dissipative and even somewhat physical, as it stems from upwinding. The dissipative character is also expected from the quadratic form in Eq. (19), although it is only formally guaranteed for positive definite $|\mathbf{J}|$. While this condition is satisfied by e.g.

the Lax-Friedrichs flux, where $|\mathbf{J}| = |\lambda|_{max}\mathbf{I}$, it does not hold for Riemann solvers in general. This is however not discouraging, as being mostly (but not always) dissipative might allow for some backscatter of turbulent kinetic energy. Further study is nevertheless required regarding this point.

5 Insights from Inviscid TGV Test Cases

High-order simulations carried out with DG over the years have indicated that the particular choice of the Riemann solver is not very important. Here, we compare the performance of different solvers and show that, for under-resolved computations, the Riemann flux choice can be quite important regarding both numerical stability and solution quality when viscosity is negligible. Discussion is based on simulations of the inviscid Taylor-Green vortex (TGV) problem, taken as representative of free turbulence (away from walls) at very high Reynolds numbers.

The TGV flow was introduced in [10] as a model problem for the analysis of transition and turbulence decay. The test problem was originally proposed for the incompressible Navier-Stokes equations in a cubic domain with triply-periodic boundary conditions. Here we adopt a modified version of the initial conditions which is suited for compressible flow solvers, as done in [11], so that the Euler equations (representing inviscid flow conditions) are solved within $[-\pi, \pi]^3$ at a baseline Mach number of 0.1. Even though the Euler equations are simulated directly, the presence of numerical diffusion is expected to make results consistent with the dissipative solution of the viscous TGV problem in the limit of zero viscosity. This is in contrast with the exact solution of the inviscid TGV where energy is conserved.

The evolution of the TGV flow at high Reynolds numbers (say, higher than 10^3) can be characterized by three distinct phases, see e.g. [16]. During the first phase dissipation effects can be neglected and vortex lines begin to fold and stretch first by pressure gradients and then via three-dimensional vortex interactions, but still through a well-organized (non-chaotic) process. In the second phase transition takes place, whereby non-linear effects intensify and small-scale energy grows rapidly through the cascade mechanism leading to a peak in kinetic energy dissipation. Finally, in the third phase the TGV flow tends to a more homogeneous state of decaying turbulence, where kinetic energy decays monotonically towards zero.

The overall behaviour described above has been captured quite well by our stable computations, with minor differences being observed upon DOF refinement. The base set of test cases addressed used Roe's original solver [17] and the local Lax-Friedrichs (LxF) flux [18], but some of the cases have also been computed with the exact Riemann solver and with the HLL and HLLC fluxes [15]. Table 1 shows the base set of cases, each column corresponding to the number of polynomial modes $m = p + 1$ used, p being the polynomial order, and each row corresponding to a different number of degrees of freedom $N_{dof} = (n_{el}\, m)^3$, in which n_{el}^3 is the total number of elements. Equispaced grids have been employed (cubic elements). Values

Table 1 Summary of test cases—crossed out numbers indicate cases that crashed

	Roe					LxF				
$m = p + 1$	4	5	6	7	8	4	5	6	7	8
n_{el}	28	23	19	16	14	28	23	~~19~~	~~16~~	~~14~~
	39	32	28	23	~~19~~	39	32	~~28~~	~~23~~	~~19~~
	56	45	39	~~32~~	~~28~~	56	45	~~39~~	~~32~~	~~28~~

in the Table's core represent the number of elements per direction, n_{el}. Crossed out numbers indicate simulations that lacked stability and crashed. All simulations have been conducted through the spectral/*hp* element code *Nektar++* [19].

The results in Table 1 indicate that Roe is more robust than LxF for the problem considered. This is counter-intuitive since the former is known to be less dissipative than the latter, at least for well resolved computations. All unstable cases yielded reasonable results (with no signs of numerical instability) until the time of crash, which took place consistently within the transitional phase of the TGV flow. This lack of robustness, found especially for the higher-order discretizations, has been carefully verified not to be related to time-step restrictions or polynomial aliasing errors. Typical CFL numbers employed (based on the acoustic wave speed) are of the order of 10^{-1} and an increased number of quadrature points ($Q = 2m$) has been used to ensure consistent integration of the cubic non-linearities of the compressible Euler equations [20], even though the flow is nearly incompressible. Tests conducted to rule out these factors consistently showed the time of crash to be practically insensitive to time-step reductions or to a further increase in the number of integration points. A much more subtle cause of crash is suggested in Sect. 6.

A comparison between energy spectra obtained from test case $m = 5, n_{el} = 23$ is given in Fig. 1. Results at two different times are shown, namely at peak dissipation ($t = 9$) and within the decay phase ($t = 18$). Different Riemann fluxes followed one of two distinct behaviours: Roe, HLLC and the exact solver yielded the expected spectrum, showing an inertial range with Kolmogorov's $-5/3$ slope at $t = 9$ followed by a (numerically induced) dissipation range; LxF and HLL yielded a spectrum with less energy at the large/intermediate scales and allowed for the formation of an 'energy bump' at the small scales. This spurious build up of energy can actually be seen in flow-field visualizations in the form of small-scale noise, see [8]. QR diagrams discussed in Sect. 6, cf. Fig. 3, confirm that the first behaviour is physically correct while the second one is somewhat far from it. An explanation for the energy bump is proposed in the next section, but at this point it is important to mention that increasing the polynomial order leads to an increase in the energy bump. For $m = 8$, even the Roe-based solution yielded one, although considerably less significant than the bump observed for LxF with $m = 5$.

The above results seem to indicate that, at least for low Mach number transitional/turbulent flows, "complete" solvers such as Roe or HLLC can already handle most of the physics, as the exact solver did not improve solution quality significantly. Results also discourage the use of more simplistic fluxes such as LxF

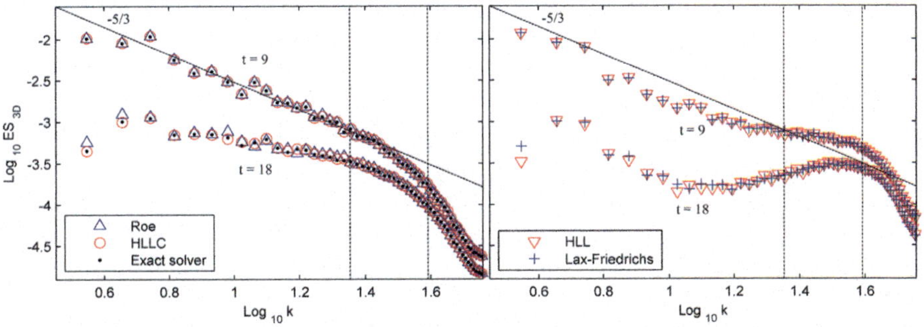

Fig. 1 Energy spectra at $t = 9$ (dissipation peak) and $t = 18$ (homogeneous decay) obtained with Roe, HLLC and the exact solver (*left*) and with LxF and HLL (*right*), from case $m = 5$, $n_{el} = 23$. The *vertical dashed lines* delimit the region where numerical dissipation is expected to begin [8]

and HLL for DG-iLES at very high Reynolds numbers. The quality of Roe-based solutions did not change much as the polynomial order was varied for a given number of DOFs, however higher-order discretizations tended to be less stable. Preliminary low-order tests conducted with $m = 3$ and $m = 2$ (not included in Table 1) provided considerably less accurate results, even though much finer grids are employed as analysis is made on a fixed DOF basis. The use of moderately high orders (e.g. $m = 4$ to 6) is therefore suggested for general practice, unless additional stabilization techniques are employed, in which case higher orders might be considered with care (to avoid strong energy bumps).

6 Dissipation Distribution and Energy Bumps

While Eq. (19) gives no obvious clue about how dissipation is distributed in Fourier space, insights can be obtained from linear dispersion-diffusion analysis [7, 8]. A simplified explanation for the energy bumps is proposed as follows. While Roe employs the correct eigenvalues when upwinding, LxF uses instead the spectral radius alone ($|u| + c$, in 1D). This results in over-upwinding for the momentum equations owing to the upwind ratio $\beta = (|u| + c)/|u| = 1 + \text{Mach}^{-1}$. We stress that β tends to infinity in the incompressible limit. Figure 2 illustrates DG's dissipation eigencurve for three ratios β when $m = 5$. There is a critical Mach number $\text{Mach}^{\star} = (\beta^{\star} - 1)^{-1}$ below which a sharp dissipation cut-off appears. Reducing the Mach number from 0.9 to 0.1 caused the eigencurve discontinuity to increase about twelve times. Further inspection showed that $\text{Mach}^{\star}$ only increases with the discretization order (e.g. $\text{Mach}^{\star} \approx 2$ for $m = 9$). The Roe flux does not have this problem as its unit upwind ratio does not change with the Mach number.

Our understanding is that a sharp upwind dissipation induces a stronger bottleneck effect [21, 22] thus promoting an energy pile-up before the cut-off wavenumber (cf. vertical dashed lines in Fig. 2). DNS experiments using hyperviscosity in place

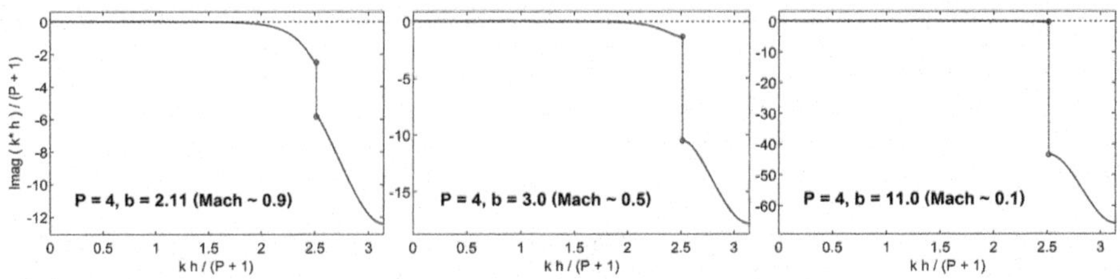

Fig. 2 Numerical diffusion in wavenumber space (LxF flux for $m = 5$) as Mach number is reduced from 0.9 to 0.1 (*left to right*). The plots show the imaginary part of the modified wavenumber k^* as a function of the actual wavenumber k, both normalized by h/m, with h being the mesh spacing and m the number of polynomial modes. Curves from linear eigensolution analysis in 1D, cf. [7]

of regular (second-order) viscosity have already demonstrated that energy bumps become more pronounced as the hyperviscosity exponent is increased [23, 24]. Another phenomenon discussed in [25] is that over-energetic small-scales can cause a more intense mixing and increase diffusion through an eddy-viscosity effect, consistent with the less energetic large/intermediate scales observed for LxF in Fig. 1.

A complementary explanation for the energy bumps observed with hyperviscosity has also been proposed in [24] and further confirmed in [26]: those emerge as the solution begins to follow a conservative dynamics, typically observed [25] when only a finite number of Fourier modes are retained (limit of increasingly sharp dissipation). In this scenario, a range of small-scale structures are said to 'thermalize' [26] and follow an independent dynamics where conservation and equipartition of energy is favoured [24]. This phenomenon has been partially confirmed in our TGV solutions through the analysis of QR diagrams [27, 28], see Fig. 3. These diagrams consist of joint PDFs of the second (Q) and third (R) invariants of the velocity gradient tensor for a given flow field, see [27], and provide an interesting statistical representation of turbulent kinematics. The 'teardrop' profile shown for the Roe case in Fig. 3 is also observed in several different turbulent flows and is regarded as one of the universal aspects of turbulent motions [28]. On the other hand, LxF profiles seemed to favour a more symmetrical distribution of kinematic states, consistent with the equipartition scenario expected from the bump-related scales.

Finally, we suggest that the lack of stability observed for higher-order simulations (especially for LxF) might be related to the sharper dissipative characteristics discussed above. This is because, the sharper the dissipation, the more the conservative dynamics will tend to overcome the dissipative one, which is expected from the Navier-Stokes equations in the limit of infinite Reynolds number (the one LES schemes should follow). It so happens that the exact (conservative) solution of the inviscid TGV problem might in fact exhibit singularities during the transitional phase, although this is still under debate in the literature [29].

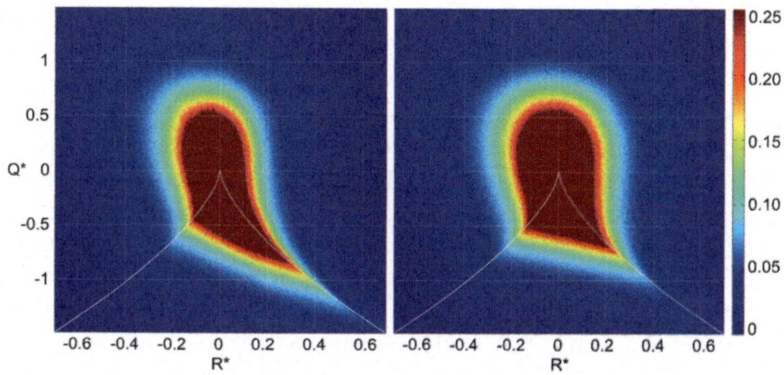

Fig. 3 QR diagrams at $t = 9$ (dissipation peak) obtained with Roe (*left*) and LxF (*right*), from case $m = 5$, $n_{el} = 23$. The *dark red colour* has been assigned to values above 1/4. The *white curve* separates rotational states (*above the curve*) from those without rotation (*under the curve*), cf. [27]

7 Concluding Remarks

The present study proposed a formal LES setting for DG-based implicit LES (iLES). This framework has been devised by analogy with the LES-PLB methodology proposed by Pope in [9], where PLB stands for 'projection onto local basis'. Through this analogy, the DG-iLES formulation was shown to be equivalent to a standard Galerkin solution of the compressible Navier-Stokes equations with extra terms related to DG's implicit turbulence model. Subsequently, we demonstrated that, for isotropic turbulence at very high Reynolds, the dissipation of 'solution energy' is primarily determined by the property-jump term of the Riemann flux employed.

In order to analyse how different fluxes performed in wavenumber space, a comprehensive set of simulations of the inviscid Taylor-Green vortex problem was assessed and their energy spectra was compared. Results showed that more sophisticated solvers (Roe and HLLC) have a better performance in terms of robustness and solution quality, yielding results very similar to those obtained with the exact Riemann solver. On the other hand, simpler fluxes (Lax-Friedrichs and HLL) showed poor accuracy and less robustness, owing to a larger number of crashes among the test cases. These results are probably also valid for the DG variants of flux reconstruction methods, given the strong connection between the two schemes [30, 31].

The main accuracy issue had to do with an 'energy bump' observed before the dissipation range of the spectra. Those have been explained in connection to sharp dissipative characteristics in wavenumber space as estimated from linear dispersion-diffusion analysis. For DG, the sharpness of the dissipation increases with the discretization order m and with the amount of upwinding. Therefore, discretizations of very high order (say, $m > 6$) have been discouraged, especially for more simplistic fluxes which do not account correctly for all the wave-speeds of the compressible formulation. For the Lax-Friedrichs solver, we showed how its flux displayed a significant over-upwind bias for the momentum equations at low Mach numbers, which resulted in strong energy bumps.

A cause of crash has also been suggested in connection to these sharp dissipation characteristics: they might be causing the inviscid TGV solution to partially follow the conservative rather than the dissipative, entropy-consistent behaviour expected in the limit of infinite Reynolds number. It has long been conjectured (but not yet proved) that the exact conservative evolution of the Taylor-Green flow might develop finite-time singularities leading to the actual collapse of the solution. As the conservative behaviour is followed in a 'truncated' Fourier solution with limited number of modes, it is believed that a very sharp dissipation might induce this behaviour to manifest at least partially. The crashes observed also highlighted that standard DG discretisations, even with consistent/over-integration, might in fact lack robustness for implicit LES at very high Reynolds numbers. This should serve as a motivation for the development (and adoption) of more robust DG formulations.

References

1. S. Kanner, P.O. Persson, Validation of a high-order large-eddy simulation solver using a vertical-axis wind turbine. AIAA J. **54**(1), 101–112 (2015)
2. A.D. Beck, T. Bolemann, D. Flad, H. Frank et al., High-order discontinuous Galerkin spectral element methods for transitional and turbulent flow simulations. Int. J. Numer. Methods Fluids **76**(8), 522–548 (2014)
3. A. Uranga, P.O. Persson, M. Drela, J. Peraire, Implicit large eddy simulation of transition to turbulence at low Reynolds numbers using a discontinuous Galerkin method. Int. J. Numer. Methods Eng. **87**(1–5), 232–261 (2011)
4. F.F. Grinstein, L.G. Margolin, W.J. Rider, *Implicit Large Eddy Simulation: Computing Turbulent Fluid Dynamics* (Cambridge University Press, Cambridge, 2007)
5. L.G. Margolin, W.J. Rider, A rationale for implicit turbulence modelling. Int. J. Numer. Methods Fluids **39**(9), 821–841 (2002)
6. R.C. Moura, S.J. Sherwin, J. Peiró, Modified equation analysis for the discontinuous Galerkin formulation. in *Spectral and High Order Methods for Partial Differential Equations*, ed. by R.M. Kirby, M. Berzins, J.S. Hesthaven (Springer, Berlin, 2015), pp. 375–383
7. R.C. Moura, S.J. Sherwin, J. Peiró, Linear dispersion-diffusion analysis and its application to under-resolved turbulence simulations using discontinuous Galerkin spectral/hp methods. J. Comput. Phys. **298**, 695–710 (2015)
8. R.C. Moura, G. Mengaldo, J. Peiró, S.J. Sherwin, On the eddy-resolving capability of high-order discontinuous Galerkin approaches to implicit LES/under-resolved DNS of Euler turbulence. J. Comput. Phys. **330**, 615–623 (2017)
9. S. Pope, Large-eddy simulation using projection onto local basis functions, in *Fluid Mechanics and the Environment: Dynamical Approaches*, ed. by J. Lumley (Springer, Berlin, 2001), pp. 239–265
10. G.I. Taylor, A.E. Green, Mechanism of the production of small eddies from large ones. Proc. R. Soc. Lond. A. **158**(895), 499–521 (1937)
11. C.W. Shu, W.S. Don, D. Gottlieb, O. Schilling et al., Numerical convergence study of nearly incompressible, inviscid Taylor-Green vortex flow. J. Sci. Comput. **24**(1), 1–27 (2005)

12. D.N. Arnold, F. Brezzi, B. Cockburn, L.D. Marini, Unified analysis of discontinuous Galerkin methods for elliptic problems. SIAM J. Numer. Anal. **93**(5), 1749–1779 (2002)
13. D.A. Kopriva, G. Gassner, On the quadrature and weak form choices in collocation type discontinuous Galerkin spectral element methods. J. Sci. Comput. **44**(2), 136–155 (2010)
14. B. Van Leer, J.L. Thomas, P.L. Roe, R.W. Newsome, A comparison of numerical flux formulas for the Euler and Navier-Stokes equations. AIAA Paper 1987–1104 (1987)
15. E.F. Toro, *Riemann Solvers and Numerical Methods for Fluid Dynamics* (Springer, Berlin 1999)
16. M.E. Brachet, D.I. Meiron, S.A. Orszag, B.G. Nickel et al., Small-scale structure of the Taylor-Green vortex. J. Fluid Mech. **130**, 411–452 (1983)
17. P.L. Roe, Approximate Riemann solvers, parameter vectors, and difference schemes. J. Comput. Phys. **43/2**, 357–372 (1981)
18. V.V. Rusanov, Calculation of interaction of non-steady shock waves with obstacles. USSR J. Comput. Math. Phys. **1**, 267–279 (1961)
19. C.D. Cantwell, D. Moxey, A. Comerford, A. Bolis et al., Nektar++: an open-source spectral/hp element framework. Comput. Phys. Commun. **192**, 205–219 (2015)
20. G. Mengaldo, D. de Grazia, D. Moxey, P.E. Vincent, S.J. Sherwin, Dealiasing techniques for high-order spectral element methods on regular and irregular grids. J. Comput. Phys. **299**, 56–81 (2015)
21. G. Falkovich, Bottleneck phenomenon in developed turbulence. Phys. Fluids **6**(4), 1411 (1994)
22. M. Coantic, J. Lasserre, On pre-dissipative 'bumps' and a Reynolds-number-dependent spectral parameterization of turbulence. Eur. J. Mech. B **18**(6), 1027–1047 (1999)
23. A.G. Lamorgese, D.A. Caughey, S.B. Pope, Direct numerical simulation of homogeneous turbulence with hyperviscosity. Phys. Fluids **17**(1), 015106 (2005)
24. U. Frisch, S. Kurien, R. Pandit, W. Pauls et al., Hyperviscosity, Galerkin truncation, and bottlenecks in turbulence. Phys. Rev. Lett. **101**(14), 144501 (2008)
25. C. Cichowlas, P. Bonaïti, F. Debbasch, M. Brachet, Effective dissipation and turbulence in spectrally truncated Euler flows. Phys. Rev. Lett. **95**(26), 264502 (2005)
26. D. Banerjee, S.S. Ray, Transition from dissipative to conservative dynamics in equations of hydrodynamics. Phys. Rev. E **90**(4), 041001 (2014)
27. M.S. Chong, A.E. Perry, B.J. Cantwell, A general classification of three-dimensional flow fields. Phys. Fluids A **2**(5), 765–777 (1990)
28. A. Tsinober, *An Informal Conceptual Introduction to Turbulence* (Springer, Berlin, 2009)
29. J.D. Gibbon, The three-dimensional Euler equations: where do we stand? Physica D **237**(14), 1894–1904 (2008)
30. D. De Grazia, G. Mengaldo, D. Moxey, P.E. Vincent, S.J. Sherwin, Connections between the discontinuous Galerkin method and high-order flux reconstruction schemes. Int. J. Numer. Methods Fluids **75**(12), 860–877 (2014)
31. G. Mengaldo, D. Grazia, P.E. Vincent, S.J. Sherwin, On the connections between discontinuous Galerkin and flux reconstruction schemes: extension to curvilinear meshes. J. Sci. Comput. **67**(3), 1272–1292 (2016)

On the Scaling of Entropy Viscosity in High Order Methods

Adeline Kornelus and Daniel Appelö

Abstract In this work, we outline the entropy viscosity method and discuss how the choice of scaling influences the size of viscosity for a simple shock problem. We present examples to illustrate the performance of the entropy viscosity method under two distinct scalings.

1 Introduction

Hyperbolic partial differential equations (PDE) are used to model various fluid flow problems. In the special case of 1-dimensional linear constant coefficient scalar hyperbolic problems, the solutions to these PDE are simply a translation of the initial data. However, for nonlinear problems the solution may deform, and as a result, shock waves can form even if the initial data is smooth [12].

In computational fluid dynamics, it is desirable that numerical methods capture shock waves and maintain a high accuracy for smooth waves. Low order methods have sufficient numerical dissipation to regularize shock waves but obtaining accurate solutions in smooth regions can become expensive. On the other hand, high order methods are capable of achieving high accuracy at a reasonable cost. Their low numerical dissipation enables such accuracy, but on the downside, it limits their ability to regularize shock waves.

Various techniques have been implemented to capture shocks while maintaining high accuracy, at least away from shocks. There are two major classes of shock capturing techniques: shock detection techniques, where we find slope limiters [12], Essentially Non-Oscillatory (ENO) and Weighted ENO (WENO) [14], and artificial

A. Kornelus (✉)
The School of Mathematical and Statistical Sciences, Arizona State University, Tempe, AZ 85287, USA
e-mail: adeline.kornelus@asu.edu

D. Appelö
The Department of Applied Mathematics, University of Colorado, Boulder, CO 80309, USA
e-mail: daniel.appelo@colorado.edu

M.L. Bittencourt et al. (eds.), *Spectral and High Order Methods for Partial Differential Equations ICOSAHOM 2016*, Lecture Notes in Computational Science and Engineering 119, DOI 10.1007/978-3-319-65870-4_11

viscosity techniques, where we find filtering [13, 15], the PDE-based viscosity method [9], the entropy viscosity method [4], among others.

In this work, we focus on the entropy viscosity method. In essence, the entropy viscosity method provides shock capturing without compromising the high accuracy away from the shock. An important advantage of this method is that it generalizes very easily to higher dimensions and unstructured grids.

As a model problem, we consider Burgers' equation

$$u_t + f(u)_x = 0, \tag{1}$$

where $f = \frac{u^2}{2}$. Physically correct solutions to (1) can be singled out by requiring that they satisfy an entropy inequality such as

$$r_{EV} = E_t + F_x \equiv \left(\frac{u^2}{2}\right)_t + \left(\frac{u^3}{3}\right)_x \leq 0. \tag{2}$$

The entropy residual, r_{EV}, is zero wherever u is smooth. If the solution u contains a shock, then the entropy residual takes the form of a negative Dirac distribution centered at the location of the shock, x_s, i.e. $r_{EV} = -C\delta(x - x_s)$. The property that the entropy residual is unbounded at a shock was first used by Guermond and Pasquetti in [4], as a way to selectively introduce viscosity. The artificial viscosity, ν, proposed in [4], defined as the minima of two viscosities

$$\nu = \min(\nu_{\max}, \nu_{EV}), \tag{3}$$

becomes the coefficient of the viscous term in the viscous Burgers' equation,

$$u_t + f(u)_x = (\nu u_x)_x. \tag{4}$$

Here, $\nu_{\max}$ is the Lax-Friedrich viscosity whose size depends on discretization and the largest eigenvalue, λ_{LF}, of the flux Jacobian, $\frac{Df(u)}{Du}$. The second viscosity ν_{EV} is proportional to the magnitude of the entropy residual (in fact, a discretization of the entropy residual) and will thus be zero (or small after discretization) away from discontinuities. In theory, the entropy residual becomes unbounded at a shock, numerically however, the entropy residual r_{EV} remains bounded with the size of the residual depending on the discretization size. As we will see below, this subtle difference has consequences for how to choose the scaling of the viscosity terms in the entropy viscosity method.

On a grid with step size h, the second viscosity ν_{EV} can be expressed as

$$\nu_{EV}(x) = \alpha_{EV} h^{\beta} |r_{EV}(x)|, \tag{5}$$

with a parameter α_{EV} that requires tuning. In recent papers on entropy viscosity method, see e.g. [3, 5–7, 16], the parameter β is chosen to be 2, but the original paper [2] uses $\beta = 1$. It is unclear to us why the later works prefer $\beta = 2$. Here, we will present analysis and computational results that suggest the original scaling $\beta = 1$ is a more natural choice. We note that the entropy residual is typically scaled by $\|E - \overline{E}\|_\infty$, with the over-bar indicating a spatial average, but as this quantity is roughly constant in the problems presented here, we omit it for brevity and reduced complexity.

The rest of the paper is organized as follows. In Sect. 2, we describe different discretizations of (4) that we consider here, in Sect. 3, we present an analysis of how the entropy viscosity ν depends on the two viscosities, ν_{EV} and $\nu_{\max}$, under different scaling for a model problem. In Sect. 4, we then conduct experiments with the entropy viscosity method where β takes on values 1 or 2 and compare the results.

2 Numerical Methods

We will consider the discretization of (4) by our conservative Hermite method [11], a standard discontinuous Galerkin (dG) method [8] and a simple finite volume type discretization [12]. For all the discretizations we let the domain $x_L \leq x \leq x_R$ be discretized by the regular grid $x_i = x_L + i\,h,\ i = 0, \ldots, n,\ h = (x_R - x_L)/n$.

The degrees of freedom for the finite volume method are cell averages centered at the grid points. For the Hermite method, the degrees of freedom are the coefficients of node centered Taylor polynomials of degree m and for the dG method, they are the $(m + 1)$ coefficients of element-wise (we take an element to be $\Omega_i = [x_{i-1}, x_i]$) expansions in Legendre polynomials. For smooth solutions the spatial accuracy of the Hermite method is $2m + 1$ and $m + 1$ for the dG method.

All three methods use the classic fourth order Runge Kutta method to evolve the semi-discretizations in a method-of-lines fashion.

In the Hermite method, we evaluate the fluxes and their derivatives at the nodes (element edges) for the four stages in the RK method. Precisely, for the first stage we compute the slope $f_1^h = \frac{1}{2}\mathcal{T}[(u_1^h)^2] - \frac{\nu}{h}\frac{du_1^h}{dx}$ for the Taylor polynomial $u_1^h = u^h$ approximating the solution at the first stage. Here $\mathcal{T}[(u_1^h)^2]$ is the truncated polynomial multiplication of u_1^h with itself and $\frac{du_1^h}{dx}$ is the derivative of the polynomial. At the next stage, the solution is $u_2^h = u^h + \frac{(\Delta t/2)}{2}\frac{df_1^h}{dx}$, the slope is $f_2^h = \frac{1}{2}\mathcal{T}[(u_2^h)^2] - \frac{\nu}{h}\frac{du_2^h}{dx}$ and so on. Once the stage slopes $f_s^h,\ s = 1, \ldots, 4$ and their spatial derivatives are known, we perform a Hermite interpolation to the element centers of the solution and the spatial derivatives of the stage slopes. These are then used to evolve the element centered Hermite interpolant of u^h to $t = t_n + \Delta t/2$. As the Hermite interpolant is of higher degree than the original Taylor polynomial, we conclude a half-step by truncating it to the appropriate degree. To advance the solution a full time step, the half-step process is repeated starting from the element centers.

To handle the artificial viscosity in the dG method, we use the approach of Bassi and Rebay [1] with a Lax-Friedrichs flux for the advective term and alternating fluxes for the viscous term. The nonlinear terms are constructed explicitly and de-aliased by over-integration [10].

For the finite volume method, we let $u_i \approx u(x_i)$ be a grid function approximating the solution and $f_{i+\frac{1}{2}} = f_{i+\frac{1}{2}}(u_i, u_{i+1})$ be an approximation to the flux at $x_{i+\frac{1}{2}}$. To compute the time derivatives, we use the spatial approximation

$$\frac{du_i}{dt} \approx \frac{f_{i+\frac{1}{2}} - f_{i-\frac{1}{2}}}{h}, \tag{6}$$

where

$$f_{i+\frac{1}{2}}(u_i, u_{i+1}) = \frac{1}{2}\left(\frac{u_i + u_{i+1}}{2}\right)^2 - \left(\frac{\nu_i + \nu_{i+1}}{2}\right)\frac{u_{i+1} - u_i}{h}. \tag{7}$$

When $\nu_i = 0$, the above discretization is linearly stable (when paired with a suitable time-stepping method) but is not non-linearly stable, and we thus add artificial viscosity to stabilize it.

For all three discretizations, we approximate the time derivative of the entropy function, E_t, by a backward difference. This approach is explicit as we use the current solution to compute E at the current time before evolving the solution in time. The residual (and hence the viscosity) is kept on each element / grid-point over each step.

To approximate the entropy flux derivative F_x using the Hermite method, we compute the derivative of the truncated polynomial multiplication $\mathcal{T}[u^h \mathcal{T}[(u^h)^2]]$ at the node. For the dG method, we evaluate the flux F on a Legendre-Gauss-Lobatto (LGL) grid and differentiate it to get an approximation for F_x. The residual on an element is taken to be the maximum of the absolute value of the residual on the LGL grid. In the finite volume method F_x is approximated by

$$\frac{dF_i}{dx} = \frac{F_{i+\frac{1}{2}} - F_{i-\frac{1}{2}}}{h}, \text{ where, } F_{i+\frac{1}{2}} = \frac{1}{3}\left(\frac{u_i + u_{i+1}}{2}\right)^3.$$

We note that more sophisticated discretizations of the entropy residual could be considered. In particular, a higher order approximation to r_{EV} would result in a higher rate of convergence for smooth solutions, but as we are mainly concerned with the scaling β, we did not pursue such discretizations here. In fact, in our experience, the results concerning the choice of scaling are not affected by the order of the accuracy of the approximation to r_{EV}. This will be discussed in Sect. 3.

We also define $\nu_{\max}$ to be the classical Lax-Friedrich viscosity, which for Burgers' equation takes the form

$$\nu_{\max} = \alpha_{\max} h \max |u|, \tag{8}$$

where the maximum is taken globally.

Finally, for the purpose of comparison we also present some results computed using the sub-cell resolution smoothness sensor of Persson and Peraire, [13]. The smoothness sensor compares the L_2 energy content of the highest (Fourier or expansion) mode with the total L_2 energy on an element and then maps its ratio (which is an indicator of the smoothness) into the size of the artificial viscosity. Precisely, if the approximate dG solution on an element is $u^h = \sum_{k=0}^{m} \hat{u}_k P_k$, with P_k being an orthogonal basis, we compute the smoothness as $s = \log_{10}(\|\hat{u}_m P_m\|^2 / \|u^h\|^2)$ and the viscosity as

$$\nu = \begin{cases} 0 & s < s_0 - \kappa, \\ \varepsilon_0 h & s > s_0 + \kappa, \\ \frac{\varepsilon_0 h}{2}\left(1 + \sin\left(\frac{\pi(s-s_0)}{2\kappa}\right)\right) & \text{otherwise.} \end{cases}$$

When applied to the Hermite method, we first project the Taylor polynomials centered at two adjacent grid-points into an orthogonal Legendre expansion on the element defined by the grid-points and then proceed as above.

3 Impact of the h-Scaling on the Selection Mechanism

To study how the selection mechanism depends on the shock speed and the size of the jump, consider a solution of the Burgers' equation consisting of a Heaviside function H with left state u_l and right state u_r, given by

$$u(x,t) = u_l + \Delta u \, H\,(x - v_s t)\,. \tag{9}$$

This corresponds to a shock of size $|\Delta u| = |u_r - u_l|$ moving with speed $v_s = 0.5(u_l + u_r)$. Solutions of the form (9) always has a negative Δu value since Lax entropy condition for Burgers' equation dictates $u_l = f'(u_l) > v_s > f'(u_r) = u_r$.

For simplicity, we use the short hand notation H for $H\,(x - v_s t)$. A direct computation

$$\begin{aligned} u_t + \left(\frac{u^2}{2}\right)_x &= \left(-\frac{u_l + u_r}{2}(\Delta u)H'\right) + \left((\Delta u)u_l H' + \frac{(\Delta u)^2}{2} H'\right) \\ &= -\Delta u \left(\frac{2u_l + \Delta u}{2}\right) H' + \Delta u \left(\frac{2u_l + \Delta u}{2}\right) H' \\ &= 0, \end{aligned}$$

shows that (9) is a solution of (1). Further, it can be shown that the entropy residual (2) for (9) is

$$r_{EV} = \frac{(\Delta u)^3}{12} H'(x - x_s) = \frac{(\Delta u)^3}{12} \delta(x - x_s). \tag{10}$$

That is, the size of the entropy residual grows with the cube of Δu.

Now, by the properties that define the Dirac delta function δ, we have

$$\int_{-\infty}^{\infty} \delta(x) dx = 1. \tag{11}$$

Thus, a consistent discretization of the Dirac delta function $\delta_0, \ldots, \delta_n$ on a grid $x_0, \ldots, x_n$ must obey the condition

$$\sum_{j=0}^{n-1} h_j \delta_j = 1, \tag{12}$$

where $h_j = x_{j+1} - x_j$. For any approximation with a finite width stencil, we must have $\delta_j \sim h_j^{-1}$ and we thus expect r_{EV} to behave like $(\Delta u)^3/h$ on a uniform grid. We therefore proceed with the analysis using the discrete approximation $r_{EV} = (\Delta u)^3/h$. Using this expression for r_{EV}, we estimate the viscosity ν by the minimum of

$$\nu_{EV} = \alpha_{EV} h^{\beta-1} |(\Delta u)^3| \text{ and } \nu_{\max} = \alpha_{\max} h \max(|u_l|, |u_r|). \tag{13}$$

The comparison between the size of ν_{EV} and $\nu_{\max}$ in various scenarios is reported in Table 1. If $\beta = 2$, then the two viscosities ν_{EV} and $\nu_{\max}$ scale as h. For a problem with multiple shocks, the homogeneity in h-scaling introduces an additional difficulty in determining α_{EV}. Should it be chosen based on the largest or smallest shock? What if new shocks appear during the course of the computation? To avoid answering these questions, we instead consider $\beta = 1$. Now $\nu_{EV} = \mathcal{O}(1)$ while $\nu_{\max} = \mathcal{O}(h)$, and the particular choice of α_{EV} is thus irrelevant since as $h \to 0$, the selection mechanism will eventually select $\nu_{\max}$ at the shocks. We will provide an example to illustrate the two-shock dilemma in Sect. 4.3.

Table 1 Size of μ_E and $\mu_{\max}$ for different size of shock speed (v_s) with respect to the size of the jump (Δu) in the entropy viscosity method

Case	ν_{EV}	$\nu_{\max}$
$\lvert\mathbf{v_s}\rvert \ll \lvert\boldsymbol{\Delta}\mathbf{u}\rvert$	$\alpha_{EV} h^{\beta-1} \lvert\Delta u\rvert^3$	$\alpha_{\max} h \lvert v_s\rvert$
$\lvert\mathbf{v_s}\rvert \approx \lvert\boldsymbol{\Delta}\mathbf{u}\rvert$	$\alpha_{EV} h^{\beta-1} \lvert\Delta u\rvert^3$	$2\alpha_{\max} h \lvert v_s\rvert$
$\lvert\mathbf{v_s}\rvert \gg \lvert\boldsymbol{\Delta}\mathbf{u}\rvert$	$\alpha_{EV} h^{\beta-1} \lvert\Delta u\rvert^3$	$0.5\alpha_{\max} h \lvert\Delta u\rvert$

4 Experiments

In this section, we describe the experiments and present a convergence study in L_2 norms, and also study the effects of the scaling in the entropy viscosity method on the convergence under grid refinement. For all the examples we solve Burgers' equation and vary the initial data. In each problem, we report the L_2-errors (the L_1-errors behaves quantitatively similar).

The solutions are obtained using the following methods: H1 and H2 refer to Hermite-entropy viscosity method for $\beta = 1$ and $\beta = 2$ respectively, DG1 and DG2 refer to dG-entropy viscosity method for $\beta = 1$ and $\beta = 2$ respectively, FV1 and FV2 refer to finite volume-entropy viscosity method with $\beta = 1$ and $\beta = 2$ respectively, DGP and HP refer to dG and Hermite method with smoothness sensor respectively.

The size of the time step is chosen close to the stability limit, which in the cases considered here results in the error being dominated by the spatial discretization.

4.1 A Single Shock

In this example, we compute the solution to (1) on the domain $D = [-1, 1]$ with the initial data imposed as the exact solution

$$u(x,t) = \begin{cases} -0.5 + v_s, & x \in [-1, v_s t), \\ 0.5 + v_s, & x \in [v_s t, 1], \end{cases} \tag{14}$$

at time $t = 0$. Here v_s is the shock speed which we choose to be either $v_s = 0$ corresponding to a stationary shock or $v_s = 0.1$ corresponding to a moving shock.

We solve until time $t = 1$ for the two different shock speeds and perform a grid refinement study using a dG method of order 5, a Hermite method of order 9, and the Finite Volume method, all using the classical fourth order Runge-Kutta time stepping. For the Hermite method, we fix $(\max|u|)\Delta t/h = 0.3$, for the dG method, the time step is set as $\Delta t/h = 0.0625$ and for the Finite Volume method, the time step is set according to $(\max|u|)\Delta t/h = 0.9$.

The L_2 norm of errors in the numerical solution u_h are plotted against the different grid sizes for different methods, see Fig. 1. In the stationary shock experiment, FV1 and FV2 use $(\alpha_{EV}, \alpha_{\max}) = (0.7, 0.5)$ and $(10, 0.5)$ respectively, DG1 and DG2 use $(\alpha_{EV}, \alpha_{\max}) = (1, 0.25)$ and $(10, 0.25)$ respectively, H1 and H2 use $(\alpha_{EV}, \alpha_{\max}) = (1, 0.4)$ and $(10, 0.4)$ respectively, DGP and HP use $(s_0, \kappa, \epsilon_0) = (-1, 2, 0.5)$ and $(\log_{10}(1/256), 1, 0.125)$ respectively.

The parameters for moving shock experiment are $(\alpha_{EV}, \alpha_{\max}) = (0.7, 0.5)$ and $(10, 0.5)$ for FV1 and FV2 respectively, $(\alpha_{EV}, \alpha_{\max}) = (1, 0.25)$ and $(10, 0.25)$ for DG1 and DG2 respectively, $(\alpha_{EV}, \alpha_{\max}) = (1, 0.4)$ and $(10, 0.4)$ for H1 and H2

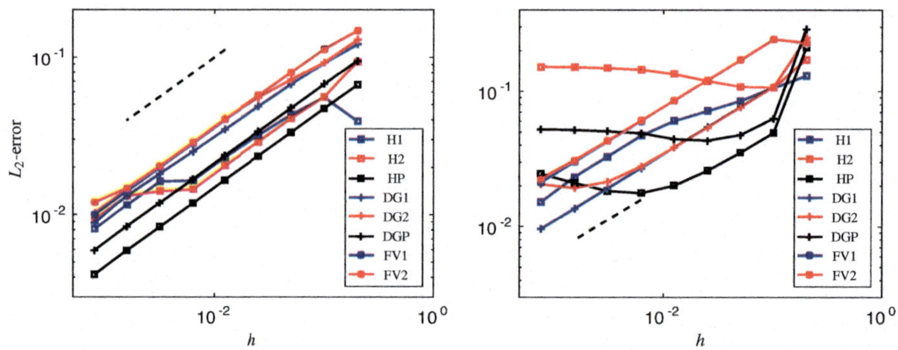

Fig. 1 Convergence of the different methods for stationary (*left*) and moving (*right*) shocks

respectively, $(s_0, \kappa, \epsilon_0) = (2\log_{10}(1/256), 1, 0.5)$ and $(\log_{10}(1/256), 1, 0.125)$ for DGP and HP respectively.

To the left in Fig. 1, we display convergence results for the stationary shock. In this case, the results indicate that all methods produce convergent solutions with roughly the same rates of convergence. The rate of convergence is limited by the smoothness of the solution but as can be seen in the same figure, the error levels are lower for the higher order methods. It is interesting to note that the smallest errors are observed for the computations using the smoothness-based sensor.

The results for the moving shock, displayed to the right in Fig. 1, are quite different. Now, for the high order methods, we observe convergence only when we use the entropy viscosity with $\beta = 1$. When we use the entropy viscosity with $\beta = 2$ or when we use the smoothness based sensor, the errors clearly saturate as the grid is refined. The errors for the low order Finite Volume method are still reduced with the grid size, independent of the scaling in the entropy viscosity method.

To understand why the convergence results obtained with $\beta = 1$ and $\beta = 2$ in the moving shock example do not agree, we study where the Lax-Friedrich viscosity $\nu_{\max}$ is activated in the vicinity of the shock. We know that when the viscosity is chosen to be just the Lax-Friedrich type viscosity, then under a suitable Courant number, the solution will converge to the correct vanishing viscosity solution of the conservation law [12].

It seems that the Lax-Friedrich viscosity is necessary in some neighborhood of the shock, and the size of this neighborhood becomes an important factor in the convergence of the solution to the moving shock problem. In Fig. 2, we plot the average (in time) of the number of elements n_m which use the Lax-Friedrich viscosity $\nu_{\max}$ as a function of total number of elements n for the stationary shock (left) and for the moving shock (right). We see that n_m is roughly constant for both $\beta = 1$ and $\beta = 2$ in the stationary shock. In the moving shock problem, n_m stays constant for $\beta = 2$ as in the stationary shock, but grows slowly for $\beta = 1$ (note the log-scale). While the growth in n_m is irrelevant in the convergence in the stationary

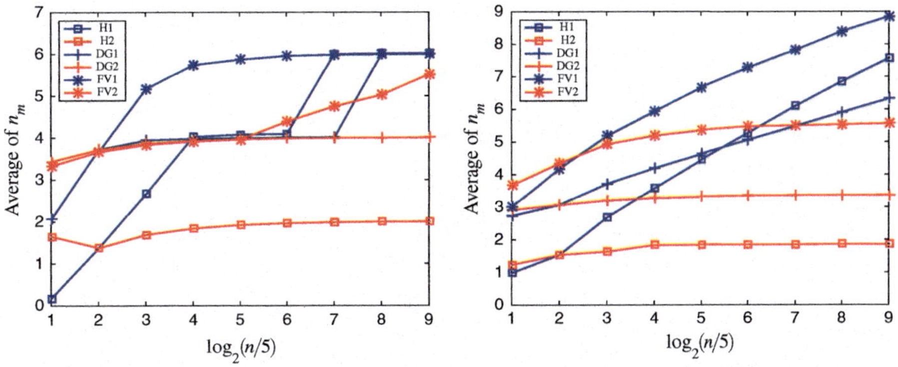

Fig. 2 Average in time of n_m, number of elements using Lax-Friedrich viscosity $\nu_{\max}$, versus the number of elements (n). *Left*: stationary shock, *right*: moving shock

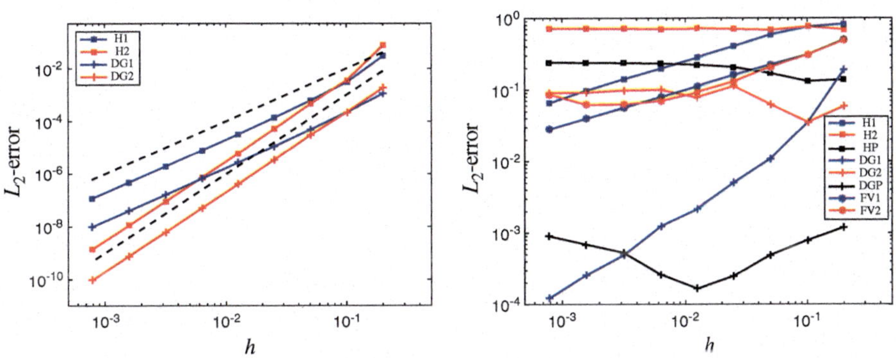

Fig. 3 Convergence of the different methods for a smooth initial data, *left*: before shock forms, *right*: after shock forms. The *dashed lines* are h^2 and h^3

shock example, it seems to play an important role in determining the convergence in the moving shock example.

4.2 *Sinusoidal to* N *Wave*

Next, we consider the smooth 2-periodic initial data

$$u(x, 0) = -\sin(\pi x) + 0.5, \tag{15}$$

which develops into a single N wave.

In Fig. 3, we present the L_2 norm of the errors at $t = 0.1$ before the shock forms (left) and at $t = 1$ after the shock forms (right). The spatial and temporal discretization of the PDE itself is performed with a high order method, so rate of convergence that we observe in Fig. 3 is limited by either the discretization of the artificial viscosity or the smoothness of the solution, whichever is more restrictive.

For this N-wave experiment, $FV1$ and $FV2$ use $(\alpha_{EV}, \alpha_{\max}) = (2, 0.5)$ and $(20, 0.5)$ respectively, $DG1$ and $DG2$ use $(\alpha_{EV}, \alpha_{\max}) = (0.1, 0.125)$ and $(1, 0.125)$ respectively, $H1$ and $H2$ use $(\alpha_{EV}, \alpha_{\max}) = (0.4, 0.4)$ and $(5, 0.4)$, DGP and HP use $(s_0, \kappa, \epsilon_0) = (2\log_{10}(1/256), 2, 0.05)$ and $(\log_{10}(1/256), 1, 0.125)$ respectively.

The discretization of the entropy residual r_{EV} is only first order due to the use of backward-Euler, so we expect the entropy-based viscosity ν_{EV} to be $(\beta + 1)$th accurate, i.e. second order when $\beta = 1$ or third order when $\beta = 2$. This analysis agrees with the convergence plot to the left in Fig. 3. To the right, we observe the same phenomena as in the moving shock example described in Sect. 4.1. We also note that the shock in this sinusoidal wave is also moving.

4.3 *Shocks of Different Size*

To complement the analysis in Sect. 3, we next consider a problem with a big shock and a small shock on the same simulation. According to the analysis, the entropy viscosity will capture the small shock when $\beta = 1$, but not when $\beta = 2$. In this setup, we start with an existing shock of size $\Delta u_1 = 0.5$ and a small sinusoidal wave that develops into an N-wave of size $\Delta u_2 = 0.2$. Thus, we consider Burgers' equation on $[-1, 5]$ with initial data

$$u(x,0) = \begin{cases} 0 & x \in [-1, -0.5), \\ -0.1\sin(2\pi x) & x \in [-0.5, 0.5), \\ 0 & x \in [0.5, 4.5), \\ -0.5 & x \in [4.5, 5], \end{cases} \tag{16}$$

and fixed boundary condition $u(-1, t) = 0$ and $u(5, t) = -0.5$.

The solution initially consists of a shock and a smooth sine wave, which are placed far away from each other so they never interact. Over time, the sinusoidal wave develops into a N-wave. In Fig. 5, we present the numerical solutions at time $t = 2$ for different grid resolutions, obtained with a Hermite method of order 9 and dG method of order 5. In these plots, we can see that the shock is resolved for both values of β, however, the N-wave comes with some overshoots when $\beta = 2$ for all the finer grid resolutions, see Fig. 5.

For this two-shock experiment, $DG1$ and $DG2$ use $(\alpha_{EV}, \alpha_{\max}) = (0.5, 0.25)$ and $(10, 0.25)$ respectively, $H1$ and $H2$ use $(\alpha_{EV}, \alpha_{\max}) = (1, 0.125)$ and $(50, 0.125)$ respectively.

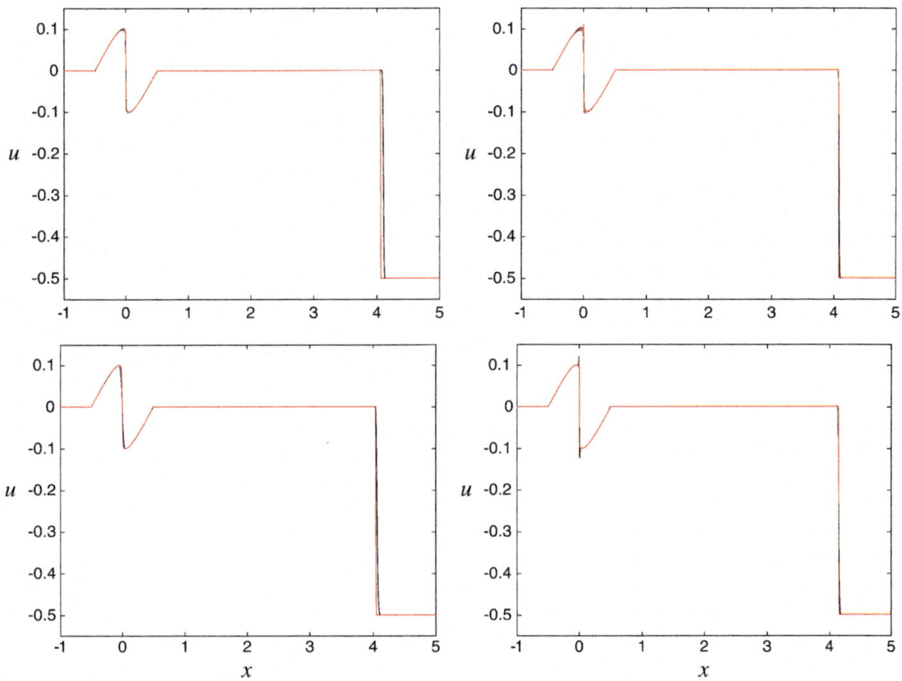

Fig. 4 Effect of the choice of scaling on a small perturbation near a larger shock. The results in the *left and right column* are for $\beta = 1$ and $\beta = 2$ respectively. The *upper figures* display the results for the dG method and the *lower figures* display the results for the Hermite method. The *black curve* is for a simulation using 320 elements and the black uses 2560

Because the magnitude of this N-wave is small, the entropy residual at the N-wave is relatively small compared to that at the existing shock. On one hand, $\beta = 1$ results are free refined, but $\beta = 2$ results do have overshoots, see Figs. 4 and 5.

5 Conclusion

In summary, we have performed a convergence study for Burgers' equation with various initial data. We demonstrated that the entropy viscosity method with $\beta = 2$ does not produce convergent results (fixing the parameters α_{EV} and $\alpha_{\max}$) in the cases where the shock is moving or more than one shock is present. Therefore, we recommend readers to use $\beta = 1$; to achieve desired accuracy or better rate of convergence, use a higher order approximation of the residual.

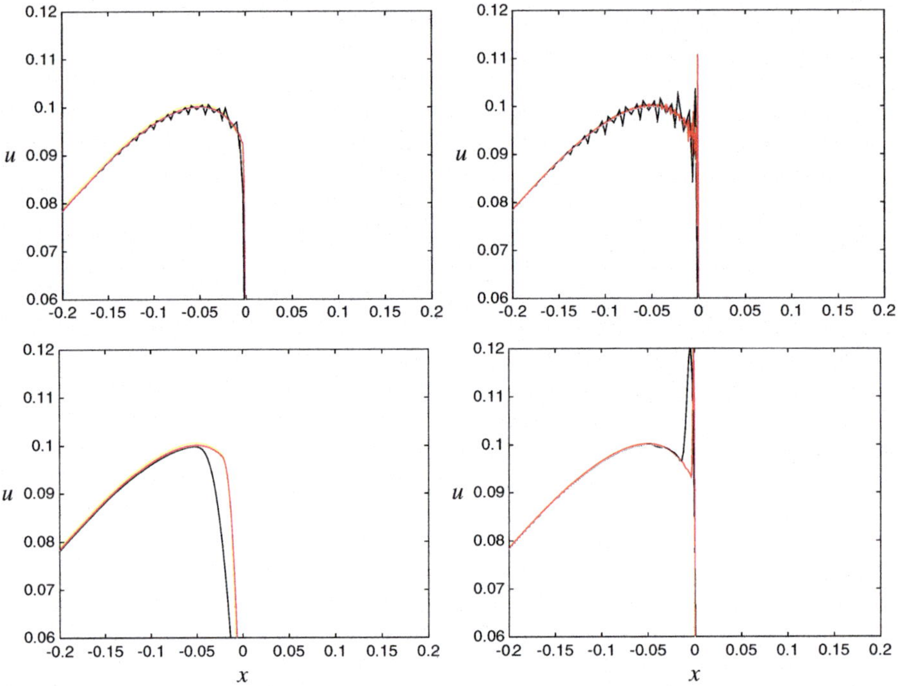

Fig. 5 Effect of the choice of scaling on a small perturbation near a larger shock. Same as in Fig. 4 but zoomed in

Acknowledgements Adeline Kornelus and Daniel Appelö are supported in part by NSF Grant DMS-1319054. Any conclusions or recommendations expressed in this paper are those of the author and do not necessarily reflect the views NSF.

References

1. F. Bassi, S. Rebay, A high-order accurate discontinuous finite element method for the numerical solution of the compressible Navier–Stokes equations. J. Comput. Phys. **131**, 267–279 (1997)
2. J.-L. Guermond, R. Pasquetti, Entropy-based nonlinear viscosity for Fourier approximations of conservation laws. C. R. Math. **346**, 801–806 (2008)
3. J.-L. Guermond, R. Pasquetti, Entropy viscosity method for high-order approximations of conservation laws, in *Spectral and High Order Methods for Partial Differential Equations* (Springer, Berlin, 2011), pp. 411–418
4. J.-L. Guermond, R. Pasquetti, Entropy viscosity method for higher-order approximations of conservation laws, in *Spectral and High Order Methods for Partial Differential Equations*. Lecture Notes in Computational Science and Engineering, vol. 76 (Springer, Berlin, 2011), pp. 411–418
5. J.-L. Guermond, R. Pasquetti, B. Popov, Entropy viscosity for conservation equations, in *V European Conference on Computational Fluid Dynamics (Eccomas CFD 2010)* (2010)

6. J.-L. Guermond, R. Pasquetti, B. Popov, Entropy viscosity method for nonlinear conservation laws. J. Comput. Phys. **230**, 4248–4267 (2011)
7. J.-L. Guermond, R. Pasquetti, B. Popov, From suitable weak solutions to entropy viscosity. J. Sci. Comput. **49**, 35–50 (2011)
8. J.S. Hesthaven, T. Warburton, *Nodal Discontinuous Galerkin Methods: Algorithms, Analysis, and Applications*, vol. 54 (Springer, New York, 2008)
9. C. Johnson, A. Szepessy, P. Hansbo, On the convergence of shock-capturing streamline diffusion finite element methods for hyperbolic conservation laws. Math. Comput. **54**, 107–129 (1990)
10. R.M. Kirby, G.E. Karniadakis, De-aliasing on non-uniform grids: algorithms and applications. J. Comput. Phys. **191**, 249–264 (2003)
11. A. Kornelus, D. Appelö, Flux-conservative Hermite methods for nonlinear conservation laws (2017). Preprints submitted to J. Sci. Comp. arXiv1703.06848
12. R. LeVeque, *Finite Volume Methods for Hyperbolic Problems*, vol. 31 (Cambridge University Press, Cambridge, 2002)
13. P. Persson, J. Peraire, Sub-cell shock capturing for discontinuous Galerkin methods, in *44-th AIAA Aerospace Sciences Meeting and Exhibit* (2006), pp. 1–13
14. C.-W. Shu, *Essentially Non-oscillatory and Weighted Essentially Non-oscillatory Schemes for Hyperbolic Conservation Laws* (Springer, Berlin, 1998)
15. H. Yee, B. Sjögreen, Development of low dissipative high order filter schemes for multiscale Navier–Stokes/MHD systems. J. Comput. Phys. **225**, 910–934 (2007)
16. V. Zingan, J.-L. Guermond, J. Morel, B. Popov, Implementation of the entropy viscosity method with the discontinuous Galerkin method. Comput. Methods Appl. Mech. Eng. **253**, 479–490 (2013)

Robust Multigrid for Cartesian Interior Penalty DG Formulations of the Poisson Equation in 3D

Jörg Stiller

Abstract We present a polynomial multigrid method for the nodal interior penalty formulation of the Poisson equation on three-dimensional Cartesian grids. Its key ingredient is a weighted overlapping Schwarz smoother operating on element-centered subdomains. The MG method reaches superior convergence rates corresponding to residual reductions of about two orders of magnitude within a single V(1,1) cycle. It is robust with respect to the mesh size and the ansatz order, at least up to $P = 32$. Rigorous exploitation of tensor-product factorization yields a computational complexity of $O(PN)$ for N unknowns, whereas numerical experiments indicate even linear runtime scaling. Moreover, by allowing adjustable subdomain overlaps and adding Krylov acceleration, the method proved feasible for anisotropic grids with element aspect ratios up to 48.

1 Introduction

Discontinuous Galerkin (DG) methods combine multiple desirable properties of finite element and finite volume methods, including geometric flexibility, variable approximation order, straightforward adaptivity and suitability for conservation laws [4, 11]. Though traditionally focused on hyperbolic systems, the need for implicit diffusion schemes and application to other problem classes, such as incompressible flow and elasticity, led to a growing interest in DG methods and related solution techniques for elliptic equations [1, 17]. This paper is concerned with fast elliptic solvers based on the multigrid (MG) method. In the context of high-order spectral element and DG methods, several approaches have been proposed: polynomial or p-MG [5, 8–10], geometric or h-MG [7, 13, 14], and algebraic MG [2, 16]. The most efficient methods reported so far [2, 18] use block smoothers that can be regarded as overlapping Schwarz methods. This work presents a hybrid Schwarz/MG method for nodal interior penalty DG formulations of Poisson

J. Stiller (✉)
Institute of Fluid Mechanics and Center of Advancing Electronics Dresden (cfaed), TU Dresden, Dresden, Germany
e-mail: joerg.stiller@tu-dresden.de

M.L. Bittencourt et al. (eds.), *Spectral and High Order Methods for Partial Differential Equations ICOSAHOM 2016*, Lecture Notes in Computational Science and Engineering 119, DOI 10.1007/978-3-319-65870-4_12

problems on 3D Cartesian grids. It extends the techniques put forward in [18, 19] and generalizes the approach to variable subdomain overlaps. The remainder of the paper is organized as follows: Section 2 briefly describes the discretization, Sect. 3 the multigrid technique, including the Schwarz smoother, and Sect. 4 presents the results of the assessment by means of numerical experiments. Section 5 concludes the paper.

2 Discontinuous Galerkin Method

We consider the Poisson equation

$$-\nabla^2 u = f \tag{1}$$

in the periodic domain $\Omega = [0, l_x] \times [0, l_y] \times [0, l_z]$.[1] For discretization, the domain is decomposed into cuboidal elements $\{\Omega^e\}$ forming a Cartesian mesh. The discrete solution u_h is sought in the function space

$$\mathcal{V}_h = \left\{ v \in L^2(\Omega) : v|_{\Omega^e} \in \mathcal{P}_P^3(\Omega^e) \quad \forall \Omega^e \subset \Omega \right\}, \tag{2}$$

where $\mathcal{P}_P^3$ is the 3D tensor product of polynomials of at most degree P. To cope with discontinuity we introduce the interior surface $\Gamma = \cup \Gamma^f$ which is composed of the element interfaces $\{\Gamma^f\}$. Let Ω^- and Ω^+ denote the elements adjacent to Γ^f and, respectively, $\mathbf{n}^-$ and $\mathbf{n}^+$ their exterior normals, and v^- and v^+ the restrictions of v to the joint face from inside the elements. Then we define the average and jump operators as

$$\{\!\{v\}\!\} = \frac{1}{2}(v^- + v^+), \quad [\![v]\!] = \mathbf{n}^- v^- + \mathbf{n}^+ v^+ . \tag{3}$$

Given these prerequisites, the interior penalty discontinuous Galerkin formulation can be stated as follows (see, e.g. [1]): Find $u_h \in \mathcal{V}_h$ such that for all $v \in \mathcal{V}_h$

$$\int_\Omega \nabla v \cdot \nabla u_h \mathrm{d}\Omega + \int_\Gamma \left([\![\nabla v]\!] \cdot \{\!\{u_h\}\!\} + \{\!\{v\}\!\} \cdot [\![\nabla u_h]\!] \right) \mathrm{d}\Gamma + \int_\Gamma \mu [\![v]\!] \cdot [\![u_h]\!] = \int_\Omega v f \mathrm{d}\Omega , \tag{4}$$

[1]Within this paper, the following symbols are used concurrently for representing the Cartesian coordinates: $\mathbf{x} = [x_i] = (x_1, x_2, x_3) = (x, y, z)$.

where μ is a piece-wise constant penalty parameter that must be chosen large enough to ensure stability.

Although, in theory, any suitable basis in $\mathcal{V}_h$ can be chosen, we restrict ourselves to nodal tensor-product bases generated from the Lagrange polynomials to the Gauss-Legendre (GL) or Gauss-Legendre-Lobatto (GLL) points, respectively. The discrete solution is expressed in Ω^e as

$$u_h(\mathbf{x})|_{\Omega^e} = u^e(\boldsymbol{\xi}^e(\mathbf{x})) = \sum_{i,j,k=0}^{P} u^e_{ijk}\varphi_i(\xi)\varphi_j(\eta)\varphi_k(\zeta)\,, \tag{5}$$

where φ_i denotes the 1D base functions and $\boldsymbol{\xi}^e(\mathbf{x})$ the transformation from Ω^e to the reference element $[-1,1]^3$. Each coefficient u^e_{ijk} is associated with one local base function, which is globalized by zero continuation outside Ω^e. Inserting these base functions in (4) for v and applying GL or GLL quadrature, according to the chosen basis, yields the discrete equations

$$\underline{A}\,\underline{u} = \underline{f} \tag{6}$$

for the solution vector $\underline{u} = [u^e_{ijk}]$. Due to the Cartesian element mesh and the tensor-product ansatz (5) the system matrix assumes the tensor-product form

$$\underline{A} = \underline{M}_z \otimes \underline{M}_y \otimes \underline{L}_x + \underline{M}_z \otimes \underline{L}_y \otimes \underline{M}_x + \underline{L}_z \otimes \underline{M}_y \otimes \underline{M}_x \tag{7}$$

where $\underline{M}_*$, $\underline{L}_*$ are the 1D mass and stiffness matrices for $* = x, y, z$. Without going into detail we remark that $\underline{M}_*$ is positive diagonal, and $\underline{L}_*$ symmetric positive semi-definite and block tridiagonal for either basis choice. The rigorous exploitation of these properties is crucial for the efficiency of the overall method.

3 Multigrid Techniques

The tensor-product structure of (8) allows for a straight-forward extension of the multigrid techniques developed in [18] for the 2D case. In the following, we examine polynomial multigrid (MG) and multigrid-preconditioned conjugate gradients (MG-CG) both using an overlapping Schwarz method for smoothing.

3.1 Schwarz Method

Schwarz methods are iterative domain decomposition techniques which improve the approximate solution by parallel or sequential subdomain solves, leading to additive or multiplicative methods, respectively. Here, we consider additive Schwarz with

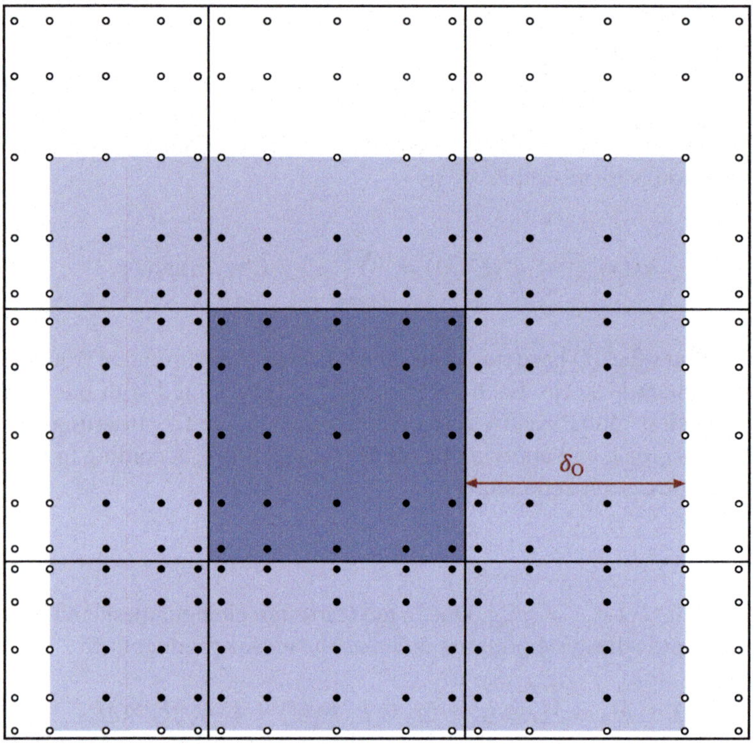

Fig. 1 Element-centered subdomain. Every subdomain consists of a core region coinciding with the embedded element (*dark*) and an overlap zone (*light shaded*). The latter represents a strip of variable width $\delta_{\rm O}$, which is adopted from the surrounding elements. The *circles* are the GL nodes for polynomial order $P = 4$. *Filled circles* indicate the unknowns that are solved for

overlapping element-centered subdomains as sketched in Fig. 1. The overlap width $\delta_{\rm O}$ can be different on each side of the embedded element, but may not exceed the width of the adjoining element. Alternatively, the overlap can be specified by prescribing the number $N_{\rm O}$ of node layers adopted from the latter.

To derive the subdomain problems, we first rewrite (8) into the residual form

$$\underline{A}\,\Delta\underline{u} = \underline{f} - \underline{A}\,\underline{\tilde{u}} = \underline{r}\,, \tag{8}$$

where $\Delta\underline{u} = \underline{u} - \underline{\tilde{u}}$ is the correction to the current approximate solution $\underline{\tilde{u}}$. For each subdomain Ω_s we define the restriction operator $\underline{R}_s$ such that $\underline{u}_s = \underline{R}_s\underline{u}$ gives the associated coefficients. Conversely, the transposed restriction operator, $\underline{R}_s^{\rm T}$ globalizes the local coefficients by adding zeros for exterior nodes. With these prerequisites the correction contributed by Ω_s is defined as the solution to the subproblem

$$\underline{A}_{ss}\Delta\underline{u}_s = \underline{r}_s\,, \tag{9}$$

where $\underline{A}_{ss} = \underline{R}_s \underline{A}\,\underline{R}_s^{\mathrm{T}}$ is the restricted system matrix and $\underline{r}_s = \underline{R}_s \underline{r}$ the restricted residual. Due to the cuboidal shape of the subdomain, the restriction operator possesses the tensor-product factorization $\underline{R}_s = \underline{R}_{s,x} \otimes \underline{R}_{s,y} \otimes \underline{R}_{s,z}$ and, as a consequence, $\underline{A}_{ss}$ inherits the tensor-structure of the full system matrix (7). Moreover it is regular and can be inverted using the fast diagonalization technique of Lynch et al. [15] to obtain

$$\underline{A}_{ss}^{-1} = (\underline{S}_z \otimes \underline{S}_y \otimes \underline{S}_x)(\underline{I} \otimes \underline{I} \otimes \underline{\Lambda}_x + \underline{I} \otimes \underline{\Lambda}_y \otimes \underline{I} + \underline{\Lambda}_z \otimes \underline{I} \otimes \underline{I})^{-1}(\underline{S}_z^{\mathrm{T}} \otimes \underline{S}_y^{\mathrm{T}} \otimes \underline{S}_x^{\mathrm{T}}),$$

where $\underline{S}_*$ is the column matrix of eigenvectors and Λ_* the diagonal matrix of eigenvalues to the generalized eigenproblem for the restricted 1D stiffness and mass matrices and $* = x, y, z$. Exploiting this structure the solution can be computed in $O(P^4)$ operations per subdomain.

One additive Schwarz iteration proceeds as follows: First, all subproblems are solved in parallel, which yields the local corrections $\Delta\underline{u}_s$. Afterwards, the global correction is computed as the weighted average

$$\Delta\underline{u} \simeq \sum_s \underline{R}_s^{\mathrm{T}} (\underline{W}_s \Delta\underline{u}_s)\,, \tag{10}$$

where $\underline{W}_s = \underline{W}_z \otimes \underline{W}_y \otimes \underline{W}_x$ is the diagonal local weighting matrix. The constituent 1D weights are computed from the hat-shaped weight function w_{H}, which is illustrated in Fig. 2. The complete definition of the weight function and alternative choices are given in [18].

3.2 Multigrid and Preconditioned Conjugate Gradient Methods

For MG we define a series of polynomial levels $\{P_l\}$ with $P_l = 2^l$ increasing from 1 at $l = 0$ to P at top level L. Correspondingly, let $\underline{u}_l$ denote the global coefficients

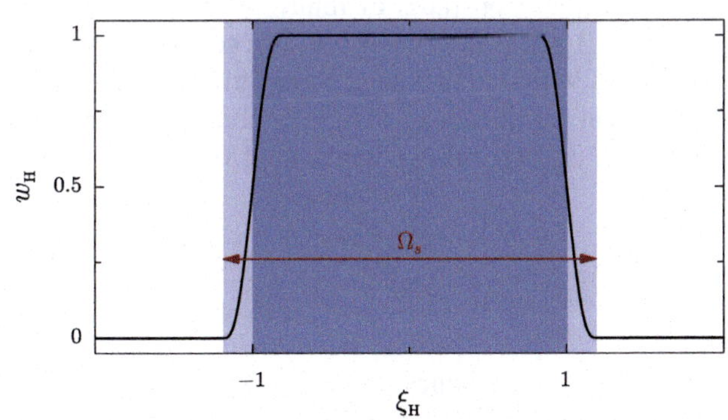

Fig. 2 Hat-shaped weight function using a piece-wise quintic transition from 0 at the subdomain boundary to 1 in the non-overlapped part of the core region. The coordinate ξ_{H} coincides with ξ inside the central element, and $\xi \mp 2$ in the left and right neighbor elements, respectively

and $\underline{A}_l$ the system matrix on level l. On the top level we have $\underline{u}_L = \underline{u}$ and $\underline{A}_L = \underline{A}$, whereas on lower levels $\underline{u}_l$ is the defect correction and $\underline{A}_l$ the counterpart of $\underline{A}$ obtained with elements of order P_l. For transferring the correction from level $l-1$ to level l we use the embedded interpolation operator $\underline{\mathcal{I}}_l$, and for restricting the residual its transpose. These ingredients allow to build the multigrid V-cycle summarized in Algorithm 1, where the SMOOTHER represents the weighted additive Schwarz method. To allow for variable V-cycles [3], the number of pre- and post-smoothing steps, $N_{S1,l}$ and $N_{S2,l}$, may change from level to level. Line 11 of Algorithm 1 defines the coarse grid solution formally by means of the pseudoinverse $\underline{A}_0^+$. In our implementation the coarse problem is solved using the conjugate gradient method. To ensure convergence, the right side is projected to the null space of $\underline{A}_0$, as proposed in [12].

Algorithm 1 Multigrid V-cycle

1: **function** MULTIGRIDCYCLE($\underline{u}, \underline{f}, \underline{N_S}$)
2: $\quad \underline{u}_L \leftarrow \underline{u}$
3: $\quad \underline{f}_L \leftarrow \underline{f}$
4: $\quad$ **for** $l = L, 1$ **step** -1 **do**
5: $\quad\quad$ **if** $l < L$ **then**
6: $\quad\quad\quad \underline{u}_l \leftarrow 0$
7: $\quad\quad$ **end if**
8: $\quad\quad \underline{u}_l \leftarrow$ SMOOTHER($\underline{u}_l, \underline{f}_l, N_{S1,l}$) ▷ Pre-smoothing
9: $\quad\quad \underline{f}_{l-1} \leftarrow \underline{\mathcal{I}}_l^{\mathrm{T}}(\underline{f}_l - \underline{A}_l \underline{u}_l)$ ▷ Residual restriction
10: $\quad$ **end for**
11: $\quad \underline{u}_0 \leftarrow \underline{A}_0^+ \underline{f}_0$ ▷ Coarse grid solution
12: $\quad$ **for** $l = 1, L$ **do**
13: $\quad\quad \underline{u}_l \leftarrow \underline{u}_l + \underline{\mathcal{I}}_l \underline{u}_{l-1}$ ▷ Correction prolongation
14: $\quad\quad \underline{u}_l \leftarrow$ SMOOTHER($\underline{u}_l, \underline{f}_l, N_{S2,l}$) ▷ Post-smoothing
15: $\quad$ **end for**
16: $\quad$ **return** $\underline{u} \leftarrow \underline{u}_L$
17: **end function**

It is well known that the robustness of multigrid method can be enhanced by Krylov acceleration [20]. Here we use the inexact preconditioned conjugate gradient method [6], which copes with the asymmetry introduced by the weighted Schwarz method without imposing significant extra cost in comparison to conventional CG. A detailed description of the algorithm is given in [18].

4 Results

For assessing robustness and efficiency, the described methods were implemented in Fortran and applied to the 2π-periodic Poisson problem with the exact solution

$$u(\mathbf{x}) = \cos(x - 3x + 2z)\sin(1 + x)\sin(1 - x)\sin(2x + x)\sin(3x - 2y + 2z).$$

To keep the test series manageable, we constrained ourselves to equidistant grids with an identical number of elements in each direction. Anisotropic meshes were realized be choosing the domain extensions as multiples of 2π, i.e. $l_* = 2\pi s_*$, which yields the aspect ratio $\Delta x : \Delta y : \Delta z = s_x : s_y : s_z$. All tests started from a random guess confined to $[-1, 1]$ and used a penalty parameter of $\mu_* = 2\mu_{\min,*}$, where $\mu_{\min,*}$ is the stability threshold, e.g., $\mu_{\min,x} = P(P+1)/\Delta x$ for the x direction [18]. The program was compiled using the Intel Fortran compiler 17.0 with optimization –O3 and run on a 3.1 GHz Intel Core i7-5557U CPU.

The primary assessment criterium is the average multigrid convergence rate

$$\rho = \sqrt[n]{\frac{r_n}{r_0}},$$

where r_n is the Euclidean norm of the residual vector after the nth cycle. Additionally we consider the number of cycles n_{10} and the average runtime per unknown τ_{10} that are required to reduce the residual by a factor of 10^{10}. These quantities follow from the convergence rate by $n_{10} = \lceil -10/\lg\rho \rceil$ and $\tau_{10} = -10\, t_{\mathrm{C}}/\lg\rho$, respectively, where t_{C} is the time required for one V-cycle.

4.1 Isotropic Meshes

First we consider the isotropic case with $s_x = s_y = s_z = 1$, such that $\Omega = [0, 2\pi]^3$. For assessing the impact of the subdomain overlap on the convergence rate and computational cost, we performed a test series for ansatz orders $P = 4$ to 32 using a degree-dependent tessellation into $N_{\mathrm{E}} = (128/P)^3$ elements. Table 1 presents the logarithmic convergence rates for 14 test cases featuring different choices for the basis functions (GLL or GL), the solution method (MG or MG-CG) and the subdomain overlap. Note that the latter was chosen identical in each direction, because of mesh isotropy. All cases employed a fixed V-cycle with one pre- and post-smoothing step. Independent of the basis and the solution method, choosing a minimal overlap of just one node ($N_{\mathrm{O}} = 1$) yields acceptable convergences rates for low order ($P = 4$), but becomes inefficient with increasing order. Using a geometrically fixed overlap of just 8 percent of the element width ($\delta_{\mathrm{O}} = 0.08\Delta x$) on every mesh level ensures robustness with respect to the ansatz order and even improves the convergence with growing P. Enlarging the overlap increases the convergence rate but also the computational cost, as will be detailed in a moment. Using Krylov acceleration tends to give faster convergence, however, this advantage melts away when increasing the overlap or the ansatz order. While these properties are consistently observed with the GLL basis, the GL results follow a less regular pattern and exhibit mostly lower convergence rates. This behavior can partly be explained by the fact that, with Gauss points, using a geometrically specified overlap width may result in zero overlapped node layers. While this phenomenon appears only at low orders, the latter are always present in the multigrid scheme, even at

Table 1 Convergence rates obtained with different multigrid methods on isotopic meshes composed of $(128/P)^3$ elements for increasing ansatz order P

					$\bar{r} = -\lg \rho$			
#	Basis	Solver	Overlap	N_O	$P = 4$	$P = 8$	$P = 16$	$P = 32$
1	GLL	MG	$N_O = 1$	$1, 1, 1, 1, 1$	1.08	0.89	0.46	0.24
2	GLL	MG-CG	$N_O = 1$	$1, 1, 1, 1, 1$	1.34	1.21	0.80	0.52
3	GLL	MG	$\delta_O = 0.08\Delta x$	$1, 1, 2, 3, 6$	1.08	1.59	2.13	2.82
4	GLL	MG-CG	$\delta_O = 0.08\Delta x$	$1, 1, 2, 3, 6$	1.34	1.85	2.24	2.81
5	GLL	MG	$\delta_O = 0.50\Delta x$	$2, 3, 5, 9, 17$	2.26	2.76	3.18	3.66
6	GLL	MG-CG	$\delta_O = 0.50\Delta x$	$2, 3, 5, 9, 17$	2.29	2.67	3.08	3.75
7	GL	MG	$N_O = 1$	$1, 1, 1, 1, 1$	0.73	0.96	0.73	0.40
8	GL	MG-CG	$N_O = 1$	$1, 1, 1, 1, 1$	1.49	1.26	1.07	0.78
9	GL	MG	$\delta_O = 0.08\Delta x$	$0, 1, 1, 3, 6$	0.73	1.15	1.70	2.20
10	GL	MG-CG	$\delta_O = 0.08\Delta x$	$0, 1, 1, 3, 6$	1.14	1.38	1.84	2.15
11	GL	MG	$\delta_O = 0.50\Delta x$	$1, 2, 4, 8, 16$	1.83	2.28	1.52	0.94
12	GL	MG-CG	$\delta_O = 0.50\Delta x$	$1, 2, 4, 8, 16$	1.90	2.34	1.88	1.37
13	GL	MG	$\delta_O = 0.09\Delta x, N_O \geq 1$	$1, 1, 2, 3, 6$	1.36	1.75	1.87	2.21
14	GL	MG-CG	$\delta_O = 0.09\Delta x, N_O \geq 1$	$1, 1, 2, 3, 6$	1.49	1.81	2.03	2.22

high ansatz orders. A remedy to this problem is to apply a lower bound of $N_{O,l} = 1$ for the nodal overlap on every mesh level l. Nevertheless, the GL-based approach remains slightly less efficient in comparison with the GLL approach. Therefore, further discussion will be constrained to the latter.

Complementary to the tabulated results, Fig. 3 depicts the number n_{10} of multigrid cycles that are required to reduce the Euclidian residual norm by ten orders of magnitude for selected cases listed in Table 1. In agreement with the above discussion, the cycle count increases considerably when using GLL MG with only one node layer overlap. Adding Krylov acceleration (MG-CG) ameliorates this drawback, especially at higher order. Yet, pure MG with a fixed geometric overlap of 8 percent is far more efficient and even attains a decreasing n_{10} with growing P. As expected, using an overlap of $\Delta x/2$ yields a further reduction of the cycle count. This advantage is, however, bought with additional computational cost related to the larger subdomain operator $\underline{A}_{ss}$. Figure 4 confirms that GLL MG with $\delta_O = 0.08\Delta x$ outpaces the other choices for all polynomial degrees but 4, where the Krylov-accelerated method (MG-CG) with one node overlap is slightly faster. Moreover, Fig. 5 illustrates the robustness of this method with respect to the mesh size. With ansatz orders up to 16, the convergence rate becomes mesh independent for $N_E \gtrsim 12^3$, whereas it still tends to improve beyond $N_E = 16^3$ for $P = 32$. It is further worth noting that the convergence rates improve with growing order, reaching an excellent $\rho \approx 6.3 \times 10^{-3}$ with $P = 16$ and even better $\rho \approx 1.6 \times 10^{-3}$ with $P = 32$. Moreover, runtimes of about 3.5 μs per degree of freedom allow to solve problems up to a million unknowns conveniently on a single core.

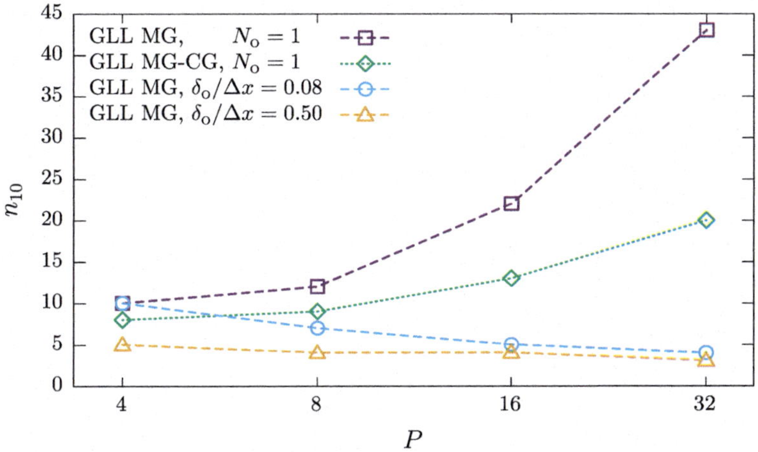

Fig. 3 Number of cycles required to reduce the residual by ten orders of magnitude for selected methods listed in Table 1

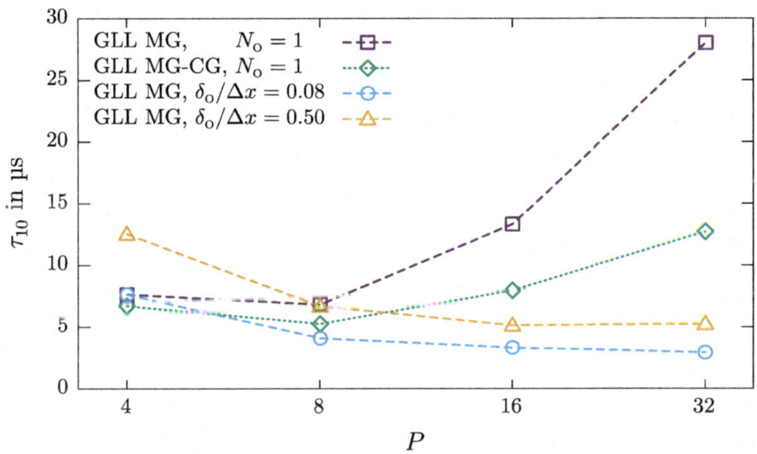

Fig. 4 Runtime per unknown required to reduce the residual by ten orders of magnitude for selected methods listed in Table 1

4.2 Anisotropic Meshes

As a second issue we investigated the suitability of the approach for anisotropic meshes. For this purpose, we defined a sequence of domains

$$\Omega = (0, 2\pi\, AR) \times (0, 2\pi \lceil AR/2 \rceil) \times (0, 2\pi),$$

with aspect ratios *AR* ranging from 1 to 48. Using a uniform tessellation featuring the same number of elements in each coordinate direction, *AR* also represents

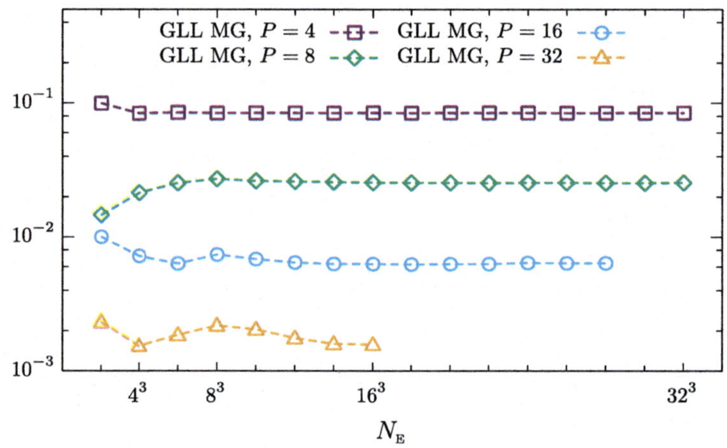

Fig. 5 Convergence rates for GLL MG with overlap $\delta_O = 0.08\Delta x$ for different orders P displayed as a function of the mesh size

Table 2 Test cases for investigating the robustness against the element aspect ratio

Case	Subdomain overlap	Smoothing steps
(0.08 rel; 1,1, fix)	$\delta_{O,x_i} = 0.08\Delta x_i$	$N_{S,l} = (1, 1)$
(0.08 rel; 1,1, var)	$\delta_{O,x_i} = 0.08\Delta x_i$	$N_{S,l} = (1, 1) \times 3^{L-l}$
(0.08 max; 1,1, var)	$\delta_{O,x_i} = \min[\max_j(0.08\Delta x_j),\ \Delta x_i]$	$N_{S,l} = (1, 1) \times 3^{L-l}$

the maximum element side aspect ratio. Thus, for example, $AR = 32$ results in $\Delta x = 2\Delta y = 32\Delta z$. In earlier 2D studies, Krylov acceleration and variable V-cycles proved helpful, though yet insufficient, for coping with anisotropy. Based on this experience, we selected methods with different overlap and smoothing strategies, which are summarized in Table 2. Methods (0.08 rel; 1,1, fix) and (0.08 rel; 1,1, var) both use a relative subdomain overlap of 8 percent, which means that the overlap width varies in each coordinate direction proportionally to the corresponding element extension. In contrast, (0.08 max; 1,1, var) sets the overlap width to 8 percent of the maximal side length, but not larger than the element width in the given direction. Additionally, the last two methods apply a variable V-cycle, which increases the number of smoothing steps by a factor of 3 with each coarser level. The performance of these methods was studied on a 8^3 tessellation using elements of order $P = 16$. Figure 6 depicts the obtained convergence rates, cycle counts and runtimes per unknown for aspect ratios up to 48. With method (0.08 rel; 1,1, fix) convergence starts to degrade at moderate aspect ratios and has already slowed by two orders of magnitude at $AR = 16$. Using a variable V-cycle improves the robustness such that a nearly constant cycle count n_{10} is maintained until $AR = 12$. From here convergence degrades more quickly, but remains superior to the previous case. Setting the overlap proportional to the largest element extension, as with method (0.08 max; 1,1, var), yields a further improvement, which becomes even

a

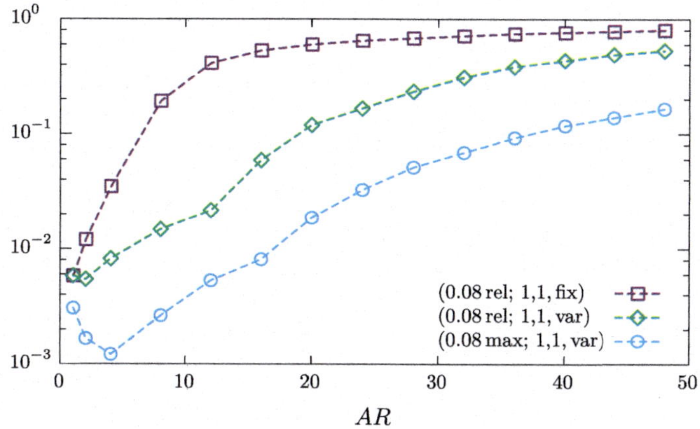

b

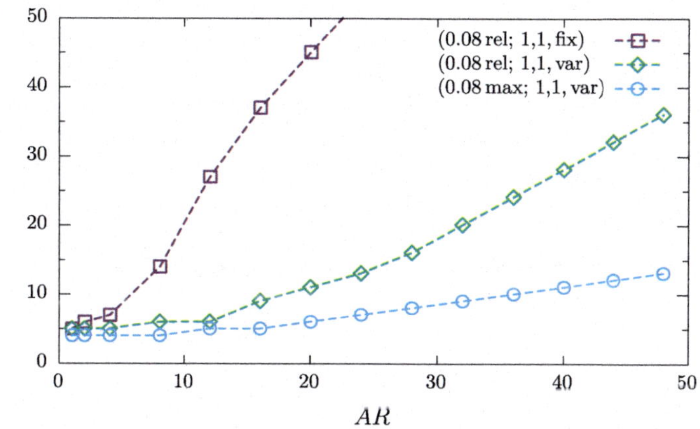

c

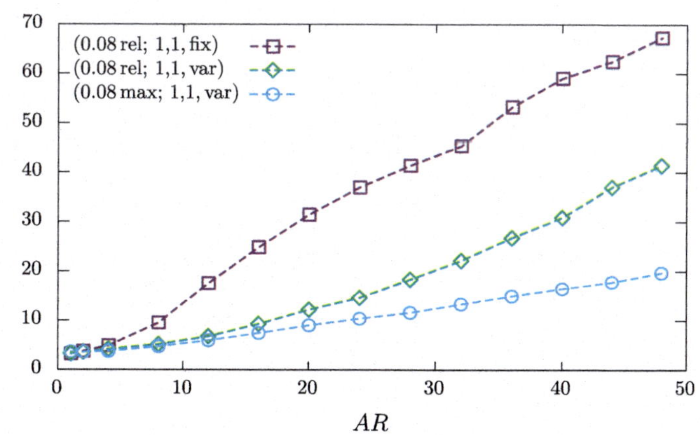

Fig. 6 (**a**) Multigrid convergence rates depending on the aspect ratio for methods listed in Table 2. (**b**) Cycle counts depending on the aspect ratio for methods listed in Table 2. (**c**) Runtime per unknown depending on the aspect ratio for methods listed in Table 2

more pronounced in the range $AR > 12$, where the overlap in the most compressed direction is already constrained by the element width. Compared to the isotropic case, the cycle count grew from 4 to 13 at $AR = 48$, whereas the serial runtime increased by factor of 5.4 to approximately 19 μs per unknown. This seems to be a good starting point, given the prospect of further acceleration, e.g. by parallelization.

5 Conclusions

We developed a polynomial multigrid method for nodal interior-penalty formulations of the Poisson equation on three-dimensional Cartesian grids. Its key ingredient is an overlapping weighted Schwarz smoother, which exploits the underlying tensor-product structure for fast solution of the subdomain problems. The method achieves excellent convergence rates and proved robust against the mesh size and ansatz orders up to at least 32. Extending the ideas put forward in [18], we showed that combining Krylov acceleration, variable smoothing and increasing the subdomain overlap proportionally to the maximum element width improves the robustness considerably and renders the approach feasible for aspect ratios up to 50. Moreover, the method is computationally efficient, allowing to solve problems with a million unknowns in a few seconds on a single CPU core.

Acknowledgements Funding by German Research Foundation (DFG) in frame of the project STI 157/4-1 is gratefully acknowledged.

References

1. D.N. Arnold, F. Brezzi, B. Cockburn, L.D. Marini, Unified analysis of discontinuous Galerkin methods for elliptic problems. SIAM J. Numer. Anal. **39**(5), 1749–1779 (2001)
2. P. Bastian, M. Blatt, R. Scheichl Algebraic multigrid for discontinuous Galerkin discretizations of heterogeneous elliptic problems. Numer. Linear Algebra Appl. **19**, 367–388 (2012)
3. J. Bramble, *Multigrid Methods*. Pitman Research Notes Mathematical Series, vol. 294 (Longman Scientific & Technical, Harlow, 1995)
4. B. Cockburn, G.E. Karniadakis, C.-W. Shu, *Discontinuous Galerkin Methods: Theory, Computation and Applications* (Springer, Berlin, Heidelberg, 2000)
5. K.J. Fidkowski, T.A. Oliver, J. Lu, D.L. Darmofal, p-multigrid solution of high-order discontinuous Galerkin discretizations of the compressible Navier-Stokes equations. J. Comput. Phys. **207**, 92–113 (2005)
6. G.H. Golub, Q. Ye Inexact preconditioned conjugate gradient method with inner-outer iteration. SIAM J. Sci. Comput. **21**(4), 1305–1320 (1999)
7. J. Gopalakrishnan, G. Kanschat, A multilevel discontinuous Galerkin method. Numer. Math. **95**, 527–550 (2003)
8. L. Haupt, J. Stiller, W. Nagel, A fast spectral element solver combining static condensation and multigrid techniques. J. Comput. Phys. **255**, 384–395 (2013)
9. B.T. Helenbrook, H.L. Atkins, D.J. Mavriplis, Analysis of p-multigrid for continuous and discontinuous finite element discretizations. AIAA Paper 2003–3989 (2003)

10. B.T. Helenbrook, H.L. Atkins, Application of p-multigrid to discontinuous Galerkin formulations of the poisson equation. AIAA J. **44**, 566–575 (2006)
11. J.S. Hesthaven, T. Warburton, *Nodal Discontinuous Galerkin Methods* (Springer, Berlin, 2008)
12. E.F. Kaasschieter, Preconditioned conjugate gradients for solving singular systems. J. Comput. Appl. Math. **24**, 265–275 (1988)
13. G. Kanschat, Multilevel methods for discontinuous Galerkin FEM on locally refined meshes. Comput. Struct. **82**, 2437–2445 (2004)
14. J.K. Kraus, S.K. Tomar, A multilevel method for discontinuous Galerkin approximation of three-dimensional anisotropic elliptic problems. Numer Linear Algebra Appl. **15**, 417–438 (2008)
15. R.E. Lynch, J.R. Rice, D.H. Thomas, Direct solution of partial difference equations by tensor product methods. Numer. Math. **6**, 185–199 (1964)
16. L.N. Olson, J.B. Schroder, Smoothed aggregation multigrid solvers for high-order discontinuous Galerkin methods for elliptic problems. J. Comput. Phys. **230**, 6959–6976 (2011)
17. B. Rivière, *Discontinuous Galerkin Methods for Solving Elliptic and Parabolic Equations: Theory and Implementation*. Frontiers in Mathematics, vol. 35 (SIAM, Philadelphia, PA, 2008)
18. J. Stiller, Robust multigrid for high-order discontinuous Galerkin methods: a fast Poisson solver suitable for high-aspect ratio Cartesian grids. J. Comput. Phys. **327**, 317–336 (2016)
19. J. Stiller, Nonuniformly weighted Schwarz smoothers for spectral element multigrid. J. Sci. Comput. **72**(1), 81–96 (2017)
20. U. Trottenberg, C.W. Oosterlee, A. Schüller, *Multigrid* (Academic, New York, 2000)

Using PGD to Solve Nonseparable Fractional Derivative Elliptic Problems

Shimin Lin, Mejdi Azaiez, and Chuanju Xu

Abstract A family of tensor-based methods called Proper Generalized Decomposition (PGD) methods have been recently introduced for the a priori construction of the solution of several partial differential equations. This strategy was tested with success to demonstrate the capability of representing the solution with a significant reduction of the calculation and storage cost. In this paper, we suggest to test the efficiency of a such approach in solving general nonseparable fractional derivative elliptic problem. We will illustrate by several numerical experiments the efficiency of PGD, especially when the mesh or the coefficients vary with high contrast ratio. Although the PGD scheme considered in this paper is based on spectral method, it is extendable to other methods such as finite element method.

1 Introduction

The study of fractional calculus has a very long history and is attracting increasing attention in recent years. The corresponding fractional differential equations provide an alternative tool for the description of memory effect and hereditary properties of various materials and processes. The fractional model has been applied in numerous diverse fields, including control theory, viscoelastic materials, chaos, electro-chemical process, etc; see, e.g., [2, 5, 9, 10]. There are three basic fractional equations: time fractional diffusion equation, space fractional diffusion equation, and time-space fractional diffusion equation. In this contribution we consider a general nonseparable fractional elliptic equation [8]. Among the issues to be addressed, solving the algebraic system arising from the discretization is an important question. It is known that finding efficient preconditioners is often a difficult task, especially when we want the condition number of the preconditioned system is independent of

S. Lin (✉) • C. Xu
School of Mathematical Sciences, Xiamen University, 361005 Xiamen, China
e-mail: linshimin@stu.xmu.edu.cn; cjxu@xmu.edu.cn

M. Azaiez
Bordeaux INP, I2M (UMR CNRS 5295), 33607 Pessac, France
e-mail: azaiez@ipb.fr

M.L. Bittencourt et al. (eds.), *Spectral and High Order Methods for Partial Differential Equations ICOSAHOM 2016*, Lecture Notes in Computational Science and Engineering 119, DOI 10.1007/978-3-319-65870-4_13

both the degree of freedom and the ration between the maximum and minimum of the coefficient. To skirt this issue we propose in this contribution to take advantage of the proper generalized decomposition (PGD) technique to approximate the solution with a reasonable storage and CPU cost.

The paper is arranged as follows. In Sect. 2, we present some notations and basic properties of fractional calculus; Then the variable coefficient fractional partial differential equation is introduced in Sect. 3; In Sect. 4 we recall the spectral method and construct the PGD method. The final section is devoted to carry out some numerical experiments to verify the efficiency of the proposed method.

2 Preliminaries

In this section, we introduce some notations and present basic results of fractional calculus, which will be used throughout the paper [1, 4]. Let c be a generic positive constant independent of any functions and of any discretization parameters. We use the expression $A \lesssim B$ (respectively, $A \gtrsim B$) to mean that $A \leq cB$ (respectively, $A \geq cB$), and use the expression $A \cong B$ to mean that $A \lesssim B \lesssim A$.

For a function $f(x)$ defined in $[a, b](-\infty < a < b < \infty)$, we denote by ${}_aI_x^s f(x)$ and ${}_xI_b^s f(x)$ the left-sided and right-sided Riemann-Liouville fractional integrals of order $s > 0$ respectively. The left-sided and right-sided Riemann-Liouville fractional derivatives of order s are denoted by ${}_aD_x^s f$ and ${}_xD_b^s f$. The notations ${}_a^CD_x^s f$ and ${}_x^CD_b^s f$ stand for the Caputo fractional derivatives. We refer to [4] for the detailed definitions of these operators.

For ease of reading, we recall below a number of properties to be used in the paper. The relation between Riemann-Liouville and Caputo derivatives is given as follows [4]:

Lemma 1 *Let $s \in [n-1, n)$ with $n \in \mathbb{N}^+$. Then it holds*

$$
{}_aD_x^s f(x) = {}_a^CD_x^s f(x) + \sum_{j=0}^{n-1} \frac{f^{(j)}(a)(x-a)^{j-s}}{\Gamma(1+j-s)};
$$

$$
{}_xD_b^s f(x) = {}_x^CD_b^s f(x) + \sum_{j=0}^{n-1} (-1)^j \frac{f^{(j)}(b)(b-x)^{j-s}}{\Gamma(1+j-s)}.
$$

We now introduce some Sobolev spaces, which will be used in the discussion that follows, Let Ω be an open set contained in $\mathbb{R}^d$, d is the space dimension. The $L^2(\Omega)$ space is defined as the space of functions which are square measurable. The associated inner product and norm are denoted respectively by

$$(u,v)_\Omega := \int_\Omega uv\,d\Omega, \quad \|u\|_{L^2(\Omega)} := (u,u)_\Omega^{\frac{1}{2}}, \quad \forall u,v \in L^2(\Omega).$$

For a nonnegative real number s, $H^s(\Omega)$ and $H_0^s(\Omega)$ denote the usual Sobolev space with norm $\|\cdot\|_{s,\Omega}$ and semi-norm $|\cdot|_{s,\Omega}$.

Given a Sobolev space X with norm $\|\cdot\|_X$, let

$$H^s(\Omega;X) := \{v; \|v(\cdot,x)\|_X \in H^s(\Omega)\},$$

endowed with the norm:

$$\|v\|_{H^s(\Omega;X)} = \big\| \|v(\cdot,x)\|_X \big\|_{s,\Omega}\,.$$

If $\Omega = \Lambda \times \Lambda$, where Λ is a finite interval in $\mathbb{R}$, we also define

$$H^{s,\gamma}(\Omega) := H^s(\Lambda; L^2(\Lambda)) \cap L^2(\Lambda; H^\gamma(\Lambda))$$

and the corresponding $H_0^{s,\gamma}(\Omega)$ that is the closure of $C_0^\infty(\Omega)$ with respect to the $H^{s,\gamma}(\Omega)$-norm:

$$\|v\|_{H^{s,\gamma}}(\Omega) := \left(\|v\|^2_{H^s(\Lambda;L^2(\Lambda))} + \|v\|^2_{L^2(\Lambda;H^\gamma(\Lambda))} \right)^{\frac{1}{2}}.$$

The $C_0^\infty(\Omega)$ is the space of infinitely differentiable functions having compact support.

The following two lemmas are useful in the construction of weak form and the analysis of well-posedness for the fractional equation considered in this paper.

Lemma 2 ([6]) *For $0 < s < 1$, if $w \in H^s(\Lambda)$, $v \in H_0^s(\Lambda)$, then*

$$\left({}_aD_x^s w, v\right)_\Lambda = \left(w, {}_xD_b^s v\right)_\Lambda, \quad \text{if } {}_aD_x^s w \in L^2(\Lambda);$$
$$\left({}_xD_b^s w, v\right)_\Lambda = \left(w, {}_aD_x^s v\right)_\Lambda, \quad \text{if } {}_xD_b^s w \in L^2(\Lambda).$$

Lemma 3 ([6]) *Let $s > 0, s \neq n - \frac{1}{2}, n \in \mathbb{N}^+$. Then for all $v \in H_0^s(\Lambda)$, we have*

$$({}_aD_x^s v, {}_xD_b^s v) \cong \cos(\pi s)\|{}_aD_x^s v\|^2_{L^2(\Lambda)} \cong \cos(\pi s)\|{}_xD_b^s v\|^2_{L^2(\Lambda)} \cong \cos(\pi s)\|v\|^2_{H^s(\Lambda)}.$$

In the following discussion, without loss of generality, we restrict the domain to be $\Lambda = [-1,1]$, $\Omega = \Lambda \times \Lambda$. To simplify the notation, we write the x direction left-sided and right-sided R-L derivatives in form ∂_{x-}^s and ∂_{x+}^s respectively, and the domain symbol may be dropped from the subscript of norm if no confusion would arise.

3 Variable Coefficient Fractional Differential Equation

For $1 < \alpha, \beta < 2$, we consider the following two dimensional variable coefficient fractional partial differential equation [8]:

$$\rho(x,y)u(x,y) + \partial_{x+}^{\frac{\alpha}{2}}\left[d_l^x(x,y)^C\partial_{x-}^{\frac{\alpha}{2}}u(x,y)\right] + \partial_{x-}^{\frac{\alpha}{2}}\left[d_r^x(x,y)^C\partial_{x+}^{\frac{\alpha}{2}}u(x,y)\right]$$
$$+ \partial_{y+}^{\frac{\beta}{2}}\left[d_l^y(x,y)^C\partial_{y-}^{\frac{\beta}{2}}u(x,y)\right] + \partial_{y-}^{\frac{\beta}{2}}\left[d_r^y(x,y)^C\partial_{y+}^{\frac{\beta}{2}}u(x,y)\right] = f(x,y), \quad (x,y) \in \Omega, \quad (1)$$

subject to the Dirichlet boundary condition

$$u|_{\partial\Omega} = 0, \quad (2)$$

where $\rho, d_l^x, d_l^y, d_r^x, d_r^y$ are positive coefficient functions.

Based on the fractional integration by parts formula given in Lemma 2, we are led to consider the following weak formulation of (1)–(2): given $f \in H_0^{\frac{\alpha}{2},\frac{\beta}{2}}(\Omega)'$, find $u \in H_0^{\frac{\alpha}{2},\frac{\beta}{2}}(\Omega)$, such that

$$\mathscr{A}(u,v) = \langle f, v\rangle_\Omega, \quad \forall v \in H_0^{\frac{\alpha}{2},\frac{\beta}{2}}(\Omega), \quad (3)$$

where $\langle\cdot,\cdot\rangle$ is the duality between $H_0^{\frac{\alpha}{2},\frac{\beta}{2}}(\Omega)'$ and $H_0^{\frac{\alpha}{2},\frac{\beta}{2}}(\Omega)$, and the *self-adjoint* bilinear form $\mathscr{A}(\cdot,\cdot)$ is defined by

$$\mathscr{A}(u,v) := (\rho(x,y)u, v) + (d_l^x(x,y)\partial_{x-}^{\frac{\alpha}{2}}u, \partial_{x-}^{\frac{\alpha}{2}}v) + (d_r^x(x,y)\partial_{x+}^{\frac{\alpha}{2}}u, \partial_{x+}^{\frac{\alpha}{2}}v)$$
$$+ (d_l^y(x,y)\partial_{y-}^{\frac{\beta}{2}}u, \partial_{y-}^{\frac{\beta}{2}}v) + (d_r^y(x,y)\partial_{y+}^{\frac{\beta}{2}}u, \partial_{y+}^{\frac{\beta}{2}}v).$$

In order to ensure the well-posedness of problem (3), we impose the following conditions on the coefficient functions: $\rho(x,y), d_j^i(x,y) \in C(\Omega)$, and

$$0 \le \check{\rho} \le \rho(x,y) \le \hat{\rho}, \quad 0 < \check{d}_j^i \le d_i^j(x,y) \le \hat{d}_j^i, \quad i = x, y,\ j = l, r. \quad (4)$$

With the condition (4), we reach the following conclusion.

Theorem 1 ([8]) *Under the assumption (4), Lemmas 1, 2, 3, there exists a unique solution $u \in H_0^{\frac{\alpha}{2},\frac{\beta}{2}}(\Omega)$ to (3). Furthermore, the solution satisfies*

$$\|u\|_{H^{\frac{\alpha}{2},\frac{\beta}{2}}(\Omega)} \lesssim \|f\|_{H^{\frac{\alpha}{2},\frac{\beta}{2}}(\Omega)'}. \quad (5)$$

4 Numerical Method

In this section we propose a method based on proper generalized decomposition approach and spectral method for numerical solutions for the underlying problem (3). The former will be used to reduce the original problem into simpler ones, which will be solved a spectral method.

We first present a monodomain spectral approximation to (3).

4.1 *Spectral-Galerkin Method (SM)*

We consider the polynomial spaces:

$$\mathbb{P}_N^0(\Lambda) := \{v \in \mathbb{P}_N(\Lambda), v(-1) = v(1) = 0\}, \ \mathbb{P}_L^0(\Omega) = \mathbb{P}_N^0(\Lambda) \times \mathbb{P}_M^0(\Lambda),$$

where the discretization parameter L represents the degree pair (N, M). The spectral method we propose uses the standard form of Galerkin's method as follows: find $u_L \in \mathbb{P}_L^0(\Omega)$, such that

$$\mathscr{A}(u_L, v_L) = \langle f, v_L \rangle_\Omega, \quad \forall v_L \in \mathbb{P}_L^0(\Omega). \tag{6}$$

To compute the inner products involving fractional derivatives in the above schema, we will make use of Jacobi-Gauss quadratures with suitable weights [6, 7]. If all the coefficients in (1) are separable, some methods like the matrix decomposition/diagonalization method can be employed to solve the discrete problem (6). If not, iterative approach is a choice. More detail can be found in [8].

4.2 *Proper Generalized Decomposition Method (PGD)*

The PGD method can be regarded as model reduction based on separation of variable (see [3] and the references therein). In the simplest form of PGD for (3), we seek an approximation $u_k(x, y)$ to the solution $u(x, y)$ in Ω in the separated form as follows:

$$u(x, y) \approx u_k(x, y) = \sum_{i=1}^{k} X_i(x) Y_i(y), \tag{7}$$

where $\forall i = 1, \cdots, k$, $X_i(x) \in \mathrm{Q}_x$, $Y_i(y) \in \mathrm{Q}_y$, Q_x and Q_y are two suitable spaces. First we consider two mappings:

- $S_k^y : \mathrm{Q}_y \to \mathrm{Q}_x$, which maps function $Y(y) \in \mathrm{Q}_y$ into a function $X(x) \in \mathrm{Q}_x$, defined by:

$$\mathscr{A}(u_{k-1} + XY, X^* Y) = \langle f, X^* Y \rangle, \quad \forall X^* \in \mathrm{Q}_x; \tag{8}$$

- $S_k^x : Q_x \to Q_y$, which maps function $X(x) \in Q_x$ into a function $Y(y) \in Q_y$, defined by:

$$\mathscr{A}(u_{k-1} + XY, XY^*) = \langle f, XY^* \rangle, \quad \forall Y^* \in Q_y. \tag{9}$$

At each enrichment step $k (k \geq 1)$, we have already computed $k-1$ first terms of approximation (7). Therefore the unknowns at the current enrichment step k are functions $X_k(x)$ and $Y_k(y)$. We now wish to compute the kth term $X_k(x)Y_k(y)$ to obtain the enriched solution

$$u_k(x,y) = u_{k-1}(x,y) + X_k(x)Y_k(y) = \sum_{i=1}^{k-1} X_i(x)Y_i(y) + X_k(x)Y_k(y).$$

The resulting problem is thus coupled and a suitable iterative scheme is required. We use the index p to denote a particular iteration, and $X_k^p(x), Y_k^p(y)$ to denote approximation of $X_k(x), Y_k(y)$ obtained at iteration p. The simplest iterative scheme is an alternating direction strategy that computes first $X_k^p(x)$ by $S_k^y(Y_k^{p-1}(y))$, and then $Y_k^p(y)$ by $S_k^x(X_k^p(x))$. An arbitrary initial non-zero $Y_k^0(y)$ is specified to start the iterative process. The non-linear iterations proceed until reaching a fixed point within a user-specified tolerance ϵ. This lead to Algorithm 1.

Algorithm 1 Progressive of PGD

1: **for** $k = 1, \cdots, k_{max}$ **do**
2: Initialize $Y_k^0(y)$
3: **for** $p = 1, \cdots, p_{max}$ **do**
4: Compute $X_k^p = S_k^y(Y_k^{p-1})$
5: Normalize X_k^p
6: Compute $Y_k^p = S_k^x(X_k^p)$
7: Check convergence of $X_k^p Y_k^p$
8: **end for**
9: Set $X_k = X_k^p$ and $Y_k = Y_k^p$
10: Set $u_k = u_{k-1} + X_k Y_k$ and check convergence
11: **end for**

Remark 1 In general, alternating directional strategy reaches criteria very fast. A slow convergence may reveal multiple or close eigenvalues. However, also in this case, a good enough couple (X_k, Y_k) is often reached in a few iterations. In practice $p_{max} = 5$ is enough.

From Algorithm 1, we find that in steps 4, 6, we only need to deal with one-dimensional problems rather than two-dimension, which is the heart of the PGD. Totally there are $2k_{max}\ p_{max}$ one-dimensional problems for computing $X_i(x)$ and $Y_i(y)$. A reasonable choice of spaces is $Q_x = H_0^{\frac{\alpha}{2}}(\Lambda), Q_y = H_0^{\frac{\beta}{2}}(\Lambda)$, since we have following lemma.

Lemma 4 *For any measurable functions $X(x) : \Lambda \to \mathbb{R}$ and $Y(y) : \Lambda \to \mathbb{R}$ such that $X(x)Y(y) \neq 0$,*

$$X(x)Y(y) \in H_0^{\frac{\alpha}{2},\frac{\beta}{2}}(\Omega) \Longleftrightarrow X(x) \in H_0^{\frac{\alpha}{2}}(\Lambda) \text{ and } Y(y) \in H_0^{\frac{\beta}{2}}(\Lambda).$$

Concerning the stopping criteria, we use

$$\frac{\|X_k^p(x)Y_k^p(y) - X_k^{p-1}(x)Y_k^{p-1}(y)\|_0}{\|X_k^{p-1}(x)Y_k^{p-1}(y)\|_0} < \epsilon \tag{10}$$

to mean convergence of the alternating direction iteration. The enrichment process itself stops (step 10) when an appropriate measure, $\varepsilon(k)$, becomes small enough, i.e., $\varepsilon(k) < \tilde{\epsilon}$. Some stopping criteria are available here:

$$\varepsilon(k) := \frac{\|X_k(x)Y_k(y)\|_{H^{\frac{\alpha}{2},\frac{\beta}{2}}(\Omega)}}{\|X_1(x)Y_1(y)\|_{H^{\frac{\alpha}{2},\frac{\beta}{2}}(\Omega)}} \tag{11}$$

or

$$\varepsilon(k) := \|\mathbf{A}\mathbf{U}_k - \mathbf{F}\|_0 \tag{12}$$

for example. Here we assume $\mathbf{AU} = \mathbf{F}$ is the linear system of (6).

Finally, the given data $f(x, y), \rho(x, y)$ and d_j^i, $i = x, y$, $j = l, r$ need to be expressed as a sum of tensor products. Otherwise, computing high-dimensional integrals would be necessary. In case the given data are not of tensor product form, we can use PGD to get an appropriate approximation of them as a sum of tensor products.

5 Numerical Experiments

We now present some numerical results to assess the efficiency of the PGD/spectral method when it is used to approximate the solution of the nonseparable fractional elliptic problem (3). As explained above the cost of the PGD is proportional to the number of iterations in the fixed point algorithm augmented by the number of enrichments we need to represent the expected solution. The objective of this section is twofold. We will numerically verify the convergence of the process for both regular and singular solutions and for each we will measure the evolution of the cost when the mesh and/or the ratio between the maximum and minimum of the data increase.

First we fix the data in (3) as follows: $\rho(x, y) = 1, d_l^x(x, y) = d_r^x(x, y) = \sin^2(\pi x) + \sin^2(\pi y) + 1, d_l^y(x, y) = d_r^y(x, y) = \cos^2(\pi x) + \cos^2(\pi y) + 1$, and

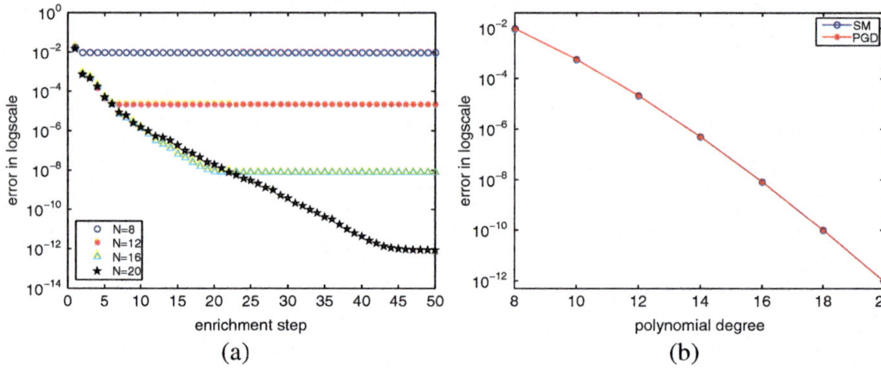

Fig. 1 Error behavior with respect to (**a**) the enrichment step k and (**b**) polynomial degree N

set $\alpha = 1.2, \beta = 1.8$. We test the method for the exact analytical solution:

$$u_{ex}(x, y) = (1 - x^2)^2 x^2 (1 - y^2)^2 sin(\pi xy).$$

In Fig. 1a we show the relationship between $H^{\frac{\alpha}{2},\frac{\beta}{2}}$—error and enrichment step in semi-log scale for several different polynomial degree N. It is observed that for N big enough the error exhibits an almost exponential decay with respect to the enrichment step number. Then we investigate the convergence behavior of the PGD/spectral method with respect to the polynomial degree. In Fig. 1b we present the $H^{\frac{\alpha}{2},\frac{\beta}{2}}$—error of solution obtained respectively by the PGD/spectral method and 2D spectral method (SM) as a function of the polynomial degree N. Clearly the both methods yield the expected spectral accuracy.

We now consider problem (3) with *singular* exact solution:

$$u_{ex}(x, y) = (1 - x^2)^2 x^\delta (1 - y^2)^2 sin(\pi xy),$$

where δ is a positive real number. The same experiments as in the previous regular case are performed. In Fig. 2a–c we show $H^{\frac{\alpha}{2},\frac{\beta}{2}}$—error decay rates as functions of the enrichment step number for different $\delta = \frac{7}{3}, \frac{13}{3}, \frac{19}{3}$. Once again the PGD/spectral method exhibits a quick convergence provided the polynomial degree is large enough, even the solution is not regular. Figure 2d plots the $H^{\frac{\alpha}{2},\frac{\beta}{2}}$—error versus polynomial degree N in log-log scale for the same values of δ, in which we observe that the convergence rates for $\delta = \frac{7}{3}, \frac{13}{3}$, and $\frac{19}{3}$ are respectively closed to $N^{-3.23}, N^{-5.23}$, and $N^{-7.23}$ as expected. This is in a quite good agreement with the regularity of the exact solution[8].

Now we investigate the impact of the polynomial degree and the coefficient variation on the performance of the PGD method. We take $\rho(x, y) = d_l^y(x, y) = d_r^y(x, y) = 1, d_l^x(x, y) = d_r^x(x, y) = a(\sin^2(\pi x) + \sin^2(\pi y)) + 1$. The ratio is then

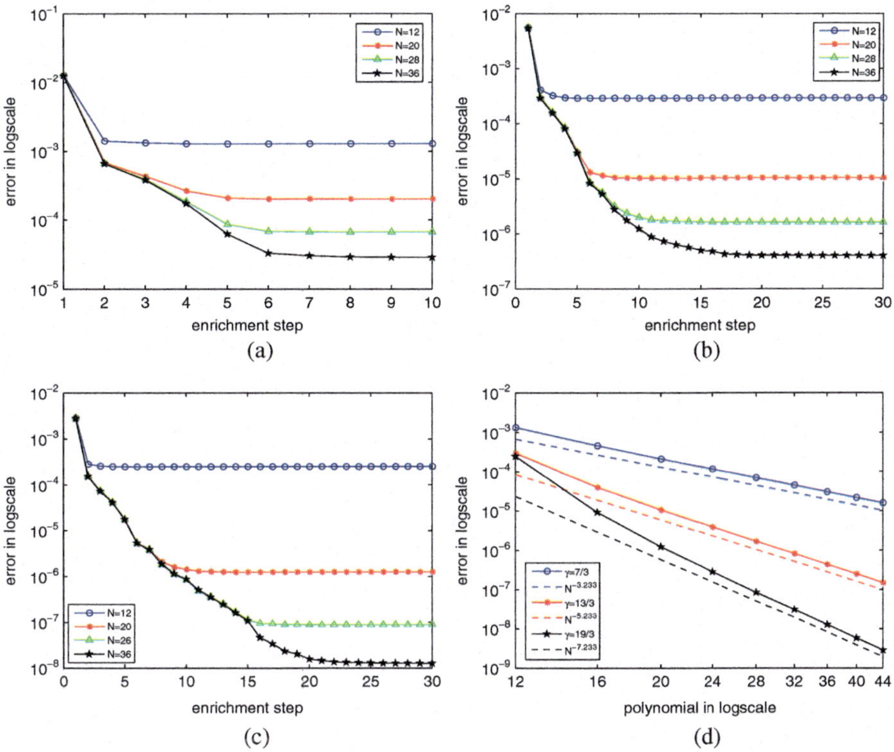

Fig. 2 Error decay versus the enrichment step (a, b, c) and polynomial degree (**d**) for singular solution. (**a**) $\delta = 7/3$. (**b**) $\delta = 13/3$. (**c**) $\delta = 19/3$

defined as

$$r := \frac{\max d_l^x(x,y)}{\min d_l^x(x,y)} = \frac{\max d_r^x(x,y)}{\min d_r^x(x,y)} = 2a + 1.$$

We solve the problem (3) with the above data and the stop criteria $\varepsilon(k) = 10^{-7}$. The number of iterations of the fixed point algorithm is fixed to 5. We list in Table 1 the minimum enrichment step number k required to reach the convergence for different N and r. From this table we observe very weak dependence (less than linearly) of the required enrichment step on the data contrast and polynomial degree N. This seems to confirm one of the key features of the PGD method, which remains true when used to approximate the solution of nonseparable fractional derivative elliptic problems.

In the last test, we measure the performance of the PGD method and the spectral method by comparing CPU time. Consider the problem (1)–(2) with $\alpha = 1.2$, $\beta = 1.8$, $\rho(x,y) = d_l^y(x,y) = d_r^y(x,y) = 1, d_l^x(x,y) = d_r^x(x,y) = \sin^2(\pi xy) + 2$, and $f = \sin(\pi x) + \sin(\pi y)$. We set the threshold $\epsilon = 10^{-7}$. The CPU time for

Table 1 Minimum enrichment step k to reach the convergence for different ratio r and polynomial degree N

N \ r	10	20	30	40	50	60
12	36	38	42	43	43	42
20	54	62	67	69	73	72
28	60	76	78	84	89	93
36	63	78	82	89	93	97

Table 2 CPU time comparison of spectral method and PGD method

N	10	20	30	40	50	60	70
SM(s)	0.0804	0.3070	0.8019	4.284485	19.0678	27.2764	159.4655
PGD(s)	0.2853	0.8354	1.0382	1.2716	1.9205	2.2717	3.0319

solving (6) by the PGD and SM are listed in Table 2. It is observed from this table that the PGD is much less expensive than the classical spectral method, especially when the polynomial degree becomes large. This is due to the fact that the condition number of the resulting algebra system for the spectral method increases quickly when the polynomial degree increases. large. It is worth emphasizing that the cost increase of the PGD is only a linear function of the polynomial degree.

6 Conclusion

In this paper, we constructed and tested a PGD/spectral method to approximate the solution of a kind of nonseparable fractional elliptic equations. The PGD has been known as a powerful model reduction technique to construct approximate solutions to some classical (integer order differential) problems. A main feature of the PGD is that its computational complexity grows only linearly with the spatial dimension, which is in contrast with the exponential growth of standard grid-based methods. The main goal of the current work is to verify the capability of the PGD when applied to nonseparable fractional elliptic equations. The latter have caused great difficulty for traditional methods due to their non-locality and low regularity. To demonstrate the efficiency of the PGD method, we have used it in the framework of spectral approximation, which has been considered as one of best methods for fractional differential equations. We provided a detailed description of the PGD/spectral method, and performed a series of numerical experiments to validate the proposed method. The numerical tests have been focused on investigating the accuracy, convergence rates of the enrichment algorithm, as well as the impact of the polynomial degree and data contrast on the convergence. The obtained numerical result seems to confirm the efficiency of the PGD method for fractional elliptic equations. Although only fractional equations of elliptic type have been considered in this paper, we believe that the PGD/spectral method can be applied to other type fractional equations such as the time fractional diffusion equation for which a suitable variational formulation exists [7]. However, it seems much care needs to be taken in constructing convergent PGD for non-elliptic type equations.

Acknowledgements This work was supported by IDEX-Bordeaux research programm and CAMPUS FRANCE PHC CAI YUANPEI Research program.
Chuanju Xu is supported by NSF of China (Grant numbers 11471274, 11421110001, 51661135011, and 91630204.

References

1. R.A. Adams, J.J.F. Fournier, *Sobolev Spaces*, vol. 140 (Academic, New York, 2003)
2. D.A. Benson, R. Schumer, M.M. Meerschaert, S.W. Wheatcraft, Fractional dispersion, Lévy motion, and the MADE tracer tests. Transp. Porous Media **42**(1), 211–240 (2001)
3. F. Chinesta, R. Keunings, A. Leygue, *The Proper Generalized Decomposition for Advanced Numerical Simulations: A Primer* (Springer Science & Business Media, New York, 2013)
4. K. Diethelm, *The Analysis of Fractional Differential Equations: An Application-Oriented Exposition Using Differential Operators of Caputo Type* (Springer, Berlin, 2010)
5. R. Gorenflo, F. Mainardi, *Fractional Calculus and Continuous-Time Finance III: The Diffusion Limit* (Birkhäuser, Basel, 2001)
6. X. Li, C. Xu, A space-time spectral method for the time fractional diffusion equation. SIAM J. Numer. Anal. **47**(3), 2108–2131 (2009)
7. X. Li, C. Xu, Existence and uniqueness of the weak solution of the space-time fractional diffusion equation and a spectral method approximation. Commun. Comput. Phys. **8**(5), 1016 (2010)
8. Z. Mao, J. Shen, Efficient spectral–Galerkin methods for fractional partial differential equations with variable coefficients. J. Comput. Phys. **307**, 243–261 (2016)
9. A. Oustaloup, P. Coiffet, Systèmes asservis linéaires d'ordre fractionnaire: théorie et pratique. Masson (1983)
10. Y.A. Rossikhin, M.V. Shitikova, Applications of fractional calculus to dynamic problems of linear and nonlinear hereditary mechanics of solids. Appl. Mech. Rev. **50**(1), 15–67 (1997)

High Order Edge Elements for Electromagnetic Waves: Remarks on Numerical Dispersion

Marcella Bonazzoli, Francesca Rapetti, Pierre-Henri Tournier, and Chiara Venturini

Abstract We recall one set of possible basis vector fields and two different sets of possible degrees of freedom, those related to "small-edges" and those defined by "moments", for the Nédélec's first family of high order edge elements. We thus address a dispersion analysis of the resulting methods, when the time-harmonic Maxwell's equation for the electric field is discretized on a simplicial mesh.

1 Introduction

Edge elements (edge-FEs) on simplices [4, 9] are perhaps the most widely used finite elements to approximate the electric field solution of the time-harmonic Maxwell's equations. They offer the simplest construction of polynomial discrete differential Whitney forms on complexes. Their associated degrees of freedom (dofs) have a very clear meaning as cochains and, thus, give a recipe for discretizing physical balance laws such as Maxwell's equations.

In the simulation of propagation phenomena, such as in seismology or in high-frequency electromagnetic transmissions, both high order accuracy and computational efficiency are mandatory. Interest thus grew for the use of high order schemes, such as *hp*-finite element or spectral element methods (see [6] for a presentation of these methods). In electromagnetism, high order extensions of Whitney forms have become an important computational tool [8]. High order

M. Bonazzoli (✉) • F. Rapetti
Laboratoire J.A. Dieudonné, CNRS & Université Côte d'Azur, Parc Valrose, 06108 Nice Cedex 02, France
e-mail: marcella.bonazzoli@unice.fr; francesca.rapetti@unice.fr

P.-H. Tournier
Laboratoire J.L. Lions, CNRS & Université Pierre et Marie Curie, B.P. 187, 4 Place Jussieu, 75252 Paris Cedex 05, France
e-mail: tournier@ljll.math.upmc.fr

C. Venturini
Università degli Studi di Verona, Via dell'Artigliere 8, 37129 Verona, Italy
e-mail: chiara.venturini_01@studenti.univr.it

M.L. Bittencourt et al. (eds.), *Spectral and High Order Methods for Partial Differential Equations ICOSAHOM 2016*, Lecture Notes in Computational Science and Engineering 119, DOI 10.1007/978-3-319-65870-4_14

edge-FEs are appreciated since they allow to reach higher accuracy at a fixed number of dofs and are compatible with a domain decomposition framework suitable for parallel computations (see a recent example of application in [2] for wave-guided transmissions). The popularity of high order finite elements for wave propagation problems is also due to the fact that they are characterized by low numerical dispersion and dissipation errors.

In this paper, the numerical dispersion of edge-FEs for a time-harmonic plane wave propagating through an "infinite" two-dimensional, simplicial mesh is investigated. Mathematical results confirm that, in the frame of edge-FEs, dispersion errors are not influenced by the set of basis functions we choose to write the discrete problem solution. Therefore, no matters which dofs we adopt, moments or circulations along small edges, dispersion errors will only depend on the mesh and on the wave direction in the mesh. Numerical results show that the dispersion relation is approximated to the order $2r$ accuracy with respect to the mesh size h, being r the polynomial degree of the edge basis functions.

2 Preliminaries on Model Problem and Dispersion Analysis

We consider the first-order two-dimensional Maxwell's equations to describe the behavior of the electric field $\mathbf{E} = (\mathrm{E}_x(\mathbf{x},t), \mathrm{E}_y(\mathbf{x},t))^t$ and magnetic field $\mathrm{H} = \mathrm{H}(\mathbf{x},t)$, where $\mathbf{x} = (x,y) \in \mathbb{R}^2$ and $t > 0$. Precisely, the fields satisfy[1]:

$$\partial_t \mathbf{E} - \mathbf{curl}\,\mathrm{H} = \mathbf{0}, \qquad \partial_t \mathrm{H} + \mathrm{curl}\,\mathbf{E} = 0. \tag{1}$$

In (1), the constitutive parameters ϵ and μ have been set to one so that the speed $c = 1/\sqrt{\epsilon\mu}$ is one. Initial and boundary conditions (or conditions at infinity) are required to specify uniquely a solution of the Maxwell's system (1). However, in a dispersion analysis we are interested in the plane wave solutions of the equations: initial and boundary conditions are then ignored. To analyze the dispersion at a single frequency, the time-dependent problem (1) is generally reformulated as a time-harmonic problem, by seeking solutions of the form:

$$\mathbf{E} = \Re(\mathbf{u}(\mathbf{x})\, e^{\mathtt{i}\,\omega\, t}), \qquad \mathrm{H} = \Re(\mathrm{v}(\mathbf{x})\, e^{\mathtt{i}\,\omega\, t}), \tag{2}$$

where $\mathtt{i}$ denotes the imaginary unit, and $\mathbf{u}(\mathbf{x})$ (resp. $\mathrm{v}(\mathbf{x})$) is a complex-valued vector (resp. scalar) function of the position $\mathbf{x}$ but not of the time t. The angular frequency ω and the wave vector $\kappa = (\kappa_x, \kappa_y)^t$ are independent of $\mathbf{x}$ and t. Substituting the

[1] In two spatial dimensions, say x, y, we denote by ∂_x (resp. ∂_y) the first order operator that associates any differentiable scalar function g with its partial derivative $\partial_x g$ (resp. $\partial_y g$) w.r.t. the variable x (resp. y). For any vector field $\mathbf{u} \in \mathbb{R}^2$, with $\mathbf{u} = (u_x, u_y)^\top$, we define $f = \mathrm{curl}\,\mathbf{u}$ such as the scalar function $f = \partial_x u_y - \partial_y u_x$; for any scalar function $v \in I\!R$, we define $\mathbf{w} = \mathbf{curl}\, v$ such as the vector field $\mathbf{w} = (\partial_y v, -\partial_x v)^\top$. Note that $\mathbf{curl}\, v = (\mathbf{grad}\, v)^\perp$, where $\mathbf{g} = \mathbf{grad}\, v = (\partial_x v, \partial_y v)^\top$.

form (2) for **E** and H in Eq. (1), we obtain the first order time-harmonic formulation of the Maxwell's system:

$$\mathrm{i}\,\omega\,\mathbf{u} - \mathbf{curl}\,\mathrm{v} = \mathbf{0}, \qquad \mathrm{i}\,\omega\,\mathrm{v} + \mathrm{curl}\,\mathbf{u} = 0. \tag{3}$$

We eliminate the field v by solving the second equation in (3) for v and replacing it into the first equation in (3), to get a second order time-harmonic formulation of the Maxwell's system for the electric field

$$\mathbf{curl}\,\mathrm{curl}\,\mathbf{u} - \omega^2\,\mathbf{u} = \mathbf{0}. \tag{4}$$

The angular frequency ω and the wave vector κ are linked by the dispersion relation $\omega = |\kappa|$ (here $c = 1$). From this relation, we can compute the phase speed $\omega/|\kappa|$, that is exactly one for the continuous problem but won't be one when a numerical scheme is used to approximate (4). To analyze the dispersion behavior of a finite element method for (4), we go through the following steps:

1. We suppose that the plane $\mathbb{R}^2$ is tiled by an "infinite" uniform triangulation τ_h and we consider edge-FEs of polynomial degree $r \geq 1$ in each triangle of τ_h.
2. For the chosen edge-FE method, we compute the equations satisfied by the degrees of freedom of the method for the given mesh.
3. We seek plane-wave solutions of the discrete equations: we thus have a matrix eigenvalue problem to solve and we select an approximated value ω_h for ω.
4. We then define the relative error $e_c = c_h/c - 1$ in the phase speed, where the grid speed c_h value depends on the computed ω_h and on the number of points per wavelength, the latter being related to the polynomial degree r of the finite element basis functions.

The discrete version of the dispersion relation shows how plane waves will propagate, namely, if they are delayed ($e_c < 0$) or accelerated ($e_c > 0$). In describing the edge-FE scheme, we shall make specific choices of the dofs for the electric variable **u**. The choice of dofs yields a particular cardinal basis function set for the finite element space. However, we will prove that the dispersion error does not depend on the fact of working with cardinal or not cardinal basis functions. Thus, the choice of the dofs has no influence of the dispersion behavior of the method, provided that the choice is conforming and unisolvent for the considered finite element space. In the next sections, we detail steps 1–4.

3 Edge-FEs of Polynomial Degree $r \geq 1$

To define high order edge reconstructions of vector fields, one can rely on [9], precisely on [9, Definition 2] for the basis functions and on [9, Definition. 4] for the needed dofs, the well-known moments. The high order basis functions for the Nédélec's first family of edge-FEs presented in [10, 11] were born instead from the

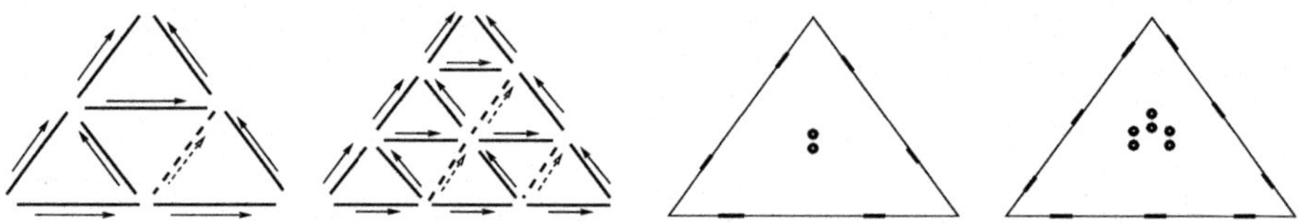

Fig. 1 The oriented small edges, in *solid line*, that support dofs (5) and allow also to list the edge-FE basis functions of degree $r = 2$ and $r = 3$, resp. (*two figures on the left*). Localization of boundary (*thick lines on edges*) and interior (*thick rounds in the triangle*) dofs (6), (7), for edge-FEs of degree $r = 2$ and $r = 3$, resp. (*two figures on the right*)

effort of addressing the localization issue, to answer to questions such as "What kind of cochains such elements should be associated with ?" or "Can the corresponding dofs be still assigned to edges, as it occurs for low order edge-FEs ?". This is the "duality" feature.[2]

To state the definition of edge-FE basis functions for $r > 1$, inside each triangle of the mesh, which will be called *big triangle*, we consider an increasing number of *small triangles* homothetic to the big one, associated with the principal lattice of the big triangle. To formalize them, let $\mathbf{k}$, boldface, be the array (k_1, k_2, k_3) of 3 integers $k_i \geq 0$, and denote by k its weight $\sum_{i=1}^{3} k_i$. The set of multi-indices $\mathbf{k}$ with three components and of weight k is denoted $\mathcal{I}(3, k)$. Given $\mathbf{k} \in \mathcal{I}(3, k)$, we set $\lambda^{\mathbf{k}} = (\lambda_1)^{k_1}(\lambda_2)^{k_2}(\lambda_3)^{k_3}$, where λ_ℓ is the barycentric coordinate w.r.t. n_ℓ in the triangle $T = \{n_1, n_2, n_3\}$. To introduce the *small edges* in T, one first considers the principal lattice of order r in T as the set

$$\mathcal{T}_r(T) = \left\{ \mathbf{x} \in T,\ \lambda_j(\mathbf{x}) \in \{0,\ \frac{1}{r},\ \frac{2}{r},\ \ldots,\ \frac{(r-1)}{r},\ 1\},\ 1 \leq j \leq 3 \right\}.$$

By connecting the points of $\mathcal{T}_r(T)$ by segments parallel to the big edges of T, one produces a partition of T into triangles (see Fig. 1): in this partition of T, the small triangles are only those homothetic to T and the *small edges* are the edges of these small triangles. We can relate a small edge to a big edge e and a multi-index $\mathbf{k}$. The *small edge* $\{\mathbf{k}, e\}$ can be found, given e and $\mathbf{k}$, as the image of e through the mapping $\tilde{\mathbf{k}}$: each mapping $\tilde{\mathbf{k}}$ is associated with a multi-index $\mathbf{k}$ as the homothety which maps a point $\mathbf{x} \in T$ onto the point of T with barycentric coordinates $\tilde{k}_i(\lambda_i(\mathbf{x})) = \frac{\lambda_i(\mathbf{x})+k_i}{k+1}$. Therefore, given a multi-index $\mathbf{k}$, we can build 3 basis functions which are associated with the small edges $\{\mathbf{k}, e\}$ of the same small triangle; in practice, to find the small edge we can say that $\{\mathbf{k}, e\}$ is parallel to the

[2]The definition of Whitney forms relies on duality features from algebraic topology. Let w^e be the Whitney edge forms, where e belongs to the set of mesh edges $\mathcal{E}$. A field v, represented by the 1-form $\sum_e \mathrm{v}_e w^e$, with $\mathrm{v}_e = \int_e \mathrm{v}$, and a curve γ, represented by the 1-chain $\sum_e \alpha_e\, e$, with $\alpha_e = \int_\gamma w^e$, are in duality via the formula $\int_\gamma \mathrm{v} = \sum_{e\in\mathcal{E}} \alpha_e \mathrm{v}_e$. What is meant by *in duality* is that $\int_\gamma \mathrm{v} = 0$ $\forall\gamma$ implies $\mathrm{v} = 0$ and the other way around. Duality stems from the property of edge elements, $\int_{e'} w^e = \delta_{e'e}$. Note that the same forms w^e are involved in the description of both fields and curves: this duality is the key to understand why Whitney elements have the expression they have [3].

big edge e, and that each component of $\mathbf{k}$ says how near the small triangle is to each vertex of the big triangle (higher is k_i nearer is the small triangle to the i-th vertex). The orientation of $\{\mathbf{k}, e\}$ is given by the orientation of e.

Definition 1 ([11], De. 3.3) The space $W^1_{h,r}(T)$ of the Nédélec's first family of edge-FEs on the triangle T, of degree $r = k + 1$ ($k \geq 0$), is spanned by the vectors $\lambda^{\mathbf{k}}\mathbf{w}^e$, with $\mathbf{k} \in \mathscr{I}(3, k)$, $e \in \mathscr{E}(T)$. For the edge $e = \{n_i, n_j\}$, we set $\mathbf{w}^e = \lambda_i \nabla \lambda_j - \lambda_j \nabla \lambda_i$.

Definition 1 yields more functions than necessary. So, in the triangle $T = \{n_1, n_2, n_3\}$, to get a basis of $W^1_{h,r}(T)$, we neglect all functions of the form $\lambda^{\mathbf{k}}\mathbf{w}^e$ such that the small edge $\{\mathbf{k}, e\}$ is interior to T and parallel to the edge $\{n_1, n_3\}$ of T (the dashed ones in Fig. 1). It is proven in [5] that circulations on small-edges (5) are $W^1_{h,r}(T)$-unisolvent dofs, for any $r \geq 1$.

Definition 2 ([11], Sec. 3.4) For $r \geq 1$ ($k \geq 0$), the functional

$$\sigma_m = \sigma_{\{\mathbf{k},e\}} : \mathbf{u} \rightarrow \frac{1}{|\mathbf{t}_e|} \int_{\{\mathbf{k},e\}} \mathbf{u} \cdot \mathbf{t}_e \qquad (m \leftrightarrow \{\mathbf{k}, e\}) \tag{5}$$

for any small edges $\{\mathbf{k}, e\}$, with $\mathbf{k} \in \mathscr{I}(3, r-1)$, $e \in \mathscr{E}(T)$ and $\mathbf{t}_e$ the vector of length $|e|$ tangent to e, is a dof for vector functions $\mathbf{w} \in W^1_{h,r}(T)$. Note that only the small edges $\{\mathbf{k}, e\}$ associated with linearly independent generators have to be considered.

In Section 1.2 of [9] we have classical $W^1_{h,r}(T)$-unisolvent dofs, called "moments", for any $r \geq 1$. By relying on the generators introduced in Definition 1, the functionals in [9] can be recast in a new more friendly form.

Definition 3 ([1], Def. 6) For $r \geq 1$ ($k \geq 0$), the functionals

$$\sigma_e : \mathbf{w} \mapsto \tfrac{1}{|e|} \textstyle\int_e (\mathbf{w} \cdot \mathbf{t}_e)\, q \qquad \forall\, q \in \mathscr{P}_{r-1}(e),\ \forall\, e \in \mathscr{E}(T), \tag{6}$$

$$\sigma_f : \mathbf{w} \mapsto \tfrac{1}{|f|} \textstyle\int_f (\mathbf{w} \cdot \mathbf{t}_{f,i})\, q, \quad \forall\, q \in \mathscr{P}_{r-2}(f) \quad (\text{here } f \equiv T), \quad \mathbf{t}_{f,i} \text{ two independent sides of } f,\ i = 1, 2 \tag{7}$$

where each $\mathbf{t}_e$, the vector of length $|e|$, is tangent to e, are the dofs for a vector function $\mathbf{w} \in W^1_{h,r}(T)$. A convenient choice for the functions q spanning the polynomial spaces over (sub)simplices e,f is given by suitable products of the barycentric coordinates associated with the nodes of the considered (sub)simplex.

Here, σ_e (resp. σ_f) denotes one of those dofs which involve the edge e (resp. the face $f \equiv T$) in their definition. For $r \geq 1$, they are in number equal to the dimension of the space $\mathscr{P}_{r-1}(e)$ (resp. twice that of $\mathscr{P}_{r-2}(T)$, therefore an additional label, that we have omitted, should be added to distinguish the one from the other. In Definition 3, only dofs which involve polynomials of non-negative degree are meaningful (e.g., σ_f is not defined for $r = 1$).

When $r > 1$, fields in Definition 1 are not cardinal functions neither for dofs in Definition 2 (see Table 2 in [10]) nor for dofs in Definition 3 (see Example 2 in [1]), namely, the matrix $\mathbf{V}$ with entries $\mathbf{V}_{ij} = \sigma_i(\mathbf{w}^j)$, $i,j = 1, \dim(W^1_{h,r}(T))$

after a suitable renumbering of dofs, is not the identity matrix for $r > 1$. Cardinal basis functions are generally required to introduce the considered FE in an existing software. They can be easily computed by considering suitable linear combinations of the basis functions of Definition 1 with coefficients given by the entries of $\mathbf{V}^{-1}$, as described in, e.g., [1]. The independence of a dof from the metric of the sub-simplex of T that supports the dof makes that the entries of $\mathbf{V}$ (and thus of $\mathbf{V}^{-1}$) can be computed once on a generic triangle T and are valid in any other triangle T' different from T, up to a suitable orientation of the edges and choice of independent sides in T'. However, working with basis functions, not necessarily cardinal functions, does not influence the dispersion analysis, as we prove in Sect. 5.

4 The Discrete Problem

The triangulation τ_h of $\mathbb{R}^2$ we shall use is the "infinite" uniform mesh generated by the translates, in the x and y directions, of the square cell $\Omega = [0,h] \times [0,h]$, that we divide into two right-triangles using the right-bottom to top-left diagonal (see Fig. 2). The cell Ω, with sides f_q, and neighboring cells Ω_q, for $q = \{R, L, T, B\}$, is composed by two triangles, as a single triangle could not be the generator of a periodic pattern.

We thus seek the approximate solution of (4) on the cell Ω that behaves like a plane wave when translated through a distance of h in the x or y direction. We are computing a very coarse, just two triangles, approximation to the problem (4) under the periodic conditions $\mathbf{u}\cdot\mathbf{t}_{|f_R} = \mathbf{u}\cdot\mathbf{t}_{|f_L}$, $\mathbf{u}\cdot\mathbf{t}_{|f_T} = \mathbf{u}\cdot\mathbf{t}_{|f_B}$ on the boundary $\partial\Omega = f_R \cup f_L \cup f_T \cup f_B$ (here $\mathbf{t}_{|f_q}$ denotes the unit tangent vector to the curve f_q). We first write the weak formulation of Eq. (4) that is suitable for finite element discretizations. We take the dot product of (4) with a vector test function $\mathbf{v}$, integrate over the domain Ω and use Stokes's theorem for the term containing the curl of $\mathbf{u}$.

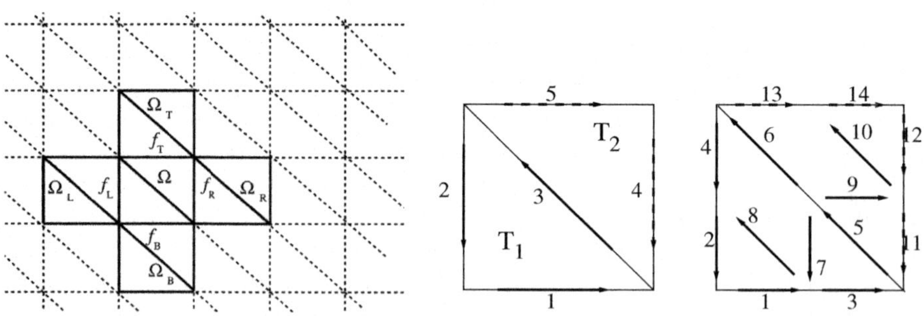

Fig. 2 The simplest periodic grid for a triangle-based FEM in $\mathbb{R}^2$ (*left*, as considered in [7]). In the two figures, *center* and *right*, *small dark/dashed arrows* indicate the oriented small edges $\{\mathbf{k}, e\}$ in Ω, with $\mathbf{k} \in \mathcal{I}(3, r-1)$ and $e \in \mathcal{E}(T_1) \cup \mathcal{E}(T_2)$, for $r = 1, 2$, respectively. An example of dof numbering is provided: dofs supported by *dashed segments* are numbered, for simplicity, at the end of the dof list as they will be expressed in terms of dofs supported by *solid segments*

We thus obtain: find $\mathbf{u} \in \mathscr{V}_P$ s.t.

$$\int_\Omega (\text{curl}\, \mathbf{u})(\text{curl}\, \mathbf{v}) - \omega^2 \int_\Omega \mathbf{u} \cdot \mathbf{v} = 0, \qquad \forall \mathbf{v} \in \mathscr{V}_0. \tag{8}$$

The space $\mathscr{V}_P$ is the set of vector fields in $H(\text{curl}, \Omega)$ satisfying the periodic conditions on $\partial\Omega$, with $H(\text{curl}, \Omega) = \{\mathbf{v} \in L^2(\Omega)^2,\ \text{curl}\, \mathbf{v} \in L^2(\Omega)\}$ and $\mathscr{V}_0 = \{\mathbf{v} \in \mathscr{X},\ (\mathbf{v} \cdot \mathbf{t})_{|\partial\Omega} = 0\}$. Problem (8) admits a unique solution $\mathbf{u} \in H(\text{curl}, \Omega)$ [8]. By introducing a finite dimensional space $\mathscr{V}_h$ which is a suitable approximation of $\mathscr{V}_P$ over Ω, the discrete equivalent of (8) reads: find $\mathbf{u}_h \in \mathscr{V}_h$ s.t.

$$\int_\Omega (\text{curl}\, \mathbf{u}_h)(\text{curl}\, \mathbf{v}) - \omega^2 \int_\Omega \mathbf{u}_h \cdot \mathbf{v} = 0, \qquad \forall \mathbf{v} \in \mathscr{V}_{0,h}. \tag{9}$$

In the present case, we consider $\mathscr{V}_h = \{\mathbf{v} \in H(\text{curl}, \Omega),\ \mathbf{v}_{|T} \in W^1_{h,r}(T),\ \forall T \in \tau_h,\ \mathbf{v} \cdot \mathbf{t}_{|f_R} = \mathbf{v} \cdot \mathbf{t}_{|f_L},\ \mathbf{v} \cdot \mathbf{t}_{|f_T} = \mathbf{v} \cdot \mathbf{t}_{|f_B}\}$ and $\mathscr{V}_{0,h} = \{\mathbf{v} \in H(\text{curl}, \Omega),\ \mathbf{v}_{|T} \in W^1_{h,r}(T),\ \forall T \in \tau_h,\ \mathbf{v} \cdot \mathbf{t}_{|\partial\Omega} = 0\}$. Problem (9) has a unique solution in $\mathscr{V}_h$ [8].

5 The Eigenvalue Problem

We need to select a basis for $\mathscr{V}_h$ to rewrite (9) as an algebraic linear system. We start by defining in each triangle $T \in \tau_h$ the vectors $\psi_m = \mathbf{w}_m$ where $\{\mathbf{w}_m\}$ is a set of basis functions for $W^1_{h,r}(T)$ as defined in Definition 1. Let D_r be the dimension of the vector space $\mathscr{V}_h$. We write the trial functions $\mathbf{u}_h \in \mathscr{V}_h$ as linear combination of basis functions as follows: $\mathbf{u}(\mathbf{x}) \approx \mathbf{u}_h(\mathbf{x}) = \sum_{l=1}^{D_r} U_j\, \mathbf{w}_{m(l)}$, with $m(j)$ the local index of the unknown with global number j. Replacing in (9) $\mathbf{u}$ by $\mathbf{u}_h$ and $\mathbf{v}$ by the $\mathbf{w}_{m(i)}$ we obtain a linear system

$$KU = \eta M\, U, \tag{10}$$

where $\eta = \omega^2$ can be seen as an eigenvalue of a generalized eigenvalue problem if also ω, and not only $\mathbf{u}_h$, is sought in (9). We have $M = \sum_T M_T$, $K = \sum_T K_T$, being $\sum_T$ the assembling procedure on the elemental matrices M_T, K_T, that are computed for each $T \in \tau_h$ as follows:

$$(M_T)_{ij} = \int_T \mathbf{w}_{m(i)} \cdot \mathbf{w}_{m(j)}, \qquad (K_T)_{ij} = \int_T \text{curl}\mathbf{w}_{m(i)}\, \text{curl}\mathbf{w}_{m(j)}.$$

Proposition 1 *If M_T (resp. K_T) is the local mass (resp. stiffness) matrix on a triangle T for the basis functions $\mathbf{w}_{m(j)}$, then the local mass (resp. stiffness) matrix $\tilde{M}_T$ (resp. $\tilde{K}_T$) for the cardinal basis functions $\tilde{\mathbf{w}}_{m(j)}$ is $\tilde{M}_T = V^{-t} M_T V^{-1}$ (resp. $\tilde{K}_T = V^{-t} K_T V^{-1}$).*

Proof Let us denote $C = V^{-1}$, where V is the square matrix of size n_{dofs} with entries $V_{\ell j} = \sigma_\ell(\mathbf{w}_j)$, being σ_ℓ the functionals given in either Definition 2 or Definition 3. For given i, j, since $\tilde{\mathbf{w}}_i = \sum_{k=1}^{n_{\text{dofs}}} C_{ki}\, \mathbf{w}_k$ and $\tilde{\mathbf{w}}_j = \sum_{\ell=1}^{n_{\text{dofs}}} C_{\ell j}\, \mathbf{w}_\ell$, we have

$$\tilde{M}_{ij} = \int_T \tilde{\mathbf{w}}_i \cdot \tilde{\mathbf{w}}_j = \sum_{k=1}^{n_{\text{dofs}}} \sum_{\ell=1}^{n_{\text{dofs}}} C_{ki} C_{\ell j} \int_T \mathbf{w}_k \cdot \mathbf{w}_\ell = \sum_{k=1}^{n_{\text{dofs}}} \sum_{\ell=1}^{n_{\text{dofs}}} (C^t)_{ik} M_{k\ell} C_{\ell j} = (C^t M C)_{ij}.$$

Proposition 2 *Working with either the basis functions $\mathbf{w}_j$ or the cardinal basis functions $\tilde{\mathbf{w}}_i$, for a set of dofs, does not modify the spectrum of problem (10).*

Proof As stated in Proposition 1, replacing the basis functions $\mathbf{w}_j$ by the cardinal basis functions $\tilde{\mathbf{w}}_j$ yields new matrices $\tilde{M}_T = \sum_T C^t M_T C$, $\tilde{K}_T = \sum_T C^t K_T C$ (with $C = V^{-1}$). Let (η, U), with $U \neq 0$, be a solution of the system $\tilde{K}_T U = \eta \tilde{M}_T U$. Then, (η, W), with $W = CU$, is a solution of the system (10). Indeed,

$$\eta = \frac{(\tilde{K}_T U, U)}{(\tilde{M}_T U, U)} = \frac{(K_T CU, CU)}{(M_T CU, CU)} = \frac{(K_T W, W)}{(M_T W, W)}.$$

For a standard dispersion analysis of (10), we seek plane-wave solutions moving in a given direction κ in Ω with periodic boundary conditions on the faces f_R, f_T (cf. Fig. 2, left). To visualize which dofs are concerned by periodic conditions on the considered fields, one can look at Fig. 2, right.

The (cardinal or not) basis functions $\psi^{\Omega_\dagger}$ with support in $\Omega_\dagger = \Omega \cup_q \Omega_q$ where $q \in \{T, B, L, R\}$ (cf. Fig. 2, left) are trial functions in $\mathscr{V}_h$; the (cardinal or not) basis functions ψ^{Ω} with support in Ω, prolongated by zero outside Ω, are test functions in $\mathscr{V}_{0,h}$. By rewriting Eq. (10), we obtain a rectangular linear system in the unknown U where

$$U = [U^{\Omega}, U^{\Omega_T}, U^{\Omega_B}, U^{\Omega_L}, U^{\Omega_R}].$$

Clearly the system on U is under-determined because the number of columns, $5[2r(r+2) - r] - 4r$, exceeds the number of rows, $[2r(r+2) - 3r]$. To reduce it into a square linear system we work in Ω and make use of the following plane wave periodicity hypothesis. Let us assume that the electric field is a plane wave of amplitude U_0 of modulus 1, i.e., we have $U^{\Omega} = U_0\, e^{i(\kappa \cdot \mathbf{x})}$, for $\mathbf{x} \in \Omega$, where $\kappa = (\kappa_x \cos\theta, \kappa_y \sin\theta)^\top$ is the wave vector and θ the incidence angle of the wave w.r.t. the x-direction. The above periodicity assumption, imposed for the unknown dof U_j supported by the face f_q of Ω, yields

$$(U_{j_q})_{|f_q} = Q^{f_q, f_p} (U_{j_p})_{|f_p}, \qquad f_q = \{f_R, f_T\},\ f_p = \{f_L, f_B\}. \tag{11}$$

In (11), $Q_{f_q f_p} = e^{\beta_{f_q}} \mathrm{I}_r$ where I_r denotes the identity matrix of size equal to the number r of dofs supported by f_q (same for f_p) and $\beta_{f_q} = \{-i\kappa_x h, -i\kappa_y h\}$. Following the numbering defined in Fig. 2, for $r = 1$, conditions (11) read

$$U_4 = Q_{f_R f_L}\, U_2, \quad \text{with } Q_{f_R f_L} = +e^{-i\kappa_x h}\, \mathrm{I}_1,$$
$$U_5 = Q_{f_T f_B}\, U_1, \quad \text{with } Q_{f_T f_B} = +e^{-i\kappa_y h}\, \mathrm{I}_1.$$

For $r = 2$, we have

$$\begin{pmatrix} U_{11} \\ U_{12} \end{pmatrix} = Q_{f_R f_L} \begin{pmatrix} U_2 \\ U_4 \end{pmatrix}, \quad \text{with } Q_{f_R f_L} = \begin{pmatrix} e^{-i\kappa_x h} & 0 \\ 0 & e^{-i\kappa_x h} \end{pmatrix} = +e^{-i\kappa_x h}\, \mathrm{I}_2,$$
$$\begin{pmatrix} U_{13} \\ U_{14} \end{pmatrix} = Q_{f_T f_B} \begin{pmatrix} U_1 \\ U_3 \end{pmatrix}, \quad \text{with } Q_{f_T f_B} = \begin{pmatrix} e^{-i\kappa_y h} & 0 \\ 0 & e^{-i\kappa_y h} \end{pmatrix} = +e^{-i\kappa_y h}\, \mathrm{I}_2.$$

When the faces linked by periodic conditions, say f_q, f_p, have not the same orientation, a minus sign has to be put in front of the quantity $e^{\beta_{f_q}}$ when defining $Q_{f_q f_p}$ in (11). Note that looking for $\mathbf{u} \in \mathcal{V}_h(\Omega)$, with thus periodicity conditions, corresponds to seeking a solution $\mathbf{u} \in \mathcal{V}_h(\Omega)$ verifying (11) on f_R and f_T, for example. We thus set $U = Q\, U_{glo}$, $K_{glo} = Q^t K Q$ and $M_{glo} = Q^t M Q$, with M, K defined as before. The rectangular matrix Q is composed of the blocks $Q_{f_q,\Omega}$, for those sets of dofs supported on f_q ($q = R$ and $q = T$) which are eliminated from U, because of the periodic boundary conditions, and of the identity matrix, otherwise. The square linear system, of size $[2r(r + 2) - 3r]$, to solve for (ξ, U_{glo}), reads

$$K_{glo} U_{glo} = \xi M_{glo} U_{glo}. \tag{12}$$

Matrices K_{glo} and M_{glo} are symmetric and positive semi-definite, thus all eigenvalues are real, of the form $\xi = \omega_h^2$, where ω_h is the angular frequency at which the wave travels in the grid. To identify which eigenvalue corresponds to the plane wave, we calculate the velocities (one for each eigenvalue) and compare them to the known value ω. Let us denote by ξ_c the eigenvalue successful in that comparison (thus the closest to ω^2): the angular frequency of the wave in the grid is given by $\omega_h = \sqrt{\xi_c}$.

6 Relative Error on the Wave Speed and Numerical Results

The grid dispersion of the wave can be given by the ratio between the velocity at which the wave travels in the grid and the physical velocity. The angular frequency of the wave in the grid is given by $\omega_h = \sqrt{\xi_c}$, therefore, the velocity at which the wave travels in the grid is given by

$$c_h = \frac{h\,\omega_h}{2\,\pi\,\delta} = \frac{h}{2\,\pi\,\delta}\sqrt{\xi_c},$$

where δ is the sampling ratio (the inverse of the number of points per wavelength L). In the case of polynomial degree $r = 1$, then $\delta = h/L$. Indeed, the size h of the mesh elements can also be interpreted as the distance between two adjacent nodes thus we have around L/h sampling nodes, that is exactly $1/\delta$. In the high-order case, $r > 1$, we have around r nodes in the length h, therefore the distance between two adjacent nodes is not h but h/r. We thus have $L/(h/r) = rL/h$ sampling nodes and

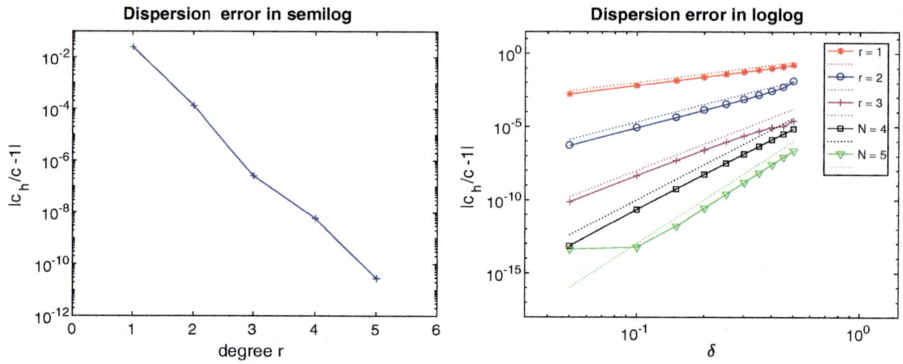

Fig. 3 Grid dispersion versus the approximation degree r ($\delta = 0.2$, $\theta = \pi/12$, *left*) and versus the sampling ratio δ ($r = 1, 2, 3, 4, 5$, $\theta = \pi/12$, *right*)

Table 1 Grid dispersion error values w.r.t. r, for $\theta = \frac{\pi}{12}$ and five nodes per wavelength ($\delta = 0.2$)

r	1	2	3	4	5
$\max_{0\le\theta\le2\pi} \lvert e_c\rvert$	2.42e−2	1.28e−4	2.52e−7	2.59e−9	2.75e−11

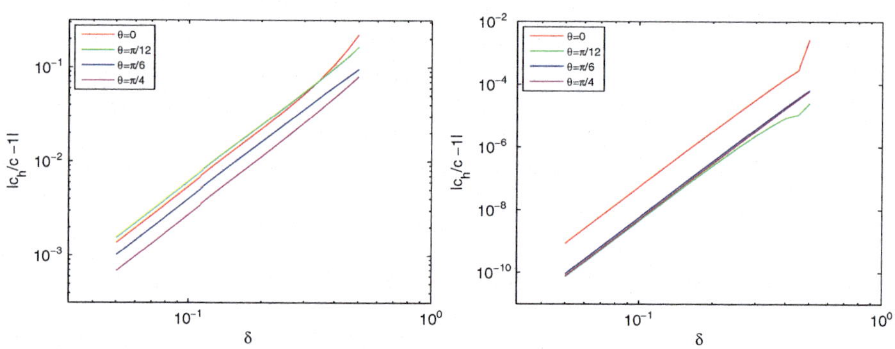

Fig. 4 Grid dispersion versus δ for different values of θ, with $r = 1$ (*left*) and $r = 3$ (*right*)

we rather set $\delta = h/(rL)$. It is convenient to measure grid dispersion as the relative error in the velocities, given by $e_c = (c_h - c)/c = (c_h/c) - 1$. The sign of the error indicates if the numerical scheme causes the waves to be delayed or accelerated. The grid dispersion error depends on the sampling ratio δ, the wave-vector κ and the degree r of the basis functions.

In Fig. 3 (left) we show the grid dispersion errors of Table 1 w.r.t. the approximation degree r of the edge-FE basis functions. The r-convergence is spectral, i.e. aligned error values in a semi-log plot. The grid dispersion as a function of the sampling ratio δ is presented in Fig. 3 (right) for the degrees $r = 1, 2, 3, 4, 5$. This convergence is algebraic, i.e., $|e_c| = O(h^p)$ with the order of convergence

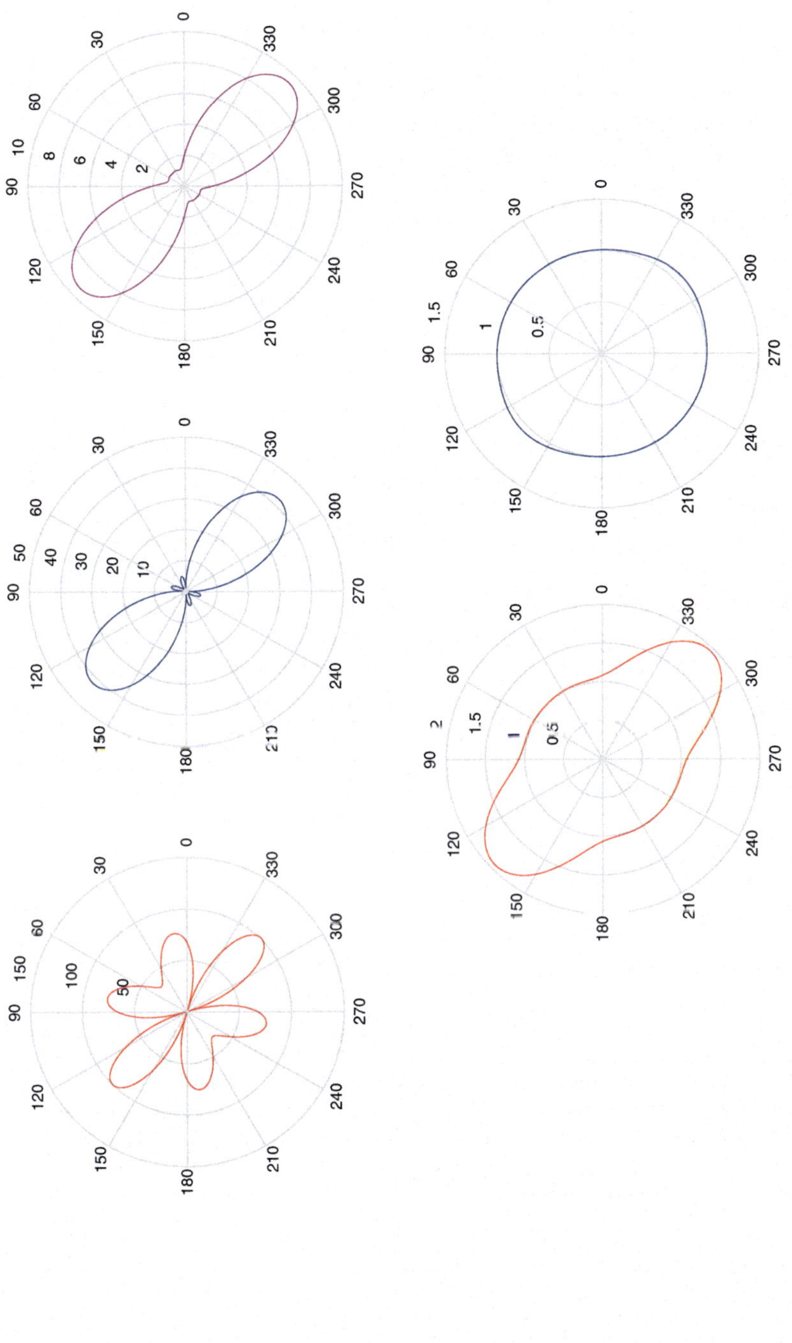

Fig. 5 Anisotropy curves: c_h/c varying the incidence angle θ, with sampling ratio $\delta = 0.2$ and approximation degree $r = 1$ (*top, left*), $r = 2$ (*top, center*), $r = 3$ (*top, right*), $r = 4$ (*bottom, left*) and $r = 5$ (*bottom, right*). For visualization purposes, the grid dispersion has been magnified by a factor $3 \times \{10^3, 10^4, 10^5, 10^6, 10^7\}$, respectively

p estimated by the slope of the represented lines. From the results reported in Fig. 3 (right), the h-convergence appears of order $p = O(2r)$, as indicated by the asymptotic lines. Here $\theta = \frac{\pi}{12}$ but similar figures can be drawn for $\theta = 0, \frac{\pi}{6}, \frac{\pi}{4}$. Indeed, as shown in Fig. 4, the lines that report on the dispersion error, for a fixed degree r, as a function of δ for different angles θ are parallel to each other. Note that the smaller is θ, the higher is $|e_c|$, for a fixed value of δ.

In Fig. 5, we show the anisotropy, namely, the dependence of the ratio c_h/c on the angle θ, introduced by the numerical scheme when $r = 1, 2, 3, 4, 5$ and $\delta = 0.2$. The case $r = 1$ in Fig. 5 is in agreement with the results presented in the literature for this mesh (see Figure 4 in [12]). The anisotropy dramatically decreases as the polynomial degree increases. Note that the direction of higher anisotropy is from top-left to bottom-right in which the mesh squares have been cut into triangles.

7 Conclusions

In these pages, we have analyzed the numerical dispersion of the high order edge-FE method applied to the second order time-harmonic Maxwell's equations on a two-dimensional simplicial mesh. We have proved that the dispersion error does not change if the basis functions are not cardinal functions for the selected dofs. This error is thus independent from the adopted dofs, provided that the choice is unisolvent and compatible with the considered FE space. The dispersion error depends both on the type of mesh and on the plane wave vector κ; it decreases algebraically as $O(\delta^p)$, with $p = O(2r)$, w.r.t. the sampling ratio δ, thus the mesh element size h, and spectrally w.r.t. the approximation degree r.

Acknowledgements The authors acknowledge the French National Research Agency (ANR) for its financial support (project MEDIMAX, ANR-13-MONU-0012). The last author warmly thanks the Università degli Studi di Verona for the possibility of studying as ERASMUS fellow at the Université Côte Azur in Nice.

References

1. M. Bonazzoli, F. Rapetti, High order finite elements in numerical electromagnetism: degrees of freedom and generators in duality. Numer. Algorithms **74**(1), 111–136 (2017)
2. M. Bonazzoli, V. Dolean, F. Rapetti, P.-H. Tournier, Parallel preconditioners for high order discretizations arising from full system modeling for brain microwave imaging. Int. J. Numer. Modell. Electron. Netw. Devices Fields. doi:10.1002/jnm.2229 [math.NA] (2017, in press)
3. A. Bossavit, *Computational Electromagnetism* (Academic, New York, 1998)
4. A. Bossavit, J.C. Vérité, A mixed FEM-BIEM method to solve eddy-current problems. IEEE Trans. Magn. **18**, 431–435 (1982)
5. S.H. Christiansen, F. Rapetti, On high order finite element spaces of differential forms. Math. Comput. **85**(298), 517–548 (2016)

6. G.E. Karniadakis, S.J. Sherwin, *Spectral/hp Element Methods for CFD* (Oxford University Press, New York, 1999)
7. I. Mazzieri, F. Rapetti, Dispersion analysis of triangle-based spectral element methods for elastic wave propagation. Numer. Algorithms **60**, 631–650 (2012)
8. P. Monk, *Finite Element Methods for Maxwell's Equations*. (Oxford Science Publications, New York, 2003)
9. J.-C. Nédélec, Mixed finite elements in R^3. Numer. Math. **35**, 315–341 (1980)
10. F. Rapetti, High order edge elements on simplicial meshes. Meth. Math. en Anal. Num. **41**(6), 1001–1020 (2007)
11. F. Rapetti, A. Bossavit, Whitney forms of higher degree. SIAM J. Numer. Anal. **47**(3), 2369–2386 (2009)
12. G.S. Warren, W.R. Scott, Numerical dispersion in the finite-element method using triangular edge elements. Microw. Opt. Technol. Lett. **9**(6), 315–319 (Wiley, New York, 1995)

A Mimetic Spectral Element Method for Non-Isotropic Diffusion Problems

B. Gervang, K. Olesen, and M. Gerritsma

Abstract We present a mimetic spectral element method for the solution of the stationary Darcy's problem. We show that the divergence constraint is satisfied exactly for both heterogeneous, non-isotropic, and deformed mesh problems.

1 Introduction

The steady, non-isotropic, heterogeneous diffusion problem for single phase flow through porous media is investigated. We are, in particular, interested in the Darcy flow problem for reservoir simulations. The governing equation is

$$-\nabla \cdot \mathbb{K}\nabla p = f \ , \tag{1}$$

where $\mathbb{K}$ is a symmetric positive definite tensor representing the permeability field of the domain, p is the pressure and f is a mass source term. In Darcy flow, $\mathbb{K}$ represents the material's preferred direction of flow when subject to a pressure gradient ∇p. We rewrite (1) as two first order equations

$$\nabla \cdot \boldsymbol{q} = f \ , \tag{2}$$

and

$$\boldsymbol{q} = -\mathbb{K}\nabla p \ , \tag{3}$$

B. Gervang (✉) • K. Olesen
Department of Engineering, Aarhus University, Inge Lehmanns Gade 10, 8000 Aarhus, Denmark
e-mail: bge@ase.au.dk; keol@eng.au.dk

M. Gerritsma
Faculty of Aerospace Engineering, Delft University of Technology, Kluyverweg 2, 2629 HS Delft, The Netherlands
e-mail: m.i.gerritsma@tudelft.nl

M.L. Bittencourt et al. (eds.), *Spectral and High Order Methods for Partial Differential Equations ICOSAHOM 2016*, Lecture Notes in Computational Science and Engineering 119, DOI 10.1007/978-3-319-65870-4_15

where $\nabla \cdot \boldsymbol{q} = f$ represents the mass conservation condition and $\boldsymbol{q} = -\mathbb{K}\nabla p$ is the constitutive relation between the pressure gradient and the velocity field, $\boldsymbol{q}$. We seek the solution of (2) and (3) subjected to boundary conditions $\boldsymbol{q} = \boldsymbol{q}_0$ on Γ_q and $p = p_0$ on Γ_p, $\partial\Omega = \Gamma_q + \Gamma_p$.

The non-isotropic heterogeneous Darcy flow problem was also analyzed in [6], where the algorithm used a support operator based on a 2nd order finite difference method. Even for very large variations of the permeability coefficient 2nd order accuracy was obtained. In [10] the hybrid discontinuous Galerkin method was used and it was shown that it is locally conservative. A mixed finite element method was used in [12], where a multiscale mortar method was used. In the present work we will use a mimetic discretization method, which ensures that the numerical solution of the discrete divergence-free velocity field, i.e. (2) is satisfied exactly, when $f = 0$, independent of the order of the expanding polynomial. It is ensured that the invariant (mass in the present system) is conserved both locally and globally. A similar method was used in [2] and [8]. The governing equations will be written in the notation of differential forms. By doing so we are associating the individual quantities in (2) and (3) with a geometrical basis, which the quantities are naturally integrated over. The conservation of mass in (2) is actually an algebraic relation between the fluxes over the surfaces of an arbitrary volume and the volumetric flux inside the volume and this relation is an inherent part of using differential forms. Instead of representing the dependent variables of (2) and (3) as normal differential equations we will make use of differential geometry to represent variables and differential equations.

2 Differential Geometry

Differential geometry deals with the geometric aspects of differential equations. Rewriting the differential equations using differential geometry we can introduce basic geometric objects as points, curves, surfaces, and volumes into the constitution of the equations and obtain exact discretization of the grad, curl, and div operators. Below we have defined a few of the concepts used in differential geometry and in the present work. However, you may consult other texts as [1, 4, 7, 11] for a much broader and deeper exposition.

2.1 *Differential k-Forms*

In contrast to vector calculus where we use vector and scalar fields, we use differential forms in differential geometry. In $\mathbb{R}^3$ we resort to four different differential forms. A 0-form, $\alpha^{(0)}$, is given by $\alpha^{(0)} = f(x, y, z)$ and is a normal scalar function. A 1-form, $\beta^{(1)}$, is given by $\beta^{(1)} = f_1(x, y, z)dx + f_2(x, y, z)dy + f_3(x, y, z)dz$. A 1-form can be integrated over smooth curves. A 2-form, $\gamma^{(2)}$, is given by

$\gamma^{(2)} = g_1(x,y,z)dy \wedge dz + g_2(x,y,z)dz \wedge dx + g_3(x,y,z)dx \wedge dy$. A 2-form can be integrated over two dimensional manifolds or surfaces. And a 3-form $\delta^{(3)}$, is given by $\delta^{(3)} = h(x,y,z)dx \wedge dy \wedge dz$. A 3-form can be integrated over volumes.

2.2 Wedge (or Exterior) Product

The wedge product introduced in the expression of two and three forms is defined as

$$\alpha \wedge \beta = \alpha \otimes \beta - \beta \otimes \alpha ,$$

where α and β are two arbitrary forms. The wedge product between k-forms and l-forms produces a $(k+l)$-form. If the space of k-forms is denoted Λ^k and the space of l-forms is given by Λ^l then

$$\wedge : \Lambda^k \times \Lambda^l \longrightarrow \Lambda^{k+l} .$$

2.3 Orientation

In n-dimensional space there are $(n+1)$ submanifolds. For $n = 3$ these submanifolds are points, curves, surfaces and volumes. There are two types of orientations, an inner orientation, which is solely connected to the geometrical object and an outer orientation, which is related to both the geometrical object and the embedding space. In Fig. 1 both inner and outer oriented k-manifolds in $\mathbb{R}^3$ are sketched.

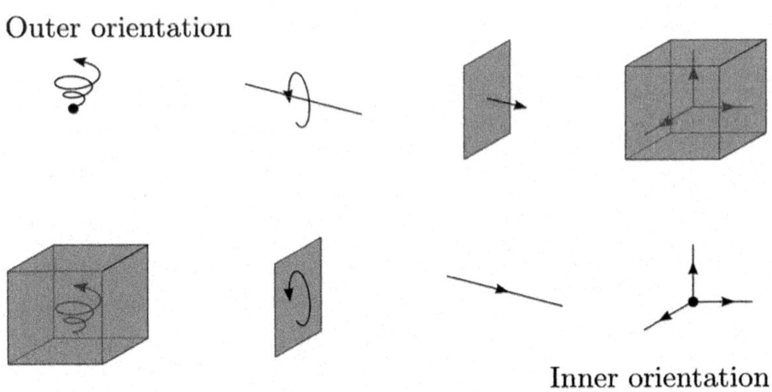

Fig. 1 Outer and inner oriented k-manifolds in $\mathbb{R}^3$ for $k = 0, 1, 2, 3$

2.4 Exterior Derivative

The exterior derivative (or exterior differential), d, is a metric free operator that generalizes the vector operators grad, curl, and div and is defined as: The exterior derivative on an n-dimensional manifold $\mathbb{M}$ is a mapping $d : \Lambda^k(\mathbb{M}) \longrightarrow \Lambda^{k+1}(\mathbb{M}), 0 \le k \le n-1$. In a local coordinate system $(x^1, \ldots, x^n)$ this map is given by

$$\begin{aligned} d\alpha^{(k)} &= d\sum_I f_I(x) dx^{i_1} \wedge \ldots \wedge dx^{i_k} \\ &= \sum_I df_I(x) \wedge dx^{i_1} \wedge \ldots \wedge dx^{i_k} \\ &= \sum_I \sum_{j=1}^{n} \frac{\partial f_I}{\partial x^j} dx^j \wedge dx^{i_1} \wedge \ldots \wedge dx^{i_k} , \end{aligned}$$

where $I = i_1, \ldots, i_k$ with $1 \le i_1 < \cdots < i_k \le n, f_I(x)$ is a continuous differentiable scalar function, $f_I(x) \in C^\infty(\mathbb{M})$ and $\frac{\partial}{\partial x^j}$ is the partial derivative with respect to x^j.

2.5 Pullback Operator

We consider two n-dimensional manifolds $\mathbb{M}$ and $\mathbb{N}$ and the mapping between them $\Phi : \mathbb{M} \rightarrow \mathbb{N}$, such that local coordinates ξ^i in $\mathbb{M}$ are mapped to local coordinates $x^i = \Phi^i(\xi^1, \cdots, \xi^n)$ in $\mathbb{N}$. Then the pullback of a k-form, $\Phi^* : \Lambda^k(\mathbb{N}) \rightarrow \Lambda^k(\mathbb{M}), k \ge 1$, is given by

$$\Phi^*(\alpha^{(k)}(\mathbf{v}_1, \cdots, \mathbf{v}_k) := \alpha^{(k)}(\Phi_*(\mathbf{v}_1), \cdots, \Phi_*(\mathbf{v}_k)) ,$$

where $\Phi_*(\mathbf{v})$ is the push forward operator of a vector field $\mathbf{v}$.

2.6 Hodge Star Operator

The Hodge star operator in an n-dimensional manifold $\mathbb{M}$ is an operator, $\star : \Lambda^k(\mathbb{M}) \rightarrow \Lambda^{n-k}(\mathbb{M})$, induced by the inner product (metric) and wedge product (orientation),

$$\alpha^{(k)} \wedge \star\beta^{(k)} := (\alpha^{(k)}, \beta^{(k)})\sigma^{(n)} ,$$

where $\sigma^{(k)}$ is the unit volume form defined in local coordinates as

$$\sigma^{(n)} := \sqrt{g} dx^1 \wedge \cdots \wedge dx^n ,$$

where g is the determinant of the metric tensor. Application of the Hodge star operator to the unit 0-form yields $\star 1 := \sigma^{(n)}$.

2.7 Using the Building Blocks

We are now in a position to rewrite (2) and (3) using differential forms. In (2) $\boldsymbol{q}$ is the velocity tensor, which in this case represents a volume flux density over a surface, and in a 3 dimensional space it is an outer oriented differential 2-form, $q^{(2)}$, given by

$$q^{(2)} = q_1\, dx^2 \wedge dx^3 + q_2\, dx^3 \wedge dx^1 + q_3\, dx^1 \wedge dx^2 ,$$

where q_i are the velocity components and $\wedge$ is the wedge product. In (2), f is a volumetric volume flux density and is an outer oriented differential 3-form, $f^{(3)}$, and is written as

$$f^{(3)} = f\, dx^1 \wedge dx^2 \wedge dx^3 .$$

In the notation of differential forms (2) is written as

$$dq^{(2)} = f^{(3)} , \tag{4}$$

where d is the exterior derivative. The pressure in (3) is a potential associated to points, and is therefore an inner-oriented differential 0-form, $\tilde{p}^{(0)} = p$, and we may rewrite (3) as

$$\star_{\mathbb{K}^{-1}} q^{(2)} = -d\tilde{p}^{(0)} , \tag{5}$$

where $\star_{\mathbb{K}^{-1}} = \mathbb{K}^{-1}\star$ with $\star$ being the Hodge star operator.

Let Φ be a map from a reference cube of size 2×2×2 in the (ξ^1, ξ^2, ξ^3) coordinate frame to an arbitrary volume in the physical coordinate frame (x^1, x^2, x^3), then (4) is pulled back to the reference frame by use of the pullback operator, $\Phi^\star$,

$$\begin{aligned} dq^{(2)} = f^{(3)} \quad &\Rightarrow \quad \Phi^* dq^{(2)} = \Phi^* f^{(3)} \\ \Rightarrow \quad d\Phi^* q^{(2)} = f^{(3)} \quad &\Rightarrow \quad d\hat{q}^{(2)} = \hat{f}^{(3)} , \end{aligned} \tag{6}$$

where the third step is valid since the pullback operator commutes with the exterior derivative, see [7]. $q^{(2)}$ is the outer oriented 2-form, which is associated to the reference basis, i.e.

$$\hat{q}^{(2)} = \hat{q}_1\, d\xi^2 \wedge d\xi^3 + \hat{q}_2\, d\xi^3 \wedge d\xi^1 + \hat{q}_3\, d\xi^1 \wedge d\xi^2 ,$$

and $\hat{f}^{(3)}$ is the outer-oriented 3-form associated to the reference basis

$$\hat{f}^{(3)} = \hat{f}\, d\xi^1 \wedge d\xi^2 \wedge d\xi^3 \,.$$

The relation in (6) states that integral values are independent of the frame that they are represented in. The constitutive relation in (5) is pulled back to the reference frame by

$$\Phi^* \star_{\mathbb{K}^{-1}} q^{(2)} = -\Phi^* d\tilde{p}^{(0)} \quad \Rightarrow \quad \Phi^* \star_{\mathbb{K}^{-1}} \left(\Phi^*\right)^{-1} \hat{q}^{(2)} = -d\hat{\tilde{p}}^{(0)} \,, \tag{7}$$

where the right hand side again is a consequence of the commuting property between the pullback operator and the exterior derivative, see also [2]. The pullback operator does, however, not commute with the Hodge star operator, and therefore, $\hat{q}^{(2)}$ must be mapped to $q^{(2)}$ by $\left(\Phi^*\right)^{-1}$, where $\star_{\mathbb{K}^{-1}}$ is applied and this term is then pulled back to the reference frame. The discretizations are based on Eqs. (6) and (7).

3 Discretization of the Equations

As in the mixed SEM, we are pairing our equations with suitable arbitrary k-forms, such that the product is a 3-form, which is naturally integrated over a volume. The mass balance in (6) is an outer oriented 3-form, so this is paired with an arbitrary inner oriented 0-form, $\varrho^{(0)}$, and (7) is an inner-oriented 1-form on both sides, so this is paired with an arbitrary outer-oriented 2-form, $\zeta^{(2)}$, i.e

$$\left\langle \hat{\tilde{\varrho}}^{(0)}, d\hat{q}^{(2)} \right\rangle_{\hat{\Omega}} = \left\langle \hat{\tilde{\varrho}}^{(0)}, \hat{f}^{(3)} \right\rangle_{\hat{\Omega}} \,, \quad \forall \hat{\tilde{\varrho}}^{(0)} \in \mathscr{P} \tag{8}$$

$$\left\langle \hat{\zeta}^{(2)}, \Phi^* \star_{\mathbb{K}^{-1}} \left(\Phi^*\right)^{-1} \hat{q}^{(2)} \right\rangle_{\hat{\Omega}} = -\left\langle \hat{\zeta}^{(2)}, d\hat{\tilde{p}}^{(0)} \right\rangle_{\hat{\Omega}} \,, \quad \forall \hat{\zeta}^{(2)} \in \mathscr{Q} \tag{9}$$

where $\hat{\Omega}$ is an arbitrary volume in the reference frame. To acquire the adjoint of $\left\langle \hat{\tilde{\varrho}}^{(0)}, d\hat{q}^{(2)} \right\rangle_{\hat{\Omega}}$ the term $\left\langle \hat{\zeta}^{(2)}, d\hat{\tilde{p}}^{(0)} \right\rangle_{\hat{\Omega}}$ in the constitutive equation is integrated by parts

$$\left\langle \hat{\zeta}^{(2)}, d\hat{\tilde{p}}^{(0)} \right\rangle_{\hat{\Omega}} = -\left\langle d\hat{\zeta}^{(2)}, \hat{\tilde{p}}^{(0)} \right\rangle_{\hat{\Omega}} + \left\langle \hat{\zeta}^{(2)}, \hat{\tilde{p}}^{(0)} \right\rangle_{\partial\hat{\Omega}} \,,$$

where $\partial\hat{\Omega}$ is the boundary of $\hat{\Omega}$. Inserting this into (9) and rearrange we obtain

$$\left\langle d\hat{\zeta}^{(2)}, \hat{\tilde{p}}^{(0)} \right\rangle_{\hat{\Omega}} - \left\langle \hat{\zeta}^{(2)}, \Phi^* \star_{\mathbb{K}^{-1}} \left(\Phi^*\right)^{-1} \hat{q}^{(2)} \right\rangle_{\hat{\Omega}} = \left\langle \hat{\zeta}^{(2)}, \hat{\tilde{p}}^{(0)} \right\rangle_{\partial\hat{\Omega}} \,. \tag{10}$$

The functionals that we want to solve are (8) and (10). Let $\partial\hat{\Omega} = \Gamma_p \cup \Gamma_q$ then we are assuming that $p^{(0)}$ is known on Γ_p, while $q^{(2)}$ is prescribed on Γ_q, and $f^{(3)}$ is given on Ω. These known values can be transformed to the reference domain by the pullback operator. Dividing $\hat{\Omega}$ into N_E non-overlapping elements $\hat{\Omega}^s$, i.e. $\hat{\Omega} = \bigcup_s \hat{\Omega}^s$, and define $\mathscr{P} \subset L^2$ and $\mathscr{Q} \subset H_0^1(\text{div})(\Omega)$ as finite dimensional subspaces, where $H_0^1(\text{div})(\Omega)$ denotes the Sobolev space of vector functions with square-integrable divergence with vanishing trace along Γ_q. We are now formulating the variational statement as: Find $\left(\hat{\tilde{p}}^h, \hat{q}_i^h\right) \in \mathscr{P} \times \mathscr{Q}$ for $i = 1, 2, 3$ such that

$$\sum_{s=1}^{N_E} \left\langle \hat{\tilde{\varrho}}^{(0),h}, d\hat{q}^{(2),h} \right\rangle_{\hat{\Omega}^s} = \sum_{s=1}^{N_E} \left\langle \hat{\tilde{\varrho}}^{(0),h}, \hat{f}^{(3),h} \right\rangle_{\hat{\Omega}^s} \quad \forall \hat{\tilde{\varrho}}^h \in \mathscr{P} \tag{11}$$

and

$$\sum_{s=1}^{N_E} \left\langle d\hat{\zeta}^{(2),h}, \hat{\tilde{p}}^{(0),h} \right\rangle_{\hat{\Omega}^s} - \sum_{s=1}^{N_E} \left(\hat{\zeta}^{(2),h}, \Phi^* \star_{\mathbb{K}^{-1}} \left(\Phi^*\right)^{-1} \hat{q}^{(2),h} \right)_{\hat{\Omega}^s}$$

$$= \left\langle \hat{\zeta}^{(2),h}, \hat{\tilde{p}}^{(0),h} \right\rangle_{(\partial\hat{\Omega})_p} \quad \forall \hat{\zeta}_i^h \in \mathscr{Q} . \tag{12}$$

4 Expansion Polynomials

The differential forms are associated to the geometry, that they are naturally integrated over, and hence it will be natural to expand these based on integral values. This can be accomplished using a combination of Lagrange polynomials given by

$$h_i(\xi) = \frac{\prod_{j=0,j\neq i}^{N} (\xi - \xi_j)}{\prod_{j=0,j\neq i}^{N} (\xi_i - \xi_j)} ,$$

and *edge polynomials* defined in [5] as

$$e_i(\xi) = -\sum_{k=0}^{i-1} \frac{dh_k(\xi)}{d\xi} d\xi , \quad i = 1, \ldots, N .$$

Just like the Lagrange polynomials have the property

$$h_i(\xi_k) = \begin{cases} 1 & \text{if } i = k \\ 0 & \text{if } i \neq k \end{cases} ,$$

the edge polynomials have the property

$$\int_{\xi_{k-1}}^{\xi_k} e_i(\xi) = \begin{cases} 1 & \text{if } i = k \\ 0 & \text{if } i \neq k \end{cases} ,$$

The Hodge star operator is represented discretely by expanding our outer and inner oriented differential forms on two different grids. Let ξ_i, $i = 0,\dots,N$ be the Gauss-Lobatto-Legendre (GLL) points of polynomial degree N and $\tilde{\xi}_i$, $i = 0, \dots, N-1$ the Gauss-Legendre (GL) points. Note that $\xi_i < \tilde{\xi}_i < \xi_{i+1}$, for $i = 0,\dots,N-1$. The Lagrange polynomials associated with the GLL points will be denoted by $h_i(\xi)$ and the Lagrange polynomials associated with the GL points will be denoted by $\tilde{h}_i(\xi)$. For more details see [3]. The edge polynomial $e_i(\xi)$ is a polynomial of degree $N-1$ and $\tilde{e}_i(\xi)$ is a polynomial of degree $N-2$ The expansion of the differential forms in (11) and (12) are expanded by

$$\hat{\tilde{p}}^{(0),h}(\xi^1,\xi^2,\xi^3) = \sum_{i=0}^{N-1}\sum_{j=0}^{N-1}\sum_{k=0}^{N-1} \tilde{p}_{i,j,k}\tilde{h}_i(\xi^1)\tilde{h}_j(\xi^2)\tilde{h}_j(\xi^3)\,, \tag{13}$$

$$\begin{aligned}\hat{q}^{(2),h}(\xi^1,\xi^2,\xi^3) &= \sum_{i=0}^{N}\sum_{j=1}^{N}\sum_{k=1}^{N} Q^1_{i,j,k}h_i(\xi^1)e_j(\xi^2)e_k(\xi^3)\\ &+\sum_{i=1}^{N}\sum_{j=0}^{N}\sum_{k=1}^{N} Q^2_{i,j,k}e_i(\xi^1)h_j(\xi^2)e_i(\xi^3) + \sum_{i=1}^{N}\sum_{j=1}^{N}\sum_{k=0}^{N} Q^3_{i,j,k}e_i(\xi^1)e_j(\xi^2)h_i(\xi^3)\,,\end{aligned} \tag{14}$$

and

$$\hat{f}^{(3),h}(\xi^1,\xi^2,\xi^3) = \sum_{i=1}^{N}\sum_{j=1}^{N}\sum_{k=1}^{N} F_{i,j,k}e_i(\xi^1)e_j(\xi^2)e_k(\xi^3)\,. \tag{15}$$

The expansion coefficients are respectively

$$\tilde{p}_{i,j,k} = p(\tilde{\xi}_i,\tilde{\xi}_j,\tilde{\xi}_k)\,,$$

$$Q^1_{i,j,k} = \int_{\xi^2_{j-1}}^{\xi^2_j}\int_{\xi^3_{k-1}}^{\xi^3_k} \hat{q}_1(\xi^1_i,\xi^2,\xi^3)\,d\xi^2\wedge d\xi^3\,,\quad Q^2_{i,j,k} = \int_{\xi^3_{k-1}}^{\xi^3_k}\int_{\xi^1_{i-1}}^{\xi^1_i} \hat{q}_2(\xi^1,\xi^2_j,\xi^3)\,d\xi^3\wedge d\xi^1\,,$$

$$Q^3_{i,j,k} = \int_{\xi^1_{i-1}}^{\xi^1_i}\int_{\xi^2_{j-1}}^{\xi^2_j} \hat{q}_3(\xi^1,\xi^2,\xi^3_k)\,d\xi^1\wedge d\xi^2\,,\quad F_{i,j,k} = \int_{\xi^1_{i-1}}^{\xi^1_i}\int_{\xi^2_{j-1}}^{\xi^2_j}\int_{\xi^3_{k-1}}^{\xi^3_k} \hat{f}(\xi^1,\xi^2,\xi^3)\,d\xi^1\wedge d\xi^2\wedge d\xi^3\,.$$

Let $\phi^h(\xi) = \sum_{i=0}^{N}\phi_i h_i(\xi)$ be a function in $\mathbb{R}$ expanded by Lagrange polynomials then the derivative is calculated by

$$\frac{d\phi^h}{d\xi} = \sum_{i=1}^{N}(\phi_i - \phi_{i-1})e_i(\xi)\,.$$

Using this, $d\hat{q}^{(2),h}$ in (11) is calculated as

$$d\hat{q}^{(2),h} = \sum_{i=1}^{N}\sum_{j=1}^{N}\sum_{k=1}^{N}\Big(Q^1_{i,j,k} - Q^1_{i-1,j,k} + Q^2_{i,j,k} - Q^2_{i,j-1,k} + Q^3_{i,j,k} - Q^3_{i,j,k-1}\Big)\, e_i(\xi^1)e_j(\xi^2)e_k(\xi^3)\,. \quad (16)$$

The pullback operators in (12) are basically just mappings of geometrical bases of different order. First $(\Phi^*)^{-1}\,\hat{q}^{(2),h}$ maps the surface basis of $\hat{q}^{(2),h}$ from the reference frame to the physical domain. This is similar to a Piola transformation known from continuum mechanics, [9], but applied to a first order tensor instead of a second order tensor

$$q^{(2)} = \frac{1}{J}\mathscr{F}\hat{q}^{(2)}\,, \quad (17)$$

with

$$\mathscr{F} = \begin{bmatrix} \frac{\partial x_1}{\partial \xi_1} & \frac{\partial x_1}{\partial \xi_2} & \frac{\partial x_1}{\partial \xi_3} \\ \frac{\partial x_2}{\partial \xi_1} & \frac{\partial x_2}{\partial \xi_2} & \frac{\partial x_2}{\partial \xi_3} \\ \frac{\partial x_3}{\partial \xi_1} & \frac{\partial x_3}{\partial \xi_2} & \frac{\partial x_3}{\partial \xi_3} \end{bmatrix},$$

$J = \det(\mathscr{F})$ and the differential forms arranged in column vectors. By applying $\star_{\mathbb{K}^{-1}}$ in (12) the surface basis is changed to a line basis, and Φ^* maps this from the physical frame to the reference frame and it is done by

$$\Phi^* \star_{\mathbb{K}^{-1}} q^{(2),h} = \mathscr{F}^T\mathbb{K}^{-1}q^{(2),h}\,,$$

where T denotes the transpose operator. The whole term is thereby given by

$$\Phi^* \star_{\mathbb{K}^{-1}} (\Phi^*)^{-1}\,\hat{q}^{(2),h} = \frac{1}{J}\mathscr{F}^T\mathbb{K}^{-1}\mathscr{F}\hat{q}^{(2),h}\,. \quad (18)$$

Gathering all terms and applying appropriate Gaussian quadratures then (11) and (12) can be written in a system of equations as

$$\begin{bmatrix} \mathbf{0} & \boldsymbol{P}_3\boldsymbol{E}_{(3,2)} \\ \boldsymbol{E}^T_{(3,2)}\boldsymbol{P}^T_3 & \boldsymbol{M}_2 \end{bmatrix}\begin{Bmatrix} \boldsymbol{\Delta}_p \\ \boldsymbol{\Delta}_Q \end{Bmatrix} = \begin{Bmatrix} \boldsymbol{P}_3\boldsymbol{\Delta}_F \\ \boldsymbol{B}\boldsymbol{\Delta}^{bc}_p \end{Bmatrix}.$$

Here $\boldsymbol{\Delta}_p$, $\boldsymbol{\Delta}_Q$ and $\boldsymbol{\Delta}_F$ contain all expansion coefficients of (13), (14) and (15), respectively, and $\boldsymbol{\Delta}^{bc}_p$ contains all known values of $p^{(0)}$ in the GL points on $(\partial\Omega)_p$. $\boldsymbol{E}_{(3,2)}$ is a representation of the expression within the bracket in (16) for the entire system, and is called an incidence matrix, see [7]. This matrix only contains the values 0,-1 and 1 and is completely determined by the connectivity of the grid. The remaining block matrices contain the wedge of the expansion polynomials of the individual terms in (11) and (12) weighed with the Gaussian integration weights. Note that $\boldsymbol{M}_2$ contains the mapping and the inverse permeability field in (18).

5 Numerical Results

5.1 Case 1: Homogeneous, Isotropic, and Cartesian Coordinate System

The method has been tested on a 2D model problem of which we have an analytic solution. For a given forcing function, $f(x, y) = 8\pi^2 sin(2\pi x)cos(2\pi y)$, the analytic solution for the pressure and velocity fields are

$$p = sin(2\pi x)cos(2\pi y), \quad q_x = -2\pi cos(2\pi x)cos(2\pi y), \quad q_y = -2\pi sin(2\pi x)sin(2\pi y).$$

The solution domain is, $\Omega = [-1, 1]^2$ and for $\mathbb{K} = 1$, four elements, and a polynomial order of $N = 12$ the residual of mass conservation is shown to the left in Fig. 2 and on the right the associated conservation of mass as function of polynomial order is shown. It is observed that the mass conservation constraint is satisfied exactly independent of polynomial order.

5.2 Case 2: Heterogeneous, Non-Isotropic, and Cartesian Coordinate System

A four element, non-isotropic case is now tested using the permeability coefficients

$$K_{11} = 4 \quad K_{12} = 3 \quad K_{21} = 3 \quad K_{22} = 20$$

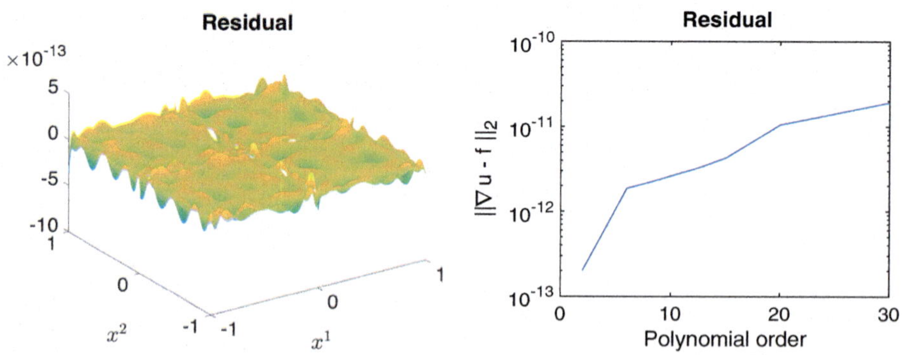

Fig. 2 Residual of mass conservation; *left*: N=12. Number of elements is 4

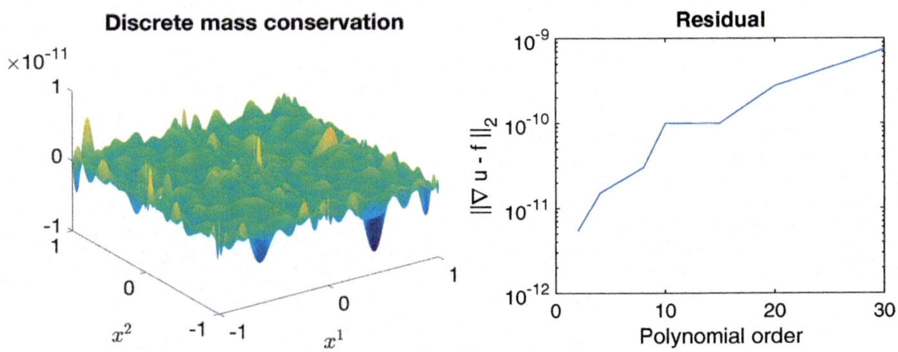

Fig. 3 Residual of mass conservation; *left*: N=15. Number of elements is 4

in each element and the pressure field $P(x, y) = sin(2\pi x)sin(2\pi y)$. This leads to the following velocity and forcing terms

$$u_x = -2\pi K_{11} cos(2\pi x)cos(2\pi y) + 2\pi K_{12} sin(2\pi x)sin(2\pi y),$$
$$u_y = -2\pi K_{21} cos(2\pi x)cos(2\pi y) + 2\pi K_{22} sin(2\pi x)sin(2\pi y),$$
$$f = 4\pi^2((K_{11} + K_{22})sin(2\pi x)cos(2\pi y) + (K_{12} + K_{21})cos(2\pi x)sin(2\pi y)).$$

The residual for mass conservation is shown in Fig. 3 and it is again observed that mass is conserved exactly.

5.3 Case 3: Heterogeneous, Non-Isotropic, and Curvilinear Coordinate System

The results for a nine element case with a very distorted grid, (see Fig. 4 for an example of a 25 elements domain), are shown in Fig. 5. It is noticed that even on a very distorted grid mass is conserved exactly. In our discrete-mass-conservation plots we observe peaks at part of the boundary and inter-element boundaries. These peaks are of the size 10^{-13} in Fig. 5 and therefore close to machine accuracy. We don't attribute them any major importance.

5.4 Case 4: The 'Tenth SPE Comparative Solution Project'

In the last numerical example we have taken the permeability field from the 'Tenth SPE Comparative Solution Project' http://www.spe.org/web/csp/.

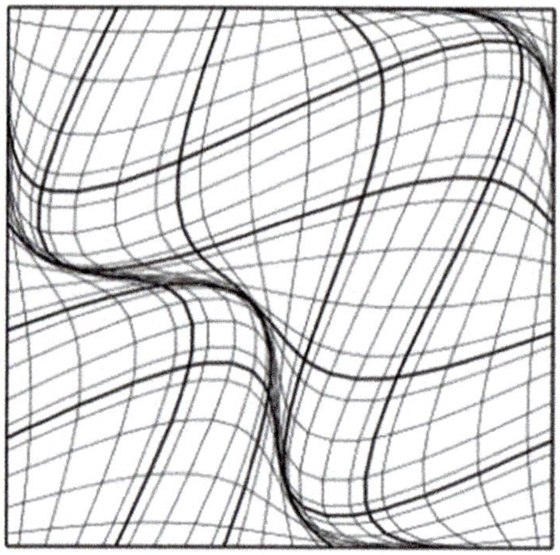

$$K11 = (1, 2, 3, 4, 5, 6, 7, 8, 9)$$

$$K12 = (0.25, 0.5, 0.75, 1, 0.5, 1, 0.5, 1, 0.5)$$

$$K21 = (0.25, 0.5, 0.75, 1, 0.5, 1, 0.5, 1, 0.5)$$

$$K22 = (9, 8, 7, 6, 5, 4, 3, 2, 1)$$

Fig. 4 Distorted grid (for a 25 elements domain) and values for the permeability coefficients for a 9 element case, which is used in the calculations

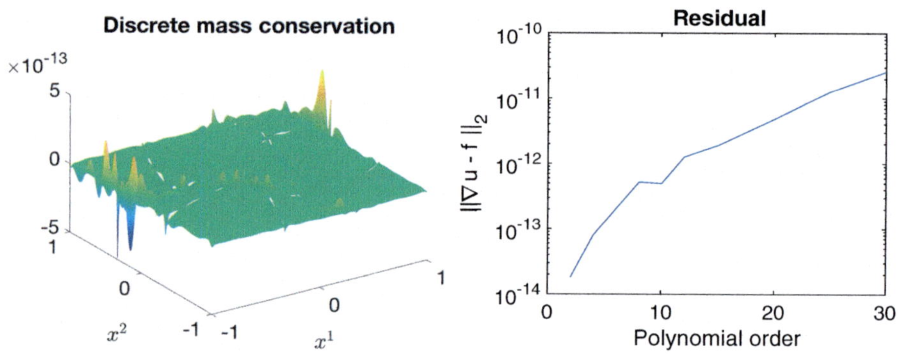

Fig. 5 Residual of mass conservation; *left*: N=15. Number of elements are 9

The original model is a 2-phase (oil and gas) model of which we only simulate the oil phase. The model has a simple 2D cross-sectional geometry with no dipping or faults. The dimensions of the model are 762 m long by 15.24 m thick. The fine scale grid is 100×20 with uniform size for each of the grid blocks. The top of the model is at 0.0 m with initial pressure at this point of 100 psia. The model is fully saturated with oil.

In Fig. 6 we again see that the mass is conserved to machine accuracy.

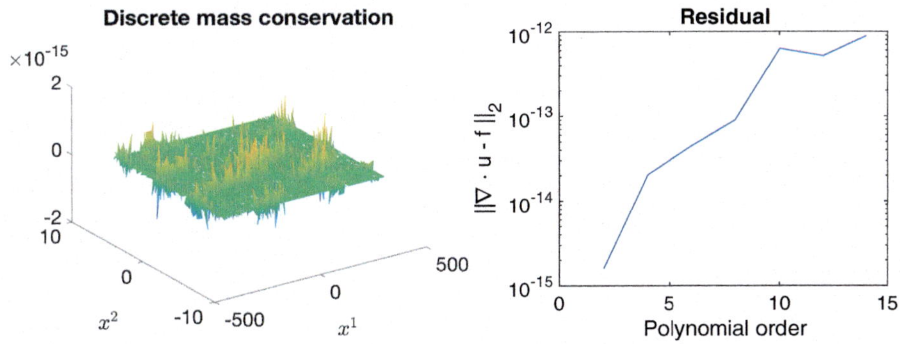

Fig. 6 Residual of mass conservation. *Left*: N=2. Permability field taken from http://www.spe.org/web/csp/

Table 1 Condition number of the algebraic system as function of the permeability field

$k_{12} = k_{21}$	k_{11}	k_{22}	Cond #
0	1	1	$3.77 * 10^3$
0	1	10	$3.74 * 10^3$
0	1	100	$3.73 * 10^3$
0	1	1000	$3.74 * 10^3$
0	1	10000	$1.20 * 10^4$
10	10	20	$0.85 * 10^3$
10	10	100	$2.45 * 10^3$
10	10	1000	$9.67 * 10^3$
10	10	10000	$2.99 * 10^4$
10	10	100000	$9.36 * 10^4$

5.5 *Condition Number of the Coefficient Matrix*

In Table 1 we analyse the condition number of the algebraic system as the permeability field changes. The condition number is shown for a 2D domain with 4 elements and a polynomial order of 12. As seen from Table 1, when the off-diagonal elements of the symmetric positive definite $\mathbb{K}$ tensor are zero, the condition number is constant as the an-isotropy increases until the an-isotropy has reach a very high value, which give rise to an increased condition number. In contrast to this, it is seen that when the off-diagonal elements are non-zero, the condition number increases as the an-isotropy increases.

The code is written in MatLab and all solutions have been obtained using the built-in direct solvers. Solution of the problem obtained from the 'Tenth SPE Comparative Solution Project' with the polynomial order $N = 4$ took 5 min on an PC using the Core processor i7-3615QM running 2.3 GHz.

6 Conclusion

A mimetic spectral element method for the Darcy's problem has been developed. It is shown that mass conservation is satisfied exactly for both isotropic and non-isotropic permeability coefficients and for regular and very distorted grids.

References

1. R. Abraham, J. Marsden, T. Ratiu, *Manifolds, Tensors Analysis, and Applications*. Applied Mathematical Sciences, vol. 75 (Springer, Berlin, 2001)
2. M. Bouman, A. Palha, J. Kreeft, M. Gerritsma, in *A Conservative Spectral Element Method for Curvilinear Domains*, ed. by J.S. Hesthaven, E.M. Rønquist. Lecture Notes in Computational Sciences and Engineering, vol. 76 (Springer, Heidelberg, 2011), pp. 111–119
3. C. Canuto, M. Hussaini, A. Quarteroni, T. Zang, *Spectral Methods, Fundamentals in Single Domains* (Springer, Berlin, 2006)
4. T. Frankel, *The Geometry of Physics* (Cambridge University Press, Cambridge, 2012)
5. M. Gerritsma, Edge Functions for Spectral Element Methods, in *Spectral and High Order Methods for Partial Differential Equations*, ed. by J.S. Hesthaven, E.M. Rønquist. Lecture Notes in Computational Science and Engineering, vol. 76 (Springer, Heidelberg, 2011), pp. 199–207
6. J.M. Hyman, M. Shashkov, S. Steinberg, The numerical solution of diffusion problems in strongly heterogeneous non-isotropic materials. J. Comput. Phys. **132**, 130–148 (1997)
7. J. Kreeft, A. Palha, M. Gerritsma, Mimetic framework on curvilinear quadrilaterals of arbitrary order. Arxiv preprint (2011)
8. A. Palha, P.P. Rebelo, R. Hiemstra, J. Kreeft, M. Gerritsma, Physics-compatible discretization techniques on single and dual grids, with application to the Poisson equation of volume forms. J. Comput. Phys. **257**, 1394–1422 (2014)
9. J.N. Reddy, *An Introduction to Continuum Mechanics*, 2nd edn. (Cambridge University Press, Cambridge, 2013)
10. A. Samii, C. Michoski, C. Dawson, A parallel and adaptive hybridized discontinuous Galerkin method for anisotropic nonhomogeneous diffusion. Comput. Methods Appl. Math. Eng. **134**, 118–139 (2016)
11. E. Tonti, Why starting from differential equations for computational physics? J. Comput. Phys. **257**, 1260–1290 (2014)
12. M. Wheeler, G. Xue, I. Yoto, A multiscale mortar multipoint flux mixed finite element method. ESIAM: M2AN **46**, 759–796 (2012)

High Order DGTD Solver for the Numerical Modeling of Nanoscale Light/Matter Interaction

Stéphane Lanteri, Claire Scheid, Maciek Klemm, and Jonathan Viquerat

Abstract Nanophotonics is the field of science and technology which aimed at establishing and using the peculiar properties of light and light/matter interactions in various nanostructures. The numerical modeling of such interactions requires to solve the system of time-domain Maxwell equations possibly coupled to appropriate models of physical dispersion in metals such as the Drude and Drude-Lorentz models. In this paper, we discuss about the development of a high order discontinuous Galerkin time-domain solver for nanophotonics applications in the linear regime. For the numerical treatment of dispersion models in metals, we have adopted an Auxiliary Differential Equation (ADE) technique leading to solve the time-domain Maxwell equations coupled to a system of ODEs. We present numerical results that demonstrate the accuracy of the proposed numerical methodology for nanstructured settings involving curvilinear geometrical features.

1 Introduction

The numerical modeling of light interaction with nanometer scale structures generally relies on the solution of the system of time-domain Maxwell equations, possibly taking into account an appropriate physical dispersion model, such as the Drude or Drude-Lorentz models, for characterizing the material properties of metallic nanostructures at optical frequencies [11]. In the computational nanophotonics literature, a large number of studies are devoted to Finite Difference Time-Domain (FDTD) type discretization methods based on Yee's scheme [18]. As a matter of

S. Lanteri (✉) • J. Viquerat
Inria Sophia Antipolis-Méditerranée Research Center, Valbonne, France
e-mail: Stephane.Lanteri@inria.fr; Jonathan.Viquerat@inria.fr

C. Scheid
Mathematics Laboratory, University of Nice-Sophia Antipolis, Nice, France
e-mail: Claire.Scheid@unice.fr

M. Klemm
Department of Electrical and Electric Engineering, University of Bristol, Bristol, UK
e-mail: M.Klemm@bristol.ac.uk

M.L. Bittencourt et al. (eds.), *Spectral and High Order Methods for Partial Differential Equations ICOSAHOM 2016*, Lecture Notes in Computational Science and Engineering 119, DOI 10.1007/978-3-319-65870-4_16

fact, the FDTD [15] method is a widely used approach for solving the systems of partial differential equations modeling nanophotonic applications. In this method, the whole computational domain is discretized using a structured (cartesian) grid. However, in spite of its flexibility and second-order accuracy in a homogeneous medium, the Yee scheme suffers from serious accuracy degradation when used to model curved objects or when treating material interfaces. During the last 20 years, numerical methods formulated on unstructured meshes have drawn a lot of attention in computational electromagnetics with the aim of dealing with irregularly shaped structures and heterogeneous media. In particular, the Discontinuous-Galerkin Time-Domain (DGTD) method has met an increased interest because these methods somehow can be seen as a crossover between Finite Element Time-Domain (FETD) methods (their accuracy depends of the order of a chosen local polynomial basis upon which the solution is represented) and Finite Volume Time-Domain (FVTD) methods (the neighboring cells are connected by numerical fluxes). Thus, DGTD methods offer a wide range of flexibility in terms of geometry (since the use of unstructured and non-conforming meshes is naturally permitted) as well as local approximation order refinement strategies, which are of useful practical interest.

In this paper, we report on our recent efforts aiming at the development of a family of high order DG-based solvers for the numerical treatment of a wide class of problems involving the interaction of light with matter at the nanoscale. Although we concentrate here on a presentation of the basic ingredients and characteristics of a DG method for time-domain nanophotonics/plasmonics applications in the linear regime assuming local dispersion effects for metallic nanostructures, we note that the present work falls within a global approach which aims at considering more general physical settings as outlined in the conclusion of the paper. The basic ingredient of a DG-based solver is a discretization method which relies on a compact stencil high order interpolation of the electromagnetic field components within each cell of an unstructured tetrahedral mesh. This piecewise polynomial numerical approximation is allowed to be discontinuous from one mesh cell to another, and the consistency of the global approximation is obtained thanks to the definition of appropriate numerical traces of the fields on a face shared by two neighboring cells. Time integration is achieved using an explicit scheme and no global mass matrix inversion is required to advance the solution at each time step. Moreover, the resulting time-domain solver is particularly well adapted to parallel computing. For the numerical treatment of dispersion models in metals, we have adopted an Auxiliary Differential Equation (ADE) technique that has already proven its effectiveness in the FDTD framework. From the mathematical point of view, this amounts to solve the time-domain Maxwell equations coupled to a system of *ordinary differential equations*. The resulting ADE-based DGTD method is detailed in [16].

2 Mathematical Modeling

Towards the general aim of being able to consider concrete physical situations relevant to nanophotonics, one of the most important features to take into account in the numerical treatment is physical dispersion. In the presence of an exterior electric field, the electrons of a given medium do not reach their equilibrium position instantaneously, giving rise to an electric polarization that itself influences the electric displacement. In the case of a linear homogeneous isotropic non-dispersive medium, there is a linear relation between the applied electric field and the polarization. However, for some range of frequencies (depending on the considered material), the dispersion phenomenon cannot be neglected, and the relation between the polarization and the applied electric field becomes complex. In practice, this is modeled by a frequency-dependent complex permittivity. Several such models for the characterization of the permittivity exist; they are established by considering the equation of motion of the electrons in the medium and making some simplifications. There are mainly two ways of handling the frequency dependent permittivity in the framework of time-domain simulations, both starting from models defined in the frequency domain. A first approach is to introduce the polarization vector as an unknown field through an auxiliary differential equation which is derived from the original model in the frequency domain by means of an inverse Fourier transform. This is called the *Direct Method* or *Auxiliary Differential Equation* (ADE) formulation. Let us note that while the new equations can be easily added to any time-domain Maxwell solver, the resulting set of differential equations is tied to the particular choice of dispersive model and will never act as a black box able to deal with other models. In the second approach, the electric field displacement is computed from the electric field through a time convolution integral and a given expression of the permittivity which formulation can be changed independently of the rest of the solver. This is called the *Recursive Convolution Method* (RCM).

In [16], an ADE formulation has been adopted. We first considered the case of Drude and Drude-Lorentz models, and further extended the proposed ADE-based DGTD method to be able to deal with a generalized dispersion model in which we make use of a Padé approximant to fit an experimental permittivity function. The numerical treatment of such a generalized dispersion model is also presented in [16]. We outline below the main characteristics of the proposed DGTD approach in the case of the Drude model. The latter is associated to a particularly simple theory that successfully accounts for the optical and thermal properties of some metals. In this model, the metal is considered as a static lattice of positive ions immersed in a free electrons gas. In the case of the Drude model, the frequency dependent permittivity is given by $\varepsilon_r(\omega) = \varepsilon_\infty - \frac{\omega_d^2}{\omega^2 + i\omega\gamma_d}$, where ε_∞ represents the core electrons contribution to the relative permittivity ε_r, γ_d is a coefficient linked to the electron/ion collisions representing the friction experienced by the electrons, and $\omega_d = \sqrt{\frac{n_e e^2}{m_e \varepsilon_0}}$ (m_e is the electron mass, e the electronic charge and n_e the electronic density) is the plasma frequency of the electrons. Considering a constant permeability and a linear homogeneous and isotropic medium, one can write the Maxwell equations as

$$\nabla \times \mathbf{H} = \frac{\partial \mathbf{D}}{\partial t}, \quad \nabla \times \mathbf{E} = -\frac{\partial \mathbf{B}}{\partial t}, \tag{1}$$

along with the constitutive relations $\mathbf{D} = \varepsilon_0 \varepsilon_\infty \mathbf{E} + \mathbf{P}$ and $\mathbf{B} = \mu_0 \mathbf{H}$, which can be combined to yield

$$\nabla \times \mathbf{E} = -\mu_0 \frac{\partial \mathbf{H}}{\partial t}, \quad \nabla \times \mathbf{H} = \varepsilon_0 \varepsilon_\infty \frac{\partial \mathbf{E}}{\partial t} + \frac{\partial \mathbf{P}}{\partial t}. \tag{2}$$

with ε_0 and μ_0 the electric permittivity and magnetic permeability of vacuum. In the frequential domain, we have that $\hat{\mathbf{D}} = \varepsilon_0 \varepsilon_r(\omega) \hat{\mathbf{E}}$, meaning that the polarization $\mathbf{P}$ is linked to the electric field through the relation $\hat{\mathbf{P}} = -\frac{\varepsilon_0 \omega_d^2}{\omega^2 + i\gamma_d \omega} \hat{\mathbf{E}}$, where $\hat{\cdot}$ denotes the Fourier transform of the time-domain field. An inverse Fourier transform gives

$$\frac{\partial^2 \mathbf{P}}{\partial t^2} + \gamma_d \frac{\partial \mathbf{P}}{\partial t} = \varepsilon_0 \omega_d^2 \mathbf{E}. \tag{3}$$

By defining the dipolar current vector $\mathbf{J}_p = \dfrac{\partial \mathbf{P}}{\partial t}$, (2)–(3) can be rewritten as

$$\mu_0 \frac{\partial \mathbf{H}}{\partial t} = -\nabla \times \mathbf{E} \ , \quad \varepsilon_0 \varepsilon_\infty \frac{\partial \mathbf{E}}{\partial t} = \nabla \times \mathbf{H} - \mathbf{J}_p \ , \quad \frac{\partial \mathbf{J}_p}{\partial t} + \gamma_d \mathbf{J}_p = \varepsilon_0 \omega_d^2 \mathbf{E}. \tag{4}$$

Recalling the definitions of the impedance and light velocity in vacuum, $Z_0 = \sqrt{\mu_0/\varepsilon_0}$ and $c_0 = 1/\sqrt{\varepsilon_0 \mu_0}$, and introducing the following substitutions, $\widetilde{\mathbf{H}} = Z_0 \mathbf{H}$, $\widetilde{\mathbf{E}} = \mathbf{E}$, $\widetilde{\mathbf{J}}_p = Z_0 \mathbf{J}_p$, $\widetilde{t} = c_0 t$, $\widetilde{\gamma}_d = \gamma_d / c_0$ and $\widetilde{\omega}_d^2 = \omega_d^2 / c_0^2$, it can be shown that system (4) can be normalized to yield

$$\frac{\partial \widetilde{\mathbf{H}}}{\partial t} = -\nabla \times \widetilde{\mathbf{E}} \ , \quad \varepsilon_\infty \frac{\partial \widetilde{\mathbf{E}}}{\partial t} = \nabla \times \widetilde{\mathbf{H}} - \widetilde{\mathbf{J}}_p \ , \quad \frac{\partial \widetilde{\mathbf{J}}_p}{\partial t} + \gamma_d \widetilde{\mathbf{J}}_p = \widetilde{\omega}_d^2 \widetilde{\mathbf{E}}, \tag{5}$$

knowing that $\mu_0 c_0 / Z_0 = 1$ and $\varepsilon_0 c_0 Z_0 = 1$. From now on, we omit the $\widetilde{X}$ notation for the normalized variables.

3 DGTD Method

The DGTD method can be considered as a finite element method where the continuity constraint at an element interface is released. While it keeps almost all the advantages of the finite element method (large spectrum of applications, complex geometries, etc.), the DGTD method has other nice properties:

- It is naturally adapted to a high order approximation of the unknown field. Moreover, one may increase the degree of the approximation in the whole mesh as easily as for spectral methods but, with a DGTD method, this can also be done locally i.e. at the mesh cell level.
- When the discretization in space is coupled to an explicit time integration method, the DG method leads to a block diagonal mass matrix independently of the form of the local approximation (e.g the type of polynomial interpolation). This is a striking difference with classical, continuous FETD formulations.
- It easily handles complex meshes. The grid may be a classical conforming finite element mesh, a non-conforming one or even a hybrid mesh made of various elements (tetrahedra, prisms, hexahedra, etc.). The DGTD method has been proven to work well with highly locally refined meshes. This property makes the DGTD method more suitable to the design of a *hp*-adaptive solution strategy (i.e. where the characteristic mesh size h and the interpolation degree p changes locally wherever it is needed).
- It is flexible with regards to the choice of the time stepping scheme. One may combine the discontinuous Galerkin spatial discretization with any global or local explicit time integration scheme, or even implicit, provided the resulting scheme is stable.
- It is naturally adapted to parallel computing. As long as an explicit time integration scheme is used, the DGTD method is easily parallelized. Moreover, the compact nature of method is in favor of high computation to communication ratio especially when the interpolation order is increased.

As in a classical finite element framework, a discontinuous Galerkin formulation relies on a weak form of the continuous problem at hand. However, due to the discontinuity of the global approximation, this variational formulation has to be defined at the element level. Then, a degree of freedom in the design of a discontinuous Galerkin scheme stems from the approximation of the boundary integral term resulting from the application of an integration by parts to the element-wise variational form. In the spirit of finite volume methods, the approximation of this boundary integral term calls for a numerical flux function which can be based on either a centered scheme or an upwind scheme, or a blend of these two schemes.

The DGTD method has been considered rather recently as an alternative to the widely used FDTD method for simulating nanoscale light/matter interaction problems [1, 12–14]. The main features of the DGTD method studied in [16] for the numerical solution of system (5) are the following:

- It is formulated on an unstructured tetrahedral mesh;
- It can deal with linear or curvilinear elements through a classical isoparametric mapping adapted to the DG framework [17];
- It relies on a high order nodal (Lagrange) interpolation of the components of $\mathbf{E}$, $\mathbf{H}$ and $\mathbf{J}_p$ within a tetrahedron;

- It offers the possibility of using a fully centered [5] or a fully upwind [6] scheme, as well as blend of the two schemes, for the evaluation of the numerical traces (also referred as numerical fluxes) of the **E** and **H** fields at inter-element boundaries;
- It can be coupled to either a second-order or fourth-order leap-frog (LF) time integration scheme [4], or to a low-storage fourth-order Runge-Kutta (LSRK) time integration scheme [2];
- It can rely on a Silver-Muller absorbing boundary condition or a CFS-PML technique for the artificial truncation of the computational domain.

Starting from the continuous Maxwell-Drude equations (5), the system of semi-discrete DG equations associated to an element τ_i of the tetrahedral mesh writes

$$\left\{\begin{aligned} \mathbb{M}_i \frac{d\overline{\mathbf{H}}_i}{dt} &= -\mathbb{K}_i \times \overline{\mathbf{E}}_i + \sum_{k\in\mathcal{V}_i} \mathbb{S}_{ik}\left(\overline{\mathbf{E}}_\star \times \mathbf{n}_{ik}\right), \\ \mathbb{M}_i^{\varepsilon\infty} \frac{d\overline{\mathbf{E}}_i}{dt} &= \mathbb{K}_i \times \overline{\mathbf{H}}_i - \sum_{k\in\mathcal{V}_i} \mathbb{S}_{ik}\left(\overline{\mathbf{H}}_\star \times \mathbf{n}_{ik}\right) - \mathbb{M}_i\overline{\mathbf{J}}_i, \\ \frac{d\overline{\mathbf{J}}_i}{dt} &= \omega_d^2 \overline{\mathbf{E}}_i - \gamma_d \overline{\mathbf{J}}_i. \end{aligned}\right. \tag{6}$$

In the above system of ODEs, $\overline{\mathbf{E}}_i$ is the vector of all the degrees of freedom of **E** in τ_i with size $3n_{p_i}$ (with similar definitions for $\overline{\mathbf{H}}_i$ and $\overline{\mathbf{J}}_i$), $\mathbb{M}_i$ and $\mathbb{M}_i^{\varepsilon\infty}$ are local mass matrices of size $3n_{p_i} \times 3n_{p_i}$, $\mathbb{K}_i$ is a local pseudo-stiffness matrix of size $3n_{p_i} \times 3n_{p_i}$, and $\mathbb{S}_{ik}$ is a local interface matrix of size $3n_{p_i} \times 3n_{p_k}$. Here, n_{p_i} denotes the number of basis functions for a polynomial interpolation $\mathbb{P}_p$ of the components of a field of degree p in τ_i. Moreover, $\overline{\mathbf{E}}_\star$ and $\overline{\mathbf{H}}_\star$ are numerical traces computed using an appropriate centered or upwind scheme. The various steps leading to the semi-discrete system (6) from the Maxwell-Drude continuous system (5) are detailed in [16].

4 Numerical Results

4.1 Scattering by a Nanosphere

Many nano-optics devices rely on the coupled plasmon resonances of metallic nanospheres, such as nano-arrays for Raman scattering [9], Fano resonators [10], or nanosphere-based biosensors [3]. We illustrate here the accuracy improvement obtained thanks to the use of high order elements (curvilinear tetrahedra) when considering the problem of the scattering of a plane wave by a gold nanosphere. The analytical solution of this problem can be computed *via* the Mie scattering theory [8]. We consider a sphere of radius $r = 50\,\text{nm}$ with Drude parameters $\varepsilon_\infty = 3.7362$, $\omega_d = 1.38710^7\,\text{GHz}$, $\gamma_d = 4.51510^4\,\text{GHz}$, and we are interested in its behavior in the [600, 1200] THz range. The incident field is a plane wave, with a sine-module Gaussian time profile, in order to provide a wide enough spectrum for the calculation. A so-called total/field (TF/SF) technique is adopted for imposing the plane wave incident field. The scatterer is enclosed by a total-field/scattered-field interface, on which the incident field is imposed. A CFS-PML layer surrounds the scattered-field region, and is terminated by an ABC condition.

We compare the results from DGTD simulations with the Mie solution of the problem. To conduct this study, we build three meshes, referred as M1, M2 and M3, corresponding to a discretization of the geometry of the sphere with an increased accuracy (the mesh characteristics and a visual representation are respectively given in Table 1 and Fig. 1). Curved elements are exploited only with the coarsest mesh (mesh M1), whereas linear elements are used for the three meshes. Results are presented in Fig. 2. One immediately notices the convergence of the results obtained on the linear meshes toward the reference solution. The $\mathbb{P}_1$-based solution on the mesh M3 almost perfectly fits the Mie prediction, at the cost of a high refinement level of the sphere surface. On the other hand, the solutions obtained with the

Table 1 Scattering by a nanosphere: characteristics of the tetrahedral meshes

	M1	M2	M3
n_s	962	1 677	10 736
n_t	4 706	8 767	61 718
h_{sphere}	2510^{-9}	10^{-9}	3.510^{-9}
n_c	764	0	0
n_r	3 942	8 767	61 718

n_s is the number of vertices, n_t the number of tetrahedra and h_{sphere} the size of the largest tetrahedron used to discretize the scatterer. For the curvilinear versions, n_c represents the number of curved tetrahedrons, whereas n_r is the number of rectilinear tetrahedrons

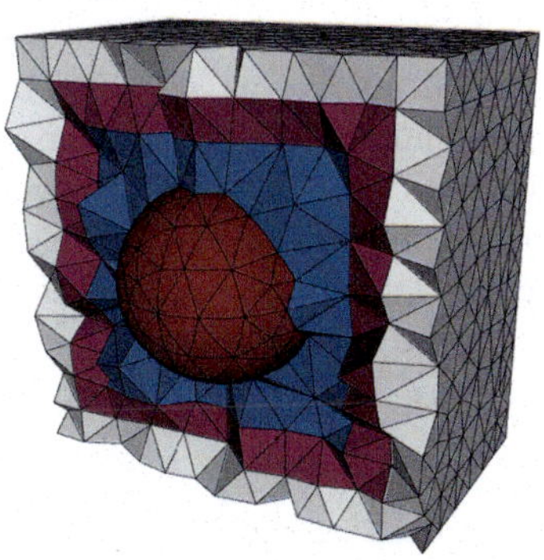

Fig. 1 Scattering by a nanosphere: mesh M1 for the cross-section calculation. The scatterer in (*red*) is enclosed by the total field region in (*blue*), delimited by the TF/SF interface on which the incident field is imposed. The scattered field region in (*purple*) is surrounded by PMLs in (*gray*)

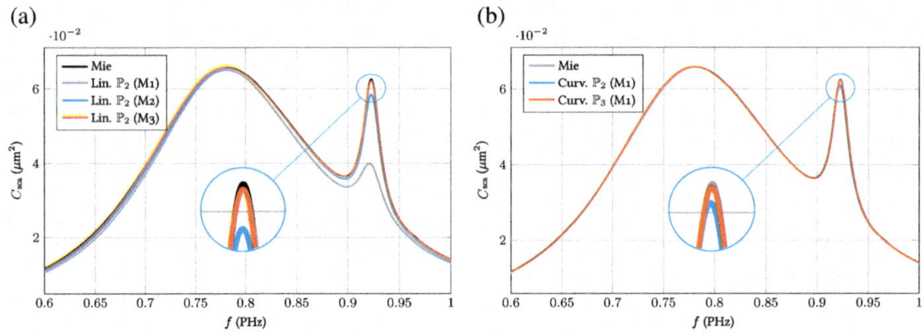

Fig. 2 Scattering by a nanosphere: scattering cross-section of a metallic sphere obtained with $\mathbb{P}_2$ and $\mathbb{P}_3$ approximations for linear and curvilinear meshes. (**a**) C_{sca} calculations with linear elements. (**b**) C_{sca} calculations with curvilinear elements

curvilinear M1 mesh are already in very good agreement with the reference solution: the $\mathbb{P}_2$ result is close, but the amplitude of the second resonance peak is still a bit undervalued. The numerical solution is improved by exploiting $\mathbb{P}_3$ approximation, yielding a relative error to the exact solution of less than 1%. Although this case corresponds to a basic but realistic nanophotonics configuration, the gains obtained in terms of CPU time and memory consumption when using curvilinear elements are very encouraging. The solution obtained with the DGTD-$\mathbb{P}_3$ method run on mesh M1 with curvilinear elements required 92 MB of memory and 884 s of CPU time.[1] In comparison, the solution obtained with the DGTD-$\mathbb{P}_2$ method run on mesh M3 with linear elements, which is of similar accuracy, required 312 MB and 6800 s. Hence, it makes the curvilinear simulation more than 3 times cheaper in terms of memory, and more than 7 times faster.

4.2 Optical Reflecting Arrays

In the past few years, important efforts have been deployed to find alternatives to on-chip, low-performance metal interconnects between devices. Because of the ever-increasing density of integrated components, intra- and inter-chip data communications have become a major bottleneck in the improvement of information processing. Given the compactness and the simple implantation of the devices, communications *via* free-space optics between nanoantenna-based arrays have recently drawn more attention [7]. Here, we focus on a specific low-loss design of dielectric reflectarray (DRA), whose geometry is based on a periodic repartition of dielectric cylinders on a metallic plate [20]. A sketch of the unit cell is presented on

[1] All the simulations are run in parallel on 16 CPUs.

Fig. 3. When illuminated in normal incidence, specific patterns of such resonators provide a constant phase gradient along the dielectric/metal interface, thus altering the phase of the incident wavefront. The gradient of phase shift generates an effective wavevector along the interface, which is able to deflect light from specular reflection. However, as can be seen on Fig. 4, the flaws of the lithographic production process can lead to discrepancies between the ideal device and the actual resonator array.

Here, we propose to exploit our DGTD solver to study the impact of the lithographic flaws on the performance of a 1D reflectarray. Efficient computations are obtained by combining high-order polynomial approximation with curvilinear meshing of the resonators, yielding accurate results on very coarse meshes.In our simulations, the silver slab is described by a simple Drude model of parameters $\varepsilon_\infty = 4.0$, $\gamma_d = 2.7310^4$ GHz and $\omega_d = 1.3810^7$ GHz. The resonators are made of a diagonally anisotropic material with relative permittivity parameters given by $\left[\varepsilon_r^x = 8.29, \varepsilon_r^y = 8.29, \varepsilon_r^z = 6.71\right]$. The slab thickness h, as well as the height d are respectively fixed to 200 and 50 nm. The defect parameter is denoted δ, and describes the impact of the lithography flaws on the cylindrical shape of

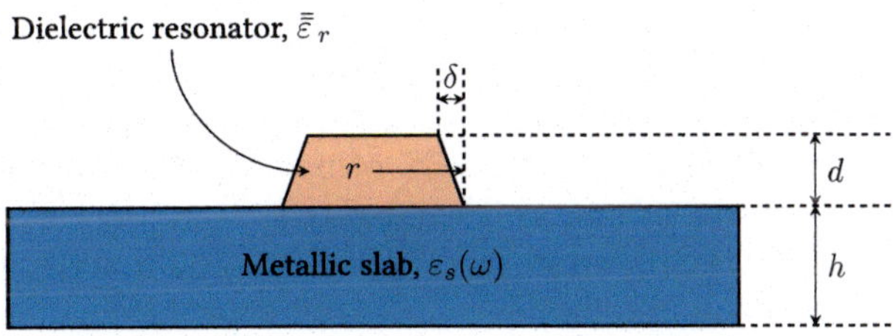

Fig. 3 Unit cell of a realistic monodimensional dielectric reflectarray composed of dielectric cylinders on a silver plate. The defect parameter δ is equal to zero for an ideal resonator. The view is a lateral cut of the cell

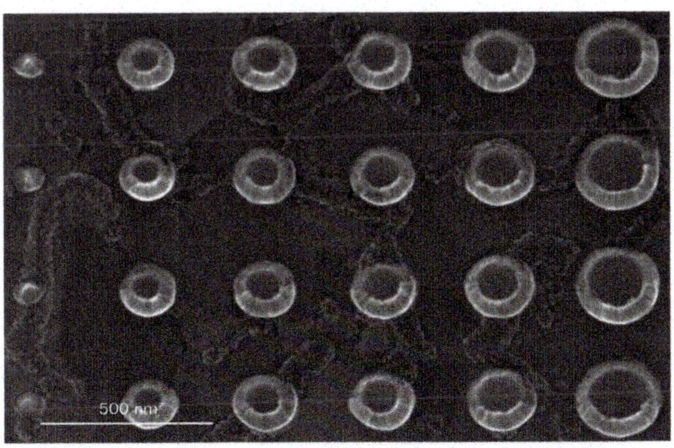

Fig. 4 6-element dielectric reflectarray produced by lithography

the resonator. The last geometric parameter is the basis radius of the resonator, denoted by r. In all computations, the devices are terminated with periodic boundary conditions in both planar directions. The incident field is a monochromatic plane wave, impinging from above in normal incidence.

The physical quantities of interest in this work are: (i) the reflection coefficient R, (ii) the reflected phase θ for single resonators, and (iii) the radar cross-section σ_{RCS} for the resonator arrays. Details about the definition and calculation of these quantities are given in [16].

We propose here to study the effects of the flaws induced by the lithography production of the dielectric resonators on its scattering regime. A single resonator with doubly periodic boundary conditions is considered. The lateral size of the periodic cell is 350 nm, the radius is fixed to $r = 85$ nm, and δ varies from 0 to 15 nm. The frequency of the incident plane wave is fixed to $f = 473.6$ THz ($\lambda = 633$ nm). The reflection coefficient and the reflected phase are computed, and plotted on Fig. 5. As can be seen, the reflected amplitude and phases are significantly blueshifted when δ is increased, which will have a major impact on the 1D dielectric array, as will be shown below. Here, we consider the 1D dielectric reflectarray presented in [19]. This array is designed to deflect normally-incident light with an angle of 19.9°, according to reflectarray theory. As before, the frequency of the incident plane wave is $f = 473.6$ THz ($\lambda = 633$ nm). The array is declined in two versions: the first one is made of ideal resonators, while the second one is composed of realistic resonators, with representative lithography flaws (see Fig. 6 for a close-up view of the array). The RCS of both arrays is computed with $\mathbb{P}_4$ polynomial approximation and quadratic tetrahedra, and plotted on Fig. 7. The ideal array provides a very good directivity toward 18.0°, with a very small parasitic lobe around 50.0°. This is confirmed by the field map of Fig. 8, where one can clearly see nearly-plane waves propagating away from the array. In this case, nearly 60% of the incident power is deflected with a non-cartesian angle. On the other

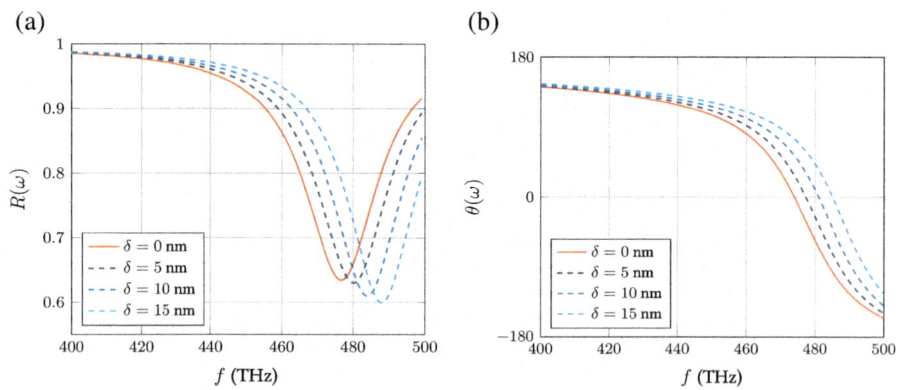

Fig. 5 Reflection coefficient and reflected phase of a single dielectric resonator with lithography defect. (**a**) Reflection coefficient. (**b**) Reflected phase

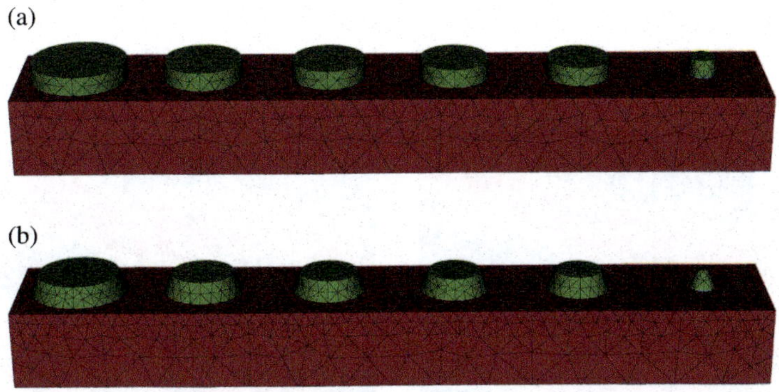

Fig. 6 Ideal and realistic 1D dielectric reflectarray meshes. The *red tetrahedra* correspond to silver, while the *green* ones are made of an anisotropic dielectric material. The device is surrounded by air and terminated by a PML above and below, and by periodic boundary conditions on the lateral sides. (**a**) Ideal reflectarray. (**b**) Realistic reflectarray, $\delta = 20\,\text{nm}$

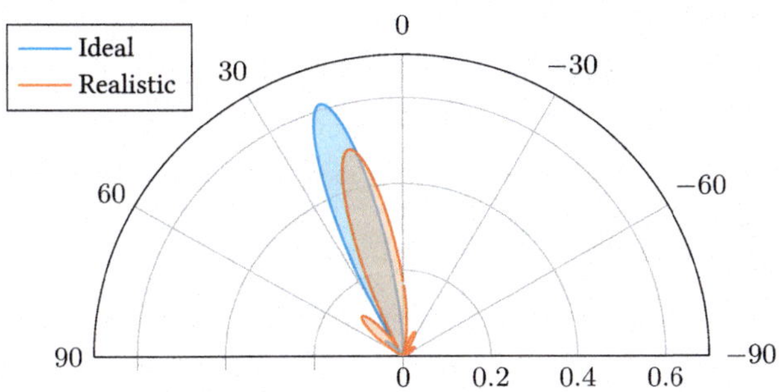

Fig. 7 Radar cross-section of ideal and realistic 1D dielectric reflectarrays at frequency f. The directivity peak in the ideal case is observed around 18.0°, while it is obtained at 14.5° for the realistic array

hand, the realistic array presents more imperfections in its directivity patterns, with numerous parasitic lobes, and a lower efficiency (around 50%). Additionally, the deflection angle is very different from what was predicted by the reflectarray theory. This results in a much less satisfying field map, where the plane wave is severely distorted. This may enlight the need to compensate these flaws at the conception level by adjusting the physical parameters of the reflectors.

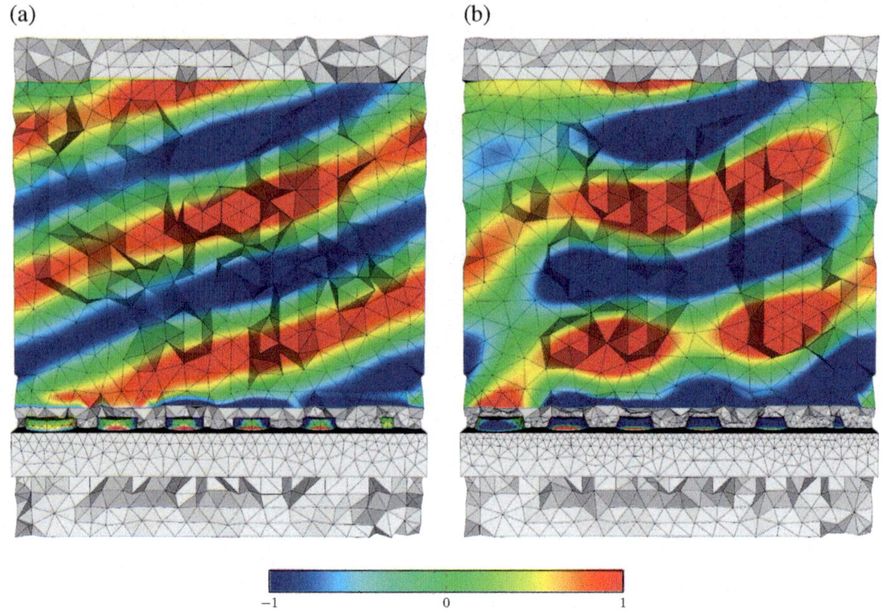

Fig. 8 Time-domain snapshot of E_y component for ideal and realistic 1D dielectric reflectarrays. Solution is obtained in established regime at $t = 0.1$ ps. Fields are scaled to $[-1, 1]$. (**a**) Ideal reflectarray. (**b**) Realistic reflectarray, $\delta = 20\,\text{nm}$

5 Conclusion

The work described here is part of a larger initiative aiming at the development of a software suite dedicated to nanophotonics/nanoplasmonics that will ultimately include DG-based solvers for both time-domain and frequency-domain problems, as well as the capabilities to numerically consider various material models in the linear and non-linear regimes, considering local and non-local (i.e. spatial) dispersion effects.

References

1. K. Busch, M. König, J. Niegemann, Discontinuous Galerkin methods in nanophotonics. Laser Photonics Rev. **5**, 1–37 (2011)
2. M. Carpenter, C. Kennedy, Fourth-order $2n$-storage Runge-Kutta schemes. Tech. rep., NASA Technical Memorandum MM-109112 (1994)
3. T. Chung, S.Y. Lee, E.Y. Song, H. Chun, B. Lee, Plasmonic nanostructures for nanoscale biosensing. Sensors **11**, 10907–10929 (2011)
4. H. Fahs, S. Lanteri, A high-order non-conforming discontinuous Galerkin method for time-domain electromagnetics. J. Comput. Appl. Math. **234**, 1088–1096 (2010)

5. L. Fezoui, S. Lanteri, S. Lohrengel, S. Piperno, Convergence and stability of a discontinuous Galerkin time-domain method for the 3D heterogeneous Maxwell equations on unstructured meshes. ESAIM: Math. Model. Numer. Anal. **39**(6), 1149–1176 (2005)
6. J. Hesthaven, T. Warburton, Nodal high-order methods on unstructured grids. I. Time-domain solution of Maxwell's equations. J. Comput. Phys. **181**(1), 186–221 (2002)
7. J. Huang, J.A. Encinar, *Reflectarray Antennas* (IEEE Press, New York, 2008)
8. H. van de Hulst, *Light Scattering by Small Particles* (Dover, New York, 1981)
9. F. Le, D.W. Brandl, Y.A. Urzhumov, H. Wang, J. Kundu, N.J. Halas, J. Aizpurua, P. Nordlander, Metallic nanoparticle arrays: a common substrate for both surface-enhanced Raman scattering and surface-enhanced infrared absorption. ACS Nano **2**, 707–718 (2008)
10. B. Luk'yanchuk, N.I. Zheludev, S.A. Maier, N.J. Halas, P. Nordlander, H. Giessen, C.T. Chong, The Fano resonance in plasmonic nanostructures and metamaterials. Nat. Mater. **9**, 707–715 (2010)
11. S. Maier, *Plasmonics - Fundamentals and Applications* (Springer, Berlin, 2007)
12. C. Matysseka, J. Niegemann, W. Hergertb, K. Busch, Computing electron energy loss spectra with the discontinuous Galerkin time-domain method. Photonics Nanostruct. **9**(4), 367–373 (2011)
13. J. Niegemann, M. König, K. Stannigel, K. Busch, Higher-order time-domain methods for the analysis of nano-photonic systems. Photonics Nanostruct. **7**, 2–11 (2009)
14. J. Niegemann, R. Diehl, K. Busch, Efficient low-storage Runge-Kutta schemes with optimized stability regions. J. Comput. Phys. **231**(2), 364–372 (2012)
15. A. Taflove, S. Hagness, *Computational Electrodynamics: The Finite-Difference Time-Domain Method*, 3rd edn. (Artech House Publishers, Boston, 2005)
16. J. Viquerat, Simulation of electromagnetic waves propagation in nano-optics with a high-order discontinuous Galerkin time-domain method. Ph.D. thesis, University of Nice-Sophia Antipolis (2015). https://tel.archives-ouvertes.fr/tel-01272010
17. J. Viquerat, C. Scheid, A 3D curvilinear discontinuous Galerkin time-domain solver for nanoscale light-matter interactions. J. Comput. Appl. Math. **289**, 37–50 (2015)
18. K. Yee, Numerical solution of initial boundary value problems involving Maxwell's equations in isotropic media. IEEE Trans. Antennas Propag. **14**(3), 302–307 (1966)
19. L. Zou, M. Lopez-Garcia, W. Withayachumnankul, C.M. Shah, A. Mitchell, M. Bhaskaran, S. Sriram, R. Oulton, M. Klemm, C. Fumeaux, Spectral and angular characteristics of dielectric resonator metasurface at optical frequencies. Appl. Phys. Lett. **105**, 191, 109 (2014)
20. L. Zou, W. Withayachumnankul, C. Shah, A. Mitchell, M. Bhaskaran, S. Sriram, C. Fumeaux, Dielectric resonator nanoantennas at visible frequencies. Opt. Express **21**, 1344–1352 (2013)

High-Order Embedded WENO Schemes

Bart S. van Lith, Jan H.M. ten Thije Boonkkamp, and Wilbert L. IJzerman

Abstract Embedded WENO schemes are a new family of weighted essentially nonoscillatory schemes that always utilise *all* adjacent smooth substencils. This results in increased control over the convex combination of lower-order interpolations. We show that more conventional WENO schemes, such as WENO-JS and WENO-Z (Borges et al., J. Comput. Phys., 2008; Jiang and Shu, J. Comput. Phys., 1996), do not exhibit this feature and as such do not always provide a desirable linear combination of smooth substencils. In a previous work, we have already developed the theory and machinery needed to construct embedded WENO methods and shown some five-point schemes (van Lith et al., J. Comput. Phys., 2016). Here, we construct a seven-point scheme and show that it too performs well using some numerical examples from the one-dimensional Euler equations.

1 Recap of WENO

Weighted essentially non-oscillatory (WENO) schemes are a class of high-order reconstruction methods commonly used in the numerical approximation of hyperbolic PDEs[6]. They are able to combine oscillation suppressing properties with a high order of convergence. As such they are particularly useful when faced with discontinuous solutions, e.g., shocks or contact discontinuities.

Consider the one-dimensional hyperbolic PDE

$$\frac{\partial u}{\partial t} + \frac{\partial f(u)}{\partial x} = 0, \tag{1}$$

B.S. van Lith (✉) • J.H.M. ten Thije Boonkkamp
CASA, Eindhoven University of Technology, PO Box 513, 5600 MB Eindhoven, The Netherlands
e-mail: b.s.v.lith@tue.nl

W.L. IJzerman
CASA, Eindhoven University of Technology, PO Box 513, 5600 MB Eindhoven, The Netherlands

Philips Lighting, High Tech Campus 48, 5656 AE Eindhoven, The Netherlands

M.L. Bittencourt et al. (eds.), *Spectral and High Order Methods for Partial Differential Equations ICOSAHOM 2016*, Lecture Notes in Computational Science and Engineering 119, DOI 10.1007/978-3-319-65870-4_17

where f is the flux function. Given some computational domain, we apply a grid $\{x_j\}_{j=1}^N$. With each grid point x_j we associate a cell, $[x_{j-\frac{1}{2}}, x_{j+\frac{1}{2}}]$ of size Δx. Averaging (1) over a cell, we obtain

$$\frac{\mathrm{d}\bar{u}_j}{\mathrm{d}t} + \frac{1}{\Delta x}\Big(f\big(u(x_{j+\frac{1}{2}}, t)\big) - f\big(u(x_{j-\frac{1}{2}}, t)\big)\Big) = 0, \tag{2}$$

where $\bar{u}_j$ is the average of u over the cell associated with x_j. Note that (2) is still exact. As per usual, we replace the fluxes at the cell boundaries by numerical fluxes $F_{j\pm\frac{1}{2}}$ and the exact average by its numerical approximation u_j, yielding

$$\frac{\mathrm{d}u_j}{\mathrm{d}t} + \frac{1}{\Delta x}\left(F_{j+\frac{1}{2}} - F_{j-\frac{1}{2}}\right) = 0. \tag{3}$$

In the following, we suppress the time-dependency. The numerical flux typically depends on the left and right limits of the function value at the cell boundary, i.e., $F_{j+\frac{1}{2}} = F(u(x^+_{j+\frac{1}{2}}), u(x^-_{j+\frac{1}{2}}))$, where the superscripts plus and minus denote a right and left limit, respectively. As such, the numerical method is completed by specifying an approximation for finding the solution value at cell boundaries.

A WENO scheme attempts to reconstruct the sought values from a stencil S of several cells. The most popular variants use a five-point stencil centred on x_j, see Fig. 1. The five-point stencil is divided into three smaller substencils, each of three points. On each of the three-point substencils, we can find a parabolic reconstruction of the solution, based on the averages over each cell. The key insight of a WENO scheme is that a suitable linear combination of the three parabolic approximations produces the fifth-order reconstruction.

If we organise the values of the large stencil into an auxiliary vector, $\mathbf{v} = (u_{j-2}, u_{j-1}, u_j, u_{j+1}, u_{j+2})^T$, we can find a vector composed of the three lower-order approximations as $C\mathbf{v}$, where C is a 3×5 matrix. Furthermore, a linear combination

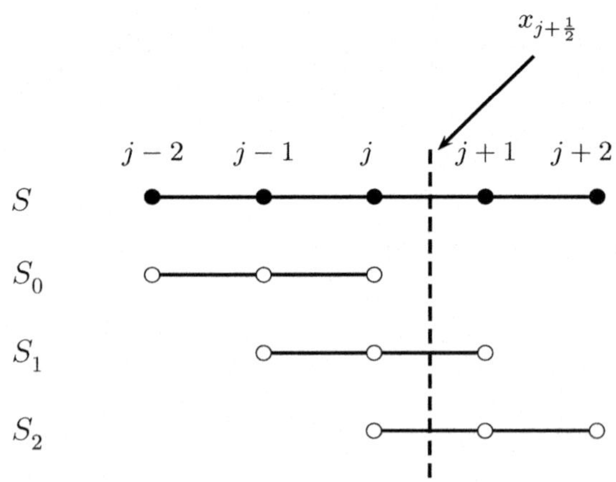

Fig. 1 Large five-point stencil and substencils that are used in the WENO reconstruction technique

of these three values gives a fifth-order approximation for smooth solutions, so that we have

$$u_{j+\frac{1}{2}}^{(\mathrm{UW5})} = \boldsymbol{\gamma}^T C \mathbf{v}, \tag{4}$$

where $\boldsymbol{\gamma}$ is the vector of linear weights. The basic scheme can then be captured in a tableau, inspired by Butcher tableaux[2], given by

$$\begin{array}{c|c} C & \boldsymbol{\gamma} \\ \hline \end{array}. \tag{5}$$

Organised in this way, the tableau contains all the coefficients involved in a WENO scheme, thus giving a concise overview of the underlying linear method. The tableau for the five-point WENO scheme looks as follows,

$$\begin{array}{ccccc|c} \frac{2}{6} & -\frac{7}{6} & \frac{11}{6} & & & \frac{1}{10} \\ & -\frac{1}{6} & \frac{5}{6} & \frac{2}{6} & & \frac{6}{10} \\ & & \frac{2}{6} & \frac{5}{6} & -\frac{1}{6} & \frac{3}{10} \\ \hline \end{array}, \tag{6}$$

where the zeros have been left out for clarity. These coefficients are related through Taylor expansions of the solution around $x_{j+\frac{1}{2}}$.

A WENO scheme attempts to suppress oscillations by replacing the linear weights, $\boldsymbol{\gamma}$, by a set of nonlinear weights that lowers the contribution of substencils containing a discontinuity. To facilitate this, smoothness indicators are introduced, the most commonly used are given by Jiang and Shu[5] as

$$\beta_k := \int_{x_{j-\frac{1}{2}}}^{x_{j+\frac{1}{2}}} \left(p_k''(x)\right)^2 \Delta x^3 + \left(p_k'(x)\right)^2 \Delta x \mathrm{d}x, \quad k = 0, 1, 2, \tag{7}$$

where p_k is the polynomial reconstruction on substencil S_k, i.e., p_k has average value u_j in cell j for each $j \in S_k$. A tedious but straightforward calculus exercise shows that

$$\beta_0 = \frac{13}{12}(u_{j-2} - 2u_{j-1} + u_j)^2 + \frac{1}{4}(u_{j-2} - 4u_{j-1} + 3u_j)^2, \tag{8a}$$

$$\beta_1 = \frac{13}{12}(u_{j-1} - 2u_j + u_{j+1})^2 + \frac{1}{4}(u_{j-1} - u_{j+1})^2, \tag{8b}$$

$$\beta_2 = \frac{13}{12}(u_j - 2u_{j+1} + u_{j+2})^2 + \frac{1}{4}(3u_j - 4u_{j+1} + u_{j+2})^2. \tag{8c}$$

A Taylor expansion around the point x_j shows that $\beta_k = \mathcal{O}(\Delta x^2)$ when the solution is smooth and $\partial_x u(x_j) = \mathcal{O}(1)$, meaning the gradient of the interpolation is independent of the grid size. However, around discontinuities we have $\partial_x u(x_j) =$

$\mathcal{O}(\frac{1}{\Delta x})$, resulting in $\beta_k = \mathcal{O}(1)$, so that the smoothness indicators allow us to find a set of nonlinear weights that pick the smoothest substencils. The WENO scheme of Jiang and Shu[5] then uses nonlinear weights ω_k $(k = 0, 1, 2)$ defined by

$$\tilde{\omega}_k^{\text{JS}} = \frac{\gamma_k}{(\beta_k + \varepsilon)^p}, \tag{9}$$

where ε is a small number to avoid division by 0 and $p > 0$. Typical values are $\varepsilon = 10^{-6}$ to 10^{-12} and $p = 2$. In any WENO scheme, the nonlinear weights are normalised to obtain a consistent method, i.e.

$$\omega_k = \frac{\tilde{\omega}_k}{\sum_{l=0}^{2} \tilde{\omega}_l}. \tag{10}$$

The JS weights satisfy $\omega_k^{\text{JS}} = \gamma_k + \mathcal{O}(\Delta x^2)$ when $\beta_k = \mathcal{O}(\Delta x^2)$. Note that together with the set of coefficients (6), smoothness indicators and unnormalised nonlinear weights, the WENO scheme is completely specified. Furthermore, if we denote by $\boldsymbol{\omega}$ the column vector of nonlinear weights, then

$$u_{j+\frac{1}{2}}^{(\text{WENO})} = \boldsymbol{\omega}^T C \mathbf{v} \tag{11}$$

is the WENO approximation.

Borges et al.[1] showed that in order to achieve fifth-order accuracy, a sufficient condition is

$$\omega_k = \gamma_k + \mathcal{O}(\Delta x^3), \tag{12}$$

in smooth regions of the solution. However, the WENO-JS scheme does not satisfy this condition, even though it does indeed provide fifth-order accuracy in smooth regions. To satisfy the sufficient condition (12), Borges et al. constructed the WENO-Z scheme, introducing a global smoothness indicator τ given by

$$\tau = |\beta_2 - \beta_0|. \tag{13}$$

The global smoothness indicator is constructed to use information from the entire stencil S. Furthermore, one can show that $\tau = \mathcal{O}(\Delta x^5)$ when the solution is smooth on the large stencil S. The WENO-Z weights are defined by

$$\tilde{\omega}_k^{\text{Z}} = \gamma_k \left(1 + \left(\frac{\tau}{\beta_k + \varepsilon} \right)^p \right), \tag{14}$$

so that the WENO-Z scheme satisfies the sufficient condition (12) for any $p \geq 1$.

1.1 A Flaw

Let us examine what WENO-JS and WENO-Z do when a discontinuity is present in substencils S_0 or S_2. Thus, there is a single nonsmooth substencil while there are two smooth substencils. In principle, it is therefore be possible to make some suitable linear combination of the smooth substencils. For instance, we might want to form a fourth-order combination or a combination that minimises dissipation.

Before we discuss these possibilities, let us investigate the ratio of the nonlinear weights of WENO-JS and WENO-Z, starting with the JS weights,

$$\frac{\omega_k^{\text{JS}}}{\omega_l^{\text{JS}}} = \frac{\tilde{\omega}_k^{\text{JS}}}{\tilde{\omega}_l^{\text{JS}}} = \frac{\gamma_k}{\gamma_l}\left(\frac{\beta_l + \varepsilon}{\beta_k + \varepsilon}\right)^p, \tag{15}$$

where we will assume for the remainder that $\varepsilon \ll \beta_k$ for all smooth S_k. For the five-point WENO scheme, we have

$$\frac{\beta_l}{\beta_k} = \begin{cases} 1 + \mathcal{O}(\Delta x^3) & \text{if } k = 0, l = 2 \text{ or } k = 2, l = 0, \\ 1 + \mathcal{O}(\Delta x^2) & \text{otherwise}, \end{cases} \tag{16}$$

which follows from the Taylor expansions of the smoothness indicators (8). For more general WENO schemes with r substencils, one can show that $\frac{\beta_l}{\beta_k} = 1 + \mathcal{O}(\Delta x^s)$, where $s \geq 2$. The WENO-JS weights therefore satisfy

$$\frac{\omega_k^{\text{JS}}}{\omega_l^{\text{JS}}} = \frac{\gamma_k}{\gamma_l}\left(1 + \mathcal{O}\left(\Delta x^s\right)\right), \tag{17}$$

with $s \geq 2$. Note however, that this relation only depends on the local smoothness of S_k and S_l.

For WENO-Z, we have a similar result, since

$$\frac{\omega_k^{Z}}{\omega_l^{Z}} = \frac{\gamma_k\left(1 + (\frac{\tau}{\beta_k})^p\right)}{\gamma_l\left(1 + (\frac{\tau}{\beta_l})^p\right)} = \frac{\gamma_k\left(\beta_l^p + \tau^p(\frac{\beta_l}{\beta_k})^p\right)}{\gamma_l\left(\beta_l^p + \tau^p\right)}. \tag{18}$$

Again using (16), we find that now independent of the value of τ,

$$\frac{\omega_k^{Z}}{\omega_l^{Z}} = \frac{\gamma_k}{\gamma_l}\left(1 + \mathcal{O}\left(\Delta x^s\right)\right), \tag{19}$$

with $s \geq 2$ provided there are no critical points. Note that the lower bound on s can be increased when the solution is smooth on the entire stencil.

Now consider the situation where substencils S_0 and S_1 are smooth with no critical points but S_2 contains a discontinuity. Both the JS and Z schemes will lead to $\frac{\omega_0}{\omega_1} = \frac{\gamma_0}{\gamma_1} + \mathcal{O}(\Delta x^2)$. This leads to $\omega_0 \approx \frac{1}{7}$ and $\omega_1 \approx \frac{6}{7}$ for both JS and Z schemes, from which the WENO approximation becomes

$$u^{(\text{WENO})}_{j+\frac{1}{2}} - u(x_{j+\frac{1}{2}}) = \frac{1}{28} u^{(3)}_j \Delta x^3 + \mathcal{O}(\Delta x^4), \tag{20}$$

by a Taylor expansion of the third-order approximations. However, if we could achieve in this situation $\omega_0 \approx \frac{1}{4}$ and $\omega_1 \approx \frac{3}{4}$, we could obtain a fourth-order approximation for the cell boundary value.

The flaw of WENO-JS and WENO-Z addressed in this work is the following: when encountering a discontinuity, both schemes revert directly to their lower-order modes, even when there are multiple adjacent smooth substencils. When multiple adjacent smooth substencils are available, control over the weights leads to direct control over the truncation error. This obviously has some advantages, for instance by increasing the local order of approximation. We have already developed a complete theory in an earlier work, where we constructed some five-point schemes [10]. Here, we will also develop and demonstrate seven-point embedded WENO schemes.

2 Embedded WENO

We propose a new class of WENO schemes, which we call embedded WENO, that allows control over the nonlinear weights for all possible locations of a discontinuity. For instance, in five-point WENO schemes there are two possible locations such that two substencils are smooth, see Fig. 2.

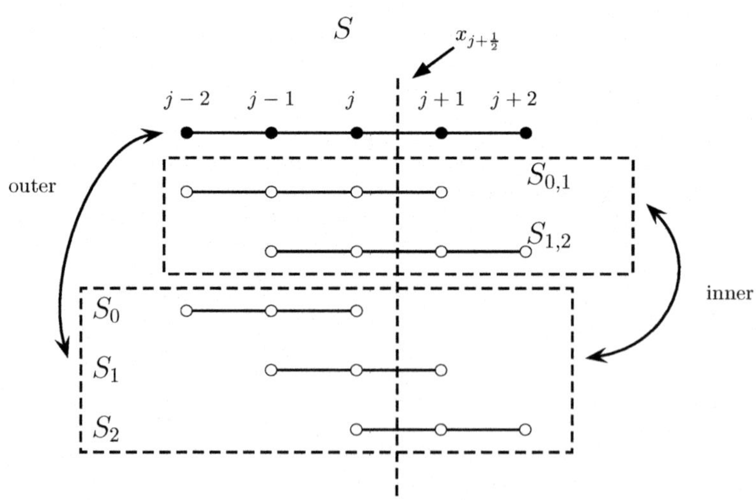

Fig. 2 Embedded WENO stencils and substencils. The four-point stencils are composed by controlling the linear combination of the two smooth substencils

Being able to control the linear combination in more situations results in more control over the numerical solution. We may, for instance, wish to have the highest possible order of convergence in all cases. Alternatively, we may wish to minimise the numerical dissipation of the scheme. We therefore need to construct nonlinear weights that converge to weights of our own choosing in these cases.

The desired weights when either S_0 or S_2 is not smooth are designated the inner weights of the scheme and are denoted $\alpha_k^{(K)}$, where K is the set of indices for the substencils that are not smooth. To reiterate, the inner weights are chosen by the user to achieve some desirable effect on the numerical solution. For instance, the inner weights optimised for order of convergence are given by $\alpha_0^{(2)} = \frac{1}{4}$, $\alpha_1^{(2)} = \frac{3}{4}$, where S_2 would contain the discontinuity. Likewise, when S_0 contains the discontinuity, the inner weights maximising order of convergence would be $\alpha_1^{(0)} = \frac{1}{2}$ and $\alpha_2^{(0)} = \frac{1}{2}$.

To escape from (17) and (19), the nonlinear weights will have to be redefined. We propose the following general form,

$$\tilde{\omega}_k = \gamma_k \left(1 + \left(\frac{|\sum_{l=0}^{r-1} a_{kl}\beta_l|}{\beta_k + \varepsilon} \right)^p \right), \tag{21}$$

where r is then number of substencils, $r = 3$ for five-point WENO schemes. The coefficients a_{kl} are called the embedding coefficients. This form of the unnormalised weights can be considered as a generalisation of the WENO-Z scheme of Borges et al. [1]. Indeed, we can choose the embedding coefficients so that (21) results in the WENO-Z scheme, i.e., $a_{k0} = 1$, $a_{k1} = 0$ and $a_{k2} = -1$ for $k = 0, 1, 2$.

Using the Taylor expansions of the smoothness indicators, we can find conditions under which $\sum_{l=0}^{r-1} a_{kl}\beta_l = \mathcal{O}(\Delta x^{r+2})$ in smooth regions. For five-point WENO schemes, we have

$$a_{k0} + a_{k1} + a_{k2} = 0, \tag{22a}$$

$$-2a_{k0} + a_{k1} - 2a_{k2} = 0, \tag{22b}$$

for $k = 0, 1, 2$. The following theorem states the conditions under which the nonlinear weights converge to the inner weights.

Theorem 1 *Let $\tilde{\omega}_k$ be the unnormalised nonlinear weights of a WENO scheme given by* (21)*. Let K be the set of indices of discontinuous substencils, assume $\varepsilon \ll \beta_k$ for all $k \notin K$, and let $\alpha_k^{(K)}$ be the desired inner weights. Let $k, l \notin K$, if the embedding coefficients then satisfy*

$$\left(\frac{\gamma_k}{\alpha_k^{(K)}} \right)^{\frac{1}{p}} \sum_{m \in K} a_{km} = \pm \left(\frac{\gamma_l}{\alpha_l^{(K)}} \right)^{\frac{1}{p}} \sum_{n \in K} a_{ln}, \tag{23}$$

the nonlinear weights will converge to the inner weights.

Proof Let k and l be indices of smooth substencils, then the ratio of their nonlinear weights is given by

$$\frac{\omega_k}{\omega_l} = \frac{\tilde{\omega}_k}{\tilde{\omega}_l} = \frac{\gamma_k \left|\sum_{m\in K} a_{km}\right|^p \frac{C}{\Delta x^2} + \mathcal{O}(1)}{\gamma_l \left|\sum_{n\in K} a_{ln}\right|^p \frac{C}{\Delta x^2} + \mathcal{O}(1)},$$

where C is a constant related to the jump in the numerical solution. Simplifying, we find

$$\frac{\omega_k}{\omega_l} = \frac{\gamma_k \left|\sum_{m\in K} a_{km}\right|^p}{\gamma_l \left|\sum_{n\in K} a_{ln}\right|^p} + \mathcal{O}(\Delta x^2).$$

Taking the limit of $\Delta x \to 0$, the $\mathcal{O}(\Delta x^2)$ term drops out. To complete the theorem, we equate this expression with $\frac{\alpha_k^{(K)}}{\alpha_l^{(K)}}$ and separate terms.

□

Remark 1 The absolute signs in (21) result in a sign freedom in the embedding coefficients. Writing the coefficients as a matrix, each row can be freely multiplied by -1. We choose the positive sign in (23) from here on out.

Remark 2 Theorem 1 only gives ratios between the embedding coefficients. As a consequence, there will always be at least one degree of freedom in the choice of the embedding coefficients. WENO-Z scheme also exhibits this degree of freedom, as one can change the weights to

$$\tilde{\omega}_k^Z = \gamma_k \left(C + \left(\frac{\tau}{\beta_k + \varepsilon} \right)^p \right),$$

for any $C > 0$ with impunity due to the normalisation. Some schemes have even used $C = 1000$, see for instance [4]. In our formulation, this is equivalent to multiplying all embedding coefficients with $\frac{1}{C}$. The larger C, the closer the weights will be to the optimal weights compared to $C = 1$. Currently, there are no guidelines on how to choose it.

For five-point WENO schemes, the theorem provides us with one equation for each four-point substencil. To give a concise overview of all the coefficients, we extend our tableaux notation to include the matrix A with elements a_{kl}, i.e.,

$$\begin{array}{c|c|c} C & \boldsymbol{\gamma} & A \\ \hline \end{array} \tag{24}$$

For example, choosing $p = 2$ and optimising for order of convergence, we find the scheme

$$
\begin{array}{ccccc|c|ccc}
\frac{2}{6} & -\frac{7}{6} & \frac{11}{6} & & & \frac{1}{10} & \frac{\sqrt{2}}{2} & 0 & -\frac{\sqrt{2}}{2} \\
 & -\frac{1}{6} & \frac{5}{6} & \frac{2}{6} & & \frac{6}{10} & \frac{1}{2} & 0 & -\frac{1}{2} \\
 & & \frac{2}{6} & \frac{5}{6} & -\frac{1}{6} & \frac{3}{10} & \frac{\sqrt{2}}{2} & 0 & -\frac{\sqrt{2}}{2} \\
\hline
\end{array}, \qquad (25)
$$

where we have chosen $a_{00} = \frac{\sqrt{2}}{2}$ to fix the coefficients.

2.1 Seven-Point Stencils

Our approach is generally valid and can also be applied to any high-order WENO scheme. As a demonstration, we present a seven-point WENO scheme, where we have taken into account all five possibilities for K: $\{0\}$, $\{0,1\}$, $\{2,3\}$, $\{3\}$ and $\{0,3\}$, where the last case represents two discontinuities on opposite sides of the large stencil S. Choosing $p = 2$, the tableau for the embedded seven-point WENO scheme is given by

$$
\begin{array}{ccccccc|c|cccc}
\frac{-3}{12} & \frac{13}{12} & -\frac{23}{12} & \frac{25}{12} & & & & \frac{1}{35} & \frac{\sqrt{2}-1}{4} & -\frac{1}{4} & \frac{1}{2} - \frac{3}{4\sqrt{2}} & \frac{1}{4\sqrt{2}} \\
 & \frac{1}{12} & -\frac{5}{12} & \frac{13}{12} & \frac{3}{12} & & & \frac{12}{35} & \frac{\sqrt{2}}{8\sqrt{3}} & \frac{\sqrt{2}-4}{8\sqrt{3}} & \frac{1-\sqrt{2}}{4\sqrt{3}} & \frac{1}{4\sqrt{3}} \\
 & & -\frac{1}{12} & \frac{7}{12} & \frac{7}{12} & -\frac{1}{12} & & \frac{18}{35} & \frac{1}{4\sqrt{3}} & \frac{1-\sqrt{2}}{4\sqrt{3}} & \frac{\sqrt{2}-4}{8\sqrt{3}} & \frac{\sqrt{2}}{8\sqrt{3}} \\
 & & & \frac{3}{12} & \frac{13}{12} & -\frac{5}{12} & \frac{1}{12} & \frac{4}{35} & \frac{1}{4\sqrt{2}} & \frac{1}{2} \quad \frac{3}{4\sqrt{2}} & \frac{1}{1} & \frac{\sqrt{2}-1}{1} \\
\hline
\end{array} \qquad (26)
$$

Note that the matrix A satisfies

$$
a_{kl} = a_{3-k,3-l}, \qquad (27)
$$

which means the embedding matrix for the left cell edge is the same as for the right cell edge. In this case too we employ the Jiang and Shu smoothness indicators together with the general form (21). We have chosen $a_{00} = \frac{\sqrt{2}-1}{4}$ to fix the coefficients. This choice was motivated by being reasonably close to unity while giving, in the authors' opinions, the most aesthetically pleasing tableau.

3 Results

To demonstrate the improvements of the embedded scheme over its standard counterpart, WENO-Z7[3], we use several numerical examples. All our examples consists of solving the Euler equations in one dimension. We solve the Euler

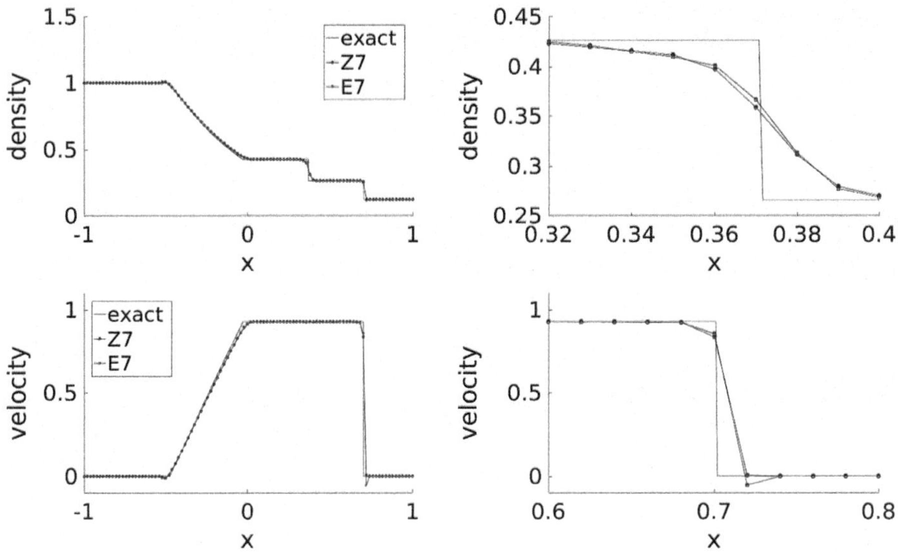

Fig. 3 Numerical solution of Sod's test. Z7 refers to the WENO-Z7 scheme while E7 to the embedded WENO7 scheme. Both numerical solutions were computed with 200 grid points and a CFL number of 0.45

equations in conjunction with an ideal gas equation of state and ratio of specific heats of 1.4.

The scheme uses a characteristic decomposition combined with global Lax-Friedrichs flux splitting and the SSPRK(5,4) integrator of Spiteri and Ruuth[8]. We use the L_1-norm at the final integration time to define a global error, the sum of absolute differences over all components and grid points.

Sod's test, see [9], is used to verify that the embedded WENO scheme performs well in relatively easy problems, see Fig. 3. Indeed, for Sod's test, the performance of the WENO-Z7 and embedded WENO7 is very similar. We define the global error as

$$e = \sum_{i=1}^{3} \sum_{j=1}^{N} |U_{ij} - U_i(T)| \Delta x, \tag{28}$$

where U is a conserved variable, T is the integration time and i is the component index. The relative decrease in the global error compared to WENO-Z7 is 0.26%, which is to say the performance of the schemes is almost equal in this test problem. However, it is interesting to note that the contact discontinuity is captured with a slightly steeper gradient.

For the Mach-3 shock entropy-wave interaction, see [7], the embedded scheme really shows its merits, see Fig. 4. Especially in the high-frequency region, the numerical solution approximates much more closely the reference

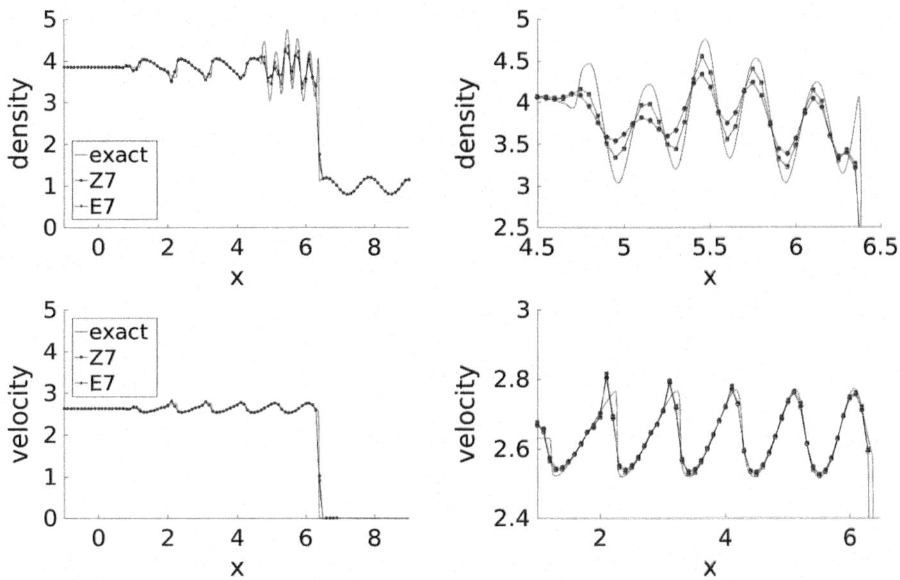

Fig. 4 Numerical solution of the Mach-3 shock entropy-wave interaction problem. Z7 refers to the WENO-Z7 scheme while E7 to the embedded WENO7 scheme. Both numerical solutions were computed with 200 grid points and a CFL number of 0.45

solution, resulting in a whopping 24.7% decrease in global error. We believe this radically better result comes from the embedded scheme having reduced numerical dissipation in the medium-to-high range of wave numbers compared to the standard one. However, we have not investigated this matter fully.

4 Conclusions

We have shown that conventional WENO schemes such as WENO-JS and WENO-Z, have a fixed ratio of the nonlinear weights dictated by the linear weights. As such, the convex combination will not always be optimal in the presence of discontinuities. We have constructed a framework that allows WENO schemes to utilise *all* adjacent smooth substencils.

The general form we proposed may be considered a generalisation of the WENO-Z scheme, which can be recovered by a special choice of embedding coefficients. Using the proposed form, we have presented order-optimised schemes for five and seven-point stencils. For the seven-point scheme we also consider two discontinuities on either side of the large stencil.

Two numerical examples from the one-dimensional incompressible Euler equations were presented for the seven-point scheme. We have demonstrated similar performance for Sod's test of the embedded scheme versus its standard counterpart, WENO-Z7. However, for a harder problem we have demonstrated vast improvement in the performance. The numerical solution exhibits both less spurious oscillations and less dissipation in high-frequency regions.

Acknowledgements This work was generously supported by Philips Lighting and the Intelligent Lighting Institute. The authors would especially like to thank Rafael Borges and Wai-Sun Don for the fruitful discussions and the generous sharing of their Matlab code.

References

1. R. Borges, M. Carmona, B. Costa, W.-S. Don, An improved weighted essentially non-oscillatory scheme for hyperbolic conservation laws. J. Comput. Phys. **227**, 3191–3211 (2008)
2. J.C. Butcher, *The Numerical Analysis of Ordinary Differential Equations* (Wiley, Chichester, 1987)
3. M. Castro, B. Costa, W.S. Don, High order weighted essentially non-oscillatory WENO-Z schemes for hyperbolic conservation laws. J. Comput. Phys. **230**(5), 1766–1792 (2011)
4. L. Fu, X. Y. Hu, N.A. Adams, A family of high-order targeted ENO schemes for compressible-fluid simulations. J. Comput. Phys. **305**, 333–359 (2016)
5. G.-S. Jiang, C.-W. Shu, Efficient implementation of weighted ENO schemes. J. Comput. Phys. **126**, 202–228 (1996)
6. C.-W. Shu, High order weighted essentially non-oscillatory schemes for convection dominated problems. SIAM Rev. **51**(1), 82–126 (2009)
7. C.-W. Shu, S. Osher, Efficient implementation of essentially non-oscillatory shock-capturing schemes, II. J. Comput. Phys. **83**(1), 32–78 (1989)
8. R.J. Spiteri, S.J. Ruuth, A new class of optimal high-order strong-stability-preserving time discretization methods. SIAM J. Numer. Anal. **40**(2), 469–491 (2002)
9. E.F. Toro, *Riemann Solvers and Numerical Methods for Fluid Dynamics* (Springer, Berlin, 1997)
10. B.S. van Lith, J.H.M. ten Thije Boonkkamp, W.L. IJzerman, Embedded WENO: a design strategy to improve existing WENO schemes. J. Comput. Phys. **330**, 529–549 (2016)

Finite Element Heterogeneous Multiscale Method for Time-Dependent Maxwell's Equations

Marlis Hochbruck and Christian Stohrer

Abstract We propose a Finite Element Heterogeneous Multiscale Method (FE-HMM) for time dependent Maxwell's equations in second-order formulation in locally periodic materials. This method can approximate the effective behavior of an electromagnetic wave traveling through a highly oscillatory material without the need to resolve the microscopic details of the material. To prove an a-priori error bound for the semi-discrete FE-HMM scheme, we need a new generalization of a Strang-type lemma for second-order hyperbolic equations. Finally, we present a numerical example that is in accordance with the theoretical results.

1 Introduction

We want to simulate electromagnetic wave propagation in a highly oscillatory material. FE-HMMs have proven to be efficient and reliable methods for many multiscale problems, see e.g. [1, 3]. Their most important advantage is that the influence of the microscopic details of the material are taken into account, whilst only a macroscopic discretization of the whole computational domain is needed. These methods were first proposed for elliptic and parabolic equations. In [2] it was proven, that the same ideas can be applied to the acoustic wave equation. This equation can be seen as an easily manageable special case of Maxwell's equations. Therefore, it is reasonable that FE-HMM can also be generalized to second-order time-dependent Maxwell's equation.

Recently, FE-HMMs for time-harmonic Maxwell's equations in rapidly oscillatory materials were presented, see [12] and [9]. There, two types of micro problems were used to approximate the effective (or upscaled or homogenized) solution. These micro problems are solved on small sampling domains such that the overall computational cost does not become infeasibly large. Here, we apply the FE-HMM scheme from [9] to second-order time-dependent Maxwell's equation. To the best

M. Hochbruck • C. Stohrer (✉)
Institute of Applied and Numerical Analysis, Karlsruhe Institute of Technology, Englerstrasse 2, 76131 Karlsruhe, Germany
e-mail: marlis.hochbruck@kit.edu; christian.stohrer@kit.edu

M.L. Bittencourt et al. (eds.), *Spectral and High Order Methods for Partial Differential Equations ICOSAHOM 2016*, Lecture Notes in Computational Science and Engineering 119, DOI 10.1007/978-3-319-65870-4_18

of our knowledge, this is the first FE-HMM scheme for this equation, while other multiscale schemes have already been proposed, see e.g. the recent article [6] and the references therein.

We consider a multiscale material with permittivity ε^η and permeability μ^η, where η denotes the characteristic microscopic length of the material. We assume that η is much smaller than the diameter of the computational domain Ω. In this article we restrict ourselves to locally periodic materials, see Definition 1 below, for simplicity. We are convinced that the Finite Element Heterogeneous Multiscale Methods (FE-HMM) presented here can be adapted to more general situations, but a rigorous justification thereof is ongoing research and beyond the scope of the current article. For a locally periodic material, η denotes the length of the microscopic oscillations in it.

The multiscale second order time-dependent Maxwell's equation is given by

$$\partial_{tt}\varepsilon^\eta(x)\mathbf{E}^\eta(t;x) + \nabla \times \big(\nu^\eta(x)(\nabla \times \mathbf{E}^\eta(t;x)\big) = \boldsymbol{f}(t;x) \quad \text{in } (0,T) \times \Omega, \tag{1}$$

where $\mathbf{E}^\eta$ is the unknown multiscale electric field and

$$\nu^\eta = (\mu^\eta)^{-1}$$

is the inverse of the magnetic permeability. To derive this equation from the standard first-order Maxwell's equations we assumed that the electric field is generated by a density free current and that the conductivity is zero (lossless material). The precise functional analytic setting, the initial and boundary conditions are given in Sect. 2, where we also recall a homogenization result derived from [18, Theorem 3.2]. In a nutshell, it states that $\mathbf{E}^\eta$ converges to the solution $\mathbf{E}^{\text{eff}}$ of an effective Maxwell's equation as the characteristic length η tends to zero. In Sect. 3 we describe how the idea of [9] can be used to build a FE-HMM for (1) to approximate $\mathbf{E}^{\text{eff}}$. All the advantages of FE-HMM schemes mentioned above carry over to the time-dependent case. We give an a-priori estimate of the difference between the FE-HMM and the effective solution in Sect. 4. This estimate is based on a improved version of the Strang-type Lemma given in [2]. To conclude this article we give a numerical example that corroborates our theoretical findings.

Notation Let $\Omega \subset \mathbb{R}^d$ be a Lipschitz domain, with $d = 2, 3$. We denote by $H^\ell(\Omega)$ the standard Sobolev spaces and set $L^2(\Omega) = H^0(\Omega)$ as usual. Vector valued function spaces are denoted in bold face, e.g. we set $\boldsymbol{H}^\ell(\Omega) := H^\ell(\Omega)^d$. We denote the corresponding scalar product and norm by $(\cdot,\cdot)_{\ell,\Omega}$, and $\|\cdot\|_{\ell,\Omega}$ respectively. The space $\boldsymbol{H}(\mathbf{curl};\Omega)$ consists of all $\boldsymbol{L}^2(\Omega)$ functions with a bounded curl. This space is a Hilbert space with respect to the scalar product

$$(\mathbf{v},\mathbf{w})_{\mathbf{curl},\Omega} = (\mathbf{v},\mathbf{w})_{0,\Omega} + (\mathbf{curl}\,\mathbf{v},\mathbf{curl}\,\mathbf{w})_{0,\Omega}.$$

We denote by $\boldsymbol{H}_0(\mathbf{curl};\Omega)$ the closure of $\boldsymbol{C}_0^\infty(\Omega)$ in $\boldsymbol{H}(\mathbf{curl};\Omega)$. This is the subspace of $\boldsymbol{H}(\mathbf{curl};\Omega)$ of functions with vanishing tangential components on the boundary $\partial\Omega$. Details about these spaces can e.g. be found in [17]. We denote likewise periodic boundary condition. For example for the centered unit cube $Y = (-1/2, 1/2)^d$, we denote by $\boldsymbol{H}_{\mathrm{per}}(\mathbf{curl}; Y)$ the closure of $\boldsymbol{C}^\infty_{\mathrm{per}}(Y)$.

2 Analytic Setting

As already mentioned in the introduction, we assume that the permittivity ε^η and the inverse permeability ν^η are locally periodic.

Definition 1 A tensor $\xi^\eta : \Omega \to \mathbb{R}^{d\times d}$ is *locally periodic* if there is a tensor $\xi : \Omega \times \mathbb{R}^d \to \mathbb{R}^{d\times d}$, which is Y-periodic ($Y = (-1/2, 1/2)^d$) in its second argument, such that $\xi^\eta(x) = \xi(x, x/\eta)$ for almost every $x \in \Omega$. We call such a function ξ *blueprint* of ξ^η.
In addition to the local periodicity we make from now on the following regularity assumptions on the tensors ε^η and ν^η:

$$\text{The blueprints of } \varepsilon^\eta \text{ and } \nu^\eta \text{ are symmetric and in } \left(C(\Omega; L^\infty_{\mathrm{per}}(Y))\right)^{d\times d}. \tag{A$_1$}$$

$$\text{The tensors } \varepsilon^\eta \text{ and } \nu^\eta \text{ are uniformly bounded and positive definite.} \tag{A$_2$}$$

Assumption (A$_2$) means that there are $0 < \alpha \le \beta$ such that for $\xi \in \{\varepsilon^\eta, \nu^\eta\}$ and almost every $x \in \Omega$

$$\alpha|z|^2 \le \xi(x)z\cdot z \quad \text{and} \quad \xi(x)z\cdot\tilde{z} \le \beta|z||\tilde{z}| \qquad \text{for all } z, \tilde{z} \in \mathbb{R}^d. \tag{A$_2'$}$$

We consider the variational formulation of (1).

$$\begin{cases} \text{Find } \mathbf{E}^\eta : (0,T) \to \boldsymbol{H}_0(\mathbf{curl}; \Omega), \text{ such that for all } \mathbf{v} \in \boldsymbol{H}_0(\mathbf{curl}; \Omega) \\ \left(\partial_{tt}\varepsilon^\eta \mathbf{E}^\eta(t), \mathbf{v}\right)_{0,\Omega} + \left(\nu^\eta \,\mathbf{curl}\,\mathbf{E}^\eta(t), \mathbf{curl}\,\mathbf{v}\right)_{0,\Omega} = \left(\mathbf{f}(t), \mathbf{v}\right)_{0,\Omega}, \\ \mathbf{E}^\eta(0) = \mathbf{E}_0, \qquad \text{and} \qquad \partial_t \mathbf{E}^\eta(0) = \mathbf{E}_0'. \end{cases} \tag{2}$$

This problem has a unique solution if, see e.g. [14, Chap. 3, Theorem 8.1],

$$\mathbf{E}_0 \in \boldsymbol{H}_0(\mathbf{curl}; \Omega), \quad \mathbf{E}_0' \in \boldsymbol{L}^2(\Omega), \quad \text{and} \quad \mathbf{f} \in \boldsymbol{L}^2(0, T; L^2(\Omega)).$$

Note that by the choice of the space $\boldsymbol{H}_0(\mathbf{curl}; \Omega)$ we use boundary conditions of a perfect electric conductor. This means that the tangential component of $\mathbf{E}^\eta$ vanishes at the boundary.

Homogenization Theory In [18] homogenization results for time-dependent first order Maxwell's equations have been proven, that answer the question how $\mathbf{E}^\eta$ behaves as $\eta \to 0$. In the case of lossless materials with no charge density, it is easy to rewrite this result in a second-order formulation. Similar results can be found in [5, 13], and [15]. Let us first introduce the involved micro problems.

Definition 2 Let $Y_\eta(x) = x + \eta Y$ be the scaled and shifted unit cell. The *first micro problem at* $x \in \Omega$ *constrained with a given* $\mathbf{v} \in \boldsymbol{H}(\mathbf{curl};\Omega)$ is defined as follows.

$$\begin{cases} \text{Find } \varphi^{\mathbf{v}}(x,\cdot) \in \varphi^{\mathbf{v}}_{\text{lin}}(x,\cdot) + H^1_{\text{per}}(Y_\eta(x)), \text{ such that } \displaystyle\int_{Y_\eta(x)} \varphi^{\mathbf{v}}(x,y)\,dy = 0 \text{ and} \\ \left(\varepsilon\left(x,\frac{\cdot}{\eta}\right)\nabla_y \varphi^{\mathbf{v}}(x,\cdot), \nabla\zeta\right)_{0,Y_\eta(x)} = 0, \qquad \text{for all } \zeta \in H^1_{\text{per}}(Y_\eta(x)), \end{cases} \tag{3}$$

where $\varphi^{\mathbf{v}}_{\text{lin}}(x,y) = \mathbf{v}(x)\cdot(y-x)$.

Definition 3 The *second micro problem at* $x \in \Omega$ *constrained with a given* $\mathbf{v} \in \boldsymbol{H}(\mathbf{curl};\Omega)$ is defined as follows.

$$\begin{cases} \text{Find } \big(\mathbf{u}^{\mathbf{v}}(x,\cdot),p\big) \in \big(\mathbf{u}^{\mathbf{v}}_{\text{lin}} + \boldsymbol{H}_{\text{per}}(\mathbf{curl};Y_\eta(x))\big) \times H^1_{\text{per}}(Y_\eta(x)), \\ \text{such that } \int_{Y_\eta(x)} \mathbf{u}^{\mathbf{v}}(x,y)\,dy = \mathbf{0},\ \int_{Y_\eta(x)} p(y)\,d\mathbf{y} = 0, \text{ and} \\ \left(\nu\left(x,\frac{\cdot}{\eta}\right)\mathbf{curl}_y\,\mathbf{u}^{\mathbf{v}}(x,\cdot), \mathbf{curl}\,\mathbf{z}\right)_{0,Y_\eta(x)} + \big(\mathbf{u}^{\mathbf{v}}(x,\cdot),\nabla q\big)_{0,Y_\eta(x)} + \big(\mathbf{z},\nabla p\big)_{0,Y_\eta(x)} = 0, \\ \text{for all } (\mathbf{z},q) \in \boldsymbol{H}_{\text{per}}(\mathbf{curl};Y_\eta(x)) \times H^1_{\text{per}}(Y_\eta(x)), \end{cases} \tag{4}$$

where $\mathbf{u}^{\mathbf{v}}_{\text{lin}}(x,y) = \mathbf{v}(x) + \frac{1}{2}\,\mathbf{curl}\,\mathbf{v}(x) \times (y-x)$.
Note that the first micro problem is the well-known elliptic cell problem of classical homogenization theory posed over the shifted sampling domain $Y_\eta(x)$ instead of the unit square Y if one chooses $\mathbf{v}$ to be a (constant) unit vector of $\mathbb{R}^d$. The second micro problem is used less frequently and related to the first one through "dual formulas", see [5, Chap. 1, Remark 5.9]. We recall the following homogenization result.

Theorem 1 (cf. [18, Theorem 3.2]) *Let* ε^η *and* ν^η *be locally periodic with blueprints* ε*, respectively* ν*, which fulfill the assumptions* (A_1) *and* (A_2)*. For* $\eta > 0$ *let* $\mathbf{E}^\eta$ *be the solution of the multiscale Maxwell's equation* (2)*. Then, as* $\eta \to 0$*,* $\mathbf{E}^\eta$ *converges weakly-$*$ in* $L^\infty(0,T;\boldsymbol{L}^2(\Omega))$ *to* $\mathbf{E}^{\text{eff}}$*, where* $\mathbf{E}^{\text{eff}}$ *is the solution of the following effective Maxwell's equation.*

$$\begin{cases} \textit{Find } \mathbf{E}^{\text{eff}} : (0,T) \to \boldsymbol{H}_0(\mathbf{curl};\Omega), \ \textit{such that for all } \mathbf{v} \in \boldsymbol{H}_0(\mathbf{curl};\Omega) \\ S^{\text{eff}}(\partial_{tt}\mathbf{E}^{\text{eff}}(t),\mathbf{v}) + B^{\text{eff}}(\mathbf{E}^{\text{eff}}(t),\mathbf{v}) = (\mathbf{f}(t),\mathbf{v})_{0,\Omega}, \\ \mathbf{E}^{\text{eff}}(0) = \mathbf{E}_0, \qquad \textit{and} \qquad \partial_t\mathbf{E}^{\text{eff}}(0) = \mathbf{E}_0'. \end{cases} \tag{5}$$

The effective scalar product S^{eff} is given by

$$S^{\text{eff}}(\mathbf{v},\mathbf{w}) = \int_\Omega \frac{1}{|Y_\eta(x)|}\left(\varepsilon\left(x,\frac{\cdot}{\eta}\right)\nabla_y\varphi^{\mathbf{v}}(x,\cdot),\nabla_y\varphi^{\mathbf{w}}(x,\cdot)\right)_{0,Y_\eta(x)} dx,$$

for all $\mathbf{v},\mathbf{w}\in \boldsymbol{H}(\mathbf{curl};\Omega)$, where $\varphi^{\mathbf{v}}$ and $\varphi^{\mathbf{w}}$ are the solutions of the first micro problem at x constrained with $\mathbf{v}$, respectively $\mathbf{w}$, see Definition 2. The effective bilinear form B^{eff} is given by

$$B^{\text{eff}}(\mathbf{v},\mathbf{w}) = \int_\Omega \frac{1}{|Y_\eta(x)|}\left(\nu\left(x,\frac{\cdot}{\eta}\right)\mathbf{curl}_y\,\mathbf{u}^{\mathbf{v}}(x,\cdot),\mathbf{curl}_y\,\mathbf{u}^{\mathbf{w}}(x,\cdot)\right)_{0,Y_\eta(x)} dx$$

for all $\mathbf{v},\mathbf{w}\in \boldsymbol{H}(\mathbf{curl};\Omega)$, where $\mathbf{u}^{\mathbf{v}}$ and $\mathbf{u}^{\mathbf{w}}$ are the solutions of the second micro problem at x constrained with $\mathbf{v}$, respectively $\mathbf{w}$, see Definition 3.
We choose to give the effective scalar product and the effective bilinear form in a non-standard version, since it reveals well the connection with our multiscale scheme defined below.

Nevertheless, we would like to mention that S^{eff} and B^{eff} could also be given with the help of an effective permittivity ε^{eff} and an effective inverse permeability ν^{eff} as

$$S^{\text{eff}}(\mathbf{v},\mathbf{w}) = (\varepsilon^{\text{eff}}\mathbf{v},\mathbf{w})_{0,\Omega} \qquad \text{and} \qquad B^{\text{eff}}(\mathbf{v},\mathbf{w}) = (\nu^{\text{eff}}\,\mathbf{curl}\,\mathbf{v},\mathbf{curl}\,\mathbf{w})_{0,\Omega}. \quad (6)$$

Explicit formulas for the effective tensors ε^{eff} and ν^{eff} in terms of the solutions of the micro problems can e.g. be found in [5, Remark 5.8]. This rewriting process has been shown in [9] for discretized versions of S^{eff} and B^{eff}, but one can follow the lines of the given proof also in the continuous case. We mention here the involved ideas. With the help of the "dual formulas" one can rewrite the effective equation as effective first order Maxwell's equations with effective electric permittivity and effective magnetic permeability. These effective equations are simplified versions of the ones given in [18]. The simplification originates by considering only lossless materials. In [18] the notion of two-scale convergence [4] was applied to Maxwell's equation to derive the convergence result.

Note, that it is well known that ε^{eff} and ν^{eff} only vary on a macroscopic length scale and that they are again uniformly bounded and positive definite. More precisely, we have that (A_2') holds for $\xi\in\{\varepsilon^{\text{eff}},\nu^{\text{eff}}\}$ with the same constants α and β. For the bilinear forms S^{eff} and B^{eff} this means, that there are $0<\lambda_S\le\Lambda_S$ and $0<\lambda_B\le\Lambda_B$, such that

$$\begin{aligned}\lambda_S\|\mathbf{v}\|^2_{0,\Omega} &\le S^{\text{eff}}(\mathbf{v},\mathbf{v}), \quad S^{\text{eff}}(\mathbf{v},\mathbf{w}) \le \Lambda_S\|\mathbf{v}\|_{0,\Omega}\|\mathbf{w}\|_{0,\Omega},\\ \lambda_B\|\mathbf{curl}\,\mathbf{v}\|^2_{0,\Omega} &\le B^{\text{eff}}(\mathbf{v},\mathbf{v}), \quad B^{\text{eff}}(\mathbf{v},\mathbf{w}) \le \Lambda_B\|\mathbf{curl}\,\mathbf{v}\|_{0,\Omega}\|\mathbf{curl}\,\mathbf{w}\|_{0,\Omega}.\end{aligned} \quad (7)$$

3 Multiscale Algorithm

As usual for FE-HMM schemes our algorithm consists of a macro and a micro solver. For the macro solver we discretize the effective equation (5) with edge elements from Nédélec's first family. Let $\mathscr{T}_H$ be a shape regular triangulation of the computational domain Ω into simplicial elements K. We let H be the largest diameter of all elements K in $\mathscr{T}_H$. Note that H can be much larger than the characteristic length η of the material. By $\boldsymbol{V}_H \subset \boldsymbol{H}_0(\mathbf{curl};\Omega)$ we denote the corresponding finite element space, for instance consisting of edge elements. The finite element discretization of (5) reads as follows.

$$\begin{cases} \text{Find } \mathbf{E}_H^{\text{eff}} : (0,T) \to \boldsymbol{V}_H, \text{ such that for all } \mathbf{v}_H \in \boldsymbol{V}_H \\ S^{\text{eff}}(\partial_{tt}\mathbf{E}_H^{\text{eff}}(t), \mathbf{v}_H) + B^{\text{eff}}(\mathbf{E}_H^{\text{eff}}(t), \mathbf{v}_H) = (\mathbf{f}(t), \mathbf{v}_H), \\ \mathbf{E}_H^{\text{eff}}(0) = \Pi_H \mathbf{E}_0, \qquad \text{and} \qquad \partial_t \mathbf{E}_H^{\text{eff}}(0) = \Pi_H \mathbf{E}_0', \end{cases} \tag{8}$$

where Π_H is a suitable $\boldsymbol{L}^2$-projection onto $\boldsymbol{V}_H$. Yet, this formulation can not be used directly, since the evaluation of S^{eff} and B^{eff} would require the exact solution of micro problems at every point $x \in \Omega$, i.e. of infinitely many micro problems.

To overcome these issues we replace S^{eff} and B^{eff} by their discretized counterparts. In this process, two discretization steps are involved. Firstly, the outer integral over the computational domain Ω is replaced by a quadrature formula: In every element $K \in \mathscr{T}_H$ we choose J quadrature nodes x_j^K and corresponding quadrature weights ω_j^K, $j = 1, \dots, J$. Then we approximate

$$\int_\Omega g(x)\,dx \approx \sum_{K \in \mathscr{T}_H} \sum_{j=1}^{J} \omega_j^K g(x_j^K) =: \sum_{K,j} \omega_j^K g(x_j^K).$$

Secondly, the micro problems are not solved analytically, but the solutions are approximated using finite elements. Therefore, we consider microscopic triangulations $\mathscr{T}_h(x)$ of the sampling domains $Y_\eta(x)$ into simplicial elements with maximal diameter h. Let $\varphi_h^{\mathbf{v}}$ be the FE solution of the first micro problem (3). This means, that $\varphi_h^{\mathbf{v}}$ is the solution of (3), where the space $H^1_{\text{per}}(Y_\eta(x))$ has been replaced with the space $W_{h,\text{per}}$ of Lagrange finite elements with periodic boundary conditions defined over $\mathscr{T}_h(x)$ of a given order. Similarly, let $\mathbf{u}_h^{\mathbf{v}}$ be the FE solution of the second micro problem (4). Here we replace additionally the space $\boldsymbol{H}_{\text{per}}(\mathbf{curl}; Y_\eta(x))$ with an edge element space $\boldsymbol{V}_{h,\text{per}}$ with periodic boundary conditions defined again over $\mathscr{T}_h(x)$. With these notations, we can define the HMM scalar product and bilinear form by

$$S_H^{\mathrm{HMM}}(\mathbf{v}_H, \mathbf{w}_H) = \sum_{K,j} \frac{\omega_j^K}{|Y_\eta|} \left(\varepsilon\left(x_j^K, \frac{\cdot}{\eta}\right) \nabla_y \varphi_h^{\mathbf{v}_H}(x_j^K, \cdot), \nabla_y \varphi_h^{\mathbf{w}_H}(x_j^K, \cdot) \right)_{0, Y_\eta(x_j^K)},$$

$$B_H^{\mathrm{HMM}}(\mathbf{v}_H, \mathbf{w}_H) = \sum_{K,j} \frac{\omega_j^K}{|Y_\eta|} \left(\nu\left(x_j^K, \frac{\cdot}{\eta}\right) \mathbf{curl}_y\, \mathbf{u}_h^{\mathbf{v}_H}(x_j^K, \cdot), \mathbf{curl}_y\, \mathbf{u}_h^{\mathbf{w}_H}(x_j^K, \cdot) \right)_{0, Y_\eta(x_j^K)}.$$

Remark 1 From the definition, it is obvious, that S^{HMM} and B^{HMM} are symmetric. Furthermore, it can be shown, that (7) holds as well for S^{HMM} and B^{HMM}, if ε^η, ν^η are sufficiently smooth and if the quadrature formula is accurate enough, with respect to the chosen macroscopic FE space $\boldsymbol{V}_H$. This is well known for FE-HMM, see [1, 3] and the references therein. For the specific case of Maxwell's equation a detailed discussion on the regularity assumptions can be found in [9]. Regarding the quadrature formula, we also refer to [7, Chap. 4].

Finally the FE-HMM scheme for second-order time-dependent Maxwell's equation can be written as follows.

$$\begin{cases} \text{Find } \mathbf{E}_H^{\mathrm{HMM}} : (0,T) \to \boldsymbol{V}_H, \text{ such that for all } \mathbf{v}_H \in \boldsymbol{V}_H \\ S_H^{\mathrm{HMM}}(\partial_{tt}\mathbf{E}_H^{\mathrm{HMM}}(t), \mathbf{v}_H) + B_H^{\mathrm{HMM}}(\mathbf{E}_H^{\mathrm{HMM}}(t), \mathbf{v}_H) = (\mathbf{f}(t), \mathbf{v}_H), \\ \quad \mathbf{E}_H^{\mathrm{HMM}}(0) = \Pi_H \mathbf{E}_0, \qquad \text{and} \qquad \partial_t \mathbf{E}_H^{\mathrm{HMM}}(0) = \Pi_H \mathbf{E}_0'. \end{cases} \tag{9}$$

Note that this FE-HMM scheme leads to a system of second-order ordinary differential equations.

For the full discretization, an appropriate time integration method has to be applied, e.g. the leap-frog or the Crank-Nicolson scheme. We refer to [8] for an error analysis for second-order Maxwell's equation for these two methods.

4 Error Analysis

FE-HMM schemes can be seen as non-conforming FE methods, since the true effective and the HMM bilinear form differ from each other. In [9] the FE-HMM for time harmonic Maxwell's equation was analyzed using the notion of T-coercivity. Since we now consider a hyperbolic time-dependent PDE we can no longer use this theory. However, the present situation is closely related to the one in [2], where a FE-HMM scheme for the scalar valued acoustic wave equation was introduced. There, a Strang-type lemma for wave equations was proven, where only the bilinear forms, but not the involved scalar products may differ from each other. Here we generalize it, such that it is applicable to our FE-HMM scheme.

Let $V \subset H \sim H' \subset V'$ be a Gelfand triple of Hilbert spaces and $W \subset V$ be a closed subset. We consider the following problem.

$$\begin{cases} \text{Find } u : (0,T) \to W, \text{ such that for all } w \in W \\ S\big(\partial_{tt} u(t), w\big) + B\big(u(t), w\big) = \big\langle f(t), w \big\rangle, \\ u(0) = u_0, \qquad \text{and} \qquad \partial_t u(0) = u_0', \end{cases} \tag{10}$$

where $S, B : W \times W \to \mathbb{R}$ are symmetric bilinear forms. S and B are assumed to be H-coercive and V-coercive, respectively, i.e., there are constants $0 < \lambda \le \Lambda$ with

$$S(v,v) \ge \lambda \|v\|_H^2, \qquad\qquad S(v,w) \le \Lambda \|v\|_H \|w\|_H, \tag{11a}$$

$$B(v,v) \ge \lambda \|v\|_V^2, \qquad\qquad B(v,w) \le \Lambda \|v\|_V \|w\|_V, \tag{11b}$$

for all $v, w \in W$. We denote the norms of bilinear forms by

$$\|B\|_V := \sup_{v,w \in W\setminus\{0\}} \frac{|B(v,w)|}{\|v\|_V \|w\|_V}, \qquad \|S\|_H := \sup_{v,w \in W\setminus\{0\}} \frac{|S(v,w)|}{\|v\|_H \|w\|_H}.$$

In the following, we will drop the explicit indication of the time dependence whenever possible, for better readability. Additionally, for the energy norm we use the abbreviation

$$\|v\|_{E(H,V)} = \|\partial_t v\|_{L^\infty(0,T;H)} + \|v\|_{L^\infty(0,T;V)} \qquad \text{for } v \in V.$$

Theorem 2 (Strang-Type Lemma for Second-Order Hyperbolic Equations) *Let $S, \tilde{S}, B, \tilde{B} : W \times W \to \mathbb{R}$ be symmetric bilinear forms satisfying* (11a) *and* (11b)*, respectively. For given $f : [0,T] \to V'$ and $u_0, u_0' \in W$, let u be the solution of* (10)*. Furthermore, let $\tilde{u}$ be the solution of* (10) *with S and B being replaced by $\tilde{S}$ and $\tilde{B}$, respectively. If $\partial_t^r u, \partial_t^r \tilde{u} \in C(0,T;V)$ for $r \in \{0,1,2\}$, then there is a constant C (depending on T and $\partial_t^r u$ for $r \in \{0,1,2\}$) such that*

$$\|u - \tilde{u}\|_{E(H,V)} \le C\big(\|S - \tilde{S}\|_H + \|B - \tilde{B}\|_V\big).$$

Proof The proof consists of three steps. The key idea is to consider the projection $\hat{u}(t) \in W$ of $u(t)$ given by

$$\tilde{B}\big(\hat{u}(t), w\big) = B\big(u(t), w\big) \qquad \text{for all } w \in W \tag{12}$$

and splitting the error into

$$e := u - \tilde{u} = \hat{e} + \tilde{e}, \qquad \text{where} \qquad \hat{e} := u - \hat{u} \qquad \text{and} \qquad \tilde{e} := \hat{u} - \tilde{u}. \tag{13}$$

(a) Due to the continuous embedding of $H^1(0,T;V)$ into the Bochner space $C([0,T];V)$, see e.g. [10, Sect. 5.9.2], we have for $v \in H^1(0,T;V)$

$$\|v\|_{L^\infty(0,T;V)} \le C\big(\|v\|_{L^2(0,T;V)} + \|\partial_t v\|_{L^2(0,T;V)}\big). \tag{14}$$

Using (14) for $v = \hat{e}$ and $v = \partial_t \hat{e}$, respectively, we obtain

$$\|e\|_{E(H,V)} \le C\big(\|\hat{e}\|_{L^2(0,T;V)} + \|\partial_t \hat{e}\|_{L^2(0,T;V)} + \|\partial_t^2 \hat{e}\|_{L^2(0,T;V)}\big) + \|\tilde{e}\|_{E(H,V)}.$$

It remains to bound $\hat{e}$ and $\tilde{e}$ defined in (13).

(b) To bound $\hat{e}$ one can follow the lines of the first paragraph of the proof of [2, Lemma 4.4]

$$\|\partial_t^r \hat{e}\|_{L^2(0,T;V)} \le C\|B - \tilde{B}\|_V \|\partial_t^r u\|_{L^2(0,T;V)}, \qquad r = 0, 1, 2.$$

(c) Bounding $\tilde{e}$ is motivated by the second part of the proof of [2, Lemma 4.4]. However, here we have to deal with the different scalar products S and $\tilde{S}$. From the definitions of the projection $\hat{u}$ in (12) and $\tilde{e}$ in (13) we obtain

$$\tilde{S}(\partial_t^2 \tilde{e}, w) + \tilde{B}(\tilde{e}, w) = \tilde{S}(\partial_t^2 \hat{u}, w) - S(\partial_t^2 u, w) \qquad \text{for all } w \in W.$$

Setting $w = \partial_t \tilde{e}$ yields

$$\frac{1}{2}\frac{d}{dt}\Big(\tilde{S}(\partial_t \tilde{e}, \partial_t \tilde{e}) + \tilde{B}(\tilde{e}, \tilde{e})\Big) = (\tilde{S} - S)(\partial_t^2 u, \partial_t \tilde{e}) - \tilde{S}(\partial_t^2 \hat{e}, \partial_t \tilde{e}).$$

By (11), we conclude

$$\frac{\lambda}{2}\frac{d}{dt}\big(\|\partial_t \tilde{e}\|_H^2 + \|\tilde{e}\|_V^2\big) \le \big(\|S - \tilde{S}\|_H \|\partial_t^2 u\|_H + \Lambda \|\partial_t^2 \hat{e}\|_H\big)\|\partial_t \tilde{e}\|_H.$$

Using the abbreviations

$$\rho = \|\partial_t \tilde{e}\|_H^2 + \|\tilde{e}\|_V^2 \quad \text{and} \quad \sigma = \|S - \tilde{S}\|_H \|\partial_t^2 u\|_H + \Lambda \|\partial_t^2 \hat{e}\|_H,$$

we find by applying Young's inequality

$$\frac{\lambda}{2}\frac{d}{dt}\rho(t) \le \sigma(t)\|\partial_t \tilde{e}(t)\|_H \le \frac{1}{2}\big(\sigma^2(t) + \rho(t)\big).$$

Gronwall's lemma yields for $0 \le t \le T$

$$\rho(t) \le \mathrm{e}^{T/\lambda}\Big(\rho(0) + \int_0^t \sigma^2(s)\, ds\Big). \tag{15}$$

The initial conditions of (10) imply $\tilde{e}(0) = -\hat{e}(0)$ and $\partial_t \tilde{e}(0) = -\partial_t \hat{e}(0)$. Using again that $H^1(0,T;V)$ is continuously embedded in $C([0,T];V)$ we have

$$\rho(0) \leq C\|\partial_t \hat{e}\|^2_{L^\infty(0,T;V)} + \|\hat{e}\|^2_{L^\infty(0,T;V)}.$$

Inserting the definition of ρ, taking square roots of the inequality (15), considering the supremum over $t \in [0,T]$, and using the bound (14) for $v = \hat{e}$ and $v = \partial_t \hat{e}$, proves the desired bound. □

Our next goal is to apply Theorem 2 to FE-HMM. To get more insight in the following a-priori error bound, we will split it into macro and HMM error. To this end we approximate the effective scalar product and the effective bilinear form, c.f. (6), using numerical integration. For $\mathbf{v}_H, \mathbf{w}_H \in \boldsymbol{V}_H$ we set

$$S_H^{\mathrm{eff}}(\mathbf{v}_H, \mathbf{w}_H) = \sum_{K,j} \omega_j^K \varepsilon^{\mathrm{eff}}(x_j^K)\mathbf{v}_H(x_j^K) \cdot \mathbf{w}_H(x_j^K),$$

$$B_H^{\mathrm{eff}}(\mathbf{v}_H, \mathbf{w}_H) = \sum_{K,j} \omega_j^K \nu^{\mathrm{eff}}(x_j^K)\, \mathbf{curl}\, \mathbf{v}_H(x_j^K) \cdot \mathbf{curl}\, \mathbf{w}_H(x_j^K),$$

and define

$$\Delta S_{\mathrm{mac}} = \|S^{\mathrm{eff}} - S_H^{\mathrm{eff}}\|_{\boldsymbol{L}^2(\Omega)}, \qquad \Delta B_{\mathrm{mac}} = \|B^{\mathrm{eff}} - B_H^{\mathrm{eff}}\|_{\boldsymbol{H}(\mathbf{curl};\Omega)},$$

$$\Delta S_{\mathrm{HMM}} = \|S_H^{\mathrm{eff}} - S_H^{\mathrm{HMM}}\|_{\boldsymbol{L}^2(\Omega)}, \qquad \Delta B_{\mathrm{HMM}} = \|B_H^{\mathrm{eff}} - B_H^{\mathrm{HMM}}\|_{\boldsymbol{H}(\mathbf{curl};\Omega)}.$$

Corollary 1 *As above, let* $\mathbf{E}^{\mathrm{eff}}$, $\mathbf{E}_H^{\mathrm{eff}}$, *and* $\mathbf{E}_H^{\mathrm{HMM}}$ *be the solution of* (5), (8), *and* (9), *respectively. Suppose that* $\partial_t^r \mathbf{E}_H^{\mathrm{eff}}, \partial_t^r \mathbf{E}_H^{\mathrm{HMM}} \in C(0,T;\boldsymbol{H}_0(\mathbf{curl};\Omega))$ *for* $r \in \{0,1,2\}$. *If* ε^η, ν^η *are sufficiently smooth and if the quadrature formulas are accurate enough, then*

$$\|\mathbf{E}^{\mathrm{eff}} - \mathbf{E}_H^{\mathrm{HMM}}\|_{E(\boldsymbol{L}^2(\Omega),\boldsymbol{H}(\mathbf{curl};\Omega))} \leq \|\mathbf{E}^{\mathrm{eff}} - \mathbf{E}_H^{\mathrm{eff}}\|_{E(\boldsymbol{L}^2(\Omega),\boldsymbol{H}(\mathbf{curl};\Omega))} + C(\Delta S_{\mathrm{mac}} + \Delta B_{\mathrm{mac}} + \Delta S_{\mathrm{HMM}} + \Delta B_{\mathrm{HMM}}). \tag{16}$$

Proof We only have to bound $\|\mathbf{E}_H^{\mathrm{eff}} - \mathbf{E}_H^{\mathrm{HMM}}\|_{E(\boldsymbol{L}^2(\Omega),\boldsymbol{H}(\mathbf{curl};\Omega))}$ due to the triangle inequality. For this we can apply Theorem 2 with $H = \boldsymbol{L}^2(\Omega)$, $V = \boldsymbol{H}_0(\mathbf{curl};\Omega)$, and $W = \boldsymbol{V}_H$. Since the bilinear forms B^{eff} and B_H^{HMM} are not W-elliptic, we consider the following modified bilinear forms

$$B(\cdot,\cdot) = B^{\mathrm{eff}}(\cdot,\cdot) + \tfrac{\lambda_S}{2}(\cdot,\cdot)_{0,\Omega}, \qquad \tilde{B}(\cdot,\cdot) = B_H^{\mathrm{HMM}}(\cdot,\cdot) + \tfrac{\lambda_S}{2}(\cdot,\cdot)_{0,\Omega},$$

$$S(\cdot,\cdot) = S^{\mathrm{eff}}(\cdot,\cdot) - \tfrac{\lambda_S}{2}(\cdot,\cdot)_{0,\Omega}, \qquad \tilde{S}(\cdot,\cdot) = S_H^{\mathrm{HMM}}(\cdot,\cdot) - \tfrac{\lambda_S}{2}(\cdot,\cdot)_{0,\Omega}.$$

The coercivity of B, S, $\tilde{B}$ and $\tilde{S}$ follows from (7) and Remark 1. Moreover, assumption (11) holds with $\lambda = \min\{\lambda_B, \lambda_S/2\}$ and $\Lambda = \lambda_S/2 + \max\{\Lambda_B, \Lambda_S\}$. With these choices we get from Theorem 2

$$\begin{aligned}\|\mathbf{E}_H^{\mathrm{eff}} - \mathbf{E}_H^{\mathrm{HMM}}\|_{E(\boldsymbol{L}^2(\Omega),\boldsymbol{H}(\mathbf{curl};\Omega))} &\le C\big(\|S-\tilde{S}\|_{\boldsymbol{L}^2(\Omega)} + \|B-\tilde{B}\|_{\boldsymbol{H}(\mathbf{curl};\Omega)}\big)\\ &= C\big(\|S^{\mathrm{eff}} - S_H^{\mathrm{eff}}\|_{\boldsymbol{L}^2(\Omega)} + \|B^{\mathrm{eff}} - B_H^{\mathrm{eff}}\|_{\boldsymbol{H}(\mathbf{curl};\Omega)}\big)\\ &\le C\big(\Delta S_{\mathrm{mac}} + \Delta B_{\mathrm{mac}} + \Delta S_{\mathrm{HMM}} + \Delta B_{\mathrm{HMM}}\big).\end{aligned}$$

□

The first term on the right hand side of (16) can be bounded by standard FE theory. E.g. for $\boldsymbol{V}_H$ being chosen as lowest order $\boldsymbol{H}(\mathbf{curl};\Omega)$-conforming edge element from Nédélec's first family, we have under appropriate regularity conditions, see [16, Theorem 3.1],

$$\|\mathbf{E}^{\mathrm{eff}} - \mathbf{E}_H^{\mathrm{eff}}\|_{E(\boldsymbol{L}^2(\Omega),\boldsymbol{H}(\mathbf{curl};\Omega))} \le C\big(\big\|\mathbf{E}_0' - \Pi_H\mathbf{E}_0'\big)\big\|_{0,\Omega} + \big\|\mathbf{E}_0 - \Pi_H\mathbf{E}_0\big\|_{\mathbf{curl},\Omega} + H\big). \tag{17}$$

Convergence rates for the differences in the scalar products and bilinear forms in terms of H and h can be found in [9].

5 Numerical Example

We present a first simple numerical example corroborating our analytical results. More involved examples will presented in a forthcoming publication. Let $\mathcal{T}_H$ be a triangulation of the computational domain $\Omega = [0,1]^2$ into uniform meshes of different mesh sizes H. Furthermore, define the function g^η by

$$g^\eta(x) = \sqrt{2} + \sin\left(2\pi\frac{x}{\eta}\right)$$

and let the electric permittivity and the inverse magnetic permeability be given by

$$\varepsilon^\eta(x_1,x_2) = \frac{g^\eta(x_1)g^\eta(x_2)}{\sqrt{2}}, \qquad \nu^\eta(x_1,x_2) = \frac{2}{g^\eta(x_1)g^\eta(x_2)},$$

with $\eta = 2^{-8} \approx 0.004$. For this particular case the effective parameters can be computed analytically and one finds $\varepsilon^{\mathrm{eff}} = \nu^{\mathrm{eff}} = 1$. We choose the source term

$$\boldsymbol{f}(t;x_1,x_2) = \begin{pmatrix} -\pi^2\sin(-\pi t)\cos(\pi x_1)\sin(\pi x_2)\\ \pi^2\sin(\pi t)\sin(\pi x_1)\cos(\pi x_2)\end{pmatrix},$$

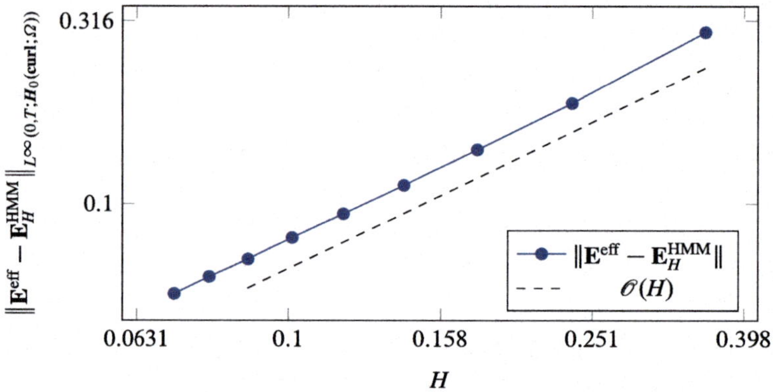

Fig. 1 Maximal difference between the effective and the FE-HMM solution, computed with first order elements. As expected we retrieve first order convergence. The experiment was conducted with `FreeFem++` [11]

such that the solution of the effective Maxwell's equation (5) is given by

$$\mathbf{E}^{\text{eff}}(t; x_1, x_2) = \begin{pmatrix} -\sin(\pi t)\cos(\pi x_1)\sin(\pi x_2) \\ \sin(-\pi t)\sin(\pi x_1)\cos(\pi x_2) \end{pmatrix}.$$

We discretize using lowest order $\boldsymbol{H}(\mathbf{curl};\Omega)$-conforming edge element from Nédélec's first family for the macro solver. For the micro solver we use Lagrange and edge elements of order one. For this particular choice it is shown in [9, Sect. 5] that we have

$$\Delta S_{\text{mac}} = \Delta B_{\text{mac}} = 0 \qquad \text{and} \qquad \Delta S_{\text{HMM}}, \, \Delta B_{\text{HMM}} \leq C\Big(\frac{h}{\eta}\Big)^2,$$

where C is independent of h and η.

In Fig. 1 we show the maximal $\boldsymbol{H}(\mathbf{curl};\Omega)$-error between $\mathbf{E}^{\text{eff}}$ and $\mathbf{E}_H^{\text{HMM}}$ for various values of H. If $r = H_1/H_2$ denotes the refinement factor between two macro meshes $\mathcal{T}_{H_1}$ and $\mathcal{T}_{H_2}$, then we use $\sqrt{r}$ as the refinement factor between the corresponding micro meshes. This simultaneous refinement strategy accounts for the different convergence orders (1 for the macro and 2 for the micro solver). As expected from the theoretical consideration above, we see that the proposed FE-HMM scheme (9) converges linearly for the above choices of the finite element spaces.

Acknowledgements We gratefully acknowledge financial support by the Deutsche Forschungsgemeinschaft (DFG) through CRC 1173 and the Klaus Tschira Stiftung. In addition we thank the anonymous referee for helpful suggestions.

References

1. A. Abdulle, The finite element heterogeneous multiscale method: a computational strategy for multiscale PDEs. GAKUTO Int. Ser. Math. Sci. Appl. **31**, 133–181 (2009)
2. A. Abdulle, M.J. Grote, Finite element heterogeneous multiscale method for the wave equation. Multiscale Model. Simul. **9**(2), 766–792 (2011)
3. A. Abdulle, W. E, B. Engquist, E. Vanden-Eijnden, The heterogeneous multiscale method. Acta Numer. **21**, 1–87 (2012)
4. G. Allaire, Homogenization and two-scale convergence. SIAM J. Math. Anal. **23**(6), 1482–1518 (1992)
5. A. Bensoussan, J.L. Lions, G. Papanicolaou, Asymptotic analysis for periodic structures, in *Studies in Mathematics and its Applications*, vol. 5 (North-Holland Publishing Co., Amsterdam/New York, 1978)
6. V.T. Chu, V.H. Hoang, High-dimensional finite elements for multiscale Maxwell-type equations. IMA J. Numer. Anal. **drx001** (2017, Online). doi:https://doi.org/10.1093/imanum/drx001
7. P.G. Ciarlet, The finite element method for elliptic problems, in *Classics in Applied Mathematics*, vol. 40 (SIAM, Philadelphia, 2002)
8. P. Ciarlet Jr., J. Zou, Fully discrete finite element approaches for time-dependent Maxwell's equations. Numer. Math. **82**(2), 193–219 (1999)
9. P. Ciarlet Jr., S. Fliss, C. Stohrer, On the approximation of electromagnetic fields by edge finite elements. Part 2: a heterogeneous multiscale method for Maxwell's equations. Comput. Math. Appl. **73**(9), 1900–1919 (2017)
10. L.C. Evans, Partial differential equations, *Graduate Studies in Mathematics*, vol. 19 (American Mathematical Society, Providence, 1998)
11. F. Hecht, New development in FreeFem++. J. Numer. Math. **20**(3–4), 25–265 (2012)
12. P. Henning, M. Ohlberger, B. Verfürth, A new heterogeneous multiscale method for time-harmonic Maxwell's equations. SIAM J. Numer. Anal. **54**(6), 3493–3522 (2016)
13. V.V. Jikov, S.M. Kozlov, O.A. Oleinik, *Homogenization of Differential Operators and Integral Functionals* (Springer, Berlin, 1994)
14. J.L. Lions, E. Magenes, Non-homogeneous boundary value problems and applications. Vol. I, *Die Grundlehren der mathematischen Wissenschaften*, vol. 181 (Springer, New York/Heidelberg, 1972)
15. P.A. Markowich, F. Poupaud, The Maxwell equation in a periodic medium: Homogenization of the energy density. Ann. Scuola Norm. Sup. Pisa Cl. Sci. (4) **23**(2), 301–324 (1996)
16. P. Monk, Analysis of a finite element method for Maxwell's equations. SIAM J. Numer. Anal. **29**(3), 714–729 (1992)
17. P. Monk, *Finite Element Methods for Maxwell's Equations* (Oxford University Press, Oxford, 2003)
18. N. Wellander, Homogenization of the Maxwell equations: Case I. Linear theory. Appl. Math. **46**(1), 29–51 (2001)

Computational Aspects of a Time Evolution Scheme for Incompressible Boussinesq Navier-Stokes in a Cylinder

Damián Castaño, María Cruz Navarro, and Henar Herrero

Abstract In this work we show some computational aspects of the implementation of a three dimensional spectral time evolution scheme for incompressible Boussinesq Navier-Stokes including rotation effects in a cylinder with a primitive variable formulation. The scheme is a second-order time-splitting method combined with pseudo-spectral Fourier Chebyshev in space. To deal with the singularity at the origin a radial expansion is considered in the diameter of the cylinder. The order expansion in the radial coordinate gets doubled. We develop a matrix processing that combines the use of the parity of the fields and the discretization functions to cancel half of the terms in the matrix reducing the radial dimension to the original one.

1 Introduction

Incompressible Boussinesq Navier-Stokes partial differential equations are of great interest to model fluid dynamics problems related with phenomena in nature or industrial processes [1, 7, 8, 20, 21]. The inclusion of the Coriolis force is relevant for most of the atmospheric events [9, 15].

The usual approach to solve these problems is numerical. Depending on the interest of the particular problem some numerical methods for solving partial differential equations are more adequate. Spectral methods have been proved to be very efficient for this task [3, 16, 18]. The method considered in this paper includes several ingredients from different works. The time splitting second order in time method was proposed by Hugues and Randriamampianina [14] and constitutes an

D. Castaño (✉)
Instituto de Matemática Aplicada a la Ciencia y la Ingeniería (IMACI), Universidad de Castilla-La Mancha, 13071 Ciudad Real, Spain

IMACI, Universidad de Castilla-La Mancha, 13071 Ciudad Real, Spain
e-mail: Damian.Castano@uclm.es

M. Cruz Navarro • H. Herrero
IMACI, Universidad de Castilla-La Mancha, 13071 Ciudad Real, Spain
e-mail: MariaCruz.Navarro@uclm.es; Henar.Herrero@uclm.es

M.L. Bittencourt et al. (eds.), *Spectral and High Order Methods for Partial Differential Equations ICOSAHOM 2016*, Lecture Notes in Computational Science and Engineering 119, DOI 10.1007/978-3-319-65870-4_19

improvement on the projection scheme proposed by Goda [11] and implemented by Gresho [12] in finite element approximations. The fractional steps consist of a predictor for the pressure, derived from the Navier-Stokes equations, a predictor for an intermediate velocity field obtained from the momentum equation by using the predicted pressure and a final projection step with a explicit evaluation of the divergence-free velocity field [16]. This scheme has been used by the authors to describe the generation of time dependent vertical vortices in cylindrical domains [4–6].

In order to avoid the singularity and the clustering at the origin a radial expansion in the diameter of the cylinder that doubles the dimension is considered, as proposed in [10, 18, 22]. We develop a matrix processing that combines the use of the parity of the fields and the discretization functions to cancel half of the terms in the matrix reducing the radial dimension to the original one.

The paper is organized as follows. Section 2 presents the mathematical formulation of the problem in a dimensionless form. Section 3 describes the numerical implementation including computational aspects and validation. Finally, in Sect. 4, conclusions are presented.

2 Formulation of the Problem

The physical setup consists of a horizontal fluid layer in a cylindrical container of radius Γ (r coordinate) and height 1 (z coordinate) in a rotating frame with a constant rotation rate Ω. At $z = 0$ the imposed temperature has a Gaussian profile which takes the value $T_{\max}$ at $r = 0$ and the value $T_{\min}$ at the outer part ($r = \Gamma$). The upper surface is at temperature $T = T_0$. We define $\triangle T_v = T_{\max} - T_0$, $\triangle T_h = T_{\max} - T_{\min}$ and $\delta = \triangle T_h / \triangle T_v$.

In the governing equations, $\mathbf{u} = (u_r, u_\phi, u_z)$ is the velocity field, T is the temperature, p is the pressure, r is the radial coordinate, and t is the time. They are expressed in dimensionless form. The domain in (r, θ, z) coordinates is $D = [0, \Gamma] \times [0, 1] \times [0, 2\pi]$.

The non-dimensional equations for Boussinesq convection with rotation are,

$$\nabla \cdot \mathbf{u} = 0, \tag{1}$$

$$\partial_t T + \mathbf{u} \cdot \nabla T = \nabla^2 T, \tag{2}$$

$$\partial_t \mathbf{u} + (\mathbf{u} \cdot \nabla)\, \mathbf{u} = Pr \left(-\nabla p + \nabla^2 \mathbf{u} + RT\mathbf{e}_z - \frac{2}{E} \mathbf{e}_z \times \mathbf{u} \right), \tag{3}$$

where the operators and fields are expressed in cylindrical coordinates. Here $\mathbf{e}_z$ is the unit vector in the z direction. The following dimensionless numbers have been introduced: the Prandtl number Pr which describes the ratio of the viscous terms to the thermal diffusivity, the Ekman number E which compares the viscous terms to the Coriolis term, and the Rayleigh number R, that characterizes the buoyancy forcing.

At $r = \Gamma$, a rigid insulating wall is considered,

$$u_r = u_\phi = u_z = \partial_r T = 0, \quad \text{on} \quad r = \Gamma. \tag{4}$$

On the top surface slip boundary conditions are considered, and the temperature is $T = T_0$, that after rescaling become,

$$\partial_z u_r = \partial_z u_\phi = u_z = T = 0, \quad \text{on} \quad z = 1, \tag{5}$$

and at the bottom slip boundary conditions are also considered

$$\partial_z u_r = \partial_z u_\phi = u_z = 0, \quad \text{on} \quad z = 0. \tag{6}$$

For temperature at the bottom, a Gaussian profile is imposed as in [17, 19],

$$T = 1 - \delta(e^{(\frac{1}{\beta})^2} - e^{(\frac{1}{\beta} - (\frac{r}{\Gamma})^2 \frac{1}{\beta})^2})/(e^{(\frac{1}{\beta})^2} - 1) \quad \text{on} \quad z = 0. \tag{7}$$

3 Numerical Method

The code presented here is inspired by the work of Mercader in [18], which is based on a time-stepping method. It considers for the temporal discretization, a second-order semi-implicit scheme, and a pseudospectral approximation for the space variables. The primitive variable formulation described in [13] is used in this code. The boundary conditions are compatible with the restriction in [18].

3.1 *Temporal Discretization and Projection Scheme*

The temporal scheme consists of the second order combination of Adams-Bashforth and backward differentiation formula (AB/BDF) stiffly stable scheme for time evolution used by Karniadakis in [16]. Equations are as follows

$$\nabla \cdot \mathbf{u}^{n+1} = 0, \tag{8}$$

$$\frac{3\mathbf{u}^{n+1} - 4\mathbf{u}^n + \mathbf{u}^{n-1}}{2\Delta t} = -2NL(\mathbf{u}^n) + NL(\mathbf{u}^{n-1}) + Pr(-\nabla p^{n+1} + \nabla^2 \mathbf{u}^{n+1} + R\Theta^{n+1} e_z), \tag{9}$$

$$\frac{3\Theta^{n+1} - 4\Theta^n + \Theta^{n-1}}{2\Delta t} = -2NL(\mathbf{u}^n, \Theta^n) + NL(\mathbf{u}^{n-1}, \Theta^{n-1}) + \nabla^2 \Theta^{n+1}, \tag{10}$$

where the superindex indicates the time step and *NL* are the nonlinear terms,

$$NL(\mathbf{u}, \Theta) = u_r \partial_r \Theta + \frac{1}{r} u_\phi \partial_\phi \Theta + u_z \partial_z \Theta, \tag{11}$$

$$NL(\mathbf{u}) = u_r \partial_r \mathbf{u} + \frac{1}{r} u_\phi \partial_\phi \mathbf{u} + u_z \partial_z \mathbf{u}, \tag{12}$$

The fractional steps consist of a predictor for the pressure, which is obtained directly from the Navier-Stokes equations with the Neumann boundary conditions; a predictor for an intermediate velocity field obtained from the momentum equation, which takes into account the predictor for the pressure obtained from the previous time, and a projection step with an explicit evaluation of the final divergence-free velocity field [16, 18].

3.2 *Spatial Discretization*

The velocity, temperature and pressure fields are written in cylindrical coordinates (r, ϕ, z). The dependence of the azimuthal component of the velocity is solved using Fourier expansions

$$\mathbf{u}(r, \phi, z) = \sum_{k=-n_\phi/2}^{n_\phi/2-1} F_k(r, z) e^{ik\phi}. \tag{13}$$

If the coefficients $F_k(r, z)$ come from a field which is real, they satisfy the following conditions

$$F_0(r, z) = f_0(r, z) \in \mathbb{R}, \tag{14}$$

$$F_{-n_\phi/2}(r, z) = f_{-n_\phi/2}(r, z) \in \mathbb{R}, \tag{15}$$

$$F_{-k}(r, z) = \bar{F}_k(r, z), \tag{16}$$

where the overbar means complex conjugate and the lowercase letter refers to real field. These conditions allow to know the entire field only with half of the Fourier coefficients and the $F_{-n_\phi/2}$ coefficient. In our calculations, as the field is real, we compute the coefficients $k = 0, 1, \ldots, n_\phi/2-1, -n_\phi/2$, and we obtain the complete field by using the condition (16),

Each Fourier coefficient is expanded by a Chebyshev collocation method, using Chebyshev polynomials and evaluating them in the Gauss-Lobatto collocation points

$$F_k(r, z) = \sum_{l=0}^{2n_r+1} \sum_{n=0}^{n_z} f_{ln} T_l(r) T_n(z), \tag{17}$$

The radial expansions are considered in the diameter of the cylinder, i.e., $r \in [-\Gamma, \Gamma]$. A transformation from the domain $[-\Gamma, \Gamma] \times [0, 1] \times [0, 2\pi]$ to $[-1, 1] \times [-1, 1] \times [0, 2\pi]$ is required due to the Chebyshev collocation implementation. We assume a radial Chebyshev expansion in $2n_r + 2$ polynomials. We introduce the expansion into equations and evaluate them in $(0, \Gamma]$ at n_r+1 Chebyshev collocation points ($r_j = \Gamma \cos(\pi j/(2n_r + 1))$ for $j = 0, \ldots, n_r + 1$) at each fixed angle. We take half of the points due to the symmetry properties of the Chebyshev polynomials as will be proven next. With this radial expansion, the singularity at the origin is avoided by ensuring that $r = 0$ is not a collocation point and preventing also the excessive clustering of points near the center [10]. This technique is known as unshifted Chebyshev polynomials of appropriate parity [2] and it has already been used in [18].

3.3 Computational Aspects

The fields u_z, p and Θ are even, i.e., $u(r, \phi, z) = u(-r, \phi + \pi, z)$, and u_r and u_ϕ are odd fields, i.e., $u(r, \phi, z) = -u(-r, \phi + \pi, z)$. For even fields, the parity follows the Fourier mode ($k = 0$ is even, $k = 1$ is odd, etc.) and for odd fields, the parity changes ($k = 0$ is odd, $k = 1$ is even, etc.). Each field expands as follows:

$$u(r, \phi, z) = \sum_{k=-n_\phi/2}^{n_\phi/2-1} F_k(r, z) e^{ik\phi}, \text{ where } F_k(r, z) = \sum_{j=0}^{n_z} \sum_{i=0}^{2n_r+1} a_{ij}^k T_i(r) T_j(z).$$

The radial derivatives are treated in a special way, being calculated by using a matrix multiplication method. Instead of using a single $(2n_r + 2) \times (2n_r + 2)$ Chebyshev differentiation matrix, two different matrices of dimension $(n_r + 1) \times (n_r + 1)$ are built, one for functions of odd parity (even Fourier coefficients of u_r and u_ϕ, and odd Fourier coefficients of u_z, p and T), and another one for functions of even parity (odd Fourier coefficients of u_r and u_ϕ, and even Fourier coefficients of u_z, p and T). In order to apply properly this method, we must take into account that the radial derivative of an even field is an odd field and vice versa, and therefore we must apply the appropriate derivation matrix depending on whether the field which is going to be derived is an even field or an odd field. This is proven in the following theorem.

Theorem 1 *Consider a field u in the discrete Fourier and Chebyshev space* $\mathbb{C}^{n_\phi} \times \mathbb{C}^{2n_r+2} \times \mathbb{C}^{n_z+1}$. *For even k, and an even field u the even columns of the inverse Chebyshev transformation matrix are null; if the field u is odd the odd columns of the inverse Chebyshev transformation matrix are null. For odd k, and an even field u the odd columns of the inverse Chebyshev transformation matrix are null; if the field u is odd the even columns of the inverse Chebyshev transformation matrix are null.*

Proof For a fixed (r_i, z_j) and even k, $F_k(r_i, z_j)$ can be written as follows in matrix notation,

$$F_k(r_i, z_j) = \Big(T_0(z_j) \ \dots \ T_{n_z}(z_j) \Big) \begin{pmatrix} a^k_{0,0} & a^k_{1,0} & \dots & a^k_{2n_r+1,0} \\ a^k_{0,1} & a^k_{1,1} & \dots & a^k_{2n_r+1,1} \\ \vdots & \vdots & \ddots & \vdots \\ a^k_{0,n_z} & a^k_{1,n_z} & \dots & a^k_{2n_r+1,n_z} \end{pmatrix} \begin{pmatrix} T_0(r_i) \\ T_1(r_i) \\ \vdots \\ T_{2n_r+1}(r_i) \end{pmatrix}$$

If we name A_k the previous central matrix that is the inverse Chebyshev transformation, then we can write in matrix form all the values $F_k(r_i, z_j)$, $i = 0, 1, .., 2n_r + 1, j = 0, \dots, n_z$,

$$T_z \cdot A_k \cdot T_{r_c} = \begin{pmatrix} F_k(r_0, z_0) & F_k(r_1, z_0) & \dots & F_k(r_{2n_r+1}, z_0) \\ \vdots & \vdots & \ddots & \vdots \\ F_k(r_0, z_{n_z}) & F_k(r_1, z_{n_z}) & \dots & F_k(r_{2n_r+1}, z_{n_z}) \end{pmatrix}$$

where T_z and T_{r_c} are the following matrices

$$T_z = \begin{pmatrix} T_0(z_0) & \dots & T_{n_z}(z_0) \\ \vdots & \ddots & \vdots \\ T_0(z_{n_z}) & \dots & T_{n_z}(z_{n_z}) \end{pmatrix}, \qquad T_{r_c} = \begin{pmatrix} T_0(r_0) & \dots & T_0(r_{2n_r+1}) \\ \vdots & \ddots & \vdots \\ T_{2n_r+1}(r_0) & \dots & T_{2n_r+1}(r_{2n_r+1}) \end{pmatrix}.$$

We are only interested in $r \in (0, 1]$, and by symmetry, r_j and r_{2n_r+1-j} ($j = 0, 1, \dots, n_r$) are symmetrical points relative to the axis. If we name

$$T_r = \begin{pmatrix} T_0(r_{n_r+1}) & \dots & T_0(r_{2n_r+1}) \\ \vdots & \ddots & \vdots \\ T_{2n_r+1}(r_{n_r+1}) & \dots & T_{2n_r+1}(r_{2n_r+1}) \end{pmatrix}, \quad T_{\bar{r}} = \begin{pmatrix} T_0(r_{n_r}) & \dots & T_0(r_0) \\ \vdots & \ddots & \vdots \\ T_{2n_r+1}(r_{n_r}) & \dots & T_{2n_r+1}(r_0) \end{pmatrix},$$

we obtain

$$T_z \cdot A_k \cdot T_r = \begin{pmatrix} F_k(r_{n_r+1}, z_0) & \dots & F_k(r_{2n_r+1}, z_0) \\ \vdots & \ddots & \vdots \\ F_k(r_{n_r+1}, z_{n_z}) & \dots & F_k(r_{2n_r+1}, z_{n_z}) \end{pmatrix}$$

and

$$T_z \cdot A_k \cdot T_r = \begin{pmatrix} F_k(r_{n_r}, z_0) & \dots & F_k(r_0, z_0) \\ \vdots & \ddots & \vdots \\ F_k(r_{n_r}, z_{n_z}) & \dots & F_k(r_0, z_{n_z}) \end{pmatrix}.$$

Taking into account the symmetry of the fields and the fact that Chebyshev polynomials with even subscript are even ($T_k(r) = T_k(-r)$, if k is even), and with odd subscript are odd ($T_k(r) = -T_k(-r)$, if k is odd), we conclude

- For even fields, $T_z \cdot A_k \cdot T_r = T_z \cdot A_k \cdot T_{\tilde{r}}$. As T_z is a regular matrix. By multiplying by T_z^{-1} to the left at both sides of the equality we get $A_k \cdot T_r = A_k \cdot T_{\tilde{r}}$. Therefore $A_k \cdot (T_r - T_{\tilde{r}}) = 0$.
- For odd fields, $T_z \cdot A_k \cdot T_r = -T_z \cdot A_k \cdot T_{\tilde{r}}$. By multiplying by T_z^{-1} to the left at both sides of the equality we get $A_k \cdot T_r = -A_k \cdot T_{\tilde{r}}$. Therefore $A_k \cdot (T_r + T_{\tilde{r}}) = 0$.

For an even field, if we separate the product by columns we obtain $A(v_i - w_i) = 0$ for $i = 1, \ldots n_{r+1}$, where v_i is the i-column of T_r and w_i is the i-column of $T_{\tilde{r}}$, and $v_i - w_i \neq 0$. Let's name $\hat{r}$ the corresponding symmetrical value of r in [-1,0). Taking into account the parity of the Chebyshev polynomials as explained before, we obtain

$$A_k \cdot \begin{pmatrix} T_0(r_i) - T_0(\hat{r}_i) \\ T_1(r_i) - T_1(\hat{r}_i) \\ \vdots \\ T_{2n_r+1}(r_i) - T_{2n_r+1}(\hat{r}_i) \end{pmatrix} = A_k \cdot \begin{pmatrix} 0 \\ 2T_1(r_i) \\ \vdots \\ 2T_{2n_r+1}(r_i) \end{pmatrix} = 0.$$

Therefore, $a^k_{1,j} T_1(r_i) + a^k_{3,j} T_3(r_i) + \ldots + a^k_{2n_r+1,j} T(r_i) = 0$ for $j = 0, .., n_z$ and $r_i \in (0, 1], i = 0, .., n_r$, The matrices of these homogeneous systems are regular, consequently the solution is zero, i.e. $a^k_{i,j} = 0$, $i = 1, 3 \ldots, 2n_r + 1, j = 0, \ldots, n_z$. This means that for an even field, the matrix A_k will have the form,

$$A_k^{e0} = \begin{pmatrix} a^k_{0,0} & 0 & a^k_{2,0} & 0 & \ldots & 0 & a^k_{2n_r,0} & 0 \\ a^k_{0,1} & 0 & a^k_{2,1} & 0 & \ldots & 0 & a^k_{2n_r,1} & 0 \\ \vdots & \vdots & \vdots & \vdots & \ddots & \vdots & \vdots & \vdots \\ a^k_{0,n_z} & 0 & a^k_{2,n_z} & 0 & \ldots & 0 & a^k_{2n_r,n_z} & 0 \end{pmatrix}$$

In a similar way, if the field is odd, the matrix A_k will be

$$A_k^{o0} = \begin{pmatrix} 0 & a^k_{1,0} & 0 & \ldots & 0 & a^k_{2n_r+1,0} \\ 0 & a^k_{1,1} & 0 & \ldots & 0 & a^k_{2n_r+1,1} \\ \vdots & \vdots & \vdots & \ddots & \vdots & \vdots \\ 0 & a^k_{1,n_z} & 0 & \ldots & 0 & a^k_{2n_r+1,n_z} \end{pmatrix}$$

The proof is analogous for odd k, with the roles of the matrices interchanged, for even fields the matrix is A_k^{o0} and for odd fields the matrix is A_k^{e0}. □

Corollary 1 *In the Fourier Chebyshev transformations of theorem 1, given the matrix T_r^e with the even Chebyshev polynomials evaluated in $r_i \in (0, 1]$, $i = 0, \ldots, n_r$ by rows, i.e., the even rows of T_r, and A_k^e the matrix A_k^{e0} eliminating the* *null columns, the transformation $T_z \cdot A_k^e \cdot T_r^e$ should be used for an even field in the case of an even k and for an odd field in the case of an odd k. Given the matrix T_r^o with the odd Chebyshev polynomials evaluated in $r_i \in (0, 1]$, $i = 0, \ldots, n_r$ by rows, that is, the odd rows of T_r, and A_k^o the matrix A_k^{o0} eliminating the null columns, the transformation $T_z \cdot A_k^o \cdot T_r^o$ should be used for an even field in the case of odd k and for an odd field in the case of an even k.*

Proof For even k and taking into account the null columns in matrices A_k^{e0} and A_k^{o0}, it is straightforward that $T_z \cdot A_k^{e0} \cdot T_r = T_z \cdot A_k^e \cdot T_r^e$, in the case of an even field, and $T_z \cdot A_k^{o0} \cdot T_r = T_z \cdot A_k^o \cdot T_r^o$, in the case of an odd field.

For odd k and taking into account the null columns in matrices A_k^{e0} and A_k^{o0}, it is straightforward that $T_z \cdot A_k^{o0} \cdot T_r = T_z \cdot A_k^o \cdot T_r^o$, in the case of an even field, and $T_z \cdot A_k^{e0} \cdot T_r = T_z \cdot A_k^e \cdot T_r^e$, in the case of an odd field. □

Corollary 2 *In the Fourier Chebyshev transformations of theorem 1 and corollary 1, given d_zT_z the matrix that results from derivating each term of T_z with respect to z, the derivative with respect to z of a field is built with the product $d_zT_z \cdot A_k^e \cdot T_r^e$ in the case of an even k for an even field and in the case of an odd k for an odd field, or with the product $T_z \cdot A_k^o \cdot T_r^o$ in the case of an odd k for an even field and in the case of an even k for an odd field.*

Proof For even k even and taking into account the null columns in matrices A_k^{e0} and A_k^{o0}, it is straightforward that $d_zT_z \cdot A_k^{e0} \cdot T_r = d_zT_z \cdot A_k^e \cdot T_r^e$, in the case of an even field, and $d_zT_z \cdot A_k^{o0} \cdot T_r = d_zT_z \cdot A_k^o \cdot T_r^o$, in the case of an odd field.

For odd k and taking into account the null columns in matrices A_k^{e0} and A_k^{o0}, it is straightforward that $d_zT_z \cdot A_k^{o0} \cdot T_r = d_zT_z \cdot A_k^o \cdot T_r^o$, in the case of an even field, and $d_zT_z \cdot A_k^{e0} \cdot T_r = d_zT_z \cdot A_k^e \cdot T_r^e$, in the case of an odd field. □

Corollary 3 *In the Fourier Chebyshev transformations of theorem 1 and corollaries 1 and 2, given d_rT_r, $d_rT_r^o$ and $d_rT_r^e$ the matrices that result from derivating each term of T_r, T_r^o and T_r^e with respect to r, the derivative with respect to r of a field is built with the product $T_z \cdot A_k^e \cdot d_rT_r^e$ in the case of an even k for an even field and in the case of an odd k for an odd field or with the product $T_z \cdot A_k^o \cdot d_rT_r^o$ in the case of an odd k for an even field and in the case of an even k for an odd field.*

Proof For even k and taking into account the null columns in matrices A_k^{e0} and A_k^{o0}, it is straightforward that $T_z \cdot A_k^{e0} \cdot d_rT_r = T_z \cdot A_k^e \cdot d_rT_r^e$, in the case of an even field, and $T_z \cdot A_k^{o0} \cdot d_rT_r = T_z \cdot A_k^o \cdot d_rT_r^o$, in the case of an odd field.

For odd k and taking into account the null columns in matrices A_k^{e0} and A_k^{o0}, it is straightforward that $T_z \cdot A_k^{o0} \cdot d_rT_r = T_z \cdot A_k^o \cdot d_rT_r^o$, in the case of an even field, and $T_z \cdot A_k^{e0} \cdot d_rT_r = T_z \cdot A_k^e \cdot d_rT_r^e$, in the case of an odd field. □

Remark 1 The elimination of half of the columns in the inverse Chebyshev transformation matrix leads to the original dimension of the radial expansion.

Remark 2 Note that when a field is differentiated once with respect to r the parity changes, therefore the previous theorem and corollaries should be applied properly in the case of successive derivatives.

4 Conclusions

In this paper we have shown some computational aspects of the implementation of a three dimensional spectral time evolution scheme for incompressible Boussinesq Navier-Stokes including rotation effects in a cylinder with a primitive variable formulation. The scheme is a second-order time-splitting method combined with pseudo-spectral Fourier Chebyshev in space. To deal with the singularity at the origin a radial expansion has been considered in the diameter of the cylinder. The order expansion in the radial coordinate gets doubled. We have developed a matrix processing that combines the use of the parity of the fields and the discretization functions to cancel half of the terms in the matrix reducing the radial dimension to the original one.

Acknowledgements This work was partially supported by the Research Grant GI20163529 (UCLM) and MTM2015-68818-R MINECO (Spanish Government), which includes RDEF funds.

References

1. C.K. Batchelor, *An Introduction to Fluid Dynamics* (Cambridge University Press, Cambridge, 1967)
2. J. Boyd, *Chebyshev and Fourier Spectral Methods* (Dover, New York, 2001)
3. C. Canuto, M.Y. Hussain, A. Quarteroni, T.A. Zang, *Spectral Methods in Fluid Dynamics* (Springer, Berlin, 1988)
4. D. Castaño, M.C. Navarro, H. Herrero, Secondary whirls in thermoconvective vortices developed in a cylindrical annulus locally heated from below. Commun. Nonlinear Sci. Numer. Simul. **28**(1–3), 201–209 (2015)
5. D. Castaño, M.C. Navarro, H. Herrero, Evolution of secondary whirls in thermoconvective vortices: strengthening, weakening and disappearance in the route to chaos. Phys. Rev. E (2016). doi:10.1103/PhysRevE.93.013117
6. D. Castaño, M.C. Navarro, H. Herrero, Double vortices and single-eyed vortices in a rotating cylinder non-homogeneously heated. Comp. and Math. with Appl. **73**, 2238–2257 (2017)
7. S. Chandrasekhar, *Hydrodynamic and Hydromagnetic Stability* (Dover Publications, New York, 1981)
8. R. Chokri, B. Brahim, Three-dimensional natural convection of molten Lithium in a differentially heated rotating cubic cavity about a vertical ridge. Powder Technol. **291**, 97–109 (2006)
9. K.A. Emanuel, *Divine Wind* (Oxford University Press, Oxford, 2005)
10. B. Fornberg, *A Practical Guide to Pseudospectral Methods* (Cambridge University Press, Cambridge, 1998)
11. K. Goda, A multistep technique with implicit difference schemes for calculating two- and three-dimensional cavity flows. J. Comput. Phys. **30**, 76–95 (1979)
12. P. Gresho, On the theory of semi-implicit projection methods for viscous incompressible flow and its implementation via finite element method that also introduces a nearly consistent mass matrix. Int. J. Numer. Meths. Fluids **11**, 587–620 (1990)
13. H. Herrero, A.M. Mancho, On pressure boundary conditions for thermoconvective problems. Int. J. Numer. Methods Fluids **39**, 391–402 (2002)
14. S. Hugues, A. Randriamampianina, An improved projection scheme applied to pseudospectral methods for the incompressible Navier-Stokes equations. Int. J. Numer. Methods Fluids **28**, 501–521 (1998)

15. C.L. Jordan, Marked changes in the characteristics of the eye of intense typhoons between the deeping and filling stages. J. Meteorol. **18**, 779–789 (1961)
16. G.E. Karniadakis, M. Israeli, S.A. Orsaq, High order splitting methods for the incompressible Navier-Stokes equations. J. Comput. Phys. **97**, 414–443 (1991)
17. A.M. Mancho, H. Herrero, J. Burguete, Primary instabilities in convective cells due to nouniform heating. Phys. Rev. E **56**, 2916–2923 (1997)
18. I. Mercader, O. Batiste, A. Alonso, An efficient spectral code for incompressible flows in cylindrical geometries. Comput. Fluids **39**, 215–224 (2010)
19. M.C. Navarro, A.M. Mancho, H. Herrero, Instabilities in Buoyant flows under localized heating. Chaos Interdisciplinary J. Nonlinear Sci. **17**, 023105-1-12 (2007)
20. Lord Rayleigh, on convective currents in a horizontal layer of fluid when the temperature is on the under side. Phil. Mag. **32**, 529–46 (1916)
21. P.C. Sinclair, The lower structure of dust devils. J. Atmos. Sci. **30**, 1599–1619 (1973)
22. L.N. Trefethen, *Spectral Methods in Matlab* (SIAM, Philadelfia, 2000)

High Order Compact Mimetic Differences and Discrete Energy Decay in 2D Wave Motions

Jose E. Castillo and Guillermo Miranda

Abstract Mimetic difference operators Div, Grad and Curl, have been constructed to provide a high order of accuracy in numerical schemes that mimic the properties of their corresponding continuum operators; hence they would be faithful to the physics. However, this faithfulness of the discrete basic operators might not be sufficient if the numerical difference scheme introduces some numerical energy increase, which would obviously result in a potentially unstable performance. We present a high order compact mimetic scheme for 2D wave motions and show that the energy of the system is also conserved in the discrete sense.

1 Introduction

A numerical model must incorporate proper physics in order to deliver acceptable numerical results. However, the numerical scheme that is used to solve the Partial Differential Equations (PDEs) also has a major impact on the quality of the results [8, 10]. Difference schemes have different requirements and both the accuracy and the performance of a model will vary based on the discretization scheme used. The majority of equations describing physical phenomena are written using the first order invariant operators: Gradient, divergence, and curl. Mimetic discrete operators are derived by constructing discrete analogs of these continuum differential operators. These discrete operators are used to build a discrete analogue of the equation modeling the physical problem. Because the discrete operators mimic the continuum ones by satisfying the same properties, in the discrete sense, numerical schemes based on these operators are more faithful to the physics of the problem under investigation. Castillo and Grone [4] have developed a methodology to construct mimetic operators, known as (CGM) operators that have the same order of approximation in the interior of the domain as at the boundary. CGM operators have been used in many applications, such as wave propagation, seismic studies, electrodynamics, and image processing, [2, 4, 6, 11, 12], with a very high

J.E. Castillo (✉) • G. Miranda
Computational Science Research Center, San Diego State University, San Diego, CA 92182, USA
e-mail: jcastillo@mail.sdsu.edu

M.L. Bittencourt et al. (eds.), *Spectral and High Order Methods for Partial Differential Equations ICOSAHOM 2016*, Lecture Notes in Computational Science and Engineering 119, DOI 10.1007/978-3-319-65870-4_20

rate of success, making the schemes based on these operators competitive with the established ones.

High-order methods are becoming more important for many applications where greater accuracy is needed than provided by conventional second-order methods. Compact finite differences allows the implementation of high-order finite differences using the shortest stencil [9]. Carpenter [3] has worked on stability for compact finite differences including boundary conditions. In [1], Abouali and Castillo presented an approach to implement high-order mimetic finite differences in a compact way so the stencils are kept at a minimum length. Cordova et al. [7] have implemented compact mimetic schemes for the acoustic wave equation comparing them to the compact finite difference schemes. We present a high order compact mimetic scheme for 2D wave motions and show that the energy of the system is also conserved in the discrete sense. The present scheme deviates from the standard derivation sequence, namely, discrete - continuous - discrete, since while keeping the original media discrete decomposition, for the time behavior, it is combined with the discrete analog of the continuous term div grad u used to account for the variation in space of the internal forces under consideration. Here, we look at the energy conservation of the mimetic methods for a model problem.

This paper is organized as follows: First, we give a brief description of mimetic finite difference operators and present their matrix representation in 1D as well as the formulas for extension to 2D using Kronecker products [4, 5]. It is followed by the compact form of the operators for a fourth order divergence and gradient mimetic operators. In Sect. 3, we present a model problem where the energy is conserved. In Sect. 4, we show that the energy is also conserved in the discrete sense when we used the CGM operators.

2 Compact Finite Difference Schemes

2.1 Mimetic Difference Operators

Mimetic discretization operators are discrete analogs of gradient G and divergence D, plus an auxiliary boundary operator B, which satisfies a discrete version of the Green–Stokes–Gauss theorem:

$$\langle Dv, u\rangle_Q + \langle v, Gu\rangle_P = \langle Bv, u\rangle_I . \tag{2.1}$$

Weighted inner products on above expression (2.1) are defined in the standard form,

$$\langle x, y\rangle_A = y^t A x. \tag{2.2}$$

As it is known, in the one-dimensional case, the discrete divergence will act on the v-values defined at $(n+1)$ nodes, $x_i = i\Delta x,\ i = 0, 1, \cdots, n$, so that the discrete v will be regarded as an $(n+1) \times 1$ matrix or an $(n+1)$–tuple.

By the same token, the discrete gradient will act on the u-values defined at both the n $1D$ cell centers $x_{i+\frac{1}{2}} = i\Delta x,\ i = 0, 1, \cdots, (n-1)$, and the two boundary nodes x_0 and x_n, so that the discrete u will be regarded as an $(n+2) \times 1$ matrix or an $(n+2)$ - tuple, and such u might be aptly symbolized as ucb, since it considers both cell centers and boundary nodes, but it is to be distinguished from a discrete u evaluated only at cell centers, naturally symbolized then by uc.

Besides G being an $(n+1) \times (n+2)$ matrix, the resulting Gu or $Gucb$ is an $(n+1) \times 1$ matrix.

While the matrix D is an $n \times (n+1)$ matrix, in the one dimensional case, it should be augmented with two rows-first and last-of zeroes, so that the matrix augmented D will be an $(n+2) \times (n+1)$. The reason for this augmentation is because the divergence is zero at the boundary and also the need to take a weighted inner product of Dv with $u = ucb$, but ucb is an $(n+2) \times 1$ matrix, and thus it will not conform to Dv, being an $n \times 1$ matrix prior to augmentation.

Here, we present the one-dimensional second-order mimetic gradient operator,

$$\mathbf{G} = \frac{1}{\Delta x} \begin{vmatrix} \frac{-8}{3} & 3 & \frac{-1}{3} & & & \\ & -1 & 1 & & & \\ & & \ddots & \ddots & & \\ & & & -1 & 1 & \\ & & & \frac{1}{3} & -3 & \frac{8}{3} \end{vmatrix}_{(n+1)\times(n+2)} \quad \text{and,} \tag{2.3}$$

the one-dimensional second-order mimetic divergence operator,

$$\mathbf{D} = \frac{1}{\Delta x} \begin{vmatrix} 0 & 0 & \cdots & 0 \\ -1 & 1 & & \\ & \ddots & \ddots & \\ & & -1 & 1 \\ 0 & 0 & \cdots & 0 \end{vmatrix}_{(n+2)\times(n+1)} . \tag{2.4}$$

2.2 2D Mimetic Operators

While working with 2D rectangular domains, $[0, a] \times [0, b]$, the corresponding grid will be made of m by n cells with mn cell centers $\left(x_{i+\frac{1}{2}}, y_{j+\frac{1}{2}}\right)$, with $i = 0, 1, \cdots, (n-1)$, $x_0 = 0$, $x_n = a$, and $j = 0, 1, \cdots, (m-1)$, $y_0 = 0$, $y_m = b$.

The 2D mimetic operators can be constructed from the 1D operators using Kronecker products. Here, e_s stands for an s-tuple filled with an appropriate number s of 1 entries and k is the order of approximation.

$$\breve{G}^k_{xy} = \begin{bmatrix} G_x(k) \\ G_y(k) \end{bmatrix} G_x(k) = \left[\hat{I}^T_n \otimes \breve{G}^k_x\right] G_y(k) = \left[\breve{G}^k_y \otimes \hat{I}^T_m\right], \tag{2.5}$$

$$\breve{D}^k_{xy} = [D_x(k) \; D_y(k)], \tag{2.6}$$

$$\text{with } D_x(k) = [\hat{I}_n \otimes \breve{D}_x^k]\ D_y(k) = [\breve{D}_y^k \otimes \hat{I}_m], \tag{2.7}$$

$$\text{with } \hat{I}_n^s = \begin{bmatrix} 0 \\ I_n \\ 0 \end{bmatrix} = e_s \otimes I_n. \tag{2.8}$$

2.3 Compact Scheme: Explicit Approach

We use an explicit approach to construct the compact mimetic finite difference schemes.

Let $\dfrac{du}{dx} = M_k u$, where M_k is a mimetic difference operator of order k. In this explicit approach, the high-order-accurate derivative is calculated as follows:

$$\left(\frac{\partial u}{\partial x}\right) = R_k D_2 u, \tag{2.9}$$

This approach eliminates the need for solving a system of linear equations. Here, $M_k = R_k D_2$, where D_2 is the second-order derivative. Here are the corresponding R matrices for the Castillo–Grone's Mimetic (CGM) gradient and divergence operators of order 4 respectively.

$$R_4^G = \begin{bmatrix} \frac{17958}{14245} & \frac{-8776}{14245} & \frac{154787}{341880} & \frac{-3415}{34188} & \frac{25}{9768} & & & & \\ \frac{-2}{35} & \frac{941}{840} & \frac{-29}{420} & \frac{1}{168} & & & & & \\ & \frac{-1}{24} & \frac{13}{12} & \frac{-1}{24} & & & & & \\ & & \frac{-1}{24} & \frac{13}{12} & \frac{-1}{24} & & & & \\ & & & \vdots & & & & & \\ & & & \frac{-1}{24} & \frac{13}{12} & \frac{-1}{24} & & & \\ & & & \frac{1}{168} & \frac{-29}{420} & \frac{941}{840} & \frac{-2}{35} & & \\ & & \frac{25}{9768} & \frac{-3415}{34188} & \frac{154787}{341880} & \frac{-8776}{14245} & \frac{17958}{14245} & & \end{bmatrix} \tag{2.10}$$

$$R_4^D = \begin{bmatrix} \frac{1045}{1142} & \frac{492}{2291} & \frac{-418}{2371} & \frac{328}{6821} & \frac{-25}{15576} & & & & \\ \frac{-1}{24} & \frac{13}{12} & \frac{-1}{24} & & & & & & \\ & \frac{-1}{24} & \frac{13}{12} & \frac{-1}{24} & & & & & \\ & & & & \vdots & & & & \\ & & & & & & \frac{-1}{24} & \frac{13}{12} & \frac{-1}{24} \\ & & & & \frac{-25}{15576} & \frac{328}{6821} & \frac{-418}{2371} & \frac{492}{2291} & \frac{1045}{1142} \end{bmatrix}. \tag{2.11}$$

3 Energy Conservation

In order to exemplify our proposed scheme, which deviates from the standard sequence described as "discretize the continuous media - derive a continuous PDE - discretize this PDE," we shall start reconstructing the preliminary discrete decomposition of an elastic membrane's surface into "surface elements dS," later to be considered as equivalent material particles subjected to Newton's Laws.

In this way, the approximations needed before the linear PDE appears in the form of the standard wave equation can be clearly seen. This PDE is not an "exact modeling" of the continuous elastic media vibrations anyway.

These considerations motivate an alternative approach, not passing to the limit of shrinking surface elements, but remaining in the context of a many body mechanical problem. A natural approximation for the variation in space of the internal tensional forces under consideration will be the discrete mimetic analog of the Laplacian of the vertical membrane displacement.

We also recall the continuous version for the time derivative of the total elastic membrane energy, so that it can be compared with our discrete version, which is derived step by step in Sect. 4.

In order to better illustrate the proposed scheme, consider the model problem provided by the mechanical vibrations of an elastic membrane on a rectangular domain. Our techniques apply to any 2D spatial domain with a closed rectifiable boundary, but we use a rectangular shape for simplicity of drawing and explanation (Fig. 1).

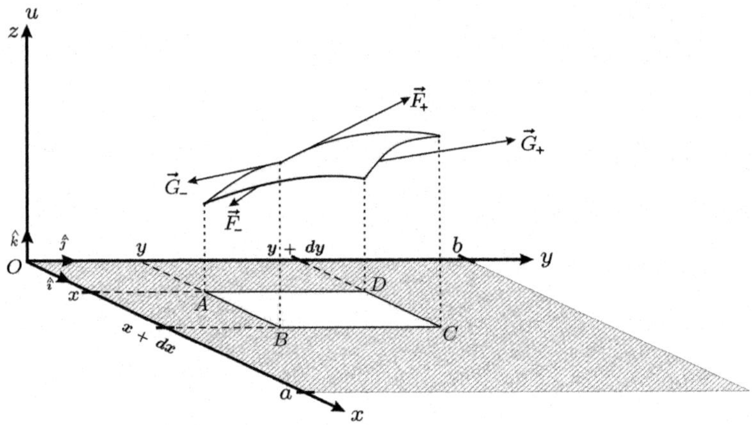

Fig. 1 Model problem: Find an elastic membrane vertical displacement $u(x, y, t)$

The membrane vector tensions $\vec{F}$, $\vec{G}$ are assumed normal to the lines of interaction and tangent to the membrane surface.

$$\vec{F}_+ = \| \vec{F}_+ \| \sin\alpha \; (x + dx, y_+^*, t)\hat{k} + \| \vec{F}_+ \| \cos \; \alpha \; (x+dx, y_+^*, t)\hat{i}; \tan \; \alpha \; (x, y, t) = \frac{\partial u}{\partial x} \; (x, y, t) \tag{3.1}$$

$$\vec{G}_+ = \| \vec{G}_+ \| \sin \; \beta \; (x_+^*, y+dy, t)\hat{k} + \| \vec{G}_+ \| \cos \; \beta \; (x_+^*, y + dy, t)\hat{j}; \tan \; \beta \; (x, y, t) = \frac{\partial u}{\partial y} \; (x, y, t) \tag{3.2}$$

Approximations: $\| \vec{F}_+ \|$ per unit length $= \| \vec{G}_+ \|$ per unit length $=$ constant $=$ To.
Now approximating

$$\sin \; \alpha \; (x, y, t) \approx \tan \; \alpha \; (x, y, t) = \frac{\partial u}{\partial x}(x, y, t), \tag{3.3}$$

and

$$\sin \; \beta \; (x, y, t) \approx \tan \; \beta \; (x, y, t) = \frac{\partial u}{\partial y}(x, y, t), \tag{3.4}$$

since

$$ds_x = \sqrt{1 + \left(\frac{\partial u}{\partial x}\right)^2} dx, \;\; \text{and} \;\; ds_y = \sqrt{1 + \left(\frac{\partial u}{\partial y}\right)^2} dy, \tag{3.5}$$

we get,

$$\mathbf{dS} = dx\,dy \begin{vmatrix} i & j & k \\ 1 & 0 & \frac{\partial u}{\partial x} \\ 0 & 1 & \frac{\partial u}{\partial y} \end{vmatrix} = \left(-\frac{\partial u}{\partial x}\, i \;-\; \frac{\partial u}{\partial y}\, j + k\right) dxdy, \tag{3.6}$$

surface element area $dS = \sqrt{1 + \left(\frac{\partial u}{\partial x}\right)^2 + \left(\frac{\partial u}{\partial y}\right)^2}\, dxdy.$

The mass element is

$$\begin{aligned} dm &= \rho(x, y, t) dS \\ &= \rho(x, y, 0) dxdy, \text{ at } t = 0, \\ &= \rho_0(x, y) dxdy \text{ (By mass conservation).} \end{aligned} \tag{3.7}$$

The vector tension $F+$ acting along the edge with length dsy, is located at some intermediate point, which is projected down to $(x + dx, y^*_+)$ in the XY Plane, and similar intermediate points are considered for the vector tensions $F-$, $G+$ and $G-$.

Resulting upwards force upon dm:

$$\text{To}\left\{\left[\sin\alpha(x + dx, y^*_+, t) - \sin\alpha(x, y^*_-, t)\right] dsy + \left[\sin\beta(x^*_+, y + dy, t) - \sin\beta(x^*_-, y, t)\right] dsx\right\} \tag{3.8}$$

Approximating $y^*_+ \approx y^*_- \approx y$ and $x^*_+ \approx x^*_- \approx x$, get with $dsy \simeq dy$ and $dsx \approx dx$; (3.9)

$$\text{To}\left\{\left[\frac{\partial^2 u}{\partial x^2}(x, y, t) dx\right] dy + \left[\frac{\partial^2 u}{\partial y^2}(x, y, t) dy\right] dx\right\} = \tag{3.10}$$

$$\rho_0(x, y,) dxdy \frac{\partial^2 u}{\partial t^2}(x, y, t). \tag{3.11}$$

The standard treatment ([13]) of the momentum equation in the presence of elastic forces is used to derive the time derivative of the kinetic energy, when applied to the vibrating membrane linearized PDE 3.11, $\left(T_o \text{ div grad} u = \rho_0 \frac{\partial^2 u}{\partial t^2}\right)$ and leads to $\frac{d}{dt} E(t) = 0 + \oint_{\partial \Box} To \frac{\partial u}{\partial t} (\text{grad}u.\hat{n}) (\tilde{x}, \tilde{y}, t)\, ds$, with $(\tilde{x}, \tilde{y})$ belonging to $\partial \Box$. Here, **n is the outward unit** normal to the boundary, $E(t) = K(t) + V(t)$, where, as usual for continuous media, the Kinetic Energy K and the Potential Elastic

Membrane Energy are given by: $K(t) = \int_{x=0}^{x=a} \int_{y=0}^{y=b} \rho_0\ (x,y) \cdot \frac{1}{2} \left(\frac{\partial u}{\partial t}\right)^2 (x,y,t)\ dxdy$ and $V(t) = \frac{To}{2} \iint\ \nabla u . \nabla u\ (x,y,t)\ dxdy.$

□

This formula for $\frac{dE}{dt}$, even though written for a rectangular boundary, is immediately seen to be valid for any domain shape with a rectifiable boundary, since the hypotheses needed to use the extended Gauss's 2D Divergence Theorem, in order to obtain the boundary integral term, remain valid for such general domains under the assumed smoothness up to the boundary of the vertical displacement gradient.

Besides, in the case of **homogeneous** boundary conditions, i.e., when $\frac{\partial u}{\partial n}$ (grad $u \cdot n) = 0$, we get the known conservation of energy in the form $\frac{d}{dt}E(t) = 0$ in the continuous elastic media model, a standard result that will be shown in the next section to hold for the time derivative of the total discrete energy.

4 Discrete Energy Conservation

In a semi-discrete approach, we consider the vertical displacement $u_{ij}(t)$ of the mn cell centers $(x_{i+\frac{1}{2}}, y_{j+\frac{1}{2}})$, and define the discrete analogs of the continuous kinetic and potential energy, using the mimetic discretisation for double integrals over a rectangular domain, and considering the weights $w_{ij} = \rho(x_{i+\frac{1}{2}}, y_{j+\frac{1}{2}})\ h_i k_j$ to be the ijth mass element undergoing mechanical vibrations, where $h_i = x_i - x_{i-1}$ with $i = 0, 1, \cdots, (n-1)$ and $k_j = y_j - y_{j-1}$ with $j = 0, 1, \cdots, (m-1)$.

For simplicity, we shall assume that ρ is constant, and that spatial grids are uniform, i.e., $h_i = h$ and $k_j = k$, $w_{ij} = \rho hk$.

$$K\text{discrete}(t) = \left(\frac{1}{2}\right)\ hk\ \rho \sum\sum\ \dot{u}_{ij}^2 \tag{4.1}$$

$$V\text{discrete}(t) = \left(\frac{1}{2}\right) hk\ (T_0) \sum\sum\ \langle \text{GRAD}\ u_{ij}, \text{GRAD} u_{ij} \rangle .$$

Instead of starting from the continuous wave equation, as is usually done to obtain some discrete scheme to be numerically solved, we consider the equation of motion for the mass element labeled ij in the form:

$$\rho hk\ \ddot{u}_{ij}(t) = T_0\ hk\ \text{DIV GRAD}\ u_{ij}(t). \tag{4.2}$$

Multiplying by $\dot{u}_{ij}\ \ (t)$ both sides of this equation: $\rho\ \ \dot{u}_{ij}\ \ \ddot{u}_{ij}\ \ (t) = T_0\ \dot{u}_{ij}\ (t)$ DIV GRAD $u_{ij}\ (t)$ Substitute t by τ and integrating from $\tau = 0$ to $\tau = t$, we get :

$$\rho\left(\frac{1}{2}\right)\left[\dot{u}_{ij}(t)^2 - \dot{u}_{ij}(0)^2\right] = \int_{\tau=0}^{\tau=t} T_0\ \dot{u}_{ij}\ (\tau)\ \text{DIV GRAD}\ u_{ij}(\tau)\ d\tau. \tag{4.3}$$

Next, carry out $hk \sum\sum$ on both sides of the previous equation, to get:

$$K\text{discrete}(t) - K\text{ discrete}(0) = hk \sum\sum \int_{\tau=0}^{\tau=t} T_0\, \dot{u}_{ij}\,(\tau)\text{DIV GRAD}\; u_{ij}\,(\tau)\; d\tau. \tag{4.4}$$

Now, interchanging the τ integration with the finite double sum operation we get,

$$K\text{ discrete}(t) - K\text{ discrete }(0) = hk \int_{\tau=0}^{\tau=t} \left\{\sum\sum T_0\, \dot{u}_{ij}\,(\tau)\text{ DIV GRAD }u_{ij}(\tau)\right\}\, d\tau. \tag{4.5}$$

Now recall that

$$\text{DIV}\left\{\dot{u}_{ij}\,(\tau)\text{ GRAD }u_{ij}\,(\tau)\right\} = \dot{u}_{ij}\,(\tau)\text{DIV GRAD}u_{ij} + \left\langle\text{GRAD}\dot{u}_{ij}\,(\tau),\ \text{GRAD }u_{ij}\,(\tau)\right\rangle. \tag{4.6}$$

Next apply $hk \sum\sum$ to both sides, using the discrete mimetic analog of Gauss's Divergence Theorem, as worked by Castillo, et al. [4–6].

$$\text{Boundary}\sum\ \dot{u}_{ij}\,(\tau)\left\langle\text{GRAD }u_{ij}\,(\tau), n_{ij}\right\rangle\, ds_{ij} = \tag{4.7}$$

$$hk \sum\sum\ \dot{u}_{ij}\,(\tau)\text{ DIV GRAD }(\tau) + hk \sum\sum \left\langle\text{GRAD }\dot{u}_{ij}\,(\tau), \text{GRAD }u_{ij}\,(\tau)\right\rangle.$$

Where, in the Boundary sum, the index pair (i,j) refers only to those mass elements that have some membrane edge projected on the boundary of the rectangular domain, $u_{ij}\ (\tau)$ would be the vertical displacement corresponding to the middle point of such edge element at time τ; ds_{ij} would be plus or minus h or k, depending upon the orientation of the edge as the boundary is circulated counterclockwise, and n_{ij} would then be the outward unit normal to that element's projected edge.

Since,

$$\dot{u}_{ij}\,(\tau)\text{ DIV GRAD}\,u_{ij}\,(\tau) = \text{DIV}\left\langle\dot{u}_{ij}\,(\tau)\text{ GRAD }u_{ij}\,(\tau)\right\rangle - \left\langle\text{GRAD }\dot{u}_{ij}\,(\tau),\ \text{GRAD }u_{ij}\,(\tau)\right\rangle, \tag{4.8}$$

then omitting the "discrete" labeling for K and V, we first obtain:

$$K(t) - K(0) = T_0 \int_{\tau=0}^{\tau=t} H(\tau)\, d\tau\ - T_0\, hk \int_{\tau=0}^{\tau=t} F(\tau)d\tau, \tag{4.9}$$

where

$$H(\tau) = \text{Boundary} \sum \dot{u}_{ij}\ (\tau)\ \langle \text{GRAD} u_{ij}\ (\tau),\ n_{ij} \rangle\ ds_{ij} \tag{4.10}$$

and

$$F\ (\tau) = \sum \sum \langle \text{GRAD}\ \dot{u}_{ij}\ (\tau),\ \text{GRAD}\ u_{ij}(\tau) \rangle,$$

where in terms H and F under integration with respect to τ, dot means $\frac{d}{d\tau}$.

Using $\frac{d}{d\tau}\left\langle \left(\frac{1}{2}\right) f^2 \right\rangle = f\frac{df}{d\tau}$, the above inner product can be put in the form:

$$\frac{d}{d\tau}\left\{ \left(\frac{1}{2}\right)\ [G_x u_{ij}\ (\tau)]^2 + \left(\frac{1}{2}\right) [G_y u_{ij}(\tau)]^2 \right\}. \tag{4.11}$$

Next, interchange the τ integration and $\sum\sum$ to get:

$$K(t) - K(0) = T_0 \int_{\tau=0}^{\tau=t} H(\tau) d\tau - T_0 \sum\sum \int_{\tau=0}^{\tau=t} \frac{d}{d\tau}\left\{ \left(\frac{1}{2}\right) \left([G_x u_{ij}(\tau)]^2 + [G_y u_{ij}(\tau)]^2 \right) \right\} d\tau, \tag{4.12}$$

thus getting:

$$K(t) - K(0) = T_0 \int_{\tau=0}^{\tau=t} H(\tau)\ d\tau\ -\ [V(t) - V(0)],\ \text{where} \tag{4.13}$$

$$V(t) = \left(\frac{T_0}{2}\right)\ hk \sum\sum \langle \text{GRAD}\ u_{ij}\ (t),\ \text{GRAD}\ u_{ij}\ (t) \rangle. \tag{4.14}$$

Finally, for the total discrete energy $E(t)$ we obtain:

$$E(t) = K\text{discrete}\ (t) + V\text{discrete}\ (t) = K\text{discrete}\ (0) + V\text{discrete}\ (0) + T_0 \int_{\tau=0}^{\tau=t} H(\tau)\ d\tau.$$

Now, differentiating with respect to t, we get the desired result for our scheme:

$$\frac{d}{dt} E(t) = T_0 H(t) = T_0\ \text{Boundary} \sum \dot{u}_{ij}\ (t)\ \langle \text{GRAD}\ u_{ij}\ (t),\ n_{ij} \rangle\ ds_{ij}. \tag{4.15}$$

In the case of the rectangular boundary taken as a model problem, and using upper case letters W, E, S and N to label the boundary edges located at $x = 0,\ x = a,\ y = 0$ and $y = b$ respectively, and by means of the discrete operators BG_x and

BG_y, the Boundary sum can be given the following concrete expression in $\frac{d}{dt}\,E(t)$:

$$\frac{dE}{dt}_{(discrete)} = -\,h\,To\,\left\langle \dot{u}^S, (BG_y u)^S\right\rangle + k\,To\,\left\langle \dot{u}^E,\ (BG_x u)^E\right\rangle$$

$$-\,h\,To\,\left\langle \dot{u}^N, (BG_y u)^N\right\rangle + k\,To\,\left\langle \dot{u}^W,\ (BG_x u)^W\right\rangle.$$

We are "integrating" (that is, summing with inner products) counter clockwise starting at $(0, 0)$ so we have used South (S), East (E), North (N) and West (W) in that order, recalling that the unit outward normals to the boundary edges would then be $(0, -1)$, $(1, 0)$, $(0, 1)$ and $(-1, 0)$ respectively, and each numerical integral is expressed by a corresponding mimetic inner product $<, >$, taken along straight boundary edges.

Hence, the discrete total energy will be conserved under homogeneous boundary conditions for the vertical displacement.

5 Conclusions

In this work, we have presented the time derivative of the total discrete Energy by employing Castillo-Grone mimetic difference Operators for the two-dimensional case in the semi-discrete version. Having shown that under homogeneous boundary conditions the discrete total Energy formulated by means of the Castillo-Grone 2D Operator G is conserved in a 2D spatial domain, we have exhibited an additional faithfulness of the High Order Castillo-Grone Mimetic Finite Differences to the underlying physics, and done so in a more complex environment. This technique can be applied to any 2D spatial domain with a closed rectifiable boundary as the consideration of a rectangular shape was used only for simplicity of drawings and explanations.

The relatively simple explicit formula derived for the ordinary time derivative of the total Energy in this bounded 2D spatial domain, could eventually lead to some stability characteristics simpler than those provided by Carpenter et al. [3]. We can consider that after Von Neumann's initial stability findings, many criteria rely upon bounding energy or amplitude related discrete expressions, sometimes exhibiting a complex dependence upon the wavelengths involved.

With our proposed scheme, which assures boundedness of $\left\langle \text{GRAD}\ u_{ij}, \text{GRAD} u_{ij}\right\rangle$ and of the discrete kinetic energy, this would amount to a global boundedness of the vertical displacement $u_{ij}\ (t)$, considering the discrete analog for the continuous relation between $u\ (x, y)$ and GRAD $u\ (x, y)$.

References

1. M. Abouali, J.E. Castillo, High-order compact Castillo-Grone's operators. Report of Computational Science Research Center at San Diego State University. CSRCR02 1–13 (2012)
2. C. Bazan, M. Abouali, J. Castillo, P. Blomgren, Mimetic finite difference methods in image processing. Comput. Appl. Math. **30**(3), 701–720 (2011)
3. M. Carpenter, D. Gottlieb, S. Abarbanel, Stable and accurate boundary treatments for compact, high-order finite-difference schemes. Appl. Numer. Math. **12**, 55–87 (1993)
4. J.E. Castillo, R. Grone, A matrix analysis to high-order approximations for divergence and gradients satisfying a global conservation law. SIAM Matrix Anal. Appl. **25**, 128–142 (2003)
5. J.E. Castillo, G.F. Miranda, *Mimetic Discretization Methods* (CRC Press, West Palm Beach, 2013)
6. J. Castillo, J. Hyman, M. Shashkov, S. Steinberg, Fourth and sixth order conservative finite difference approximations of the divergence and gradient. Appl. Numer. Math. **37**(1–2), 171–187 (2001)
7. L.J. Cordova, O. Rojas, B. Otero, J.E. Castillo, Compact finite difference modeling of 2-D acoustic wave propagation. J. Comput. Appl. Math. (2015). http://www.sciencedirect.com/science/article/pii/S0377042715000618. Available online 19 February 2015
8. K. Hoffmann, S. Chiang, *Computational Fluid Dynamics*, vol. 2, 4th ed. (Engineering Education System Book, Wichita, KS, 2000)
9. S. Lele, Compact finite difference schemes with spectral-like resolution. J. Comput. Phys. **103**, 16–42 (1992)
10. R. LeVeque, *Finite Difference Methods for Ordinary and Partial Differential Equations* (SIAM, Philadelphia, 2007)
11. J. de la Puente, M. Ferrer, M. Hanzich, J.E. Castillo, J.M. Cela, Mimetic seismic wave modeling including topography on deformed staggered grids. Geophysics **79**(3), T125-T141 (2014)
12. O. Rojas, B. Otero, J.E. Castillo, S.M. Day, Low dispersive modeling of Rayleigh waves on partly-staggered grids. Int. J. Comput. Geom. Appl. **18**(1), 29–43 (2014)
13. R. Aberayatne, *Continuum Mechanics*, vol. II (Massachusetts Institute of Technology, Cambridge, MA, 2012)

A Spectral Mimetic Least-Squares Method for Generalized Convection-Diffusion Problems

Rasmus O. Hjort and Bo Gervang

Abstract We present a spectral mimetic least-squares method for a model convection-diffusion problem, which preserves conservation properties. The problem is solved using differential geometry where the topological part and the constitutive part have been separated. It is shown that the topological part is solved exactly independent of the order of the spectral expansion. The mimetic method incorporates the Lie derivative for the convective term, by means of Cartans homotopy formula, see for example Abraham et al. (1988) (Manifolds, Tensor Analysis, and Applications, Springer, New York). The spectral mimetic least-squares method is compared to a more classic spectral least-squares method. It is shown that both schemes lead to spectral convergence.

1 Introduction

We consider a general convection-diffusion problem in 2D:

$$\nabla \cdot (\mathbf{u}\phi) + \Delta\phi = f \qquad in\ \Omega, \tag{1}$$

where ϕ is the potential, f the source term, $\mathbf{u}$ a known divergence free vector field and a homogeneous Dirichlet boundary condition:

$$\phi = 0 \qquad on\ \partial\Omega. \tag{2}$$

The method presented is based on a combination of mimetic methods, presented in [3, 6] and [10] and least-squares spectral element methods, [2] and [12]. Recent work combining the two methods can be found in [8]. The method is derived using basic components from differential geometry, which leads to conservation of invariants both locally and globally of the system. Using the least-squares principles lead to a symmetric positive definite matrix when the problem is discretized. This

R.O. Hjort (✉) • B. Gervang
Department of Engineering, Aarhus University, Lehmanns Gade 10, 8000 Aarhus C, Denmark
e-mail: R.O.Hjort@live.dk; bge@ase.au.dk

M.L. Bittencourt et al. (eds.), *Spectral and High Order Methods for Partial Differential Equations ICOSAHOM 2016*, Lecture Notes in Computational Science and Engineering 119, DOI 10.1007/978-3-319-65870-4_21

is in contrast to a normal Galerkin method, which only leads to a symmetric matrix for the diffusion parts. The least-squares methods also give rise to a symmetric matrix for the convective term. The paper is structured as follows: Section 2 is an introduction to differential geometry, while Sect. 3 establishes the mimetic least-squares formulation and Sect. 4 presents the mimetic discretization of the governing equations. Finally Sect. 5 shows the numerical results.

2 Differential Geometry

In this section a short presentation of differential geometry is given. Additional information can be found in [1] and [11].

In differential geometry the unknowns are presented by forms instead of vector and scalar fields as addressed in vector calculus. Variables associated with points, such as the temperature, are represented by a 0-form while variables associated with a volume are represented by 3-forms, e.g. the density. 1-forms and 2-forms can likewise represent variables associated with lines and surfaces. Furthermore forms have geometric orientation, which makes it possible to further distinguish different variables. Outer orientated 2-forms represent variables working through surfaces, e.g. heat flux, while inner orientated 2-forms represent variables working on a surface e.g. describing vorticity.

Generalising the definitions of 0-forms, 1-forms, 2-forms and 3-forms, the general k-form is denoted $\omega^{(k)} \in \Lambda^k(\Omega_n)$ on the n-dimensional domain Ω_n, for $0 \leq k \leq n$. $\Lambda^k(\Omega_n)$ is the space of k-forms on Ω_n, i.e. the collection of all k-linear, antisymmetric mappings of vectors belonging to the n-dimensional tangent vector space V:

$$\omega^{(k)} : \underbrace{V \times \ldots \times V}_{k-\text{times}} \rightarrow \mathbb{R}. \tag{3}$$

Differential geometry also introduces the wedge product between k-forms and m-forms, which produces a (k+m)-form: $\wedge : \Lambda^k(\Omega_n) \times \Lambda^m(\Omega_n) \rightarrow \Lambda^{k+m}(\Omega_n)$. The wedge product is skew-symmetric such that: $\alpha^{(k)} \wedge \beta^{(m)} = (-1)^{km}\beta^{(m)} \wedge \alpha^{(k)}$.

Instead of using three different operators to represent curl, divergence and gradient, differential forms are equipped with an operator representing all three operators; the exterior derivative, d. The exterior derivative operates on k-forms and maps them into (k+1)-forms: $d : \Lambda^k(\Omega_n) \rightarrow \Lambda^{k+1}(\Omega_n)$. The exterior derivative can be defined by means of the Stokes theorem [4]:

$$\int_{\Omega_{k+1}} d\omega^{(k)} = \int_{\partial\Omega_{k+1}} \omega^{(k)}. \tag{4}$$

Since the exterior derivative is constructed using the boundary operator, the discrete version of the exterior derivative can be performed exactly.

The interior product is the inverse operation of the exterior derivative and is the mapping: $\iota_Y : \Lambda^k(\Omega_n) \to \Lambda^{k-1}(\Omega_n)$ for some vector field $Y \in \Omega_n$ and $0 \le k \le n$, defined as:

$$\iota_Y \alpha^{(k)}(X_2, \cdots, X_k) := \alpha^{(k)}(Y, X_2, \cdots, X_k) \quad \forall X_i, Y \in V \tag{5}$$

The Lie-derivative represents how forms change when they are altered by some vector field $\mathbf{v} \in \Omega_n$ and is the mapping: $\mathcal{L}_{\mathbf{v}} : \Lambda^k(\Omega_n) \to \Lambda^k(\Omega_n)$, see [11] and [13]. The Lie-derivative can be seen as the convection operator for differential geometry and is defined by applying Cartan's formula:

$$\mathcal{L}_{\mathbf{v}}\alpha^{(k)} = \iota_{\mathbf{v}} d\alpha^{(k)} + d\iota_{\mathbf{v}}\alpha^{(k)}. \tag{6}$$

The Hodge-star operator is a map between inner and outer oriented forms such that $\star : \Lambda^k(\Omega^n) \to \tilde{\Lambda}^{n-k}(\Omega^n)$, where $\sim$ denotes the space of opposite oriented forms. In this paper, the Hodge-star operator is used to construct the constitutive relations such that the approximation takes place here.

3 Mimetic Least-Squares Formulation

3.1 Mimetic Method

The mimetic formulation uses differential geometry as its cornerstone. The variable ϕ in (1), is represented by the inner oriented 0-form $\tilde{\phi}^{(0)}$. The Laplace operator working on a 0 form is constructed using the exterior derivative and Hodge star operator $\Delta \to d \star d$, which results in a 2-form. Since the diffusion term is represented by a 2-form, it is natural to also represent the source term with a 2-form, $f^{(2)}$. The term $\nabla \cdot (\mathbf{u}\phi)$ represents convection ϕ, which is naturally constructed using the Lie-derivative. One way of implementing this is to consider the 2-form $\star\phi^{(0)}$ for the convective term resulting in the following equation:

$$\mathcal{L}_{\mathbf{u}} \star \tilde{\phi}^{(0)} + d \star d\tilde{\phi}^{(0)} = f^{(2)}. \tag{7}$$

Using Cartans homotopy formula, (6), reduces the convective term to only one term, since $d \circ d \equiv 0$. This leads to the following equation:

$$d\iota_{\mathbf{u}} \star \tilde{\phi}^{(0)} + d \star d\tilde{\phi}^{(0)} = f^{(2)}. \tag{8}$$

This allows us to define the outer oriented 1-form $q^{(1)}$ in relation to the potential:

$$q^{(1)} = \iota_{\mathbf{u}} \star \tilde{\phi}^{(0)} + \star d\tilde{\phi}^{(0)}, \tag{9}$$

which can be interpreted as the total flux of the potential, i.e. the sum of diffusive and convective fluxes. A solution to the problem in (1), can then be obtained by solving a conservation equation and a constitutive relation. The conservation equation can be solved exactly while the approximation is introduced in the constitutive equation:

$$\nabla \cdot (\mathbf{u}\phi) + \Delta\phi = f \Leftrightarrow \begin{cases} dq^{(1)} = f^{(2)} \\ q^{(1)} = \iota_{\mathbf{u}} \star \tilde{\phi}^{(0)} + \star d\tilde{\phi}^{(0)}. \end{cases} \tag{10}$$

3.2 Mimetic Least-Squares Method

To establish the least-squares method we construct the functional $\mathcal{J}$, which squares the residual of (10):

$$\mathcal{J}(\tilde{\phi}^{(0)}, q^{(1)}; f^{(2)}) := \frac{1}{2}\left(\|dq^{(1)} - f^{(2)}\|_0^2 + \|q^{(1)} - \iota_{\mathbf{u}} \star \tilde{\phi}^{(0)} - \star d\tilde{\phi}^{(0)}\|_0^2 \right). \tag{11}$$

The least-squares method is a minimisation problem where the functional $\mathcal{J}$ is minimised by setting the derivative of $\mathcal{J}$ to zero. If we define $\tilde{G}_0^{(0)}$ as the space of all inner oriented 0-forms, satisfying the boundary conditions in (2), and $V^{(1)}$ as the space of all outer oriented 1-forms, then the variational formulation is obtained as: Find $\tilde{\phi}^{(0)} \in \tilde{G}_0^{(0)}$ and $q^{(1)} \in V^{(1)}$ such that:

$$\begin{aligned} (dp^{(1)}, dq^{(1)} - f^{(2)}) &= 0 \\ (p^{(1)} - \iota_{\mathbf{u}} \star \tilde{\varsigma}^{(0)} - \star d\tilde{\varsigma}^{(0)}, q^{(1)} - \iota_{\mathbf{u}} \star \tilde{\phi}^{(0)} - \star d\tilde{\phi}^{(0)}) &= 0 \end{aligned} \qquad \forall \tilde{\varsigma}^{(0)} \in \tilde{G}^{(0)}, p^{(1)} \in V^{(1)}. \tag{12}$$

4 Mimetic Spectral Discretization

In this section the discretization of the system is presented. The unknowns will be expanded using Lagrange polynomials [9] and edge polynomials [7]. The discrete representation of forms is obtained by mapping them onto discrete function spaces.

4.1 Discrete Representation of Forms

Consider the one dimensional domain $\Omega_1 := [-1, 1]$ on which $N+1$ Gauss-Lobatto-Legendre (GLL) nodes are defined: $-1 = x_0 < \cdots < x_N = 1$. On this grid we approximate the outer oriented forms, so that the outer oriented 0-form $a^{(0)}$ can then be represented by the approximate function $a^h(x)$ defined as:

$$a^h(x) = \sum_{i=0}^{N} a_i h_i(x), \tag{13}$$

where $h_i(x)$ are the Lagrange polynomials of order N, defined as $h_i(x_j) = \delta_{ij}$. The expansion coefficients are then equal to the 0-form evaluated in the nodes: $a_i = a^{(0)}(x_i)$. Now we consider the 1-form $b^{(1)}(x)$ on Ω_1, which can be represented by the function:

$$b^h(x) = \sum_{i=0}^{N-1} b_i e_i(x), \tag{14}$$

where $e_i(x)$ are the edge functions mentioned earlier, which have the property $\int_{x_i}^{x_{i+1}} e_j(x)dx = \delta_{ij}$. The expansion coefficients in (14) represent the integral values of the form, $b_i = \int_{x_i}^{x_{i+1}} b^{(1)}$.

The derivative of a function can also be taken by using the edge function, since the edge functions are defined in terms of derivative of the nodal expansion [7]:

$$\frac{d}{dx} a^h(x) = \sum_{i=0}^{N} a_i h_i'(x) = \sum_{i=0}^{N-1} (a_{i+1} - a_i) e_i(x). \tag{15}$$

Knowing the discrete representation of forms defined in 1D, we can to proceed to 2D by using the tensor product rule. On Fig. 1 the domain $[-1, 1]^2$ has been discretized using two grids, one using GLL nodes and one using GL nodes. We use the red grid where the nodes are denoted with squares to represent the outer oriented forms. The dual grid is used to represent the inner oriented forms. On the dual grid

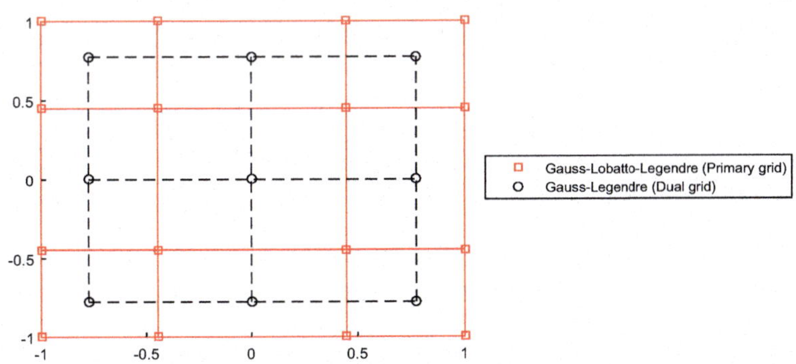

Fig. 1 Double grid configuration showing the 2D domain $[-1, 1]^2$ discretized with $N = 3$. The *red nodes and lines* denotes grid with GLL nodes whereas the *black* denotes the GL nodes

the expansion functions are denoted $\tilde{h}(x)$ and $\tilde{e}(x)$, which are derived using the GL nodes. The primary grid consists of $(N+1)^2$ grid points, (x_i, y_j) for $i, j = 0 : N$. The dual grid is constructed using $(N)^2$ grid points denoted $(\tilde{x}_i, \tilde{y}_j)$ for $i, j = 0 : N\text{-}1$.

The inner oriented 0-form $\tilde{\phi}^{(0)}$ is then be expanded using the dual grid:

$$\phi^h(x,y) = \sum_{i=0}^{N-1} \sum_{j=0}^{N-1} \phi_{ij} \tilde{h}_i(x) \tilde{h}_j(y), \tag{16}$$

The nodal expansion has the property that $\phi^h(\tilde{x}_i, \tilde{y}_j) = \phi_{ij} = \tilde{\phi}^{(0)}(\tilde{x}_i, \tilde{y}_j)$. We now introduce the discrete function space G^h, which is spanned by the basis of $\phi^h(x,y)$, i.e. $\phi^h(x,y) \in G^h$.

The outer oriented 1-form $q^{(1)}$ can be described as the flux of the potential across a line segment on the primary grid, which we represent in the finite setting as:

$$q^h(x,y) = \begin{cases} q_h^x(x,y) \\ q_h^y(x,y) \end{cases} = \begin{cases} \displaystyle\sum_{i=0}^{N} \sum_{j=0}^{N-1} q_{i,j}^x h_i(x) e_j(y) \\ \displaystyle\sum_{i=0}^{N-1} \sum_{j=0}^{N} q_{i,j}^y e_i(x) h_j(y). \end{cases} \tag{17}$$

In this situation the finite discretization of the form results in a vector field with a component in each direction and the basis of q^h is used to define the space V^h. As in the one dimensional case the expansion coefficients represent the integral values of the form:

$$q_{i,j}^x = \int_{x_i} \int_{y_j}^{y_{j+1}} q^{(1)} \qquad q_{i,j}^y = \int_{x_i}^{x_{i+1}} \int_{y_j} q^{(1)}. \tag{18}$$

In this case the expansion coefficients can then be interpreted as the flux over the line it is associated with. The same grid is also used to discretize outer oriented 2-forms so the 2-form $\rho^{(2)}$ can be approximated by $\rho^h(x,y) \in S^h$, which we define as:

$$\rho^h(x,y) = \sum_{i=0}^{N-1} \sum_{j=0}^{N-1} \rho_{ij} e_i(x) e_j(y). \tag{19}$$

Again the basis used here to define $\rho^h(x,y)$ span the space S^h. Since ρ^h represents a 2-form, the expansion coefficients relate to the integrated values of the form, over the associated area:

$$\rho_{i,j} = \int_{x_i}^{x_{i+1}} \int_{x_j}^{x_{j+1}} \rho^{(2)}. \tag{20}$$

This could be described as the expansion coefficient representing the mass, the form then expresses the mass density. The expression in (19) can also be written by introducing a matrix $\mathbb{A}_\rho$:

$$\bar{\rho}^{GL} = \mathbb{A}_\rho \bar{\rho}, \tag{21}$$

where $\bar{\rho}^{GL}$ contains the function $\rho^h(x, y)$ evaluated in the GL nodes and $\bar{\rho}$ is the expansion coefficients of ρ^h: $\bar{\rho} = [\rho_{0,0}, \cdots, \rho_{N-1,N-1}]^T$. Applying the discrete Hodge star operator on ρ^h leads to an inner oriented 0-from, i.e. the map $S^h \to G^h$. This is constructed by setting the values in the vector $\bar{\rho}^{GL}$ equal to the expansion coefficients for a 0-form on the dual grid. The matrix $\mathbb{A}_\rho$ can then be seen as the map from $S^h \to G^h$. The reverse map, $G^h \to S^h$, is then described by $\mathbb{A}_\rho^{-1}$.

4.2 Discrete Operators

First we consider the exterior derivative, which represents either the gradient, divergence or curl operator from vector calculus. The exterior derivative working on a 0-form results in a 1-form, so if we consider the 0-form from (16) and apply the same principles as used in (15) we get:

$$d\phi^h(x,y) = \begin{cases} \dfrac{d\phi^h(x,y)}{dx} = \displaystyle\sum_{i=0}^{N-2}\sum_{j=0}^{N-1} \left(\phi_{i+1,j} - \phi_{i,j}\right) \tilde{e}_i(x)\tilde{h}_j(y) \\ \dfrac{d\phi^h(x,y)}{dy} = \displaystyle\sum_{i=0}^{N-1}\sum_{j=0}^{N-2} \left(\phi_{i,j+1} - \phi_{i,j}\right) \tilde{h}_i(x)\tilde{e}_j(y). \end{cases} \tag{22}$$

In the least-squares minimisation problem in (12) we also apply the exterior derivative on an outer orientated 1-form $q^{(1)}$. In this case the discrete exterior derivative represents $d : V^h \to S^h$. The exterior derivative is applied to q^h using the same approach as in (15):

$$dq^h(x,y) = \sum_{i=0}^{N-1}\sum_{j=0}^{N-1} \left(q^x_{i+1,j} - q^x_{i,j} + q^y_{i,j+1} - q^y_{i,j}\right) e_i(x)e_j(y). \tag{23}$$

It is here seen that the basis of dq^h is the same as the one used in (19), which is a natural result since both dq^h and ρ^h is in S^h. Since the basis is the same, $dq^h = \rho^h$ can be expressed independently of the basis so we get an exact operation that can be represented by use of an incidence matrix:

$$\bar{\rho} = \mathbb{E}^{2,1}\bar{q}, \tag{24}$$

where $\bar{q} = [q^x_{0,0}, \cdots, q^x_{N-1,N}, q^y_{0,0} \cdots, q^y_{N,N-1}]^T$. The incidence matrix contains only 1, -1 and 0 and is non-square. For $N = 2$ the corresponding incidence matrix would be defined as:

$$\mathbb{E}^{2,1} = \begin{pmatrix} -1 & 1 & 0 & 0 & 0 & -1 & 0 & 1 & 0 & 0 & 0 \\ 0 & -1 & 1 & 0 & 0 & 0 & -1 & 0 & 1 & 0 & 0 \\ 0 & 0 & -1 & 1 & 0 & 0 & 0 & -1 & 0 & 1 & 0 \\ 0 & 0 & 0 & -1 & 1 & 0 & 0 & 0 & -1 & 0 & 1 \end{pmatrix}. \tag{25}$$

In order to compute the inner products used to establish the least-squares functional in (12), we need to construct the discrete version of this operation. The inner product on the finite space G^h can be defined by considering ϕ^h and $\varphi^h \in G^h$:

$$(\phi^h, \varphi^h) = \int_{-1}^{1} \int_{-1}^{1} \varphi^h(x,y)\phi^h(x,y)dxdy. \tag{26}$$

Integration is performed by applying Gauss-Lobatto integration so the integration is exact for polynomials of up to degree $2N - 1$, see [5]. By evaluating the expansion polynomials in the points described by the Gauss-Lobatto quadrature and using the weights from the quadrature, it is possible to construct a matrix $\mathbb{M}_G$ such that the inner product can be defined as:

$$(\phi^h, \varphi^h) = \bar{\phi}^T \mathbb{M}_G \bar{\varphi}, \tag{27}$$

where $\bar{\phi}$ and $\bar{\varphi}$ are column vectors containing the expansion coefficients of. The inner product on V^h can be constructed in a similar manner using the matrix $\mathbb{M}_V$. On S^h we denote the resulting matrix $\mathbb{M}_S$. For $\phi^h \in G^h$ and $\rho^h \in S^h$ we define the inner product:

$$(\phi^h, \rho^h) = \int_{-1}^{1} \int_{-1}^{1} \phi^h(x,y)\rho^h(x,y)dxdy = \bar{\phi}^T \mathbb{M}_{GS} \bar{\rho}. \tag{28}$$

We now construct the adjoint gradient of $p^h \in G^h$, which enables us to invoke the boundary condition in a weak sense. This operator is a map $\star d : G^h \to V^h$, and is constructed using integration by parts:

$$(\star dp^h, v^h) = -(p^h, dv^h) + \int_{\partial\Omega} p^h(v^h \cdot n)d\Omega \quad \forall v^h \in V^h. \tag{29}$$

Consider the space $G^h_0 \subset G^h$, which contains all functions within G^h that satisfies the boundary condition given in (2), then for $\phi^h \in G^h_0$ the last term in (29) vanishes such that:

$$
\begin{aligned}
(-\star\, d\phi^h, v^h) = (\phi^h, dv^h) &= \bar{\phi}^T \mathbb{M}_{GS} \mathbb{E}^{2,1} \bar{v} \\
&= \bar{\phi}^T \mathbb{M}_{GS}\, \mathbb{E}^{2,1}\, \mathbb{M}_V^{-1} \mathbb{M}_V \bar{v} \\
&= (\underbrace{\mathbb{M}_V^{-1} \mathbb{E}^{2,1T} \mathbb{M}_{GS}^T \bar{\phi}}_{-\star d\phi^h}\,)^T \mathbb{M}_V \bar{v},
\end{aligned}
\tag{30}
$$

with $\bar{v}$ containing the expansion coefficients of v^h.

Now consider the convective flux represented by $\gamma^{(1)} \in V^h$, which is expanded in the same way as the 1-form in (16). This flux is defined by applying the interior derivative on a 2-form, which leads to a 1-form. If $\mathbf{u} = u_x \frac{\partial}{\partial x} + u_y \frac{\partial}{\partial y}$, then applying the interior product on ρ^h leads to:

$$
\iota_{\mathbf{u}} \rho^h = \sum_{i=0}^{N-1} \sum_{j=0}^{N-1} \rho_{ij} (u_x \varepsilon_i(x) e_j(y) - u_y e_i(x) \varepsilon_j(y)), \tag{31}
$$

since $u_x \frac{\partial}{\partial x} e_i(x) = u_x \frac{\partial}{\partial x} \varepsilon_i(x) dx = u_x \varepsilon_i(x)$ and $u_x \frac{\partial}{\partial x} e_j(y) = 0$.

Instead of calculating the expansion coefficients by integrating $\gamma^{(1)}$ as it was done in (18), we now integrate $\iota_{\mathbf{u}} \rho^{(2)}$:

$$
\gamma_{i,j}^x = \int_{x_i} \int_{y_j}^{y_{j+1}} \iota_{\mathbf{u}} \rho^{(2)} \qquad \gamma_{i,j}^y = \int_{x_i}^{x_{i+1}} \int_{y_j} \iota_{\mathbf{u}} \rho^{(2)}. \tag{32}
$$

The integrals above are calculated using the property of the edge functions, which means that $\int_{x_i}^{x_{i+1}} e_j(x) = \delta_{ij}$. Using this property the following relations are obtained:

$$
\gamma_{ij}^x = u_x(x_i, y_j) \sum_{k=0}^{N-1} \rho_{kj} \varepsilon_k(x_i) \qquad \gamma_{ij}^y = u_y(x_i, y_j) \sum_{l=0}^{N-1} \rho_{il} \varepsilon_l(y_j). \tag{33}
$$

If we then construct matrices $\mathbb{A}_x$ and $\mathbb{A}_y$ such that the following relations can be established: $\bar{\gamma}^x = \mathbb{A}_x \bar{\rho}$ and $\bar{\gamma}^y = \mathbb{A}_y \bar{\rho}$, where $\bar{\gamma}^x$, $\bar{\gamma}^y$ and $\bar{\rho}$ are column vectors containing the expansion coefficients. Then we can construct the matrix $\mathbb{E}_A$ such that:

$$
\bar{\gamma} = \begin{bmatrix} \mathbb{A}_x \\ \mathbb{A}_y \end{bmatrix} \bar{\rho} = \mathbb{E}_A \bar{\rho}, \tag{34}
$$

which describes the finite representation of the mapping $S^h \to V^h$.

4.3 Coefficient Matrix

Now we establish the matrix vector system, which can be solved in order to obtained a solution to (1):

$$\mathbb{A}\bar{x} = \bar{b}, \tag{35}$$

where $\bar{x}$ is a column vector containing all the degrees of freedom in the system defined as $\bar{x} = [\bar{\phi}, \bar{q}]^T$. In order to construct the coefficient matrix $\mathbb{A}$ we use the coefficient matrices introduced above to construct the terms from (12):

$$\begin{aligned}
(\iota_{\mathbf{u}} \star \varsigma^h, \iota_{\mathbf{u}} \star \phi^h) &\Rightarrow \mathbb{A}_1 = (\mathbb{A}_\rho^{-1})^T \mathbb{E}_A^T \mathbb{M}_V \mathbb{E}_A \mathbb{A}_\rho^{-1} \\
(\iota_{\mathbf{u}} \star \varsigma^h, \star d\phi^h) &\Rightarrow \mathbb{A}_2 = -(\mathbb{A}_\rho^{-1})^T \mathbb{E}_A^T \mathbb{M}_V \mathbb{M}_V^{-1} \mathbb{E}^{2,1T} \mathbb{M}_{GS}^T \\
(\iota_{\mathbf{u}} \star \varsigma^h, q^h) &\Rightarrow \mathbb{A}_3 = (\mathbb{A}_\rho^{-1})^T \mathbb{E}_A^T \mathbb{M}_V \\
(\star d\varsigma^h, \star d\phi^h) &\Rightarrow \mathbb{A}_4 = \mathbb{M}_{GS}\, \mathbb{E}^{2,1} \mathbb{M}_V^{-1} \mathbb{M}_V \mathbb{M}_V^{-1} \mathbb{E}^{2,1T} \mathbb{M}_{GS}^T \\
(p^h, q^h) &\Rightarrow \mathbb{A}_5 = \mathbb{M}_V \\
(\star d\varsigma^h, q^h) &\Rightarrow \mathbb{A}_6 = -\mathbb{M}_{GS}\, \mathbb{E}^{2,1} \mathbb{M}_V^{-1} \mathbb{M}_V = -\mathbb{M}_{GS}\, \mathbb{E}^{2,1} \\
(dp^h, dq^h) &\Rightarrow \mathbb{A}_7 = \mathbb{E}^{2,1^T} \mathbb{M}_S \mathbb{E}^{2,1}.
\end{aligned}$$

for $\forall \varsigma^h \in G_0^h$ and $p^h \in V^h$. These terms are collected in $\mathbb{A}$ in the following way:

$$\mathbb{A} = \begin{pmatrix} \mathbb{A}_1 - \mathbb{A}_2 - \mathbb{A}_2^T + \mathbb{A}_4 & -\mathbb{A}_3 + \mathbb{A}_6 \\ -\mathbb{A}_3^T + \mathbb{A}_6^T & \mathbb{A}_5 + \mathbb{A}_7 \end{pmatrix}, \qquad \bar{b} = \begin{pmatrix} 0 \\ \mathbb{E}^{2,1T} \mathbb{M}_S \bar{f} \end{pmatrix}.$$

Here $\bar{f}$ is a column vector containing the source function values integrated over each area on the primary grid, corresponding to the expansion coefficients in (20).

5 Numerical Results

The method has been tested on model problems in order to observe convergence and conservation properties. The spectral mimetic least-squares (MLS) method is compared to a traditional spectral least-sqaures (LS) method. The model problem solved here has the analytical solution of (1) defined as:

$$\phi_{exact}(x, y) = (-1 + x^2)(-1 + y^2)\, \sin(\frac{1}{2}\pi x). \tag{36}$$

The velocity field is defined as $\mathbf{u} = [1, 1]$, and the source term is found by taking $f(x, y) = \nabla \cdot (\mathbf{u}\phi_{exact}) + \Delta\phi_{exact}$.

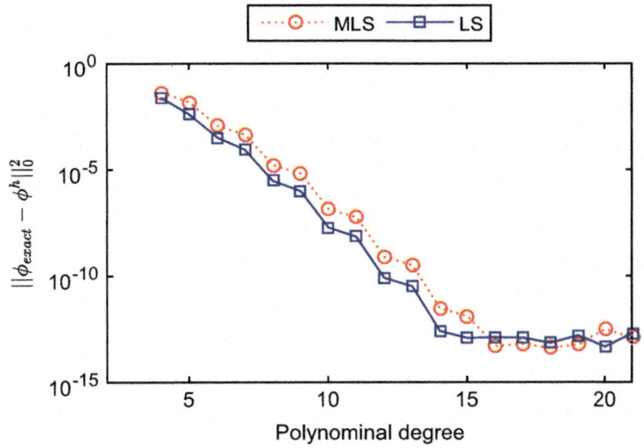

Fig. 2 Convergence comparison between MLS and LS for the model problem

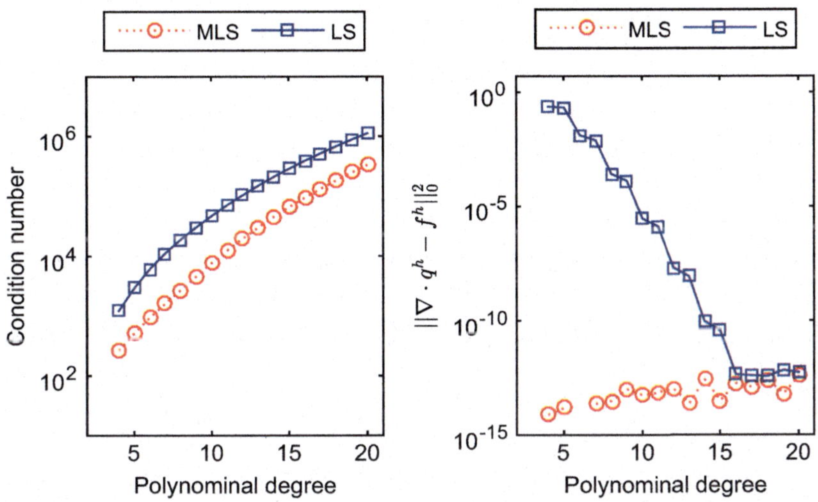

Fig. 3 *Left*: Condition number of the coefficient matrix used to solve the system. *Right*: Conservation equation error for the two methods

In Fig. 2 the error of the numerical solution compared to the analytical solution is shown. It is seen that both methods give rise to exponential convergence. The MLS method achieves the same convergence rate as the standard LS method. At around a polynomial order of 15 the LS method has converged and at $N = 16$ the MLS also has converged.

In Fig. 3 the right plot shows the residual of the conservation part of (10). It is seen that for the LS method the error decreases exponentially for increasing polynomial order which is expected. For the MLS method the error is independent of the polynomial order and is satisfied to around machine precision for even the coarsest grid. The small increase in residual is due to increase in condition number.

The left plot in Fig. 3 shows how the condition number varies as a function of the polynomial order. Here, it is seen that the MLS method has a smaller condition number for all polynomial orders.

6 Conclusion

In this paper we present a spectral mimetic least-squares method. We show that by encapsulating the underlying geometric properties in the problem, we are able to discretize the convection-diffusion problem such that the invariant is conserved both globally and locally. The topological part of the problem can be satisfied to machine precision. By using a least-squares approach the method results in a positive definite symmetric coefficient matrix which can be solved using well established iterative solvers.

References

1. R. Abraham, J.E. Marsden, T. Ratiu, *Manifolds, Tensor Analysis, and Applications*. Applied Mathematical Sciences (Springer, New York, 1988)
2. P. Bochev, M. Gunzburger, On least-squares finite element methods for the Poisson equation and their connection to the Dirichlet and Kelvin principles. SIAM J. Numer. Anal. **43**(1), 340–362 (2005). ISSN: 0036–1429
3. P. Bochev, J.M. Hyman, Principles of mimetic discretizations of differential operators, in *Compatible Spatial Discretizations*, ed. by D.N. Arnold et al. (Springer, New York, 2006)
4. W.L. Burke, *Applied Differential Geometry* (Cambridge University Press, Cambridge, 1985)
5. C. Canuto et al., *Spectral Methods: Fundamentals in Single Domains*, 1st edn. Scientific Computation (Springer, Berlin, 2007)
6. M. Desbrun et al. Discrete exterior calculus. ArXiv Mathematics e-prints (2005)
7. M. Gerritsma, Edge functions for spectral element methods, in *Spectral and High Order Methods for Partial Differential Equations: Selected Papers from the ICOSAHOM '09 Conference, June 22–26, Trondheim, Norway*, ed. by S. Jan Hesthaven, M. Einar Rønquist (Springer, Berlin, 2011), pp. 199–207
8. M. Gerritsmam, P. Bochev, A spectral mimetic least-squares method for the Stokes equations with no-slip boundary condition. Comput. Math. Appl. **71**(11), 2285–2300 (2016)
9. G. Karniadakis, S. Sherwin, *Spectral/hp Element Methods for Computational Fluid Dynamics*, 2nd edn. (Oxford University Press, Oxford, 2005), pp. 1–650
10. C. Mattiussi, An analysis of finite volume, finite element, and finite difference methods using some concepts from algebraic topology. J. Comput. Phys. **133**(2), 289–309 (1997)
11. A. McInerney, *First Steps in Differential Geometry: Riemannian, Contact, Symplectic*, 1st edn. Undergraduate Texts in Mathematics (Springer, New York, 2013)
12. M.M.J. Proot, M. Gerritsma, A least-squares spectral element formulation for the Stokes problem. J. Sci. Comput. **17**, 285–296, (2002)
13. L.W. Tu, *An Introduction to Manifolds*, 2nd edn. Universitext (Springer, New York, 2011)

Krylov Subspace Spectral Methods with Coarse-Grid Residual Correction for Solving Time-Dependent, Variable-Coefficient PDEs

Haley Dozier and James V. Lambers

Abstract Krylov Supspace Spectral (KSS) methods provide an efficient approach to the solution of time-dependent, variable-coefficient partial differential equations by using an interpolating polynomial with frequency-dependent interpolation points to approximate a solution operator for each Fourier coefficient. KSS methods are high-order accurate time-stepping methods that also scale effectively to higher spatial resolution. In this paper, we will demonstrate the effectiveness of using coarse-grid residual correction, generalized to the time-dependent case, to improve the accuracy and efficiency of KSS methods. Numerical experiments demonstrate the effectiveness of this correction.

1 Introduction

Consider a time-dependent, variable-coefficient PDE, such as

$$u_t + Lu = 0, \quad 0 < x < 2\pi, \quad t > 0 \tag{1}$$

$$u(x,0) = f(x), \qquad 0 < x < 2\pi, \tag{2}$$

where L is a second order, self-adjoint, positive definite differential operator, such as a Sturm-Liouville operator. This type of problem often poses difficulties for both implicit and explicit time-stepping methods due to the lack of scalability of these methods caused by stiffness. That is, unless the chosen time-step is sufficiently small, the computed solutions might exhibit nonphysical behavior with large input sizes [7].

Krylov Subspace Spectral (KSS) methods [9] are designed specifically for solving time-dependent, variable-coefficient problems. The main idea behind KSS

H. Dozier (✉) • J.V. Lambers
Department of Mathematics, The University of Southern Mississippi, 118 College Drive #5045, Hattiesburg, MS 39406, USA
e-mail: Haley.Dozier@usm.edu; James.Lambers@usm.edu

M.L. Bittencourt et al. (eds.), *Spectral and High Order Methods for Partial Differential Equations ICOSAHOM 2016*, Lecture Notes in Computational Science and Engineering 119, DOI 10.1007/978-3-319-65870-4_22

methods is to use an interpolating polynomial with frequency-dependent interpolation points to approximate the solution operator for each Fourier coefficient. As a result, KSS methods exhibit a high order of accuracy and stability. The dilemma is that this approach is only practical when applied to the high frequency components of the solution, and so a less efficient approach, such as standard Krylov projection, must be used on the low frequency components.

We have found that with the addition of coarse-grid residual correction, we can eliminate low frequency components of the error by restricting the problem to a coarser grid, and then using KSS methods on that coarser grid. In this paper, an overview of KSS methods in its current form will be given, as well as a description of how the addition of coarse-grid residual correction can be added to improve the accuracy of KSS. Numerical results will demonstrate this improvement.

The outline of this paper is as follows, Sect. 2 reviews KSS methods. Section 3 will discuss the addition of coarse-grid residual correction to KSS. Numerical results will be presented in Sect. 4, and conclusions will be given in Sect. 5.

2 Krylov Subspace Spectral Methods

We start by examining the parabolic PDE $u_t + Lu = 0$ on $(0, 2\pi)$ with periodic boundary conditions $u(0,t) = u(2\pi, t)$. The solution of this PDE can be represented by the Fourier series

$$u(x,t) = \frac{1}{\sqrt{2\pi}} \sum_{\omega=-\infty}^{\infty} e^{i\omega x}\hat{u}(\omega,t). \tag{3}$$

To find the Fourier coefficients of the solution of the solution at time t^{n+1}, we can represent them using the standard inner product on $(0, 2\pi)$,

$$\hat{u}(\omega, t_{n+1}) = \left\langle \frac{1}{\sqrt{2\pi}} e^{i\omega x}, e^{-L\Delta t} u(x, t_n) \right\rangle, \tag{4}$$

where $e^{-L\Delta t}$ is the exact solution operator.

The main idea behind KSS methods, as first described in [11], is to independently approximate all Fourier coefficients of the solution using an approximation of the exact solution operator that is tailored to each Fourier coefficient. To approximate the exact solution operator, spatially discretizing the right hand side of (4) leads to

$$\hat{u}(\omega, t_{n+1}) \approx \left(\frac{\Delta x}{\sqrt{2\pi}} e^{i\omega \mathbf{x}} \right)^H \left(e^{-L_N \Delta t} \mathbf{u}(x, t_n) \right) \tag{5}$$

where L_N is a $N \times N$ symmetric positive definite matrix obtained from spatial discretization of L using finite differences, and $\mathbf{x}$ is a vector of equally spaced points in $[0, 2\pi)$ with spacing $\Delta x = 2\pi/N$. If we let $\mathbf{u} = \frac{\Delta x}{\sqrt{2\pi}} e^{i\omega \mathbf{x}}$, $\mathbf{v} = u(\mathbf{x}, t_n)$, and $\phi(L_N) = e^{-L_N \Delta t}$, then we can represent the right side of (5) by the bilinear form

$$\mathbf{u}^H \phi(L_N) \mathbf{v}. \tag{6}$$

The matrix L_N is symmetric positive definite, and therefore has positive, real eigenvalues $b = \lambda_1 \geq \lambda_2 \geq \ldots \geq \lambda_N = a$, and orthonormal eigenvectors $\mathbf{q}_j$ where $j = 1, \ldots, N$. Then the spectral decomposition of (6) is

$$\mathbf{u}^H \phi(L_N) \mathbf{v} = \sum_{j=1}^{N} \phi(\lambda_j) \mathbf{u}^H \mathbf{q}_j \mathbf{q}_j^H \mathbf{v}. \tag{7}$$

From here, we define the measure $\alpha(\lambda)$ by

$$\alpha(\lambda) = \begin{cases} 0 & \text{if } \lambda < a \\ \sum_{j=i}^{N} \mathbf{u}^H \mathbf{q}_j \mathbf{q}_j^H \mathbf{v} & \text{if } \lambda_i \leq \lambda \leq \lambda_{i-1} \\ \sum_{j=1}^{N} \mathbf{u}^H \mathbf{q}_j \mathbf{q}_j^H \mathbf{v} & \text{if } b \leq \lambda. \end{cases} \tag{8}$$

Now, as shown in [5], the bilinear form in (6) can be expressed as a Riemann-Stieltjes integral

$$\mathbf{u}^H \phi(L_N) \mathbf{v} = \int_u^b \phi(\lambda)\, d\alpha(\lambda). \tag{9}$$

To approximate this integral, Gaussian quadrature is used because it has a high degree of accuracy, and the weights are guaranteed to be positive if the measure $\alpha(\lambda)$ is positive and increasing [5]. After applying Gaussian quadrature to (9) the following approximation can be obtained

$$\int_a^b \phi(\lambda)\, d\alpha(\lambda) = \sum_{j=1}^{K} \phi(\lambda_j) w_j + error \tag{10}$$

where the nodes are λ_j and weights are w_j, for $j = 1, \ldots, K$. This quadrature rule is exact for polynomials of degree up to $2K - 1$ [5].

In the case where $\mathbf{u} = \mathbf{v}$, the nodes and weights for Gaussian quadrature can be obtained using the symmetric Lanczos algorithm applied to L_N using initial vector $\mathbf{u}$. In the case where $\mathbf{u} \neq \mathbf{v}$, the weights for Gaussian quadrature, w_j, are not always guaranteed to be positive real numbers. This occurrence can destabilize the quadrature rule as shown in [1]. In this case, we consider a block approach:

$$[\mathbf{u}\ \mathbf{v}]^H \phi(L_N) [\mathbf{u}\ \mathbf{v}]. \tag{11}$$

We can represent this matrix as the Riemann-Stieltjes integral

$$\int_a^b \phi(\lambda)\, d\mu(\lambda) = \begin{bmatrix} \mathbf{u}^H \phi(L_N)\mathbf{u} & \mathbf{u}^H \phi(L_N)\mathbf{v} \\ \mathbf{v}^H \phi(L_N)\mathbf{u} & \mathbf{v}^H \phi(L_N)\mathbf{v} \end{bmatrix}, \tag{12}$$

where $\mu(\lambda)$ is a 2×2 matrix with entries of the form $\alpha(\lambda)$ from (8) [5]. Then a quadrature rule approximates the integral (12) as follows:

$$\int_a^b \phi(\lambda)\, d\mu(\lambda) \approx \sum_{j=1}^{2K} \phi(\lambda_j)\mathbf{v}_j\mathbf{v}_j^H + error. \tag{13}$$

where each λ_j is a scalar and each $\mathbf{v}_j$ is a 2-vector.

The block Lanczos algorithm applied to L_N using initial block $[\mathbf{u}\ \mathbf{v}]$ yields the nodes and weights for block Gaussian quadrature [6]. Specifically, block Lanczos produces the block tridiagonal matrix with 2×2 blocks

$$\mathscr{T}_K = \begin{bmatrix} M_1 & B_1^T & & \\ B_1 & M_2 & B_2^T & \\ & \ddots & \ddots & \ddots \\ & & B_{K-1} & M_K \end{bmatrix}, \tag{14}$$

where each B_j is upper triangular. The eigenvalues of $\mathscr{T}_K$ are used as the nodes λ_j in (13), and $\mathbf{v}_j\mathbf{v}_j^H$ are the matrix-valued "weights", where $\mathbf{v}_j$ consists of the first two components of the normalized eigenvector corresponding to λ_j.

A time step of block KSS, as seen in [11], proceeds as follows. First, we define

$$R_0(\omega) = [\hat{\mathbf{e}}_\omega\ \mathbf{u}^n] \tag{15}$$

where $\hat{\mathbf{e}}_\omega$ is a discretization of $\frac{\Delta x}{\sqrt{2\pi}} e^{i\omega\mathbf{x}}$ on a uniform N-point grid, and $\mathbf{u}^n$ is the computed solution at time t_n (these are $\mathbf{u}$ and $\mathbf{v}$ in (11) above). The QR factorization of (15) leads to

$$R_0(\omega) = X_1(\omega)B_0(\omega) \tag{16}$$

with

$$X_1(\omega) = \begin{bmatrix} \hat{\mathbf{e}}_\omega & \frac{\mathbf{u}_\omega^n}{||\mathbf{u}_\omega^n||^2} \end{bmatrix}, \quad B_0(\omega) = \begin{bmatrix} 1 & \hat{\mathbf{e}}_\omega^H\mathbf{u}^n \\ 0 & ||\mathbf{u}_\omega^n||^2 \end{bmatrix},$$

where

$$\mathbf{u}_\omega^n = \mathbf{u}^n - \hat{\mathbf{e}}_\omega\hat{\mathbf{e}}^H\mathbf{u}^n = \mathbf{u}^n - \hat{\mathbf{e}}_\omega\hat{u}(\omega, t_n). \tag{17}$$

Block Lanczos is then applied to the discretized operator, L_N, with initial block $X_1(\omega)$. From block Lanczos, we obtain our M_j and B_j so that we can produce the matrix $\mathscr{T}_K(\omega)$ with the same form as (14), the entries of which depend on ω. Then, each Fourier coefficient of the solution at time t_{n+1} can be approximated by

$$[\hat{\mathbf{u}}^{n+1}]_\omega = [B_0^H(\omega)E_{12}^H e^{-\mathscr{T}_K(\omega)\Delta t} E_{12}B_0(\omega)]_{12}, \quad E_{12} = [\,\mathbf{e}_1\ \mathbf{e}_2\,] \tag{18}$$

By applying an inverse Fast Fourier Transform (FFT) to the vector of Fourier coefficients, we obtain the vector $\mathbf{u}^{n+1}$, which approximates the solution $u(x, t_{n+1})$. In [11] it was shown that this algorithm has local temporal accuracy of $O(\Delta t^{2K-1})$ for the parabolic problem and in [10] it was shown to have local temporal accuracy $O(\Delta t^{4K-2})$ for the second-order wave equation.

To improve the efficiency of block KSS methods, asymptotic analysis of block Lanczos iteration was performed in [3, 14]. It was shown that at high frequencies, the eigenvalue problem for $\mathscr{T}_K(\omega)$ approximately decouples, so that the Gaussian quadrature nodes could instead be estimated by performing "non-block" Lanczos on L_N with initial vectors $\hat{\mathbf{e}}_\omega$ and $\mathbf{u}^n$, which yields *frequency-dependent* and *frequency-independent nodes*, respectively.

This improves efficiency for two reasons. First, the frequency-independent nodes need only be computed once per time step, and shared by all quadrature rules of the form (13) for each ω. Second, the entries of the Jacobi matrix obtained by applying Lanczos with to L_N initial vector $\hat{\mathbf{e}}_\omega$ can easily be estimated in terms of the coefficients of the underlying differential operator L.

With these enhancements taken into account, the following algorithm from [3] describes a time step of KSS on $[t_n, t_{n+1}]$ to solve $u_t + Lu = 0$, on an N-point uniform grid, with periodic boundary conditions, and $O(\Delta t^{2K-1})$ accuracy in time.

1. Perform K iterations of Lanczos on L_N with initial vector $\mathbf{u}^n$ to obtain Jacobi matrix T_K. The eigenvalues $\lambda_1, \ldots, \lambda_K$ of T_K are the frequency-independent nodes.
2. For each $\omega = -N/2 + 1, \ldots, N/2$, compute the frequency-dependent nodes $\lambda_{1,\omega}, \ldots, \lambda_{K,\omega}$ from analytically computed estimates of the entries of $T_K(\omega)$, obtained through K iterations of Lanczos on L_N with initial vector $\hat{\mathbf{e}}_\omega$ [3, 14].
3. For each $\omega = -N/2 + 1, \ldots, N/2$, compute the polynomial interpolant

$$p_{2K-1,\omega}(\lambda) = \sum_{j=0}^{2K-1} c_{j,\omega}\lambda^j$$

of $\phi(\lambda) = e^{-\lambda\Delta t}$, with interpolation points $\lambda_1, \ldots, \lambda_K, \lambda_{1,\omega}, \ldots, \lambda_{K,\omega}$.
4. Each Fourier coefficient of the solution at time t_{n+1} is then computed as follows:

$$[\mathbf{u}^{n+1}]_\omega = \sum_{j=0}^{2K-1} c_{j,\omega}\hat{\mathbf{e}}_\omega^H L^j \mathbf{u}^n. \tag{19}$$

FFTs are used to compute Fourier coefficients of $L^j\mathbf{u}^n$, but by performing Newton interpolation in step 3 with the frequency-independent nodes listed first, the number of FFTs can be reduced from $2K$ in (19) to K [3].
5. Perform an inverse FFT to obtain the solution at time t_{n+1}.

3 KSS with Coarse-Grid Residual Correction

The "smoothing" property [2] is a phenomenon that many iterative methods possess. This property describes a method in which the rapid decrease in error in earlier iterations is due to the elimination of higher frequency error. Methods with the smoothing property are often not as effective at eliminating low frequency error.

Similarly, due to the work of Cibotarica, Lambers, and Palchak in [3, 14], KSS methods are already highly effective at computing the high frequency components of the solution, but since the asymptotic analysis in these works applied only to high-frequency components, they are not as effective at eliminating low frequency error.

Solving a PDE on a coarse grid is an effective way to eliminate low-frequency error in linear systems that arise from the spatial discretization of elliptic partial differential equations. To apply this multigrid-inspired technique to a time-dependent problem of the form $u_t + Lu = 0$, we first define the residual as $R = u_t + Lu$, and then solve a non-homogeneous version of the PDE to estimate the error for correction.

We therefore need three functions to implement coarse grid residual correction generalized to the time-dependent PDE:

- a function to restrict the problem to a coarser grid
- a function to discretize the spatial differential operator on the coarse grid, and
- a function to interpolate back to the fine grid

The Krylov Subspace Spectral method with Coarse-Grid Residual Correction (KSS-CG) proceeds as follows, during *each* time step:

1. use KSS as described in Sect. 2 to compute an initial solution and residual,
2. use a FFT to restrict the residual to a coarse grid,
3. compute a correction on the coarse grid by solving the same PDE, but with the residual as a source term,
4. use a FFT to transfer the correction to the fine grid, and
5. add the correction to the initial solution from step 1.

It should be noted that since the test cases we use have homogeneous Neumann boundary conditions instead of periodic, we will be using a discrete cosine transform that employs FFTs.

In [4] it is shown why multigrid methods are ineffective for nonelliptic problems such as the Helmholtz equation, as any choice of a relaxation parameter results in an amplification of some modes. In [12], KSS methods were applied to the Helmholtz equation, and difficulties arose due to a singularity in the integrand $\phi(\lambda) = 1/\lambda$ in (9) resulting from the indefiniteness of the underlying matrix.

In both cases, the PDE is being solved on the entire (spatial) domain through the solution of a single system of linear equations; by contrast, KSS-CG is not solving a nonelliptic PDE on the entire (space-time) domain simultaneously. As described above, the algorithm is essentially a time-stepping method, with residual correction performed after each time step on a coarser spatial grid.

3.1 *Using KSS-CG to Solve a Parabolic PDE*

Consider the parabolic PDE

$$u_t + Lu = 0, \quad (0, 2\pi) \times (0, \infty), \tag{20}$$

$$u(x, 0) = f(x), \quad 0 < x < 2\pi, \tag{21}$$

with either periodic or homogeneous boundary conditions. In this section, we restrict ourselves to one space dimension for concreteness; a 2-D problem is considered in the results section.

As previously stated, multigrid can be used to improve the accuracy of iterative methods that have the smoothing property. After KSS is applied during a single time step as described in Sect. 2, we are left with a relatively smooth error. To perform residual correction, first the solution computed from KSS is used to find the residual, $R(x, t) = u_t(x, t) + Lu(x, t)$. This entails using the time derivative of the solution operator, in this case $S(t) = e^{-Lt}$, to compute u_t. That is, the same Gaussian quadrature rules are used as with computing the solution itself, but with the integrand $f(\lambda) = -\lambda e^{-\lambda t}$.

To restrict the residual to a coarse grid, the low-frequency components of its discrete Fourier transform are extracted. Once the residual is restricted to the coarse grid, the differential operator L must also be restricted to the coarse grid. Then, the non-homogeneous equation

$$e_t + Le = R(x, t) \tag{22}$$

must be solved where e is the error, and the initial condition is $e(x, 0) = e_0 = 0$. It follows that

$$e(x, t) = \int_0^t e^{-L(t-s)} R(x, s)\, ds. \tag{23}$$

If we use Gaussian quadrature to approximate the integral in (23), we obtain the error estimate

$$e(x, \Delta t) = \int_0^{\Delta t} e^{-L(\Delta t - s)} R(x, s) ds \approx \sum_{k=1}^{m} w_k e^{-L(\Delta t - s_k)} R(x, s_k) \tag{24}$$

where the s_k are the Gauss-Legendre points, transformed to the interval $[0, \Delta t]$, and the w_k are the weights transformed to the same interval. To correct the solution, the newly obtained error estimate can be interpolated back to the fine grid by padding its discrete Fourier transform with zeros.

A straightforward modification of the above algorithm to perform multiple coarse-grid corrections would have the drawback that with each correction, the total number of quadrature nodes in time would increase substantially, because the residual of each term in each correction would have to be evaluated at m times, where m is the number of nodes used in the quadrature rule in (24). To avoid the resulting increase in computational expense, future work will focus on coarsening in both space and time to make multiple corrections practical.

3.2 Using KSS-CG to Solve a Hyperbolic Problem

Consider the hyperbolic PDE

$$u_{tt} = Lu, \quad (0, 2\pi) \times (0, \infty), \tag{25}$$

$$u(x, 0) = f(x), \quad u_t(x, 0) = g(x), \quad 0 < x < 2\pi. \tag{26}$$

The solution operator for this problem can be expressed as a matrix of functions of the operator L:

$$\begin{bmatrix} u(x, t + \Delta t) \\ u_t(x, t + \Delta t) \end{bmatrix} = \begin{bmatrix} \cos(\sqrt{-L}\Delta t) & \frac{1}{\sqrt{-L}} \sin(\sqrt{-L}\Delta t) \\ -\sqrt{-L} \sin(\sqrt{-L}\Delta t) & \cos(\sqrt{-L}\Delta t) \end{bmatrix} \begin{bmatrix} u(x, t) \\ u_t(x, t) \end{bmatrix}. \tag{27}$$

The entries of the propagator matrix in (27) indicate which functions are the integrands in the Riemann-Stieltjes integrals that are used to compute the Fourier coefficients of the solution [13].

The residual, R, computed at each time step is

$$R = u_{tt} - Lu,$$

where the second time derivative of the solution operator from Sect. 2 is used to compute u_{tt}. That is, the second derivatives of the matrix functions in (27) with respect to Δt are used as integrands in the required Riemann-Stieltjes integrals. Then, the error used to update the solution is obtained by solving

$$e_{tt} = Le + R(x,t)$$

which yields

$$e(x,t_{n+1}) = \int_0^{\Delta t} \frac{1}{\sqrt{-L}} \sin(\sqrt{-L}(\Delta t - s))R(x,s)\,ds. \tag{28}$$

This error estimate and its time derivative are then interpolated back to the fine grid by padding their discrete Fourier transforms with zeros, as in the parabolic case.

4 Numerical Results

In this section, the effectiveness of KSS with coarse grid residual correction (KSS-CG) will be demonstrated. The following approaches will be compared:

- KSS method as described in Sect. 2.
- KSS method with coarse grid residual correction, as described in Sect. 3, using 2 Gaussian quadrature nodes (KSS-CG2). This will only be done in the parabolic case.
- KSS method with coarse grid residual correction, as described in Sect. 3, using 4 Gaussian quadrature nodes (KSS-CG5). This will only be done in the hyperbolic case.
- KSS method with coarse grid residual correction using 3 Gaussian quadrature nodes (KSS-CG3)
- Krylov projection as described in [8] (KP)
- KSS-EPI method as described in [3].

The errors reported are relative errors with respect to an "exact solution" using the MATLAB ODE solver `ode15s`, computed using the smallest allowable time step. For each test problem we use grid sizes of $N = 50, 150$ grid points per dimension to demonstrate how increased spatial resolution will affect the performance of each method. For KSS-CG, a grid with $N = 25$ grid points per dimension is used for residual correction.

4.1 Parabolic Problem

We first compare all five methods when solving the 2-D parabolic problem

$$u_t = \alpha\Delta u + (1 - 3u_0^2)u, \tag{29}$$

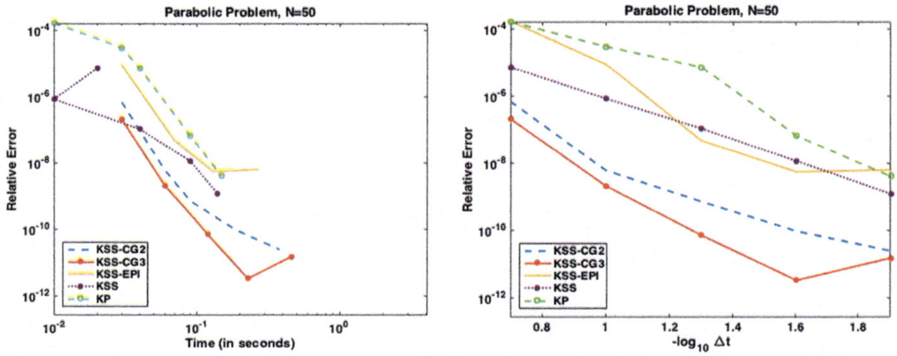

Fig. 1 Time of each timestep for each method vs error with grid sizes $N = 50$ points per dimension. The *blue dashed curve* represents KSS-CG using 2 Gaussian quadrature nodes, the *red solid and star curve* represents KSS-CG using 3 Gaussian quadrature nodes, the *yellow dash-dot curve* represents KSS-EPI, the *purple dot-star curve* represents the KSS method with filtering (but without correction), and the *green dash-circle curve* represent standard Krylov Projection

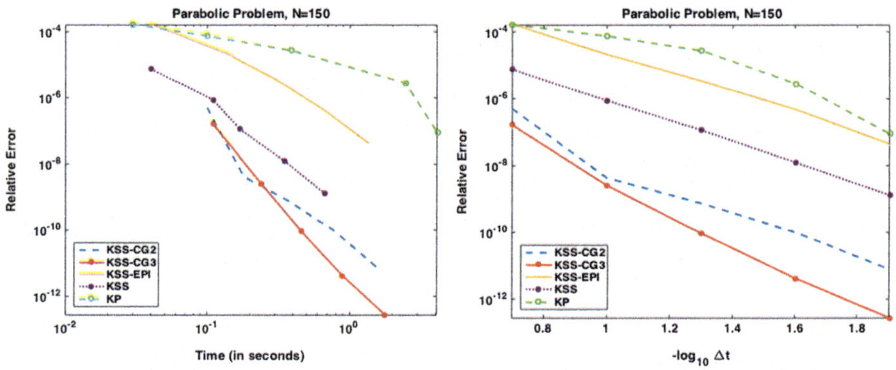

Fig. 2 Time of each timestep for each method vs error with grid sizes $N = 150$ points per dimension. The *blue dashed curve* represents KSS-CG using 2 Gaussian quadrature nodes, the *red solid and star curve* represents KSS-CG using 3 Gaussian quadrature nodes, the *yellow dash-dot curve* represents KSS-EPI, the *purple dot-star curve* represents the KSS method with filtering (but without correction), and the *green dash-circle curve* represent standard Krylov Projection

on the rectangle $[0, 1]^2$ and for $0 < t < 0.2$, with initial condition

$$u(x, y, 0) = u_0(x, y) = 0.4 + 0.1 \cos(2\pi x) \cos(5\pi y) \tag{30}$$

and homogeneous Neumann boundary conditions.

Figures 1 and 2 show the error vs. time performance and error vs. time step performance for each approach used, with grid sizes $N = 50$ and 150 points per dimension, respectively. From these plots we can see that the errors for both the KSS-CG methods are smaller that the errors for any other method. On the larger grid size, seen in Fig. 2, the difference in efficiency between each method is more pronounced. KSS-CG3 had the smallest relative error for each time-step, followed closely by KSS-CG2. Standard KSS and KSS-EPI methods demonstrated less computational time per time-step although comparatively the percentage increase in computational time between grid sizes for KSS is much larger than the percentage increase for both KSS-CG methods. This implies that for even larger grid sizes, KSS-CG may be more efficient than the other tested methods, as also observed in [3], though further numerical experiments would have to be performed to validate this theory. It is also important to note that although all KSS methods used are third-order accurate, KSS-CG achieved fourth-order accuracy.

4.2 Hyperbolic Problem

We now compare the performance of these methods when solving the hyperbolic problem

$$u_{tt} = \alpha \Delta u + (1 - 3u_0^2)u, \tag{31}$$

on the rectangle $[0, 1]^2$ and for $0 < t < 2$, with initial conditions

$$u(x, y, 0) = u_0(x, y), \quad u_t(x, y, 0) = 1,$$

where $u_0(x, y)$ is as defined in (30). For this problem, all KSS methods used are sixth-order accurate, and therefore we use KSS-CG with 3 and 4 Gaussian quadrature nodes for the correction.

We can see from the graph of the smaller grid size in Fig. 3 that all methods yield similar computational time in the last time-step, yet KSS-CG3 and KSS-CG4

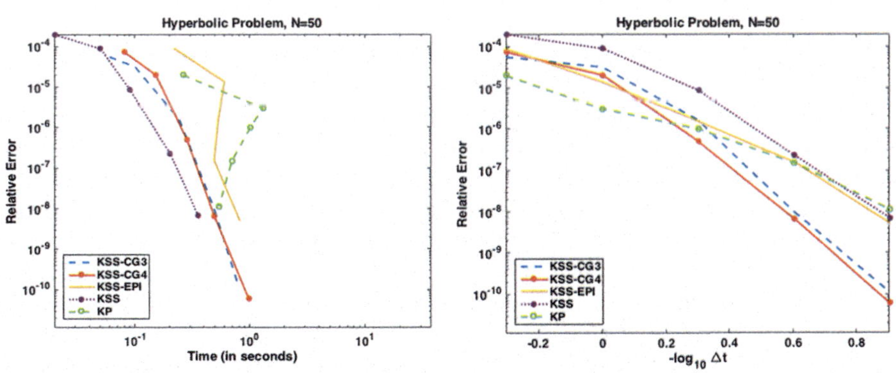

Fig. 3 Time of each timestep for each method vs error with grid size $N = 50$ points per dimension. The *blue dashed curve* represents KSS-CG using 3 Gaussian quadrature nodes, the *red solid and star curve* represents KSS-CG using 5 Gaussian quadrature nodes, the *yellow dash-dot curve* represents KSS-EPI, the *purple dot-star curve* represents the KSS method with filtering (but without correction), and the *green dash-circle curve* represent standard Krylov Projection

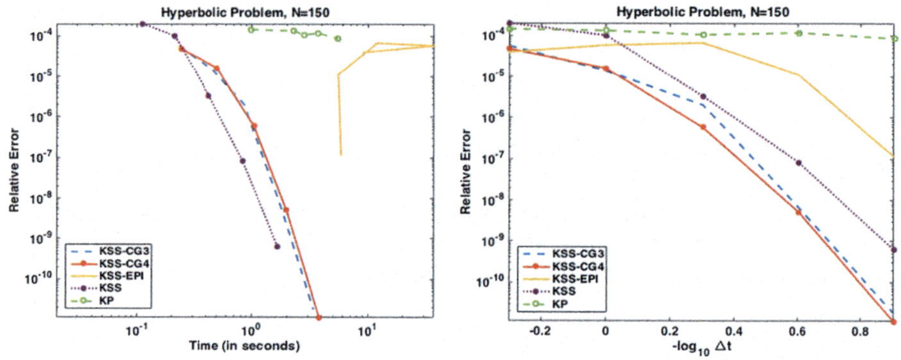

Fig. 4 Time of each timestep for each method vs error with grid size $N = 150$ points per dimension. The *blue dashed curve* represents KSS-CG using 3 Gaussian quadrature nodes, the *red solid and star curve* represents KSS-CG using 5 Gaussian quadrature nodes, the *yellow dash-dot curve* represents KSS-EPI, the *purple dot-star curve* represents the KSS method with filtering (but without correction), and the *green dash-circle curve* represent standard Krylov Projection

yield substantially higher accuracy. As the grid size increases from $N = 50$ grid points per dimension to $N = 150$, the accuracy of Krylov projection and KSS-EPI decreased while the accuracy of KSS and both KSS-CG methods increased, as can be seen in Fig. 4. Most significantly, the KSS-CG methods are far more efficient and scalable than Krylov Projection or KSS-EPI. As in the parabolic case, both KSS-CG methods exhibited higher-order accuracy: eighth-order instead of sixth over the smaller time steps.

5 Conclusion

It has been demonstrated that with coarse-grid residual correction, KSS methods become more accurate when solving time-dependent variable-coefficient PDEs, and also achieve a higher order of temporal accuracy. This represents a significant step forward in the evolution of KSS methods, as previous versions used the less efficient approaches of explicitly performing block Lanczos [14] or Krylov projection [3] to compute low-frequency components. Further optimization of KSS-CG will be needed to obtain faster computation time with sustained accuracy.

Future work on the combination of coarse-grid residual correction and KSS is needed to fully explore the effectiveness of this method. Topics for further research include generalizing KSS-CG to solve a wider variety of problems, including nonlinear PDEs in combination with EPI methods as in [3]. Future work must also focus on further grid coarsening, as in this paper only the next coarsest grid was used. An efficient way to use any number of corrections must be developed to fully realize the potential of KSS-CG; this would involve coarsening in time as well as space.

References

1. K. Atkinson, *An Introduction to Numerical Analysis*, 2nd ed. (Wiley, Hoboken, NJ, 1989)
2. W. Briggs, V.E. Henson, S.F. McCormick, *A Multigrid Tutorial* (Society for Industrial and Applied Mathematics, Philidelphia, PA, 2000)
3. A. Cibotarica, J.V. Lambers, E.M. Palchak. Solution of nonlinear time-dependent pde through componentwise approximation of matrix functions. J. Comput. Phys. **321**, 1120–1143 (2016)
4. O.G. Ernst, M.J. Gander, Why it is difficult to Solve Helmholtz problems with classical iterative methods, in *Numerical Analysis of Multiscale Problems*. Lecture Notes in Computational Science and Engineering, vol. 83 (Springer, Berlin/Heidelberg, 2011), pp. 325–363
5. G.H. Golub, G. Meurant, *Matrices, Moments and Quadrature with Applications* (Princeton University Press, Princeton, 2010)
6. G.H. Golub, R. Underwood, The block lanczos method for computing eigenvalues, in *Mathematical Software III* (Academic press, Cambridge, MA, 1977), pp. 361–377
7. B. Gustafsson, H.O. Kreiss, J. Oliger, *Time-Dependent Problems and Difference Methods* (Wiley, New York, 1995)
8. M. Hochbruck, C. Lubich: Approximations to the matrix exponential operator. SIAM J. Numer. Anal. **34**, 1911–1925 (1996)
9. J.V. Lambers, Krylov subspace methods for variable-coefficient initial-boundary value problems, Ph.D. thesis, SC/CM Program, Stanford University, 2003
10. J.V. Lambers, An explicit, stable, high-order spectral method for the wave equation based on block Gaussian quadrature. IAENG J. Appl. Math. **38**, 233–248 (2008)
11. J.V. Lambers, Enhancement of Krylov subspace spectral methods by block lanczos iteration. Electron. Trans. Numer. Anal. **31**, 86–109 (2008)
12. J.V. Lambers, A multigrid block Krylov subspace spectral method for variable-coefficient elliptic PDE. IAENG J. Appl. Math. **39**, 236–246 (2009)
13. J.V. Lambers, A spectral time-domain method for computational electrodynamics. Adv. Appl. Math. Mech. **1**, 781–798 (2009)
14. E.M. Palchak, A. Cibotarica, J.V. Lambers, Solution of time-dependent pde through rapid estimation of block gaussian quadrature nodes. Linear Algebra Appl. **468**, 233–259 (2015)

Extension of the Velocity-Correction Scheme to General Coordinate Systems

Douglas Serson, Julio R. Meneghini, and Spencer J. Sherwin

Abstract The velocity-correction scheme is a time-integration method for the incompressible Navier-Stokes equations, and is a common choice in the context of spectral/hp methods. Although the spectral/hp discretization allows the representation of complex geometries, in some cases the use of a coordinate transformation is desirable, since it may lead to symmetries which allow a more efficient solution of the equations. One example of this occurs when the transformed geometry has a homogeneous direction, in which case a Fourier expansion can be applied in this direction, reducing the computational cost. In this paper, we revisit two recently proposed forms of extending the velocity-correction scheme to general coordinate systems, the first treating the mapping terms explicitly and the second treating them semi-implicitly. We then present some numerical examples illustrating the properties and applicability of these methods, including new tests focusing on the time-accuracy of these schemes.

1 Introduction

The velocity-correction scheme [5, 8] is a widely used time-integration method for the unsteady incompressible Navier-Stokes equations, having being applied in conjunction with finite volume [12], finite element [5], spectral-Legendre [12] and spectral/hp [8] discretizations. In particular, this method is commonly used in conjunction with spectral/hp discretizations, for example in the Nektar++ package [2]. This method has the advantage of allowing the pressure and the velocity to be solved separately, leading to an efficient solution.

D. Serson (✉) • S.J. Sherwin
Department of Aeronautics, Imperial College London, South Kensington Campus, London SW7 2AZ, UK
e-mail: d.serson14@imperial.ac.uk; s.sherwin@imperial.ac.uk

J.R. Meneghini
NDF, Escola Politécnica, Universidade de São Paulo, Av. Prof. Mello Moraes, 2231, São Paulo 05508-030, Brazil
e-mail: jmeneg@usp.br

M.L. Bittencourt et al. (eds.), *Spectral and High Order Methods for Partial Differential Equations ICOSAHOM 2016*, Lecture Notes in Computational Science and Engineering 119, DOI 10.1007/978-3-319-65870-4_23

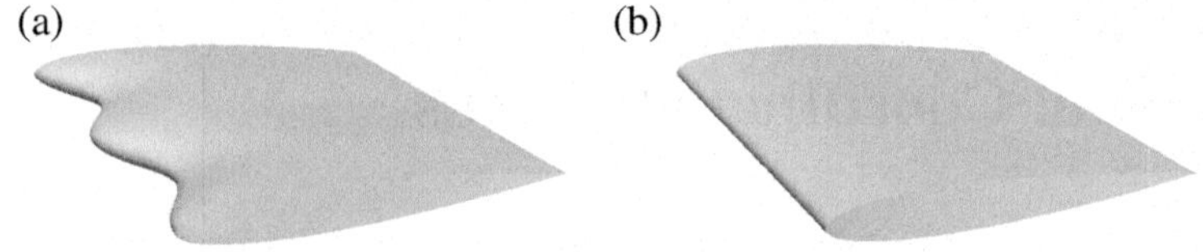

Fig. 1 Example of how a complex geometry in the physical Cartesian coordinate system can be mapped into a simpler geometry in a different coordinate system. (**a**) Cartesian system. (**b**) Transformed system

Although the spectral/hp discretization allows us to consider arbitrary geometries, in some problems it is desirable to solve the equations in a coordinate system other than the typical Cartesian frame of reference. For example, if the transformation creates a homogeneous direction, we can employ what is called the Fourier-spectral/hp element method [6]. In this approach, the homogeneous direction is discretized by a Fourier expansion, leading to a more efficient solution that can compensate for the extra computational costs of solving the Navier-Stokes equations in the general coordinates. This idea is illustrated in Fig. 1, which shows how we can obtain a simpler representation of a complex geometry by changing the coordinate system. Another situation where using a coordinate transformation can be advantageous occurs when we have a moving geometry. In this case, the coordinate transformation removes the need to deform the computational mesh, which is usually an expensive operation in spectral/hp methods.

Although being able to employ the velocity-correction scheme in general coordinates would be desirable, until recently this method had not been extended to account to general coordinate transformations. As far as the authors are aware, only specialised situations had been considered, like constant-Jacobian time-dependent transformation [11], and constant-Jacobian time-independent mappings [4]. Considering other approximations of the Navier-Stokes equations, a method for accounting for general coordinate transformations in the context of pseudo-spectral methods was proposed [3], using iterative procedures to solve for the pressure and velocity fields.

In a recent paper [13], we proposed two methods for including coordinate transformations in the velocity-correction scheme. The first one is a generalization of the approach of [4, 11], with the mapping being treated explicitly. On the other hand, the second method is a modified version of the iterative procedure employed by Carlson et al. [3], with the pressure and viscous terms of the mapping being treated implicitly. Neither of these new methods are restricted to constant-Jacobian transformation, and both of them can be used with time-dependent transformations.

In this paper, we revisit the methods proposed in [13], presenting also numerical tests focused on demonstrating the spatial and time-accuracy of these schemes. The paper is organized as follows. Section 2 describes the numerical methods. Then, Sect. 3 presents results from simulations employing them. Finally, Sect. 4 contains the conclusions of the paper.

2 Numerical Formulation

In this section the numerical methods that allow the velocity-correction scheme to be applied to general coordinate systems are briefly described. A more detailed presentation of the formulation can be found in [13]. We start by considering the velocity-correction scheme in a Cartesian coordinate frame, and then proceed to extending it to a general coordinate system.

In the original velocity-correction scheme [5, 8], we are interested in solving the incompressible Naiver-Stokes equations, which assuming a unity density can be written as

$$\begin{aligned} \frac{\partial \mathbf{u}}{\partial t} &= \mathbf{N}(\mathbf{u}) - \mathbf{G}(p) + \nu \mathbf{L}(\mathbf{u}), \\ \nabla \cdot \mathbf{u} &= 0, \end{aligned} \tag{1}$$

where $\mathbf{u}$ is the velocity, p is the pressure and ν is the kinematic viscosity. In this equation, $\mathbf{N}(\mathbf{u}) = -(\mathbf{u} \cdot \nabla)\mathbf{u}$ are the non-linear convective terms, $\mathbf{G}(p) = \nabla p$ is the pressure gradient and $\nu \mathbf{L}(\mathbf{u}) = \nu \nabla^2 \mathbf{u}$ are the linear viscous terms.

We begin by approximating $\mathbf{N}(\mathbf{u})$ by an explicit polynomial extrapolation

$$\mathbf{N}(\mathbf{u}(t^{n+1})) \approx \mathbf{N}^* = \sum_{q=0}^{J_e-1} \beta_q \mathbf{N}(\mathbf{u}^{n-q}) \tag{2}$$

and the time derivative by a backward differentiation formula (BDF)

$$\frac{\partial \mathbf{u}}{\partial t}(t^{n+1}) \approx \frac{\gamma_0 \mathbf{u}^{n+1} - \mathbf{u}^+}{\Delta t} = \frac{\gamma_0 \mathbf{u}^{n+1} - \sum_{q=0}^{J_i-1} \alpha_q \mathbf{u}^{n-q}}{\Delta t}. \tag{3}$$

We note that the extrapolation from Eq. (2) can also be applied to other operators.

Then, the time-integration is performed in two steps. In the first step, we solve for the pressure assuming an intermediate velocity field which satisfies the continuity equation, while the second step corresponds to a viscous correction. The first step consists of the system

$$\begin{cases} \frac{\gamma_0 \bar{\mathbf{u}}^{n+1} - \mathbf{u}^+}{\Delta t} + \nabla p^{n+1} + \nu \mathbf{Q}^* - \mathbf{N}^* = 0 & \text{in } \Omega, \\ \nabla \cdot \bar{\mathbf{u}}^{n+1} = 0 & \text{in } \Omega, \\ \bar{\mathbf{u}}^{n+1} \cdot \mathbf{n} = \mathbf{u}_{\mathfrak{D}} \cdot \mathbf{n} & \text{on } \Gamma_{\mathfrak{D}}, \end{cases} \tag{4}$$

where $\bar{\mathbf{u}}^{n+1}$ is the intermediate divergence-free velocity, which does not need to be calculated during the solution, and $\mathbf{Q} = \nabla \times \nabla \times \mathbf{u}$ is a valid form of the viscous terms. The reason for using $\mathbf{Q}$ instead of $\mathbf{L}$ in this equation is that the latter leads to spurious boundary conditions [5, 8].

Equation (4) can be solved by dotting it with $\nabla\phi$ and integrating to obtain the weak form. After applying some identities, the resulting equation for the pressure is

$$\int_{\Omega} \nabla p^{n+1} \cdot \nabla \phi \, d\Omega = \int_{\Omega} \phi \nabla \cdot \left(-\frac{\hat{\mathbf{u}}}{\Delta t} \right) d\Omega + \int_{\Gamma} \phi \left[\frac{\hat{\mathbf{u}} - \gamma_0 \bar{\mathbf{u}}^{n+1}}{\Delta t} - \nu \mathbf{Q}^* \right] \cdot \mathbf{n} \, dS, \tag{5}$$

with $\hat{\mathbf{u}} = \mathbf{u}^+ + \Delta t \mathbf{N}^*$. This is a Poisson equation, with the second term in the right hand side representing the high order pressure boundary conditions.

Having obtained the pressure, the second step of the scheme consists in solving for the velocity using the Helmholtz equation

$$\frac{\gamma_0 \mathbf{u}^{n+1} - \hat{\hat{\mathbf{u}}}}{\Delta t} = \nu \mathbf{L}(\mathbf{u}^{n+1}) \tag{6}$$

with the velocity boundary conditions from the specific problem applied to $\mathbf{u}^{n+1}$, and where $\hat{\hat{\mathbf{u}}} = \hat{\mathbf{u}} - \nabla p^{n+1} \Delta t$.

We now consider the problem in a general coordinate system. We will denote the usual Cartesian system by $(\bar{x}, \bar{y}, \bar{z})$ and the transformed system by (x, y, z). In the transformed system, we can obtain the appropriate form of the incompressible Navier-Stokes equations using tensor calculus [10], where the resulting equations can be represented by

$$\begin{aligned} \frac{\partial \mathbf{u}}{\partial t} &= \bar{\mathbf{N}}(\mathbf{u}) - \bar{\mathbf{G}}(p) + \nu \bar{\mathbf{L}}(\mathbf{u}), \\ D(\mathbf{u}) &= 0, \end{aligned} \tag{7}$$

where:

$$\begin{aligned} \bar{\mathbf{N}}(\mathbf{u}) &= -u^j u^i_{,j} + V^j u^i_{,j} - u^j V^i_{,j}, \\ \bar{\mathbf{G}}(p) &= g^{ij} p_{,j}, \\ \nu \bar{\mathbf{L}}(\mathbf{u}) &= \nu g^{jk} u^i_{,jk}, \\ D(\mathbf{u}) &= \frac{1}{J} \nabla \cdot (J u^i), \end{aligned} \tag{8}$$

with g^{ij} representing the inverse of the metric tensor, u^i the components of the vector $\mathbf{u}$, J the Jacobian of the transformation to the Cartesian system, and a subscript after a comma denotes the covariant derivative. The term $V^j = -\frac{\partial x^j}{\partial t}$ represents the velocity of the coordinate system, and therefore is only relevant for time-dependent transformations. For the fundamentals of tensor calculus leading to the derivation of equation (7), the reader is referred to [1]. To simplify the notation, the ∇ operator will be assumed to correspond to the usual Cartesian operation representing the partial derivatives.

In the following, we will present two approaches to solving Eq. (7). In the first all extra terms arising from the transformation are treated explicitly, while in the second the convective terms are treated explicitly and the pressure and viscous terms implicitly.

2.1 Explicit Formulation

This section describes an approach to solve Eq. (7) treating all the mapping terms explicitly, which can be seen as a generalization of a method previously presented in the literature for constant Jacobian transformations [4, 11]. For this explicit formulation, we rewrite Eq. (7) as

$$\begin{aligned} \frac{\partial \mathbf{u}}{\partial t} &= \mathbf{N}(\mathbf{u}) - \frac{\nabla p}{J} + \nu \mathbf{L}(\mathbf{u}) + \mathbf{A}(\mathbf{u}, p), \\ D(\mathbf{u}) &= 0, \end{aligned} \tag{9}$$

where $\mathbf{N}(\mathbf{u})$ and $\mathbf{L}(\mathbf{u})$ are the usual convective and viscous operators, and

$$\mathbf{A}(\mathbf{u}, p) = \left[\bar{\mathbf{N}}(\mathbf{u}) - \mathbf{N}(\mathbf{u})\right] + \left[-\bar{\mathbf{G}}(p) + \frac{\nabla p}{J}\right] + \nu\left[\bar{\mathbf{L}}(\mathbf{u}) - \mathbf{L}(\mathbf{u})\right] \tag{10}$$

is a forcing term that imposes the coordinate transformation and can clearly be interpreted as the difference between the Cartesian and transformed expressions.

Since the incompressibility condition is now different, we also need to modify the decomposition of the viscous terms which lead to the operator $\mathbf{Q}$. Following a similar idea from that leading to $\mathbf{Q}$, we propose replacing $\mathbf{Q}$ by $\bar{\mathbf{Q}}_e = \nabla(\frac{\mathbf{u}}{J} \cdot \nabla J) + \nabla \times \nabla \times \mathbf{u}$. The resulting equation for the pressure is then

$$\begin{aligned} \int_\Omega \nabla p^{n+1} \cdot \nabla \phi \, d\Omega = &\int_\Omega \phi \nabla \cdot \left[-\frac{J\hat{\mathbf{u}}}{\Delta t} + \nu \left(\nabla \left(\frac{\mathbf{u}}{J} \cdot \nabla J \right) \right)^* \right] + \phi \nu \nabla J \cdot (\nabla \times \nabla \times \mathbf{u})^* \, d\Omega \\ &+ \int_\Gamma \phi J \left[\frac{\hat{\mathbf{u}} - \gamma_0 \bar{\mathbf{u}}^{n+1}}{\Delta t} - \nu \bar{\mathbf{Q}}_e^* \right] \cdot \mathbf{n} \, dS \end{aligned} \tag{11}$$

where this time $\hat{\mathbf{u}} = \mathbf{u}^+ + \Delta t(\mathbf{N}^* + \mathbf{A}^*)$ and the terms of the form $(.)^*$ are calculated according to the extrapolation procedure from Eq. (2). This is still a Poisson equation, with modified forcing terms and boundary conditions.

In the second step of the solution, the only modification required is in the definition of $\hat{\hat{\mathbf{u}}}$, which is now $\hat{\hat{\mathbf{u}}} = \hat{\mathbf{u}} - \frac{\nabla p^{n+1}}{J} \Delta t$. Therefore, only slight modifications are required to the original velocity-correction scheme in order to implement this explicit formulation. This characteristic is one of the advantages of the method, since an existing solver can be easily adapted to include the coordinate transformation.

2.2 Semi-Implicit Formulation

This section describes an approach to solve the equations where the mapping terms arising from the convective part of the equation are treated explicitly, while the pressure and viscous terms are treated implicitly, maintaining the characteristics of the original splitting scheme. This is a modified version of the method used by Carlson et al.[3], with the difference being that our approach allows us to readily obtain the appropriate pressure boundary conditions.

The main idea of this method is to follow the original velocity-correction scheme, replacing the operators $\mathbf{N}$, $\mathbf{G}$ and $\mathbf{L}$ by their analogous counterparts $\bar{\mathbf{N}}$, $\bar{\mathbf{G}}$ and $\bar{\mathbf{L}}$. However, instead of directly solving the implicit equations involving the generalized operators, they are solved iteratively. This is done because inverting those operators would lead to strong couplings (e.g. coupling between different velocity components and between Fourier modes) which would make the method very expensive computationally. Another change required is in the $\bar{\mathbf{Q}}$ operator, which should be replaced by $\bar{\mathbf{Q}}_i = \varepsilon^{imn}\varepsilon^{ljk}g_{nl}g_{kp}u^p_{,jm}$, where g_{ij} is the metric tensor and $\varepsilon^{ijk} = g^{-1/2}\epsilon^{ijk}$, with ϵ^{ijk} being the permutation symbol, is a generalization of the permutation symbol.

The first step is solved using the following iteration:

$$\nabla p^{n+1}_{s+1} = \nabla p^{n+1}_{s} + J\left[\frac{\mathbf{u}^{+} - \gamma_0\bar{\mathbf{u}}^{n+1}}{\Delta t} - \nu\bar{\mathbf{Q}}^{*}_{i} + \bar{\mathbf{N}}^{*} - \bar{\mathbf{G}}(p^{n+1}_{s})\right], \tag{12}$$

where s is the iteration counter.

Dotting equation (12) with $\nabla\phi$ and integrating to obtain the weak form, and after using the identities $\nabla\cdot(J\bar{\mathbf{u}}^{n+1}) = 0$ and $D(\mathbf{Q}) = 0$, the equation becomes

$$\begin{aligned}\int_{\Omega}\nabla p^{n+1}_{s+1}\cdot\nabla\phi\, d\Omega = &\int_{\Omega}\phi\left[JD\left(\frac{-\hat{\mathbf{u}}}{\Delta t}\right) + JD(\bar{\mathbf{G}}(p^{n+1}_{s})) - \nabla^2 p^{n+1}_{s}\right] d\Omega\\ &+\int_{\Gamma}\phi\left[J\left(\frac{\hat{\mathbf{u}} - \gamma_0\bar{\mathbf{u}}^{n+1}}{\Delta t}\right) - \nu J\bar{\mathbf{Q}}^{*}_{i} - J\bar{\mathbf{G}}(p^{n+1}_{s}) + \nabla p^{n+1}_{s}\right]\cdot\mathbf{n}\, dS,\end{aligned} \tag{13}$$

where $\hat{\mathbf{u}} = \mathbf{u}^{+} + \Delta t\bar{\mathbf{N}}^{*}$. This Poisson equation needs to be solved at each iteration; however, most of the terms in the right-hand-side are not modified during the iterative procedure, and therefore only need to be computed once per time-step.

Similarly, the velocity system of the second step can be solved using the following iterative procedure:

$$\frac{\gamma_0\mathbf{u}^{n+1}_{s+1}}{\Delta t} - \nu\mathbf{L}(\mathbf{u}^{n+1}_{s+1}) = \frac{\hat{\mathbf{u}}}{\Delta t} - \bar{\mathbf{G}}(p^{n+1}) + \nu\bar{\mathbf{L}}(\mathbf{u}^{n+1}_{s}) - \nu\mathbf{L}(\mathbf{u}^{n+1}_{s}), \tag{14}$$

where each iteration consists of solving a Helmholtz equation for each velocity component. Also, we note that the iterative procedures of equations (12) and (14) can be modified to include a relaxation parameter, making them more robust.

3 Test Cases

This section presents results of tests employing the previous methods for three different flows, the first two considering flows with analytic solutions in order to demonstrate the accuracy of the schemes, and the last one illustrating a possible practical application. The first case considers the two-dimensional Taylor-Green vortex, which represents a decaying vortex in a periodic square. This is an unsteady flow, allowing us to assert the time accuracy of our methods, but due to the periodic boundary conditions the high order pressure boundary conditions are not tested. The second case is the Kovasznay flow, which is a steady solution of the two-dimensional Navier-Stokes equations. The last simulation presented is the two-dimensional flow around two circular cylinders in tandem, with the upstream cylinder subject to forced oscillations, while the downstream cylinder is held fixed. We also tested the case of a uniform flow with a time-dependent transformation. In this case the flow remained uniform, demonstrating that our treatment of time-dependent mappings does not introduce spurious structures.

All simulations employ the spatial discretization of the spectral/hp method [7] implemented in Nektar++ [2]. However, we note that the methods proposed here do not depend on this particular choice of discretization.

3.1 Two-Dimensional Taylor-Green Vortex

The two-dimensional Taylor-Green vortex is a decaying vortex in a periodic domain, with analytic solution

$$\begin{aligned} u &= cos(x)sin(y)e^{-2\nu t}, \\ v &= -sin(x)cos(y)e^{-2\nu t}, \\ p &= \frac{1}{4}\left(sin\left(2x - \frac{\pi}{2}\right) + sin\left(2y - \frac{\pi}{2}\right)\right)e^{-4\nu t}. \end{aligned} \tag{15}$$

We solved this problem in a computational domain extending from 0 to 2π in both x and y, using a uniform mesh consisting of 16 quadrilateral elements. Periodic boundary conditions were used, with the initial conditions obtained from (15) with $t = 0$. The kinematic viscosity was $\nu = \frac{1}{100}$ and the final time of the simulations was $t = 1$.

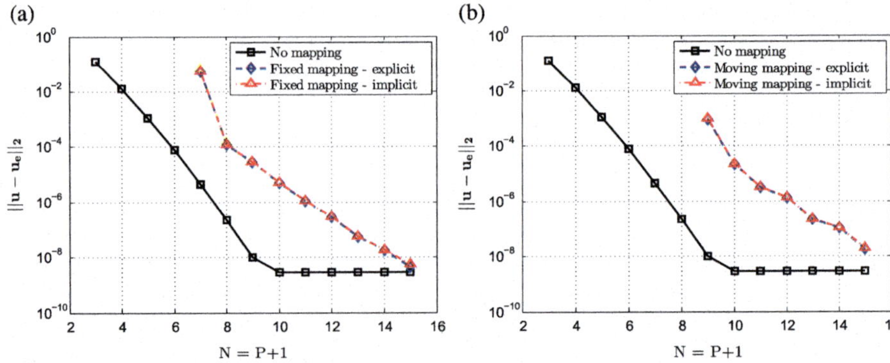

Fig. 2 Convergence with the polynomial order of simulations for the two-dimensional Taylor-Green vortex. (**a**) Fixed transformation. (**b**) Time-dependent transformation

We considered transformations of the form

$$\bar{x} = x - \frac{h}{2}cos(\frac{2\pi y}{\lambda})cos(\frac{2\pi x}{\lambda}) - \frac{A}{2}cos(2\pi t\omega), \tag{16}$$

using $\lambda = \pi$ and $\omega = 1.0$. Three configurations were studied: no mapping ($h = 0$ and $A = 0$), fixed mapping ($h = 0.15$ and $A = 0.0$) and time-dependent mapping ($h = 0.15$ and $A = 0.1$). For the simulations with a transformation, both the explicit and semi-implicit formulations were considered.

Figure 2 shows the convergence with the polynomial order for simulations using a second-order time integration with $\Delta t = 10^{-5}$. We notice that the coordinate transformation may increase the level of error, what is likely caused by the exact solution in this case being more difficult to represent using polynomials in the transformed domain. However, it is reasonable to expect that in practical applications this preferential representation in the Cartesian system will not occur in general. Therefore, the most important characteristic of these results is that the exponential convergence of the method is preserved.

Figure 3 presents the temporal convergence of the method, for a polynomial order $P = 11$. Once again, the error levels are higher, but the order of the schemes is maintained.

3.2 *Kovasznay Flow*

The Kovasznay flow [9] consists in a steady analytical solution for the two-dimensional Navier-Stokes equations with a periodic direction, which can be viewed as a representation of the flow behind a two-dimensional grid. The exact solution is

$$u = 1 - e^{kx}\cos(2\pi y), \quad v = \frac{k}{2\pi}e^{kx}\sin(2\pi y), \quad p = \frac{1}{2}\left(1 - e^{2kx}\right), \tag{17}$$

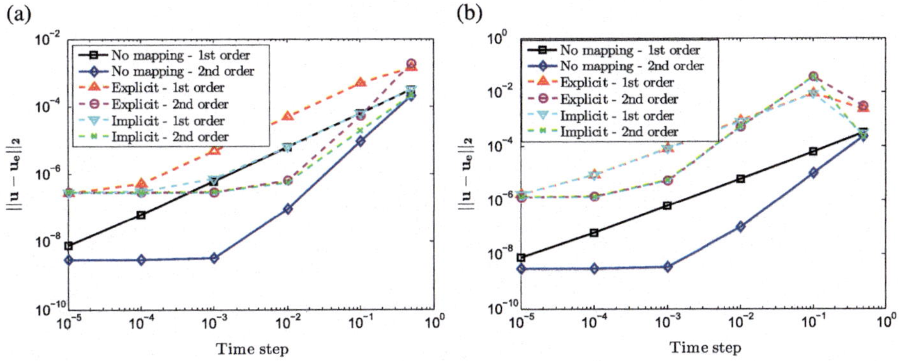

Fig. 3 Convergence with the time step of simulations for the two-dimensional Taylor-Green vortex. (**a**) Fixed transformation. (**b**) Time-dependent transformation

where the constant k is defined as

$$k = \frac{1}{2\nu} - \sqrt{\frac{1}{4\nu^2} + 4\pi^2}. \tag{18}$$

We solved this problem in a computational domain extending from -0.5 to 1 in x and from -0.5 to 1.5 in y, using a uniform mesh consisting of 12 quadrilateral elements. A uniform flow was used as initial condition, and the time integration was performed until $t = 20$. Dirichlet conditions were used in all boundaries, except for the inflow where the high order conditions were used for the pressure. Also, all simulations employed $\nu = \frac{1}{40}$.

We considered a transformation of the form

$$\bar{x} = x - (1-x)\frac{h}{2}cos(\frac{2\pi y}{\lambda}), \quad \bar{y} = y + \frac{A}{2}cos(2\pi t\omega), \tag{19}$$

using $\lambda = 1.0$ and $\omega = 2.0$. Once again, three configurations were studied: no mapping ($h = 0$ and $A = 0$), fixed mapping ($h = 0.05$ and $A = 0.0$) and time-dependent mapping ($h = 0.05$ and $A = 0.2$), with both the explicit and semi-implicit formulations considered.

Figure 4 shows the convergence with the polynomial order for simulations using a second-order time integration with $\Delta t = 10^{-5}$. Both methods maintain the exponential convergence with respect to P. Also, we notice that for the time-dependent transformation, the error saturates at a higher level. This happens because in this case the solution in the computational domain is not steady, and therefore there are also time integration errors. The convergence with Δt for this case is presented in Fig. 5 (with $P = 11$), demonstrating that the temporal convergence is consistent with the original scheme.

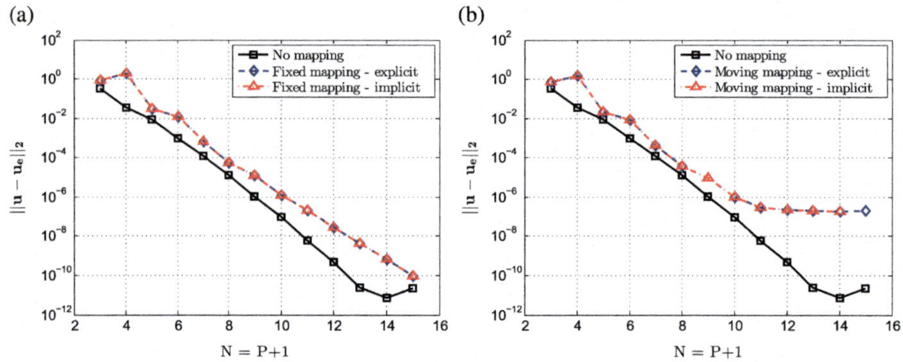

Fig. 4 Convergence with the polynomial order of simulations for the Kovasznay flow. (**a**) Fixed transformation. (**b**) Time-dependent transformation

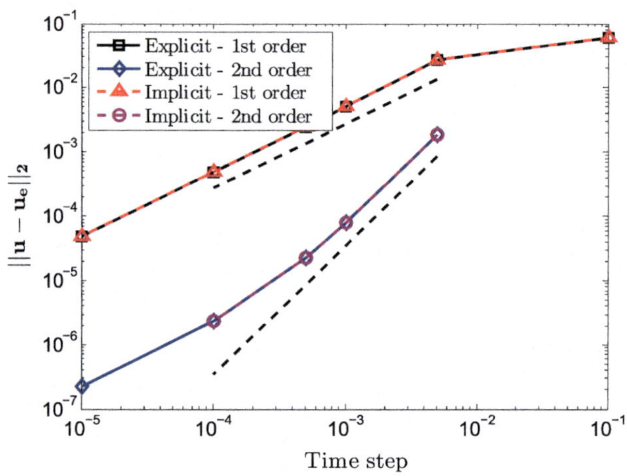

Fig. 5 Convergence with the time step of simulations for the Kovasznay flow with a time-dependent transformation. The *dashed black lines* are reference slopes

3.3 Flow Around Moving Cylinders

To demonstrate the possibility of using the techniques presented in this paper to problems involving moving bodies, a simulation of the two-dimensional flow around a pair of moving circular cylinders was performed, with Reynolds number $Re = 100$. The centre-to-centre distance is 3 diameters, the downstream cylinder is held fixed, and a forced oscillation in the y direction with non-dimensional frequency 0.3 and amplitude of 0.75 diameter was imposed on the upstream cylinder. Instead of using a moving mesh to solve this problem, a fixed mesh where the displacement of the cylinders is zero was used, with a mapping accounting for the movement of the upstream cylinder. Before each time-step, the displacements on the boundaries were used as boundary conditions to solve a Laplace equation, leading to a global representation of the mapping which was used to solve the equations with the semi-implicit method. Figure 6 shows instantaneous contours of vorticity for this case, exhibiting a behaviour that is compatible with what is expected for this flow. The effect of the transformation is illustrated in Fig. 7, showing the computational mesh in the region close to the cylinders, along with the corresponding representations in the physical domain at two different time instants.

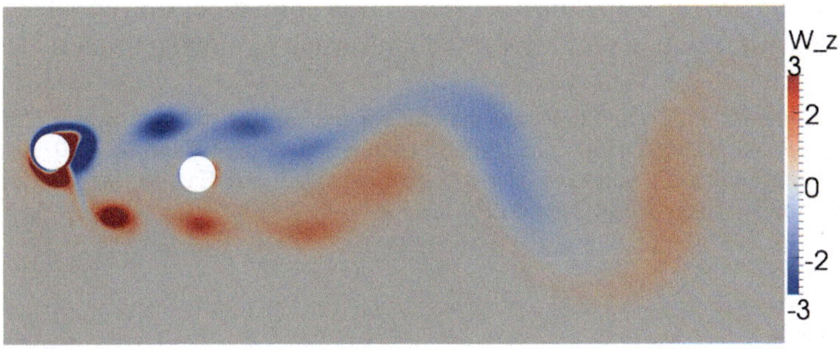

Fig. 6 Instantaneous contours of vorticity for flow around two circular cylinder in tandem with $Re = 100$. The downstream cylinder is fixed, while the upstream cylinder oscillates with frequency 0.3 and amplitude 0.75

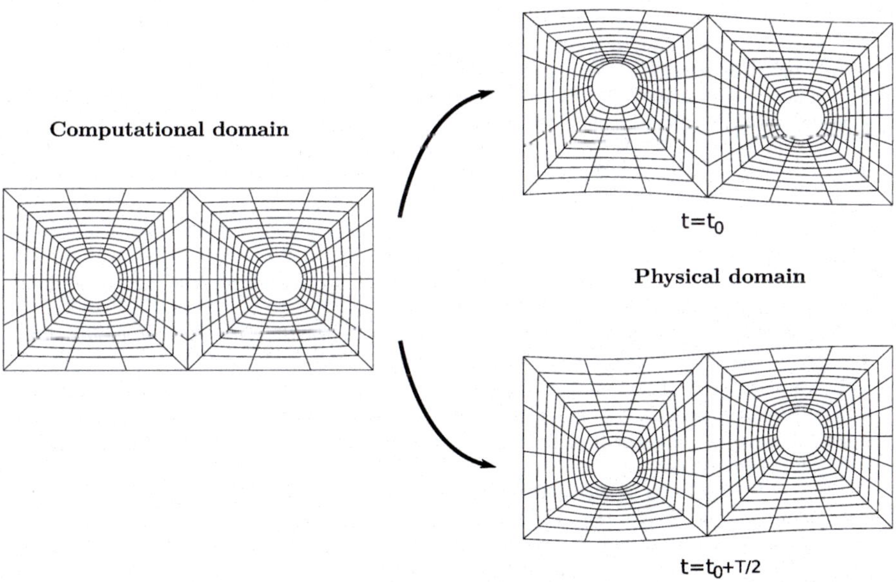

Fig. 7 Detail of computational mesh used in the simulation of the flow around two circular cylinders, and the same mesh after applying the coordinate transformation at two different time instants

4 Conclusions

In this paper, we revisited two methods recently proposed to introduce coordinate transformations in the velocity-correction scheme. Using numerical tests, we show that these methods maintain the exponential convergence of the underlying spatial discretization, and also the order of accuracy of the time integration scheme being employed. These methods are useful when they lead to a simplification of the geometry, allowing for efficient numerical techniques such as a Fourier expansion to be employed. Also, by using time-dependent transformations it is possible to solve fluid-structure interaction problems without resorting to moving meshes.

Acknowledgements D.S. and J.R.M. are grateful for the support received from CNPq (grants 231787/2013-8 and 312755/2014-7) and FAPESP (grants 2012/23493-0 and 2014/50279-4). S.J.S. would like to acknowledge support under the Royal Academy of Engineering Research Chair Scheme and support under EPSRC grant EP/K037536/1.

References

1. R. Aris, *Vectors, Tensors, and the Basic Equations of Fluid Mechanics* (Dover Publications, Inc., Mineola, NY, 1989)
2. C. Cantwell, D. Moxey, A. Comerford, A. Bolis, G. Rocco, G. Mengaldo, D.D. Grazia, S. Yakovlev, J.E. Lombard, D. Ekelschot, B. Jordi, H. Xu, Y. Mohamied, C. Eskilsson, B. Nelson, P. Vos, C. Biotto, R. Kirby, S. Sherwin, Nektar++: an open-source spectral/hp element framework. Comput. Phys. Commun. **192**, 205–219 (2015)
3. H.A. Carlson, G. Berkooz, J.L. Lumley, Direct numerical simulation of flow in a channel with complex, time-dependent wall geometries: a pseudospectral method. J. Comput. Phys. **121**, 155–175 (1995)
4. R.M. Darekar, S.J. Sherwin, Flow past a square-section cylinder with a wavy stagnation face. J. Fluid Mech. **426**, 263–295 (2001)
5. J.L. Guermond, J. Shen, Velocity-correction projection methods for incompressible flows. SIAM J. Numer. Anal. **41**(1), 112–134 (2003)
6. G.E. Karniadakis, Spectral element-Fourier methods for incompressible turbulent flows. Comput. Methods Appl. Mech. Eng. **80**(1–3), 367–380 (1990)
7. G.E. Karniadakis, S.J. Sherwin, *Spectral/hp Element Methods for Computational Fluid Dynamics*, 2nd edn. (Oxford University Press, Oxford, 2005)
8. G.E. Karniadakis, M. Israeli, S.A. Orszag, High-order splitting methods for the incompressible Navier-Stokes equations. J. Comput. Phys. **97**(2), 414–443 (1991)
9. L.I.G. Kovasznay, Laminar flow behind a two-dimensional grid. Math. Proc. Camb. Philos. Soc. **44**(1), 58–62 (1948)
10. H. Luo, T.R. Bewley, On the contravariant form of the Navier-Stokes equations in time-dependent curvilinear coordinate systems. J. Comput. Phys. **199**, 355–375 (2004)
11. D.J. Newman, G.E. Karniadakis, A direct numerical simulation study of flow past a freely vibrating cable. J. Fluid Mech. **344**, 95–136 (1997)
12. A. Poux, S. Glockner, E. Ahusborde, M. Azaïez, Open boundary conditions for the velocity-correction scheme of the Navier-Stokes equations. Comput. Fluids **70**, 29–43 (2012)
13. D. Serson, J.R. Meneghini, S.J. Sherwin, Velocity-correction schemes for the incompressible Navier-Stokes equations in general coordinate systems. J. Comput. Phys. **316**, 243–254 (2016)

A Parallel High-Order CENO Finite-Volume Scheme with AMR for Three-Dimensional Ideal MHD Flows

Lucie Freret, Clinton P.T. Groth, and Hans De Sterck

Abstract A highly-scalable and efficient parallel high-order finite-volume method with local solution-dependent adaptive mesh refinement (AMR) is described for the solution of steady plasma flows governed by the equations of ideal magnetohydrodyamics (MHD) on three-dimensional multi-block body-fitted hexahedral meshes, including cubed-sphere grids based on cubic-gnomonic projections. The approach combines a family of robust and accurate high-order central essentially non-oscillatory (CENO) spatial discretization schemes with a block-based anisotropic AMR scheme. The CENO scheme is a hybrid approach that avoids some of the complexities associated with essentially non-oscillatory (ENO) and weighted ENO schemes and is therefore well suited for application to meshes having irregular and unstructured topologies. The anisotropic AMR method uses a binary tree and hierarchical data structure to permit local refinement of the grid in preferred directions as directed by appropriately selected refinement criteria. Applications will be discussed for several steady MHD problems and the computational performance of the proposed high-order method for the efficient and accurate simulation of a range of plasma flows is demonstrated.

1 Introduction and Motivation

Physics-based space weather modeling [6, 7, 14] is a challenging problem that requires accurate numerical modeling for both disparate spatial and temporal scales. Accurate solutions can be achieved by using either high-order schemes or an adaptive mesh refinement (AMR) technique. A combination of both approaches would appear to be particularly desirable [15].

L. Freret (✉) • C.P.T. Groth
Institute for Aerospace Studies, University of Toronto, Toronto, ON, Canada M3H5T6
e-mail: lfreret@utias.utoronto.ca; groth@utias.utoronto.ca

H. De Sterck
School of Mathematical Sciences, Monash University, Melbourne, Australia
e-mail: hans.desterck@monash.edu

M.L. Bittencourt et al. (eds.), *Spectral and High Order Methods for Partial Differential Equations ICOSAHOM 2016*, Lecture Notes in Computational Science and Engineering 119, DOI 10.1007/978-3-319-65870-4_24

The high-order central essentially non-oscillatory (CENO) finite-volume scheme from Ivan et al. [16, 18] uses a hybrid reconstruction approach based on a fixed central stencil. An unlimited high-order k-exact reconstruction is performed in the cells where the solution is well resolved while the scheme reverts to a low-order limited linear approach for cells with under-resolved/discontinuous solution content. Switching in the hybrid procedure is determined by a smoothness indicator. The CENO high-order scheme has been successfully applied to a broad range of flows on multi-block body-fitted meshes including non-viscous flows [18], viscous flows [16], large-eddy simulation (LES) of turbulent premixed flames [24] and magnetohydrodyamics (MHD) problems [18, 23]. The efficiency of the CENO scheme has also been assessed on cubed-sphere meshes [18] and extended to unstructured meshes for laminar viscous flows [5] and turbulent reactive flows [4].

Block-based AMR approaches [2, 3, 14, 21] are very attractive since they are naturally suitable for parallel implementation and lead to highly scalable methods while requiring an overall light data structure to compute the block connectivity. The multi-block AMR scheme considered here is based on the previous work by Gao and Groth [11] for reacting flows with isotropic refinement. This numerical scheme has also been applied to the solution of complex flow problems such as non-premixed laminar and turbulent flames [10, 12, 20] as well as turbulent multi-phase rocket core flows [22], MHD simulations [17, 18, 23], and micron-scale flows [13, 19]. The isotropic AMR scheme was originally extended to allow for anisotropic refinement by Williamschen and Groth [26] for non-viscous flows. More recently, Freret and Groth [9] reformulated the anisotropic AMR scheme using a non-uniform treatment of the cells (both interior and ghost or halo cells) within a given block. It directly makes use of the neighboring cells as the ghost cells, even those at different levels of refinement as found at grid resolution changes. The resulting anisotropic AMR multi-block scheme is better suited for high-order finite-volume schemes.

The focus of this study is the extension of the enhanced anisotropic AMR algorithm of Freret and Groth [9] for use in conjunction with the fourth-order CENO finite-volume scheme (the former permits the use of efficient high-order solution transfer operators) and the subsequent application of the combined method to the prediction of steady-state solutions of the ideal MHD equations. For this application, the solenoidal constraint on the magnetic field is controlled using the generalized Lagrange multiplier (GLM) proposed by Dedner et al. [8, 18, 23]. The ideal MHD equations and the GLM formulation are described in Sect. 2. In Sect. 3, a brief outline of the high-order CENO scheme is provided. The proposed anisotropic AMR block-based method is reviewed in Sect. 4 with the necessary extension for use with the high-order spatial discretization scheme. Finally, three-dimensional (3D) numerical results are presented in Sect. 5, including an accuracy demonstration of the high-order CENO reconstruction procedure for a known function and numerical results for two steady-state flow problems on cubed-sphere grids. Numerical results for both non-magnetized and magnetized flows are used to evaluate the grid convergence of the proposed fourth-order CENO scheme for uniformly and anisotropically refined meshes and compare the convergence behavior to that of the second-order limited method described by Ivan et al. [17].

The latter was originally developed for use with the isotropic AMR of Gao and Groth [11] and has been extended for the purpose of this study to non-uniform block-based anisotropic AMR.

2 Ideal Magnetohydrodynamics Equations

Solution of the hyperbolic system of ideal MHD equations is considered here using a high-order Godunov-type finite-volume scheme with a GLM formulation [8] which couples the divergence constraint, $\nabla \cdot \mathbf{B} = 0$, with the induction equation through the introduction of the potential, ψ. The system of conservation laws for which numerical solutions are sought may be expressed in weak conservation form as

$$\frac{\partial \mathbf{U}}{\partial t} + \nabla \cdot \mathbf{F} = \mathbf{S} + \mathbf{Q}, \tag{1}$$

where $\mathbf{U}$ is the vector of conserved variables, $\mathbf{F}$ is the solution flux dyad, and $\mathbf{S}$ and $\mathbf{Q}$ are volumetric source terms. The solution vector, $\mathbf{U}$, has the form

$$\mathbf{U} = [\,\rho,\ \rho\mathbf{V},\ \mathbf{B},\ \rho e,\ \psi\,]^T, \tag{2}$$

where ρ is the plasma density, $\mathbf{V}$ the velocity field, $\mathbf{B}$ the magnetic field, ρe is the total energy and ψ is the so-called generalized Lagrange multiplier variable associated with the GLM $\nabla \cdot \mathbf{B}$ treatment. The flux dyad, $\mathbf{F}$, is given by

$$\mathbf{F} = \begin{bmatrix} \rho\mathbf{V} \\ \rho\mathbf{V}\mathbf{V} + (p + \frac{\mathbf{B}\cdot\mathbf{B}}{2})\mathbf{I} - \mathbf{B}\mathbf{B} \\ \mathbf{V}\mathbf{B} - \mathbf{B}\mathbf{V} + \psi\mathbf{I} \\ (\rho e + p + \frac{\mathbf{B}\cdot\mathbf{B}}{2})\mathbf{V} - (\mathbf{V}\cdot\mathbf{B})\mathbf{B} \\ c_h^2\mathbf{B} \end{bmatrix}. \tag{3}$$

The specific total plasma energy is $e = p/(\rho(\gamma - 1)) + V^2/2 + B^2/(2\rho)$, where p is the molecular pressure, V is the magnitude of the fluid velocity, and B is the magnitude of the magnetic field. The numerical source term, $\mathbf{S}$, is due to the GLM-MHD formulation and has the form

$$\mathbf{S} = [0, \mathbf{0}, \mathbf{0}, 0, -\frac{c_h^2}{c_p^2}\psi]^T, \tag{4}$$

in which the coefficients c_p and c_h control the relative rates of dissipation and transport of ψ, as well as the corresponding advection speed of the $\nabla \cdot \mathbf{B}$ cleaning mechanism, respectively. The ideal gas equation of state, $p = \rho RT$, is assumed, where T is the gas temperature and $R = 1/\gamma$ is the gas constant. For a polytropic gas (thermally and calorically perfect), the ratio of plasma specific heats, γ, is a constant, and the specific heats are given by $C_v = 1/(\gamma - 1)$ and $C_p = \gamma/(\gamma - 1)$. The source vector, $\mathbf{Q}$, appearing in Eq. (1) generally represents different volumetric sources arising from the physical modelling of various space plasma flows, such as gravitational forces.

2.1 Semi-Discrete Finite-Volume Formulation

The semi-discrete form of the preceding upwind finite-volume scheme applied to Eq. (1) for hexahedral computational cell (i, j, k) of a three-dimensional grid is

$$\frac{\mathrm{d}\overline{\mathbf{U}}_{ijk}}{\mathrm{d}t} = -\frac{1}{V_{ijk}} \sum_{f=1}^{6} \sum_{m=1}^{N_g} (\tilde{\omega}\mathbf{F} \cdot \mathbf{n})_{i,j,k,f,m} + (\overline{\mathbf{S}})_{ijk} + (\overline{\mathbf{Q}})_{ijk} = (\overline{\mathbf{R}})_{ijk}(\overline{\mathbf{U}}) , \tag{5}$$

where N_g is the number of Gauss quadrature points and $\mathbf{n}$ is the local normal of the face f at each of the N_g Gauss quadrature points. The hexahedral cells are contained within logically Cartesian blocks that form a multi-block body-fitted mesh with general unstructured connectivity between blocks. The total number of Gauss integration points, N_g, at which the numerical flux is evaluated is chosen as the minimum required to preserve the targeted rate of convergence for solution accuracy. In this work, standard tensor-product quadrature consisting of four Gauss quadrature points are used for the cell faces, providing a fourth-order accurate spatial discretization. The latter is the target accuracy for the high-order scheme considered here.

The numerical fluxes, $\mathbf{F} \cdot \mathbf{n}$, at each Gauss quadrature point on each face of a cell (i, j, k) are determined from the solution of a Riemann problem. Given the left and right interface solution values, $\mathbf{U}_l$ and $\mathbf{U}_r$, an upwind numerical flux is evaluated by solving a Riemann problem in the direction defined by the normal to the face. The values of $\mathbf{U}_l$ and $\mathbf{U}_r$ are determined by performing the CENO reconstruction as detailed in the next section. The contributions of the volumetric sources $\overline{\mathbf{S}}_{ijk}$, $\overline{\mathbf{Q}}_{ijk}$ are evaluated to fourth-order accuracy by again using a standard tensor-product Gauss quadrature with twenty-seven points for the volumetric integration. In the present computational studies, the Lax-Friedrichs approximate Riemann solver and fourth-order accurate Runge-Kutta explicit time-marching scheme have been used. Steady-state solutions are obtained using the latter by integrating the solution forward in time until a steady result is achieved.

3 High-Order CENO Finite-Volume Scheme

The hybrid CENO finite-volume method for conservation laws originally proposed by Ivan and Groth [16] is used to discretize the governing equations on a hexahedral computational grid. The hybrid CENO procedure uses the multidimensional unlimited k-exact reconstruction of Barth [1] in smooth regions and reverts to a limited piecewise-linear reconstruction algorithm in regions deemed as non-smooth or under-resolved by a solution smoothness indicator, thus providing monotone solutions near discontinuities.

In the present study, only smooth flows are considered reducing the CENO procedure to an unlimited fourth-order reconstruction. The Kth-order Taylor series polynomial expansion of the spatial distribution of a scalar solution quantity, U_{ijk}, within a cell with index ijk about the cell-centroid $(x_{ijk}, y_{ijk}.z_{ijk})$ can be expressed as:

$$U_{ijk}(x, y, z) = \sum_{p_1=0}^{K} \sum_{p_2=0}^{K} \sum_{\substack{p_3=0 \\ (p_1+p_2+p_3 \le K)}}^{K} (x - x_{ijk})^{p_1} (y - y_{ijk})^{p_2} (z - z_{ijk})^{p_3} D_{p_1 p_2 p_3} . \tag{6}$$

The coefficients, $D_{p_1 p_2 p_3}$, of the Taylor polynomials are referred to as the unknown derivatives and their number is equal to 20 for the target fourth-order accurate ($K = 3$ piecewise cubic) reconstruction. They are obtained by solving a constrained least-squares problem as detailed in Ivan et al. [16]. In order to obtain an exactly determined or overdetermined set of linear equations, a stencil including the two nearest rings of neighbours is used whatever is the mesh discretization size in the neighbouring cells. In particular, $5 \times 5 \times 5$ cells are used in a region with uniform resolution, and for regions with resolution changes or where the grid connectivity is irregular (such as at cubed-sphere sector edges), more or less numbers of cells may be used.

Both Householder QR factorization and singular value orthogonal decomposition (SVD) can be used to solve the weighted least-squares problem associated with the CENO reconstruction [16]. The latter is exploited here. The SVD approach permits the computation of a pseudo-inverse matrix after which the solution of the least-squares problem is then given by a simple matrix-vector product. The use of a single fixed stencil, the same for all dependent variables, allows the pseudo-inverse matrix to be stored and re-used in the reconstruction of all variables, thereby avoiding the repeated evaluation of the pseudo inverse. This was found to reduce significantly the computational costs of performing the CENO reconstruction without requiring substantial additional storage [16]. Additionally, there are conventionally issues with k-exact reconstruction related to conditioning and/or invertibility that generally increase with the order of the scheme as well as can be very dependent on mesh features, such as cell size, aspect ratio, and topology. However, a rather simple column-scaling procedure is applied here to the least-squares problem which significantly improves the conditioning, makes it virtually independent of the mesh, and affords robust and reliable solutions to the least-squares problem [16].

4 Parallel Anisotropic Block-Based AMR

A flexible block-based hierarchical binary tree data structure is used in conjunction with the spatial discretization procedure described in Sect. 2 to facilitate automatic solution-directed anisotropic mesh adaptation on body-fitted multi-block mesh. Figure 1 shows the resulting binary tree after several refinements of an initial mesh consisting of a single block. A binary tree is used rather than the usual octree used in isotropic methods, as the refinement decisions are made separately for each coordinate direction in the anisotropic AMR approach applied herein [9, 26].

The anisotropic AMR framework of Freret and Groth [9], based on extensions to the previous work by Williamschen and Groth [26], is well suited and readily allows the use of high-order spatial discretization by adopting a non-uniform representation of the cells within each block. An example of a non-uniform block obtained from a multi-block structure is shown in Fig. 2. In this treatment, the neighboring cells are used directly as the ghost cells, even those at different levels of refinement as found at grid resolution changes. This non-uniform treatment presents many advantages as outlined by Freret and Groth [9]. In particular, high-order

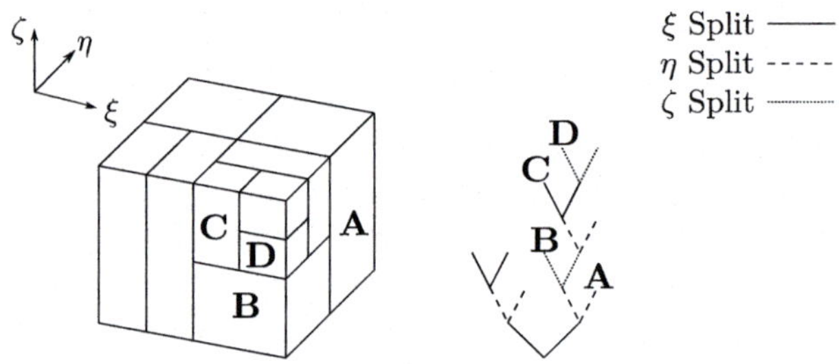

Fig. 1 3D binary tree and the corresponding blocks after several anisotropic refinements

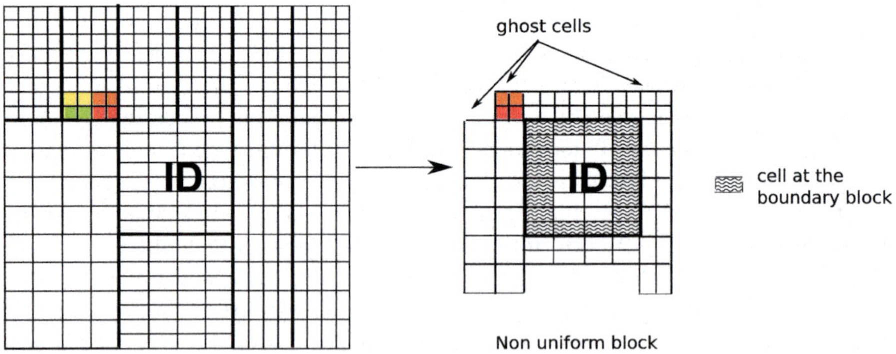

Fig. 2 Example of a non-uniform structured mesh block (*right*) obtained from a block-based anisotropic AMR grid mesh (*left*). This block is called non-uniform because its ghost cells may have different resolution from the interior cells

restriction and prolongation operators are not required to evaluate the solution within ghost cells.

Mesh adaptation is accomplished by refining and coarsening grid blocks. Each refinement produces new blocks called "children" from a "parent" block and the children can be refined further. This refinement process can be reversed in regions that are deemed over-resolved and two, four or eight children can coarsen or merge into a single parent block. In the present work we use a refinement criteria based on the gradient of a given quantity. This quantity can be a test function as in Sect. 5.1 or the fluid density as used in Sects. 5.2 and 5.3. The refinement criteria is a three-component vector such that the mesh can be refined in an anisotropic way.

A high-order accurate solution transfer from the coarse cell to the fine cells is required to distribute the average solution quantity among offspring with high-order accuracy. The high-order reconstruction polynomials of all solution variables on the coarse cell are readily integrated over the domain of each new fine cell having a volume, V_{fine}, and the resulting integrated average values of a solution quantity within the fine cells, $\bar{u}_{fine}$, is given by

$$\bar{u}_{fine} = \frac{1}{V_{fine}} \iiint_{V_{fine}} u_{coarse}(\mathbf{X}) \mathrm{d}V = \frac{1}{V_{fine}} \sum_{m=1}^{N_g} \omega_m \, u_{coarse}(\mathbf{X_m}) \,, \tag{7}$$

where the volume integral is computed exactly for the reconstruction polynomial with an appropriate-order tensor-product Gauss quadrature volumetric integration technique ($N_g = 27$ quadrature points are used for fourth-order spatial accuracy [16]). Here, ω_m and $\mathbf{X_m}$ are the fine-cell Gauss weights and quadrature points.

5 Numerical Results

To validate the proposed fourth-order CENO finite-volume method for use in combination with the anisotropic AMR strategy outlined in Sect. 4, 3D numerical results are now considered, including a demonstration of the accuracy of the high-order CENO scheme for reconstruction of a known function and numerical predictions for steady-state flow problems on cubed-sphere grids. For the latter, numerical results for both non-magnetized and magnetized flows are used to evaluate the grid convergence of the CENO method when anisotropic AMR is applied. Additionally, the computational efficiency of the fourth-order CENO method is also compared to that of the second-order method described previously by Ivan et al. [17] and also extended herein for use in conjunction with the anisotropic AMR scheme.

5.1 Function Reconstruction on a Cubed-Sphere Grid

To demonstrate the accuracy of the CENO reconstruction applied in conjunction with anisotropic AMR, numerical results for the reconstruction of a smooth

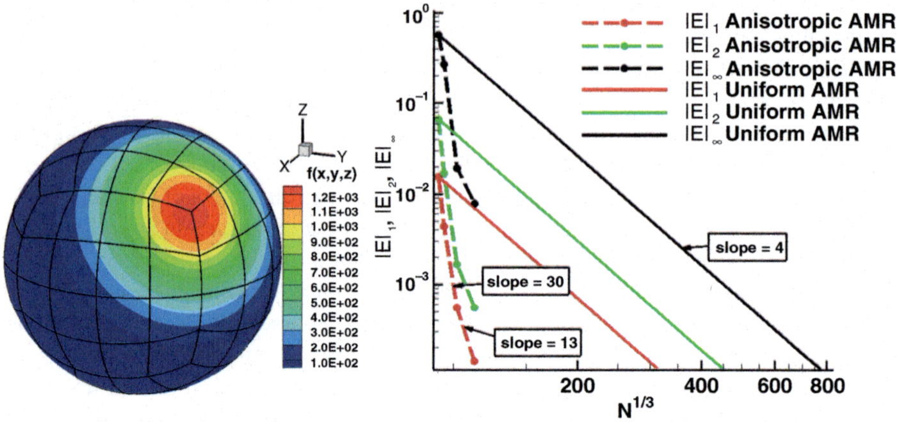

Fig. 3 (*left*) Contours of the test function f, (*right*) L_1, L_2, L_∞ error norms for the fourth-order CENO scheme with anisotropic AMR (*dashed lines*) compared to the error norms with fourth-order scheme with uniform refinements (*solid lines*)

continuous function are examined. This initial numerical test proceeds by first computing accurate cell averages for the function to be reconstructed and then using these cell averages to compute high-order polynomial reconstructions in the cells and finally computing the error between the original function and the polynomial reconstruction by high-accuracy numerical integration over each cell. The order of convergence of this error measures the order of accuracy of the CENO reconstruction. For the particularly case of interest, reconstruction of the function

$$f(x, y, z) = (1 - R + R^2)e^{x+y+z}, \tag{8}$$

is considered on the spherical computational domain defined by two concentric spheres with inner and outer radius $R_i = 1$ and $R_o = 3$, where R denotes the radius. As depicted in Fig. 3-left, this function exhibits a large smooth variation spanning several orders of magnitude that is oriented along the line connecting two diametrically-opposed cubed-sphere corners, where the function maximum and minimum occur. The computational meshes used in this grid convergence study range in size from 786,432 to 50,331,648 cells using from 96 to 6144 solution blocks. As shown in Fig. 3-right, the L_1, L_2, and L_∞ error norms obtained for the reconstruction procedure show that the CENO scheme achieves the theoretical fourth-order convergence accuracy when uniform refinements are applied (solid lines). The improved accuracy exhibited by the use of the anisotropic AMR translates into significant savings in terms of computational cell number for a targeted solution error. For example, to achieve $L_1 = 10^{-3}$ solution error, the fourth-order CENO method requires about 5,832,000 cells which is more than 5 times the mesh requirements when anisotropic AMR is applied. Moreover it is worth mentioning that the refinement criteria is based on the gradient of f defined in Eq. (8) and user-defined thresholds for refinement and coarsening. Refinement strategies based on estimated solution error [25] are not applied here and therefore the mesh refinement does not guarantee a specific target solution accuracy. For this reason, the slopes of the error norms with AMR vary between 13 and 30, as shown in Fig. 3-right.

5.2 *Steady Supersonic Outflow of Non-Magnetized Plasma*

To assess the accuracy of the finite-volume scheme on cubed-sphere grids, numerical convergence studies for a spherically symmetric expanding supersonic non-magnetized plasma flow have been performed and are considered next. The accuracy of the fourth-order CENO scheme for a series of uniform and anisotropic refined AMR meshes was determined and is compared here to similar results obtained using the corresponding second-order scheme [17]. The computational domain of the steady supersonic outflow of interest is defined by inner and outer spheres of radius $R_i = 1$ and $R_o = 4$ respectively. For boundary data, the exact solution is imposed on the inner sphere: $\rho_i = 10$, $V_{r,i} = 4.5$, $\mathbf{V}_{||,\mathbf{i}=\mathbf{0}}$ and $p_i = 26$. An outflow supersonic boundary condition is imposed at R_o. As described by Ivan et al. [17], the analytical solution of this flow problem can be obtained in spherical coordinates as the solution of the equation

$$C_3 - \frac{1}{r^2 V_r \left[(C_2 - V_R^2)^{\frac{1}{\gamma-1}} \right]} = 0 \,, \tag{9}$$

where C_2 and C_3 are constants depending on the inflow conditions.

The L_1, L_2 and L_∞ norms of the error in the predicted solution density obtained on a series of grids are given in Fig. 4. These convergence results show that

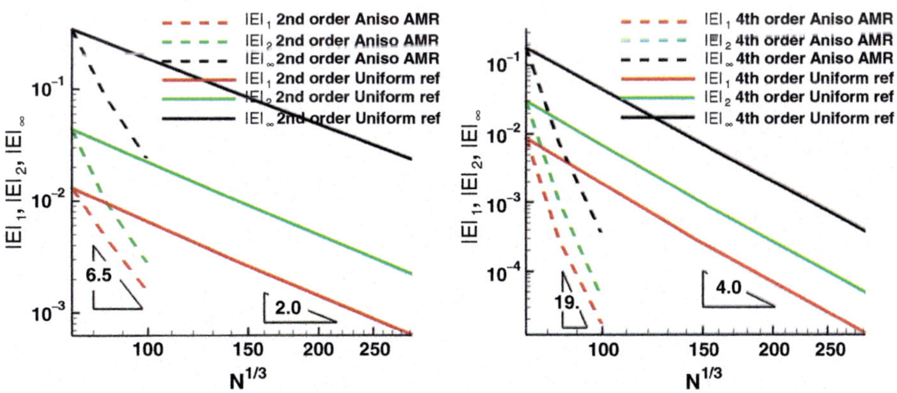

Fig. 4 L_1, L_2, L_∞ error norms for second-order (*left*) and fourth-order CENO schemes (*right*) with anisotropic AMR (*dashed lines*) compared to successive uniform refinements (*solid lines*)

the expected second-order (Fig. 4-left) and fourth-order (Fig. 4-right) theoretical accuracies are achieved in all error norms as the mesh is uniformly refined (solid lines). For anisotropic refinement of the mesh via the AMR strategy, the effective convergence rate approaches 6.5 for the second-order scheme and 19 for the CENO fourth-order scheme. As noted previously [9, 26], the solution varies only along the radial direction and the anisotropic AMR exploits this feature by refining only in the radial direction, thus avoiding the introduction of an unnecessary large number of computational cells. When the CENO scheme is used (Fig. 4-right), for an error target of $L_1 = 10^{-4}$, the memory requirement of the anisotropic AMR is only 12% of the memory requirement of the uniform refinements. For the second-order scheme (Fig. 4-left), for an error target $L_1 = 3 \times 10^{-3}$, the mesh saving of the anisotropic AMR strategy is around 73% compared to the uniform refinements. Finally, the mesh saving between second-order and fourth-order schemes with uniform refinements is about 90% for a target error of $L_1 = 10^{-3}$.

5.3 *Steady Supersonic Outflow of Magnetized Plasma*

As a final example, steady supersonic outflow of a magnetized plasma on a spherical domain is considered. The exact solution for this case is given by

$$\mathbf{U}(x, y, z) = \left[r^{-\frac{5}{2}}, \frac{x}{\sqrt{r}}, \frac{y}{\sqrt{r}}, \frac{z}{\sqrt{r}} + \kappa r^{\frac{5}{2}}, \frac{x}{r^3}, \frac{y}{r^3}, \frac{z}{r^3} + \kappa, r^{-\frac{5}{2}} \right]^T, \tag{10}$$

where $\kappa = 0.017$ is a perturbation parameter chosen such that the solution has significant latitudinal variation [17]. In Eq. (1), the source term $\mathbf{Q}$ can be written as

$$\mathbf{Q} = \begin{bmatrix} 0 \\ \frac{1}{2} x r^{-\frac{5}{2}} (r^{-1} - 5r^{-2} - \kappa z) \\ \frac{1}{2} y r^{-\frac{5}{2}} (r^{-1} - 5r^{-2} - \kappa z) \\ \frac{1}{2} z r^{-\frac{5}{2}} (r^{-1} - 5r^{-2} - \kappa z) + \frac{5}{2} r^{-\frac{1}{2}} \kappa (1 + \kappa r z) + \kappa r^{-\frac{1}{2}} \\ 0 \\ \frac{1}{2} r^{-2} + \kappa z (3.5 r^{-1} + 2\kappa z) + \frac{(\kappa r)^2}{2} (7 + 5\kappa r z) \end{bmatrix}.$$

The computational domain used for the outflowing plasma flow problem is defined by inner and outer spheres of radius $R_i = 2$ and $R_o = 3.5$. The inflow boundary conditions are specified at R_i based on the exact solution and outflow boundary conditions are applied at R_o.

The L_1, L_2 and L_∞ norms of the error in the predicted solution density at cells centroids were obtained on a series of grids and are given in Fig. 5. As the mesh is uniformly refined, the theoretical fourth-order accuracy is achieved for the CENO scheme. When anisotropic AMR is applied the slopes of the L_1, L_2 and L_∞ norms

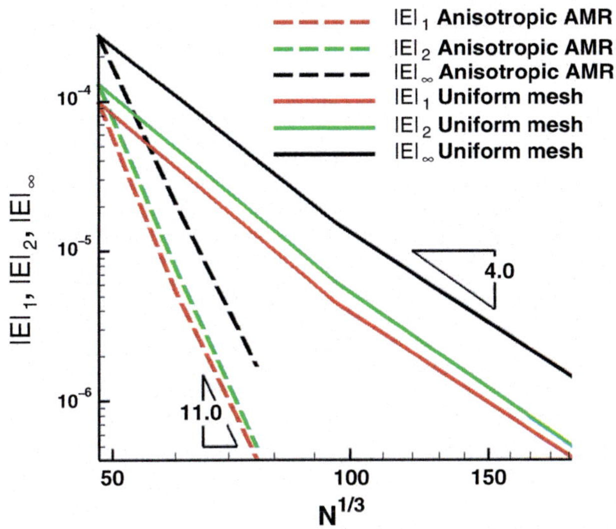

Fig. 5 L_1, L_2, L_∞ error norms of the solution density for the fourth-order CENO scheme with anisotropic AMR (*dashed lines*) compared to successive uniform refinements (*solid lines*)

are close to the value of 11. In terms of mesh size reduction, to achieve the error norm $L_1 = 10^{-6}$, the anisotropic AMR scheme uses only 9.2% of the number of cells of the uniform CENO scheme.

6 Conclusions

A fourth-order CENO finite-volume scheme has been extended for use with an efficient anisotropic block-based AMR method. High-order solutions on adapted anisotropic AMR grids have been obtained for three test problems on 3D cubed-sphere grids. The predicted results have been compared to those obtained using the high-order solution with uniform refinement as well as those of the associated second-order scheme, in order to assess the efficiency of the proposed approach. It is shown that high accurate solutions have been obtained with a reduced computational effort and significant reductions in mesh size. A natural future extension will be to consider the application to 3D unsteady MHD flows with both smooth solution content and shocks.

Acknowledgements This work was supported by the Canadian Space Agency and by the Natural Sciences and Engineering Research Council (NSERC) of Canada. In particular, the authors would like to acknowledge the financial support received from the Canadian Space Agency through the Geospace Observatory Canada program. Computational resources for performing all of the calculations reported herein were provided by the SciNet High Performance Computing

Consortium at the University of Toronto and Compute/Calcul Canada through funding from the Canada Foundation for Innovation (CFI) and the Province of Ontario, Canada.

References

1. T.J. Barth, Recent developments in high order k-exact reconstruction on unstructured meshes. Paper 93-0668, AIAA (1993)
2. J.B. Bell, M.J. Berger, J.S. Saltzman, M. Welcome, A three-dimensional adaptive mesh refinement for hyperbolic conservation laws. SIAM J. Sci. Comput. **15**(1), 127–138 (1994)
3. M.J. Berger, J. Oliger, Adaptive mesh refinement for hyperbolic partial differential equations. J. Comput. Phys. **53**, 484–512 (1984)
4. M.R.J. Charest, C.P.T. Groth, A high-order central ENO finite-volume scheme for three-dimensional turbulent reactive flows on unstructured mesh. Paper 2013–2567, AIAA (2013)
5. M.R.J. Charest, C.P.T. Groth, P.Q. Gauthier, A high-order central ENO finite-volume scheme for three-dimensional low-speed viscous flows on unstructured mesh. Commun. Comput. Phys. **17**(3), 615–656 (2015)
6. C.R. Clauer, T.I. Gombosi, D.L. De Zeeuw, A.J. Ridley, K.G. Powell, B. van Leer, Q.F. Stout, C.P.T. Groth, T.E. Holzer, High performance computer methods applied to predictive space weather simulations. IEEE Trans. Plasma Sci. **28**(6), 1931–1937 (2000)
7. D.L. De Zeeuw, T.I. Gombosi, C.P.T. Groth, K.G. Powell, Q.F. Stout, An adaptive MHD method for global space weather simulations. IEEE Trans. Plasma Sci. **28**(6), 1956–1965 (2000)
8. A. Dedner, F. Kemm, D. Kröner, C.D. Munz, T. Schnitzer, M. Wesenberg, Hyperbolic divergence cleaning for the MHD equations. J. Comput. Phys. **175**, 645–673 (2002)
9. L. Freret, C.P.T. Groth, Anisotropic non-uniform block-based adaptive mesh refinement for three-dimensional inviscid and viscous flows. Paper 2015–2613, AIAA (2015)
10. X. Gao, A parallel solution-adaptive method for turbulent non-premixed combusting flows, Ph.D. thesis, University of Toronto, 2008
11. X. Gao, C.P.T. Groth, A parallel solution-adaptive method for three-dimensional turbulent non-premixed combusting flows. J. Comput. Phys. **229**(5), 3250–3275 (2010)
12. X. Gao, S.A. Northrup, C.P.T. Groth, Parallel solution-adaptive method for two-dimensional non-premixed combusting flows. Prog. Comput. Fluid Dyn. **11**(2), 76–95 (2011)
13. C.P.T. Groth, J.G. McDonald, Towards physically-realizable and hyperbolic moment closures for kinetic theory. Contin. Mech. Thermodyn. **21**(6), 467–493 (2009)
14. C.P.T. Groth, D.L. De Zeeuw, T.I. Gombosi, K.G. Powell, Global three-dimensional MHD simulation of a space weather event: CME formation, interplanetary propagation, and interaction with the magnetosphere. J. Geophys. Res. **105**(A11), 25053–25078 (2000)
15. L. Ivan, Development of high-order CENO finite-volume schemes with block-based adaptive mesh refinement, Ph.D. thesis, University of Toronto, 2011
16. L. Ivan, C.P.T. Groth, High-order solution-adaptive central essentially non-oscillatory (CENO) method for viscous flows. J. Comput. Phys. **257**, 830–862 (2014)
17. L. Ivan, H. De Sterck, S.A. Northrup, C.P.T. Groth, Hyperbolic conservation laws on three-dimensional cubed-sphere grids: a parallel solution-adaptive simulation framework. J. Comput. Phys. **255**, 205–227 (2013)
18. L. Ivan, H. De Sterck, A. Susanto, C.P.T. Groth, High-order central ENO finite-volume scheme for hyperbolic conservation laws on three-dimensional cubed-sphere grids. J. Comput. Phys. **282**, 157–182 (2015)
19. J.G. McDonald, J.S. Sachdev, C.P.T. Groth, Application of Gaussian moment closure to micro-scale flows with moving and embedded boundaries. AIAA J. **51**(9), 1839–1857 (2014)
20. S.A. Northrup, C.P.T. Groth, Solution of laminar diffusion flames using a parallel adaptive mesh refinement algorithm. Paper 2005-0547, AIAA (2005)

21. J.J. Quirk, An adaptive grid algorithm for computational shock hydrodynamics, Ph.D. thesis, Cranfield Institute of Technology, 1991
22. J.S. Sachdev, C.P.T. Groth, J.J. Gottlieb, A parallel solution-adaptive scheme for predicting multi-phase core flows in solid propellant rocket motors. Int. J. Comput. Fluid Dyn. **19**(2), 159–177 (2005)
23. A. Susanto, L. Ivan, H.D. Sterck, C.P.T. Groth, High-order central ENO finite-volume scheme for ideal MHD. J. Comput. Phys. **250**, 141–164 (2013)
24. L. Tobaldini Neto, C.P.T. Groth, A high-order finite-volume scheme for large-eddy simulation of turbulent premixed flames. Paper 2014–1024, AIAA (2014)
25. D.A. Venditti, D.L. Darmofal, Anisotropic grid adaptation for functional outputs: application to two-dimensional viscous flows. J. Comput. Phys. **187**, 22–46 (2003)
26. M.J. Williamschen, C.P.T. Groth, Parallel anisotropic block-based adaptive mesh refinement algorithm for three-dimensional flows. Paper 2013–2442, AIAA (2013)

GPU Acceleration of Hermite Methods for the Simulation of Wave Propagation

Arturo Vargas, Jesse Chan, Thomas Hagstrom, and Timothy Warburton

Abstract The Hermite methods of Goodrich, Hagstrom, and Lorenz (2006) use Hermite interpolation to construct high order numerical methods for hyperbolic initial value problems. The structure of the method has several favorable features for parallel computing. In this work, we propose algorithms that take advantage of the many-core architecture of Graphics Processing Units. The algorithm exploits the compact stencil of Hermite methods and uses data structures that allow for efficient data load and stores. Additionally the highly localized evolution operator of Hermite methods allows us to combine multi-stage time-stepping methods within the new algorithms incurring minimal accesses of global memory. Using a scalar linear wave equation, we study the algorithm by considering Hermite interpolation and evolution as individual kernels and alternatively combined them into a monolithic kernel. For both approaches we demonstrate strategies to increase performance. Our numerical experiments show that although a two kernel approach allows for better performance on the hardware, a monolithic kernel can offer a comparable time to solution with less global memory usage.

1 Introduction

Wave simulation is essential to many fields of study. For example, in geophysics the numerical solution to the acoustic wave equation is central to various imaging algorithms such as Reverse Time Migration [2] and Full Waveform Inversion [16].

A. Vargas (✉) • J. Chan
Rice University, Houston, TX, USA
e-mail: arturo.vargas@rice.edu; jesse.chan@caam.rice.edu

T. Hagstrom
Southern Methodist University, Dallas, TX, USA

T. Warburton
Virginia Tech, Blacksburg, VA, USA

M.L. Bittencourt et al. (eds.), *Spectral and High Order Methods for Partial Differential Equations ICOSAHOM 2016*, Lecture Notes in Computational Science and Engineering 119, DOI 10.1007/978-3-319-65870-4_25

In the context of electromagnetism, numerical simulations of Maxwell's equations are employed in the design of new products such as radars and antennae [14]. The need to resolve high frequency waves over long periods of time makes these simulations challenging. High order numerical methods can be more efficient than lower order methods for such simulations as they minimize dispersion and offer high convergence rates for smooth solutions [8].

The Hermite methods introduced by Goodrich et al. [5], are a class of cell based numerical methods which reconstruct a polynomial-based solution at each cell through Hermite interpolation. The methods can be constructed to achieve high order accuracy making them well suited for high frequency wave simulations. An advantageous feature of these methods is their high computation to communication ratio, making them ideal for parallel computing [3]. High performance implementations of Hermite methods have been carried out in [1, 6] for aero-acoustics and compressible flows in which numerical experiments demonstrated favorable results in terms of scalability on CPU-based clusters.

Recent trends in processor design has resulted in multi-core processors with wide single instruction multiple data (SIMD) vector units. Each SIMD group has access to a relatively small shared memory cache and each SIMD lane has a small number of fast registers. Typical GPUs are further equipped with large bandwidth, high latency, global shared memory storage. To achieve high performance on GPUs fine-grained parallelism must be exposed with minimal communication between computing units. Examples of numerical algorithms that have demonstrated utility of the GPU can be found in [9–12]

Hermite methods were first implemented on a GPU by Dye in [4], wherein strategies for two-dimensional equations were presented. Building on the work of Dye, we introduce strategies for three-dimensional linear equations. In Sect. 2 we provide a brief overview of Hermite methods. Section 3 introduces our strategies for tailoring Hermite methods onto the GPU and lastly Sect. 4 studies performance with respect to the GPU hardware.

2 Overview of Hermite Methods

To highlight key concepts of Hermite methods, we consider the three-dimensional advection equation

$$\frac{\partial u}{\partial t} = -\frac{\partial u}{\partial x_1} - \frac{\partial u}{\partial x_2} - \frac{\partial u}{\partial x_3}. \tag{1}$$

Hermite methods represent the solution of an initial value boundary problem on a grid constructed through tensor products of one-dimensional grids. We denote the mth node for the kth dimension as x_{k,m_k} and for simplicity consider periodic grids. The degrees of freedom of the method, at time step $t_n = t_0 + n\Delta t$, are represented over each node in the form of the tensor product of the function value and first N

(scaled) derivatives in each dimension

$$p^{i}_{m_1,m_2,m_3}(t_n) \approx \frac{h^{|i|}}{i!} D^i u\left(x_{1,m_1}, x_{2,m_2}, x_{3,m_3}, t_n\right). \tag{2}$$

Here h denotes the spacing between the nodes, D denotes the derivative operator, and $i = (i_1, i_2, i_3)$ denotes the multi-index with i_j ranging from 0 to N.

To represent the solution on each cell of the grid a staggered (dual) grid is introduced. The cell midpoints of the primary grid make up the dual grid. The dual grid facilitates the construction of tensor polynomials (Hermite interpolants)

$$Rp_{m_1+\frac{1}{2},\ldots,m_3+\frac{1}{2}} = \sum_{j_1=0}^{2N+1} \cdots \sum_{j_3=0}^{2N+1} b_{j_1,\ldots,j_3} \left(\frac{x_1 - x_{1,m_1+\frac{1}{2}}}{h_{x_1}}\right)^{j_1} \cdots \left(\frac{x_3 - x_{3,m_3+\frac{1}{2}}}{h_{x_3}}\right)^{j_3}, \tag{3}$$

which interpolate the function value and derivatives at each cell's vertices. The coefficients of the tensor product polynomial are the approximation of the function value and the derivatives at the midpoint of the cell.

Evolution from t_n to $t_{n+\frac{1}{2}}$ is carried out independently on each cell by the use of a q-order temporal Taylor series expansion centered at t_n of the tensor product polynomial

$$TRp = \sum_{j_1=0}^{2N+1} \cdots \sum_{j_3=0}^{2N+1} \sum_{s=0}^{q} b_{j_1,\ldots,j_3,s} \left(\frac{x_1 - x_{1,m_1+\frac{1}{2}}}{h_{x_1}}\right)^{j_1} \cdots \left(\frac{x_3 - x_{3,m_3+\frac{1}{2}}}{h_{x_3}}\right)^{j_3} \left(\frac{t - t_n}{\Delta t}\right)^{s}. \tag{4}$$

The scalar Δt denotes the size of a full time step. For $s = 0$ the coefficients of equation 4 are simply the coefficients from the Hermite interpolant (Eq. 3). The time-stepping scheme of Hermite methods, Hermite-Taylor, expresses the values of unknown coefficients in terms of known coefficients by applying the Cauchy-Kowalweski recursion to the PDE. For brevity, we omit the derivation and provide the resulting recursion for the three-dimensional advection equation

$$b_{j_1 j_2 j_3, s+1} = -\frac{j_1+1}{s+1}\frac{\Delta t}{h_{x_1}} b_{j_1+1 j_2 j_3,s} - \frac{j_2+1}{s+1}\frac{\Delta t}{h_{x_2}} b_{j_1 j_2+1 j_3,s} - \frac{j_3+1}{s+1}\frac{\Delta t}{h_{x_3}} b_{j_1 j_2 j_3+1,s}. \tag{5}$$

With the determined coefficients, the function value and derivatives for the midpoint are computed by evaluating the series at $t_{n+1/2}$. To complete a full time step the process is repeated on the dual grid to approximate the solution on the primary grid.

Hermite methods converge at a rate of $O(h^{2N+1})$ for smooth solutions, and are stable as long as the waves do not propagate from the cell boundaries to the cell center in a half step. A significant feature of the method's stability is that the result is independent of order. We refer the reader to [5, 7, 15] for further details on the methods.

By exploiting the compact stencil (the vertices of a cell) and local evolution of the method, we expose opportunities for parallelism. At a coarse level the polynomial reconstruction and evolution can be performed independently for each cell. At a finer level many of the operations can be carried out as one-dimensional matrix-vector multiplications. Because of the two levels of parallelism we demonstrate how the method can be mapped on to the many-core architecture of the GPU.

3 Implementing Hermite Methods on Graphics Processing Units

To simplify the performance analysis, we first implement the interpolation and evolution procedure as separate kernels. The drawback of this approach is the additional temporary memory required to store the reconstructed polynomial and additional memory transfers. In an effort to minimize global memory usage we implement a monolithic kernel performing both the interpolation and evolution.

Computation on the GPU is performed on a predefined grid of compute units. Following NVIDIA's nomenclature each unit of the grid is referred to as a thread. Threads are grouped to form thread blocks. The hardware provides a similar hierarchy for memory. Threads are provided with a small amount of exclusive memory, threads in a thread block share block exclusive memory (shared memory), and lastly the entire compute grid shares global memory. Moving data between the CPU and GPU is accomplished through the use of global memory which acts as a general buffer. We refer the reader to [13] for a detailed overview on GPU computing. All numerical experiments in this work are written in the Open Concurrent Compute Abstraction (OCCA) API [10] allowing for portability across hardware. Numerical experiments are carried out using an NVIDIA GTX 980 GPU in single precision using OCCA generated Compute Unified Device Architecture (CUDA) code. The hardware has theoretical peak bandwidth of 224 GB/sec and floating point performance of 4612 GFLOP/sec.

3.1 *Hermite Interpolation on the GPU*

The fundamental data structure used throughout our implementation is the tensor. For example the tensor

$$\mathbf{u}[m_3][m_2][m_1][n_3][n_2][n_1],$$

Algorithm 1 Polynomial reconstruction in the x_1 dimension

```
procedure ReconstructionInx1(H_x1, u_loc, Ru)
    for tz=0:2N+1 do
        for ty=0:2N+1 do
            for tx=0:2N+1 do
                c=0
                for k=0:2N+1 do
                    c += H_x1 [tx][k] u_loc[tz][ty][k]
                end for
                Ru[tz][ty][tx]=c;
            end for
        end for
    end for
end procedure
```

Algorithm 2 Polynomial reconstruction

```
procedure PolynomialReconstruction(H_x1, H_x2, H_x3, u_loc, Ru)
    Ru = H_x1 u_loc
    u_loc = H_x2 Ru
    Ru = H_x3 u_loc
end procedure
```

is used to store the function value and derivatives at each node of a three-dimensional grid. The three innermost indices correspond to a grid point on the grid and the outermost indices catalog the corresponding tensor product of function value and derivatives.

In three dimensions, polynomial reconstruction at a node on the dual grid, is accomplished by interpolating the function value and derivatives from vertices of the encapsulating cell. This requires reading $(N+1)^3$ degrees of freedom per vertex, for a total of eight vertices in three dimensions.

To facilitate the interpolation procedure a one-dimensional Hermite interpolation operator, $\mathbf{H}$ (see [15] for details on construction), is pre-computed enabling dimension-by-dimension reconstruction of the polynomial. In this kind of reconstruction, the degrees of freedom of the encapsulating cell are stored in a local rank 3 tensor, $\mathbf{u_{loc}}$. The one-dimensional operator, $\mathbf{H}$, is then applied to the degrees of freedom of nodes parallel to the x_1 dimension as a series of matrix-vector multiplications. Next, the operator is applied to the degrees of freedom of nodes parallel to the x_2 dimension, and lastly to the degrees of freedom of nodes parallel to the x_3 dimension. For clarity we define $\mathbf{H}_{x_1}$, $\mathbf{H}_{x_2}$, and $\mathbf{H}_{x_3}$ as operators to be applied in the x_1, x_2, and x_3 dimensions respectively. Algorithm 1 presents the application of the interpolation operator to nodes parallel to the x_1 dimension using nested for loops. Applying the operator in the x_2, and x_3 dimensions is performed analogously. The complete reconstruction procedure for a single polynomial is listed as Algorithm 2.

Our GPU implementation exposes two levels of parallelism: coarse parallelism, in which threads in a block collectively reconstructs polynomials, and fine-grain

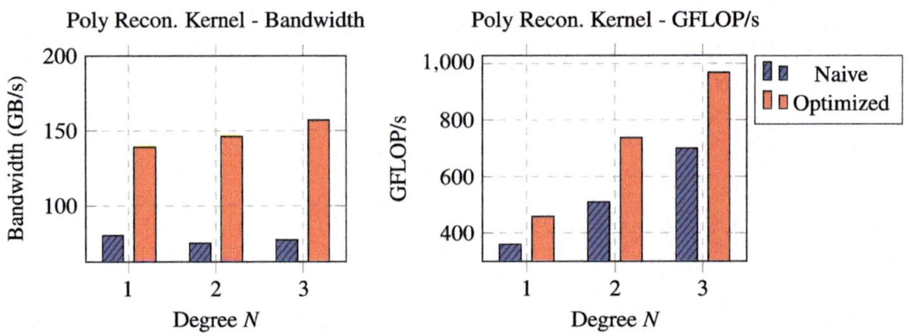

Fig. 1 Performance of the interpolation kernel. The optimized kernel assigned the construction of 16, 10, and 4 interpolants to each block threads for $N = 1, 2, 3$, reconstructing order 3, 5, and 7 polynomials respectively

parallelism in which threads apply the interpolation operator as a series of matrix-vector multiplications. The reconstruction is carried out locally by moving the necessary degrees of freedom to shared memory.

To minimize and reuse global memory reads we apply a similar register rolling technique as used in Finite Difference Time Domain methods [11]. Hermite methods can mimic this technique by having a block of threads reuse a subset of shared memory. This is accomplished by setting up a two-dimensional grid of thread blocks. A single block of threads moves the bottom four vertices of a cell to shared memory. The block of threads then applies the interpolation operators $\mathbf{H}_{x_1}$ and $\mathbf{H}_{x_2}$. As it progresses along the x_3-dimension it stores the next four vertices of the cell in shared memory and applies $\mathbf{H}_{x_1}$ and $\mathbf{H}_{x_2}$ to the newly added degrees of freedom. As there are now degrees of freedom for eight vertices, the $\mathbf{H}_{x_3}$ operator is then applied to the degrees of freedom parallel to the x_3 dimension and the result is stored in a rank 6 tensor similar to the initial degrees of freedom. The block of threads then shifts forward to the next set of four nodes and repeats the polynomial reconstruction.

We add a tunable parameter: the number of polynomials reconstructed along the x_1-dimension per block of threads. This can further reduce the total amount of global memory reads as neighboring cells share nodes on the interface. Figure 1 reports the performance for the polynomial reconstruction kernel under a naive implementation, with no reuse of existing memory reads, and the optimized kernel with reuse of global memory reads. Figure 2 visualizes the relationship between global throughput and bandwidth. As we increased data reuse we observed higher bandwidth. Additionally for the reconstruction of order 3, and 5 polynomials, $N = 1, 2$ respectively, there was a reduction in throughput in the optimized kernel suggesting caches are being exploited.

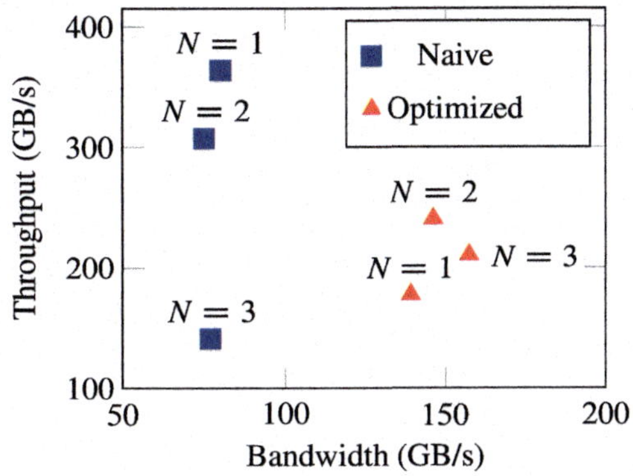

Fig. 2 Plotting bandwidth against throughput shows that assigning more cells per block reduces global throughput for orders $N = 1, 2$, which reconstruct polynomials of orders 3, and 5, while increasing global bandwidth

3.2 *Hermite-Taylor Methods on the GPU*

With the polynomial reconstruction procedure described in the previous section completed we may now advance the solution using the Hermite-Taylor method. For each reconstructed polynomial, the procedure can be performed locally using a rank 3 tensor to store the coefficients

$$(\mathbf{Ru})[n_3][n_2][n_1],$$

where n_3, n_2, n_1 range from $0, \cdots, 2N+1$, corresponding to the order of spatial derivative in each spatial dimension. Differentiating the reconstructed polynomial with respect to a spatial dimension is achieved by applying the following derivative matrix

$$\mathbf{D}_{ij} = \begin{cases} \frac{i+1}{h} & , \quad j = i+1 \\ 0 & , \quad \text{otherwise} \end{cases} \qquad 0 \leq i,j \leq 2N+2,$$

to the reconstructed polynomial. For convenience $\mathbf{D}_{x_1}$ will represent an operator to be applied to the reconstructed polynomial with respect to the x_1 dimension. In a similar manner the operators $\mathbf{D}_{x_2}$, and $\mathbf{D}_{x_3}$ will represent an operator to be applied to the reconstructed polynomial with respect to the x_2, and x_3 dimensions. For example application of the derivative matrix along the x_1 dimension to the reconstructed polynomial is illustrated in Algorithm 3 using nested for loops. Differentiating the reconstructed polynomial in the remaining dimensions is accomplished analogously. With the compact notation the Hermite-Taylor algorithm can be reduced to a q-stage loop as listed in Algorithm 4. Carrying out $q = d(2N+1)$ stages, in d dimensions, allows for the largest possible time step. Taking $q < d(2N+1)$ corresponds to a lower order temporal approximation and may require a smaller time-step in order to maintain expected order of convergence and stability. Similar to the polynomial reconstruction kernel, we consider two levels of parallelism: a

Algorithm 3 Differentiation in the x_1-dimension

procedure DIFFERENTIATIONINx_1($\mathbf{D}_{x_1}$,**Ru**,$\mathbf{Ru}_x$)
  for $tz = 0, 2N + 1$ **do**
    for $ty = 0, 2N + 1$ **do**
      for $tx = 0, 2N + 1$ **do**
        if $tx < 2N + 1$ **then**
          $p_{x_1} = \frac{(tx+1)}{h_x}\mathbf{Ru}[tz][ty][tx + 1]$
        else
          $p_{x_1} = 0$
        end if
        $\mathbf{Ru}_x[tz][ty][tx] = p_{x_1}$
      end for
    end for
  end for
end procedure

Algorithm 4 Hermite-Taylor evolution

procedure HERMITE-TAYLOREVOLUTION($\mathbf{D}_{x_1}$,$\mathbf{D}_{x_2}$,$\mathbf{D}_{x_3}$,**Ru**)
  $\hat{\mathbf{w}} = \mathbf{Ru}$
  for $k = q, q - 1, \ldots, 1$ **do**
    $\hat{\mathbf{w}} = \mathbf{Ru} - \frac{\Delta t}{k}(\mathbf{D}_{x_1}\hat{\mathbf{w}} + \mathbf{D}_{x_2}\hat{\mathbf{w}} + \mathbf{D}_{x_3}\hat{\mathbf{w}})$
  end for
  $\mathbf{Ru} = \hat{\mathbf{w}}$
end procedure

coarse level in which each block of threads carries out the Hermite-Taylor scheme for a number of cells and a fine-grained level in which threads to carry out the computation. Numerical experiments demonstrated that increasing the number of stages, q, in the scheme increases computational intensity. Peak performances were observed when assigning a block of threads to evolve the solution at 16, 10, and 2 cells for orders N=1, 2, and 3 respectively. Performance results are reported in Fig. 3.

3.3 A Monolithic Kernel

A two kernel approach allows for fine tuning of each individual procedure at the cost of storing the coefficients for the reconstructed polynomial. In the interest of minimizing global storage we combine the polynomial reconstruction and evolution procedures to a single monolithic kernel. We repeat previous experiments carried out in Sect. 3.2 and observe the relationship between number of stages in the Hermite-Taylor scheme and performance. Figure 4 reports the performance for the monolithic kernels.

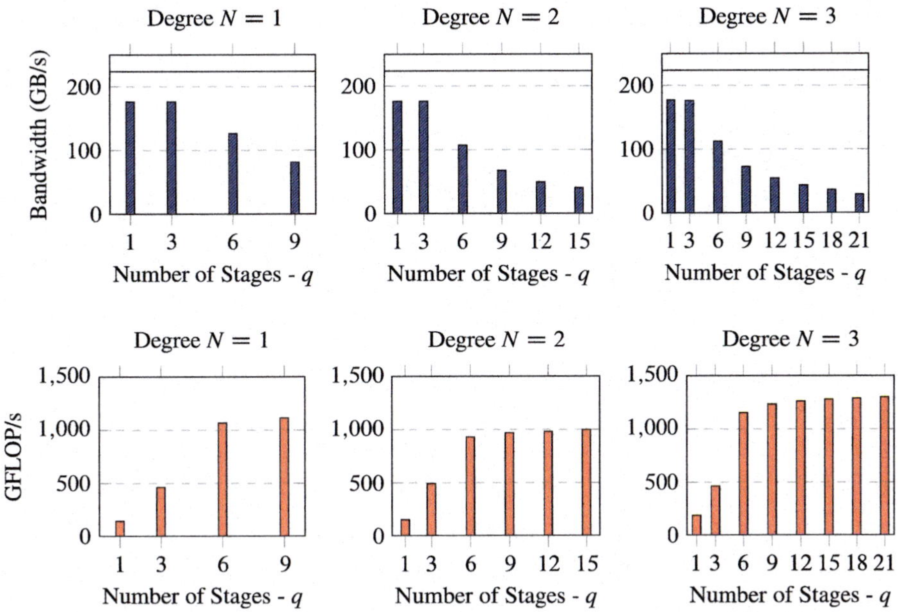

Fig. 3 Performance of the Hermite-Taylor kernel, the kernel assigns the evolution at 16, 10, and 2 cells per block of threads. An Nth degree method reconstructs local order $2N + 1$ polynomials. As the order of the temporal expansion increases the kernel becomes more compute intensive. The solid line on the bandwidth plot denotes the peak theoretical bandwidth of the device

4 Roofline Analysis and Time to Solution

The Roofline model relates flops, bandwidth, and hardware [17]. It provides an upper bound on the rate of floating point operations based on the arithmetic intensity of a given kernel. Arithmetic intensity is defined as

$$\text{arithmetic intensity} = \frac{\text{FLOPs performed}}{\text{bytes loaded}}.$$

Pairing the arithmetic intensity and the physical capabilities of the hardware allows the roofline model to present a theoretical ceiling on performance for a given kernel. Theoretical achievable performance is defined as

$$\min(\text{arithmetic intensity} \times \text{peak bandwidth}, \ \text{peak GFLOP/s}).$$

Figure 5 profiles the Hermite kernels in this work with respect to the Roofline model and reports the computational efficiency. The Hermite-Taylor and monolithic kernels are profiled using a $q = d(2N + 1)$ stage loop. Typically there are two types of computational bottle necks, bandwidth or compute. Kernels which are bandwidth

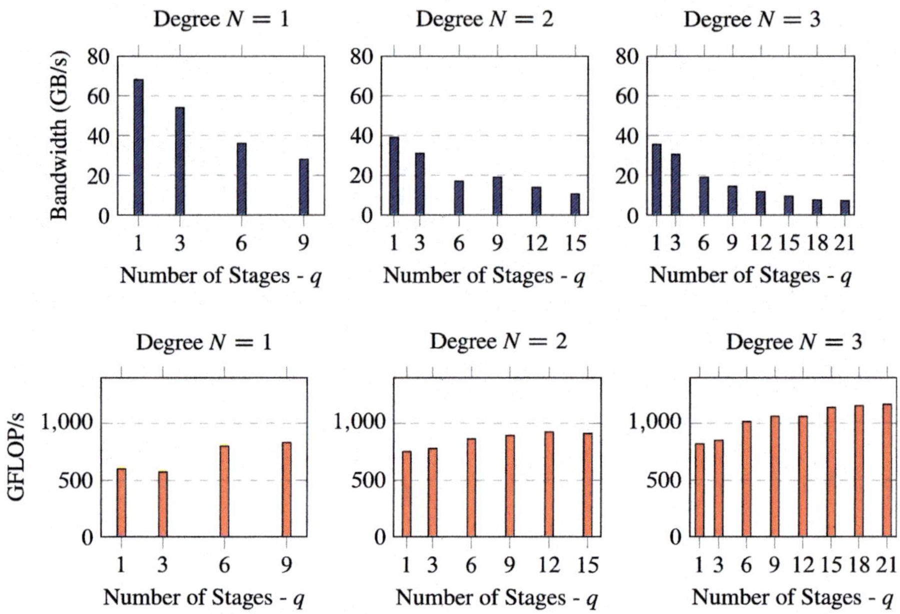

Fig. 4 Performance of the monolithic kernel. An Nth degree method reconstructs local order $2N+1$ polynomials. As the order of the temporal expansion increases the kernel becomes more compute intensive. Peak performances were found when assigning 12, 10, 2 cells per block of threads for orders $N = 1, 2, 3$ respectively

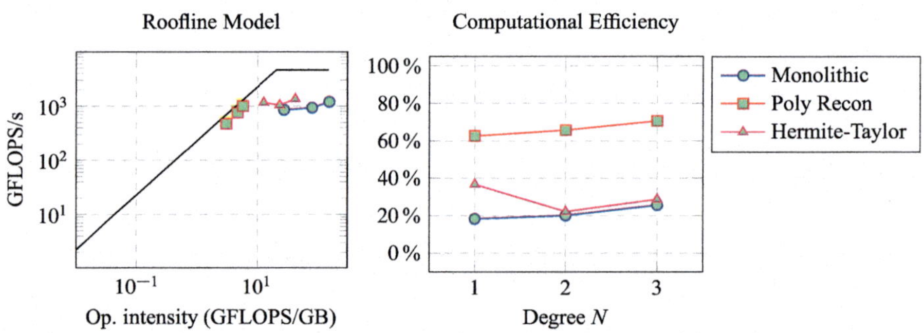

Fig. 5 Roofline performance analysis for the various Hermite method kernels

limited are constrained by a device's ability to read and write to global memory. Compute bound kernels are limited by the device's ability to perform floating point operations. The Roofline model places kernels limited by bandwidth on the bottom left while compute bound kernels are found on the top right. We observe that our kernels have a higher compute intensity and are closer to being compute bound. This is largely due to the high number of stages in the Hermite-Taylor scheme.

Table 1 Comparison of time to solution

	$N = 1$	$N = 2$	$N = 3$
Advection: monolithic kernel	2.09 s	13.77 s	36.17 s
Advection: two kernels	2.42 s	13.77 s	37.23 s

The initial condition is propagated for 200 time-steps on a fixed grid of 150 grid points in each dimension. A degree N Hermite method converges at a rate of $O(h^{2N+1})$

Reducing the number of stages reduces the floating point intensity. Noticeably the interpolation and evolution kernels achieve a higher hardware efficiency.

Although separate kernels for the interpolation and evolution lead to better hardware efficiency, computational experiments have demonstrated that both approaches lead to comparable times to solution. The monolithic kernel has the advantage of requiring less reads/writes to global memory in comparison to the two kernel approach. Table 1 reports a comparison of time to solution for the advection equation on a fixed grid with 150 grid points in each dimension propagated for 200 time-steps. The caveat is that the local variables must be able to fit in shared memory when using a monolithic kernel. We carried out similar experiments for the acoustic wave equations and have found that peak performances were found by reducing the number of cells per block relative to the advection equation. The additional variables increases the use of hardware resources.

5 Conclusion

This work examines the use of a GPU as a kernel accelerator for Hermite methods. Hermite methods consist of two main components, the reconstruction of a polynomial of order $2N + 1$ and evolution via a space-time expansion. We presented two strategies in which to exploit the many-core architecture of the GPU. The first considered separate kernels for the polynomial reconstruction and evolution while the second considered a monolithic kernel. We demonstrated that separate kernels for the polynomial reconstruction and evolution make better use of the hardware capabilities but the fewer global memory read/writes of a single monolithic kernel enables for a comparable time to solution with less global memory usage. Future work will examine optimization strategies in the case of spatially varying coefficients and the employment of multiple GPUs.

Acknowledgements TH was supported in part by NSF Grant DMS-1418871. TW and JC were supported in part by NSF Grant DMS-1216674. Any opinions, findings, and conclusions or recommendations expressed in this material are those of the authors and do not necessarily reflect the views of the National Science Foundation.

References

1. D. Appelö, M. Inkman, T. Hagstrom, T. Colonius, Recent progress on Hermite methods in aeroacoustics, in 17th AIAA/CEAS Aeroacoustics Conference. AIAA, 2011
2. E. Baysal, D.D. Kosloff, J.W. Sherwood, Reverse time migration. Geophysics **48**, 1514–1524 (1983)
3. X. Chen, Numerical and analytical studies of electromagnetic waves: hermite methods, supercontinuum generation, and multiple poles in the SEM, Doctoral Thesis, University of New Mexico, 2012
4. E.T. Dye, Performance analysis and optimization of hermite methods on NVIDIA GPUs using CUDA, Master Thesis, The University of New Mexico, 2015
5. J. Goodrich, T. Hagstrom, J. Lorenz, Hermite methods for hyperbolic initial-boundary value problems. Math. Comput. **75**, 595–630 (2006)
6. T. Hagstrom, D. Appelö. Experiments with Hermite methods for simulating compressible flows: Runge-Kutta time-stepping and absorbing layers, in 13th AIAA/CEAS Aeroacoustics Conference. AIAA, 2007
7. T. Hagstrom, D. Appelö, 2015. Solving PDEs with hermite interpolation, in *Spectral and High Order Methods for Partial Differential Equations ICOSAHOM 2014* (Springer International Publishing, Cham, 2014), pp. 31–49
8. J.S. Hesthaven, T. Warburton, *Nodal Discontinuous Galerkin Methods: Algorithms, Analysis, and Applications* (Springer Science & Business Media, New York, 2007)
9. A. Klöckner, T. Warburton, J. Bridge, J.S. Hesthaven, Nodal discontinuous Galerkin methods on graphics processors. J. Comput. Phys. **228**, 7863–7882 (2009)
10. D. Medina. OKL: a unified language for parallel architectures, Doctoral Thesis, Rice University, 2015
11. P. Micikevicius. 3D finite difference computation on GPUs using CUDA, in *Proceedings of 2nd workshop on general purpose processing on graphics processing units*, ACM, 2009, pp. 79–84
12. A. Modave, A. St-Cyr, T. Warburton, GPU performance analysis of a nodal discontinuous Galerkin method for acoustic and elastic models. Comput. Geosci. **91**, 64–76 (2006)
13. J. Sanders, E. Kandrot, *CUDA by Example: An Introduction to General-Purpose GPU Programming* (Addison-Wesley Professional, Boston, MA, 2010)
14. A. Taflove, S.C. Hagness, *Computational Electrodynamics: The Finite-Difference Time-Domain Method*, 2nd edn. (Artech House, Norwood, MA, 1995)
15. A. Vargas, J. Chan, T. Hagstrom, T. Warburton, Variations on Hermite methods for wave propagation. arXiv:1509.08012 (2015, arXiv preprint)
16. J. Virieux, S. Operto, An overview of full-waveform inversion in exploration geophysics. Geophysics **74**, WCC1–WCC2 (2009)
17. S. Williams, A. Waterman, D. Patterson, Roofline: an insightful visual performance model for multicore architectures. Commun. ACM **52**, 65–76 (2009)

Helically Reduced Wave Equations and Binary Neutron Stars

Stephen R. Lau and Richard H. Price

Abstract We describe ongoing work towards construction—via multidomain, modal, spectral methods—of helically symmetric spacetimes representing binary neutron stars. In particular, we focus on the influence of both the helically reduced wave operator and boundary conditions on the self-consistent field method, a widely used iterative scheme for the construction of stellar models.

1 Introduction and Preliminaries

This section both gives an overview of our research program and describes the particular problem analyzed in this article. The next section describes the spectral methods that we adopt to solve it, while the final section presents numerical results.

1.1 Overview

Studies of binary stars in Newtonian theory have a long history; see, for example, [1]. Equilibrium configurations were first studied numerically in the 1980s, in particular via Hachisu's generalization [2] of the self-consistent field (SCF) method. The SCF method was developed for and applied to single stars by Ostriker and Mark [3] in the late 1960s. It is essentially a fixed-point procedure in which the material and gravitational fields are updated in succession.

S.R. Lau (✉)
Mathematics and Statistics, University of New Mexico, Albuquerque, NM 87131, USA
e-mail: lau@math.unm.edu

R.H. Price
Department of Physics, University of Massachusetts at Dartmouth, Dartmouth, MA 02747-2300, USA
e-mail: rprice.physics@gmail.com

M.L. Bittencourt et al. (eds.), *Spectral and High Order Methods for Partial Differential Equations ICOSAHOM 2016*, Lecture Notes in Computational Science and Engineering 119, DOI 10.1007/978-3-319-65870-4_26

Our long-term goal is the numerical construction—via multidomain, modal, spectral methods—of helically symmetric spacetimes representing binary neutron stars. To maintain helical symmetry, such spacetimes must contain equal amounts of outgoing and incoming gravitational radiation, and therefore correspond to standing-wave equilibrium configurations. Such configurations are admittedly unphysical. Indeed, as energy is lost through gravitational radiation, an isolated binary configuration undergoes inspiral and eventual merger. Nonetheless, standing-wave spacetimes might serve as useful approximations, perhaps yielding excellent trial data for solving the initial value constraints of general relativity (GR).

Our task of numerically constructing helically symmetric solutions to the matter/gravity field equations of GR is daunting, and our progress is step-by-step. In the steps taken so far, we have developed new spectral-tau methods, with a focus on sparsity and preconditioning for problems beyond one dimension. We have investigated random-matrix based preconditioning for multidomain methods [4]. Furthermore, we have developed a strategy [5, 6] for treating the a priori unknown location of a stellar surface; this method should have applications beyond binary neutron stars.

Two intermediate steps are the Newtonian [6] and linearized-gravity binary-star problems. A third [7] is based on perturbative expansion of the Einstein equations, with spacetime geometry viewed as a small deviation of flat (no curvature) spacetime. For scenarios beyond the Newtonian problem, we are often interested in outgoing radiation conditions, despite their ultimate unsuitability for the full GR standing-wave problem. Indeed, these conditions are physical, and their use precludes the need for the two solutions (one with incoming conditions, one with outgoing conditions) per SCF iteration required for standing-wave configurations.

In this paper we consider a generalization of the Newtonian problem, with the Poisson equation for the gravitational potential replaced by the helically reduced wave equation (HRWE). The partial differential operator [8, 9] in this equation is mixed-type (its symbol is elliptic near the rotation axis and hyperbolic far from it) and it plays a key role in the helical reduction of the Einstein equations formulated in [10]. Our focus here is on the formulation of the SCF method in the presence of the helically reduced wave operator and outgoing radiation conditions. We also investigate formulation of the method in the construction of standing-wave models. Our immediate objective is to understand whether adoption of outgoing conditions in our theoretical framework [7] leads to a viable SCF scheme.

1.2 HRWE and Binary Neutron Stars: The Continuum Problem

Following the presentation in [6], this section describes our continuum PDE problem: construction of a comoving neutron star binary. Ref. [6] considers the

Newtonian problem, with the gravitational potential determined by the Poisson equation. Here we replace the Poisson equation with the HRWE, and consider outgoing radiation boundary conditions. Moreover, the solutions we construct here have reflection symmetry only across the orbital plane, whereas solutions to the Newtonian problem are reflection symmetric across two coordinate planes. We believe we are the first to contend with the ramifications of this symmetry loss for the SCF algorithm.

We label one star I and the other II. The stellar extents are determined by a density $\rho(\mathbf{x})$ which is non-vanishing on the interior $U = \{\mathbf{x} : \rho(\mathbf{x}) > 0\}$ of a compact set closure(U). The set $U = U_I \cup U_{\mathit{II}}$ is itself the union of two disjoint sets, one for each star. The boundaries ∂U_I and ∂U_{II} are the stellar surfaces and are a priori unknown. The strength of gravity G and constants $K_{I,\mathit{II}}$ appearing in the stellar equation of state for star I, II remain fixed throughout. Moreover, this work only considers a polytropic equation of state with index $n = 1$ (admittedly the simplest case; see [6]).

1.2.1 Problem Statement

A binary pair (stars I and II with known equations of state) is determined by three fixations: the masses $M_{I,\mathit{II}}$ of the stars and the separation d. (Rather than d, we might fix the angular velocity Ω; for point particles d and Ω are related by Kepler's law.) In addition to these *physical* choices, three *gauge* fixations are required to locate the binary within the orbital plane, our $y = 0$ plane. Since we view the orbital plane as coordinatized by Cartesian coordinates which co-rotate with the binary, the two stars (more precisely, their centers of mass) should remain fixed.

Fixation of d along with the three gauge choices is equivalent to fixation of the four center-of-mass coordinates: $C_{x,I}$, $C_{z,I}$, $C_{x,\mathit{II}}$, $C_{z,\mathit{II}}$. (The conditions $C_{y,I} = 0 = C_{y,\mathit{II}}$ are automatically enforced in our problem by reflection symmetry across the orbital $y = 0$ plane.) However, we are only able to enforce *three* fixations among these quantities. Indeed, as seen below, for our SCF scheme each such fixation corresponds to an auxiliary variable, and we have only five auxiliary variables at our disposal, with two reserved for fixation of the masses. In any case, these fixations amount to integral conditions, and thus are different from pointwise conditions (see the MEUDON and HACHISU conditions in [6]) developed in the literature for Newtonian binaries.

Our domain is $\mathscr{D} \equiv \{\mathbf{x} : |\mathbf{x}| \le r_{\text{out}}\}$, and we consider the following unknowns: (i) the stellar enthalpy $h(\mathbf{x}) \equiv 2K_{I,\mathit{II}}\rho(\mathbf{x})$ and the scalars (ii,iii,iv,v,iv) $\kappa_{I,\mathit{II}}$, Ω^2, ℓ_x, ℓ_z. The stellar surfaces are described by envelope functions $r_{I,\mathit{II}}(\boldsymbol{\theta}_{I,\mathit{II}})$ for $\partial U_{I,\mathit{II}}$. Here $\boldsymbol{\theta}_{I,\mathit{II}}$ are direction cosines relative to star I, II. The stellar surfaces are zero sets of the enthalpy: $h(r_{I,II}(\boldsymbol{\theta}_{I,II})\boldsymbol{\theta}_{I,II}) = 0$. The unknowns obey the following equations:

$$L\Psi = 4\pi G\rho(\mathbf{x}), \qquad \mathscr{B}(\Psi) = 0 \tag{1a}$$

$$\kappa_{I,II} = h(R_\beta^{-1}\mathbf{x}) + \Psi(\mathbf{x}) - \tfrac{1}{2}\Omega^2\varpi^2(\mathbf{x}) \text{ for } \mathbf{x} \in U_I, U_{II} \tag{1b}$$

$$M_{I,II} = \mu_{I,II}, \qquad \sqrt{(C_{x,II} - C_{x,I})^2 + (C_{z,II} - C_{z,I})^2} = d$$

$$\frac{\mu_I C_{x,I} + \mu_{II} C_{x,II}}{\mu_I + \mu_{II}} = x_{\text{center}}, \qquad \frac{\mu_I C_{z,I} + \mu_{II} C_{z,II}}{\mu_I + \mu_{II}} = z_{\text{center}}, \tag{1c}$$

with $\mu_{I,II}, d, x_{\text{center}}, z_{\text{center}}$ chosen constants. Respectively, (ℓ_x, ℓ_z) and $(x_{\text{center}}, z_{\text{center}})$ are the *rotation center* and *mass center*; these points need not coincide numerically. Relative to the mass center, R_β is a rotation of the orbital plane by an angle β; its presence ensures that the stars remain on the z-axis. $L = \nabla^2 - \Omega_0^2(x\partial_z - z\partial_x)^2$ is the helical reduction of the wave operator; it features the Laplacian ∇^2 and a squared angular momentum term proportional to a fiducial rotation rate Ω_0. Moreover, $\mathscr{B}(\Psi) = 0$ refers to either nonlocal outgoing or incoming boundary conditions, $\varpi^2(\mathbf{x}) := (x-\ell_x)^2+(z-\ell_z)^2$ is the squared distance of a point from the rotation axis, and Ω is the binary rotation rate. Constancy of $\kappa_{I,II}$ reflects a balance of chemical, gravitational, and rotational potential.

For the Newtonian problem Ω_0, β, and ℓ_x all vanish. Moreover, in this case both the outgoing/incoming boundary conditions collapse to the same nonlocal conditions associated with the Laplacian. "Standing-wave conditions" correspond to averaging $\frac{1}{2}(\Psi_{\text{out}} + \Psi_{\text{inc}})$ of the outgoing and incoming solutions. With such averaging, the enthalpy satisfies the same symmetries as for the Newtonian problem, in particular reflection symmetry across the $x = 0$ plane. This symmetry ensures $C_{x,I} = 0 = C_{x,II}$. When constructing standing-wave configurations, we therefore drop ℓ_x as a variable, drop the equation in (1c) involving x_{center}, and enforce $\beta = 0$.

A crucial point is that the conditions in (1c) correspond to integral expressions stemming from the constraint in (1b) with $\beta = 0$. For example, to fix the mass $M_I = \int_{V_I} \rho(\mathbf{x})d\mathbf{x}$, where V_I is the volume of star I determined by $r_I(\boldsymbol{\theta}_I)$, we enforce

$$\mu_I = \frac{1}{2K_I}\int_{V_I}\left[\kappa_I + \tfrac{1}{2}\Omega^2(\ell_x^2 + \ell_z^2) - \Psi(\mathbf{x}) + \tfrac{1}{2}\Omega^2(x^2 + z^2) - \Omega^2\ell_x x - \Omega^2\ell_z z\right]d\mathbf{x}. \tag{2}$$

In practice, the integration is distributed over the quantities $1, \Psi, x^2, z^2, x, z$ in order to achieve a constraint of the form

$$\mu_I = \left[\kappa_I + \tfrac{1}{2}\Omega^2(\ell_x^2 + \ell_z^2)\right]a_1 + a_\Psi + \Omega^2(a_{x^2} + a_{z^2} + \ell_x a_x + \ell_z a_z), \tag{3}$$

where $a_1, a_\Psi, a_{x^2}, a_{z^2}, a_x, a_z$ are precomputed integrals. This constraint is then viewed as one equation on the unknowns κ_I (and κ_{II}), Ω^2, ℓ_x, and ℓ_z. When imposing the conditions involving center-of-mass components, we use, for example,

$$C_{x,I} = \frac{1}{2K_I\mu_I}\int_{V_I} x\big[\kappa_I + \tfrac{1}{2}\Omega^2(\ell_x^2+\ell_z^2) - \Psi(\mathbf{x}) + \tfrac{1}{2}\Omega^2(x^2+z^2) - \Omega^2\ell_x x - \Omega^2\ell_z z\big]d\mathbf{x}. \quad (4)$$

The integration is distributed over the quantities $x, x\Psi, x^3, xz^2, x^2, xz$, thereby giving

$$C_{x,I} = \big[\kappa_I + \tfrac{1}{2}\Omega^2(\ell_x^2+\ell_z^2)\big]b_x + b_{x\Psi} + \Omega^2(b_{x^3} + b_{xz^2} + \ell_x b_{x^2} + \ell_z b_{xz}), \quad (5)$$

where $b_x, b_{x\Psi}, b_{x^3}, b_{xz^2}, b_{x^2}, b_{xz}$ are precomputed integrals. Similar expressions correspond to μ_{II}, $C_{x,II}$, and $C_{z,I,II}$. With all of these, we view (1c) as a system of five nonlinear equations for the "constants" $\kappa_{I,II}, \Omega^2, \ell_x, \ell_z$. The precise form of these five nonlinear equations are determined by the gravitational potential Ψ and the envelope functions $r_{I,II}(\boldsymbol{\theta}_{I,II})$ (which in turn define the integration regions $V_{I,II}$).

1.2.2 Self-Consistent Field Method

Algorithm 1 summarizes our procedure for constructing comoving binaries. Essentially, we use the SCF method [3] of Ostriker and Mark with stabilization, through the auxiliary equations (1c), of a type first considered by Hachisu [2, 11, 12]. The method is a fixed-point iteration, and the Broyden method [13] may be used to accelerate its convergence. We will report on such acceleration elsewhere.

Step 5, 6, and 7 in the algorithm merit further comment. Update of the enthalpy h_0 via the generalized potential equation (1b) must occur in a region surrounding, indeed larger, than each star. As a result, on this region the enthalpy takes both positive and negative values. Relative to a chosen coordinate center of star I, II,

---

Algorithm 1 BINARY SCF ITERATION Inputs are the enthalpy h and the scalars $\kappa_{I,II}$,Ω^2, ℓ_x, ℓ_z. Outputs are updates of the same objects

1: Enforce symmetry (or symmetries) on h. This step is only necessary in the numerical implementation.
2: Find the zero sets (stellar surfaces) $r_{I,II}(\boldsymbol{\theta}_{I,II})$ of the enthalpy h, thereby determining from h the *non-negative* density ρ.
3: Solve the problem $L\Psi = 4\pi G\rho(\mathbf{x})$, $\mathscr{B}(\Phi) = 0$.
4: With $r_{I,II}(\boldsymbol{\theta}_{I,II})$ and the new Ψ, get updated scalars $\kappa_{I,II}$, Ω^2,ℓ_x,ℓ_z through solution of the auxiliary equations (1c).
5: Obtain a provisional enthalpy $h_0 = \kappa_{I,II} - \Psi + \frac{1}{2}\Omega^2\varpi^2$ via (1b). As described in the text, this step involves negative values of the enthalpy.
6: Find the stellar surfaces $r_{I,II}(\boldsymbol{\theta}_{I,II})$ of h_0, and use these to compute the individual stellar centers-of-mass of the provisional density ρ_0. These define the angle β.
7: Rotate the enthalpy configuration via $h(R_\beta^{-1}\mathbf{x}) = h_0(\mathbf{x})$.

---

the provisional surfaces $r_{I,II}(\boldsymbol{\theta}_{I,II})$ are then determined by $h_0\big(r_{I,II}(\boldsymbol{\theta}_{I,II})\boldsymbol{\theta}_{I,II}\big) = 0$. Numerically, each stellar surface corresponds to a discretization $\boldsymbol{\theta}_{jk}$ (where j,k respectively range over longitudinal and latitudinal points) of the unit sphere with corresponding radial values $r_{jk} = r(\boldsymbol{\theta}_{jk})$. The values r_{jk} are found via one-dimensional root-finding.

Computation of β is straightforward:

$$\tan\beta_{I,II} = \Big[\int_{V^0_{I,II}} \rho_0(\mathbf{x})x d\mathbf{x} - \mu_{I,II}x_{\text{center}}\Big]\Big/\Big[\int_{V^0_{I,II}} \rho_0(\mathbf{x})z d\mathbf{x} - \mu_{I,II}z_{\text{center}}\Big], \tag{6}$$

where $V^0_{I,II}$ is the star I,II region where the provisional density ρ_0 is nonnegative. Consistency of our algorithm demands that $\beta_I = \beta_{II}$ (up to numerical error) and that this angle converges to a fixed value with increased resolution. The final step of rotating the provisional enthalpy is actually technically involved, due the way [5] we represent stellar surfaces via our modal tau approach. Indeed, a stellar surface "lives" in the overlap between two spherical shells, and on the overlap of these shells our numerical solution is double valued (with the unique point-space physical solution drawn from the inner/outer shell depending on whether a physical point lies inside/outside of the star). Although the details will be presented elsewhere, the rotation described in step 7 involves extrapolation of Chebyshev series outside of the standard interval. The distance from an evaluation point (at which an extrapolation takes place) to the standard interval is always many orders of magnitude smaller than 2 (the standard-interval length), and such extrapolation does not cause troubles.

Algorithm 1 is simpler for construction of standing-wave configurations. As mentioned, in this case ℓ_x is not solved for in step 4. Moreover, h_0 is h in step 5, so that steps 6 and 7 are omitted. However, for standing waves step 3 refers to two solves (one with outgoing conditions, one with incoming conditions) and averaging.

2 Sparse Modal Tau Methods

This section (i) surveys key features of our 3d sparse modal-tau method for solving the continuum problem (1) and (ii) presents a toy 2d model meant to further elucidate some of the ideas behind our 3d method. We consider a "2-center domain" $\mathscr{D}$ that is decomposed, as depicted in Fig. 1, into an arrangement of Cartesian blocks, cylindrical shells, and spherical shells. Respectively, on these subdomains we choose the following classical basis functions: triple products of Chebyshev polynomials, products of trigonometric functions (sines and cosines) with double products of Chebyshev polynomials, and products of spherical harmonics with Chebyshev polynomials. However, the actual variables in our problem are the expansion coefficients associated with series in these basis functions.

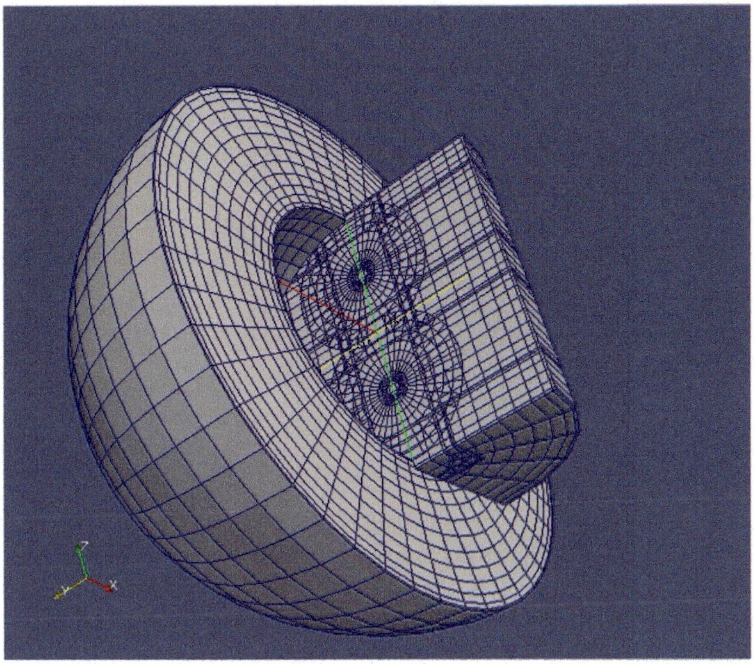

Fig. 1 2-center domain decomposition. The figure depicts all subdomains in the decomposition, although for the sake of visualization the outer radius of the outer shell is smaller than usual

2.1 *Survey of Key Features for 3D*

The domain $\mathscr{D}$ is split into a collection of (overlapping and conforming) subdomains; again see Fig. 1. For binary problems such domain decompositions were pioneered by Pfeiffer et al. [14]. The minimal configuration involves 15 subdomains: blocks $B^{1,2,3,4,5}$; cylindrical shells $C^{1,2,3,4,5}$; inner spherical shells $S_{I}^{1,2}$ around star I; inner spherical shells $S_{II}^{1,2}$ around star II; and an outer spherical shell S_{out}^{1}. In particular, the stellar surface $\partial U_I \in S_I^1 \cap S_I^2$, and the block B^2 fills in the central "hole" of S_I^1. The subdomains S_{II}^1, S_{II}^2, B^4 play the same role for star II. The blocks B^1, B^3, B^5 cover the remaining portion of the z-axis joining the stars; and this axis is wrapped by the five cylinders. Finally, this whole configuration is surrounded by S_{out}^{1}.

The SCF procedure described in the last section involves the following details.

- *Sparse modal representation of the HRWE.* Reference [9] describes modal approximation of (1a) on blocks, cylinders, and shells.
- *"Gluing" of conforming and overlapping subdomains.* Reference [9] examines how the gluing of subdomains is reflected in the overall linear system corresponding to representation of (1a) on all of $\mathscr{D}$.
- *Preconditioning of the global solve over $\mathscr{D}$.* The details are found in [9].
- *Low-rank treatment of stellar surfaces through tau-conditions.* References [5, 6] describe this relatively new feature of our modal approach.

The toy 2d model below conveys a sense of these details for the first three bullets.

2.2 Toy Model: 2D HRWE

Much of the following comes from [4]. Our 2d domain is a disk $\mathscr{D} = \{\mathbf{x} \in \mathbb{R}^2 : 0 \leq |\mathbf{x}| < r_{\max}\}$. Figure 2 shows decomposition of $\mathscr{D}$ into a central square and two annuli. The square S fills the disk center, thereby avoiding troublesome coordinate issues. Compare this situation with the region around one of the stars in Fig. 1.

2.2.1 Modal Representation of the 2D HRWE

The angle on $\mathscr{D}$ is $\varphi = \phi - \Omega_0 t$, where t, r, ϕ are the time-spherical polar coordinates associated with an inertial frame. Therefore, our Cartesian coordinates $x = r\cos\varphi$, $y = r\sin\varphi$ are not fixed in an inertial frame, rather in a frame that rotates at angular velocity Ω_0 with respect to an inertial frame. Moreover, as is appropriate for a "two-center problem", we will assume that the *rotation center* $x, y = -a, -b$ is not the coordinate center $x, y = 0, 0$ of $\mathscr{D}$; see Fig. 2. In Cartesian coordinates the wave

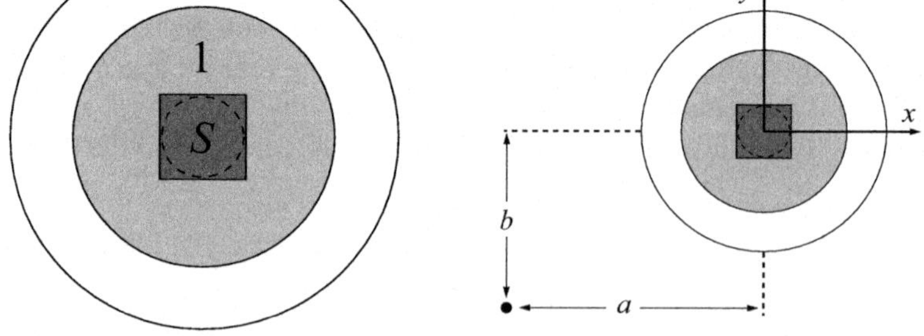

Fig. 2 Domain $\mathscr{D}$. The *dashed smallest circle* is the inner boundary of annulus 1. (**a**) Domain decomposition. (**b**) Rotational center for HRWE

operator, $L = \nabla^2 - \partial_t^2$, subject to helical reduction, takes the form

$$L \equiv \partial_x^2 + \partial_y^2 - \Omega_0^2[(x+a)\partial_y - (y+b)\partial_x]^2. \tag{7}$$

The problem to solve is therefore

$$L\Psi = g,\ \mathbf{x} \in \mathscr{D}; \quad \Psi = h,\ \mathbf{x} \in \partial\mathscr{D}, \tag{8}$$

where and g and h are prescribed sources.

We approximate (8) via modal methods with "integration preconditioning" [8, 9, 15], also referred to as *integration sparsification*. The first step is to express the HRWE on the square S and on a generic annulus in analytical forms that facilitate the integration sparsification technique [8, 9, 15]. First, on S we write

$$\begin{aligned} L = {} & \partial_x^2[1 - \Omega_0^2(y+b)^2] + \partial_y^2[1 - \Omega_0^2(x+a)^2] \\ & - \Omega_0^2[\partial_x(x+a) + \partial_y(y+b) - 2\partial_x\partial_y(x+a)(y+b)], \end{aligned} \tag{9}$$

of course here viewing, for example, $\partial_x x = 1 + x\partial_x$. On each annulus we consider (7) as $r^2 L\Psi = r^2 g$. Then, in terms of the angular functions $F(\varphi) = a\sin\varphi - b\cos\varphi$, $G(\varphi) = a\cos\varphi + b\sin\varphi$, we have [8]

$$\begin{aligned} r^2 L = {} & \partial_r^2 r^2(1 - \Omega_0^2 F^2) + \partial_r r[-3 + \Omega_0^2(2F^2 + G^2 + rG)] + \partial_\varphi^2[1 - \Omega_0^2(r+G)^2] \\ & + \Omega_0^2\partial_\varphi rF - 2\Omega_0^2\partial_\varphi\partial_r rF(r+G) + 1 - \Omega_0^2(G^2 + rG). \end{aligned} \tag{10}$$

In Eqs. (9,10) all derivatives have been pulled to the left; therefore each equation has a form ready for integration sparsification. For example, starting with the form (10), we replace all physical space operators by their matrix counterparts in the space of modal coefficients (e.g., $\partial_r \to D_r$, $r \to A_r$, $F(\varphi) \to \mathsf{F}$). All derivative matrices then appear at the left. Therefore, these differentiations can be "undone" via the left action of (banded) integration matrices [15]. This action is achieved with the "sparsifier" $\mathscr{B} = B_\varphi^2 \otimes B_{r[2]}^2$, where B_φ^2 (Fourier) and $B_{r[2]}^2$ (Chebyshev) are modal-basis matrices representing double integration. The subscript [2] indicates that the first two rows of $B_{r[2]}^2$ are zeroed out. The result is a sparse matrix; see (62) of [8].

2.2.2 Block Jacobi Preconditioner

We now describe the block Jacobi preconditioner used when iteratively solving the above equations with GMRES. On S we let $\mathscr{AL}$ denote L and on each annulus we let $\mathscr{AL}$ denote $r^2 L$. With $\mathscr{B}$ representing integration sparsification on all subdomains, the linear system approximating the problem (8) has the form

$$\mathscr{M}\widetilde{\boldsymbol{\Psi}} = \mathscr{B}\mathscr{A}\mathbf{g}, \tag{11}$$

where $\mathscr{M} = \mathscr{B}\mathscr{A}\mathscr{L}$ and $\widetilde{\boldsymbol{\Psi}}$ is the concatenation $(\widetilde{\boldsymbol{\Psi}}^S, \widetilde{\boldsymbol{\Psi}}^1, \widetilde{\boldsymbol{\Psi}}^2)^T$ of modal coefficients from each subdomain. In (11) rows in both $\mathscr{M}$ and $\mathscr{B}\mathscr{A}\mathbf{g}$ that would have otherwise been empty (due to empty rows in $\mathscr{B}$) have been filled with the tau conditions responsible for both the Dirichlet boundary conditions and the "gluing" of S to 1 and 1 to 2. The gluing conditions involve use of Dirichlet and Neumann vectors; see [8].

The *sparse* matrix $\mathscr{M}$ and preconditioner $\mathscr{G}$ we use have the structure

$$\mathscr{M} = \begin{pmatrix} \mathscr{M}_{SS} & \mathscr{M}_{S1} & \\ \mathscr{M}_{1S} & \mathscr{M}_{11} & \mathscr{M}_{12} \\ & \mathscr{M}_{21} & \mathscr{M}_{22} \end{pmatrix}, \qquad \mathscr{G} = \begin{pmatrix} \mathscr{G}_{SS} & & \\ & \mathscr{G}_{11} & \\ & & \mathscr{G}_{22} \end{pmatrix}, \tag{12}$$

where $\mathscr{G}_{SS} \equiv \mathscr{M}_{SS}^{-1}$ is stored as a precomputed *PLU* factorization. However, $\mathscr{G}_{11} \neq \mathscr{M}_{11}^{-1}$ and $\mathscr{G}_{22} \neq \mathscr{M}_{22}^{-1}$, but $\mathscr{G}_{11}$ and $\mathscr{G}_{22}$ are themselves block diagonal. On an annulus the operator (10) is not diagonal in Fourier space as it mixes Fourier modes.

To explain the construction of $\mathscr{G}_{11}$, we let $\alpha, \beta, \cdots$ denote the global indices associated with the overall set of concatenated equations (11), so that $\widetilde{\boldsymbol{\Psi}}$ has components $\widetilde{\boldsymbol{\Psi}}(\alpha)$. The unknowns for the central square are modal coefficients for a double Chebyshev expansion. For the square let $N_x + 1$ be the number of x modes, and $N_y + 1$ the number of y modes. Then the integer $\sigma_1 \equiv (N_x + 1)(N_y + 1)$ is the "shift" for annulus 1. The components $\widetilde{\boldsymbol{\Psi}}(\alpha)$ which belong to S correspond to $0 \leq \alpha \leq \sigma_1 - 1$, since the indexing of α starts at 0. The block matrix $\mathscr{M}_{SS}$ operating on these modal coefficients has elements $\mathscr{M}_{SS}(I, J)$ with $0 \leq I, J \leq \sigma_1 - 1$.

Let N_r^1 and N_φ^1 specify the number of radial and angular modes on annulus 1. The components $\widetilde{\boldsymbol{\Psi}}(\alpha)$ for annulus 1 then have $\sigma_1 \leq \alpha \leq \sigma_1 + (N_r^1 + 1)(N_\varphi^1 + 1) - 1$. Moreover, the block $\mathscr{M}_{11}$ has components $\mathscr{M}_{11}(I, J) = \mathscr{M}(\sigma_1 + I, \sigma_1 + J)$, with $0 \leq I, J \leq (N_r^1 + 1)(N_\varphi^1 + 1) - 1$. The direct product representation on annulus 1 is chosen such that the two-index modal coefficients are [8] $\widetilde{\Psi}^1_{qj} = \widetilde{\boldsymbol{\Psi}}(\sigma_1 + (N_r^1 + 1)q + j)$, with q the mode index dual to φ and j the mode index dual to r. Therefore, we may likewise write $I = (N_r^1 + 1)q + j$ and $J = (N_r^1 + 1)p + k$, with $0 \leq q, p \leq N_\varphi^1$ and $0 \leq j, k \leq N_r^1$. Therefore, $\mathscr{M}_{11}(I, J) = \mathscr{M}_{11}\big(q(N_r^1 + 1) + j, p(N_r^1 + 1) + k\big)$.

We fix $\bar{\mathscr{M}}_{11} \simeq \mathscr{M}_{11}$ by $\bar{\mathscr{M}}_{11}(I, J) = \mathscr{M}_{11}(I, J)$ for (I, J) pairs with $p = q$ (the Fourier block diagonal), and $\bar{\mathscr{M}}_{11}(I, J) = 0$ for pairs with $p \neq q$. We then set $\mathscr{G}_{11} \equiv \bar{\mathscr{M}}_{11}^{-1}$, with application of $\mathscr{G}_{11}$ performed via *PLU* factorizations of the Fourier blocks comprising $\bar{\mathscr{M}}_{11}$. $\mathscr{G}_{22}$ is defined similarly. Analogous block Jacobi preconditioning *drastically* reduces iteration counts for 3d subdomain GMRES solves [9].

We comment on the scalability of our preconditioner. Our comments also pertain to the neutron star problem. For the number of annuli fixed at two our preconditioner scales with increased modal resolution; the GMRES iteration count appears nearly independent of the subdomain resolutions [4]. Of course, our preconditioner would not scale with increased number of annuli. However, this is a limit we do not take in practice. Indeed, for the neutron star problem we have worked with the fixed 15 subdomain configuration, although our software does allow for increasing the number of spherical shells and cylinders. For the binary configuration we have also employed the simple additive Schwarz method to address the subdomain coupling.

3 Numerical Results

This section considers construction of comoving binaries, presenting convergence studies for the following: (i) the problem (1) with $\mathscr{B}(\Psi) = 0$ specifying outgoing radiation conditions and (ii) the same problem, but now with MEUDON pointwise conditions (described in [6]) and $\mathscr{B}(\Psi) = 0$ specifying "standing wave conditions". For (i) we enforce reflection symmetry only across the $y = 0$ plane, whereas for (ii) we enforce symmetry across both the $x = 0$ and $y = 0$ planes. The MEUDON conditions fix maximum enthalpy values at prescribed points $(0, 0, z_{I,II})$ on the z-axis joining the stars. With these conditions, $\beta = 0$ and we do not use (1c).

We consider the equal-mass case for simplicity; Ref. [6] documents our approach applied to the unequal-mass Newtonian case. We adopt nearly the same domain and truncation sequence used in [6]. Since that reference gives details, here we only summarize and point out differences. The 2-center domain $\mathscr{D}$ has $r_{\text{out}} = 10$, and our seed configuration for the coarsest resolution is the superposition of two spherically symmetric Lane-Emden stars with central densities $\rho_{c,I,II} = 1.0$ and radii $R_{I,II} = 0.4375$. These choices change the inner shells listed in Table 1 of [6]; here $0.01 \leq S^1_{I,II} \leq 0.48$ and $0.39 \leq S^2_{I,II} \leq 1$. We fix Ω_0 and the initial Ω by Kepler's law.

We pick five resolution levels, the fifth for error computations. For levels 1–5 we have respectively chosen 30,25,20,15,15 SCF iterations. These "hand" choices yield Table 1, but automatic stopping criteria are possible. Each solution is the initial guess for the next level. To compute errors, we interpolate the numerical solution onto reference grids: one covering the shell S^1_{out} and another the block B^4. We also interpolate the surface $r_I(\boldsymbol{\theta}_I)$ onto a reference grid. Table 1 lists relative L_2 errors computed thusly for cases (i) and (ii). Tables 2 and 3 are for case (i). For each SCF iteration we compute each star's drift angle β with (6). Table 2 lists the last β for each level; these suggest convergence. Table 3 indicates that (ℓ_x, ℓ_z) convergences to $(x_{\text{center}}, z_{\text{center}}) = (0, 0)$. The rotation and mass centers coincide to numerical error.

Table 1 Convergence study errors

	Truncation	Shell error (i)	Block error (i)	Surface error (i)	Shell error (ii)	Block error (ii)	Surface error (ii)
Level 1	29,752	1.6906e-05	1.9689e-05	6.1152e-07	1.6767e-05	1.9675e-05	6.0416e-07
Level 2	117,103	7.8765e-07	8.0692e-07	2.1870e-08	8.0465e-07	8.2401e-07	2.1930e-08
Level 3	239,496	1.3715e-08	1.4422e-08	4.0022e-10	1.3928e-08	1.4647e-08	4.0017e-10
Level 4	411,112	1.7467e-10	2.8126e-10	5.6591e-13	1.7383e-10	2.7984e-10	4.8059e-13
Level 5	474,464	–	–	–	–	–	–

Here we list errors for cases (i,ii) described in the text

Table 2 Angle offset β

	β from star I	β from star II
Level 1	−1.205652769721e-05	−1.205652769354e-05
Level 2	−1.209687570418e-05	−1.209687570300e-05
Level 3	−1.209687347907e-05	−1.209687347671e-05
Level 4	−1.209687342227e-05	−1.209687342098e-05
Level 5	−1.209687342235e-05	−1.209687342129e-05

This table corresponds to case (i)

Table 3 Convergence of the rotation center

	ℓ_x	ℓ_z
Level 1	8.785744527481e-05	1.911555336618e-05
Level 2	7.840454591376e-09	−3.001365628128e-08
Level 3	−3.260309822213e-12	4.365263771623e-10
Level 4	8.850781768913e-12	−6.603800320517e-13
Level 5	−5.606252946658e-13	6.642903812566e-12

This table corresponds to case (i)

4 Conclusions

We have considered construction of comoving binary stars, with the Poisson equation of Newtonian theory replaced by the inhomogeneous HRWE. The HRWE features a mixed-type operator L which also appears in the helically reduced matter/field equations of GR (and in post-Minkowski approximations thereof [7]). Replacement of the Poisson equation by the HRWE in the binary model gives rise to the numerical challenges confronted here.

As part of the binary problem (1) the HRWE wave equation is associated with radiation conditions, and for either outgoing or incoming conditions it gives rise to solutions with less symmetry than those encountered for the Newtonian problem. This symmetry loss leads to wrinkles in applying the SCF iteration for binary construction. With the reflection symmetry across the $x = 0$ plane relaxed, each star may wander off the z-axis. When solving the Newtonian problem *without* enforcement of symmetry across $x = 0$ (the solution has reflection symmetry across this plane), this wander arises only from numerical error. However, with the HRWE it appears fundamental, necessitating introduction of the rotation matrix R_β in (1b). An important next step is to explain the drift β in terms of radiation reaction.

We have also compared the construction of binaries (with the Poisson/HRWE replacement) subject to outgoing radiation conditions (and less symmetry) to standing-wave configurations (with the same symmetries as Newtonian binaries). While the standing-wave conditions are less physical, this scenario does not require introduction of the angular drift β. Although construction of a standing-wave binary requires twice as many solves of the HRWE as construction of an outgoing-wave binary (two per SCF iteration rather than one), our current code computes the standing-wave configuration faster than the outgoing one. The reason is that for the standing-wave scenario the auxiliary system of equations for the parameters relies on pointwise conditions (MEUDON conditions), whereas the integral conditions described in this article are expensive to evaluate. Our current evaluation of such integrals relies on poorly organized interpolation; we could surely speed up these computations.

Acknowledgements We gratefully acknowledge support by NSF grant no. DMS-1216866.

References

1. S. Chandrasekhar, *An Introduction to the Study of Stellar Structure* (Dover, New York, 1967)
2. I. Hachisu, A versatile method for obtaining structures of rapidly rotating stars. Astrophys. J. Suppl. Ser. **61**, 479–507 (1986)
3. J.P. Ostriker, J.W.-K. Mark, Astrophys. J. **151**, 1075–1088 (1968)
4. M. Beroiz, T. Hagstrom, S.R. Lau, R.H. Price, Multidomain, sparse, spectral-tau method for helically symmetric flow. Comput. Fluids **102**, 250–265 (2014)
5. S.R. Lau, R.H. Price, Sparse modal tau-method for helical binary neutron stars, in *Proceedings of Spectral and High Order Methods for Partial Differential Equations ICOSAHOM 2014*. Lecture Notes in Computational Science and Engineering, vol. 106 (Springer, Cham, 2015), pp. 315–323
6. S.R. Lau, Stellar surface as low-rank modification in iterative methods for binary neutron stars. J. Comput. Phys. **348**, 460–481 (2017)
7. S.R. Lau, Second-order formalism for helically symmetric spacetimes describing binary neutron stars (2017, in preparation)
8. S.R. Lau, R.H. Price, Multidomain spectral method for the helically reduced wave equation. J. Comput. Phys. **227**, 1126–1161 (2007). We regret an error in Eq. (42). The correct expressions are $\nu^{\pm} = \left[T_0'(\pm 1), T_1'(\pm 1), T_2'(\pm 1), T_3'(\pm 1), T_4'(\pm 1), \cdots\right] = \left[0, 1, \pm 4, 9, \pm 16, \cdots\right]$. The right-hand side of the second equation of (69) is also off by a sign
9. S.R. Lau, R.H. Price, Sparse spectral-tau method for the three-dimensional helically reduced wave equation on two-center domains. J. Comput. Phys. **231**(2), 7695–7714 (2012)
10. C. Beetle, B. Bromley, N. Hernández, R. H. Price, Periodic standing-wave approximation: Post-Minkowski computations. Phys. Rev. D **76**, 084016 (2007)
11. I. Hachisu, A versatile method for obtaining structures of rapidly rotating stars. II. Three-dimensional self-consistent field method. Astrophys. J. Suppl. Ser. **62**, 461–499 (1986)
12. I. Hachisu, Y. Eriguchi, K. Nomoto, Fate of merging double white dwarfs. II. Numerical method. Astrophys. J. **311**, 214–225 (1986)
13. C.T. Kelley, *Iterative Methods for Linear and Nonlinear Equations* (SIAM, Philadelphia, 1995)
14. H.P. Pfeiffer, L.E. Kidder, M.A. Scheel, S.A. Teukolsky, A multidomain spectral method for solving elliptic equations. Comput. Phys. Commun. **152**(3), 253–273 (2003)
15. E.A. Coutsias, T. Hagstrom, J.S. Hesthaven, D. Torres, Integration preconditioners for differential operators in spectral τ-methods. Houst. J. Math. (Special Issue), 21–38 (1996)

Detecting Discontinuities Over Triangular Meshes Using Multiwavelets

Mathea J. Vuik and Jennifer K. Ryan

Abstract It is well known that solutions to nonlinear hyperbolic PDEs develop discontinuities in time. The generation of spurious oscillations in such regions can be prevented by applying a limiter in the troubled zones. In earlier work, we constructed a multiwavelet troubled-cell indicator for one and (tensor-product) two dimensions (SIAM J. Sci. Comput. 38(1):A84–A104, 2016). In this paper, we investigate multiwavelet troubled-cell indicators on structured triangular meshes. One indicator uses a problem-dependent parameter; the other indicator is combined with outlier detection.

1 Introduction

It is well known that solutions to nonlinear hyperbolic PDEs develop discontinuities in time. The generation of spurious oscillations in such regions can be prevented by applying a limiter in the troubled zones. In [18–20], two different multiwavelet troubled-cell indicators were introduced, one based on a parameter, the other combined with outlier detection. In these papers, we focused on one-dimensional and tensor product two-dimensional meshes. Here, the use of multiwavelets on triangular meshes is investigated [17, 21]. We again consider two approaches to troubled-cell indication: one based on a parameter, the other combined with outlier detection. We demonstrate the performance of the indicators on a test problem based on the two-dimensional linear advection equation, using the vertex-based limiter in the identified troubled cells [10].

The outline of this paper is as follows: in Sect. 2, the triangular mesh is defined, and information about barycentric coordinates is given. The multiresolution analysis is described in Sect. 3. The multiwavelet troubled-cell indicators are defined in

M.J. Vuik
VORtech, Post Box 260, 2600AG Delft, The Netherlands
e-mail: thea.vuik@vortech.nl

J.K. Ryan (✉)
School of Mathematics, University of East Anglia, Norwich NR4 7TJ, UK
e-mail: Jennifer.Ryan@uea.ac.uk

M.L. Bittencourt et al. (eds.), *Spectral and High Order Methods for Partial Differential Equations ICOSAHOM 2016*, Lecture Notes in Computational Science and Engineering 119, DOI 10.1007/978-3-319-65870-4_27

Sect. 4. Preliminary results are shown in Sect. 5, and some concluding remarks are given in Sect. 6.

2 Structured Triangular Mesh and Barycentric Coordinates

In this section, the definition of a structured triangular mesh on a rectangular domain $\Omega \in \mathbb{R}^2$ is given, following the notation in [5, 17]. To compute the multiwavelet decomposition at a later time, the relation between the mesh on the finest level n and level $n-1$ is explained. Furthermore, several properties of the barycentric coordinate system are given.

Definition 1 Let i and j be space indices in the x- and y-direction, respectively, and let M account for the orientation of the triangle: $M = 1$ corresponds to triangles with the right angle located in the bottom-left corner, $M = 2$ belongs to the triangles with right angles in the upper-right corner. The uniform triangulation of a rectangular domain $\Omega \in \mathbb{R}^2$ on level n consists of 2^{2n+1} elements, and is expressed as

$$\mathscr{T}^n = \{T^n_{(i,j,M)}\}^{M=1,2}_{i,j=0,\ldots,2^n-1} = \{T^n_\lambda\}_\lambda,$$

with $\lambda = (i,j,M)$, $i,j = 0,\ldots,2^n-1$, $M = 1,2$.

The triangulation on level $n-1$ is obtained by uniting four triangles on level n:

$$T^{n-1}_{(i,j,1)} = T^n_{(2i,2j,2)} \cup T^n_{(2i,2j,1)} \cup T^n_{(2i+1,2j,1)}, \cup T^n_{(2i,2j+1,1)},$$
$$T^{n-1}_{(i,j,2)} = T^n_{(2i+1,2j+1,1)} \cup T^n_{(2i+1,2j+1,2)} \cup T^n_{(2i,2j+1,2)} \cup T^n_{(2i+1,2j,2)},$$

$i,j = 0,\ldots,2^{n-1}-1$, see Fig. 1.

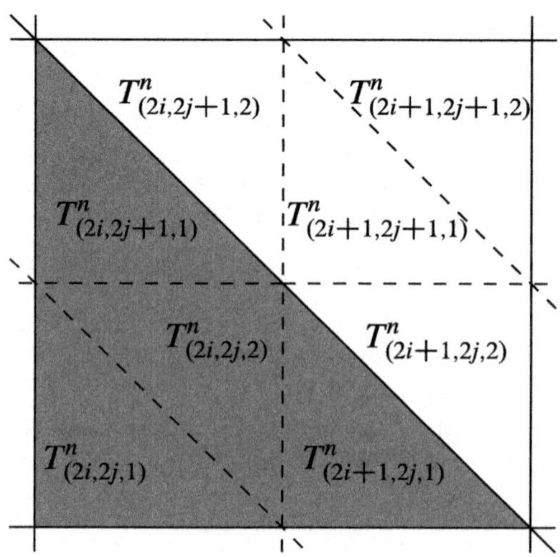

Fig. 1 Triangulation $\mathscr{T}^n$ of a rectangular domain $\Omega \in \mathbb{R}^2$. *Solid lines* correspond to the elements $T^{n-1}_{(i,j,1)}$ and $T^{n-1}_{(i,j,2)}$ on level $n-1$

Points inside a triangle are efficiently expressed using barycentric coordinates.

Definition 2 Let triangle T be defined by its vertices $\mathbf{P}_i = (x_i, y_i)^\top$, $i = 1, 2, 3$. Every point $\mathbf{P} = (x, y)^\top$ can be expressed in terms of the barycentric coordinates $\boldsymbol{\tau} = (\tau_1, \tau_2, \tau_3)^\top$ with respect to triangle T as follows: $\mathbf{P} = (\mathbf{P}_1\mathbf{P}_2\mathbf{P}_3)\boldsymbol{\tau}$.
The barycentric coordinates are uniquely given by requiring $|\boldsymbol{\tau}| = \tau_1 + \tau_2 + \tau_3 = 1$. If $\mathbf{P}$ is located inside T, then $\tau_i \geq 0, i = 1, 2, 3$.

Integrals on a triangle can be transformed to barycentric coordinates as follows:

$$\iint_T f(x, y)dxdy = 2|T| \int_0^1 \int_0^{1-\tau_1} f(x(\tau_1, \tau_2), y(\tau_1, \tau_2))d\tau_2 d\tau_1, \tag{1}$$

where $|T|$ is the area of T [17, 21].

The transformation from original coordinates to barycentric coordinates equals

$$\begin{pmatrix} \tau_1 \\ \tau_2 \\ \tau_3 \end{pmatrix} = \begin{pmatrix} x_1 & x_2 & x_3 \\ y_1 & y_2 & y_3 \\ 1 & 1 & 1 \end{pmatrix}^{-1} \begin{pmatrix} x \\ y \\ 1 \end{pmatrix}.$$

With this expression, it is possible to relate the barycentric coordinates on different triangles which will be necessary when discussing multiwavelets. If $\mathbf{P}$ has barycentric coordinates $\boldsymbol{\tau}$ relative to triangle T (defined by $\{(x_i, y_i), i = 1, 2, 3\}$), then the barycentric coordinates $\boldsymbol{\tau}'$ with respect to T' (defined by $\{(x_i', y_i'), i = 1, 2, 3\}$) can be calculated using $\boldsymbol{\tau}' = M_{T\to T'}\boldsymbol{\tau}$, where

$$M_{T\to T'} = \begin{pmatrix} x_1' & x_2' & x_3' \\ y_1' & y_2' & y_3' \\ 1 & 1 & 1 \end{pmatrix}^{-1} \begin{pmatrix} x_1 & x_2 & x_3 \\ y_1 & y_2 & y_3 \\ 1 & 1 & 1 \end{pmatrix}. \tag{2}$$

The right matrix transforms $\boldsymbol{\tau}$ to $\mathbf{P}$, and the left matrix computes $\boldsymbol{\tau}'$ from $\mathbf{P}$ [17, 21].

Finally, the midpoint subdivision of a triangle $T_\lambda^{n-1} = T_{\lambda_0}^n \cup T_{\lambda_1}^n \cup T_{\lambda_2}^n \cup T_{\lambda_3}^n$ can easily be described in barycentric coordinates, see Fig. 2 [17, 21].

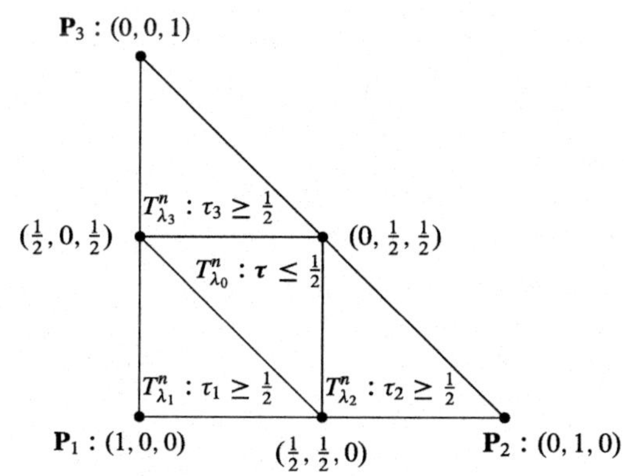

Fig. 2 Midpoint subdivision of triangle T_λ^{n-1}. Coordinates are in barycentric form (τ_1, τ_2, τ_3)

3 Multiresolution Analysis

In this section, the multiresolution analysis for a triangular mesh is presented, together with the formulae for multiwavelet decomposition [17]. For the reconstruction procedure, we refer to [17]. The scaling functions and multiwavelets are constructed for the so-called *base triangle*, T_B, which has vertices $\mathbf{P}_1 = (0,0)$, $\mathbf{P}_2 = (1,0)$, and $\mathbf{P}_3 = (0,1)$, and subdivision $T_B = T_0 \cup T_1 \cup T_2 \cup T_3$ (numbering similar to Fig. 2). The extension to general triangles is given as well.

3.1 Scaling-Function Space

In this section, the orthonormal scaling-function basis is constructed for the base triangle, using barycentric coordinates [21]. The scaling-function space on T_B is defined as $V^{k+1}(T_B) = \mathbb{P}^k(T_B)$, which means that the space is spanned by polynomials of total degree less than or equal to k on T_B. The standard monomial basis for $V^{k+1}(T_B)$ consists of N_k functions $\{1, x, y, x^2, xy, y^2, \dots\}$. For the base triangle, the coordinates (x, y) transform to

$$x = \tau_1 x_1 + \tau_2 x_2 + \tau_3 x_3 = \tau_2,$$
$$y = \tau_1 y_1 + \tau_2 y_2 + \tau_3 y_3 = \tau_3 = 1 - \tau_1 - \tau_2$$

in barycentric coordinates. This means that the monomial basis is equivalent to the set $\{1, \tau_2, 1-\tau_1-\tau_2, \tau_2^2, (1-\tau_1-\tau_2)\tau_2, (1-\tau_1-\tau_2)^2, \dots\}$ in the barycentric coordinate system. Orthonormality of this basis is achieved by the application of the Gram-Schmidt procedure with respect to the inner product

$$\langle f, g\rangle = \int_0^1 \int_0^{1-\tau_1} f(\tau_1,\tau_2) g(\tau_1,\tau_2) d\tau_2 d\tau_1,$$

together with normalization. This results in the orthonormal scaling functions ϕ_{ℓ,T_B}, $\ell = 1,\dots,N_k$. The first six functions (corresponding to $k \leq 2$) are given in [17, 21].

The scaling-function space on level n is defined as the space of piecewise polynomials of total degree less than or equal to k on every triangle $T_\lambda^n \in \mathcal{T}^n$:

$$V_n^{k+1} = \{f : f \in \mathbb{P}^k(T_\lambda^n), \quad \forall T_\lambda^n \in \mathcal{T}^n\}. \tag{3}$$

The orthonormal basis for V_n^{k+1} can be found by substituting the correct barycentric coordinates (translation) and scaling the functions ϕ_{ℓ,T_B} [21]. Let $\boldsymbol{\tau}$ be the barycentric coordinates with respect to T_B, and let $\boldsymbol{\tau}'$ be the corresponding barycentric coordinates with respect to $T_\lambda^n \in \mathcal{T}^n$. The space V_n^{k+1} is spanned by $2^{2n+1} \cdot N_k$ functions that are obtained from ϕ_{ℓ,T_B} using

$$\phi_{\ell\lambda}^n(\tau_1', \tau_2', \tau_3') = \sqrt{\frac{1}{2|T_\lambda^n|}}\phi_{\ell,T_B}(\tau_1, \tau_2, \tau_3). \tag{4}$$

The orthogonal projection of an arbitrary function $f \in L^2(\Omega)$ onto V_n^{k+1} is given by

$$P_n^{k+1}f(\mathbf{x}) = \sum_{T_\lambda^n \in \mathcal{T}^n} \sum_{\ell=1}^{N_k} s_{\ell\lambda}^n \phi_{\ell\lambda}^n(\boldsymbol{\tau}),$$

which is the single-scale decomposition of f on level n. The scaling-function coefficients are given by $s_{\ell\lambda}^n = \langle f, \phi_{\ell\lambda}^n \rangle$. Note that if $f \in V_n^{k+1}$, then $P_n^{k+1}f = f$.

3.2 *Nodal DG Approximation and Scaling-Function Expansion*

Although it is possible to use modal DG based on a PKD-polynomial basis on triangular meshes [12], it is more convenient to use the nodal form of the DG method for this mesh type [2, 3, 8].

The DG approximation space, V_h, is equal to the scaling-function space on level n (Eq. (3)). This means that it is possible to express the nodal DG approximation as a scaling-function approximation in level n. Since $u_h \in V_h = V_n^{k+1}$, we know that $u_h = P_n^{k+1}u_h$. Therefore, we can write the global nodal DG approximation as

$$u_h(\mathbf{x}) = \sum_{T_\lambda^n \in \mathcal{T}^n} \sum_{i=1}^{N_k} u_h(\mathbf{x}^i)\ell_i(\mathbf{x}) = \sum_{T_\lambda^n \in \mathcal{T}^n} \sum_{\ell=1}^{N_k} s_{\ell\lambda}^n \phi_{\ell\lambda}^n(\boldsymbol{\tau}).$$

Knowing the values $u_h(\mathbf{x}^i)$, we can efficiently compute the scaling-function coefficients by a matrix-vector multiplication. Define the vectors $\mathbf{s}_\lambda^n = (s_{1\lambda}^n, \ldots, s_{N_k\lambda}^n)^\top$, $\mathbf{u}_h = (u_h(\mathbf{x}^1), \ldots, u_h(\mathbf{x}^{N_k}))^\top$, and a Vandermonde matrix by $V_{mi} = \phi_{i\lambda}^n(\boldsymbol{\tau}(\mathbf{x}^m))$, then $\mathbf{V}\mathbf{s}_\lambda^n = \mathbf{u}_h$ and $\mathbf{V}^{-1}\mathbf{u}_h = \mathbf{s}_\lambda^n$.

This procedure is very similar to the transformation from nodal to modal DG. This is because the scaling-function basis for V_n^{k+1} is closely related to the modal DG basis, which is given by the so-called *PKD polynomials* [4, 9]. The difference between both bases is the reference triangle that is used [8].

3.3 *Multiwavelets*

In addition to the scaling-function space, the multiwavelet space should be defined. This is done by computing the orthogonal complement: $V_{n-1}^{k+1} \oplus W_{n-1}^{k+1} = V_n^{k+1}$, such that $W_{n-1}^{k+1} \perp V_{n-1}^{k+1}$, $W_{n-1}^{k+1} \subset V_n^{k+1}$. In Algorithm 6.1 in [17], the procedure to compute the multiwavelets for the base triangle is given, in a manner very similar to Alpert's construction in one dimension [1, 21]. The execution of this algorithm leads to the multiwavelets as provided in [15].

Similar to Eq. (4), the multiwavelets on triangle $T_\lambda^n \in \mathscr{T}^n$ are equal to

$$\psi_{\ell\lambda}^{m,n}(\tau_1', \tau_2', \tau_3') = \sqrt{\frac{1}{2|T_\lambda^n|}}\psi_\ell^m(\tau_1, \tau_2, \tau_3), \quad m = 1, 2, 3, \quad \ell = 1, \dots, N_k.$$

In [7], a similar multiwavelet basis is constructed, but normalization is done in the L^∞-norm instead of the L^2-norm.

3.4 Multiwavelet Decomposition

In Sect. 3.2, the relation between the DG approximation and the scaling-function coefficients on level n was given. In this section, the scaling-function expansion on level n is decomposed to a multiwavelet expansion on level $n-1$ [21], using the same notation as in [17]. The full decomposition is derived in [7, 14, 15].

In the following, the scaling-function basis of $\mathbb{P}^k(T_\lambda^{n-1})$ is written in terms of a vector $\boldsymbol{\phi}_\lambda^{n-1} = (\phi_{1\lambda}^{n-1}, \dots, \phi_{N_k\lambda}^{n-1})^\top$. Because $V_{n-1}^{k+1} \subset V_n^{k+1}$, we can express $\boldsymbol{\phi}_\lambda^{n-1}$ in terms of $\boldsymbol{\phi}_{\lambda_i}^n$, $i = 0, 1, 2, 3$, using the local numbering $T_\lambda^{n-1} = T_{\lambda_0}^n \cup T_{\lambda_1}^n \cup T_{\lambda_2}^n \cup T_{\lambda_3}^n$ (Fig. 2). This means that

$$\boldsymbol{\phi}_\lambda^{n-1} = H_0\boldsymbol{\phi}_{\lambda_0}^n + H_1\boldsymbol{\phi}_{\lambda_1}^n + H_2\boldsymbol{\phi}_{\lambda_2}^n + H_3\boldsymbol{\phi}_{\lambda_3}^n. \tag{5a}$$

The $N_k \times N_k$ matrices H_i are similar to the QMF coefficients in the one-dimensional case [16, 17], and are defined as ($i = 0, 1, 2, 3,\ p, q = 1, \dots, N_k$)

$$(H_i)_{p,q} = \langle \phi_{p\lambda}^{n-1}, \phi_{q\lambda_i}^n \rangle = \iint_{T_{\lambda_i}^n} \phi_{p\lambda}^{n-1}(x, y)\phi_{q\lambda_i}^n(x, y)dxdy,$$

using that $\phi_{q\lambda_i}^n$ is only nonzero in $T_{\lambda_i}^n$. We transform to barycentric coordinates $\boldsymbol{\tau}$ based on the vertices of $T_{\lambda_i}^n$. Using Eqs. (1), (2) and (4), this yields

$$\begin{aligned}
(H_i)_{p,q} &= 2|T_{\lambda_i}^n|\sqrt{\frac{1}{2|T_\lambda^{n-1}|}}\sqrt{\frac{1}{2|T_{\lambda_i}^n|}}\int_0^1\int_0^{1-\tau_1} \phi_p(M_{T_{\lambda_i}^n \to T_\lambda^{n-1}}\boldsymbol{\tau})\phi_q(\boldsymbol{\tau})d\tau_2 d\tau_1 \\
&= \sqrt{\frac{|T_{\lambda_i}^n|}{|T_\lambda^{n-1}|}}\int_0^1\int_0^{1-\tau_1} \phi_p(M_{T_{\lambda_i}^n \to T_\lambda^{n-1}}\boldsymbol{\tau})\phi_q(\boldsymbol{\tau})d\tau_2 d\tau_1 \\
&= \frac{1}{2}\int_0^1\int_0^{1-\tau_1} \phi_p(M_{T_{\lambda_i}^n \to T_\lambda^{n-1}}\boldsymbol{\tau})\phi_q(\boldsymbol{\tau})d\tau_2 d\tau_1,
\end{aligned}$$

since $|T^n_{\lambda_i}| = |T^{n-1}_\lambda|/4$. For a structured triangular mesh, the matrices H_i do not depend on the mesh size [15].

Similarly, the multiwavelet basis is written as $\boldsymbol{\psi}^{m,n-1}_\lambda = (\psi^{m,n-1}_{1\lambda}, \ldots, \psi^{m,n-1}_{N_k\lambda})^\top$, $m = 1, 2, 3$. Because $W^{k+1}_{n-1} \subset V^{k+1}_n$, the vectors of multiwavelets can be written as

$$\boldsymbol{\psi}^{m,n-1}_\lambda = G_{m,0}\boldsymbol{\phi}^n_{\lambda_0} + G_{m,1}\boldsymbol{\phi}^n_{\lambda_1} + G_{m,2}\boldsymbol{\phi}^n_{\lambda_2} + G_{m,3}\boldsymbol{\phi}^n_{\lambda_3}, \quad \text{for } m = 1, 2, 3, \tag{5b}$$

with $(G_{m,i})_{p,q} = \langle \psi^{m,n-1}_{p\lambda}, \phi^n_{q\lambda_i}\rangle$, $i = 0, 1, 2, 3,\ p, q = 1, \ldots, N_k$. The matrices $G_{m,i}$ are computed similarly to the matrices H_i.

From Eq. (5) and the fact that $\mathbf{s}^{n-1}_\lambda = \langle f, \boldsymbol{\phi}^{n-1}_\lambda\rangle$, $\mathbf{d}^{m,n-1}_\lambda = \langle f, \boldsymbol{\psi}^{m,n-1}_\lambda\rangle$, it follows that we can decompose the scaling-function coefficients on level n to scaling-function and multiwavelet coefficients on level $n-1$ as follows:

$$\mathbf{s}^{n-1}_\lambda = H_0\mathbf{s}^n_{\lambda_0} + H_1\mathbf{s}^n_{\lambda_1} + H_2\mathbf{s}^n_{\lambda_2} + H_3\mathbf{s}^n_{\lambda_3}, \tag{6a}$$

$$\mathbf{d}^{1,n-1}_\lambda = G_{1,0}\mathbf{s}^n_{\lambda_0} + G_{1,1}\mathbf{s}^n_{\lambda_1} + G_{1,2}\mathbf{s}^n_{\lambda_2} + G_{1,3}\mathbf{s}^n_{\lambda_3}, \tag{6b}$$

$$\mathbf{d}^{2,n-1}_\lambda = G_{2,0}\mathbf{s}^n_{\lambda_0} + G_{2,1}\mathbf{s}^n_{\lambda_1} + G_{2,2}\mathbf{s}^n_{\lambda_2} + G_{2,3}\mathbf{s}^n_{\lambda_3}, \tag{6c}$$

$$\mathbf{d}^{3,n-1}_\lambda = G_{3,0}\mathbf{s}^n_{\lambda_0} + G_{3,1}\mathbf{s}^n_{\lambda_1} + G_{3,2}\mathbf{s}^n_{\lambda_2} + G_{3,3}\mathbf{s}^n_{\lambda_3}, \tag{6d}$$

which is called the *multiwavelet decomposition* from level n to level $n-1$.

4 Multiwavelet Troubled-Cell Indicator

In this section, multiwavelet troubled-cell indicators are defined for triangular meshes [17]. Here, the number of multiwavelet coefficients is increased by a renumbering technique [17, 19]. This leads to the multiwavelet coefficients $\tilde{d}^{m,n-1}_{\ell\lambda}$, where $\ell = 1, \ldots, N_k$, $m = 1, 2, 3$, and λ belongs to the triangles in level n (instead of level $n-1$).

4.1 Parameter-Based Indicator

The parameter-based multiwavelet troubled-cell indicator is defined similarly to the indicator for the one-dimensional and tensor-product two-dimensional case [17, 19]. The major difference lies in the number of coefficients that is needed for accurate detection. In the one-dimensional or tensor-product two-dimensional case, knowledge of the jump relation at element boundaries made it possible to use one coefficient per element for detection. In the triangular case, however, such a relation has not yet been proven, neither theoretically, nor numerically. Therefore, we will use all multiwavelet coefficients for detection: triangle T_λ^n is detected as troubled if for any $m = 1, 2, 3$, $\ell = 1, \ldots, N_k$:

$$\left|\tilde{d}_{\ell\lambda}^{m,n-1}\right| > C \cdot \max_{T_\lambda^n \in \mathcal{T}^n} \left\{ \left|\tilde{d}_{\ell\lambda}^{m,n-1}\right| \right\},$$

where $C \in [0, 1]$ is a parameter that defines the strictness of the indicator. The parameter C is problem-dependent: it depends on the strength of different shocks in the domain. This limits the applicability of this troubled-cell indicator. Therefore, an outlier-detection approach is also considered.

4.2 *Outlier-Detection Approach*

In this section, a troubled-cell indication technique for the multiwavelet coefficients on a structured triangular mesh is proposed that is based on outlier detection. In this way, a problem-dependent parameter is not needed.

A triangle is detected as troubled if it is detected in either the x- or the y-direction, using the one-dimensional approach [20]. Regions with triangles in the x-direction are split into local regions of size 16, as is visualized in Fig. 3, and a similar approach is followed for regions in the y-direction. The resulting outlier-detection approach is given in Algorithm 1. Note that this approach is closely related to the outlier-detection algorithm for a rectangular tensor-product mesh [20].

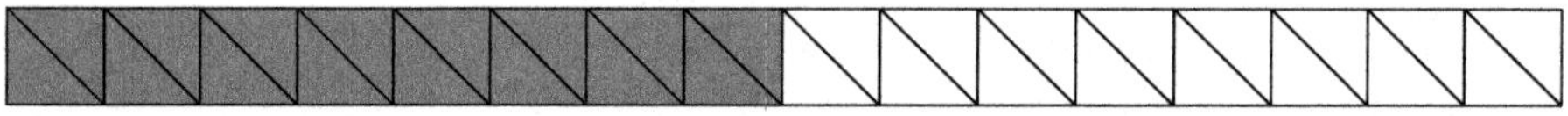

Fig. 3 Split of a 32-triangle region in the x-direction into two local regions of size 16

Algorithm 1 Outlier-detection algorithm for multiwavelet coefficients on triangular meshes, using local vectors

for all $\ell = 1, \ldots, N_k$ **do**
  for all $m = 1, 2, 3$ **do**
    for all $i = 0, \ldots, 2^n - 1$ **do**
      Form troubled-cell indication vector $\mathbf{D}_{\ell i}^{m,n-1}$ consisting of multiwavelet coefficients $\tilde{d}_{\ell\lambda}^{m,n-1}$, where $\lambda = (i, j, M)$, with $j = 0, \ldots, 2^n - 1$, $M = 1, 2$ (definition 1).
      Apply Algorithm 2 in [20].
    end for
    for all $j = 0, \ldots, 2^n - 1$ **do**
      Form troubled-cell indication vector $\mathbf{D}_{\ell j}^{m,n-1}$ consisting of multiwavelet coefficients $\tilde{d}_{\ell\lambda}^{m,n-1}$, where $\lambda = (i, j, M)$, with $i = 0, \ldots, 2^n - 1$, $M = 1, 2$ (definition 1).
      Apply Algorithm 2 in [20].
    end for
  end for
end for
Label an element as troubled if it is detected in any application of Algorithm 2 in [20].

5 Numerical Results

In this section, preliminary numerical results are shown for which the multiwavelet troubled-cell indicator has been tested [17]. The test is done for an example based on the linear advection equation on $[0, 1] \times [0, 1]$, given by $u_t + \nabla \cdot (\mathbf{v}u) = 0$. Here, $\mathbf{v} = (v_1, v_2)^\top$ is the velocity vector, and $u = u(x, y, t)$ is the unknown quantity to be resolved. We use a diagonally-directed velocity: $\mathbf{v} = \sqrt{2}/2 \cdot (1, 1)^\top$. The following discontinuous initial condition is used:

$$u_0(x, y) = \begin{cases} 1, \text{ if } (x - 0.5)^2 + (y - 0.5)^2 \leq 0.1, \\ 0, \text{ else,} \end{cases}$$

together with periodic boundary conditions. The exact solution of this boundary-value problem is equal to $u(x, y, t) = u_0(x - v_1 t, y - v_2 t)$. This means that the initial function should be recovered at the final time $T = \sqrt{2}$ [11].

The multiwavelet troubled-cell indicator is applied both using the parameter C and with the outlier-detection approach. For the tests, the Matlab code of Hesthaven and Warburton is used [8], which is extended to the advection equation together with the vertex-based limiter by Raees et al. [13].

This section only shows the results for one specific initial condition. For more different test problems, we refer to [17].

5.1 Multiwavelet Coefficients of Initial Condition

In order to investigate the information gleaned from multiwavelets on structured triangular meshes, in this section we first study multiwavelet coefficients of the initial condition in a DG basis (Fig. 4). Clearly, the multiwavelet coefficients can be used to distinguish between smooth and nonsmooth regions. However, a clear meaning of the coefficients (as is the case in one and two dimensions) is difficult to establish.

5.2 Detection at Final Time

In this section, the approximations and detected troubled-cells are shown at the final time $T = \sqrt{2}$. Note that the exact solution equals the initial condition at this time.

The multiwavelet troubled-cell indicator is applied using either the parameter-based method or the outlier-detection approach (for which a problem-dependent parameter is not necessary). For the parameter-based method, the vertex-based limiter is applied only in the detected elements [10]. The outlier-detection scheme, however, turned out to only be stable if the limiter is also applied to all boundary

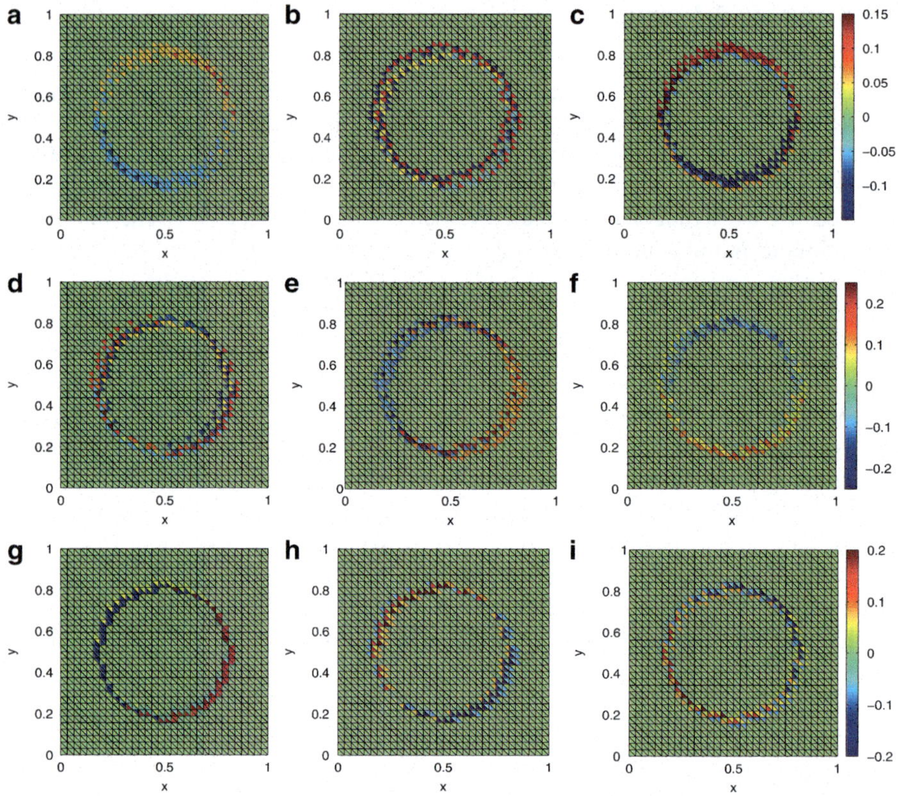

Fig. 4 Multiwavelet coefficients of initial condition, structured triangular mesh based on 32 × 32 rectangles, $k = 1$. **(a)** $d_{1\lambda}^{1,n-1}$, **(b)** $d_{2\lambda}^{1,n-1}$, **(c)** $d_{3\lambda}^{1,n-1}$, **(d)** $d_{1\lambda}^{2,n-1}$, **(e)** $d_{2\lambda}^{2,n-1}$, **(f)** $d_{3\lambda}^{2,n-1}$, **(g)** $d_{1\lambda}^{3,n-1}$, **(h)** $d_{2\lambda}^{3,n-1}$, **(i)** $d_{3\lambda}^{3,n-1}$

elements. These elements are not always detected by the outlier scheme and are therefore not marked as such in the figures. Similar boundary problems were also observed in [6], where it was proposed to either use an adaptive mesh with more triangles near the boundary or ignore the boundary triangles for certain resolution levels.

The results are shown in Fig. 5. The parameter-based method detects the discontinuities accurately if a suitable value for the parameter C is chosen. The outlier-detection method detects more elements near the circle wave but is not as sharp as we expect compared to results for the quadrilateral mesh case [20]. Inspection of the multiwavelet coefficients at the final time reveals that the discontinuous region is spread out wide, and therefore, the local region of size 16 is too small to contain both continuous and discontinuous regions. At certain locations, all coefficients in a local vector belong to a discontinuous region, and therefore, the fences are wide

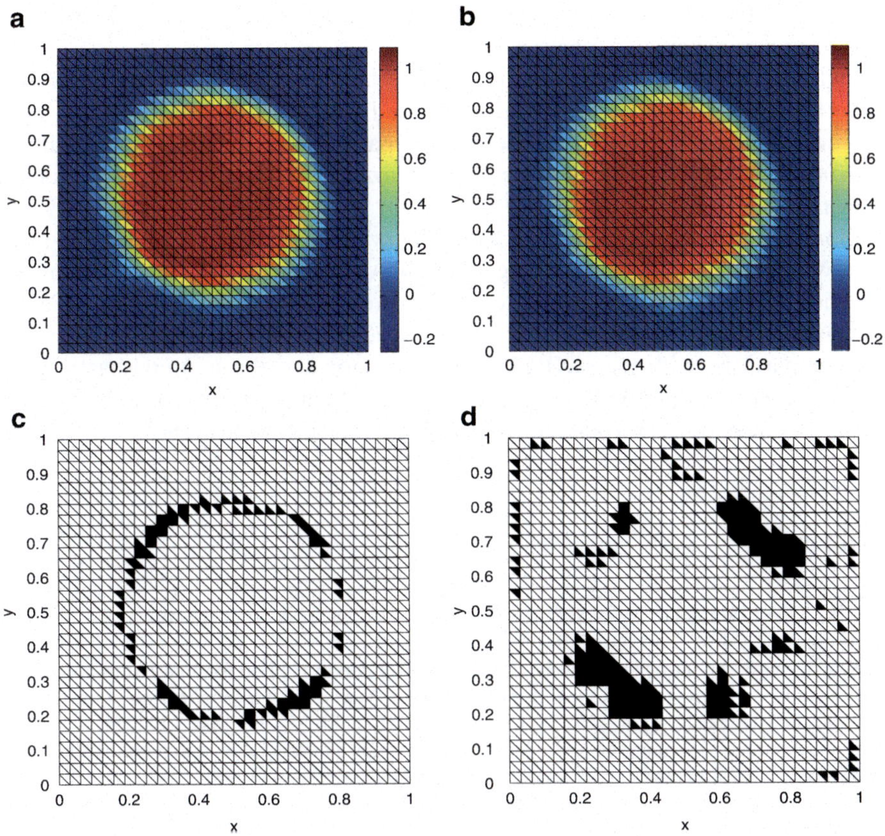

Fig. 5 Final-time approximations and corresponding detected troubled cells, using the parameter-based multiwavelet troubled-cell indicator or outlier detection on the multiwavelet coefficients, $T = \sqrt{2}$, structured triangular mesh based on 32×32 rectangles, $k = 1$. (**a**) $C = 0.9$, (**b**) Outlier, (**c**) $C = 0.9$, (**d**) Outlier

enough such that no elements are detected. Further research is needed to understand which outlier-detection strategy should be used.

6 Conclusion

In this paper, the use of multiwavelets for troubled-cell indication on structured triangular meshes has been investigated. Inspection of the multiwavelet coefficients reveals that they are very useful to detect nonsmooth regions in the underlying function. Two different troubled-cell indicators were introduced: one indicator that uses a problem-depending parameter, and another indicator that applies outlier detection to the multiwavelet coefficients. By using outlier detection, a problem-dependent parameter is no longer needed.

Preliminary results have been shown for a test based on the two-dimensional linear advection equation. The parameter-based troubled-cell indicator detects the correct features if a suitable choice for the parameter is made. For the outlier-detection method, it seems as if the optimal size of the local vectors is no longer equal to 16.

More research should be done to recognize which multiwavelet coefficient measures which feature of the underlying function. Also, an improvement of the outlier-detection strategy is needed to detect the correct regions after time integration. Furthermore, tests for nonlinear PDEs such as the two-dimensional Euler equations, and comparisons with the KXRCF shock detector and the minmod-based TVB indicator should be performed to thoroughly test the applicability of multiwavelets and outlier detection for troubled-cell indication on triangular meshes.

References

1. B.K. Alpert, A class of bases in L^2 for the sparse representation of integral operators. SIAM J. Math. Anal. **24**(1), 246–262 (1993)
2. B. Cockburn, An introduction to the discontinuous galerkin method for convection-dominated problems, in *Advanced Numerical Approximation of Nonlinear Hyperbolic Equations*. Lecture Notes in Mathematics, vol. 1697 (Springer, Berlin/Heidelberg, 1998), pp. 150–268
3. B. Cockburn, C.-W. Shu, The runge-kutta discontinuous galerkin method for conservation laws v: multidimensional systems. J. Comput. Phys. **141**(2), 199–224 (1998)
4. M. Dubiner, Spectral methods on triangles and other domains. J. Sci. Comput. **6**(4), 345–390 (1991)
5. M.A. Fortes, M. Moncayo, Multiresolution analysis and supercompact multiwavelets for surfaces. Math. Comput. Simul. **81**(10), 2129–2149 (2011)
6. M.A. Fortes, M.L. Rodríguez, Non-uniform multiresolution analysis for surfaces and applications. Appl. Numer. Math. **75**, 123–135 (2014)
7. N. Gerhard, S. Müller, Adaptive multiresolution discontinuous Galerkin schemes for conservation laws: multi-dimensional case. Comput. Appl. Math. **35**, 321–349 (2016)
8. J.S. Hesthaven, T. Warburton, *Nodal Discontinuous Galerkin Methods: Algorithms, Analysis, and Applications*. Texts in Applied Mathematics, vol. 54 (Springer, New York, 2008)
9. T. Koornwinder, Two-variable analogues of the classical orthogonal polynomials, in *Theory and Application of Special Functions* ed. by R. Askey (Elsevier BV, Amsterdam, 1975), pp. 435–495
10. D. Kuzmin, A vertex-based hierarchical slope limiter for p-adaptive discontinuous Galerkin methods. J. Comput. Appl. Math. **233**(12), 3077–3085 (2010)
11. R.J. LeVeque, in *Finite Volume Methods for Hyperbolic Problems*, 6th edn. Cambridge Texts in Applied Mathematics (Cambridge University Press, New York, 2002)
12. A. Meister, S. Ortleb, T. Sonar, M. Wirz, An extended discontinuous galerkin and spectral difference method with modal filtering. ZAMM - J. Appl. Math. Mech. Z. Angew. Math. Mech. **93**(6–7), 459–464 (2013)
13. F. Raees, D.R. van der Heul, C. Vuik, A mass-conserving level-set method for simulation of multiphase flow in geometrically complicated domains. Int. J. Numer. Methods Fluids **81**(7), 399–425 (2016)

14. A.B. Shelton, A multi-resolution discontinuous Galerkin method for unsteady compressible flows, PhD thesis, Georgia Institute of Technology, 2008
15. F. Sieglar, Konstruktion von multiwavelets auf dreiecksgittern, Bachelor's thesis, Rheinisch-Westfälischen Technischen Hochschule Aachen, 2013
16. M.J. Vuik, Limiting and shock detection for discontinuous Galerkin solutions using multiwavelets, Master's thesis, Delft University of Technology, 2012
17. M.J. Vuik, The use of multiwavelets and outlier detection for troubled-cell indication in discontinuous Galerkin methods, PhD thesis, Delft University of Technology, 2017
18. M.J. Vuik, J.K. Ryan, Multiwavelet troubled-cell indicator for discontinuity detection of discontinuous Galerkin schemes. J. Comput. Phys. **270**, 138–160 (2014)
19. M.J. Vuik, J.K. Ryan, Multiwavelets and jumps in DG approximations, in *Spectral and High Order Methods for Partial Differential Equations—ICOSAHOM 2014*, ed. by R.M. Kirby, M. Berzins, J.S. Hesthaven. Lecture Notes in Computational Science and Engineering, vol. 106 (Springer International Publishing, Switzerland, 2015)
20. M.J. Vuik, J.K. Ryan, Automated parameters for troubled-cell indicators using outlier detection. SIAM J. Sci. Comput. **38**(1), A84–A104 (2016)
21. T.P.Y. Yu, K. Kolarov, W. Lynch, Barysymmetric multiwavelets on triangle. Technical report 1997-006, Interval Research Corporation (1997)

Solution of Wave Equation in Rods Using the Wavelet-Galerkin Method for Space Discretization

Rodrigo B. Burgos, Marco A. Cetale Santos, and Raul R. e Silva

Abstract The use of multiresolution techniques and wavelets has become increasingly popular in the development of numerical schemes for the solution of partial differential equations (PDEs) in the last three decades. Therefore, the use of wavelets scale functions as a basis in computational analysis holds some promise due to their compact support, orthogonality, localization and multiresolution properties. The present work discusses an alternative to the usual finite difference (FDM) approach to the acoustic wave equation modeling by using a space discretization scheme based on the Galerkin Method. The combination of this method with wavelet analysis using scale functions results in the Wavelet Galerkin Method (WGM) which has been adapted for the direct solution of the wave differential equation in a meshless formulation. This paper presents an extension of previous works which dealt with linear elasticity problems. This work also introduces Deslauriers-Dubuc scaling functions (also known as Interpolets) as interpolating functions in a Galerkin approach considering wave propagation problems. Examples in 1-D were formulated using a central difference (second order) scheme for time differentiation. Encouraging results were obtained when compared with the FDM using the same time steps. The main improvement in the presented formulation was the recognition of a different dispersion pattern when comparing FDM and WGM results using the same space and time grid.

R.B. Burgos (✉)
Department of Structures and Foundations, UERJ, Rua S. Francisco Xavier, 524, Rio de Janeiro, RJ, Brazil
e-mail: rburgos@eng.uerj.br

M.A. Cetale Santos
Department of Geology and Geophysics, UFF, Av. Gen. Milton Tavares de Souza, s/n, Niterói, RJ, Brazil
e-mail: marcocetale@id.uff.br

R.R. e Silva
Department of Civil Engineering, PUC-Rio, Rua Marquês de São Vicente, 225, Rio de Janeiro, RJ, Brazil
e-mail: raul@puc-rio.br

M.L. Bittencourt et al. (eds.), *Spectral and High Order Methods for Partial Differential Equations ICOSAHOM 2016*, Lecture Notes in Computational Science and Engineering 119, DOI 10.1007/978-3-319-65870-4_28

1 Introduction

Among the various techniques available for the solution of the partial differential equation (PDE) that describes wave propagation (wave equation), the Finite Difference Method (FDM) [1] is by far the most employed one, being used frequently as a standard for the validation of new methods. As a disadvantage, the FDM is known for requiring excessive refining in terms of space and time discretization.

The use of wavelet-based numerical methods has become popular in the last three decades, especially for problems with local high gradients and singularities. Wavelets have many properties that are quite useful for representing solutions of PDEs, such as orthogonality, compact support and a certain number of vanishing moments (exact representation of polynomials). These characteristics allow the efficient and stable calculation of functions with high gradients or singularities at different levels of resolution [2].

A complete basis of wavelets can be generated through dilation and translation of a mother scaling function. Although many applications use only the wavelet filter coefficients of the multiresolution analysis, there are some which explicitly require the values of the basis functions and their derivatives, such as the Wavelet Finite Element Method (WFEM) [3].

Compactly supported wavelets have a finite number of derivatives which can be highly oscillatory, which makes numerical evaluation of integrals of their inner products difficult and unstable. Those integrals are known as connection coefficients and they are employed in the calculation of stiffness and mass matrices in the Wavelet-Galerkin Method (WGM). Due to some properties of wavelet functions, these coefficients can be obtained by solving an eigenvalue problem using filter coefficients.

Working with dyadically refined grids, Deslauriers and Dubuc [4] obtained a new family of scaling functions and wavelets with interpolating properties, later called Interpolets. Their filter coefficients are obtained from the autocorrelation of the Daubechies' coefficients [5]. As a consequence, interpolets are symmetric, which is especially useful in numerical analysis. The use of interpolets instead of Daubechies' wavelets considerably improves the method's accuracy [6, 7].

In this work, the Wavelet-Galerkin Method has been adapted for the direct solution of the acoustic wave equation in a meshless formulation. Accuracy can be improved by increasing either the level of resolution or the order of the wavelet used. Normally, the former works better than the latter, since increasing the order of the functions involved can lead to ill conditioned systems.

As a preliminary study, the formulation of an interpolet-based Galerkin scheme was demonstrated for a one-dimensional wave propagation problem. Some examples were formulated and results compared with the standard Finite Differences Method (FDM).

2 Wavelet Theory and Interpolets

Multiresolution analysis using orthogonal, compactly supported wavelets has become increasingly popular in numerical simulation. Wavelets are localized in both frequency and space, which allows the analysis of local variations of the problem at various levels of resolution.

The following expression, known as the two-scale relation, is a recursive relation and is essential to defining wavelets on spaces of functions. In Eq. (1), N is the order and a_k are the filter coefficients of the wavelet scaling function. The limits for the index k depend on the wavelet family.

$$\varphi(x) = \sum_{k=1-N}^{N-1} a_k \varphi\,(2x - k) = \sum_{k=1-N}^{N-1} a_k \varphi_k(2x) \tag{1}$$

The basic characteristics of interpolating wavelets require that the mother scaling function satisfies the following condition [8]:

$$\varphi(k) = \delta_{0,k} = \begin{cases} 1, & k = 0 \\ 0, & k \neq 0 \end{cases} \quad k \in \mathbb{Z} \tag{2}$$

Working with dyadically refined grids, Deslauriers and Dubuc [4] obtained a new family of scaling functions and wavelets with interpolating properties. The Deslauriers-Dubuc interpolating function of order N is given by an autocorrelation of the Daubechies' scaling filter coefficients (h_m) of the same order (with $N/2$ vanishing moments). Its support is given by $[1 - N; N - 1]$, it has even symmetry and is capable of representing polynomials of order up to $N - 1$ (i.e. N vanishing moments).

$$a_k = \sum_{m=0}^{N-1} h_m h_{m-k} \tag{3}$$

Interpolets satisfy the same requirements as other wavelets, especially the two-scale relation, which is fundamental for their use as interpolating functions in numerical methods. Figure 1 shows the interpolet IN6 (autocorrelation of DB6). Its symmetry and interpolating properties are evident. Its support is given by $[-5; 5]$ and there is only one integer abscissa which evaluates to a non-zero value.

The numerical solution of differential equations is one of the possible applications of the wavelet theory. The Wavelet-Galerkin Method (WGM) results from the use of wavelet scaling functions as the interpolating basis in a traditional Galerkin scheme. In the following sections, the WGM will be applied to solve the typical DE for acoustic wave propagation.

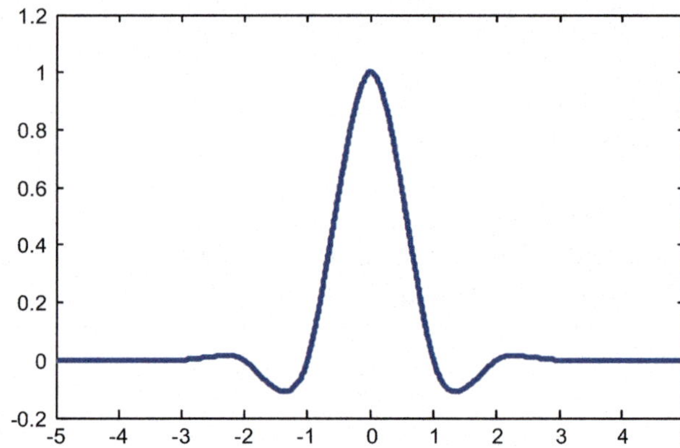

Fig. 1 Interpolet IN6 scaling function

3 Wave Propagation Using the Wavelet-Galerkin Method

The partial differential equation (PDE) which rules the one dimensional wave propagation is:

$$\alpha^2 \frac{\partial^2 u\,(x,t)}{\partial x^2} = \frac{\partial^2 u\,(x,t)}{\partial t^2} \tag{4}$$

In Eq. (4), u is the horizontal displacement and α is the medium velocity. The problem at hand will be solved in an acoustic homogeneous domain using Dirichlet boundary conditions. Assuming that the displacement u is approximated at the lowest level of discretization (level 0) by a series of interpolating scale functions using a normalized (non dimensional) coordinate ξ, the following may be written:

$$u\,(\xi) = \sum_{k=2-N}^{N-1} d_k \varphi\,(\xi - k) \tag{5}$$

Coordinate $\xi = x/L$ is used in order to allow the domain to be considered as a unit length rod ([0 1]). The Galerkin method consists in substituting the expression above in the differential equation and forcing the approximation error to be orthogonal to a test result which is formulated using the same interpolating functions [9].

$$\alpha^2 \left\{ \int_0^1 \mathbf{\Phi}^{\mathrm{T}} \mathbf{\Phi}'' d\xi \right\} \mathbf{d} = \left\{ \int_0^1 \mathbf{\Phi}^{\mathrm{T}} \mathbf{\Phi}\; d\xi \right\} \ddot{\mathbf{d}} \tag{6}$$

In Eq. (6), Φ is a vector consisting of translations of wavelet scaling functions and **d** is a vector of coefficients for interpolating the displacement function. Using this approach, the PDE can be rewritten at a specific time t as a system of linear equations, which in matrix form is:

$$\mathbf{m}\,\ddot{\mathbf{d}}_t + \mathbf{k}\mathbf{d}_t = \mathbf{0} \tag{7}$$

In this expression, **m** represents the mass matrix and **k** is the stiffness matrix of the model, which in normalized coordinates (ξ) within the interval [0; 1] are given by:

$$\begin{aligned} k_{i,j} &= -\alpha^2 \int_0^1 \varphi_i(\xi)\, \varphi_j''(\xi)\, d\xi = -\alpha^2 \Lambda_{i,j}^{0,2} \\ m_{i,j} &= \int_0^1 \varphi_i(\xi)\, \varphi_j(\xi)\, d\xi = \Lambda_{i,j}^{0,0} \end{aligned} \tag{8}$$

The so-called connection coefficients Λ appear in the expressions above. Wavelet dilation and translation properties allow the calculation of connection coefficients to be summarized by the solution of an eigenvalue problem based only on filter coefficients [10].

$$\begin{aligned} &\left(\mathbf{P} - \tfrac{1}{2^{d_1+d_2-1}}\mathbf{I}\right) \mathbf{\Lambda}^{\mathbf{d_1,d_2}} = \mathbf{0} \\ &\mathbf{P} = \left[a_{r-2i}a_{s-2j} + a_{r-2i+1}a_{s-2j+1}\right]_{ij,r,s=(2-N)\ldots(N-1)} \end{aligned} \tag{9}$$

Since the expression above leads to an infinite number of solutions, there is the need for a normalization rule that provides a unique eigenvector. This unique solution comes with the inclusion of the so-called moment equation, derived from the wavelet scaling functions property of exact polynomial representation [11]. For the two-dimensional modeling, a Kronecker product appears when using 2-D functions, resulting in:

$$\begin{aligned} \mathbf{m} &= \mathbf{\Lambda}^{0,0} \otimes \mathbf{\Lambda}^{0,0} \\ \mathbf{k} &= -\alpha^2 \left(\mathbf{\Lambda}^{0,2} \otimes \mathbf{\Lambda}^{0,0} + \mathbf{\Lambda}^{0,0} \otimes \mathbf{\Lambda}^{0,2}\right) \end{aligned} \tag{10}$$

As in the FDM, it becomes necessary to solve the system of equations at discrete time intervals. There are several effective direct integration methods, among which the most intuitive one is the Central Difference Method:

$$\ddot{\mathbf{d}}_t \cong \frac{\mathbf{d}_{t+\Delta t} - 2\,\mathbf{d}_t + \mathbf{d}_{t-\Delta t}}{(\Delta t)^2} \tag{11}$$

Substituting the expression of the acceleration obtained by the Central Difference Method and solving for the next time step at $t + \Delta t$:

$$\mathbf{m}\,\mathbf{d}_{t+1} = \mathbf{m}\,(2\mathbf{d}_t - \mathbf{d}_{t-1}) - (\Delta t)^2 \mathbf{k}\mathbf{d}_t \tag{12}$$

Stability of the Central Difference Method is conditioned to the choice of the time step, whose upper bound is obtained from a generalized eigenvalue problem.

$$\begin{gathered}\left(\mathbf{k}-\omega^2\mathbf{m}\right)\mathbf{d}=\mathbf{0}\rightarrow\left(\mathbf{X}-\omega^2\,\mathbf{I}\right)\mathbf{d}=\mathbf{0}\\ \Delta t_{\max}=\frac{\sqrt{2}}{\omega_{\max}}\end{gathered} \tag{13}$$

Matrix **m** might not be invertible. In this case, boundary conditions shall be imposed by using Lagrange multipliers. This procedure is used commonly in meshless methods [12] and leads to a square matrix which can be useful for some system solvers.

$$\begin{bmatrix}\mathbf{m} & \mathbf{g}^{\mathrm{T}}\\ \mathbf{g} & \mathbf{0}\end{bmatrix}\begin{Bmatrix}\mathbf{d}_{t+1}\\ \boldsymbol{\lambda}\end{Bmatrix}=\begin{Bmatrix}\mathbf{m}\left(2\mathbf{d}_t-\mathbf{d}_{t-1}\right)-(\Delta t)^2\mathbf{k}\mathbf{d}_t\\ \mathbf{0}\end{Bmatrix} \tag{14}$$

In the expression above, the matrix **g** is associated with boundary conditions and λ is a vector of Lagrange multipliers which is not used in the solution. The main difference in relation to the FDM is that the unknowns in vector **d** are the interpolating coefficients of the basis functions instead of nodal displacements. In fact, there is no need to establish nodal coordinates.

When dealing with one-dimensional problems, most wavelets (including Daubechies and Interpolets) present a mass matrix whose rank is one unit less than its size. This means that only one boundary condition needs to be imposed for the system to have a solution.

4 Numerical Results

To validate the formulation, a one dimensional example was implemented, consisting in applying a ricker source at the midpoint of a pinned, unit length rod. The propagation was modeled by the FDM using 265 points and $\Delta t = 0.3$ ms, with fourth and second order discretization in space and time, respectively. This time step was obtained using the upper bound described in the previous section. For this example, the lowest central frequency of the source that produces a numerical dispersion is $\omega = 80$ Hz. The same source central frequency, spatial discretization and time step were used in the WGM example, with no visible dispersion in the results. The spatial discretization in the WGM was adjusted in terms of function order and level of resolution in order to give the same number of degrees of freedom. In this example, this was achieved using IN4 and level $j = 8$. Figure 2 shows the response at time $t = 0.45$ s for both methods using a source central frequency of $\omega = 40$ Hz. There is no visible numerical dispersion in either case.

Figure 3 shows the response using a source central frequency of $\omega = 80$ Hz, which produces numerical dispersion only in the FDM model.

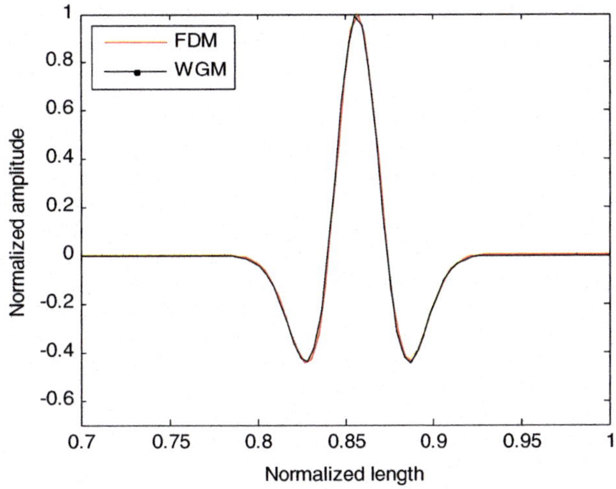

Fig. 2 FDM and WGM results for $\omega = 40$ Hz

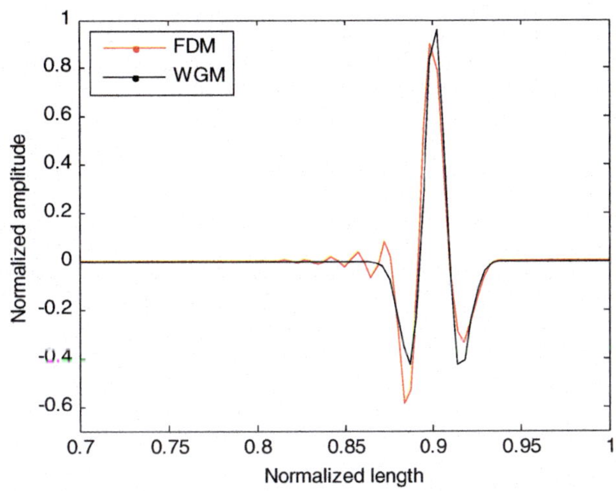

Fig. 3 FDM and WGM results for $\omega = 80$ Hz

As a second example, the rod is made by two different materials and the dispersion in the case of the FDM is even greater, as shown in Fig. 4. As expected, the change in velocity introduces additional errors in the FDM model. These errors are not present in the WGM model. Figure 5 shows the time record of a point at a normalized length of 0.4. It's clear that numerical dispersion becomes visible after the reflection.

Only for the purpose of testing the ability of the method for two-dimensional applications, a simple two-material model with the same velocity profile in depth as the 1-D model was tested and a snapshot taken at time $t = 0.52$ s is shown in Fig. 6.

The time step used of $\Delta t = 2.3$ ms was obtained employing the same approach given in (13). A smaller time step, $\Delta t = 2.0$ ms, was used in the FDM model and led to numerical instability, as indicated in Fig. 7. This finding shows that the FDM requires smaller time steps than the WGM.

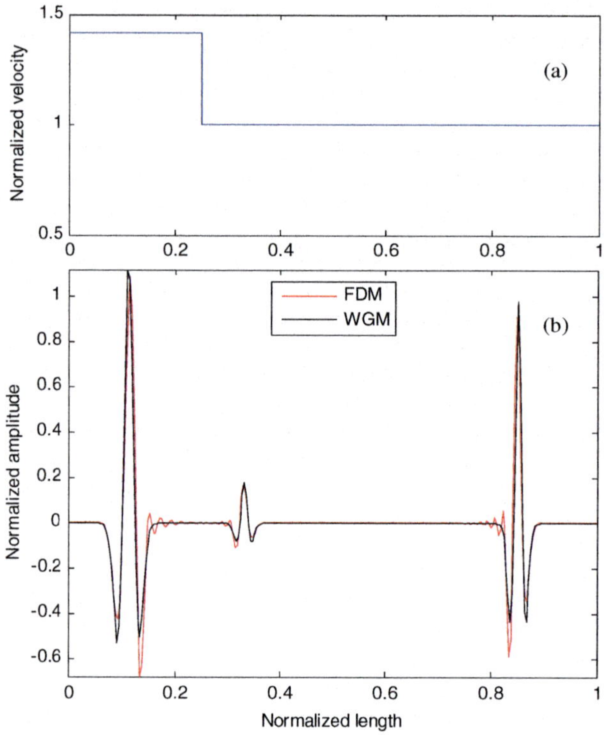

Fig. 4 (**a**) velocity profile; (**b**) snapshot for comparison at time t = 0.45 s

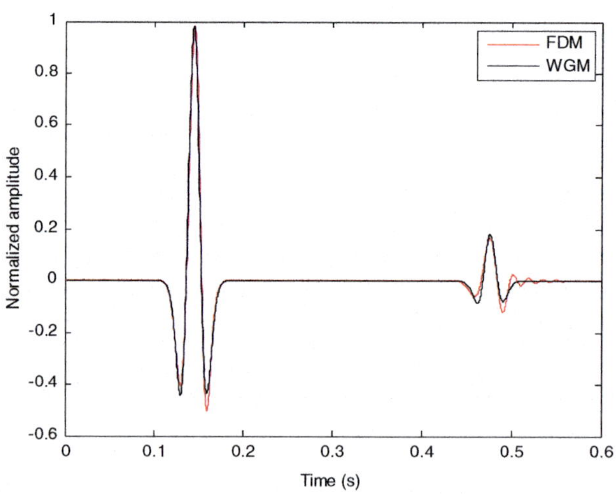

Fig. 5 Records of normalized displacement at point 0.4

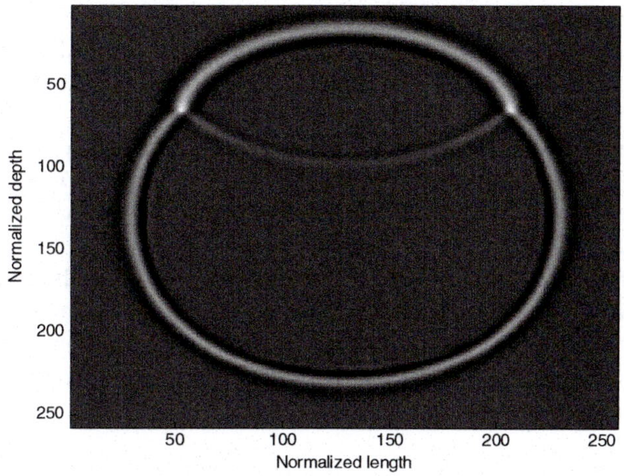

Fig. 6 Propagation snapshot at time t = 0.52 s for the 2-D model

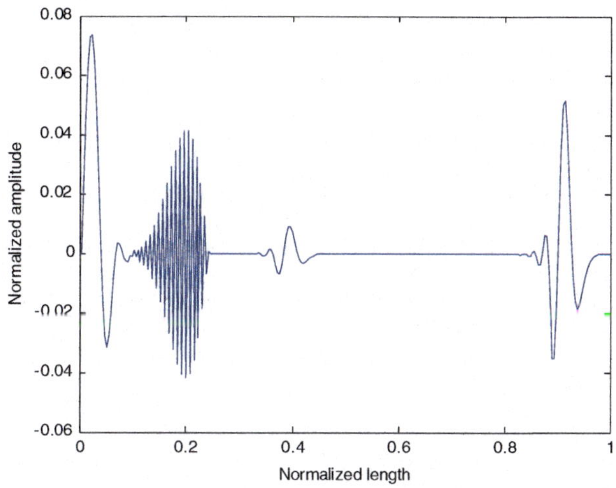

Fig. 7 Snapshot of FDM model at time t = 0.52 s showing numerical instability

5 Conclusions

This work presented the formulation and validation of the Wavelet-Galerkin Method (WGM) using Deslauriers-Dubuc Interpolets. These preliminary results are promising, but the simplicity of the studied models has to be taken into account. The main improvement in the presented formulation was the recognition of a different dispersion pattern when comparing FDM and WGM results using the same space and time grid. Both methods used second order time discretization and the FDM used fourth order space discretization, which shows that comparisons were made against a robust numerical scheme. The different dispersion pattern is probably due to the convergent nature of Galerkin type methods, which generally have dispersion errors of twice their truncation errors, while FD methods have dispersion errors on the order of their truncation errors.

All matrices involved can be stored and operated in a sparse form, since most of their components are null, thus saving computer resources. Due to the compact support of wavelets, the sparseness of matrices increases along with the level of resolution.

In future works, models with greater complexity will be analyzed and different families of wavelets will be explored. The extension of the method to irregular geometries in two-dimensional problems is still a challenge, but one potential advantage is the possibility of implementing absorbing boundary conditions analytically with the use of Lagrange Multipliers.

Acknowledgments Authors would like to thank CNPq, PETROBRAS and FAPERJ for their financial support.

References

1. K.R. Kelly, R.W. Ward, S. Treitel, R.M. Alford, Synthetic seismograms: a finite-difference approach. Geophysics **41**, 2–27 (1976)
2. S. Qian, J. Weiss, Wavelets and the numerical solution of partial differential equations. J. Comput. Phys. **106**, 155–175 (1992)
3. X. Chen, S. Yang, J. Ma, Z. He, The construction of wavelet finite element and its application. Finite Elem. Anal. Des. **40**, 541–554 (2004)
4. G. Deslauriers, S. Dubuc, Symmetric iterative interpolation processes. Constr. Approx. **5**, 49–68 (1989)
5. I. Daubechies, Orthonormal bases of compactly supported wavelets. Commun Pure Appl Math **41**, 909–996 (1988)
6. R.B. Burgos, M.A. Cetale Santos, R.R. Silva, Analysis of beams and thin plates using the Wavelet-Galerkin method. Int J Eng Technol **7**, 261–266 (2015)
7. A.J.M. Ferreira, L.M. Castro, S. Bertoluzza, Analysis of plates on Winkler foundation by wavelet collocation. Meccanica **46**(4), 865–873 (2011)
8. Z. Shi, D.J. Kouri, G.W. Wei, D.K. Hoffman, Generalized symmetric interpolating wavelets. Comput. Phys. Commun. **119**, 194–218 (1999)
9. X. Du, J.C. Bancroft, in *Proceedings of the SEG Int'l Exposition and 74th Annual Meeting,* 2-D Wave Equation Modeling and Migration By a New Finite Difference Scheme Based on the Galerkin Method, (Denver, USA, 2004)
10. X. Zhou, W. Zhang, The evaluation of connection coefficients on an interval. Commun Nonlinear Sci Numer Simul **3**, 252–255 (1998)
11. R.B. Burgos, M.A. Cetale Santos, R.R. Silva, Deslauriers-Dubuc interpolating wavelet beam finite element. Finite Elem. Anal. Des. **75**, 71–77 (2013)
12. V.P. Nguyen, T. Rabczuk, S. Bordas, M. Duflot, Meshless methods: a review and computer implementation aspects. Math. Comput. Simul. **79**, 763–813 (2008)

On High Order Entropy Conservative Numerical Flux for Multiscale Gas Dynamics and MHD Simulations

Björn Sjögreen and H.C. Yee

Abstract The Sjögreen and Yee (On skew-symmetric splitting and entropy conservation schemes for the Euler equations, in Proceedings of ENUMATH09, June 29-July 2, Uppsala University, Sweden, 2009) high order entropy conservative numerical method for compressible gas dynamics is extended to include discontinuities and also extended to the ideal magnetohydrodynamics (MHD). The basic idea is based on Tadmor's (Acta Numer 12:451–512, 2003) original work for the Euler gas dynamics. For the MHD four formulations of the MHD formulations are considered: (a) the conservative MHD, (b) the Godunov/Powell non-conservative form, (c) the Janhunen MHD with magnetic field source terms (Janhunen, J Comput Phys 160:649–661, 2000), and (d) a MHD with source terms by Brackbill and Barnes (J Comput Phys 35:426–430, 1980). Three forms of the high order entropy numerical fluxes in the finite difference framework are constructed. They are based on the extension of the low order form by Chandrashekar and Klingenberg (SIAM J Numer Anal 54:1313–1340, 2016), and two forms with modifications of the Winters and Gassner (J Comput Phys 304:72–108, 2016) numerical fluxes. For flows containing discontinuities and multiscale turbulence fluctuations the Yee and Sjogreen (High order filter methods for wide range of compressible flow speeds, in *Proceedings of the ICOSAHOM09*, Trondheim, Norway, June 22–26, 2009) and Kotov et al. (Commun Comput Phys 19:273–300, 2016; J Comput Phys 307:189–202, 2016) high order nonlinear filter approach are extended to include the high order entropy conservative numerical fluxes as the base scheme.

B. Sjögreen (✉)
MultiD Analyses AB, Odinsgatan 28, SE-411 03 Goteborg, Sweden
e-mail: bjorn.sjogreen@multid.se

H.C. Yee
NASA Ames Research Center, Mountain View, CA, USA
e-mail: Helen.M.Yee@nasa.gov

M.L. Bittencourt et al. (eds.), *Spectral and High Order Methods for Partial Differential Equations ICOSAHOM 2016*, Lecture Notes in Computational Science and Engineering 119, DOI 10.1007/978-3-319-65870-4_29

1 Introduction

Stability and accuracy in high order numerical method development for multiscale turbulent flows consist of conflicting requirements. On one hand, stable methods for the subject flow require an appropriate amount of numerical dissipation to achieve stability. On the other hand, numerical accuracy requirement in direct numerical simulations (DNS) and large eddy simulations (LES) cannot tolerate numerical dissipation and high dispersive error. Moreover, due to the CPU intensive computations of such flows, practical efficient methods require low dispersive and low dissipative error with coarse grid computation which are at the same time applicable for non-periodic boundaries in generalized geometries. A prime candidate for an efficient high order spatial discretization with non-dissipative and low dispersion error is the high order central schemes and their low dispersion counterparts of Linders and Nordström [10] to be used as the primary spatial discretizations at almost everywhere in the computed flow field. After the completion of every full time step of the non-dissipative (and low dispersion) spatial scheme, to suppress spurious oscillations at discontinuities and high frequency oscillations of long time integrations of highly coupled nonlinear governing equation sets, the computed solution is nonlinearly filtered by the dissipative portion of a high order shock-capturing scheme accompanied by a smart flow sensor developed by Yee and Sjögreen [20] and Kotov et al. [8, 9]. The smart flow sensor provides the locations and the estimated strength of the necessary numerical dissipation needed at these locations and leaves the rest of the flow field free of shock-capturing dissipations. In this paper, we only concentrate on the high order central schemes.

The aforementioned nonlinear filter scheme with adaptive numerical dissipation control in high order shock-capturing schemes and their hybrid cousins have shown excellent performance for certain turbulent test cases. For more practical 3D test cases of DNS and LES of compressible shock-free turbulence, low speed turbulence with shocklets, and supersonic turbulence for non-periodic boundaries in curvilinear geometries, some improvement in numerical stability is needed without resorting to added numerical dissipation that can interfere with the accuracy of numerical simulations. The skew-symmetric splitting of the inviscid flux derivatives for central schemes can help with numerical stability. See Arakawa [1], Blaisdell et al. [2], Yee et al. [21], Ducros et al. [5], Yee and Sjögreen [19, 20], Sjögreen and Yee [12], Kotov et al.[8, 9] for some discussions and performance of the combined approach for DNS and LES applications. For their skew-symmetric splitting extension to the ideal MHD, see Sjögreen and Yee [13]. Entropy conservative schemes [4, 15, 17] is another class of methods that might have better stability properties than straightforward centered discretizations. Here, entropy conservative schemes refer to conservative schemes satisfying a discrete entropy equation. This is the subject of our current study.

Objective and Outline In this paper we will develop and test entropy conservative schemes for the equations of gas dynamics and equations of the ideal MHD for compressible flows. Their stability for problems with smooth solutions and problems with discontinuities will be investigated. An additional aspect is that there are several slightly different, but equivalent, ways to formulate the equations of MHD. We show in [14] that the formulation of the MHD equations have a strong effect on the stability of non-dissipative approximations. Specifically, four formulations of the MHD are considered: (a) the conservative MHD, (b) the Godunov/Powell non-conservative form, (c) the Janhunen MHD with magnetic field source terms [7], and (d) the MHD source term of [3]. Three formulations of the high order entropy numerical fluxes in the finite difference framework are constructed. They are based on the extension of the low order form of Chandrashekar and Klingenberg [4], and two forms with modifications of the Winters and Gassner [17] numerical fluxes. For flows containing discontinuities and multiscale turbulence fluctuations the Yee and Sjogreen [20] and Kotov et al. [8, 9] high order nonlinear filter approach is extended to include the high order entropy conservative numerical fluxes as the base scheme. Due to page limitations, many details of the development and extensive numerical testing with more representative test cases among the different formulations and the different governing equation sets will be reported in [14] for journal publication.

2 High Order Entropy Conservative Schemes for Gas Dynamics

Entropy conservative schemes were introduced in the 1980s. See, e.g., [15]. These schemes are in conservation form, and admit a discrete conservation law for the entropy. We consider a conservation law in one space dimension

$$\mathbf{u}_t + \mathbf{f}(\mathbf{u})_x = \mathbf{0}. \tag{1}$$

An entropy, $E(\mathbf{u})$, and an entropy flux, $F(\mathbf{u})$, are two functions satisfying $E_{\mathbf{u}}^T A(\mathbf{u}) = F_{\mathbf{u}}^T$, where $E_{\mathbf{u}}$ denotes the gradient of E with respect to $\mathbf{u}$, and $A(\mathbf{u})$ denotes the matrix $\mathbf{f}_{\mathbf{u}}$. Furthermore, $E(\mathbf{u})$ is assumed to be a convex function. The entropy variables are defined by $\mathbf{v} = E_{\mathbf{u}}(\mathbf{u})$. Multiplying (1) by $\mathbf{v}^T$ gives the entropy equation

$$\mathbf{v}^T \mathbf{u}_t + \mathbf{v}^T A \mathbf{u}_x = E(u)_t + F_{\mathbf{u}}^T u_x = E(u)_t + F(u)_x = 0.$$

The entropy flux potential, defined by

$$\psi = \mathbf{v}^T \mathbf{f} - F$$

has the property that $\mathbf{f} = \psi_{\mathbf{v}}$.

The following construction, proved in [12], defines a high order entropy conservation scheme.

Theorem 1 *Let the coefficients* $\alpha_k^{(p)}$ *be determined such that*

$$D_p u_j = \sum_{k=-p}^{p} \alpha_k^{(p)} \frac{u_{j+k} - u_{j-k}}{2k\Delta x} \tag{2}$$

is the standard 2*pth order accurate centered difference operator approximating the first derivative. The semi-discrete approximation of a system of conservation laws given by*

$$\Delta x \frac{d}{dt}\mathbf{u}_j + \sum_{k=1}^{p} \frac{\alpha_k^{(p)}}{k}(\mathbf{g}_{j+k/2}^{(k)} - \mathbf{g}_{j-k/2}^{(k)}) = \mathbf{0}, \tag{3}$$

where $\mathbf{g}_{j+k/2}^{(k)}$ *satisfies*

$$(\mathbf{v}_{j+k} - \mathbf{v}_j)^T \mathbf{g}_{j+k/2}^{(k)} = \psi_{j+k} - \psi_j \tag{4}$$

and where the kth flux differences approximate the flux derivative to second order with a truncation error of even powers of $k\Delta x$*,*

$$\mathbf{g}_{j+k/2}^{(k)} - \mathbf{g}_{j-k/2}^{(k)} = k\Delta x \mathbf{f}_x + k^3 \Delta x^3 \boldsymbol{\phi}_1 + k^5 \Delta x^5 \boldsymbol{\phi}_2 + \dots, \tag{5}$$

is 2*pth order accurate, and admits a discrete entropy equation*

$$\Delta x \frac{d}{dt} E_j + \sum_{k=1}^{p} \frac{\alpha_k^{(p)}}{k}(H_{j+k/2}^{(k)} - H_{j-k/2}^{(k)}) = 0, \tag{6}$$

where $H_{j+k/2}^{(k)} = \frac{1}{2}[(\mathbf{v}_{j+k} + \mathbf{v}_j)^T \mathbf{g}_{j+k/2}^{(k)} - (\psi_{j+k} + \psi_j)]$*. Both (3) and (6) can be cast in conservation form, because*

$$a_{j+k/2} - a_{j-k/2} = \Delta_+ (\sum_{m=0}^{k-1} a_{j-k/2+m})$$

for any arbitrary grid function $a_{j+k/2}$ *that satisfies* $a_{j+k/2-k} = a_{j-k/2}$*.*

For a scalar conservation law the simple choice $g_{j+k/2}^{(k)} = (\psi_{j+k} - \psi_j)/(v_{j+k} - v_j)$ satisfies both (4) and (5). For the one dimensional Euler system [16] defined entropy conserving fluxes based on integration in phase space. An alternative approach, which was used in [12], is to write ψ as a function of the entropy variables and determine functions φ_i consistent with the gradient of ψ and satisfying

$$(\psi_{j+k} - \psi_j) = \varphi_1[(v_1)_{j+k} - (v_1)_j] + \ldots + \varphi_3[(v_3)_{j+k} - (v_3)_j].$$

The definition $\mathbf{g}^{(k)}_{j+k/2} = (\varphi_1, \varphi_2, \varphi_3)$ determines an entropy conservative method.

3 MHD Formulations

The equations of magnetohydrodynamics(MHD) is the system of conservation laws

$$\mathbf{u}_t + \mathbf{f}_x + \mathbf{g}_y + \mathbf{h}_z + \mathbf{e}\,\mathrm{div}\,\mathbf{B} = \mathbf{0}, \tag{7}$$

where the unknown field vector is

$$\mathbf{u} = (\rho,\ u,\ v,\ w,\ e,\ B^{(x)},\ B^{(y)},\ B^{(z)})$$

where ρ is density, $(u,\ v,\ w)$ is the velocity vector, e the total energy and $(B^{(x)},\ B^{(y)},\ B^{(z)})$ the magnetic field components. The pressure is

$$p = (\gamma - 1)(e - \frac{1}{2}\rho|\mathbf{w}|^2 - \frac{1}{2}|\mathbf{B}|^2),$$

where $\mathbf{w} = (u,\ v,\ w)$ and $\mathbf{B} = (B^{(x)},\ B^{(y)},\ B^{(z)})$. The x-direction flux is given by

$$\mathbf{f} = \begin{pmatrix} \rho u \\ \rho u^2 + p + \frac{1}{2}|\mathbf{B}|^2 - B^{(x)}B^{(x)} \\ \rho uv - B^{(x)}B^{(y)} \\ \rho uw - B^{(x)}B^{(z)} \\ u(e + p + \frac{1}{2}|\mathbf{B}|^2) - B^{(x)}\mathbf{u}^T\mathbf{B} \\ 0 \\ uB^{(y)} - vB^{(x)} \\ uB^{(z)} - wB^{(x)} \end{pmatrix},$$

and similar expressions hold for $\mathbf{g}$ and $\mathbf{h}$. In the last term on the left hand side of (7)

$$\mathbf{e} = \mathbf{e}_G = (0,\ B^{(x)},\ B^{(y)},\ B^{(z)},\ \mathbf{w}^T\mathbf{B},\ u,\ v,\ w)^T \tag{8}$$

multiplies $\mathrm{div}\,\mathbf{B}$. This term could be removed, because $\mathrm{div}\,\mathbf{B} = 0$ from a physical standpoint. If we keep the $\mathrm{div}\,\mathbf{B}$ term, the non-conservative system (7) has an entropy, as first shown by Godnov [6]. Some variants of (7) are to replace $\mathbf{e}$ by either

$$\mathbf{e}_J = (0,\ 0,\ 0,\ 0,\ 0,\ u,\ v,\ w)$$

as suggested in [7] or by

$$\mathbf{e}_B = (0,\ B^{(x)},\ B^{(y)},\ B^{(z)},\ 0,\ 0,\ 0,\ 0)$$

as suggested in [3]. The conservative form of (7) is obtained by setting $\mathbf{e} = \mathbf{0}$.

4 High Order Entropy Conservative Numerical Fluxes for MHD

Consider (7) in one space dimension, for simplicity,

$$\mathbf{u}_t + \mathbf{f}_x + \mathbf{e}B^{(x)}_x = \mathbf{0} \tag{9}$$

and define the entropy

$$E = -\frac{\rho s}{\gamma - 1} \qquad s = \ln p\rho^{-\gamma} \tag{10}$$

and the entropy flux $F = uE$. The entropy variables are defined by $\mathbf{v} = \nabla_{\mathbf{u}} E$. Written out explicitly for (9),

$$\mathbf{v} = \left(\frac{\gamma - s}{\gamma - 1} - \frac{\rho|\mathbf{w}|^2}{2p}, \frac{\rho u}{p}, \frac{\rho v}{p}, \frac{\rho w}{p}, -\frac{\rho}{p}, \frac{\rho B^{(x)}}{p}, \frac{\rho B^{(y)}}{p}, \frac{\rho B^{(z)}}{p} \right).$$

Multiplying the entropy variables by the Jacobian of the flux function gives after some lengthy algebra,

$$\mathbf{v}^T \frac{\partial \mathbf{f}}{\partial \mathbf{u}} = \nabla_{\mathbf{u}} F - (\mathbf{v}^T \mathbf{e}_G)(0\ 0\ 0\ 0\ 0\ 1\ 0\ 0)^T. \tag{11}$$

Note that $\mathbf{e}_G$ is defined in (8). The entropy conservation law $E_t + F_x = 0$ follows if we multiply (9) by the entropy variables,

$$\mathbf{v}^T \mathbf{u}_t + \mathbf{v}^T \mathbf{f}_x + \mathbf{v}^T \mathbf{e} B^{(x)}_x = \mathbf{0}$$

and use (11) on the term $\mathbf{v}^T \mathbf{f}_x = \mathbf{v}^T \frac{\partial \mathbf{f}}{\partial \mathbf{u}} \mathbf{u}_x$. Note that only the value of $\mathbf{v}^T \mathbf{e}$ matters for entropy conservation. Either choice $\mathbf{e}_G$, $\mathbf{e}_J$, or $\mathbf{e}_B$ for $\mathbf{e}$ in (9) would work equally well since $\mathbf{v}^T \mathbf{e}_G = \mathbf{v}^T \mathbf{e}_J = \mathbf{v}^T \mathbf{e}_B$.

Denote $\psi = \mathbf{v}^T\mathbf{f} - F$ and note that

$$\psi_x + \mathbf{v}^T\mathbf{e}B_x^{(x)} = \mathbf{v}_x^T\mathbf{f}, \tag{12}$$

since by using (11),

$$\psi_x = \mathbf{v}_x^T\mathbf{f} + \mathbf{v}^T\mathbf{f}_x - F_x = \mathbf{v}_x^T\mathbf{f} + \mathbf{v}^T\mathbf{f}_\mathbf{u}\mathbf{u}_x - F_x = \mathbf{v}_x^T\mathbf{f} + F_x - \mathbf{v}^T\mathbf{e}_G B_x^{(x)} - F_x.$$

It was proved in [4], for the case $\mathbf{e} = \mathbf{e}_G$, that entropy conservation holds for the semi-discrete second order accurate scheme,

$$\frac{d}{dt}\mathbf{u}_j + \frac{1}{\Delta x}(\mathbf{h}_{j+1/2} - \mathbf{h}_{j-1/2}) + \frac{1}{2}\mathbf{e}_j(B_{j+1}^{(x)} - B_{j-1}^{(x)}) = \mathbf{0}, \tag{13}$$

if the numerical flux $\mathbf{h}_{j+1/2} = \mathbf{h}(\mathbf{u}_{j+1}, \mathbf{u}_j)$ satisfies the discrete counterpart of (12),

$$(\mathbf{v}_{j+1} - \mathbf{v}_j)^T\mathbf{h}_{j+1/2} = \psi_{j+1} - \psi_j + \frac{1}{2}(\mathbf{v}_{j+1}^T\mathbf{e}_{j+1} + \mathbf{v}_j^T\mathbf{e}_j)(B_{j+1}^{(x)} - B_j^{(x)}). \tag{14}$$

Furthermore, when the numerical flux of scheme (13) satisfies (14), the computed solution satisfies the semi-discrete entropy conservation law

$$\frac{d}{dt}\mathbf{E}_j + \frac{1}{\Delta x}(H_{j+1/2} - H_{j-1/2}) = 0,$$

where

$$H_{j+1/2} = \frac{1}{2}(\mathbf{v}_{j+1} + \mathbf{v}_j)^T\mathbf{h}_{j+1/2} - \frac{1}{2}(\mathbf{v}_{j+1}^T\mathbf{f}_{j+1} + \mathbf{v}_j^T\mathbf{f}_j) + \frac{1}{2}(F_{j+1} + F_j) - \frac{1}{4}((\mathbf{v}^T\mathbf{e})_{j+1} - (\mathbf{v}^T\mathbf{e})_j)(B_{j+1}^{(x)} - B_j^{(x)}). \tag{15}$$

Note that since $\mathbf{v}^T\mathbf{e}$ only appears in (14) evaluated at the grid points, the choices $\mathbf{e}_G$, $\mathbf{e}_J$, or $\mathbf{e}_B$ for $\mathbf{e}$ can be used interchangeably. The scheme was formulated somewhat differently in [4], but is consistent with our description.

We can generalize this to any order by using the construction in Theorem 1.

Theorem 2 *The semi-discrete scheme*

$$\frac{d}{dt}\mathbf{u}_j + \frac{1}{\Delta x}\sum_{k=1}^{p}\frac{\alpha_k^{(p)}}{k}(\mathbf{h}_{j+k/2}^{(k)} - \mathbf{h}_{j-k/2}^{(k)}) + \frac{1}{2}\mathbf{e}_j D_p B_j^{(x)} = \mathbf{0}, \tag{16}$$

where the functions $\mathbf{h}_{j+k/2}^{(k)}$ *satisfy*

$$(\mathbf{v}_{j+k} - \mathbf{v}_j)^T\mathbf{h}_{j+k/2}^{(k)} = \psi_{j+k} - \psi_j + \frac{1}{2}(\mathbf{v}_{j+k}^T\mathbf{e}_{j+k} + \mathbf{v}_j^T\mathbf{e}_j)(B_{j+k}^{(x)} - B_j^{(x)}) \tag{17}$$

and where the flux difference $\mathbf{h}^{(k)}_{j+k/2} - \mathbf{h}^{(k)}_{j-k/2}$ *satisfies (5), is 2pth order accurate, and admits a discrete entropy equation,*

$$\Delta x \frac{d}{dt} E_j + \sum_{k=1}^{p} \frac{\alpha_k^{(p)}}{k} (H^{(k)}_{j+k/2} - H^{(k)}_{j-k/2}) = 0, \tag{18}$$

with numerical entropy flux

$$H^{(k)}_{j+k/2} = \frac{1}{2}(\mathbf{v}_{j+k} + \mathbf{v}_j)^T \mathbf{h}^{(k)}_{j+k/2} - \frac{1}{2}(\mathbf{v}^T_{j+k}\mathbf{f}_{j+k} + \mathbf{v}^T_j \mathbf{f}_j) + \frac{1}{2}(F_{j+k} + F_j) - \frac{1}{4}((\mathbf{v}^T\mathbf{e})_{j+k} - (\mathbf{v}^T\mathbf{e})_j)(B^{(x)}_{j+k} - B^{(x)}_j). \tag{19}$$

The standard centered $2p$th operator D_p is defined in (2).

Construction of entropy conserving numerical fluxes is done by enforcing (14), which is one single constraint on the eight components of $\mathbf{h}^{(k)}_{j+1/2}$. In principle, this could be done by expanding the right hand side of (14) in the elements of the vector $\mathbf{v}_{j+1} - \mathbf{v}_j$. In practice it is easier to expand both sides of (14) in a parameter vector, $\mathbf{z}$. Different choices of parameter vectors are possible, and lead to different numerical fluxes. Two different entropy conserving MHD schemes were recently developed. The scheme developed in Winters and Gassner [17] uses

$$\mathbf{z}_1 = (\sqrt{\frac{\rho}{p}},\ \sqrt{\frac{\rho}{p}}u,\ \sqrt{\frac{\rho}{p}}v,\ \sqrt{\frac{\rho}{p}}w,\ \sqrt{p\rho},\ B^{(x)},\ B^{(y)},\ B^{(z)}) \tag{20}$$

while

$$\mathbf{z}_2 = (\frac{\rho}{p},\ u,\ v,\ w,\ \rho,\ B^{(x)},\ B^{(y)},\ B^{(z)}) \tag{21}$$

is used in Chandrashekar and Klingenberg [4].

The scheme in Winters and Gassner [17] uses the slightly different form

$$\frac{d}{dt}\mathbf{u}_j + \frac{1}{\Delta x}(\mathbf{h}_{j+1/2} - \mathbf{h}_{j-1/2}) + \frac{1}{2}\mathbf{e}_{j+1/2}(B^{(x)}_{j+1} - B^{(x)}_j) + \frac{1}{2}\mathbf{e}_{j-1/2}(B^{(x)}_j - B^{(x)}_{j-1}) = \mathbf{0}, \tag{22}$$

for the numerical scheme. The intermediate $\mathbf{e}_{j+1/2}$ has to be determined along with the numerical flux function. Here the scheme for the ideal MHD system we will work with is (13) (and (16) for higher order extension), since it works without modification for any choice for $\mathbf{e}$. Numerical experiment indicated that these two choices are more accurate and stable than the Winters & Gassner forms.

The procedure to define the numerical flux is a lengthy, but straightforward, expansion of both sides of (14) in the components of $\mathbf{z}_{j+1} - \mathbf{z}_j$. Equating the coefficients in front of the $\mathbf{z}$-difference components gives equations for the components of the numerical flux function; see [17] for details.

For example, the numerical flux satisfying (14), derived by use of (21), is the following

$$\mathbf{h}_{j+1/2} = \begin{pmatrix} \rho^{ln}\{u\} \\ \rho^{ln}\{u\}^2 + \frac{\{\rho\}}{\{\rho/p\}} + \frac{1}{2}\{(B^{(x)})^2 + (B^{(y)})^2 + (B^{(z)})^2\} - \{B^{(x)}\}^2 \\ \rho^{ln}\{u\}\{v\} - \{B^{(x)}\}\{B^{(y)}\} \\ \rho^{ln}\{u\}\{w\} - \{B^{(x)}\}\{B^{(z)}\} \\ h_5 \\ 0 \\ \frac{1}{\{\rho/p\}}(\{\rho u/p\}\{B^{(y)}\} - \{\rho v/p\}\{B^{(x)}\}) \\ \frac{1}{\{\rho/p\}}(\{\rho u/p\}\{B^{(z)}\} - \{\rho w/p\}\{B^{(x)}\}) \end{pmatrix},$$

where h_5 is the longer expression

$$\begin{aligned} h_5 = \frac{1}{\gamma - 1}\frac{\rho^{ln}}{(\rho/p)^{ln}}\{u\} + \hat{p}_1\{u\} - \frac{1}{2}\rho^{ln}\{u\}(\{u^2\} + \{v^2\} + \{w^2\}) \\ + \rho^{ln}\{u\}(\{u\}^2 + \{v\}^2 + \{w\}^2) + \\ \hat{u}_2(\{B^{(x)}\}^2 + \{B^{(y)}\}^2 + \{B^{(z)}\}^2) - \{B^{(x)}\}(\hat{u}_2\{B^{(x)}\} + \hat{v}_2\{B^{(y)}\} + \hat{w}_2\{B^{(z)}\}) \end{aligned} \quad (23)$$

where the arithmetic average is denoted $\{\rho\} = (\rho_{j+1} + \rho_j)/2$, and where

$$\hat{p}_1 = \frac{\{\rho\}}{\{\rho/p\}} \qquad \rho^{ln} = \frac{\log \rho_{j+1} - \log \rho_j}{\rho_{j+1} - \rho_j}$$

and

$$\hat{u}_2 = \frac{\{u\rho/p\}}{\{\rho/p\}} \qquad \hat{v}_2 = \frac{\{v\rho/p\}}{\{\rho/p\}} \qquad \hat{w}_2 = \frac{\{w\rho/p\}}{\{\rho/p\}}.$$

5 Numerical Results

This section shows numerical results for one test case for the gas dynamics containing shock waves, and one test case for the ideal MHD containing shock waves to illustrate the performance of the proposed methods for problems with shocks. The aforementioned nonlinear filter approach of Yee & Sjögreen is utilized. The nonlinear filter postprocesses the solution after each time step, applying the dissipative part of a high order shock capturing scheme. The filter is on conservative form, and consistent to high formal order of accuracy. For all test cases we use the classical fourth-order Runge-Kutta time discretization. Here, for illustration purposes, only one smart flow sensor (among the many variants indicated in [20] and Kotov et al. [8, 9]) is chosen for the numerical experiment for the nonlinear filter approach. It is the third-order B-spline wavelet flow sensor developed in Sjögreen and Yee [11]. No grid refinement results are indicated as the grid shown of these well-known test cases are the commonly used grid size for scheme comparison. Many details of the development and extensive numerical testing with more representative test cases among the different formulations and the different governing equation will be reported in Sjögreen and Yee [14]. In this expanded version, 2D and 3D test cases for smooth flows, problems with shock waves, shock-free turbulence and turbulence with weak and strong shocks will be included. Our studies show the gain in stability by the high order entropy conservative numerical fluxes over the purely high order central base scheme.

5.1 Gas Dynamics Test Case with Shocks: Shu-Osher Problem

The Shu-Osher problem is a one-dimensional Mach 3 shock moving into an oscillatory density. A highly oscillatory flow field (1D turbulent flow) develops behind the shock wave. The problem is defined for the one dimensional Euler equations with $\gamma = 1.4$ and initial data

$$(\rho,\ u,\ p) = \begin{cases} (3.857143,\ 2.629369,\ 10.33333), & x < -4 \\ (1 + 0.2\sin 5x,\ 0,\ 1), & x \geq 4 \end{cases} \tag{24}$$

on the domain $-5 \leq x \leq 5$. For this gas dynamics computation, the MHD equation solver is used with magnetic fields set to zero. The Yee & Sjögreen nonlinear filter scheme using the high order entropy conservative numerical flux as the base scheme is employed for the numerical experiment. The coarse grid has 201 points, corresponding to discretization size $\Delta x = 0.05$. The exact left and right hand solutions are imposed at the boundaries at a number of ghost points that is sufficiently large that no boundary modification of the scheme is needed. The base scheme is used to advance the solution one full time step by a Runge-Kutta time discretization. After each full time step the computed solution is nonlinearly filtered by the nonlinear dissipation part of a WENO scheme. The nonlinear numerical dissipation is multiplied with sensors designed to activate it only in the neighborhood of shocks. In the computations shown here, a wavelet sensor was used with two wavelet levels and a cut-off smoothness exponent 0.5. Figure 1 compares the density at time 1.8 computed by the Jiang & Shu seventh-order WENO (WENO7) scheme (to the left) and computed by the eighth-order entropy conservative scheme as the base scheme and then the computed solution

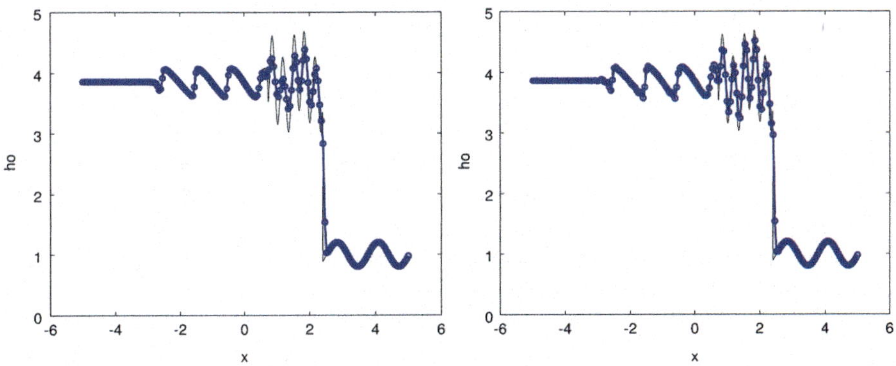

Fig. 1 1D Osher-Shu test case: Density at time 1.8 for WENO7 (*left*) and C08*ECCK*+WENO7fi (*right*)

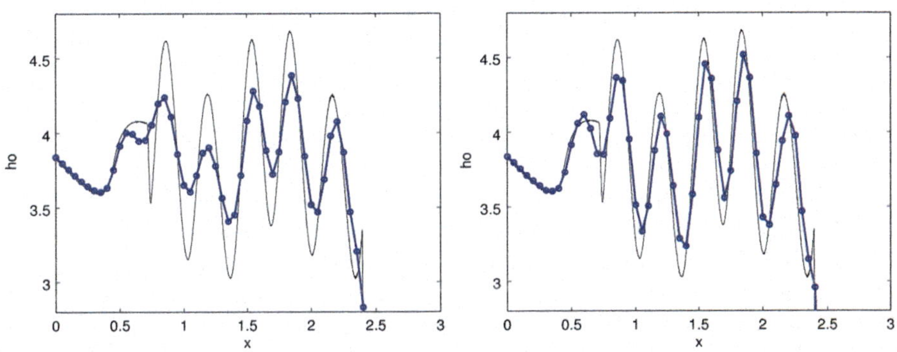

Fig. 2 1D Osher-Shu test case: Close up of the oscillations in density at time 1.8 for WENO7 (*left*) and C08*ECCK*+WENO7fi (*right*)

is nonlinearly filtered by the dissipative portion of WENO7 using the wavelet flow sensor to control the numerical dissipation (C08*ECCK*+WENO7fi) (to the right). See [18–20] for a description of the nonlinear filter scheme. Only result by the CK form of the entropy numerical flux is shown. For this test case, other forms of the entropy numerical flux behaves similarly.

Figure 2 shows a close up of the oscillatory regions of the plots in Fig. 1. The oscillations are visibly better resolved by the filter scheme. See Yee and Sjögreen[20], Sjögreen and Yee [12], Kotov et al. [8, 9] for some discussions and performance of the combined approach for DNS and LES gas dynamics applications. For extension of skew-symmetric splitting to the equations of ideal MHD, see Sjögreen and Yee [13].

5.2 MHD Test Case with Shocks: Orzag-Tang Vortex

The entropy conservative scheme with numerical flux function satisfying (14) was applied to (7). All three variants with $\mathbf{e}$ equal to $\mathbf{e}_G$, $\mathbf{e}_J$, and $\mathbf{e}_B$ were implemented. Both choices of parameter vector (20) and (21) were implemented, leading to six different schemes. The schemes were implemented with sixth-order of accuracy. The results are compared with the conservative formulation $\mathbf{e}_0$ ($\mathbf{e} = 0$). This is a problem where shock waves appear. Computed solutions exhibit oscillatory solutions without a shock-capturing dissipation at the discontinuities. In addition, centered base schemes in conjunction with skew-symmetric splitting, entropy conservative numerical fluxes for the base scheme, all give unphysical oscillations around the shock waves.

We will here use the entropy conservative numerical flux as a base scheme in the nonlinear filter scheme described in [18–20].

The Orzag-Tang vortex starts from initial data

$$\rho = 25/9 \;\; (u, v, w) = (-\sin y, \sin x, 0) \;\; p = 5/3, \tag{25}$$

$$(B^{(x)}, B^{(y)}, B^{(z)}) = (-\sin y, \sin 2x, 0) \tag{26}$$

and is solved in two space dimensions on a domain of size $2\pi \times 2\pi$ with periodic boundary conditions. Results will be displayed as contour levels of density of the solution at the time 3.14 together with contours of div $\mathbf{B}$ at the same time. Furthermore, a logscale plot of the norm of div $\mathbf{B}$ will also be given. The computational domain used 100×100 grid points.

Figure 3 shows the result obtained by sixth order base scheme (16) using (20) and post processed with dissipation from the fifth order WENO scheme (C06*ECSY*+WENO5fi), which had the smallest div $\mathbf{B}$ error of the six variants of entropy conserving base schemes. The remaining cases are not shown, since their plotted results do not differ significantly from Fig. 3.

Figures 4, 5, and 6 show the standard sixth-order centered base scheme with the source term choices $\mathbf{e} = \mathbf{0}$, $\mathbf{e} = \mathbf{e}_G$, and $\mathbf{e} = \mathbf{e}_J$, respectively. The centered base

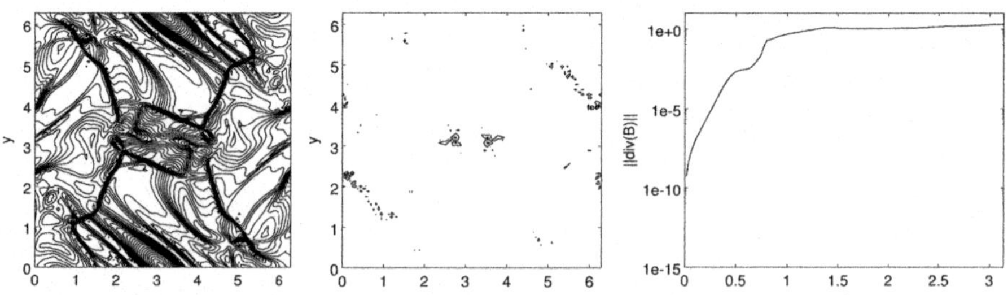

Fig. 3 2D Orzag-Tang vortex test case: C06*ECSY*+WENO5fi+e_G, density contours (*left*), div $\mathbf{B}$ contours (*middle*), norm of div $\mathbf{B}$ vs. time (*right*)

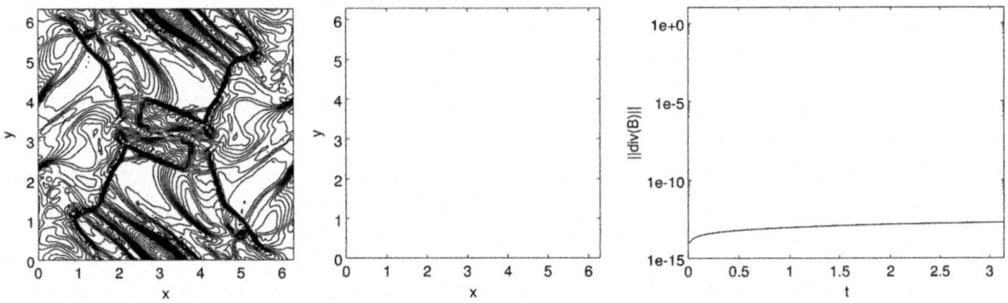

Fig. 4 C06+WENO5fi+e_0, density contours (*left*), div **B** contours (*middle*), norm of div **B** vs. time (*right*)

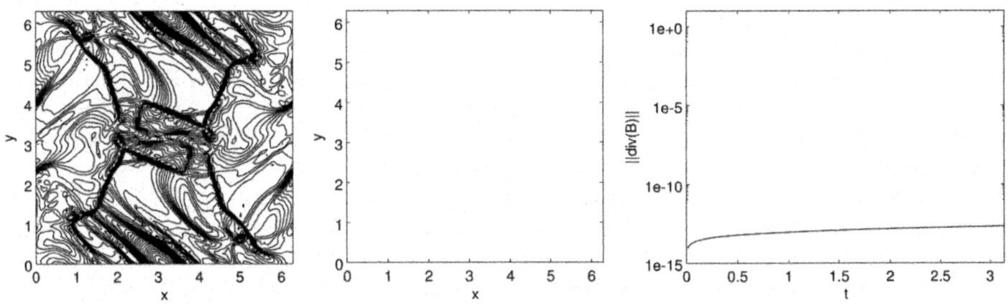

Fig. 5 C06+WENO5fi+e_G, density contours (*left*), div **B** contours (*middle*), norm of div **B** vs. time (*right*)

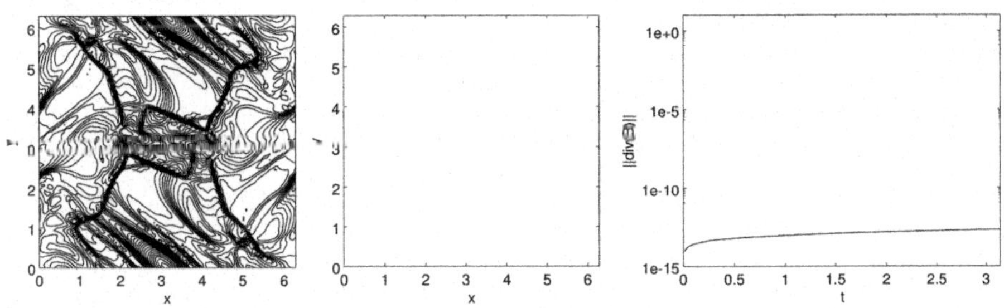

Fig. 6 C06+WENO5fi+e_J, density contours (*left*), div **B** contours (*middle*), norm of div **B** vs. time (*right*)

scheme preserves the discretized div **B** perfectly, and the nonlinear filter was not applied to the three magnetic field components of the equations, leading to an error-free discrete div **B** for the overall computation. The entropy conserving schemes do not have the perfect div **B** preservation property. In addition, the nonlinear filter dissipation will introduce errors in div **B** if it is applied to all components of the equations.

The above computations with order of accuracy eight instead of six are reported in [14], where more details and illustrations are given.

6 Conclusions

Entropy conserving schemes for the equations of MHD were implemented and some new variants of these were developed.

For problems where shock waves are present, the entropy conserving schemes exhibit oscillations. The nonlinear filter method by Yee and Sjögreen was demonstrated to maintain stability and give highly accurate computed solutions when using an entropy conservative scheme as base scheme.

References

1. A. Arakawa, Computational design for long-term numerical integration of the equations of fluid motion: two-dimensional incompressible flow, Part I. J. Comput. Phys. **1**, 119–143 (1966)
2. G.A. Blaisdell, E.T. Spyropoulos, J.H. Qin, The effect of the formulation of nonlinear terms on aliasing errors in spectral methods. Appl. Numer. Math. **21**, 207–219 (1996)
3. J.U. Brackbill, D.C. Barnes, The effect of nonzero $\nabla \cdot \mathbf{B}$ on the numerical solution of the magnetohydrodynamics equations. J. Comput. Phys. **35**, 426–430 (1980)
4. P. Chandrashekar, C. Klingenberg, Entropy stable finite volume scheme for ideal compressible MHD on 2-D Cartesian Meshes. SIAM J. Numer. Anal. **54**, 1313–1340 (2016)
5. F. Ducros, F. Laporte, T. Soulères, V. Guinot, P. Moinat, B. Caruelle, High-order fluxes for conservative skew-symmetric-like schemes in structured meshes: application to compressible flows. J. Comput. Phys. **161**, 114–139 (2000)
6. S.K. Godunov, The symmetric form of magnetohydrodynamics equation. Num. Meth. Mech. Cont. Media **1**, 26–34 (1972)
7. P. Janhunen, A positive conservative method for MHD based on HLL and Roe methods. J. Comput. Phys. **160**, 649–661 (2000)
8. D.V. Kotov, H.C. Yee, A.A. Wray, B. Sjögreen, A.G. Kritsuk, Numerical disipation control in high order shock-capturing schemes for LES of low speed flows. J. Comput. Phys. **307**, 189–202 (2016)
9. D.V. Kotov, H.C. Yee, A.A. Wray, B. Sjögreen, High order numerical methods for dynamic SGS model of turbulent flows with shocks. Commun. Comput. Phys. **19**, 273–300 (2016)
10. K. Linders, J. Nordström, Uniformly best wavenumber approximations by spatial central difference operators. J. Comput. Phys. **300**, 695–709 (2015)
11. B. Sjögreen, H.C. Yee, Multiresolution wavelet based adaptive numerical dissipation control for high order methods. J. Sci. Comput. **20**, 211–255 (2004)
12. B. Sjögreen, H.C. Yee, On skew-symmetric splitting and entropy conservation schemes for the Euler equations, in *Proceedings of ENUMATH09, June 29- July 2*, Uppsala University, Sweden (2009)
13. B. Sjögreen, H.C. Yee, D. Kotov, Skew-symmetric splitting and stability of high order central schemes. J. Phys. **837**, 012019 (2017)
14. B. Sjögreen, H.C. Yee, Construction of high order entropy conserving numerical flux for gas dynamics and MHD turbulent simulations. J. Comput. Phys. (2016, submitted)
15. E. Tadmor, Numerical viscosity and the entropy condition for conservative difference schemes. Math. Comput. **43**, 369–381 (1984)
16. E. Tadmor, Entropy stability theory for difference approximations of nonlinear conservation laws and related time-dependent problems. Acta Numer. **12**, 451–512 (2003)
17. A.R. Winters, G.J. Gassner, Affordable, entropy conserving and entropy stable flux functions for the ideal MHD equations. J. Comput. Phys. **304**, 72–108 (2016)

18. H.C. Yee, B. Sjögreen, Efficient low dissipative high order schemes for multiscale MHD flows, II: minimization of $\nabla \cdot B$ numerical error. J. Sci. Comput. **29**, 115–164 (2006)
19. H.C. Yee, B. Sjögreen, Development of low dissipative high order filter schemes for multiscale Navier-Stokes MHD systems. J. Comput. Phys. **225** 910–934 (2007)
20. H.C. Yee, B. Sjögreen, High order filter methods for wide range of compressible flow speeds, in *Proceedings of the ICOSAHOM09*, Trondheim, Norway, June 22–26, 2009
21. H.C. Yee, M. Vinokur, M.J. Djomehri, Entropy splitting and numerical dissipation. J. Comp. Phys. **162**, 33–81 (2000)

A Fast Direct Solver for the Advection-Diffusion Equation Using Low-Rank Approximation of the Green's Function

Jonathan R. Bull

Abstract We describe a new direct solution method for the advection-diffusion equation at high Reynolds number on simple bounded two-dimensional domains. The key step is to treat advection explicitly, leading to a new class of time integration schemes based on Green's functions. As a proof of concept a first-order Euler scheme is presented. We compare the accuracy and computational cost of the new scheme to existing solution techniques. Low-rank approximation of the Green's function is found to reduce cost without loss of accuracy. Stabilisation via numerical dissipation is required for high Reynolds number problems on coarse grids. Linear scaling of computational cost is achieved in 1D and 2D. This work is a building block for constructing fast direct solvers and preconditioners for the Navier-Stokes equations.

1 Introduction

Today's most powerful supercomputers make use of large numbers of parallel processors and threads. However, some of the core solution algorithms in computational physics codes are not well suited to massively parallel execution and linear scaling is generally lost. For example, Newton-Krylov (NK) solvers require a global inner product calculation every iteration, causing a performance bottleneck [1]. In this context, hierarchical algorithms such as the Fast Multipole Method (FMM) [2] and $\mathscr{H}^2$ matrices [3] have considerable potential. They are increasingly being employed as fast tunable solvers and preconditioners for elliptic and parabolic PDEs and BIEs [4–8]. In particular, when solving the discrete problem $\mathbf{A}\mathbf{x} = \mathbf{b}$ with N degrees of freedom, hierarchical low-rank approximation (HLRA) methods compute a highly compressed approximation to the inverse matrix A^{-1}. HLRA solves the problem to a chosen accuracy in $\mathscr{O}(N \log N)$ or $\mathscr{O}(N)$ operations. Furthermore, this method is competitive with multigrid in terms of parallel scaling [6, 9].

J.R. Bull (✉)
Division of Scientific Computing, Uppsala University, Polacksbacken, Uppsala, Sweden
e-mail: jonathan.bull@it.uu.se

M.L. Bittencourt et al. (eds.), *Spectral and High Order Methods for Partial Differential Equations ICOSAHOM 2016*, Lecture Notes in Computational Science and Engineering 119, DOI 10.1007/978-3-319-65870-4_30

HLRA compression is effective when the Green's function of the PDE/BIE decays with increasing distance between two points in space. In fluid flows where advection dominates diffusion this is not the case. However, in Bull et al. [10] it was shown that the linear advection-diffusion equation could also be solved by low-rank approximation of the inverse operator. By discretising in time such that the advective term was explicit and the diffusion term implicit, it was transformed into a forced heat equation with a rapidly decaying Green's function. We describe and analyse the direct solution method in greater detail and present simple and effective low-rank approximations. High-order time accuracy may be achieved by using a high-order integral approximation. This is left as a topic for future work. The paper is organised as follows. In Sect. 2 numerical methods for the solution of the advection-diffusion equation are presented. A simple method for imposing Dirichlet boundary conditions is described. The cost and accuracy of low-rank approximation schemes is analysed. In Sect. 3 the 1D and 2D numerical tests are presented. Comparisons are made with Matlab's backslash operator and the exponential integrator method. Finally in Sect. 4 conclusions are drawn and future work proposed.

2 Numerical Methods

We consider the linear advection-diffusion equation in 1, 2 or 3 dimensions:

$$\begin{aligned} \partial u/\partial t + \mathbf{a} \cdot \nabla u - \nu \nabla^2 u &= 0 \quad \text{in } \Omega, \\ u(\mathbf{x}, 0) &= u_0 \quad \text{in } \Omega, \\ u(\mathbf{x}, t) &= u_D \quad \text{on } d\Omega, \end{aligned} \tag{1}$$

where u is a smooth scalar field, $\mathbf{a}$ is the constant advection velocity and ν is the diffusion coefficient (constant). For simplicity of analysis $\mathbf{a}$ is chosen to be constant. We are interested in advection-dominated (stiff) problems for which $|\mathbf{a}| >> \nu$. The advection term is placed on the right-hand side (RHS), making it in effect a forcing term:

$$\partial u/\partial t - \nu \nabla^2 u = -\mathbf{a} \cdot \nabla u. \tag{2}$$

2.1 IMEX Schemes

A standard way to solve (2) is with a first-order implicit-explicit (IMEX) scheme:

$$\frac{u^{n+1} - u^n}{\Delta t} - \nu \nabla^2 u^{n+1} = -|\mathbf{a}| \cdot \nabla u^n, \tag{3}$$

Applying some suitable spatial discretisation, we would obtain a linear system of the form

$$(I - \Delta t A)\mathbf{u}^{n+1} = (I - \Delta t B)\mathbf{u}^n, \tag{4}$$

where B and A are the advective and diffusive operators (including boundary conditions) respectively. An overview of IMEX schemes is given in [11]. We solve the linear system using Matlab's backslash operator, which defaults to a Cholesky decomposition or LDL factorisation since the matrix is Hermitian.

2.2 *Exponential Integrator Method*

For an equation containing a constant-in-time linear operator A and a nonlinear operator B, exponential integration obtains the exact solution as:

$$u(\mathbf{x}, t) = e^{-tA} u(\mathbf{x}, 0) + \int_0^t e^{-(t-t')A} B(u(\mathbf{x}, t'), t') dt'. \tag{5}$$

In the current context, A is diffusion and B is the negative advection term plus any initial and boundary terms. An explicit Euler approximation of the time integral over a timestep Δt leads to the first-order solver

$$u(\mathbf{x}, t^{n+1}) = e^{-\Delta t A} u(\mathbf{x}, t^n) + \Delta t \phi_1(-\Delta t A) B(u(\mathbf{x}, t^n), t^n)), \tag{6}$$

where ϕ_1 is the first ϕ function defined by $\phi_1(z) = (e^z - 1)/z$. By approximating ϕ_1 at t^n we have a simpler form with only one exponential matrix:

$$u(\mathbf{x}, t^{n+1}) = e^{-\Delta t A}(u(\mathbf{x}, t^n) + \Delta t B(u(\mathbf{x}, t^n), t^n)). \tag{7}$$

2.3 *Space-Time Integration with Green's Function*

Alternatively, (2) can be solved exactly by integrating initial, boundary and forcing terms over the domain Ω and time interval t:

$$\begin{aligned} u(\mathbf{x}, t) &= \textstyle\int_0^t \int_\Omega G(\mathbf{x}, \mathbf{x}', t, t') f(\mathbf{x}', t', u(\mathbf{x}', t')) d\mathbf{x}' dt' \\ &+ \textstyle\int_\Omega G(\mathbf{x}, \mathbf{x}', t, 0) u(\mathbf{x}', 0) dx' - \int_0^t \int_{d\Omega} \nabla G(\mathbf{x}, \mathbf{x}', t, t') \cdot \mathbf{n}\, u_D(\mathbf{x}', t') d\mathbf{x}' dt', \end{aligned} \tag{8}$$

where $f(x, t) = -\mathbf{a} \cdot \nabla(u(x, t))$, G is the Green's function of the left-hand side (LHS) operator in (2) and u_D are Dirichlet boundary conditions. Specific cases of G are given below. The time integral of G is evaluated numerically in this paper. It can also be integrated analytically in the 2D and 3D cases, which, combined with a high-order quadrature rule or Runge-Kutta scheme for the RHS terms, leads to a high-order accurate time integration scheme. The integral equation (8) is solved over a single timestep from t^n to $t^{n+1} = t^n + \Delta t$. The first-order explicit Euler scheme is written:

$$u^{n+1} = \int_{\Omega} G(\Delta t)[u^n + \Delta t f(u^n)] d\mathbf{x}' + \text{boundary terms.} \tag{9}$$

Consider the solution of (8) on a finite periodic 1D domain $\Omega : x \in [0, L]$. The boundary integral in (8) is left out and the Green's function is the fundamental solution of the forced heat equation:

$$G(x, x', t, t') = H(t - t')(4\pi\nu(t - t'))^{-d/2} \exp\left(-\frac{(x - x')^2}{4\nu(t - t')}\right), \tag{10}$$

where $H(t - t')$ is the Heaviside step function. G is defined to be a Dirac delta function at $t = t'$ and also in the infinite-Reynolds number limit as $\nu \to 0$. It is necessary that G is sufficiently compact with respect to L, i.e. $G(x, x \pm L) \approx 0$, so that the tails do not overlap. For advection-dominated problems this is easily satisfied.

We now partition the domain into N intervals of uniform size $\Delta x = L/N$. Using a piecewise-constant midpoint approximation of the integral with first-order upwind differencing for the advection term:

$$\begin{aligned} u_i^{n+1} &= M_{ij}(u_j^n + \Delta t f_j^n), \quad i = 1, \ldots, N, \\ &= \textstyle\sum_{j=1}^{N} \frac{\kappa \Delta x}{\sqrt{\pi}} \exp(-(|j - i|\kappa \Delta x)^2)[(1 - c_A)u_j^n + c_A u_{j-1}^n], \end{aligned} \tag{11}$$

where M is the Green's matrix, the advective CFL number $c_A = a\Delta t/\Delta x$ and the effective wavenumber $\kappa = (4\nu\Delta t)^{-1/2}$. Periodicity is imposed by defining j such that if $j - i > N/2$ then $j = j - N$ and if $j - i < N/2$ then $j = j + N$, thus making M circulant.

The method extends easily to 2D structured grids on the rectangular domain $V = x \in [0, L] \times y \in [0, L]$. In the periodic setting the exact solution is given by a triple integration over the two space and one time dimensions:

$$\begin{aligned} u(x, y, t) &= \textstyle\int_0^t \int_V G(x, x', y, y', t, t') f(x', y', t') dx' dy' dt' \\ &\quad + \textstyle\int_V G(x, x', y, y', t, 0) u(x', y', 0) dx' dy', \end{aligned} \tag{12}$$

$$G(x, x', y, y', t, t') = H(t - t') 4\pi\nu(t - t') \exp\left(-\tfrac{r^2}{4\nu(t-t')}\right), \tag{13}$$

where $r = (x-x')^2 + (y-y')^2$. On a uniform $N \times N$ grid with explicit Euler in time:

$$u_{ij}^{n+1} = M_{ijkl}(u_{kl}^n + \Delta t f_{kl}), \quad i, j = 1, \ldots, N, \tag{14}$$

$$= \sum_{k=i-N/2}^{i+N/2} \sum_{l=j-N/2}^{j+N/2} \frac{\kappa^2 \Delta x \Delta y}{\pi}$$

$$\exp\left[-\kappa^2((|k-i|\Delta x)^2 + (|l-j|\Delta y)^2)\right](u_{kl}^n + \Delta t f_{kl}^n). \tag{15}$$

Although quadruple indexing has been used here, the solver is implemented as a matrix-vector product of dimensions $[N^2, N^2] \times (N^2)$ so u_{ij} is referred to as a vector and M_{ijkl} as a matrix.

There are three stability criteria for this method. Firstly, the timestep Δt has to satisfy any stability criteria set by the forcing terms; in this case $c_A \leq 1$ due to the upwind scheme. This can be shown by von Neumann analysis but space does not permit its inclusion. Secondly, the maximum resolvable numerical wavenumber is the Nyquist wavenumber N. We cap the wavenumber κ, resulting in a modified matrix M:

$$\kappa = \min((4\nu\Delta t)^{-1/2}, \kappa_{max}), \kappa_{max} = N. \tag{16}$$

It can be considered as stabilisation via a modified viscosity $\nu_{eff} = (4N^2\Delta t)^{-1}$. Thirdly, on a periodic domain the method is A-stable if (in 1D) $\sum_{j=1}^{N} M_{ij} \leq 1 \ \forall \ i$ (and similarly in 2D). This is satisfied by using the midpoint rule of integration and using wavenumber limiting.

2.4 Dirichlet Boundary Conditions

On a non-periodic domain, Dirichlet boundary conditions can be imposed by the method of images. There is now no restriction on the compactness of the kernel. The method of images removes the boundary integral but the volume integrals are allowed to extend outside the boundaries. In 1D, the domain $\Omega = [0 : L]$ is extended by L to the left and right and the Dirichlet boundary data are defined as constant fields in these 'ghost' regions. Homogeneous Neumann conditions are imposed by defining a new Dirichlet condition u_D at each timestep equal to the solution at the boundary. Now the solution is given by:

$$u_i^{n+1} = M_{ij}(u_j^n + \Delta t f_j^n) + B_i^L u_D + B_i^R u_D. \tag{17}$$

The two 'boundary influence vectors' B^L and B^R are simply summations of the part of M centred on a point i that lies outside the respective boundary. In effect, the domain of integration of the ith point is extended from $[0, L]$ to $[\min(0, x_i - L/2), \max(L, x_i + L/2)]$. To save effort B^L and B^R can be computed beforehand and stored, as with the matrix M. Because M is circulant, B^L and B^R can be calculated

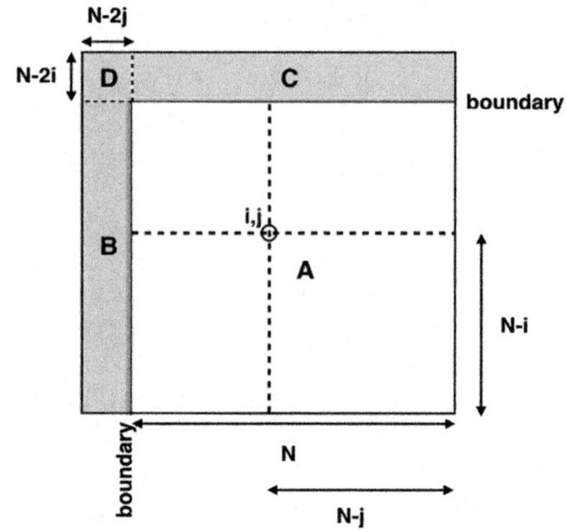

Fig. 1 Method of integration on 2D non-periodic domain

from the first row: $B_i^L = \sum_{j=i+1}^{N/2} M_{1,j}, \quad i = 1, \ldots, N, B_i^R = B_{N-i+1}^L$. Boundary conditions are imposed on the advection term separately via the finite difference stencil.

The method of images in 2D involves integration over an area outside the domain as shown in Fig. 1. Let there be homogeneous Dirichlet conditions u_D on the left and top boundaries. At a point (x_i, y_j) in the square domain of size $L \times L$, the solution is given by a double integral over the area $[\min(0, x_i - L/2), \max(L, x_i + L/2)] \times [\min(0, y_j - L/2), \max(L, y_j + L/2)]$. The portions of this area lying outside the domain (shaded grey and labelled B, C and D in Fig. 1) contribute to the boundary influence vectors. On the outflow boundaries (right and bottom) a zero Neumann condition is imposed. The solution is given by

$$u_{ij}^{n+1} = M_{ijkl}(u_{kl}^n + \Delta t f_{kl}) + B_{ij}^L u_D + B_{ij}^T u_D + E_{kl} B_{kl}^L g_{ij}^R + E_{kl} B_{kl}^T g_{ij}^B, \tag{18a}$$

$$B_{ij}^L = \sum_{k=1}^N \sum_{l=2j}^N M_{ijkl} + \frac{1}{2} \sum_{k=2i}^N \sum_{l=2j}^N M_{ijkl}, \tag{18b}$$

$$B_{ij}^T = \sum_{k=2i}^N \sum_{l=1}^N M_{ijkl} + \frac{1}{2} \sum_{k=2i}^N \sum_{l=2j}^N M_{ijkl}, \tag{18c}$$

where E is the reversal matrix of size N^2 and g^R and g^B are vectors containing the outflow boundary conditions (method described below). The first term in (18b) corresponds to the area labelled B in Fig. 1. The first term in (18c) corresponds to C. The second terms in (18b) and (18c) together represent D as an average of the contributions from the left and top boundaries.

On the right outflow boundary a zero Neumann condition is imposed as an inhomogeneous Dirichlet condition by extending the solution on the line $x = L$ horizontally to the right. Likewise the solution on the bottom boundary is extended vertically downwards. The outflow boundary value vectors g_{ij}^R and g_{ij}^B are defined such that at a point (i, j) in the domain and time t^{n+1}, $g_{ij}^R = u_{N,j}^n$ and $g_{ij}^B = u_{i,N}^n$.

2.5 Low-Rank Approximation

The Green's function or integral kernel G decays quickly and to save on storage, very small entries in the full-rank matrix M are neglected without loss of accuracy:

$$m_{ij} = \begin{cases} G(x_i, x_j), & G(x_i, x_j) \geq \epsilon, \\ 0, & \text{otherwise}, \end{cases} \tag{19}$$

where ϵ is machine epsilon. When solving the advection-diffusion equation the timestep scales with $\mathcal{O}(N^{-1})$ due to the CFL condition. The kernel's exponent $(-r^2/4\nu\Delta t)$ therefore scales at $\mathcal{O}((N^{-2})/(N^{-1})) = \mathcal{O}(N^{-1})$ and the kernel's variance, $\sigma = \sqrt{2\nu\Delta t}$, scales with $\mathcal{O}(N^{-1/2})$. By thresholding M at ϵ (or indeed any constant value), the bandwidth is restricted to a multiple of $\sigma N = \mathcal{O}(N^{1/2})$. Therefore the storage scales with $N^{1.5}$, as does the computational cost of a matrix-vector product. When solving over a fixed time period ($\mathcal{O}(N)$ timesteps) the total cost scales as $\mathcal{O}(N^{2.5})$. If the modified-wavenumber stabilisation (16) is applied, $\sigma = \mathcal{O}(N^{-1})$ and we achieve ideal scaling of $\mathcal{O}(N)$ for a matrix-vector product. However, this comes at the expense of reduced accuracy.

The rank of M can be reduced further albeit at the expense of accuracy. For preconditioning applications in particular, the required precision is not high and considerable savings might be found. We define a thresholded low-rank matrix M^T:

$$m^T_{ij} = \begin{cases} m_{ij}, & m_{ij} \geq t, \\ 0, & \text{otherwise}. \end{cases} \tag{20}$$

A threshold value of $t = 1 \times 10^{-5}$, used in the numerical tests below, selects entries within about 2σ of the diagonal. We expect the computational cost and storage also to scale with $N^{1.5}$ but with a lower constant than for M.

One can also specify a bandwidth to define a low-rank matrix. Let $P < N/2$ be the low-rank matrix bandwidth. The entries of the low-rank matrix M^B are given by

$$m^B_{ij} = \begin{cases} m_{ij}, & i - P \leq j \leq i + P, \\ 0, & \text{otherwise}. \end{cases} \tag{21}$$

Choosing P as a multiple of the variance σ is equivalent to choosing $t =$ constant. To obtain ideal scaling of $\mathcal{O}(N)$, P must be independent of N. We use $P = \text{round}(\kappa^2/N) = \text{round}(1/2N\sigma^2) = \mathcal{O}(\nu^{-1})$ to define the matrix M^B used in the numerical tests. When $P > N/2$, M^B is identical to M, otherwise it is a low-rank approximation.

As with M, if the modified numerical wavenumber stabilisation (16) is applied to M^T and M^B, a cost scaling of $\mathcal{O}(N)$ is achieved. On non-periodic domains the boundary influence vectors are also low-rank to preserve the scaling.

In 2D the thresholded low-rank matrix M^{T2} is constructed according to

$$m_{ijkl}^{T2} = \begin{cases} m_{ijkl}, & m_{ijkl} > 1.e-5, \\ 0, & \text{otherwise}, \end{cases} \tag{22}$$

and the specified-bandwidth low-rank matrix M^{B2} is given by

$$m_{ijkl}^{B2} = \begin{cases} m_{ijkl}, & (k-i)^2 + (l-j)^2 \leq P^2, \quad P = \text{round}(1/2N\sigma^2), \\ 0, & \text{otherwise}. \end{cases} \tag{23}$$

2.6 Accuracy of Low-Rank Approximation

We analyse the accuracy of the 1D periodic low-rank Green's solver relative to the full-rank solver. Low-rank approximation is by the specified-bandwidth method and no modified-wavenumber stabilisation is applied. Since it is possible to define the same matrix by careful choice of a threshold or a bandwidth this analysis applies to both low-rank approximation methods above. The Green's matrix M for the 1D periodic equation is circulant and can be treated as a wide finite difference stencil. Defining a vector $Z : \{z_0 = M_{jj}, z_{-1} = M_{j,j-1}, z_1 = M_{j,j+1}, \ldots, z_{-N/2} = M_{j,j-N/2}, z_{N/2} = M_{j,j+N/2}\}$ for any j, and using the symmetry of M, the nth eigenvalue of M is given by Leveque [12]:

$$\begin{aligned} \lambda_n^M &= z_0 + z_{-1}e^{-2\pi in\Delta x} + z_1 e^{2\pi in\Delta x} + \ldots + z_{-N/2}e^{-N\pi in\Delta x} + z_{N/2}e^{N\pi in\Delta x} \\ &= z_0 + \textstyle\sum_{k=1}^{N/2} z_k \cos(2k\pi n\Delta x). \end{aligned} \tag{24}$$

The scheme is stable since $|\lambda_n^M| \leq 1$. The eigenvalues of the specified-bandwidth low-rank matrix M^B are given by

$$\lambda_n^{M^B} = z_0 + \sum_{k=1}^{P} z_k \cos(2k\pi n\Delta x), \quad P < N/2. \tag{25}$$

We see that the eigenvalues of the low-rank approximation are a truncated Fourier expansion of Z. Furthermore, stability of the low-rank scheme is guaranteed by stability of the full-rank scheme. Let us assume that N is large enough that $z_{N/2} \approx 0$ to machine precision, i.e. the full eigenspectrum of M is resolved. The kernel is relatively compact so this is easily satisfied for moderate N. Using Parseval's identity the truncation error η incurred by low-rank approximation is [13]:

$$\eta = \|\lambda_n^M - \lambda_n^{M^B}\|_{L^2}^2 = 2\pi \sum_{k>P}^{N/2} |z_k|^2. \tag{26}$$

The truncation error is bounded because the sum converges since $z_k \leq 1, k = 0, \ldots, N/2$. Consider now a series of low-rank matrices: $M^{B_k}, k = 1, \ldots, N/2$. The error between successive entries in the series is

$$\eta_k = \|\lambda_n^{M^{B_{k+1}}} - \lambda_n^{M^{B_k}}\|_{L^2}^2 = 2\pi |z_{k+1}|^2. \tag{27}$$

This demonstrates that, for fixed Δx and Δt, the low-rank approximation M^B converges spectrally to the full-rank matrix M in the L^2 norm as P is increased. For a suitably smooth initial condition, this implies spectral convergence of the solution given by the low-rank scheme to that of the full-rank scheme (11). Numerical tests (not shown) verify this result and also show that for low ratio a/ν the kernel is wide and the cost savings by low-rank approximation could be significant. As a/ν increases the kernel becomes closer to a Dirac function and low-rank approximation becomes less important.

3 Results

The full-rank and low-rank Green's solvers were implemented in Matlab. For comparison two other solvers were tested: the Matlab `backslash` operator for the solution of the sparse linear system (4), and a first-order exponential integrator (7). Exponential integration was implemented by using the Matlab `expm` operator on the matrix $-\Delta t A$. This function uses a scaling and squaring algorithm [14].

We present the results of numerical tests in 1D and 2D at high Reynolds numbers. The conditions were such that modified-wavenumber stabilisation was employed in all cases. Low-Reynolds number tests (no stabilisation) were also conducted, confirming the scaling analyses in Sect. 2.5, but there is insufficient space in this paper to present them. All tests were run in serial on a 2015 MacBook Pro with a 3.1 GHz Intel Core i7.

3.1 1D Solvers

A Gaussian initial condition was specified that decayed to zero well within the domain: $u = \exp(-(x-0.5)^2/\nu)$ with $\nu = 1.e-6$. The mesh resolutions were $N = 100, 200, \ldots, 6400$. The initial condition was under-resolved and approximated the Dirac function. In the non-periodic case, a zero Dirichlet boundary condition was specified on the left (inflow) and a zero Neumann condition on the right (outflow). The domain length was $L = 1$, the advective constant was $a = 1.0$ and simulations were run for 10 timesteps. The advective CFL number $c_A = 0.5$ for the exponential integrator (it was found to be unstable at $c_A = 1$) and $c_A = 1$ for the others. In the current tests $\kappa > N$ so the modified-wavenumber stabilisation (16) is switched on.

The L_2 norms of the solution errors with respect to the exact solution are plotted in Fig. 2a, b. The Green's solvers obtain larger errors than the reference solvers due to the modification of κ. The Green's and exponential integrator solvers converge linearly at low N, trending towards quadratic convergence at large N. The backslash solver converges quadratically and obtains the smallest the error magnitude. Figure 2c, d shows the solver CPU times for 10 timesteps excluding matrix assembly. The backslash solver is much faster in the non-periodic case than the periodic case due to a faster method being chosen by the backslash algorithm for the circulant LHS matrix. All other solvers obtain linear scaling and the Green's solvers are marginally the fastest.

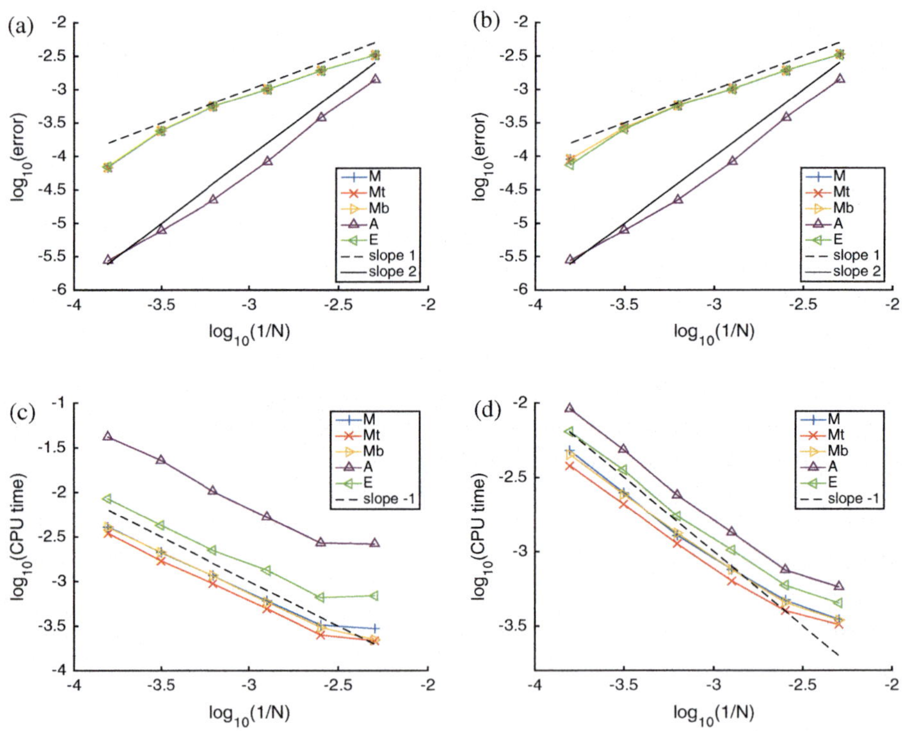

Fig. 2 Error convergence and CPU times of 1D first-order solvers. (**a**) Periodic error, (**b**) Non-periodic error, (**c**) Periodic time, (**d**) Non-periodic time. M = full-rank Green's solver, A = backslash solver, E = exponential integrator, Mt, Mb = low-rank Green's solvers

3.2 2D Solvers

In the 2D tests, the domain was $x, y = [0 : 1], [0 : 1]$ and the initial condition was a Gaussian distribution $u_0(x, y) = -0.5 + \exp(((x-0.5)^2 + (y-0.5)^2)/\nu)$. The exact solution is $u(x, y, t) = -0.5 + \frac{1}{4t+1}\exp(((x-0.5)^2 + (y-0.5)^2)/\nu(4t+1))$. The advection vector was $\mathbf{a} = (\sqrt{2}\sqrt{2})^T$ and the diffusion coefficient was $\nu = 1.e-4$. Five resolutions were used: $N = \{25, 50, 75, 100, 125\}$. The simulations proceeded for 10 timesteps with advective CFL number $c_A = 1$ for the direct solvers and $c_A = 0.5$ for the others (they were found to be unstable at $c_A = 1$). In the non-periodic cases the Dirichlet boundary condition $u = -0.5$ is imposed on the left and top, and zero Neumann boundaries on the right and bottom. Figure 3a, b plots the L_2 norm of the 2D solution errors with respect to the analytical solution. All schemes obtain linear convergence and identical error magnitudes. Figure 3c, d shows the 2D solver CPU times, excluding matrix assembly. All solvers demonstrate roughly linear scaling except the backslash operator, which tends towards $\mathscr{O}(N^{1.5})$ and is also slowest. The fastest solver is the low-rank Green's solver M^{T2}.

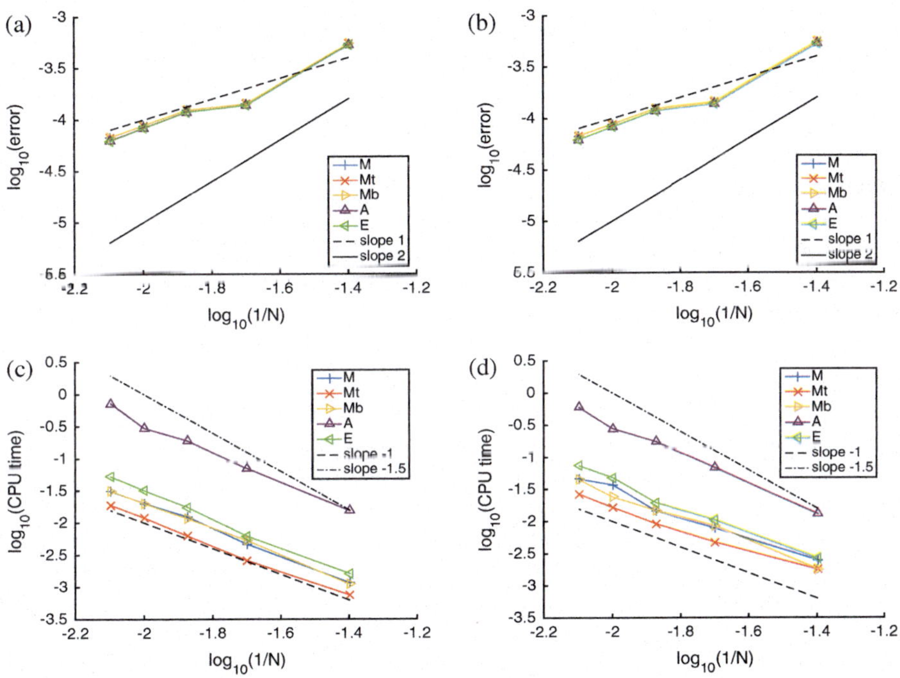

Fig. 3 Error convergence and CPU times of 2D solvers. (**a**) Periodic error, (**b**) Non-periodic error, (**c**) Periodic time, (**d**) Non-periodic time. M = full-rank Green's solver, A = backslash solver, E = exponential integrator, Mt, Mb = low-rank Green's solvers

4 Conclusions

A direct solution method for the advection-diffusion based on a new class of time integration schemes has been presented. The method is similar to the exponential integrator method but makes use of the Green's function G instead of the matrix exponential to propagate initial and boundary values through time and space. As with IMEX schemes and exponential integrators, the timestep is not limited by the linear term, which may reduce stiffness in some problems. Furthermore, existing low-rank approximation techniques can be applied to the discrete Green's matrix including hierarchical (HLRA) techniques. These are a promising route to achieving good parallel scaling on modern HPC systems.

By applying the new scheme to the linear advection-diffusion equation it became a forced heat equation for which G is a Gaussian kernel. Dirichlet boundary conditions were imposed by the method of images which is suitable for simple domains. In more complex domains it will be necessary to use a boundary integral method instead. When discretised in space G is amenable to low-rank approximation. Low-rank matrices were defined by setting a constant threshold value (matrix M^T) or by restricting the bandwidth to be a function only of ν.

Stabilisation was required when the resolution was too low to resolve G (Reynolds number too high). In that case, G was modified by restricting the numerical wavenumber κ. This led to cost scaling of $\mathcal{O}(N)$ for the full- and low-rank solvers, but reduced accuracy and order of convergence. The overall cost of the stabilised method for a fixed time interval was N^2 due to the timestep scaling with N^{-1}.

In 1D tests, the Green's solvers and exponential integrator attained first-order accuracy, tending towards second-order at large N. The backslash solver attained second-order accuracy but was also the most costly. The exponential integrator was competitive with the Green's solvers in terms of cost and accuracy in the range of N considered. In the 2D tests, all solvers obtained first-order accuracy and cost scaled linearly. The fastest computation times were obtained by the threshold-value low-rank matrix and Matlab's backslash operator was the slowest by up to an order of magnitude. The exponential integrator was competitive with the Green's solvers.

These preliminary results indicate that the new time integration scheme coupled with threshold-value low-rank approximation is competitive with, if not faster than, the reference solvers without sacrificing on accuracy. Limitations of the new method are that it is low-order accurate and requires modification to ensure stability when the Green's function is under-resolved. The next planned steps are (a) to investigate improved high-Reynolds-number stabilisation methods; (b) to extend the time integration scheme to higher-order accuracy via analytic integration of G coupled with a Runge-Kutta scheme for forcing and boundary terms; (c) to develop schemes for nonlinear advection; (d) to develop a boundary integral method for imposing boundary conditions on general domains; (e) to use HLRA on the Green's matrix. The Green's solver is intended to be used as a fast tunable-accuracy preconditioner/solver for large-scale CFD simulations. It is also suitable for other stiff time-dependent PDEs with a constant-in-time linear term.

The author wishes to thank his colleagues Per Lötstedt and Sverker Holmgren for useful discussions.

References

1. L. McInnes, B. Smith, H. Zhang, R. Mills, Hierarchical Krylov and nested Krylov methods for extreme-scale computing. Parallel Comput. **40**(1):17–31 (2014)
2. L. Greengard, V. Rokhlin, A fast algorithm for particle simulations. J. Comput. Phys. **73**, 325–348 (1987)
3. W. Hackbusch, B. Khoromskij, S. Sauter, On H^2-matrices, in *Lectures on applied mathematics* ed. by H. Bungartz, R. Hoppe, C. Zenger (Springer, Berlin/Heidelberg, 2000)
4. M. Bebendorf, Hierarchical LU decomposition-based preconditioners for BEM. Computing **74**(3), 225–247 (2005)
5. B. Engquist, L. Ying, Sweeping preconditioner for the Helmholtz equation: hierarchical matrix representation. Multiscale Model. Simul. **9**(2), 686–710 (2011)
6. R. Yokota, J. Pestana, H. Ibeid, D. Keyes, Fast multipole preconditioners for sparse matrices arising from elliptic equations (2013). ArXiv:1308.3339
7. S. Ambikasaran, E. Darve, The inverse fast multipole method (2014). ArXiv:1407.1572
8. R. Yokota, H. Ibeid, D. Keyes, Fast multipole method as a matrix-free hierarchical low-rank approximation (2016). ArXiv:1602.02244 [cs.NA]
9. A. Gholami, D. Malhotra, H. Sundar, G. Biros, FFT, FMM or multigrid? a comparative study of state-of-the-art Poisson solvers in the unit cube (2014). ArXiv:1408.6497
10. J. Bull, A direct solver for the advection-diffusion equation using Green's functions and low-rank approximation, in *Proceedings of ECCOMAS '16* (2016)
11. U. Ascher, S. Ruuth, B. Wetton, Implicit-explicit methods for time-dependent partial differential equations. SIAM J. Numer. Anal. **32**(3), 797–823 (1995)
12. R. Leveque, *Finite Difference Methods for Ordinary and Partial Differential Equations* (SIAM, Philadelphia, 2007)
13. D. Gottlieb, J. Hesthaven, Spectral methods for hyperbolic problems. J. Comput. Appl. Math. **128**, 83–131 (2001)
14. A. Al-Mohy, N. Higham, A new scaling and squaring algorithm for the matrix exponential. SIAM J. Matrix Anal. Appl. **31**(3), 970–989 (2009)

High Order in Space and Time Schemes Through an Approximate Lax-Wendroff Procedure

A. Baeza, P. Mulet, and D. Zorío

Abstract This paper deals with the scheme proposed by the authors in Zorío, Baeza and Mulet (J Sci Comput 71(1):246–273, 2017). This scheme is an alternative to the techniques proposed in Qiu and Shu (SIAM J Sci Comput 24(6):2185–2198, 2003) to obtain high-order accurate schemes using Weighted Essentially Non Oscillatory finite differences and approximating the flux derivatives required by the Cauchy-Kovalevskaya procedure by simple centered finite differences. We analyse how errors in first-order terms near discontinuities propagate through both versions of the Cauchy-Kovalevskaya procedure. We propose a fluctuation control, for which the approximation of the first-order derivative to be used in the Cauchy-Kovalevskaya procedure is obtained from a Weighted Essentially Non Oscillatory (WENO) interpolation of flux derivatives, instead of the usual finite difference of WENO flux reconstructions. The numerical results that we obtain confirm the benefits of this fluctuation control.

1 Introduction

This paper takes as starting point the scheme proposed in [8] as an alternative version of the Cauchy-Kovalevskaya procedure proposed by Qiu and Shu in [6]. This procedure consists in the replacement of the exact flux derivatives by accurate enough approximations. This replacement makes the implementation much simpler and the computational cost is reduced with respect to the original scheme, maintaining the global order of the method. The novelty in this work is the development of a fluctuation control method which avoids the propagation of large terms at the discretization of the high order derivatives. The method is applied to a Shu-Osher finite-difference spatial discretization [7].

A. Baeza (✉) • P. Mulet • D. Zorío
Departament de Matemàtiques, Universitat de València, València, Spain
e-mail: antonio.baeza@uv.es; mulet@uv.es; david.zorio@uv.es

M.L. Bittencourt et al. (eds.), *Spectral and High Order Methods for Partial Differential Equations ICOSAHOM 2016*, Lecture Notes in Computational Science and Engineering 119, DOI 10.1007/978-3-319-65870-4_31

The paper is organized as follows: in Sect. 2 we review the numerical scheme originally proposed by Qiu and Shu in [6]; in Sect. 3, we present the approximate Lax-Wendroff approach, based on using central difference approximations for the time derivatives of the flux; Sect. 4 stands for the fluctuation control, a novel technique to avoid the excessive propagation of diffusion around a discontinuity; in Sect. 5 some numerical experiments are presented; finally, some conclusions are drawn in Sect. 6.

2 Numerical Scheme

For completeness sake we include here the description and basic results of the scheme proposed in [8].

We consider a system of m hyperbolic conservation laws in d dimensions

$$u_t + \sum_{i=1}^{d} f^i(u)_{x_i} = 0.$$

For the sake of simplicity, we start with the one-dimensional scalar case ($d = m = 1$). For the solution $u(x,t)$ of $u_t + f(u)_x = 0$ on a fixed spatial grid (x_i) with spacing $h = x_{i+1} - x_i$ and some time t_n from a temporal grid with spacing $\delta = \Delta t = t_{n+1} - t_n > 0$, proportional to h, $\delta = \tau h$, where τ is dictated by stability restrictions (CFL condition) we use the following notation for time derivatives of u and $f(u)$:

$$\begin{aligned} u_{i,n}^{(\ell)} &= \frac{\partial^\ell u(x_i, t_n)}{\partial t^\ell}, \\ f_{i,n}^{(\ell)} &= \frac{\partial^\ell f(u)(x_i, t_n)}{\partial t^\ell}. \end{aligned}$$

Our goal is to obtain an R-th order accurate numerical scheme, i.e., a scheme with a local truncation error of order $R+1$, based on the Taylor expansion of the solution u from time t_n to time t_{n+1}:

$$u_i^{n+1} = \sum_{l=0}^{R} \frac{\Delta t^\ell}{\ell!} u_{i,n}^{(\ell)} + \mathcal{O}(\Delta t^{R+1}).$$

To achieve this we aim to define, by recursion on ℓ, corresponding approximations

$$\begin{aligned} \widetilde{u}_{i,n}^{(\ell)} &= u_{i,n}^{(\ell)} + \mathcal{O}(h^{R+1-\ell}), \\ \widetilde{f}_{i,n}^{(\ell)} &= f_{i,n}^{(\ell)} + \mathcal{O}(h^{R-\ell}), \end{aligned}$$

assuming (for a local truncation error analysis) that $\widetilde{u}_{i,n}^{0} = u_{i,n}^{(0)} = u(x_i, t_n)$.

The fact that u solves the system of conservation laws implies that the time derivatives $u_{i,n}^{(\ell)}$, $1 \leq \ell \leq R$, can be written in terms of the first spatial derivative of some function of $u_{i,n}^{(j)}, j < l$, by using the chain rule on f, which can be written as

$$f_{i,n}^{(\ell-1)} = F_{l-1}(u_i^n, u_{i,n}^{(\ell)}, \ldots, u_{i,n}^{(\ell-1)}) \tag{1}$$

for some function F_{l-1}, and following the Cauchy-Kowalewski (or Lax-Wendroff for second order) procedure:

$$\frac{\partial^\ell u}{\partial t^\ell} = \frac{\partial^{\ell-1}}{\partial t^{\ell-1}}(u_t) = -\frac{\partial^{\ell-1}}{\partial t^{\ell-1}}(f(u)_x) = -\left[\frac{\partial^{\ell-1} f(u)}{\partial t^{\ell-1}}\right]_x, \tag{2}$$

Specifically, to approximate the first time derivative, $u_t = -f(u)_x$, we use the Shu-Osher finite difference scheme [7] with upwinded Weighted Essentially Non-Oscillatory (WENO) spatial reconstructions [4] of order $2r-1$ in the flux function, with $r = \lceil \frac{R+1}{2} \rceil$:

$$u_{i,n}^{(1)} = u_t(x_i, t_n) = -[f(u)]_x(x_i, t_n) = -\frac{\hat{f}^n_{i+\frac{1}{2}} - \hat{f}^n_{i-\frac{1}{2}}}{h} + \mathcal{O}(h^{2r-1}). \tag{3}$$

Much cheaper centered differences are used instead for the next derivatives. We expound the general procedure for a third order accurate scheme ($R = 3$) for a scalar one-dimensional conservation law. Assume we have numerical data, $\{\widetilde{u}_i^n\}_{i=0}^{M-1}$, which approximates $u(\cdot, t_n)$ and want to compute an approximation for $u(\cdot, t_{n+1})$ at the same nodes, namely, $\{\widetilde{u}_i^{n+1}\}_{i=0}^{M-1}$.

First, we compute an approximation of u_t by the procedure stated above:

$$\widetilde{u}_{i,n}^{(1)} = -\frac{\hat{f}^n_{i+\frac{1}{2}} - \hat{f}^n_{i-\frac{1}{2}}}{h}, \tag{4}$$

with

$$\hat{f}^n_{i+\frac{1}{2}} = \hat{f}(\widetilde{u}^n_{i-r+1}, \ldots, \widetilde{u}^n_{i+r})$$

being the numerical fluxes, which are obtained through upwind WENO spatial reconstructions of order $2r-1$.

Once the corresponding nodal data is obtained for the approximated values of u_t, we compute

$$u_{tt} = [u_t]_t = [-f(u)_x]_t = -[f(u)_t]_x = -[f'(u)u_t]_x,$$

where $f'(u)u_t$ is now an approximately known expression for the required nodes. We use then a second order centered difference in order to obtain the approximation:

$$\widetilde{u}_{i,n}^{(2)} = -\frac{\widetilde{f}_{i+1,n}^{(1)} - \widetilde{f}_{i-1,n}^{(1)}}{2h}, \tag{5}$$

where

$$\widetilde{f}_{i,n}^{(1)} = F_1(\widetilde{u}_{i,n}^{(0)}, \widetilde{u}_{i,n}^{(1)}) = f'(\widetilde{u}_{i,n}^{(0)})\widetilde{u}_{i,n}^{(1)},$$

Finally, we approximate the third derivative:

$$u_{ttt} = [u_t]_{tt} = [-f(u)_x]_{tt} = -[f(u)_{tt}]_x = -\left(f''(u)u_t^2 + f'(u)u_{tt}\right)_x, \tag{6}$$

where again the function $f''(u)u_t^2 + f'(u)u_{tt}$ is approximately known at the nodes and therefore u_{ttt} can be computed by second order accurate centered differences (note that in case of Eq. (6) it would be required only a first order accurate approximation; however, the order of centered approximations is always even):

$$\widetilde{u}_{i,n}^{(3)} = -\frac{\widetilde{f}_{i+1,n}^{(2)} - \widetilde{f}_{i-1,n}^{(2)}}{2h}, \tag{7}$$

where

$$\widetilde{f}_{i,n}^{(2)} = F_2(\widetilde{u}_{i,n}^{(0)}, \widetilde{u}_{i,n}^{(1)}, \widetilde{u}_{i,n}^{(2)}) = f''(\widetilde{u}_{i,n}^{(0)}) \cdot (\widetilde{u}_{i,n}^{(1)})^2 + f'(\widetilde{u}_{i,n}^{(0)}) \cdot (\widetilde{u}_{i,n}^{(2)})^2.$$

Once all the needed data is obtained, we advance in time by replacing the terms of the third order Taylor expansion in time of $u(\cdot, t_{n+1})$ by their corresponding nodal approximations, namely, replace the exact derivatives $u_{i,n}^{(\ell)}$, $1 \leq \ell \leq 3$, by the numerical approximations $\widetilde{u}_{i,n}^{(\ell)}$, $1 \leq \ell \leq 3$, obtained in (4), (5) and (7), respectively:

$$\widetilde{u}_i^{n+1} = \widetilde{u}_i^n + \Delta t \widetilde{u}_{i,n}^{(1)} + \frac{\Delta t^2}{2}\widetilde{u}_{i,n}^{(2)} + \frac{\Delta t^3}{6}\widetilde{u}_{i,n}^{(3)}.$$

As we shall see, the above example can be extended to arbitrarily high order time schemes through the computation of the suitable high order central differences of the nodal values

$$\widetilde{f}_{i,n}^{(\ell)} = F_l(\widetilde{u}_{i,n}^{(0)}, \widetilde{u}_{i,n}^{(1)}, \ldots, \widetilde{u}_{i,n}^{(\ell)}) = f_{i,n}^{(\ell)} + \mathcal{O}(h^{R-\ell+1}).$$

The generalization to multiple dimensions is straightforward, since now the Cauchy-Kowalewski procedure, being based on the fact that $u_t = -\nabla \cdot f(u)$, yields

$$\frac{\partial^\ell u}{\partial t^\ell} = -\nabla \cdot \left(\frac{\partial^{\ell-1} f(u)}{\partial t^{\ell-1}}\right) = -\sum_{i=1}^{d} \frac{\partial}{\partial x_i}\left(\frac{\partial^{\ell-1} f^i(u)}{\partial t^{\ell-1}}\right)$$

and that the spatial reconstruction procedures are done separately for each dimension. For the case of the systems of equations, the time derivatives are now computed through tensor products of the corresponding derivatives of the Jacobian of the fluxes. The general procedure for systems and multiple dimensions is thus easily generalizable and further details about the procedure can be found in [6, 8]. Closed formulas to explicitly compute the above expressions can be found in the literature, such as the Faà di Bruno formula [3].

3 The Approximate Lax-Wendroff Procedure

As reported by the authors of [6], the computation of the exact nodal values of $f^{(k)}$ can be very expensive as k increases, since the number of required operations may increase exponentially. Moreover, implementing it is costly and requires large symbolic computations for each equation. We now present an alternative, which is much less expensive for large k and less dependent on the equation, in the sense that its only requirement is the knowledge of the flux function. This procedure also works in the multidimensional case and in the case of systems as well (by working componentwise). This technique is based on the observation that approximations $\widetilde{f}^{(l-1)} \approx f^{(l-1)}$ can be easily obtained by finite differences, rather than using the exact expression F_{l-1} in (1).

Let us introduce some notation for a one-dimensional system, that we assume for the sake of simplicity. For a function $u\colon \mathbb{R} \to \mathbb{R}^m$, we denote the discretization of the function on the grid defined by a base point a and grid space h by

$$G_{a,h}(u)\colon \mathbb{Z} \to \mathbb{R}^m, \quad G_{a,h}(u)_i = u(a+ih).$$

The symbol $\Delta_h^{p,q}$ denotes the centered finite differences operator that approximates p-th order derivatives to order $2q$ on grids with spacing h. For any u sufficiently differentiable, it satisfies:

$$\Delta_h^{p,q} G_{a,h}(u) = u^{(p)}(a) + \alpha^{p,q} u^{(p+2q)}(a) h^{2q} + \mathcal{O}(h^{2q+2}). \tag{8}$$

We aim to define approximations $\widetilde{u}_{i,n}^{(k)} \approx u_{i,n}^{(k)}$, $k = 0, \ldots, R$, recursively. We start the recursion with

$$\begin{aligned}\widetilde{u}_{i,n}^{(0)} &= u_i^n, \\ \widetilde{u}_{i,n}^{(1)} &= -\frac{\hat{f}_{i+\frac{1}{2}}^n - \hat{f}_{i-\frac{1}{2}}^n}{h},\end{aligned} \tag{9}$$

where $\hat{f}_{i+\frac{1}{2}}^n$ are computed from the known data (u_i^n) by applying upwind WENO reconstructions (see [2, 4, 7] for further details).

Associated to fixed h, i, n, once obtained $\widetilde{u}_{i,n}^{(l)}$, $l = 0, \ldots, k$, in the recursive process we define the k-th degree approximated Taylor polynomial $T_k[h, i, n]$ by

$$T_k[h, i, n](\rho) = \sum_{l=0}^{k} \frac{\widetilde{u}_{i,n}^{(l)}}{l!} \rho^l.$$

By recursion, for $k = 1, \ldots, R-1$, we define

$$\begin{aligned}\widetilde{f}_{i,n}^{(k)} &= \Delta_\delta^{k, \lceil \frac{R-k}{2} \rceil} \Big(G_{0,\delta}\big(f(T_k[h, i, n])\big)\Big), \\ \widetilde{u}_{i,n}^{(k+1)} &= -\Delta_h^{1, \lceil \frac{R-k}{2} \rceil} \widetilde{f}_{i+\cdot,n}^{(k)},\end{aligned} \tag{10}$$

where we denote by $\widetilde{f}_{i+\cdot,n}^{(k)}$ the vector given by the elements $(\widetilde{f}_{i+\cdot,n}^{(k)})_j = \widetilde{f}_{i+j,n}^{(k)}$ and $\delta = \Delta t$. With all these ingredients, the proposed scheme is:

$$u_i^{n+1} = u_i^n + \sum_{l=1}^{R} \frac{\Delta t^l}{l!} \widetilde{u}_{i,n}^{(l)}. \tag{11}$$

It can be proven that the method resulting from this construction is $R-th$ order accurate and can be written in conservation form, see [8]

4 Fluctuation Control

Now we focus on the computation of the approximate nodal values of the first order time derivative. Typically, one would simply take the approximations obtained through the upwinded reconstruction procedure in the Shu-Osher's finite difference approach, that is,

$$\widetilde{u}_{j,n}^{(1)} = -\frac{\hat{f}_{j+\frac{1}{2},n} - \hat{f}_{j-\frac{1}{2},n}}{h}. \tag{12}$$

However, taking directly these values as the first derivative used to compute the next derivatives through (11) can produce wrong results if the data is not smooth, as routinely happens in hyperbolic systems. In fact, it will include $\mathcal{O}(h^{-1})$ terms wherever there is a discontinuity, which we will call from now on *fluctuations*. These terms will appear provided $\hat{f}_{j-\frac{1}{2},n}$ and $\hat{f}_{j+\frac{1}{2},n}$ come from different sides of a discontinuity (or some of them has mixed information of both sides due to a previous flux splitting procedure to reconstruct the interface values), since in that case $\hat{f}_{j+\frac{1}{2},n} - \hat{f}_{j-\frac{1}{2},n} = \mathcal{O}(1)$.

In practice, this implies that the k-th derivative, $1 \leq k \leq R$, will have terms of magnitude $\mathcal{O}(h^{-k})$, therefore, the term which appears on the Taylor expansion term, which is multiplied by $\frac{\Delta t^k}{k!}$, a term of magnitude $\mathcal{O}(h^k)$, will be ultimately $\mathcal{O}(1)$. This may result in undesired diffusion, oscillations or even a complete failure of the scheme in some cases.

Our proposal is to compute an alternative approximation of $\widetilde{u}_{j,n}^{(1)}$ as described in Sect. 4.1 and replace (12) by this new approximation for the recursive computation in (10) only, maintaining (12) for its use in (11), thus ensuring the proper upwinding.

4.1 Central WENO Reconstructions

Let us assume that our spatial scheme is $(2r-1)$-th order accurate WENO. After having performed all the operations for the reconstruction of the numerical fluxes at the interfaces, the stencil of points that is used in order to approximate the derivative at the node x_i is the following set of $2r+1$ points:

$$\{x_{i-r}, \ldots, x_i, \ldots, x_{i+r}\}, \tag{13}$$

whose corresponding flux values, $f_j = f(u_j)$, are

$$\{f_{i-r}, \ldots, f_i, \ldots, f_{i+r}\}.$$

The procedure that we next expound only uses information from the stencil

$$\mathcal{S}_{i+r-1}^{2r-1} := \{i-r+1, \ldots, i, \ldots, i+r-1\}, \tag{14}$$

thus ignoring the flux values f_{i-r}, f_{i+r} at the edges of the stencil in (13).

For fixed i, let q_k^r be the interpolating polynomial of degree $\leq r-1$ such that $q_k^r(x_j) = f_j, j \in \mathcal{S}_{i+k} := \{i+k-r+1, \ldots, i+k,\}$, $0 \leq k \leq r-1$. After the previous discussion, our goal is to obtain an approximation of the flux derivative $f(u)_x(x_i)$ from the stencil $\mathcal{S}_{i+r-1}^{2r-1}$ which is $(2r-1)$-th order accurate if the nodes in the stencil lie within a smoothness region for u or is $\mathcal{O}(1)$ otherwise. We use WENO techniques to achieve this purpose.

The following lemma is easily established.

Lemma 1 *There exists a set of constants* $\{c_k^r\}_{k=1}^r$ *satisfying* $0 < c_k^r < 1$, *for* $0 \leq k \leq r-1$, $\sum_{k=0}^{r-1} c_k^r = 1$, *such that*

$$\sum_{k=0}^{r-1} c_k^r (q_k^r)'(x_i) = (q_{r-1}^{2r-1})'(x_i).$$

If $f_j = f(u(x_j, t_n))$, for smooth enough u and fixed t_n, then

$$(q_k^r)'(x_i) = f(u)_x(x_i, t_n) + d_k^r(x_i)h^{r-1} + \mathcal{O}(h^r), k = 0, \ldots, r-1, \tag{15}$$

$$(q_{r-1}^{2r-1})'(x_i) = f(u)_x(x_i, t_n) + d_{r-1}^{2r-1}(x_i)h^{2r-2} + \mathcal{O}(h^{2r-1}). \tag{16}$$

for continuously differentiable d_k^r, d_{r-1}^{2r-1}. The goal is to obtain the accuracy in (16) by a suitable nonlinear convex combination of (15),

$$\sum_{k=0}^{r-1} w_k^r (q_k^r)'(x_i) = f(u)_x(x_i, t) + \widetilde{d}_{r-1}^{2r-1}(x_i)h^{2r-2} + \mathcal{O}(h^{2r-1}), \tag{17}$$

where $w_k^r = c_k^r(1 + \mathcal{O}(h^{r-1}))$ if the whole stencil $x_{i-r+1}, \ldots, x_{i+r-1}$ lies within a smoothness region for u and $w_k^r = \mathcal{O}(h^{r-1})$ if the k-th stencil crosses a discontinuity and there is at least another stencil which does not. To this aim we follow the WENO idea [4, 5]. From now on we drop the superscript r in q_k^r.

Furthermore, we need the approximation in (17) to be in conservation form. To achieve this we use the polynomial p_k of degree $r-1$ satisfying

$$\frac{1}{h}\int_{x_{j-\frac{1}{2}}}^{x_{j+\frac{1}{2}}} p_k(x)dx = f_j, \quad i-r+1+k \leq j \leq i+k, \quad 0 \leq k \leq r-1,$$

and $\widetilde{p}_k(x)$ a primitive of it. It can be seen that the polynomial

$$\widetilde{q}_k(x) = \frac{\widetilde{p}_k(x+\frac{h}{2}) - \widetilde{p}_k(x-\frac{h}{2})}{h},$$

has degree $\leq r-1$ and that $\widetilde{q}_k(x_j) = f_j, j = i-r+1+k, \ldots, i+k$, and therefore $\widetilde{q}_k(x)$ must coincide with q_k. Thus

$$q_k'(x_j) = \frac{(\widetilde{p}_k)'(x_{j+\frac{1}{2}}) - (\widetilde{p}_k)'(x_{j-\frac{1}{2}})}{h} = \frac{p_k(x_{j+\frac{1}{2}}) - p_k(x_{j-\frac{1}{2}})}{h}.$$

Now, let us define the following Jiang-Shu smoothness indicators using the definition of p_k:

$$I_k = \sum_{\ell=1}^{r-1} \int_{x_{i-\frac{1}{2}}}^{x_{i+\frac{1}{2}}} h^{2\ell-1} p_k^{(\ell)}(x)^2 dx, \quad 0 \leq k \leq r-1 \tag{18}$$

so that we can define the weights as follows:

$$\omega_k = \frac{\alpha_k}{\sum_{l=1}^{r} \alpha_l}, \qquad \alpha_k = \frac{c_k}{(I_k + \varepsilon)^m}, \tag{19}$$

with $\varepsilon > 0$ a small positive quantity, possibly depending on h. Following the techniques in [1], since $p_k^{(\ell)} - p_j^{(\ell)} = \mathcal{O}(h^{r-\ell})$ at regions of smoothness, whereas $\int_{x_{i-\frac{1}{2}}}^{x_{i+\frac{1}{2}}} (p_k')^2 dx = \mathcal{O}(h^{-2})$ if the corresponding stencil $\mathcal{S}_k^r$ crosses a discontinuity, the smoothness indicators satisfy $I_k - I_j = \mathcal{O}(h^{r+1})$ at regions of smoothness and $I_k \not\to 0$ if the k-th stencil crosses a discontinuity. Therefore, the definition (19) satisfies the requirements mentioned above in order to achieve maximal order even at smooth extrema, provided that the parameter $\varepsilon > 0$, besides avoiding divisions by zero, is chosen as $\varepsilon = \lambda h^2$, with $\lambda \sim f(u)_x$, and that the exponent m in (19) makes the weight $\omega_k = \mathcal{O}(h^{r-1})$ wherever there is a discontinuity at that stencil. Since one wants to attain the maximal possible order in such case, which corresponds to the value interpolated from a smooth substencil, which is $\mathcal{O}(h^r)$, then it suffices to set $m = \lceil \frac{r}{2} \rceil$. Finally, we define $\approx{u}_{i,n}^{(1)}$, the smoothened approximation of $u_t(x_i, t_n)$ that replaces $\widetilde{u}_{i,n}^{(1)}$ in (10) as the result of the following convex combination:

$$\overset{\approx}{u}_{i,n}^{(1)} = -\sum_{k=1}^{r} \omega_k q_k'(x_i).$$

5 Numerical Experiments

In this section we present some 2D experiments with Euler equations involving comparisons of the fifth order both in space ($r = 3$) and time ($R = 2r - 1 = 5$) exact and approximate Lax-Wendroff schemes, together with the results obtained using the third order TVD Runge-Kutta time discretization. From now on we will refer as WENO[]-LW[] to the exact Lax-Wendroff procedure, WENO[]-LWA[] to the approximate Lax-Wendroff procedure, WENO[]-LWF[] if a fluctuation control is used in the exact procedure, WENO[]-LWAF[] if the fluctuation control comes together with the approximate procedure and WENO[]-RK[] when a Runge-Kutta method is used. In each case, the first bracket stands for the value of the spatial accuracy order and the second one for the time accuracy order.

5.1 Smooth Solution

In order to test the accuracy of our scheme in the general scenario of a multidimensional system of conservation laws, we perform a test using the 2D Euler equations with smooth initial conditions, given by

$$\begin{aligned} u_0(x,y) &= (\rho(x,y), v^x(x,y), v^y(x,y), E(x,y)) \\ &= \left(\frac{3}{4}+\frac{1}{2}\cos(\pi(x+y)), \frac{1}{4}+\frac{1}{2}\cos(\pi(x+y)),\right. \\ &\left.\frac{1}{4}+\frac{1}{2}\sin(\pi(x+y)), \frac{3}{4}+\frac{1}{2}\sin(\pi(x+y))\right), \end{aligned}$$

where $x \in \Omega = [-1,1] \times [-1,1]$, with periodic boundary conditions.

Numerical solutions for resolutions $n \times n$, for $n = 10 \cdot 2^k$, $1 \leq k \leq 5$ are compared with a reference solution computed using a WENO5 spatial scheme and the third order Runge-Kutta TVD method in a finer mesh, with $n = 2560$ and $\Delta t = h^{\frac{5}{3}}$, obtaining the results shown in Tables 1, 2, 3 at the time $t = 0.025$ for CFL $= 0.5$. We can thus see that our scheme achieves the desired accuracy, being the results obtained through the approximate Lax-Wendroff procedure almost the same as those obtained using the exact version. The version with fluctuation control also yields the desired accuracy.

Table 1 Error table for 2D Euler equation, $t = 0.025$. WENO5-LW5

n	Error $\|\cdot\|_1$	Order $\|\cdot\|_1$	Error $\|\cdot\|_\infty$	Order $\|\cdot\|_\infty$
40	1.80E−5	–	2.74E−4	–
80	1.09E−6	4.05	1.80E−5	3.93
160	3.89E−8	4.80	7.36E−7	4.61
320	1.29E−9	4.92	2.49E−8	4.88
640	4.11E−11	4.97	8.07E−10	4.95
1280	1.23E−12	5.06	2.43E−11	5.06

Table 2 Error table for 2D Euler equation, $t = 0.025$. WENO5-LWA5

n	Error $\|\cdot\|_1$	Order $\|\cdot\|_1$	Error $\|\cdot\|_\infty$	Order $\|\cdot\|_\infty$
40	1.80E−5	–	2.74E−4	–
80	1.09E−6	4.05	1.80E−5	3.93
160	3.89E−8	4.80	7.36E−7	4.61
320	1.29E−9	4.92	2.49E−8	4.88
640	4.11E−11	4.97	8.07E−10	4.95
1280	1.23E−12	5.06	2.43E−11	5.06

Table 3 Error table for 2D Euler equation, $t = 0.025$. WENO5-LWAF5

n	Error $\|\cdot\|_1$	Order $\|\cdot\|_1$	Error $\|\cdot\|_\infty$	Order $\|\cdot\|_\infty$
40	2.63E−5	–	2.97E−4	–
80	1.58E−6	4.06	2.01E−5	3.89
160	6.66E−8	4.57	1.06E−6	4.24
320	2.33E−9	4.84	4.08E−8	4.70
640	7.60E−11	4.94	1.34E−9	4.93
1280	2.35E−12	5.02	4.06E−11	5.04

5.2 *Double Mach Reflection*

This experiment uses the Euler equations to model a vertical right-going Mach 10 shock colliding with an equilateral triangle. By symmetry, this is equivalent to a collision with a ramp with a slope of 30 degrees with respect to the horizontal line.

For the sake of simplicity, we consider the equivalent problem of an oblique shock whose vertical angle is $\frac{\pi}{6}$ rad in the rectangle $\Omega = [0, 4] \times [0, 1]$. The initial conditions of the problem are

$$u_0(x, y) = \begin{cases} C_1 & y \leq \frac{1}{4} + \tan(\frac{\pi}{6})x, \\ C_2 & y > \frac{1}{4} + \tan(\frac{\pi}{6})x, \end{cases}$$

where

$$C_1 = (\rho_1, v_1^x, v_1^y, E_1)^T = (8, 8.25\cos(\frac{\pi}{6}), -8.25\sin(\frac{\pi}{6}), 563.5)^T,$$
$$C_2 = (\rho_2, v_2^x, v_2^y, E_2)^T = (1.4, 0, 0, 2.5)^T.$$

We impose inflow boundary conditions, with value C_1, at the left side, $\{0\} \times [0, 1]$, outflow boundary conditions both at $[0, \frac{1}{4}] \times \{0\}$ and $\{4\} \times [0, 1]$, reflecting boundary conditions at $]\frac{1}{4}, 4] \times \{0\}$ and inflow boundary conditions at the upper side, $[0, 4] \times \{1\}$, which mimics the shock at its actual traveling speed:

$$u(x, 1, t) = \begin{cases} C_1 & x \leq \frac{1}{4} + \frac{1+20t}{\sqrt{3}}, \\ C_2 & x > \frac{1}{4} + \frac{1+20t}{\sqrt{3}}. \end{cases}$$

We run different simulations until $t = 0.2$ at a resolution of 2048×512 points for CFL $= 0.4$ and a different combination of techniques, involving WENO5-RK3, WENO5-LW5 and WENO5-LWA5. The results are presented in Fig. 1 as Schlieren plots of the turbulence zone. It can be concluded that the results obtained through the exact and approximate Lax-Wendroff techniques are quite similar, and that the

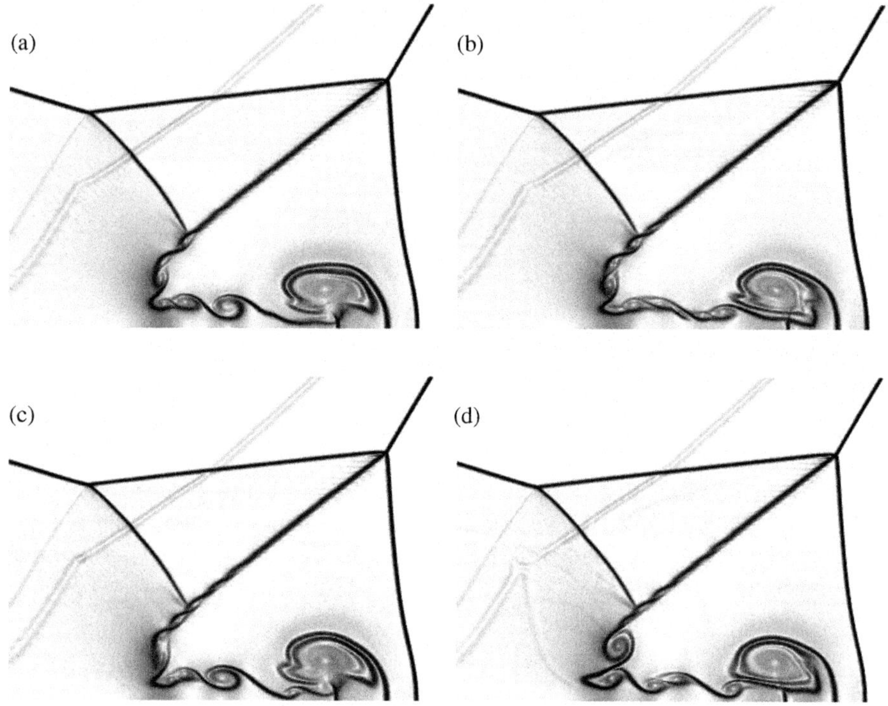

Fig. 1 Double Mach reflection results. Density field. (**a**) WENO5-RK3. (**b**) WENO5-LW5. (**c**) WENO5-LWA5. (**d**) WENO5-LWAF5

Table 4 Performance table

Method	Efficiency
WENO5-LW5	1.44
WENO5-LWA5	1.54
WENO5-LWF5	1.33
WENO5-LWAF5	1.44

results obtained through the technique with fluctuation control provides a slightly sharper profile.

Finally, in order to illustrate that the LW techniques are more efficient than the RK time discretization, we show a performance test involving the computational time required by each technique by running the Double Mach Reflection problem for the resolution 200 × 50. The results are shown in Table 4, where the field "Efficiency" stands for $\frac{t_{\mathrm{RK3}}}{t_{\mathrm{LW*}}}$. It can be seen that the fifth order Lax-Wendroff technique outperforms the third order accurate Runge-Kutta scheme.

On the other hand, we see that the version with approximate fluxes has a better performance than the main formulation, since less computations are required for high order derivatives. On the other hand, if the fluctuation control is used then the performance is lower, however, the combination of the approximate fluxes with the fluctuation control yields a fifth order accurate with approximately the same efficiency than the original formulation, but providing better results.

6 Conclusions

In this paper we have used an arbitrarily high order Lax-Wendroff-type time scheme through an approximate formulation of the scheme proposed by Qiu and Shu, which was developed in [8] and does not require symbolic computations to implement it, unlike those proposed by the aforementioned authors in [6]. The novelty of this work is the development of a fluctuation control in order to avoid the propagation of first-order errors around the discontinuities inherent to these schemes.

The results obtained in the numerical experiments are satisfactory, and show that the approximate procedure yields essentially the same results than the exact version with a much lower implementation cost and being less computationally expensive. On the other hand, the version with fluctuation control, albeit increasing the computational cost, produces numerical solutions with better resolution.

Acknowledgements This research was partially supported by Spanish MINECO project MTM2014-54388.

References

1. F. Aràndiga, A. Baeza, A.M. Belda, P. Mulet, Analysis of WENO schemes for full and global accuracy. SIAM J. Numer. Anal. **49**(2) 893–915 (2011)
2. R. Donat, A. Marquina, Capturing shock reflections: an improved flux formula. J. Comput. Phys. **125**, 42–58 (1996)
3. C.F. Faà di Bruno, Note sur une nouvelle formule de calcul différentiel. Q. J. Math. **1**, 359–360 (1857)
4. G.S. Jiang, C.W. Shu, Efficient implementation of weighted ENO schemes. J. Comput. Phys. **126**, 202–228 (1996)
5. X.-D. Liu, S. Osher, T. Chan, Weighted essentially non-oscillatory schemes. J. Comput. Phys. **115** 200–212 (1994)
6. J. Qiu, C.W. Shu, Finite difference WENO schemes with Lax-Wendroff-type time discretizations. SIAM J. Sci. Comput. **24**(6), 2185–2198 (2003)
7. C.W. Shu, S. Osher, Efficient implementation of essentially non-oscillatory shock-capturing schemes, II. J. Comput. Phys. **83**(1), 32–78 (1989)
8. D. Zorío, A. Baeza, P. Mulet, An Approximate Lax-Wendroff-type procedure for high order accurate schemes for hyperbolic conservation laws. J. Sci. Comput. **71**(1), 246–273 (2017)

On Thin Plate Spline Interpolation

M. Löhndorf and J.M. Melenk

Abstract We present a simple, PDE-based proof of the result (Math Comput 70(234):719–737, 2001) by Johnson that the error estimates of Duchon (RAIRO Anal Numér 12(4):325–334, 1978) for thin plate spline interpolation can be improved by $h^{1/2}$. We illustrate that $\mathscr{H}$-matrix techniques can successfully be employed to solve very large thin plate spline interpolation problems.

1 Introduction and Main Results

Interpolation with so-called thin plate splines (also known as surface splines, D^m-splines, or polyharmonic splines) is a classical topic in spline theory. It is concerned with the following interpolation problem (1): Given a (sufficiently smooth) function f and points $x_i \in \mathbb{R}^d$, $i = 1, \ldots, N$, find the minimizer If of the problem

$$\text{minimize} \qquad |v|_{H^m(\mathbb{R}^d)} \qquad \text{under the constraint } v(x_i) = f(x_i), \quad i = 1, \ldots, N. \tag{1}$$

Here, the seminorm $|v|_{H^m(\mathbb{R}^d)}$ is induced by the bilinear form

$$\langle v, w \rangle_m := \sum_{|\alpha|=m} \frac{m!}{\alpha!} \int_{\mathbb{R}^d} D^\alpha v D^\alpha w \, dx. \tag{2}$$

For $m > d/2$ and under very mild conditions on the point distribution, a unique minimizer If exists. The name "thin plate splines" originates from the fact that in the simplest case $m = d = 2$, If can be represented in terms of translates of the fundamental solution of the biharmonic equation. For general m the interpolant If

M. Löhndorf (✉)
Kapsch TrafficCom, Am Europlatz 2, A-1120 Wien, Austria

J.M. Melenk
Technische Universität Wien, A-1040 Wien, Austria
e-mail: melenk@tuwien.ac.at

M.L. Bittencourt et al. (eds.), *Spectral and High Order Methods for Partial Differential Equations ICOSAHOM 2016*, Lecture Notes in Computational Science and Engineering 119, DOI 10.1007/978-3-319-65870-4_32

can be expressed in terms fundamental solutions of Δ^m: There are constants $c_i \in \mathbb{R}$, $i = 1, \ldots, N$, and a polynomial $\pi \in \mathbb{P}_{m-1}$ of degree $m-1$ such that (with the Euclidean norm $\|\cdot\|_2$ on $\mathbb{R}^d$)

$$If(x) = \sum_{i=1}^{N} c_i \phi_m(\|x - x_i\|_2) + \pi_{m-1}(x), \qquad \sum_{i=1}^{N} c_i q(x_i) = 0 \qquad \forall q \in \mathbb{P}_{m-1}, \quad (3)$$

where ϕ_m is given explicitly by

$$\phi_m(r) = \begin{cases} r^{2m-d} \log r & d \text{ even} \\ r^{2m-d} & d \text{ odd}. \end{cases} \quad (4)$$

The representation (3) allows one to reformulate (1) as the problem of finding the coefficients c_i and the polynomial π_{m-1} so that the (constrained) interpolation problem (3) is solved. The classical error analysis for (1) is formulated in terms fill-distance: For a bounded domain $\Omega \subset \mathbb{R}^d$ and points $X_N = \{x_i \mid i = 1, \ldots, N\} \subset \Omega$, the *fill distance* $h(X_N)$ is given by

$$h(X_N) := \sup_{x \in \Omega} \inf_{i=1,\ldots,N} \|x - x_i\|_2. \quad (5)$$

Starting with the seminal papers by Duchon [11, 12] the error $f - If$ on Ω is controlled in terms of h and the regularity properties of f (on Ω):

Proposition 1 ([11, Prop. 3]) *Let $\Omega \subset \mathbb{R}^d$ be a bounded Lipschitz domain. Let $m > d/2$, $k \in \mathbb{N}$, $p \in [2, \infty]$ be such that $H^m(\Omega) \subset W^{k,p}(\Omega)$. Then, there are constants h_0, C_1, $C_2 > 0$ depending only on Ω, m, d such that for any collection $X_N = \{x_1, \ldots, x_N\} \subset \Omega$ with fill distance $h := h(X_N) \leq h_0$*

$$\sum_{|\alpha|=k} \|D^\alpha(f - If)\|_{L^p(\Omega)} \leq C_1 h^{m-k-d/2+d/p} |E^\Omega f - If|_{H^m(\mathbb{R}^d)} \leq C_2 h^{m-k-d/2+d/p} |f|_{H^m(\Omega)};$$

here, $E^\Omega f$ denotes the minimum norm extension of f defined in (8).

In Proposition 1 and throughout the present note, we will use the standard notation for Sobolev spaces $W^{s,p}$ and Besov spaces $B^s_{2,q}$; we refer to [26] for their definition. Interpolation space will always be understood by the so-called "real method" (also known as "K-method") as described, e.g., in [26, 27]. We will use extensively that the scales of Sobolev and Besov spaces are interpolation spaces. We will also use the notation $|\nabla^j f|^2 = \sum_{|\alpha|=j} \frac{j!}{\alpha!} |D^\alpha f|^2$.

It is worth noting that the interpolation operator I is a projection so that $I(f - If) = 0$. Proposition 1 applied to the function $f - If$ therefore yields

Corollary 1 *Under the assumptions of Proposition 1 there holds*

$$\sum_{|\alpha|=k} \|D^\alpha(f - If)\|_{L^p(\Omega)} \leq C_2 h^{m-k-d/2+d/p} |f - If|_{H^m(\Omega)}.$$

A natural question in connection with Proposition 1 is whether the convergence rate can be improved by requiring additional regularity of f. It turns out that boundary effects limits this. We mention that a doubling of the convergence rate is possible by imposing certain homogeneous boundary conditions on high order derivatives as shown in [22] and, more abstractly, in [24]. If this highly fortuitous setting is not given, then only a small further gain is possible as shown by Johnson, [16, 18]. For example, he showed that a gain of $h^{1/2}$ is possible if $f \in B_{2,1}^{m+1/2}(\Omega)$ and $\partial\Omega$ is sufficiently smooth. The purpose the present note is to give a short and simple proof of this result using different tools, namely, those from elliptic PDE theory. The techniques also open the door to reducing the smoothness assumptions on $\partial\Omega$ in [16, 18] to Lipschitz continuity as discussed in more detail in Remark 2. Our main result therefore is a simpler proof of:

Proposition 2 ([16]) *Let $\Omega \subset \mathbb{R}^d$ be a bounded Lipschitz domain with sufficiently smooth boundary. Then there are constants h_0, C_1, $C_\delta > 0$ that depend solely on Ω, m, d, and δ such that for any collection $X = \{x_1, \ldots, x_N\} \subset \Omega$ with fill distance $h := h(X_N) \leq h_0$ there holds*

$$|E^\Omega f - If|_{H^m(\mathbb{R}^d)} \leq C_1 h^{1/2} \|f\|_{B_{2,1}^{m+1/2}(\Omega)}, \tag{6}$$

$$|E^\Omega f - If|_{H^m(\mathbb{R}^d)} \leq C_\delta h^\delta \|f\|_{H^{m+\delta}(\Omega)}, \quad 0 \leq \delta < 1/2. \tag{7}$$

In particular, therefore, the estimates of [11, Prop. 3] (i.e., Proposition 1) can be improved by $h^{1/2}$ for $f \in B_{2,1}^{m+1/2}(\Omega)$ and by h^δ for $f \in H^{m+\delta}(\Omega)$.

Remark 1 A common route to error estimates for $f - If$ is via the so-called "power function" $P(x)$. Indeed, classical pointwise estimates take the form $|f(x) - If(x)| < P(x)|E^\Omega f - If|_{H^m(\mathbb{R}^d)}$ (cf., e.g., [8, Prop. 5.3], [29, Thm. 11.4]) and P is subsequently estimated in terms of the fill distance h. Thus, Proposition 2 allows for improving estimates in this setting. ∎

We close this section by referring the reader to the monographs [8, 29] as well as [17] for further details on the approximation properties of radial basis functions, in particular, thin plate splines.

2 Proof of Proposition 2

2.1 Tools

The precise formulation of the minimization problem (1) is based on the classical *Beppo-Levi space* $\mathrm{BL}^m(\mathbb{R}^d)$, which is defined as

$$\mathrm{BL}^m(\mathbb{R}^d) := \{u \in \mathscr{D}' \mid \nabla^m u \in L^2(\mathbb{R}^d)\}.$$

We refer to [10] and [29, Sec. 10.5] for more properties of the space $\mathrm{BL}^m(\mathbb{R}^d)$; in particular, $C_0^\infty(\mathbb{R}^d)$ is dense in $\mathrm{BL}^m(\mathbb{R}^d)$ (see [29, Thm. 10.40] for the precise notion). We also need the *minimum norm extension* $E^\Omega : H^m(\Omega) \to \mathrm{BL}^m(\mathbb{R}^d)$ given by

$$E^\Omega U = \arg\min\{|u|_{H^m(\mathbb{R}^d)} \mid u \in \mathrm{BL}^m(\mathbb{R}^d), \quad u|_\Omega = U\}. \tag{8}$$

The minimization property in (8) implies the orthogonality

$$\langle E^\Omega U, v\rangle_m = 0 \qquad \forall v \in \{v \in \mathrm{BL}^m(\mathbb{R}^d) \mid v|_\Omega = 0\}. \tag{9}$$

The connection with elliptic PDE theory arises from the fact that $E^\Omega U$ satisfies an elliptic PDE in $\Omega^c := \mathbb{R}^d \setminus \overline{\Omega}$:

$$\Delta^m E^\Omega U = 0 \qquad \text{in} \Omega^c. \tag{10}$$

It will be convenient to decompose $B(u, v) := \langle u, v\rangle_m = \sum_{|\alpha|=m} \frac{m!}{\alpha!} \int_{\mathbb{R}^d} D^\alpha u\, D^\alpha v$ as $B(u, v) = B_\Omega(u, v) + B_{\Omega^c}(u, v)$, where

$$B_\Omega(u, v) := \sum_{|\alpha|=m} \frac{m!}{\alpha!} \int_\Omega D^\alpha u\, D^\alpha v, \quad B_{\Omega^c}(u, v) := \sum_{|\alpha|=m} \frac{m!}{\alpha!} \int_{\Omega^c} D^\alpha u\, D^\alpha v.$$

The trace mapping is continuous $H^{1/2+\varepsilon}(\Omega) \to H^\varepsilon(\partial\Omega)$ for $\varepsilon \in (0, 1/2]$; however, the limiting case $\varepsilon = 0$ is not true; it is true if the Sobolev space $H^{1/2}(\Omega)$ is replaced with the slightly smaller Besov space $B^{1/2}_{2,1}(\Omega)$:

Lemma 1 (Trace Theorem) *Let $\Omega \subset \mathbb{R}^d$ be a Lipschitz domain, $k \in \mathbb{N}_0$. Then there exists $C > 0$ such that the multiplicative estimate $\|u\|^2_{L^2(\partial\Omega)} \leq C\|u\|_{L^2(\Omega)}\|u\|_{H^1(\Omega)}$ holds as well as*

$$\|u\|_{L^2(\partial\Omega)} \leq C\|u\|_{B^{1/2}_{2,1}(\Omega)}, \qquad \|\nabla^k u\|_{L^2(\partial\Omega)} \leq C\|u\|_{B^{k+1/2}_{2,1}(\Omega)}. \tag{11}$$

Proof The case $k \geq 1$ in (11) follows immediately from the case $k = 0$. The case $k = 0$ is discussed in [27, Thm. 2.9.3] for the case of a half-space. The generalization to Lipschitz domains can be found, for example, in [1, Lemma 1.10]. □

2.2 An Interpolation Argument

The following technical result, which is of independent interest, will be used to reduce regularity assumptions to $B^{m+1/2}_{2,1}(\Omega)$.

Lemma 2 *Let $X_1 \subset X_0$ be two Banach spaces with continuous embedding. Let $q \in [1,\infty]$, $\theta \in (0,1)$. Define (by the real method of interpolation) $X_\theta := (X_0, X_1)_{\theta,q}$ for $\theta \in (0,1)$. Let $0 < \theta_1 < \theta_2 < \cdots < \theta_n < 1$ be fixed and assume that $\ell \in X_0'$ satisfies for some C_0, C_1, $\varepsilon > 0$*

$$|\ell(f)| \le C_0 \|f\|_{X_0} \qquad \forall f \in X_0,$$

$$|\ell(f)| \le C_1 \left[\sum_{i=1}^{n} \varepsilon^{\theta_i} \|f\|_{X_{\theta_i}} + \varepsilon \|f\|_{X_1} \right] \qquad \forall f \in X_1.$$

Then there exists a constant $C > 0$ that is independent of ε such that

$$|\ell(f)| \le C\varepsilon^{\theta_1} \|f\|_{X_{\theta_1}} \qquad \forall f \in X_{\theta_1}.$$

Proof We start with the special case $n = 1$ and abbreviate $\theta = \theta_1$. Let $f \in X_\theta$. By definition of the K-functional we may choose $\widetilde{f} \in X_1$ with

$$\|f - \widetilde{f}\|_{X_0} + \varepsilon \|\widetilde{f}\|_{X_1} \le 2K(\varepsilon, f). \tag{12}$$

Using the linearity of ℓ, we can bound

$$|\ell(f)| = |\ell(f - \widetilde{f}) + \ell(\widetilde{f})| \le C_0 \|f - \widetilde{f}\|_{X_0} + C_1 \left[\varepsilon^\theta \|\widetilde{f}\|_{X_\theta} + \varepsilon \|\widetilde{f}\|_{X_1} \right]$$

$$\overset{(12)}{\le} CK(\varepsilon, f) + \varepsilon^\theta \|\widetilde{f}\|_{X_\theta} \le CK(\varepsilon, f) + \varepsilon^\theta \|f - \widetilde{f}\|_{X_\theta} + \varepsilon^\theta \|f\|_{X_\theta}.$$

We now use the bound $\|f - \widetilde{f}\|_{X_\theta} \le 3K(\varepsilon, f)$ from [7, eqn. (2.8)] and then $K(\varepsilon, f) \le C\varepsilon^\theta \|f\|_{X_\theta}$ (see, e.g., [27, Thm. 1.3.3]) to conclude

$$|\ell(f)| \le C\varepsilon^\theta \|f\|_{X_\theta}.$$

We now consider the general case $n > 1$. We choose $\widetilde{f}$ as in (12) and proceed as above to get

$$|\ell(f)| = |\ell(f - \widetilde{f}) + l(\widetilde{f})|$$

$$\le C_0 \|f - \widetilde{f}\|_{X_0} + C_1 \left[\varepsilon^{\theta_1} \|\widetilde{f}\|_{X_{\theta_1}} + \sum_{i=2}^{n} \varepsilon^{\theta_i} \|\widetilde{f}\|_{X_{\theta_i}} + \varepsilon \|\widetilde{f}\|_{X_1} \right]. \tag{13}$$

In order to treat the terms involving $\|\widetilde{f}\|_{X_{\theta_i}}$ for $i \geq 2$, we use the reiteration theorem to infer $X_{\theta_i} = (X_{\theta_1}, X_1)_{s_i,q}$, where $s_i \in (0, 1)$ is given by

$$\theta_i = \theta_1(1 - s_i) + s_i.$$

Next, the interpolation inequality $\|\widetilde{f}\|_{X_{\theta_i}} \leq C\|\widetilde{f}\|_{X_{\theta_1}}^{1-s_i}\|\widetilde{f}\|_{X_1}^{s_i}$ together with the elementary bound $ab \leq a^p/p + b^q/q$ (for $a, b > 0$, $p, q > 1$ with $1/p + 1/q = 1$) gives

$$\begin{aligned}\varepsilon^{\theta_i}\|\widetilde{f}\|_{X_{\theta_i}} &\leq C\varepsilon^{\theta_i - s_i}\|\widetilde{f}\|_{X_{\theta_1}}^{1-s_i}\varepsilon^{s_i}\|\widetilde{f}\|_{X_1}^{s_i} \leq C\left[\varepsilon^{(\theta_i - s_i)/(1-s_i)}\|\widetilde{f}\|_{X_{\theta_1}} + \varepsilon\|\widetilde{f}\|_{X_1}\right] \\ &= C\left[\varepsilon^{\theta_1}\|\widetilde{f}\|_{X_{\theta_1}} + \varepsilon\|\widetilde{f}\|_{X_1}\right].\end{aligned}$$

Inserting this result in (13), we get together with (12)

$$|\ell(f)| \leq C\left[K(\varepsilon, f) + \varepsilon^{\theta_1}\|\widetilde{f}\|_{X_{\theta_1}}\right].$$

Reasoning as in the case $n = 1$ now allows us to conclude the argument. □

2.3 *Elliptic Regularity*

Lemma 3 *Let $\Omega \subset \mathbb{R}^d$ be a bounded Lipschitz domain with a* smooth *boundary. Let $m \in \mathbb{N}$ and $k \in \mathbb{N}_0$. Then there is $C_{\Omega,m,k}$ depending only on Ω, m, k such that the following is true: If $g \in H^{-m+k}(\Omega)$ and u is the (variational) solution of the Dirichlet problem*

$$\Delta^m u = g \quad \text{in}\Omega, \qquad u = \partial_n u = \cdots \partial_n^{m-1} u = 0 \quad \text{on}\partial\Omega,$$

then $u \in H^{m+k}(\Omega)$ with the a priori bound

$$\|u\|_{H^{m+k}(\Omega)} \leq C_{\Omega,m,k}\|g\|_{H^{-m+k}(\Omega)}.$$

Proof This regularity result is a special case of a more general result for the regularity of solutions of elliptic systems, [2, 3]. Self-contained proofs of this result can also be found, for example, in [30, Sec. 20] and in [19, Chap. 2, Thm. 8.2]. □
The minimum norm extension $E^\Omega : H^m(\Omega) \to \mathrm{BL}^m(\mathbb{R}^d)$ satisfies

$$|E^\Omega f|_{H^m(\mathbb{R}^d)} \leq C_\Omega\|f\|_{H^m(\Omega)}. \tag{14}$$

However, for smooth $\partial\Omega$, it has additional mapping properties:

Corollary 2 *Let Ω be a bounded Lipschitz domain with a smooth boundary and let $\overline{\Omega}$ be contained in the (open) ball $B_R(0)$ of radius R centered at 0. For each $j \in \{0, \ldots, m\}$ there is a constant $C_{j,\Omega} > 0$ depending only on j, Ω, and R such that the following is true for the minimum norm extension $E^\Omega : H^m(\Omega) \to \mathrm{BL}^m(\mathbb{R}^d)$: It is also a bounded linear map $H^{m+j}(\Omega) \to H^{m+j}(B_R(0) \setminus \overline{\Omega})$ and, with γ_0^c denoting the trace operator for $B_R(0) \setminus \overline{\Omega}$,*

$$\|\gamma_0^c(\nabla^{m+j} E^\Omega f)\|_{L^2(\partial\Omega)} \leq C_{j,\Omega} \|f\|_{B_{2,1}^{m+j+1/2}(\Omega)}. \tag{15}$$

Proof We write $\widetilde{\Omega} := B_R(0) \setminus \overline{\Omega}$. The operator E^Ω is clearly a bounded linear map $E^\Omega : H^m(\Omega) \to H^m(\widetilde{\Omega})$. From Lemma 3, we also see that E^Ω maps $H^{2m}(\Omega)$ boundedly into $H^{2m}(\widetilde{\Omega})$: We denote by E the universal extension operator of [25, Chap. VI, 3], which we may choose such that $\operatorname{supp} Ef \subset B_R(0)$. Next, we write $E^\Omega f$ in the form $E^\Omega f = Ef + u$, where $Ef \in H^{2m}(\widetilde{\Omega})$ (since $f \in H^{2m}(\Omega)$) and u solves the differential equation

$$\Delta^m u = -\Delta^m Ef \in L^2(\widetilde{\Omega}) \quad \text{in} \widetilde{\Omega}, \qquad u = \partial_n u = \cdots = \partial_n^{m-1} u = 0 \qquad \text{on} \partial\widetilde{\Omega}.$$

Lemma 3 then gives $u \in H^{2m}(\widetilde{\Omega})$ with the a priori estimate $\|u\|_{H^{2m}(\widetilde{\Omega})} \leq C\|\Delta^m Ef\|_{L^2(\widetilde{\Omega})} \leq C\|Ef\|_{H^{2m}(\widetilde{\Omega})} \leq C\|f\|_{H^{2m}(\Omega)}$. We have thus obtained

$$\|E^\Omega f\|_{H^m(\widetilde{\Omega})} \leq C\|f\|_{H^m(\Omega)}, \qquad \|E^\Omega f\|_{H^{2m}(\widetilde{\Omega})} \leq C\|f\|_{H^{2m}(\Omega)}. \tag{16}$$

An interpolation argument then gives us

$$\|E^\Omega f\|_{B_{2,1}^{m+1/2+j}(\widetilde{\Omega})} \leq C\|f\|_{B_{2,1}^{m+j+1/2}(\Omega)}, \qquad j = 0, \ldots, m-1.$$

By the trace theorem (Lemma 1), we arrive at $\|\gamma_0^c \nabla^{j+m} E^\Omega f\|_{L^2(\partial\Omega)} \leq C\|f\|_{B_{2,1}^{m+j+1/2}(\Omega)}$ for $j = 0, \ldots, m-1$. □

2.4 PDE-Based Proof of Proposition 2

Lemma 4 *Let Ω be a Lipschitz domain. Then*

$$|E^\Omega f - If|_m \leq C_\Omega |f - If|_{H^m(\Omega)}. \tag{17}$$

Proof We exploit that $\Delta^m(E^\Omega f - If) = 0$ in Ω^c. To that end, let again E be the universal extension of operator of [25, Chap. VI, 3]. We write $E^\Omega f - If = E(f - If) + \delta$ for some $\delta \in \mathrm{BL}^m(\mathbb{R}^d)$ with $\delta|_\Omega = 0$. We get

$$|E^\Omega f - If|_m^2 = B_\Omega(f - If, f - If) + B_{\Omega^c}(E^\Omega f - If, E(f - If) + \delta) \tag{18}$$

$$= |f - If|_{H^m(\Omega)}^2 + B_{\Omega^c}(E^\Omega f - If, E(f - If)), \tag{19}$$

where, in the step from (18) to (19) we used integration by parts, the property $\Delta^m(E^\Omega f - If) = 0$ on Ω^c, and $\delta|_\Omega \equiv 0$; the integration by parts does not produce any terms "at infinity" since $C_0^\infty(\mathbb{R}^d)$ is dense in $\mathrm{BL}^m(\mathbb{R}^d)$ (in the sense described in [29, Thm. 10.40]) and thus δ can be approximated by such compactly supported functions. From (19) and the continuity of E we infer

$$|E^\Omega f - If|_m \le C_\Omega \|f - If\|_{H^m(\Omega)}. \tag{20}$$

In the estimate (20), the full norm on the right-hand side can be reduced to a seminorm with the aid of the Deny-Lions Lemma and fact that I reproduces polynomials of degree $m - 1$. Thus, (17) is proved. □

The solution If of the minimization problem (1) satisfies the orthogonality condition

$$\langle E^\Omega f - If, If \rangle_m = 0 \tag{21}$$

since $E^\Omega f - If \in \mathrm{BL}^m(\mathbb{R}^d)$ and $(E^\Omega f - If)(x_i) = f(x_i) - If(x_i) = 0$, $i = 1, \dots, N$. Therefore,

$$\begin{aligned}\langle E^\Omega f - If, E^\Omega f - If \rangle_m &= \langle E^\Omega f - If, E^\Omega f \rangle_m \\ &= B_\Omega(f - If, f) + B_{\Omega^c}(E^\Omega f - If, E^\Omega f).\end{aligned} \tag{22}$$

These last two terms are treated separately in Lemmas 5, 6. Inserting (23), (25) in (22) we get

$$|E^\Omega f - If|_{H^m(\mathbb{R}^d)}^2 \le C h^{1/2} \|f\|_{B_{2,1}^{m+1/2}(\Omega)} |f - If|_{H^m(\Omega)},$$

which readily implies (6) of Proposition 2. The bound (7) follows from (6) and an interpolation argument since the reiteration theorem asserts for $0 < \delta < 1/2$ that $H^{m+\delta}(\Omega) = (H^m(\Omega), B_{2,1}^{m+1/2}(\Omega))_{2\delta,2}$ and $|E^\Omega f - If|_{H^m(\mathbb{R}^d)} \le C\|f\|_{H^m(\Omega)}$, which follows from combining (21) and (14).

Lemma 5 *Let Ω be a Lipschitz domain. Then:*

$$|B_\Omega(f - If, f)| \le C_\Omega h^{1/2} |f - If|_{H^m(\Omega)} \|f\|_{B_{2,1}^{m+1/2}(\Omega)}. \tag{23}$$

Proof Let $\widetilde{f} \in H^{m+1}(\Omega)$. Integration by parts once gives

$$\left|B_\Omega(f - If, \widetilde{f})\right| \lesssim \tag{24}$$
$$\|\nabla^{m-1}(f - If)\|_{L^2(\partial\Omega)}\|\nabla^m\widetilde{f}\|_{L^2(\partial\Omega)} + \|\nabla^{m-1}(f - If)\|_{L^2(\Omega)}\|\nabla^{m+1}\widetilde{f}\|_{L^2(\Omega)}.$$

The multiplicative trace inequality $\|z\|^2_{L^2(\partial\Omega)} \lesssim \|z\|_{L^2(\Omega)}\|z\|_{H^1(\Omega)}$, Corollary 1 with $k = m - 1$, and the trace estimate $\|\nabla^m z\|_{L^2(\partial\Omega)} \lesssim \|z\|_{B^{m+1/2}_{2,1}(\Omega)}$ yield

$$\left|B_\Omega(f - If, \widetilde{f})\right| \lesssim$$
$$\left[\|\nabla^{m-1}(f - If)\|^{1/2}_{L^2(\Omega)}\|f - If\|^{1/2}_{H^m(\Omega)}\right]\|\nabla^m\widetilde{f}\|_{L^2(\partial\Omega)} + \|\nabla^{m-1}(f-If)\|_{L^2(\Omega)}\|\nabla^{m+1}\widetilde{f}\|_{L^2(\Omega)}$$
$$\lesssim \left[h^{1/2}|f - If|_{H^m(\Omega)}\|\nabla^m\widetilde{f}\|_{L^2(\partial\Omega)} + h|f - If|_{H^m(\Omega)}\|\nabla^{m+1}\widetilde{f}\|_{L^2(\Omega)}\right]$$
$$\lesssim \left[h^{1/2}\|\widetilde{f}\|_{B^{m+1/2}_{2,1}(\Omega)} + h\|\widetilde{f}\|_{H^{m+1}(\Omega)}\right]|f - If|_{H^m(\Omega)}.$$

We conclude that the linear functional $\widetilde{f} \mapsto B_\Omega(f - If, \widetilde{f})$ satisfies

$$|B_\Omega(f - If, \widetilde{f})| \le C|f - If|_{H^m(\Omega)}\|\widetilde{f}\|_{H^m(\Omega)},$$
$$|B_\Omega(f - If, \widetilde{f})| \le C|f - If|_{H^m(\Omega)}\left[h^{1/2}\|\widetilde{f}\|_{B^{m+1/2}_{2,1}(\Omega)} + h\|\widetilde{f}\|_{H^{m+1}(\Omega)}\right];$$

since $B^{m+1/2}_{2,1}(\Omega) = (H^m(\Omega), H^{m+1}(\Omega))_{1/2,1}$ Lemma 2 implies the estimate (23).

□

We now turn to the second part of (24). The key step is to observe that the minimum norm extension $E^\Omega f$ satisfies the homogeneous differential equation $\Delta^m E^\Omega f = 0$ in Ω^c.

Lemma 6 *Let Ω be a bounded Lipschitz domain with a sufficiently smooth boundary. Then:*

$$\left|B_{\Omega^c}(E^\Omega f - If, E^\Omega f)\right| \le C_\Omega h^{1/2}|f - If|_{H^m(\Omega)}\|f\|_{B^{m+1/2}_{2,1}(\Omega)}. \tag{25}$$

Proof Let $\widetilde{f} \in H^{2m}(\Omega)$. By Corollary 2, we have $E^\Omega\widetilde{f} \in H^{2m}(B_R(0) \cap \Omega^c)$ for every $R > 0$ sufficiently large. Furthermore, $\Delta^m E^\Omega\widetilde{f} = 0$ in Ω^c. Next, m-fold integration by parts yields

$$\left|B_{\Omega^c}(E^\Omega f - If, E^\Omega\widetilde{f})\right| \lesssim \sum_{j=1}^{m} \|\nabla^{m-j}(E^\Omega f - If)\|_{L^2(\partial\Omega)}\|\gamma_0^c\nabla^{m+j-1}E^\Omega\widetilde{f}\|_{L^2(\partial\Omega)}. \tag{26}$$

The integration by parts does not produce any terms "at infinity" since $C_0^\infty(\mathbb{R}^d)$ is dense in $\mathrm{BL}^m(\mathbb{R}^d)$ (in the sense described in [29, Thm. 10.40]) and thus $E^\Omega f - If \in \mathrm{BL}^m(\mathbb{R}^d)$ can be approximated by such compactly supported functions.

Since $\nabla^j E^\Omega f = \nabla^j f$ on $\partial\Omega$ for $j = 0, \ldots, m-1$, we use again the multiplicative trace inequality and Corollary 1 to get

$$\left|B_{\Omega^c}(E^\Omega f - If, E^\Omega \widetilde{f})\right| \leq C|f - If|_{H^m(\Omega)} \sum_{j=1}^{m} h^{-1/2+j} \|\gamma_0^c \nabla^{m+j-1} E^\Omega \widetilde{f}\|_{L^2(\partial\Omega)}$$

$$\overset{(15)}{\leq} C|f - If|_{H^m(\Omega)} \sum_{j=1}^{m} h^{-1/2+j} \|\widetilde{f}\|_{B_{2,1}^{m+j-1/2}(\Omega)}. \qquad (27)$$

We reduce the regularity requirement on $\widetilde{f}$ by applying Lemma 2 to $\widetilde{f} \mapsto B_{\Omega^c}(E^\Omega f - If, E^\Omega \widetilde{f})$: We observe that the reiteration theorem of interpolation allows us to identify

$$B_{2,1}^{m+j-1/2}(\Omega) = (H^m(\Omega), B_{2,1}^{2m-1/2}(\Omega))_{\theta_j,1}, \qquad \theta_j = \frac{j-1/2}{m-1/2};$$

hence, we get (25) from an application of Lemma 2 with $X_0 = H^m(\Omega)$, $X_1 = B_{2,1}^{2m-1/2}(\Omega)$ and $\varepsilon = h^{m-1/2}$ since we have additionally the stability bound $|B_{\Omega^c}(E^\Omega f - If, E^\Omega \widetilde{f})| \leq C|f - If|_{H^m(\Omega)} \|\widetilde{f}\|_{H^m(\Omega)}$ by Lemma 4 and (16). □

Remark 2 (Generalization to Lipschitz Domains) The proof Proposition 2 relies on three ingredients: (a) integration by parts arguments to treat B_Ω, (b) the approximation properties given in [11] of the thin plate spline interpolation operator I, and (c) regularity properties of $u := E^\Omega f$. Ingredients a) and b) are already formulated for Lipschitz domains. However, the regularity properties of $u = E^\Omega f$ are delicate in their generalization to the case of Lipschitz domains. We note that u solves in Ω^c the Dirichlet problem

$$\Delta^m u = 0 \quad \text{in} \Omega^c, \qquad \partial_n^{j-1} u|_{\partial\Omega} = \partial_n^{j-1} f|_{\partial\Omega}, \qquad j = 1, \ldots, m-1.$$

For such problems, a shift theorem by $1/2$ is shown in [23, Thm. 2] (see also [9, 28]) in the sense that for smooth f (in fact, $f \in B^{m+1/2}(\Omega)$ is sufficient), one can control $\|\nabla^j u\|_{L^2(\partial\Omega)}$ for $j = 0, \ldots, m$. This together with careful integration by parts arguments as in [23] for the treatment of B_{Ω^c} allow for an extension of the proof of Proposition 2 to Lipschitz domain and will be given in [20]. ■

3 Numerical Example

We illustrate Proposition 2 for the case $m = d = 2$, i.e., the classical thin plate splines. We employ uniformly distributed nodes on two geometries, the unit square $\Omega_1 = (0,1)^2$ and the L-shaped domain $\Omega_2 = (-1/2, 1/2)^2 \setminus [0, 1/2]^2$. As usual, we denote $r : x \mapsto \|x\|_2$. We interpolate 4 functions with different characters: the functions $r^{1.05}$ and $r^{2.76}$, which are, for any $\varepsilon > 0$, in $H^{2.05-\varepsilon}$ and $H^{3.76-\varepsilon}$, respectively, and the smooth functions $\exp(xy)$ and $F(x,y)$, where the so-called Franke function F is given by

$$\begin{aligned}F(x,y) =& 0.75\exp(-0.25((9x-2)^2 + (9y-2)^2)\\ &+ 0.75\exp(-(9x+1)^2/49 - 0.1(9y+1)^2)+\\ &0.5\exp(-0.25((9x-7)^2 + (9y-3)^2) - 0.2\exp(-(9x-4)^2 - (9y-7)^2).\end{aligned}$$

The results are presented in Fig. 1 and corroborate the assertions of Proposition 2, which read, for $m = 2$, $\|f - If\|_{L^\infty(\Omega)} \leq Ch^{1+\delta}\|f\|_{H^{2+\delta}(\Omega)}$ with $\delta \in [0, 1/2)$ and $\|f - If\|_{L^\infty(\Omega)} \leq Ch^{3/2}\|f\|_{B^{5/2}_{2,1}(\Omega)}$. These numerical results were first presented at the conference [21].

3.1 $\mathscr{H}$-Matrix Techniques for Solving the TPS Interpolation Problem

The numerical solution of the thin plate interpolation problem is numerically challenging since the system matrix is fully populated. Nevertheless, several approaches for fast solution techniques exist. For example, the matrix-vector multiplication

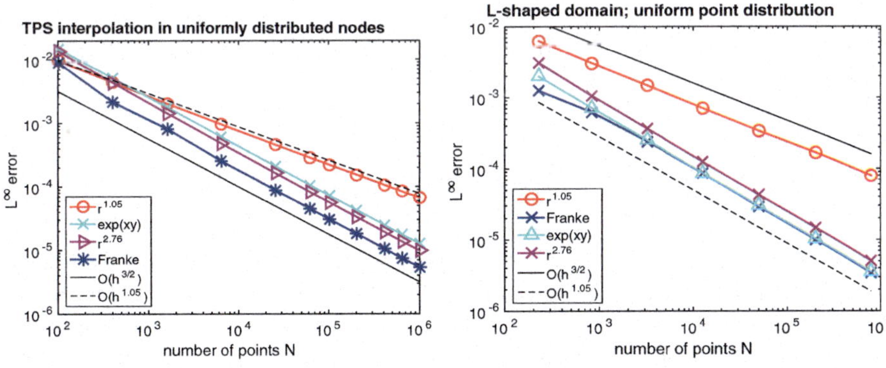

Fig. 1 Convergence of TPS interpolation. *Left*: square Ω_1. *Right*: L-shaped domain Ω_2

can be realized in log-linear complexity using techniques from fast multipole methods. This leads to efficient solution strategies based on Krylov subspace methods provided suitable preconditioners are available. We refer to [29, Sec. 15], [8, Sec. 7.3] as starting points for a literature discussion. For our calculations, we employed related techniques based on the concept of $\mathcal{H}$-matrices, [14, 15]. $\mathcal{H}$-matrices come with an (approximate) factorization that can either be used as a solver (if the approximation is sufficiently accurate) or as a preconditioner in an iterative environment. The latter use has been advocated, in a different context, for example, in [4, 13].

For the case $m = 2 = d$, the interpolation problem (3) results in a linear system of equations of the form

$$\begin{pmatrix} \mathbf{P}^\top & 0 \\ \mathbf{G} & \mathbf{P} \end{pmatrix} \begin{pmatrix} \mathbf{c} \\ \lambda \end{pmatrix} = \begin{pmatrix} 0 \\ \mathbf{f} \end{pmatrix}, \qquad \mathbf{G}_{ij} = \phi_2(\|x_i - x_j\|_2), \quad i,j = 1,\dots,N. \tag{28}$$

The matrix $\mathbf{P}^{N\times 3}$ is obtained by selecting a basis $\{b_1, b_2, b_3\}$ of $\mathbb{P}_1$ (e.g., $\{1, x, y\}$) and setting $\mathbf{P}_{i,j} = b_j(x_i)$. The vector $\mathbf{f} \in \mathbb{R}^N$ collects the values $f(x_i)$, the vector $\mathbf{c} \in \mathbb{R}^N$ the sought coefficients c_i, and the vector $\lambda \in \mathbb{R}^3$ is the Lagrange multiplier for the constrained problem (3). The function $\phi_2(z) = z^2 \log z$ is smooth for $z > 0$. Lemma 7 below shows that the function $(x, y) \mapsto \phi_2(\|x - y\|_2)$ can be approximated by a polynomial, which is in particular a *separable* function, i.e. a short sum of products of functions of x and y, only. This in turn implies that the fully populated matrix $\mathbf{G}$ can in fact be approximated as a blockwise low-rank matrix, in particular in the form of an $\mathcal{H}$-matrix, [14, 15].

By forming a Schur complement, the linear system of (28) can be transformed to SPD form. To that end, we select three points and rearrange the problem (28) as

$$\begin{pmatrix} P_1^\top & 0 & P_2^\top \\ G_{11} & P_1 & G_{12} \\ G_{21} & P_2 & G_{22} \end{pmatrix} \begin{pmatrix} \mathbf{c}_1 \\ \lambda \\ \mathbf{c}_2 \end{pmatrix} = \begin{pmatrix} 0 \\ \mathbf{f}_1 \\ \mathbf{f}_2 \end{pmatrix} \qquad G_{11} \in \mathbb{R}^{3\times 3}, \quad G_{22} \in \mathbb{R}^{(N-3)\times(N-3)},$$

where the vectors $\mathbf{c}_1, \mathbf{f}_1 \in \mathbb{R}^3$ and $\mathbf{c}_2, \mathbf{f}_2 \in \mathbb{R}^{N-3}$ result from the permutations. The Schur complement

$$\mathbf{S} := G_{22} - \begin{pmatrix} G_{21} & P_2 \end{pmatrix} \begin{pmatrix} P_1^\top & 0 \\ G_{11} & P_1 \end{pmatrix}^{-1} \begin{pmatrix} P_2^\top \\ G_{12} \end{pmatrix}$$

is SPD. We computed an (approximate) Cholesky factorization of $\mathbf{S}$ using the library HLib [5]. This factorization can be employed as a preconditioner for a CG iteration. The $\mathcal{H}$-matrix structure of $\mathbf{S}$ was ensured by so-called geometric clustering of the interpolation points. Specifically, we used this hierarchical structure to set up G_{22} by approximating its entries with the Chebyshev interpolant as described in Lemma 7. In the interest of efficiency, the thus obtained $\mathcal{H}$-matrix approximation of G_{22} was further modified by using SVD-based compression of blocks as well as coarsing of the block structure (these tools are provided by HLib). The matrix $\mathbf{S}$ is a rank-3 update of the matrix G_{22}, which can also be realized in HLib.

Lemma 7 *Let $\eta > 0$ be given. For any (closed) axiparallel boxes σ, $\tau \subset \mathbb{R}^2$ and a polynomial degree $p \in \mathbb{N}_0$ denote by $I_p^{Cheb} : C(\sigma \times \tau) \to \mathbb{Q}_p$ the tensor product Chebyshev interpolation operator associated with $\sigma \times \tau$. Then there are constants C, $b > 0$ depending only on η such that under the condition* $\max\{\operatorname{diam}(\sigma), \operatorname{diam}(\tau)\} \leq \eta \operatorname{dist}(\sigma, \tau)$ *there holds*

$$\sup_{(x,y)\in\sigma\times\tau} |\phi_2(\|x-y\|_2) - I_p^{Cheb}\phi_2(\|x-y\|)| \leq C|\operatorname{dist}(\sigma,\tau)|^2 \left(1 + |\log \operatorname{dist}(\sigma,\tau)|\right) e^{-bp}.$$

Proof The proof follows with the tool developed in [6]. Consider $Q := \prod_{i=1}^n [a_i, b_i] \subset \mathbb{R}^n$ and a function $f \in C(Q; \mathbb{C})$. Denote by Λ_p the Lebesgue constant for univariate Chebyshev interpolation (note that $\Lambda_p = O(\log p)$). Introduce, for each $x \in Q$ and each $i \in \{1, \ldots, n\}$, the univariate function $f_{x,i} : [-1, 1] \to \mathbb{C}$ by $f_{x,i}(t) := f(x_1, \ldots, x_{i-1}, (a_i + b_i)/2 + t(b_i - a_i)/2, x_{i+1}, \ldots, x_n)$. Then, standard tensor product arguments [6, Lemma 3.3] show that the tensor product Chebyshev interpolation error is bounded by

$$\|f - I_p^{Cheb} f\|_{L^\infty(Q)} \leq (1 + \Lambda_p)\Lambda_p^{n-1} \sum_{i=1}^n \sup_{x\in Q} \inf_{\pi\in\mathbb{P}_p} \|f_{x,i} - \pi\|_{L^\infty(-1,1)}.$$

The best approximation problems $\inf_{\pi\in\mathbb{P}_p} \|f_{x,i} - \pi\|_{L^\infty(-1,1)}$ in turn lead to exponentially small (in p) errors, provided the holomorphic extensions of the functions $f_{x,i}$ can be controlled. We show this for the case $f(x_1, x_2, x_3, x_4) = \phi_2(\|(x_1, x_2) - (x_3, x_4)\|_2)$ under consideration here. Note that $f_{x,1}(t) = \phi_2(\|\mathfrak{d} - t\mathfrak{p}\|_2)$, where $\mathfrak{d} = ((a_1 + b_1)/2 - x_3, x_2 - x_4)^\top$ and $\mathfrak{p} = ((a_1 - b_1)/2, 0)^\top$. Note $\|\mathfrak{d}\|_2 \leq (1 + \eta) \operatorname{dist}(\sigma, \tau)$ and $\|\mathfrak{p}\|_2 \leq 1/2 \max\{\operatorname{diam}(\sigma), \operatorname{diam}(\tau)\} \leq \eta/2 \operatorname{dist}(\sigma, \tau)$. As shown in [6, Lemma 3.6, proof of Thm. 3.13], the holomorphic extension of the function $\mathfrak{n} : t \mapsto \|\mathfrak{d} - t\mathfrak{p}\|_2$ is holomorphic on $U_r := \cup_{t\in[-1,1]} B_r(t)$ with $r = \operatorname{dist}(\sigma, \tau)/\|\mathfrak{p}\|_2 \geq 2/\eta$ and maps into the left half plane $\mathbb{C}_+ = \{z \in \mathbb{C} \mid \operatorname{Re} z > 0\}$. We note that $\sup_{z\in U_r} |\mathfrak{n}(z)| \leq \|\mathfrak{d}\|_2 + r\|\mathfrak{p}\|_2 \leq (2 + \eta) \operatorname{dist}(\sigma, \tau)$. In view of $\phi_2(z) = z^2 \log z$, we conclude $\sup_{z\in U_r} |f_{x,i}(z)| \leq C(\operatorname{dist}(\sigma, \tau))^2(1 + |\log \operatorname{dist}(\sigma, \tau)|)$ for a constant $C > 0$ that depends solely on η. We finish the proof by observing that there is $\rho > 1$ (depending only on r and thus on η) such that U_r contains the Bernstein ellipse $\mathscr{E}_\rho$ (see [6, Lemma 3.12]). A classical polynomial approximation result (see, e.g., [6, Lemma 3.11]) concludes the proof. □

3.2 Edge Effects and Concentrating Points at the Boundary

The convergence behavior of thin plate splines is limited by edge effects. Above, we mentioned that imposing certain boundary conditions on f mitigates this effect. An alternative is to suitably concentrate points near $\partial\Omega$. Without proof, we announce the following result:

Proposition 3 *Assume that the points x_i, $i = 1, \ldots, N$, satisfy for a $\delta > 0$ sufficiently small*

$$\forall x \in \Omega: \qquad \inf_{i=1,\ldots,N} \operatorname{dist}(x, x_i) \leq \delta \min\{h_{min} + \operatorname{dist}(x, \partial\Omega), h\}. \tag{29}$$

Then, for $f \in H^{m+1}(\Omega)$ there holds $|f - If|_{H^m(\Omega)} \leq C\left(h_{min}^{1/2} + h\right)|f|_{H^{m+1}(\Omega)}$.

Inserting the result of Proposition 3 in the estimates of Proposition 1 shows that a factor $h_{min}^{1/2} + h$ can be gained in the convergence estimates. Figure 2 presents numerical examples for the square Ω_1 and the functions given in Sect. 3. We selected $h_{min} = h^2$ and distributed the points so as ensure the condition

$$\forall i: \qquad \min_{j:\, j \neq i} \|x_i - x_j\|_2 \gtrsim \min\{h_{min} + \operatorname{dist}(x, \partial\Omega), h\}.$$

For the present case $d = 2$, it can then be shown that the number of points N is $O(h^{-2})$, which is also illustrated in Fig. 2.

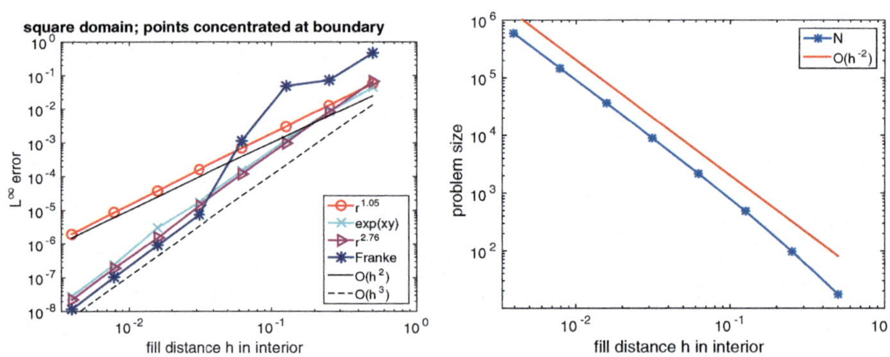

Fig. 2 Concentrating points near $\partial\Omega$. *Left*: convergence. *Right*: problem size versus fill distance in the interior

References

1. V. Adolphsson, J. Pipher, The inhomogeneous Dirichlet problem for Δ^2 in Lipschitz domains. J. Funct. Anal. **159**, 137–190 (1998)
2. S. Agmon, A. Douglis, L. Nirenberg, Estimates near the boundary for solutions of elliptic partial differential equations satisfying general boundary conditions I. Commun. Pure Appl. Math. **12**, 623–727 (1959)
3. S. Agmon, A. Douglis, L. Nirenberg, Estimates near the boundary for solutions of elliptic partial differential equations satisfying general boundary conditions II. Commun. Pure Appl. Math. **17**, 35–92 (1964)
4. M. Bebendorf, Hierarchical LU decomposition-based preconditioners for BEM. Computing **74**(3), 225–247 (2005)
5. S. Börm, L. Grasedyck, H-Lib - a library for $\mathscr{H}$- and $\mathscr{H}^2$-matrices. Available at http://www.hlib.org,1999
6. S. Börm and J.M. Melenk, Approximation of the high frequency Helmholtz kernel by nested directional interpolation: error analysis, Numer. Math. **137**, 1–24 (2017)
7. J. Bramble, R. Scott, Simultaneous approximation in scales of Banach spaces. Math. Comput. **32**, 947–954 (1978)
8. M.D. Buhmann, *Radial Basis Functions: Theory and Implementations*. Cambridge Monographs on Applied and Computational Mathematics, vol. 12 (Cambridge University Press, Cambridge, 2003)
9. B.E.J. Dahlberg, C.E. Kenig, J. Pipher, G.C. Verchota, Area integral estimates for higher order elliptic equations and systems. Ann. Inst. Fourier Grenoble **47**(5), 1425–1561 (1997)
10. J. Deny, J.L. Lions, Les espaces du type de Beppo Levi. Ann. Inst. Fourier Grenoble **5**, 305–370 (1955)
11. J. Duchon, Sur l'erreur d'interpolation des fonctions de plusieurs variables par les D^m-splines. RAIRO Anal. Numér. **12**(4), 325–334 (1978)
12. J. Duchon, Interpolation des fonctions de deux variables suivant le principe de la flexion des plaques minces. RAIRO Analyse Numér. **10**(R-3), 5–12 (1976)
13. L. Grasedyck, Adaptive recompression of $\mathscr{H}$-matrices for BEM. Computing **74**(3), 205–223 (2005)
14. W. Hackbusch, A sparse matrix arithmetic based on $\mathscr{H}$-matrices. Part I: introduction to $\mathscr{H}$-matrices. Computing **62**, 89–108 (1999)
15. W. Hackbusch, *Hierarchical Matrices: Algorithms and Analysis*. Springer Series in Computational Mathematics, vol. 49 (Springer, Heidelberg, 2015)
16. M.J. Johnson, The L_2-approximation order of surface spline interpolation. Math. Comput. **70**(234), 719–737 (2001) [electronic]
17. M. Johnson, An error analysis for radial basis function interpolation. Numer. Math. **98**, 675–694 (2004)
18. M.J. Johnson, The L_p-approximation order of surface spline interpolation for $1 \leq p \leq 2$. Constr. Approx. **20**(2), 303–324 (2004)
19. J.L. Lions, E. Magenes, *Non-Homogeneous Boundary Value Problems and Applications* (Springer, Berlin, 1972)
20. M. Löhndorf, J.M. Melenk, Approximation properties of thin plates splines on Lipschitz domain. (in preparation)
21. J.M. Melenk, on approximation in meshless methods and thin plate spline interpolation, in *Third Conference on* Meshfree Methods for PDEs. Held in Bonn, Sept. 12–15 (2005) [M. Griebel, M.A. Schweitzer, organizers]
22. J.M. Melenk, T. Gutzmer, Approximation orders for natural splines in arbitrary dimensions. Math. Comput. **70**, 699–703 (2001)
23. J. Pipher, G.C. Verchota, Dilation invariant estimates and the boundary Gårding inequality for higher order elliptic operators. Ann. Math. (2) **142**(1), 1–38 (1995)

24. R. Schaback, Improved error bounds for scattered data interpolation by radial basis functions. Math. Comput. **68**(225), 201–216 (1999)
25. E.M. Stein, *Singular Integrals and Differentiability Properties of Functions* (Princeton University Press, Princeton, 1970)
26. L. Tartar, *An Introduction to Sobolev Spaces and Interpolation Spaces*. Lecture Notes of the Unione Matematica Italiana, vol. 3 (Springer, Berlin, 2007)
27. H. Triebel, *Interpolation Theory, Function Spaces, Differential Operators*, 2nd edn. (Johann Ambrosius Barth, Heidelberg, 1995)
28. G. Verchota, The Dirichlet problem for the polyharmonic equation in Lipschitz domains. Indiana Univ. Math. J. **39**(3), 671–702 (1990)
29. H. Wendland, *Scattered Data Approximation*. Cambridge Monographs on Applied and Computational Mathematics, vol. 17 (Cambridge University Press, Cambridge, 2005)
30. J. Wloka, Partielle Differentialgleichungen. Teubner (1982)

Efficient Equilibrated Flux Reconstruction in High Order Raviart-Thomas Space for Discontinuous Galerkin Methods

Igor Mozolevski and Edson Luiz Valmorbida

Abstract We develop an efficient and computationally cheap method of equilibrated fluxes reconstruction for high-order dG solutions to elliptic problems using a specific computational basis in high order Raviart-Thomas space. The computational basis is designed in such a way that coordinates of equilibrated fluxes with respect to this basis can be easy calculated from the moments of the numerical fluxes of dG method. Some applications of this method in implementation of a posteriori error estimators for elliptic boundary value problems are considered.

1 Introduction

Equilibrated fluxes reconstruction in Raviart-Thomas space is used in finite element methods for development of fully computable (not involving unknown constants), efficient and reliable a posteriori error estimates for elliptic, convection-diffusion and parabolic problems, see e.g. [5, 7, 8, 12–14, 19, 22]. As an another important application the equilibrated velocity recuperation from a discontinuous Galerkin solution to the Darcy equation in the multiphase flow in heterogeneous porous media should be mentioned, see [11, 15]. One of the attractive properties of a posteriori error estimates, based on the equilibrated fluxes technique, is the robustness with respect to the order of polynomial approximation (cf. [5, 13]), whereas the efficiency of residual type estimates can decrease with the degree (cf. [4, 20]).

Owing to the local conservation properties, the discontinuous Galerkin (dG) finite element methods allow easy flux reconstruction in Raviart-Thomas space by locally prescribing the numerical flux moments as the degrees of freedom,

I. Mozolevski (✉)
Federal University of Santa Catarina, Florianópolis, Brazil
e-mail: igor.mozolevski@ufsc.br

E.L. Valmorbida
Federal University of Technology - Paraná, Paraná, Brazil
e-mail: edsonvalmorbida@utfpr.edu.br

M.L. Bittencourt et al. (eds.), *Spectral and High Order Methods for Partial Differential Equations ICOSAHOM 2016*, Lecture Notes in Computational Science and Engineering 119, DOI 10.1007/978-3-319-65870-4_33

cf. [9]. Such an approach offers cheap and efficient computational algorithm for implementation of flux reconstruction in lowest order Raviart-Thomas space.

Nevertheless, reconstruction of equilibrated fluxes from the prescribed moments of discrete numerical fluxes in higher order Raviart-Thomas space can be computationally involved procedure. In this article we introduce a specific modal basis in high order Raviart-Thomas space such that calculation of reconstructed flux coefficients from the prescribed moments is extremely easy owing to orthogonal properties of edge elements of the basis. Using this tool we develop an efficient and computationally cheap method of equilibrated fluxes reconstruction from high-order dG solutions to elliptic problems. To demonstrate the potential of the method we consider an application to adaptive mesh refinement, where the method is used for equilibrated fluxes calculation needed for the a posteriori error estimator.

2 Modal Basis in High Order Raviart-Thomas Space

Let Ω be a polygonal domain in $\mathbb{R}^2$. Let us denote by $H^k(\Omega)$ the Sobolev space of order $k \in \mathbb{N}_0$. The space of vector functions $\mathbf{u} \in [L^2(\Omega)]^2$ with weak divergence $\nabla \cdot \mathbf{u}$ in $L^2(\Omega)$ is denoted by $H(\text{div}, \Omega)$. The reader is referred to [1] and [3] among others, where standard properties of the Sobolev and $H(\text{div})$ spaces are exposed.

For discrete approximation of $H(\text{div}\ )$ spaces let us define in Ω a shape regular family $\mathscr{T}_h$ of triangular meshes (see e.g. [6]), where $h = \max_{T \in \mathscr{T}_h} h(T)$ denotes the mesh size and $h(T)$ is the diameter of the mesh element T. We denote the set of all mesh edges as $\mathscr{E}$ and decompose it in the set $\mathscr{E}^i$ of all interior edges (interfaces between adjacent mesh elements) and the set of all boundary faces $\mathscr{E}^\partial$. Next we define vector field $\mathbf{n}_{\mathscr{E}} : \mathscr{E} \to \mathbb{R}^2$ of edge normals, where $\mathbf{n}_{\mathscr{E}}(E) = \mathbf{n}_E$ is the fixed unit vector orthogonal to E which coincides with the external normal to $\partial\Omega$ on the boundary edges. We also denote as $\mathbf{n}_T$ the external normal to ∂T for any mesh element T. For $E \in \mathscr{E}$ we denote by $T_E = \{T \in \mathscr{T}_h \ : \ E \subset \partial T\}$ the set of all mesh elements sharing the edge E.

For any triangle $T \in \mathscr{T}_h$ the local Raviart-Thomas space is defined by

$$\mathbb{RT}^k(T) = [\mathbb{P}_k(T)]^2 + \begin{pmatrix} x \\ y \end{pmatrix} \mathbb{P}_k(T), \tag{1}$$

where $\mathbb{P}_k(T)$ denotes the space of polynomials in T of degree less than or equal to $k \in \mathbb{N}_0$. For $\mathbf{u} \in \mathbb{RT}^k(T)$ the degrees of freedom are given by

$$\int_{\partial T} (\mathbf{u} \cdot \mathbf{n}_T)p, \quad \forall p \in \mathbb{P}_k(\partial T); \tag{2}$$

$$\int_T \mathbf{u} \cdot \mathbf{q}, \quad \forall \mathbf{q} \in [\mathbb{P}_{k-1}(T)]^2 \quad \text{if } k \geq 1. \tag{3}$$

Associated with the triangulation $\mathscr{T}_h$ the global Raviart-Thomas finite element space is defined as

$$\mathbb{RT}^k(\mathscr{T}_h) = \{\mathbf{u}_h \in H(\text{div}, \Omega) \mid \mathbf{u}_h|_T \in \mathbb{RT}^k(T), \ \forall T \in \mathscr{T}_h\}. \tag{4}$$

The computational implementation of $\mathbb{RT}^0$ elements is typically included in finite element software packages and was carefully discussed in [2]. Here we aim at introducing a computational basis in high order Raviart-Thomas space $\mathbb{RT}^k(\mathscr{T}_h)$ such that the coordinates of an element in this basis can be easily calculated from its degrees of freedom (2)–(3). We start with the definition of the basis in the master element and then extend the definition to any $T \in \mathscr{T}_h$ using Piola transformation.

Let us consider the reference triangle $\hat{T}$ with vertexes

$$\hat{\mathbf{v}}_1 = (-1,-1)', \hat{\mathbf{v}}_2 = (1,-1)', \hat{\mathbf{v}}_3 = (-1,1)';$$

$$\hat{T} = \{\hat{\mathbf{v}}^1, \hat{\mathbf{v}}^2, \hat{\mathbf{v}}^3\} = \{(r,s)' \mid r,s \geq -1; \ \ r+s \leq 0\}.$$

For any $T \in \mathscr{T}_h$, $T = \{\mathbf{v}^1, \mathbf{v}^2, \mathbf{v}^3\}$, $\mathbf{v}^1 = (x_1, y_1)', \mathbf{v}^2 = (x_2, y_2)', \mathbf{v}^3 = (x_3, y_3)'$ we fix the canonical affine application $\pi_T : \hat{T} \to T$ as:

$$\pi_T(r,s) = -\frac{r+s}{2}\mathbf{v}^1 + \frac{r+1}{2}\mathbf{v}^2 + \frac{s+1}{2}\mathbf{v}^3 = (x(r,s), y(r,s))'. \tag{5}$$

The Piola transformation corresponding to π_T is defined for $\hat{\mathbf{u}} \in [L^2(\hat{T})]^2$ by

$$P_T\hat{\mathbf{u}}(x,y) = \frac{1}{|\det J_T|} J_T \hat{\mathbf{u}}(\pi_T(r,s)), \tag{6}$$

where J_T denotes the Jacobian matrix of π_T.

Lemma 1 (Properties of Piola Transformation, See e.g. [3]) *For any $\mathbf{u} \in H(\text{div}, T)$ and $v \in H^1(T)$ we have*

$$\int_T (\nabla \cdot \mathbf{u}) v = \int_{\hat{T}} (\hat{\nabla} \cdot \hat{\mathbf{u}}) \hat{v}; \quad \int_T \mathbf{u} \cdot \nabla v = \int_{\hat{T}} \hat{\mathbf{u}} \cdot \hat{\nabla} \hat{v}; \quad \int_{\partial T} \mathbf{u} \cdot \mathbf{n}_T v = \int_{\partial \hat{T}} \hat{\mathbf{u}} \cdot \hat{\mathbf{n}}_{\hat{T}} \hat{v},$$

where $\hat{\mathbf{u}} = P_T^{-1}\mathbf{u}$ and $\hat{v} = \pi_T^{-1} v$.

Now we are ready to formulate the theorem that provides a construction of the basis.

Theorem 1 *In the local Raviart-Thomas space $\mathbb{RT}^k(T)$, $k \in \mathbb{N}_0$, $T \in \mathscr{T}_h$, $T = \{\mathbf{v}^1, \mathbf{v}^2, \mathbf{v}^3\}$ there exists a basis $\{\Phi_{i,l}^{\partial T}, \Psi_m^T\}$, $i = 1,2,3$, $l = 1, \ldots, k+1$, $m = 1, \ldots, 2M$, $M = \frac{k(k+1)}{2}$ such that*

(B1) $\Phi_{i,l}^{\partial T} \cdot \mathbf{n}_{i'}^T|_{E_{i'}} = \delta_{i,i'} \mathbb{L}_l^i$, $\ i, i' \in \{1,2,3\}$, $\ l \in \{1, \ldots, k+1\}$, *where E_i denotes the triangle's edge opposite to the vertex $\mathbf{v}^i$, $\mathbf{n}_i^T$ is the unit normal to*

the edge E_i external to T and $\{\mathbb{L}_l^i\}_{l=0}^k$ is the orthonormal system of Legendre polynomials in $L^2(E_i)$.

(B2) Ψ_m^T, $m = 1, \ldots, 2M$ *form basis in* $L^2([\mathbb{P}_{k-1}(\hat{T})]^2)$ *and*

$$\Psi_m^T \cdot \mathbf{n}_i|_{E_i} = 0, \quad i \in \{1,2,3\},\ m \in \{1,\ldots,2M\};$$

Proof Following [16] let us consider in the master element $\hat{T}$ vector functions

$$\mathbf{e}_1(r,s) = \frac{1}{2}\begin{pmatrix} r+1 \\ s+1 \end{pmatrix}, \quad \mathbf{e}_2(r,s) = \frac{1}{2}\begin{pmatrix} r-1 \\ s+1 \end{pmatrix}, \quad \mathbf{e}_3(r,s) = \frac{1}{2}\begin{pmatrix} r+1 \\ s-1 \end{pmatrix}, \tag{7}$$

$$\mathbf{t}_1(r,s) = \frac{s+1}{2}\begin{pmatrix} r+1 \\ s-1 \end{pmatrix}, \quad \mathbf{t}_2(r,s) = \frac{r+1}{2}\begin{pmatrix} r-1 \\ s+1 \end{pmatrix}. \tag{8}$$

We define

$$\hat{\Phi}_{i,l}(r,s) = \mathbb{L}_{l-1}(s)\mathbf{e}_i(r,s),\ i = 1,2, \quad \hat{\Phi}_{3l}(r,s) = \mathbb{L}_{l-1}(r)\mathbf{e}_3(r,s),\ l \in \{1,\ldots,k+1\},$$

where $\{\mathbb{L}_n\}_{n=0}^k$ are (normalized) Legendre polynomials that form an orthonormal system in $L^2([-1,1])$. Since $\hat{\Phi}_{i,l}$ satisfies the property

$$\mathbf{e}_1 \cdot \mathbf{n}_j|_{\hat{E}_j} = \frac{\sqrt{2}}{2}\delta_{1j}, \quad \mathbf{e}_i \cdot \mathbf{n}_j|_{\hat{E}_j} = \delta_{ij}, \quad i \in \{2,3\}, j \in \{1,2,3\}$$

we obtain

$$\hat{\Phi}_{i,l'}^{\partial\hat{T}} \cdot \mathbf{n}_{i'}^{\hat{T}}|_{\hat{E}_{i'}} = \delta_{i,i'}\mathbb{L}_l^i.$$

Note that

$$\int_{\hat{E}_i} (\hat{\Phi}_{i,l'}^{\partial\hat{T}} \cdot \hat{\mathbf{n}}_i)(\hat{\Phi}_{i,l}^{\partial\hat{T}} \cdot \hat{\mathbf{n}}_i) = \delta_{ll'}, \quad l,l' \in \{1,\ldots,k+1\}, i \in \{1,2,3\}. \tag{9}$$

Next we define $\hat{\Psi}_m = \hat{p}_m(r,s)\mathbf{t}_1(r,s)$, $m = 1,\ldots,M$ and $\hat{\Psi}_m = \hat{p}_{m-M}(r,s)\mathbf{t}_2(r,s)$, $m = M+1,\ldots,2M$, where polynomials $\hat{p}_m$ form the orthonormal Dubiner basis in $\mathbb{P}_{k-1}(\hat{T})$, cf. [10]. Since

$$\mathbf{t}_j(r,s) \cdot \mathbf{n}_i|_{\hat{E}_i} = 0,\ j \in \{1,2\}, i \in \{1,2,3\},$$

we have $\hat{\Psi}_m \cdot \mathbf{n}_i|_{\hat{E}_i} = 0$. Using the Piola transformation the respective basis in $\mathbb{RT}^k(T), T \in \mathscr{T}_h$ is defined as

$$\mathbf{\Phi}_{i,l}^{\partial T} = P \circ \Phi_{il} \circ \pi_T^{-1},\ \Psi_m^T = P \circ \hat{\Psi}_m \circ \pi_T^{-1}$$

and the required properties (B1)–(B2) follow directly from Lemma 1. □

It should be noted that the nodal computational basis with Lagrangian property on the mesh edges was introduced in [16]. Nevertheless, such basis functions do not have the orthogonal property necessary for efficient recuperation of the basis coefficients from prescribed degrees of freedom of a finite element in $\mathbb{RT}$ space.

Next we will demonstrate how to recuperate the coefficients with respect to this basis from the degrees of freedom of a finite element in $\mathbb{RT}^k$ space.

Let us consider in $[\mathbb{P}_{k-1}(T)]^2$ a basis $\{\mathbf{P}_m^T\}_{m=1}^{2M}$,

$$\mathbf{P}_m^T = \begin{pmatrix} p_m \\ 0 \end{pmatrix}, \ m = 1, \ldots, M; \quad \mathbf{P}_m^T = \begin{pmatrix} 0 \\ p_{M-m} \end{pmatrix} \ j = M+1, \ldots, 2M;$$

where polynomials p_m form the Dubiner basis in $\mathbb{P}_{k-1}(T)$.

Lemma 2 *Assume that for* $\mathbf{u}_h \in \mathbb{RT}^k(\mathscr{T}_h)$

$$\mathbf{u}_T = \mathbf{u}_h|_T = \sum_{i',l'} c_{i',l'}^{\partial T} \Phi_{i',l'}^{\partial T} + \sum_{m'} c_{m'}^T \Psi_{m'}^T \tag{10}$$

be local representation with respect to the basis $\{\Phi_{i,l}^{\partial T}, \Psi_m^T\}$ *in* $\mathbb{RT}^k(T)$.

Let

$$\mu_{i,l}^{\partial T}(\mathbf{u}_T) = \int_{E_i} (\mathbf{u}_T \cdot \mathbf{n}_i) \mathbb{L}_l^i, \quad i \in \{1,2,3\}, l \in 1, \ldots, k+1; \tag{11}$$

$$\mu_m^T(\mathbf{u}_T) = \int_T \mathbf{u}_T \cdot \mathbf{P}_m, \quad m \in \{1, \ldots, 2M\}, \tag{12}$$

be the degrees of freedom of $\mathbf{u}_T$. *Then*

$$c_{i,l}^{\partial T} = \mu_{i,l}^{\partial T}(\mathbf{u}_T), \ i \in \{1,2,3\}, l \in 1, \ldots, k+1 \text{ and } \quad \mathbf{c}^T = G_T^{-1} \mathbf{F}^T, \tag{13}$$

where $\mathbf{c}^T = (c_1^T, \ldots, c_{2M}^T)'$,

$$G_T = \left[\int_T \Psi_i^T \cdot \mathbf{P}_j^T \right]_{2M \times 2M}, \quad \mathbf{F}_T = \left[\mu_m^T(\mathbf{u}) - \sum_{i,l} \mu_{i,l}^{\partial T} \int_T \Phi_{i,l}^{\partial T} \cdot \mathbf{P}_m^T \right]_{2M \times 1}.$$

Proof From the edge moments (11) of (10) we have:

$$\begin{aligned}\mu_{i,l}^{\partial T}(\mathbf{u}_T) &= \sum_{i',l'} c_{i',l'}^{\partial T} \int_{E_i} (\Phi_{i',l'}^{\partial T} \cdot \mathbf{n}_i) \mathbb{L}_l^i + \sum_{m'} c_{m'}^T \int_{E_i} (\Psi_{m'}^T \cdot \mathbf{n}_i) \mathbb{L}_l^i \\ &= \sum_{l'} c_{i,l'}^{\partial T} \int_{E_i} \mathbb{L}_{l'}^i \mathbb{L}_l^i = c_{i,l}^{\partial T}\end{aligned}$$

owing to (B1), (B2) and orthogonality of the Legendre polynomials (9). Similarly from the element's moments (12) we obtain:

$$\mu_m^T(\mathbf{u}_T) = \sum_{i',l'} c_{i',l'}^{\partial T} \int_T \Phi_{i',l'}^{\partial T} \cdot \mathbf{P}_m^T + \sum_{m'} c_{m'}^T \int_T \Psi_{m'}^T \cdot \mathbf{P}_{m'}^T,$$

that is $G_T \mathbf{c}^T = \mathbf{F}_T$. □

Note 1 Let us note that Lemma 2 provides extremely cheap method for the flux recuperation from the moments: in each element of the mesh we only need to solve a small linear system.

Note 2 Since $\int_T \Psi_{m'}^T \cdot \Psi_m^T = |J_T| \int_{\hat{T}} P_T \Psi_{m'}^T \cdot P_T \Psi_m^T = \int_{\hat{T}} (J_T J_T')/|J_T| \hat{\Psi}_{m'} \cdot \hat{\Psi}_m^T$, we immediately obtain $c_m^T = \mu_m^T(\mathbf{u}_T) - \sum_{i,l} \mu_{i,l}^{\partial T}(\mathbf{u}_T) \int_T \Phi_{i,l}^{\partial T} \cdot \Psi_m^T, \quad m \in 1, \ldots, 2M$ for triangles where $(J_T J_T')/|J_T| = Id$. This situation occurs for rectangular equilateral elements for example, so for such structured triangular meshes the flux reconstruction can be obtained directly from the moments and does not require a solution of the local systems.

3 Equilibrated Flux Reconstruction for Discontinuous Galerkin Method

Let us present an application of the computational basis introduced in previous section to equilibrated fluxes reconstruction from discrete gradient of discontinuous Galerkin approximation to a solution of elliptic boundary value problem.

We consider in Ω the model problem:

$$\begin{aligned} -\nabla \cdot (D \nabla u) &= f \quad \text{in } \Omega, \\ u &= g \quad \text{on } \partial\Omega. \end{aligned} \tag{14}$$

Here the diffusion coefficient $D > 0$ is supposed to be constant in Ω, $f \in L^2(\Omega)$ and $g \in H^{3/2}(\partial\Omega)$.

For a shape regular family $\mathcal{T}_h$ of triangular meshes in Ω we introduce the (discontinuous) finite element spaces $\mathcal{V}_h^k$ as:

$$\mathcal{V}_h^k := \{v_h \in L^2(\Omega) : \ v_h|_T \in \mathbb{P}_k(T), \ \forall T \in \mathcal{T}_h\}.$$

Symmetric version of the interior penalty dG method is formulated as:
find $u_h \in \mathcal{V}_h^k$ such that

$$B_h(u_h, v_h) = F(v_h), \ \forall v_h \in \mathcal{V}_h^k, \tag{15}$$

where

$$B_h(u_h, v_h) = \int_{\mathscr{T}_h} D\nabla_h u_h \cdot \nabla_h v_h - \int_{\mathscr{E}} \{\{\mathbf{n}_{\mathscr{E}} \cdot D\nabla_h u_h\}\}[\![v_h]\!]$$

$$+ \int_{\mathscr{E}} \left(-\{\{\mathbf{n}_{\mathscr{E}} \cdot D\nabla_h v_h\}\} + \gamma_{\mathscr{E}}[\![v_h]\!]\right)[\![u_h]\!],$$

$$F(v_h) = \int_{\Omega} f v_h + \int_{\mathscr{E}^{\partial}} \left(-\{\{\mathbf{n} \cdot D\nabla_h v_h\}\} + \gamma_{\mathscr{E}}[\![v_h]\!]\right) g.$$

Here we are using standard definition (see e.g. [9]) for discrete gradient, mean value and jump at the edges; the penalty parameter $\gamma_{\mathscr{E}}|_E = 2.5D(k+1)^2 h_E^{-1}$ is considered to be reasonable [17] for the stabilization. The number of unknowns (degrees of freedom) of linear system (15) is equal to

$$N_{DOF} = \dim V_h = N(\mathscr{T}_h) \cdot dim\mathbb{P}_k = N(\mathscr{T}_h)\frac{(k+1)(k+2)}{2},$$

where $N(\mathscr{T}_h)$ denotes the number of elements of $\mathscr{T}_h$.

For $v \in H^1(\mathscr{T}_h) + \mathscr{V}_h^k$ we define the energy norm associated with the dG method by

$$|||v_h|||_{dG} = \left(\|D^{\frac{1}{2}} \nabla_h v_h\|^2_{L^2(\Omega)} + \int_{\mathscr{E}} \gamma_{\mathscr{E}}[\![v_h]\!]^2 \right)^{\frac{1}{2}}. \tag{16}$$

For the energy norm of the error in dG approximation the following a priori estimate holds true, see e.g. [18].

Theorem 2 *Let $u \in H^{k+1}(\Omega)$ be a weak solution to* (14) *and $u_h \in \mathscr{V}_h^k$ be the dG finite element approximation of u. Then the estimate*

$$|||u - u_h|||_{dG} \leq Ch^k \|u\|_{H^{k+1}(\Omega)} \tag{17}$$

holds with a constant $C > 0$ independent of h.

We call the weak solution $u \in H^1(\Omega)$ the potential and the vector-valued function $\sigma(u) = -D\nabla u \in H(\mathrm{div}\,, \Omega)$ the flux. Similarly, we call the dG solution $u_h \in \mathscr{V}_h^k$ the discrete potential and define the discrete flux as $\sigma_h(u) = -D\nabla_h u_h$.

Definition 1 A vector field $\mathbf{s}_h \in H(\mathrm{div}\,, \Omega)$ is called equilibrated up to order $l \in \mathbb{N}_0$ if $\nabla \cdot \mathbf{s}_h - \pi_h^l(f) = 0$, where $\pi_h^l : L^2(\Omega) \to \mathscr{V}_h^l$ denotes the orthogonal projection operator.

Following [14], let us consider the flux $\mathbf{t}_h^{k-1}(u_h) \in \mathbb{RT}^{k-1}(\mathcal{T}_h)$ with the degrees of freedom of (11) and (12) locally prescribed by :

$$k \geq 1: \ \mu_{i,l}^{\partial T}(\mathbf{t}_h^{k-1}(u_h)) = \int_{E_i} \left(- \{\{\mathbf{n} \cdot D\nabla_h u_h\}\} + \gamma_E [\![u_h]\!]_g \right) \mathbb{L}_l^i,$$

$$E_i \in \partial T, i \in \{1,2,3\}, l \in 1, \ldots, k; \tag{18}$$

$$k \geq 2: \ \mu_m^T(\mathbf{t}_h^{k-1}(u_h)) = -\int_T D\nabla_h u_h \cdot \mathbf{P}_m \tag{19}$$

$$+ \sum_i \int_{E_i} \chi_e(E_i) D(\mathbf{P}_m \cdot \mathbf{n}_i) [\![u_h]\!]_g,$$

$$m = 1, \ldots, 2M,$$

where

$$[\![u_h]\!]_g = \begin{cases} [\![u_h]\!]|_E, & \text{for } E \in \mathcal{E}^i, \\ u_h|_E - g, & \text{for } E \in \mathcal{E}^D. \end{cases}$$

Note that this definition is correct since the numerical fluxes of dG method are uniquely defined at the edges of the mesh. Moreover, we will refer to the following lemma.

Lemma 3 *Let u be a weak solution to* (14) *and $u_h \in \mathcal{V}_h^k$ be the dG finite element approximation of u. The flux $\mathbf{t}_h^{k-1}(u_h)$, defined by* (18)–(19)*, is equilibrated up to order $k-1$ and there exists a constant $C > 0$, independent of h, such that*

$$\|D^{\frac{1}{2}} \nabla u - D^{-\frac{1}{2}} \mathbf{t}_h^{k-1}(u_h)\|_{L^2(\Omega)} \leq C |\!|\!| u - u_h |\!|\!|_{dG}. \tag{20}$$

See [14] for the proof and more details.

For computational reconstruction of $\mathbf{t}_h^{k-1}(u_h)$ from the prescribed moments (18) and (19) we use the algorithm presented in Lemma 2.

4 Numerical Examples

To demonstrate a potential of the suggested algorithm we consider an application to adaptive mesh refinement in dG approximation of elliptic boundary value problems, where the equilibrated fluxes are used in a posteriori error estimator in the energy norm (16).

We consider the error estimator introduced in [13]

$$\eta^2(u_h) = \sum_{T \in \mathcal{T}_h} \eta^2(T) = \sum_{T \in \mathcal{T}_h} ((\eta_{\mathcal{O}}(T) + \eta_{\nabla}(T))^2 + \eta_{\mathcal{H}}^2(T))$$

where $\eta_{\mathscr{O}}(T) = \|f - \nabla \cdot \mathbf{t}_h(u_h)\|_{L^2(T)}$ is the oscillations term, $\eta_{\nabla}(T) = \|D^{-\frac{1}{2}}(\sigma_h(u_h) - \mathbf{t}_h(u_h))\|_{L^2(T)}$ measures the deviation of the discrete flux $\sigma_h(u_h)$ from $\mathbf{H}(div, \Omega)$ and $\eta_{\mathscr{H}}(T) = \|D^{\frac{1}{2}}(\nabla_h(u_h) - \nabla_h(u_h^O))\|_{L^2(T)}$ measures the deviation of u_h from $H^1(\Omega)$. Here $u_h^O \in H^1(\Omega)$ is Oswald interpolator of u_h and the equilibrated flux $\mathbf{t}_h^{k-1}(u_h)$ is reconstructed by prescription from (18)–(19) in the computational basis $\{\Phi_{i,l}^{\partial T}, \Psi_m^T\}$ in $\mathbb{RT}^k(T)$. This type of estimator has proven to be reliable, efficient and robust with respect to polynomial order of approximation space, see [5, 13]. The quality of the error estimator η is assessed in terms of the effectivity index $\mathscr{I}_\eta = \frac{\eta(u_h)}{\|D^{\frac{1}{2}}(\nabla u - \nabla_h u_h)\|_{L^2(\Omega)}}$ evaluated on sequences of uniformly and adaptively refined meshes.

4.1 Test Case 1: Uniform Mesh Refinement

Firstly we confirm the order of approximation of the exact flux by the reconstructed flux for a smooth solution to elliptic problem. So let us consider the model problem (14) in $\Omega = (0, 1)^2$ with $D = 1$, the right-hand side and the homogeneous Dirichlet boundary condition corresponding to the exact solution $u(x, y) = \sin(\pi x)\sin(\pi y)$. The model problem was solved numerically using the dG method of order $k = 1, 2, 3, 4$ on a sequence of meshes obtained by successive uniform bisection from the initial unstructured mesh of 48 elements. We have used five mesh refinement levels $N_r = 1, 2, 3, 4, 5$ with 192, 768, 3072, 12,288, 49,152 elements. Let us denote by $e(u_h) = \|D^{-\frac{1}{2}}(\sigma(u) - \sigma_h(u_h))\|_{L^2(\Omega)}$ the error in dG approximation of the exact flux by the discrete flux, let $e(\mathbf{t}_h) = \|D^{-\frac{1}{2}}(\sigma(u) - \mathbf{t}_h(u_h))\|_{L^2(\Omega}$ denotes the error in approximation of the exact flux by the reconstructed Raviart-Thomas flux and let $e(\nabla \cdot \mathbf{t}_h) = \|f - \nabla \cdot \mathbf{t}_h(u_h)\|_{L^2(\Omega)}$ be the error in equilibration of the reconstructed flux. In Table 1 we show the errors and the convergence rates of the dG method and of the equilibrated flux, reconstructed from the discrete solution using the suggested computational basis. We observe that the dG method accurately approximate the exact solution to the problem and exhibits the optimal order of convergence k predicted in the Theorem 2. The error in approximation of the exact flux by reconstructed equilibrated flux is also of order k, which is the optimal order of approximation of the exact flux by dG method and the optimal order of the projection of the exact flux on the Raviart-Thomas space $\mathbb{RT}^{k-1}$. Finally, the order of convergence of $\nabla \cdot \mathbf{t}_h \to f$ is $k+1$, that is the flux $\mathbf{t}_h$ is equilibrated up to order $k-1$.

4.2 Test Case 2: Adaptive Mesh Refinement

Inspired by propagation of saturation front in a two-phase flow in porous media, we consider here the model problem (14) in $\Omega = (0, 1)^2$ with the solution $u(x, y) = x(x-1)y(y-1)\arctan\left(60\sqrt{(x - 5/4)^2 + (y + 1/4)^2)} - 1\right)$, that exhibits

Table 1 The errors $e(u_h)$, $e(\mathbf{t}_h)$, $\mathbf{t}_h(u_h)$ and convergence rates $o(e(u_h)$, $o(e(\mathbf{t}_h))$, $o(\mathbf{t}_h(u_h))$ for the approximation orders $k = 1, 2, 3, 4$ calculated on the refined meshes for refinement levels $N_r = 1, 2, 3, 4, 5$

k	N_r	$e(u_h)$	$o(e(u_h))$	$e(\mathbf{t}_h)$	$o(e(\mathbf{t}_h))$	$e(\nabla \cdot \mathbf{t}_h)$	$o(e(\nabla \cdot \mathbf{t}_h))$
1	1	2.779×10^{-1}	2.951	3.664×10^{-1}	2.605	6.013×10^{-2}	5.544
	2	1.394×10^{-1}	0.995	1.879×10^{-1}	0.964	1.521×10^{-2}	1.983
	3	6.984×10^{-2}	0.997	9.425×10^{-2}	0.995	3.813×10^{-3}	1.996
	4	3.495×10^{-2}	0.999	4.711×10^{-2}	1.000	9.540×10^{-4}	1.999
	5	1.749×10^{-2}	0.999	2.354×10^{-2}	1.001	2.385×10^{-4}	2.000
2	1	2.480×10^{-2}	5.058	3.184×10^{-2}	4.275	3.725×10^{-3}	8.005
	2	6.227×10^{-3}	1.994	7.819×10^{-3}	2.026	4.685×10^{-4}	2.991
	3	1.561×10^{-3}	1.996	1.944×10^{-3}	2.008	5.864×10^{-5}	2.998
	4	3.910×10^{-4}	1.998	4.852×10^{-4}	2.003	7.332×10^{-6}	3.000
	5	9.783×10^{-5}	1.999	1.212×10^{-4}	2.001	9.165×10^{-7}	3.000
3	1	1.011×10^{-3}	7.457	1.294×10^{-3}	6.806	2.743×10^{-4}	9.262
	2	1.294×10^{-4}	2.965	1.648×10^{-4}	2.973	1.697×10^{-5}	4.015
	3	1.623×10^{-5}	2.995	2.050×10^{-5}	3.007	1.059×10^{-6}	4.003
	4	2.029×10^{-6}	3.000	2.547×10^{-6}	3.008	6.613×10^{-8}	4.001
	5	2.536×10^{-7}	3.001	3.172×10^{-7}	3.005	4.133×10^{-9}	4.000
4	1	6.619×10^{-5}	8.177	8.910×10^{-5}	7.072	8.448×10^{-6}	11.581
	2	4.042×10^{-6}	4.033	5.414×10^{-6}	4.041	2.801×10^{-7}	4.915
	3	2.514×10^{-7}	4.007	3.371×10^{-7}	4.005	8.856×10^{-9}	4.983
	4	1.570×10^{-8}	4.001	2.109×10^{-8}	3.999	2.775×10^{-10}	4.996
	5	9.812×10^{-10}	4.000	1.319×10^{-9}	3.999	1.161×10^{-11}	4.579

a steep front in the interior of the domain. This example is commonly used for testing adaptive refinement algorithms for elliptic equations, c.f. [21]. Here we also consider $D = 1$, and the right-hand side and the homogeneous Dirichlet boundary condition corresponding to the exact solution.

Meshes are adapted from the same initial mesh, using a refinement strategy based on the method proposed by Dörfler, whereby the elements in a minimal set $\mathcal{M} \subset \mathcal{T}_h$, such that $\sum_{T \in \mathcal{M}} \eta(T) \geq \theta \sum_{T \in \mathcal{T}_h} \eta(T)$, are refined. Elements are refined using the longest edge bisection technique and additional refinements of the mesh are considered in order to eliminate hanging nodes.

Figure 1 displays in the first column the energy norm of the error and the convergence order calculated for uniform and adaptive mesh refinements with $\theta = 0.25$, $\theta = 0.5$ and $\theta = 0.75$ in the Dörfler marking as a function of DOF (degrees of freedom of dG method), on a logarithmic scale for approximation order $k = 1$ to $k = 4$. The second column shows the respective effectivity indices and the third column presents the adaptively refined meshes corresponding to the error ≈ 0.01 for $\theta = 0.25$. We can see that for given order of dG method the energy norm of the errors are very close for all values of the parameter in the Dörfler marking and asymptotically exhibit the optimal convergence rates. The effectivity indices

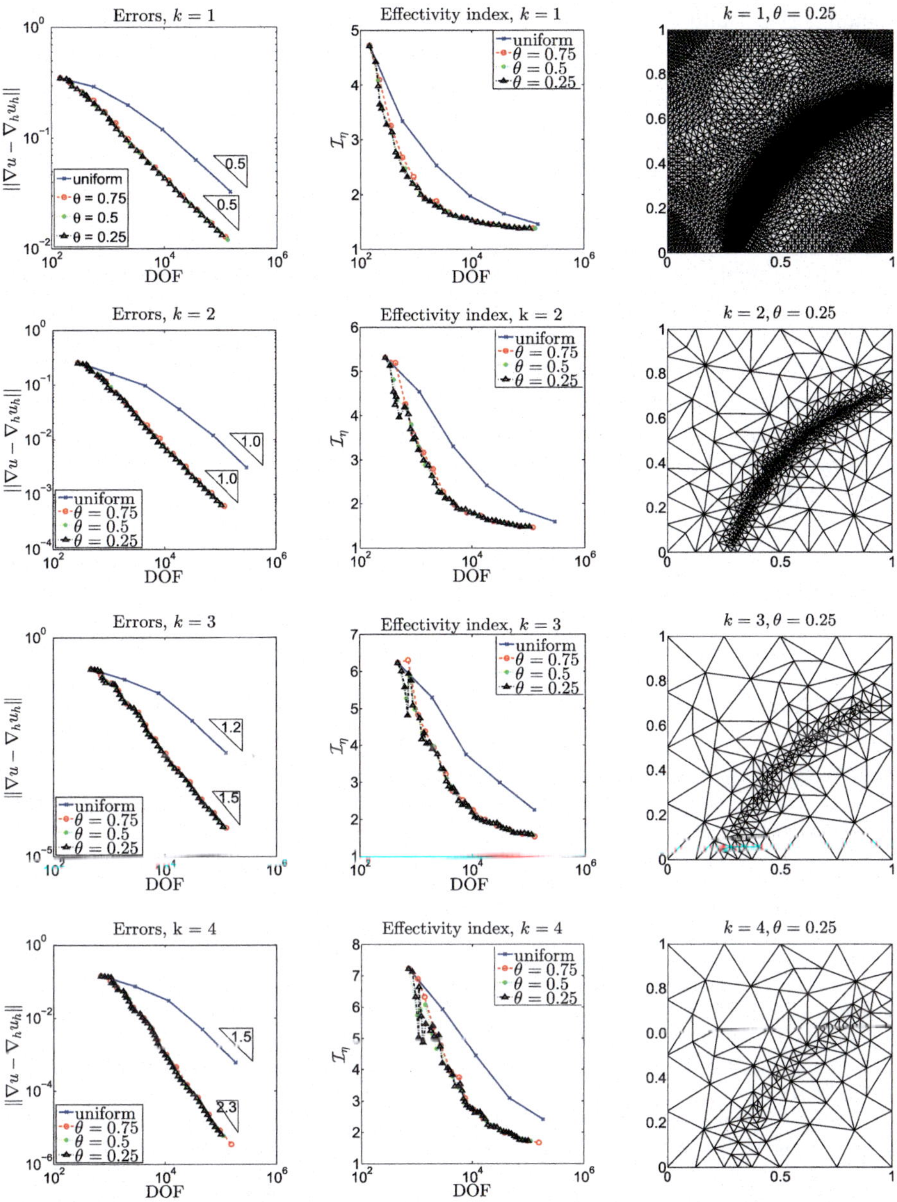

Fig. 1 The energy norm of error, the convergence order (*left column*) and the effectivity index (*middle column*) as a function of the degrees of freedom (DOF) on a logarithmic scale for various θ in the Dörfler marking and for various orders of dG method. *Right column*: adaptively refined meshes, corresponding to the error ≈ 0.01 in the energy norm, with DOF number $N_{DOF} = 104991$ for $k = 1$, $N_{DOF} = 7692$ for $k = 2$, $N_{DOF} = 4270$ for $k = 3$ and $N_{DOF} = 3930$ for $k = 4$

remain above 1 and are asymptotically close to 1 even the order of approximation increases. We observe also that the number of DOF necessary to achieve the same global approximation error decreases with increasing polynomial degree k.

5 Conclusions

A specific modal computational basis, in which the coordinates of the equilibrated fluxes can be easily calculated from the numerical fluxes of dG method, is designed for high order Raviart-Thomas space. Optimal convergence of equilibrated fluxes and the robustness of the reconstruction method with respect to the order of dG method are confirmed numerically. Adaptive mesh refinement, guided by the equilibrated error estimator calculated in this basis, exhibits robust effectivity index and provides final meshes with less DOF for the same error tolerance for higher orders of the dG method.

Acknowledgements The authors gratefully acknowledge the support by CNPq, Brazil, grant 477935/2013-3.

References

1. R.A. Adams, *Sobolev Spaces* (Academic, New York, 1975)
2. C. Bahriawati, C. Carstensen, Three matlab implementations of the lowest-order Raviart-Thomas MFEM with a posteriori error control. Comput. Methods Appl. Math. **5**(4), 333–361 (2005)
3. D. Boffi, F. Brezzi, M. Fortin, *Mixed Finite Element Methods and Applications*. Springer Series in Computational Mathematics, vol. 44 (Springer, Heidelberg, 2013)
4. D. Braess, V. Pillwein, J. Schöberl, Equilibrated residual error estimates are p-robust. Comput. Methods Appl. Mech. Eng. **198**(13–14), 1189–1197 (2009)
5. D. Braess, T. Fraunholz, R.H.W. Hoppe, An equilibrated a posteriori error estimator for the interior penalty discontinuous Galerkin method. SIAM J. Numer. Anal. **52**(4), 2121–2136 (2014)
6. S.C. Brenner, L.R. Scott, *The Mathematical Theory of Finite Element Methods* (Springer, Berlin, 1994)
7. E. Creusé, S. Nicaise, A posteriori error estimator based on gradient recovery by averaging for discontinuous Galerkin methods. J. Comput. Appl. Math. **234**(10), 2903–2915 (2010)
8. E. Creusé, S. Nicaise, A posteriori error estimator based on gradient recovery by averaging for convection-diffusion-reaction problems approximated by discontinuous Galerkin methods. IMA J. Numer. Anal. **33**(1), 212–241 (2013)
9. D.A. Di Pietro, A. Ern, *Mathematical Aspects of Discontinuous Galerkin Methods*. Mathématiques & Applications, vol. 69 (Springer, Berlin, 2011)
10. M. Dubiner, Spectral methods on triangles and other domains. J. Sci. Comput. **6**(4), 345–390 (1991)
11. A. Erm, I. Mozolevski, Discontinuous Galerkin method for two-component liquid gas porous media flows. Comput. Geosci. **16**, 677–690 (2012)

12. A. Ern, M. Vohralík, Flux reconstruction and a posteriori error estimation for discontinuous Galerkin methods on general nonmatching grids. C. R. Math. Acad. Sci. Paris **347**(7–8), 441–444 (2009)
13. A. Ern, M. Vohralík, Polynomial-degree-robust a posteriori estimates in a unified setting for conforming, nonconforming, discontinuous Galerkin, and mixed discretizations. SIAM J. Numer. Anal. **53**(2), 1058–1081 (2015)
14. A. Ern, S. Nicaise, M. Vohralík, An accurate $\mathbf{H}(div)$ flux reconstruction for discontinuous Galerkin approximations of elliptic problems. C. R. Math. Acad. Sci. Paris **345**(12), 709–712 (2007)
15. A. Ern, I. Mozolevski, L. Schuh, Accurate velocity reconstruction for discontinuous Galerkin approximations of two-phase porous media flows. C. R. Math. Acad. Sci. Paris **347**(9–10), 551–554 (2009)
16. V. Ervin, Computational bases for RT_k and BDM_k on triangles. Comput. Math. Appl. **64**(8), 2765–2774 (2012)
17. J.S. Hesthaven, T. Warburton, *Nodal Discontinuous Galerkin Methods*. Texts in Applied Mathematics, vol. 54 (Springer, New York, 2008). Algorithms, analysis, and applications
18. P. Houston, C. Schwab, E. Süli, Discontinuous hp-finite element methods for advection-diffusion problems. SIAM J. Numer. Anal. **39**(6), 2133–2163 (2002)
19. K.Y. Kim, A posteriori error estimators for locally conservative methods of nonlinear elliptic problems. Appl. Numer. Math. **57**(9), 1065–1080 (2007)
20. J.M. Melenk, B.I. Wohlmuth, On residual-based a posteriori error estimation in *hp*-FEM. Adv. Comput. Math. **15**(1–4), 311–331 (2002). A posteriori error estimation and adaptive computational methods
21. W.F. Mitchell, A collection of 2D elliptic problems for testing adaptive grid refinement algorithms. Appl. Math. Comput. **220**, 350–364 (2013)
22. I. Mozolevski, S. Prudhomme, Goal-oriented error estimation based on equilibrated-flux reconstruction for finite element approximations of elliptic problems. Comput. Methods Appl. Mech. Eng. **288**, 127–145 (2015)

Numerical Experiments on a Nonlinear Wave Equation with Singular Solutions

Thomas Hagstrom

Abstract We use a Fourier pseudospectral method to compute solutions to the Cauchy problem for a nonlinear variational wave equation originally proposed as a model for the dynamics of nematic liquid crystals. The solution is known to form singularities in finite time; in particular space and time derivatives become unbounded. Beyond the singularity time, both conservative and dissipative Hölder continuous weak solutions exist. We present results with energy-conserving discretizations as well as with a vanishing viscosity sequence, noting marked differences between the computed solutions after the solution loses regularity.

1 Introduction

We consider the second order nonlinear wave equation

$$\frac{\partial^2 u}{\partial t^2} = c(u)\frac{\partial}{\partial x}\left(c(u)\frac{\partial u}{\partial x}\right), \tag{1}$$

supplemented by initial data $u(x, 0)$, $\frac{\partial u}{\partial t}(x, 0)$. Our experiments below deal with the specific case

$$c^2(u) = \alpha \cos^2 u + \beta \sin^2 u, \tag{2}$$

which has been proposed as a simplified model for liquid crystals [15]; in this case $\mathbf{e}_x \cos u + \mathbf{e}_y \sin u$ represents the so-called director field, with $\mathbf{e}_x$ the unit vector in the direction of wave propagation and $\mathbf{e}_y$ the unit vector in an orthogonal direction, the orientations assumed to be confined to a plane. For smooth solutions we have

T. Hagstrom (✉)
Southern Methodist University, Dallas, TX, USA
e-mail: thagstrom@smu.edu

M.L. Bittencourt et al. (eds.), *Spectral and High Order Methods for Partial Differential Equations ICOSAHOM 2016*, Lecture Notes in Computational Science and Engineering 119, DOI 10.1007/978-3-319-65870-4_34

that the energy:

$$E(t) = \frac{1}{2} \int \left[\left(\frac{\partial u}{\partial t} \right)^2 + c^2(u) \left(\frac{\partial u}{\partial x} \right)^2 \right] dx \tag{3}$$

is conserved.

In recent years an interesting mathematical literature has developed devoted to the analysis of solutions to (1). Pertinent facts include:

1. Solutions of the Cauchy problem with smooth data can develop singularities in finite time [8]. Typical isolated singularities involve propagating solutions with cusps, $u \propto (x - x_s)^{2/3}$, $(t - t_s)^{2/3}$, with (x_s, t_s) space-time coordinates of a point where the solution is not smooth, but slightly stronger singularities are possible [4] and in general one can only prove that weak solutions are Hölder continuous with exponent $1/2$.
2. Unique conservative global weak solutions exist (e.g. [3, 12]). However, dissipative weak solutions can also be constructed (e.g. [2]), but global existence results for dissipative solutions are only available under assumptions which do not hold in our case [18].

There have also been a few publications proposing numerical methods. In [1, 13] the authors develop approximations to a first order reformulation of the system in Riemann-like variables, which is also the backbone of much of the analytical work. In particular in [1] both conservative and dissipative discontinuous Galerkin discretizations are developed, with the dissipation provided by a smoothness indicator and a numerical flux. On the other hand the second order system is directly discretized using a discontinuous Galerkin method in [17]. Again they propose both a conservative scheme and a scheme with artificial dissipation. A convergence theory is only given in [13], and then under assumptions which do not hold in our case.

Our goals in this work are to study the convergence of conservative pseudospectral discretizations of (1) and also to examine the convergence in the limit of vanishing viscosity of resolved approximations to a related viscous equation.

2 Energy-Conserving Pseudospectral Discretizations

In an attempt to compute the unique conservative weak solution we considered 2π-periodic Cauchy data and experimented with two energy-conserving Fourier pseudospectral discretizations. Note that the regularity properties of the solution itself are sufficient to guarantee the uniform convergence of its Fourier series, so it is not unreasonable to hope that the implicit regularization associated with truncating the series would be sufficient to produce a convergent numerical scheme.

Suppose S is any skew-symmetric discrete derivative operator. Given a discrete solution vector, U^h, with

$$U_j^h(t) \approx u(x_j, t)$$

consider the semidiscretized system:

$$\frac{d^2 U_j^h}{dt^2} = S_{jk} c^2(U_k^h) S_{kl} U_l^h - c(U_j^h) c'(U_j^h) \left(S_{jk} U_k^h\right)^2, \tag{4}$$

where summations over k and l are implied. We first note that the semidiscretization is consistent with strong solutions to (1) since one can rewrite

$$c(u)\frac{\partial}{\partial x}\left(c(u)\frac{\partial u}{\partial x}\right) = \frac{\partial}{\partial x}\left(c^2(u)\frac{\partial u}{\partial x}\right) - c(u)c'(u)\left(\frac{\partial u}{\partial x}\right)^2.$$

Second, introducing the discrete energy

$$E^h(t) = \frac{1}{2}\sum_j \left(\frac{dU_j^h}{dt}\right)^2 + c^2(U_j^h)\left(S_{jk} U_k^h\right)^2, \tag{5}$$

we calculate using the skew-symmetry of S

$$\begin{aligned}
\frac{dE^h}{dt} &= \sum_j \frac{dU_j^h}{dt}\frac{d^2 U_j^h}{dt^2} + \left(S_{jk}\frac{dU_k^h}{dt}\right)\left(c^2(U_j^h) S_{jl} U_l^h\right) + \frac{dU_j^h}{dt}\left(c(U_j^h)c'(U_j^h)\left(S_{jk}U_k^h\right)^2\right) \\
&= \sum_j \frac{dU_j^h}{dt}\left(S_{jk}c^2(U_j^h)S_{kl}U_l^h - c(U_j^h)c'(U_j^h)\left(S_{jk}U_k^h\right)^2\right) \\
&\quad - \frac{dU_j^h}{dt}\left(S_{jk}c^2(U_j^h)S_{kl}U_l^h - c(U_j^h)c'(U_j^h)\left(S_{jk}U_k^h\right)^2\right) \\
&= 0.
\end{aligned}$$

Thus the discrete energy is conserved:

$$E^h(t) = E^h(0). \tag{6}$$

Here we try two different Fourier pseudospectral approximations, S. Assuming a uniform grid

$$x_j = \frac{2\pi j}{N}, \quad j = 0, \ldots, N-1,$$

we denote by $\mathscr{F}^h$ the discrete Fourier transform and by $\mathscr{F}^{h,*}$ its inverse. Set

$$S_U = \mathscr{F}^{h,*}\text{diag}\,(ik)\,\mathscr{F}^h.$$

Assuming N even we have $k = -N/2, \ldots, N/2 - 1$. We will refer to S_U as the unfiltered derivative operator. We also consider a filtered derivative operator employing a spectral vanishing viscosity filter (e.g. [11, Ch. 9])

$$S_F = \mathscr{F}^{h,*}\text{diag}\left(ike^{-\alpha\left(\frac{2k}{N}\right)^m}\right)\mathscr{F}^h.$$

(In our experiments we set $\alpha = m = 36$ as suggested in [14].) It is easy to check that both S_F and S_U are skew-symmetric.

In time we employ an energy-conserving discretization method for the semidiscrete system. Precisely we use a symmetric eighth order multistep method taken from [10, Ch. XIV], using the implementation `gni_lmm2.f` made available by Hairer [9]:

$$\begin{aligned} U^h(t+\Delta t) &= U^h(t) + U^h(t-6\Delta t) - U^h(t-7\Delta t) \\ &\quad + \frac{\Delta t^2}{8640}\big(13207(G(t)+G(t-6\Delta t)) - 8934(G(t-\Delta t)+G(t-5\Delta t)) \\ &\quad + 42873(G(t-2\Delta t) + G(t-4\Delta t)) - 33812G(t-3\Delta t)\big). \end{aligned}$$

Here G denotes the right-hand side of (4). For energy computations we use the finite difference approximation to the time derivative provided in the code.

Remark 1 The formulation above applies to any skew-symmetric derivative approximation, so, for example, S could be chosen to be any central difference operator. Note that the spectral vanishing viscosity filter does not lead to a dissipative method as it is only applied to the derivative computation.

In our experiments we choose $\alpha = \frac{1}{2}$, $\beta = \frac{3}{2}$, matching the cases considered in [1, 13, 17]. Our initial condition, also similar to those considered in the cited references except for the assumption of periodicity, is:

$$u(x,0) = \frac{\pi}{4} + e^{-5(1-\cos x)}, \quad \frac{\partial u}{\partial t}(x,0) = -c(u(x,0))\frac{\partial u}{\partial x}(x,0). \tag{7}$$

As in their simulations, we expect a singularity to form and continue the simulation to $t = 6\pi$. We consider a sequence of grids with $\Delta x = \frac{2\pi}{N}$, $N = 16{,}384$, $N = 32{,}768$ and $N = 65{,}536$, and $\Delta t = \frac{\Delta x}{20}$. Experiments with different time steps indicate that the error due to the spatial discretization is significantly larger than the error due to the temporal discretization.

2.1 Results

We first plot in Fig. 1 solutions and their energy spectrum,

$$S(k) = |\hat{u}(k)|^2 + |\hat{u}(-k)|^2 \tag{8}$$

with the finest discretization, $N = 65{,}536$, at $t = \pi$. This is before the development of any singularities. Here the graphs are indistinguishable and a glance at the spectrum shows they are very well-resolved.

Next consider the solutions with $N = 65{,}536$ at $t = 3\pi$, some time after the singularity has occurred. (The time of its formation is roughly $t = 4$.) We see in Fig. 2 that they are now diverging from one another. Moreover, the decay of the energy spectrum is far less steep. A linear fit using modes with $k \leq 10^3$ yields slopes of -3.5 for the unfiltered solution and -3.6 for the filtered solution. Assuming

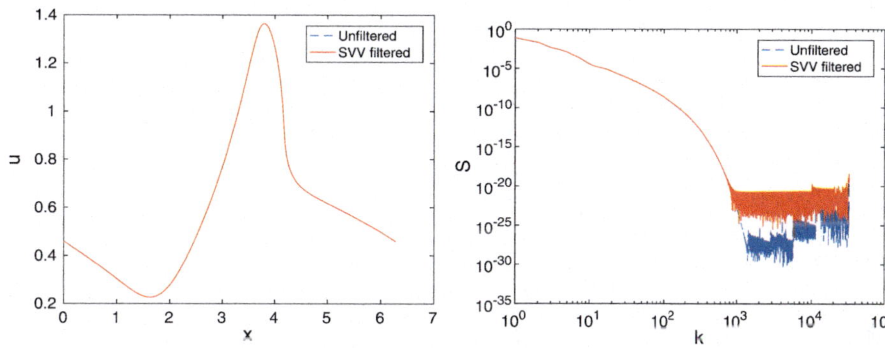

Fig. 1 Conservative solutions and energy spectrum at $t = \pi$

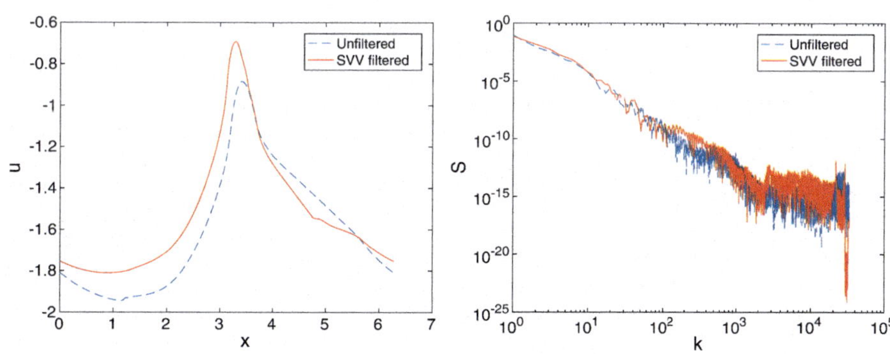

Fig. 2 Conservative solutions and energy spectrum at $t = 3\pi$

$u \in C^{2/3}$, consistent with the propagating singularities discussed in [4], we only have the bound

$$S(k) \leq Ck^{-4/3}, \tag{9}$$

so the observed decay is faster than expected.

To further assess the accuracy of these solutions we compare them in Fig. 3 at different resolutions: $N = 16{,}384$, $N = 32{,}768$ and $N = 65{,}536$. We see that both the unfiltered and filtered solutions appear to converge with grid refinement. Given the differences between the two solutions we are left with the question of which, if any, represent the unique conservative weak solution described mathematically in [3, 12].

Finally consider the solutions with $N = 65{,}536$ at $t = 6\pi$. We see in Fig. 4 that now they are completely different. A linear fit of the energy spectrum using modes with $k \leq 10^3$ at this time again yields slopes of -3.5 for the unfiltered solution and -3.6 for the filtered solution.

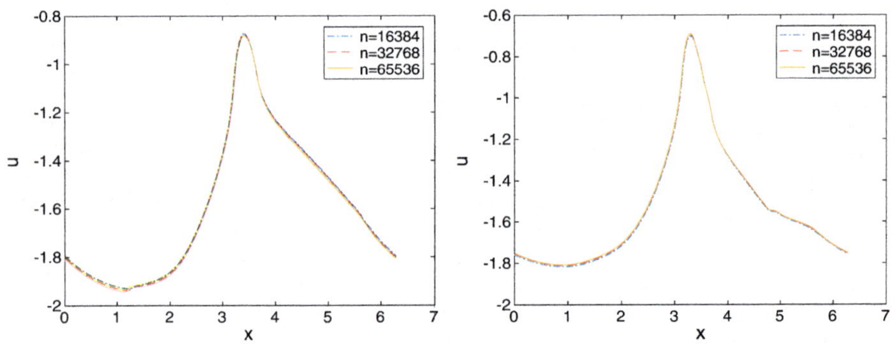

Fig. 3 Grid convergence of conservative solutions at $t = 3\pi$

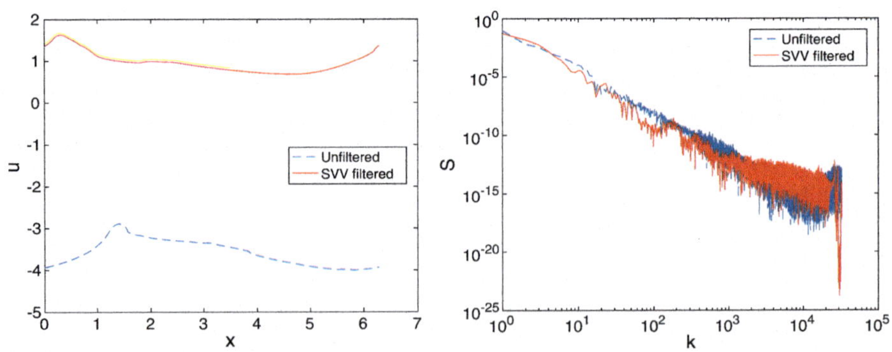

Fig. 4 Conservative solutions and energy spectrum at $t = 6\pi$

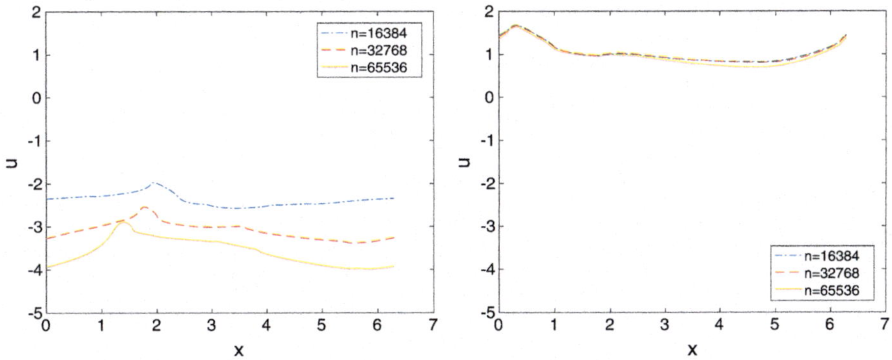

Fig. 5 Grid convergence of conservative solutions at $t = 6\pi$

Again, as proven in [3], there is a unique conservative weak solution, therefore at most one of these numerical solutions can be an accurate approximation to a conservative weak solution. We consider the question of whether either appears converged at these discretization levels. To that end we plot in Fig. 5 the solutions for each discretization of the derivative operator with $N = 16{,}384$, $N = 32{,}768$ and $N = 65{,}536$. For solutions computed with the unfiltered derivative we see large differences for the three choices of N. With the filtered derivative, on the other hand, solutions for different values of N match better, though the solution with the finest grid exhibits larger deviations from those at the coarser grid levels. At earlier times, such as $t = 5\pi$, there is apparent grid convergence in the filtered case while the unfiltered solutions are still quite different. However, as observed at $t = 3\pi$, it is possible for both conservative solutions to appear converged but to differ from one another. Thus we see that the filtering operation does enable grid convergence at these discretization levels over longer time intervals. As such it produces a potentially accurate conservative solution well beyond the formation of a singularity. However, further hard analysis is required to see if any of our conservative solutions converge to the conservative weak solution at later times.

3 Vanishing Viscosity Models

As is well-known, admissible weak solutions for first order systems of conservation laws can be obtained by adding generic viscosity terms and taking the limit as the viscosity parameter approaches zero (e.g. [6]). Here we experiment with the addition of a small viscous perturbation to (1):

$$\frac{\partial^2 u}{\partial t^2} = c(u)\frac{\partial}{\partial x}\left(c(u)\frac{\partial u}{\partial x}\right) + \nu\frac{\partial^3 u}{\partial x^2 \partial t}. \tag{10}$$

Now the energy equality reads

$$E(t) = E(0) - \nu \int_0^t \int \left(\frac{\partial^2 u}{\partial x \partial t} \right)^2 dxdt. \tag{11}$$

For singular solutions of the type described in [4, 8] the integral term in (11) quantifying the energy dissipation would be unbounded for any $\nu > 0$. Thus it is possible that in the limit $\nu \to 0$ this integral will not vanish, distinguishing a dissipative weak solution of the original problem. We note that our approach differs from those in [1, 17] as they introduce numerical dissipation which depends on the discretization parameters and on the solution.

Here we semidiscretize the system in a first order form in time, using the second order Fourier differentiation operator

$$S_2 = \mathscr{F}^{h,*} \text{diag}\left(-k^2\right) \mathscr{F}^h,$$

$$\frac{dV_j^h}{dt} = S_{U,jk} c^2(U_k^h) S_{U,kl} U_l^h - c(U_j^h) c'(U_j^h) \left(S_{U,jk} U_k^h\right)^2 + \nu S_{2,jk} V_k^h,$$

$$\frac{dU_j^h}{dt} = V_j^h.$$

To march in time we then use the standard fourth order Runge-Kutta method. Note that we expect these simulations to be reasonably well-resolved and so decided not to employ the filtered first derivative operator, S_F. We consider coarser meshes: $N = 4096$, $N = 8192$ and $N = 16{,}384$ for $\nu = 10^{-4}$ and $\nu = 10^{-5}$, and $N = 8192$, $N = 16{,}384$ and $N = 32{,}768$ for $\nu = 10^{-6}$. Studies of grid convergence indicate that these resolutions are sufficient. As our time stepping scheme is lower order we choose smaller time steps than in the conservative case, $\Delta t = \frac{\Delta x}{100}$. Experiments with different time steps suggest that as in the previous case the spatial discretization errors are dominant.

3.1 Results

We begin by comparing the energy evolution for the conservative discretizations and the viscous simulations at the finest resolutions available. The results, shown in Fig. 6, indicate a sharp energy loss occurring just prior to $t = 4$, which we believe coincides with the development of a singularity in the inviscid solution. Most important, in our view, is the fact that the total energy dissipation in this epoch is fairly insensitive to the value of the viscosity parameter. This is suggestive of a vanishing viscosity limit which involves energy dissipation.

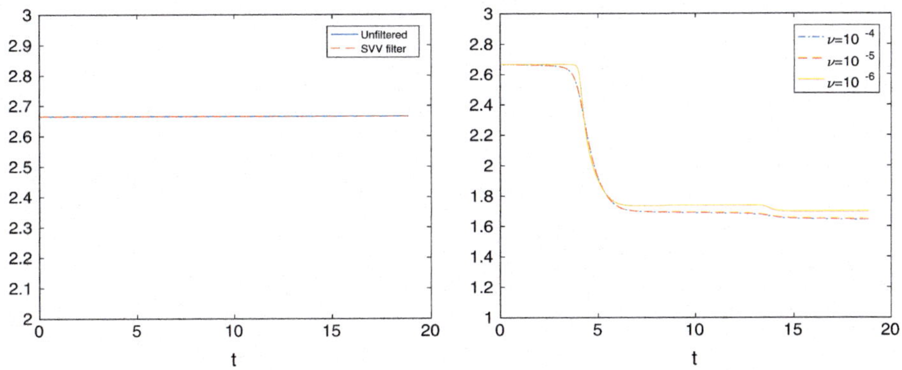

Fig. 6 Energy evolution for the conservative and viscous simulations

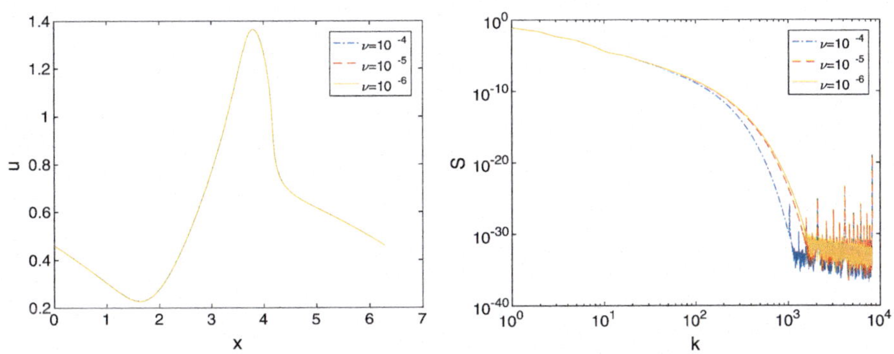

Fig. 7 Viscous solutions and energy spectrum at $t = \pi$

Beginning at $t = \pi$, we see in Fig. 7 that the viscous solutions are nearly identical at early times and are very well-resolved. Moreover, as shown in Fig. 8, they are nearly identical to the conservative solutions. (Note that throughout this section we will compare viscous solutions to conservative solutions obtained using the spectral vanishing viscosity filter based on the remarks above.)

Moving on to $t = 3\pi$, we again see in Fig. 9 excellent agreement between the solutions obtained with various values of the viscosity parameter, and the energy spectrum indicates that these are well-resolved. On the other hand we already observe a large deviation between viscous and conservative solutions; see Fig. 10.

Finally at $t = 6\pi$ we again see evidence of convergence as $\nu \to 0$ of the viscous solutions, which appear to be sufficiently well-resolved. They also differ by a large amount from the conservative solution (Figs. 11 and 12).

In Fig. 13 we consider the grid convergence of the viscous solutions with $\nu = 10^{-5}$ and $\nu = 10^{-6}$. Again this indicates that the chosen resolutions are adequate.

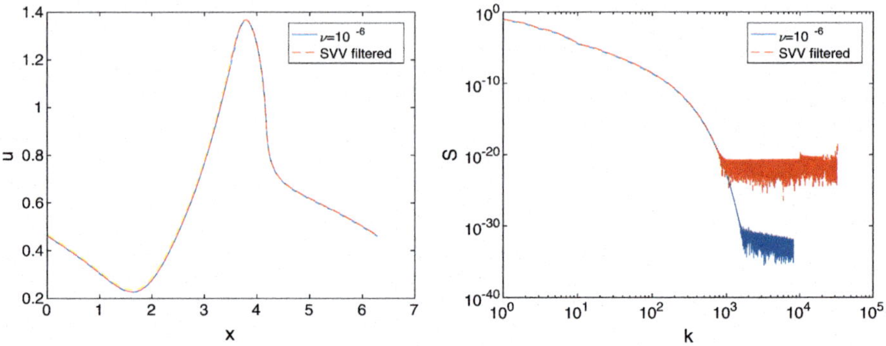

Fig. 8 Comparison of conservative and viscous solutions at $t = \pi$

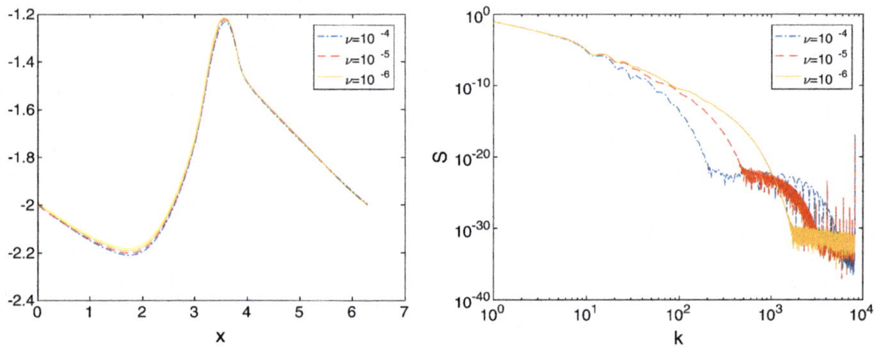

Fig. 9 Viscous solutions and energy spectrum at $t = 3\pi$

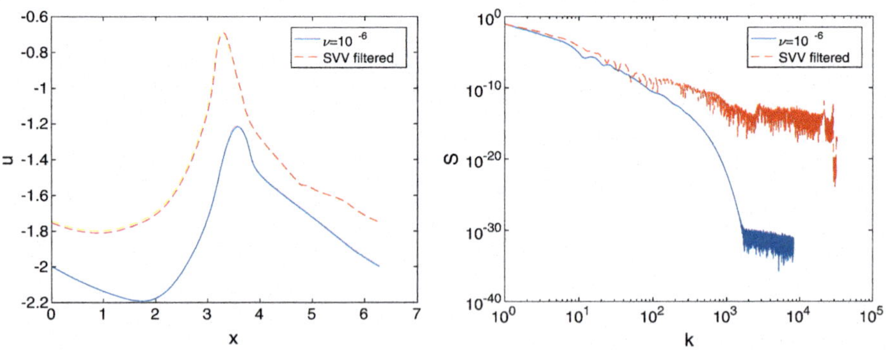

Fig. 10 Comparison of conservative and viscous solutions at $t = 3\pi$

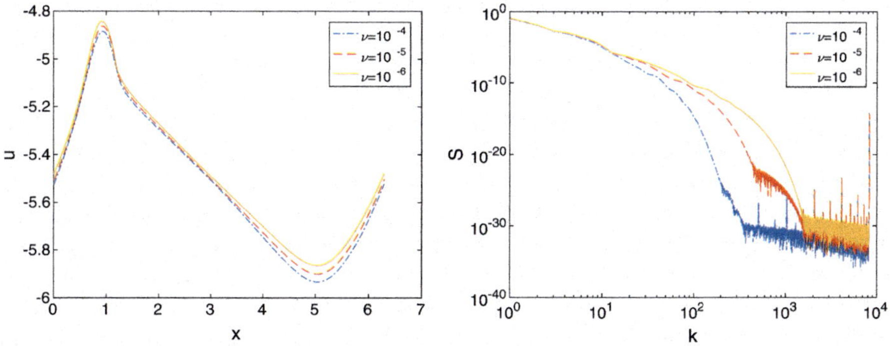

Fig. 11 Viscous solutions and energy spectrum at $t = 6\pi$

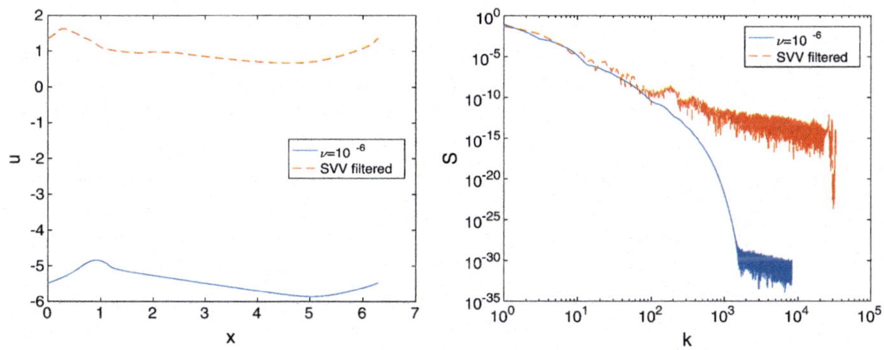

Fig. 12 Comparison of conservative and viscous solutions at $t = 6\pi$

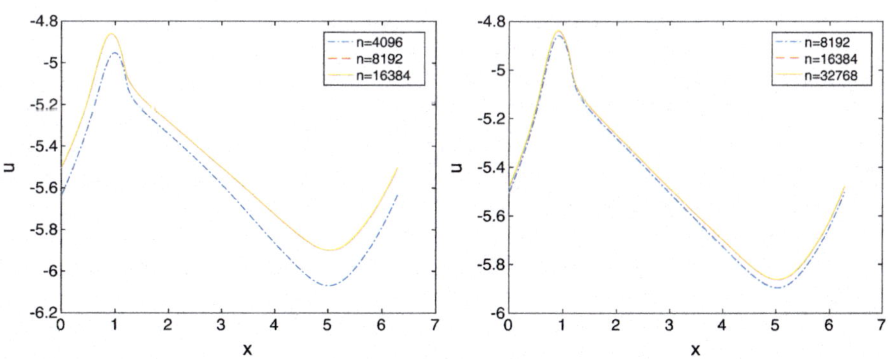

Fig. 13 Grid convergence of viscous solutions at $t = 6\pi$

4 Conclusions and Future Work

In conclusion, we have constructed and implemented conservative Fourier pseudospectral discretizations of the nonlinear variational wave equation introduced by Hunter and Saxton [15] to model the dynamics of director fields. Depending on our construction of the discrete derivative operator—more precisely depending on the use of a spectral vanishing viscosity filter—apparent grid convergence is observed at some times beyond the time where a singularity develops. The filtered solution displays grid convergence at later times than the unfiltered one. However, earlier on both solutions are apparently converged but disagree with one another. Thus an open question is what additional criteria, beyond energy conservation, must a numerical scheme satisfy to guarantee convergence to the unique conservative weak solution.

In addition we pursue numerically a vanishing viscosity limit. Considering the nonlinear wave equation (1) perturbed by a generic viscous term (10) we observe, using well-resolved Fourier pseudospectral discretizations, apparent convergence to a nonconservative solution as the viscosity parameter ν is decreased. This leads to the mathematical question of whether the vanishing viscosity limit distinguishes a unique dissipative weak solution and if this solution is to be preferred as a model of the physics.

Acknowledgements This work was supported in part by NSF Grant DMS-1418871. Any opinions, findings, and conclusions or recommendations expressed in this material are those of the author and do not necessarily reflect the views of the National Science Foundation. We also acknowledge the use of FFT routines adapted from [7] as well as OpenMP-enabled FFTs adapted from [16] and written by Burkardt [5].

References

1. P. Aursand. U. Koley, Local discontinuous Galerkin schemes for a nonlinear variational wave equation modeling liquid crystals. Preprint (2014)
2. A. Bressan, T. Huang, Representation of dissipative solutions to a nonlinear variational wave equation. Commun. Math. Sci. **14**, 31–53 (2016)
3. A. Bressan, G. Chen, Q. Zhang, Unique conservative solutions to a variational wave equation. Arch. Ration. Mech. Anal. **217**, 1069–1101 (2015)
4. A. Bressan, T. Huang, F. Yu, Structurally stable singularities for a nonlinear wave equation. Bull. Inst. Math. Acad. Sinica **10**, 449–478 (2015)
5. J. Burkardt, http://www.sc.fsu.edu/~jburkardt/
6. C. Dafermos, *Hyperbolic Conservation Laws in Continuum Physics*, 3rd edn. (Springer, Berlin, 2010)
7. B. Fornberg, *A Practical Guide to Pseudospectral Methods* (Cambridge University Press, Cambridge, 1998)
8. R. Glassey, J. Hunter, Y. Zheng, Singularities of a variational wave equation. J. Differ. Equ. **129**, 49–78 (1996)
9. E. Hairer, http://www.unige.ch/~hairer/software.html
10. E. Hairer, C. Lubich, G. Wanner, *Geometric Numerical Integration* (Springer, Berlin, 2002)

11. J. Hesthaven, S. Gottlieb, D. Gottlieb, *Spectral Methods for Time-Dependent Problems* (Cambridge University Press, Cambridge, 2007)
12. H. Holden, X. Raynaud, Global semigroup of conservative solutions of the nonlinear variational wave equation. Arch. Ration. Mech. Anal. **201**, 871–964 (2011)
13. H. Holden, K. Karlsen, N. Risebro, A convergent finite-difference method for a nonlinear variational wave equation. IMA J. Numer. Anal. **29**, 539–572 (2009)
14. T. Hou, Blow-up or no blow-up? a unified computational and analytic approach to 3D incompressible Euler and Navier-Stokes equations. Acta Numer. **18**, 277–346 (2009)
15. J. Hunter, R. Saxton, Dynamics of director fields. SIAM J. Appl. Math. **29**, 1498–1521 (1991)
16. W. Petersen, P. Arbenz, *Introduction to Parallel Computing - A Practical Guide with Examples in C* (Oxford University Press, Oxford, 2004)
17. N. Yi, H. Liu, An energy conserving discontinuous Galerkin method for a nonlinear variational wave equation. Commun. Comput. Phys. (2016, Preprint)
18. P. Zhang, Y. Zheng, Weak solutions to a nonlinear variational wave equations with general data. Ann. Inst. H. Poincaré Non-Linéaire **22**, 207–226 (2005)

Numerical Solution of the Viscous Flow Past a Cylinder with a Non-global Yet Spectrally Convergent Meshless Collocation Method

Francisco Bernal, Alfa R.H. Heryudono, and Elisabeth Larsson

Abstract The flow of a viscous fluid past a cylinder is a classical problem in fluid-structure interaction and a benchmark for numerical methods in computational fluid dynamics. We solve it with the recently introduced radial basis function-based partition of unity method (RBF-PUM), which is a spectrally convergent collocation meshless scheme well suited to this kind of problem. The resulting discrete system of nonlinear equations is tackled with a trust-region algorithm, whose performance is much enhanced by the analytic Jacobian which is provided alongside. Preliminary results up to $Re = 60$ with just 1292 nodes are shown.

1 Introduction

The steady flow of a viscous fluid past a fixed, perpendicular cylinder is one of the simplest nontrivial problems in fluid dynamics. Moreover, it furnishes a relevant model for fluid-structure interaction (for instance, bridge pillars or the drillpipe of an oil platform, see Fig. 1). The problem was already tackled by Stokes [1], who was unable to find an analytical solution; a situation which—to the best of our knowledge—persists today. Consequently, it has been mostly studied numerically, with a special interest in the structure of the wake (the recirculating region immediately downstream the cylinder, shown in Fig. 1) and the drag coefficient, as the Reynolds number (*Re*) grows.

F. Bernal (✉)
CMAP, École Polytechnique, Paris, France
e-mail: francisco.bernal@polytechnique.edu

A.R.H. Heryudono
Department of Mathematics, University of Massachusetts Dartmouth, North Dartmouth, MA, USA
e-mail: aheryudono@umassd.edu

E. Larsson
Department of Information Technology, Uppsala University, Box 337, SE-751 05 Uppsala, Sweden
e-mail: elisabeth.larsson@it.uu.se

M.L. Bittencourt et al. (eds.), *Spectral and High Order Methods for Partial Differential Equations ICOSAHOM 2016*, Lecture Notes in Computational Science and Engineering 119, DOI 10.1007/978-3-319-65870-4_35

Fig. 1 (*Left*) Highly symmetric flow past a cylinder. (*Right*) An oil rig. *Credit*: VaderSS from Wikipedia

The physical problem becomes unstable from $Re \approx 40$ onwards, when small perturbations in the symmetry of the incoming flow lead to an asymmetric vortex structure in the wake (the well-known Von Karman vortex street). In highly controlled laboratory conditions, it is possible to delay the onset of the physical instability until a higher value of Re, resulting in the symmetric, but unstable, pattern shown on Fig. 1 (left).

From a numerical point of view, it is certainly possible to enforce symmetric conditions to arbitrary Re (unlike with the physical problem). However, as Re grows, the discretized equations become more and more ill-conditioned and it is challenging to keep the numerical simulation stable. (The roundoff errors appearing during the iterations of the nonlinear mathematical problem act in a similar way as flow perturbations and are prone to pick up the physical instability.) Another source of difficulty is derived from the fact that the computational domain must be very large compared with the area of interest (the wake), in order to enforce the far field boundary conditions (BCs) at a finite distance. These and other aspects are discussed in detail by Fornberg, who solved the flow problem up to $Re = 600$ [2] (with Re based on the cylinder diameter).

In this paper, we present a novel approach based on three ingredients. First, instead of the streamfunction/vorticity formulation, we follow [3] in using the natural variables and apply a transformation of the unbounded domain into a finite rectangle (Sect. 2). Second, we discretize the resulting equations according to the recently introduced Radial Basis Functions-based Partition of Unity method (RBF-PUM) [4]. RBF-PUM has all the advantages of RBF collocation—such as spectral accuracy for smooth functions and flexibility in the choice of discretization—while being able to tackle much larger problems (Sect. 3). Third, in Sect. 4 the analytical Jacobian of the nonlinear algebraic system is derived, which is critical for convergence. This idea was introduced in [5], allowing highly nonlinear elliptic problems to be solved with RBF meshless methods. Preliminary results up to $Re = 60$ are presented and briefly discussed in Sect. 5, and Sect. 6 concludes the paper with pointers to future work.

2 Transformed Navier-Stokes Equations

A fixed, infinite circular cylinder of radius $a > 0$ is immersed in a fluid of kinematic viscosity $\nu > 0$, which flows steadily perpendicularly to the cylinder with far-field velocity $U > 0$ and far-field pressure P_0. The Reynolds number (based on the diameter) is $Re = 2aU/\nu$. By symmetry, the problem is two-dimensional, with the obstacle being the circular section. Let (x, y) be a Cartesian dimensionless frame (with $a = 1$) centred at the axis; (r, φ) polar coordinates (where $\varphi = 0$ marks the direction of advance of the flow); $\mathbf{u} = (u, v)$ the dimensionless velocity field (with $U = 1$); P the pressure and $p = P - P_0$. The steady Navier-Stokes equations are then given by

$$\frac{Re}{2}(\mathbf{u}\cdot\nabla)\mathbf{u} = \nabla^2\mathbf{u} - \nabla p, \qquad \nabla\cdot\mathbf{u} = 0.$$

They are supplemented with five boundary conditions:

$$u(r=\infty,\varphi) = 1, \qquad v(r=\infty,\varphi) = 0, \qquad \text{(unperturbed flow)}$$
$$u(r=1,\varphi) = 0, \qquad v(r=1,\varphi) = 0, \qquad \text{(non-slip condition)}$$
$$p(r=\infty,\varphi) = 0. \qquad \text{(unperturbed pressure)}$$

Instead of taking a large finite domain and enforcing the BCs far away from the cylinder, the infinite domain is compressed into the rectangle $[0, 1] \times [0, 2\pi]$ via the following transformation [3]:

$$\xi = 1 - 1/r \qquad (\text{ such that } \xi(r=1) = 0 \text{ and } \xi(r=\infty) = 1).$$

Moreover, $\partial/\partial r = (1-\xi)^2\partial/\partial\xi$, so that the unit vectors point in the same direction: $\mathbf{i}_r = \mathbf{i}_\xi$. Denoting as $\dot{\xi}$ and $\dot{\varphi}$ the components of the fluid velocity in the new coordinates, the velocity field is transformed as

$$\mathbf{u} = \dot{\xi}\mathbf{i}_r + (1-\xi)\dot{\varphi}\mathbf{i}_\varphi.$$

(Note that the dot notation is intuitive for velocities and useful so as not to overload the notation, but recall that the problem is steady and involves no time derivatives.) We also introduce the following notation: $\partial^2_{\xi\varphi}\dot{\varphi} = \partial^2\dot{\varphi}/\partial\xi\partial\varphi$, $\partial_\varphi p = \partial p/\partial\varphi$, etc. The Navier-Stokes equations in the variables (ξ, φ) are

$$\mathcal{W}_1 = \mathcal{W}_2 = \mathcal{W}_3 = 0, \qquad (1)$$

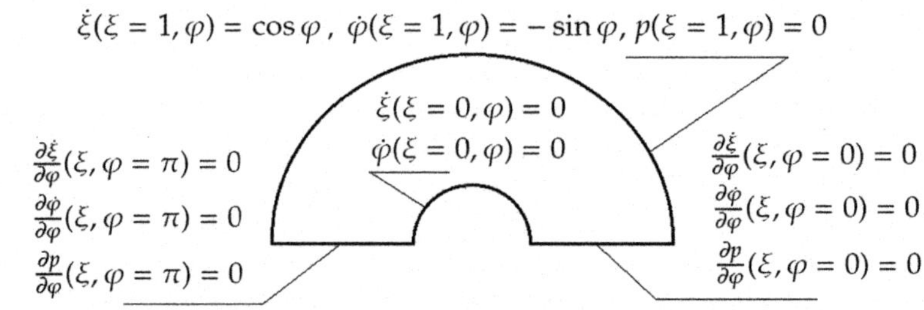

Fig. 2 Transformed BCs. The infinite angular section $(1 \leq r < \infty, 0 \leq \varphi \leq \pi)$ sketched above has been compressed into the rectangle $[0, 1] \times [0, \pi]$

where:

$$\begin{aligned} \mathcal{W}_1\big(\xi, \dot{\xi}, \dot{\varphi}, \partial_\xi \dot{\xi}, \partial_\varphi \dot{\xi}, \partial_\varphi \dot{\varphi}, \partial_\xi p, \partial^2_{\xi\xi} \dot{\xi}, \partial^2_{\varphi\varphi} \dot{\xi}\big) &= \frac{Re}{2}\left[(1-\xi)\dot{\xi}\partial_\xi \dot{\xi} + \dot{\varphi}\partial_\varphi \dot{\xi} - \dot{\varphi}^2\right] + \\ (1-\xi)\partial_\xi p - (1-\xi)^3 \partial^2_{\xi\xi} \dot{\xi} - (1-\xi)\partial^2_{\varphi\varphi} \dot{\xi} &+ (1-\xi)^2 \partial_\xi \dot{\xi} + 2(1-\xi)\partial_\varphi \dot{\varphi} + (1-\xi)\dot{\xi}, \\ \mathcal{W}_2\big(\xi, \dot{\xi}, \dot{\varphi}, \partial_\varphi \dot{\xi}, \partial_\varphi \dot{\varphi}, \partial_\xi \dot{\varphi}, \partial_\varphi p, \partial^2_{\xi\xi} \dot{\varphi}, \partial^2_{\varphi\varphi} \dot{\varphi}\big) &= \frac{Re}{2}\left[(1-\xi)\dot{\xi}\partial_\xi \dot{\varphi} + \dot{\varphi}\partial_\varphi \dot{\varphi} + \dot{\xi}\dot{\varphi}\right] + \\ \partial_\varphi p - (1-\xi)^3 \partial^2_{\xi\xi} \dot{\varphi} - (1-\xi)\partial^2_{\varphi\varphi} \dot{\varphi} &+ (1-\xi)^2 \partial_\xi \dot{\varphi} - 2(1-\xi)\partial_\varphi \dot{\xi} + (1-\xi)\dot{\varphi}, \\ \mathcal{W}_3\big(\xi, \dot{\xi}, \partial_\xi \dot{\xi}, \partial_\varphi \dot{\varphi}\big) &= (1-\xi)\partial_\xi \dot{\xi} + \partial_\varphi \dot{\varphi} + \dot{\xi}. \end{aligned} \tag{2}$$

Remark The last two terms in [3, formula (7)] are seemingly wrong.

Moreover, since the problem is symmetric along the x axis, only $0 \leq \varphi \leq \pi$ needs to be considered. (The boundary conditions along the x axis are now of reflecting type to enforce the symmetry.) The transformed BCs are sketched in Fig. 2.

3 Meshless Discretization Using the RBF-PUM

In this section we briefly review the formulation described in detail in [4].

Kansa's Method Let $\mathbf{q} = (\xi, \varphi) =\in \Omega \subset \mathbb{R}^2$, where $\Omega = [0, 1] \times [0, \pi]$, and let the *pointset* $\{\mathbf{q}_1, \ldots, \mathbf{q}_N\}$ be a discretization of Ω and its boundary $\partial\Omega$ into N scattered, distinct points (called *nodes*). A Radial Basis Function (RBF) approximation to the pressure is the *RBF interpolant*

$$p(\xi, \varphi) = p(\mathbf{q}) = \sum_{i=1}^{N} \lambda_i \phi_i(||\mathbf{q} - \mathbf{q}_i||). \tag{3}$$

Remark For notational convenience, we use the symbols $\dot{\xi}, \dot{\varphi}$, and p both for the exact solution of equation (1) and for their RBF interpolants.

Above, $||\cdot||$ is the Euclidean norm, and $\phi_i(\mathbf{q})$ is the chosen RBF, which also contains a *shape parameter* $\epsilon > 0$. For instance, the Gaussian RBF

$$\phi_i(\mathbf{q}) = \exp\left[-(\epsilon||\mathbf{q}-\mathbf{q}_i||)^2\right].$$

The RBF coefficients $\lambda_1, \ldots, \lambda_N$ can be found by collocation. Requesting that $p(\mathbf{q})$ in (3) interpolates the nodal pressures $p(\mathbf{q}_1), \ldots, p(\mathbf{q}_N)$ leads to

$$\begin{pmatrix} p_1 \\ \vdots \\ p_N \end{pmatrix} = \begin{pmatrix} \phi_1(\mathbf{q}_1) & \ldots & \phi_N(\mathbf{q}_1) \\ \vdots & \ddots & \vdots \\ \phi_1(\mathbf{q}_N) & \ldots & \phi_N(\mathbf{q}_N) \end{pmatrix} \begin{pmatrix} \lambda_1 \\ \vdots \\ \lambda_N \end{pmatrix} \Rightarrow \boldsymbol{\lambda} = [\phi]^{-1}\mathbf{p},$$

where we have introduced the notation of *nodal vectors and matrices*. Moreover, given $f_i : \Omega \mapsto \mathbb{R}$, $1 \le i \le N$, then $\mathbf{f}(\mathbf{q}) \in \mathbb{R}^N = [f_1(\mathbf{q}), \ldots, f_N(\mathbf{q}_N)]$. Note that (with a fixed ϵ) $[\phi]$ is symmetric since $[\phi]_{ij} = \phi_j(||\mathbf{q}_i - \mathbf{q}_j||) = \phi_i(||\mathbf{q}_j - \mathbf{q}_i||) = [\phi]_{ji}$. This allows to express the RBF interpolant in terms of the (unknown) nodal pressures rather than RBF coefficients,

$$p(\mathbf{q}) = \boldsymbol{\phi}^T(\mathbf{q})\boldsymbol{\lambda} = \boldsymbol{\phi}^T(\mathbf{q})[\phi]^{-1}\mathbf{p} = \boldsymbol{\psi}^T(\mathbf{q})\mathbf{p}, \tag{4}$$

where

$$\boldsymbol{\psi}(\mathbf{q}) = [\phi]^{-1}\boldsymbol{\phi}(\mathbf{q}) \Rightarrow \psi_i(\mathbf{q}_j) = \delta_{ij} \text{ (Kronecker's delta)}.$$

The functions $\psi_i(\mathbf{q})$ are the *cardinal basis functions* of the RBF ϕ and the pointset.

Linear boundary value problems (BVPs) can readily be solved as follows. Let the PDE be defined by the interior operator $\mathcal{L}^{PDE}p = f$ and the BCs by the boundary operator $\mathcal{L}^{BC}p = g$. For notational convenience, let the entire BVP be described by $\mathcal{L} = h$, where $h(\mathbf{q})$ and $\mathcal{L}_{\mathbf{q}}$ are

$$\mathcal{L} = \begin{cases} \mathcal{L}^{PDE}, & \text{if } \mathbf{q} \in \Omega/\partial\Omega, \\ \mathcal{L}^{BC}, & \text{if } \mathbf{q} \in \partial\Omega, \end{cases} \qquad h(\mathbf{q}) = \begin{cases} f(\mathbf{q}), & \text{if } \mathbf{q} \in \Omega/\partial\Omega, \\ g(\mathbf{q}), & \text{if } \mathbf{q} \in \partial\Omega. \end{cases}$$

Applying $\mathcal{L}$ on the RBF interpolant of p yields the square linear system

$$\mathcal{L}p(\mathbf{q}) = \left[\mathcal{L}\phi_1(\mathbf{q}), \ldots, \mathcal{L}\phi_N(\mathbf{q})\right]^T [\phi]^{-1}\mathbf{p} = \mathbf{h}, \tag{5}$$

whose solution is $\mathbf{p}$, and $p(\mathbf{q})$ can be reconstructed on $\Omega \cup \partial\Omega$ by (4).

The just described algorithm is easy to code, meshfree, geometrically flexible and spectrally convergent for smooth problems (see [5]). A drawback is that the last property comes at the expense of a fully populated matrix in (5). For this reason, it is often thought that Kansa's method loses many of its advantages after a few thousand

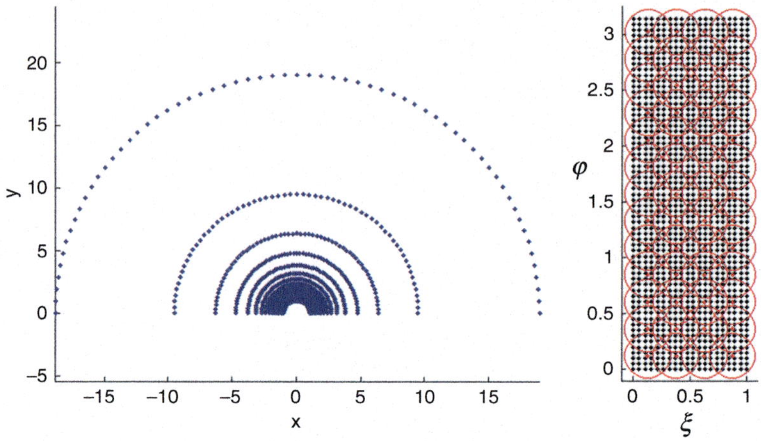

Fig. 3 PURBF discretization in the numerical example in Sect. 5. (*Left*) Nodes in real space. (*Right*) Nodes in the (ξ, φ) frame, with a cover overlaid

nodes. The RBF-PUM pushes N beyond that *without loss of performance*. It does so by "embedding" Kansa's method into a higher level of discretization, in order to attain a sparse matrix.

RBF-PUM Let $\{\Omega_j\}_{j=1}^{M}$ such that $\Omega \subset \cup_{j=1}^{N} \Omega_j$ be an open *cover* of Ω satisfying a pointwise overlap condition and

$$\forall \mathbf{q} \in \Omega, \qquad \Xi(\mathbf{q}) = \{j \,|\, \mathbf{q} \in \Omega_j\} \text{ and } \#\Xi(\mathbf{q}) \leq K,$$

where # is the cardinal of the set, and K a constant independent of M. Further, let $\{w_j(\mathbf{q})\}_{j=1}^{M}$ be a partition of unity on Ω (i.e. $\forall \mathbf{q} \in \Omega, \sum_{j=1}^{M} w_j(\mathbf{q}) = 1$) subordinate to the cover. (Fig. 3 shows an illustrative pointset and cover.) For w_j being $\mathcal{C}^2$, this can be attained with Shepard's method, using circular patches Ω_j and Wendland's compactly supported RBF, $\phi_W(\mathbf{q})$:

$$w_j(\mathbf{q}) = \frac{\phi_W(\mathbf{q})}{\sum\limits_{k \in \Xi(\mathbf{q})} \phi_W(\mathbf{q})} \text{ where } \phi_W(\mathbf{q}) = \begin{cases} (1 - ||\mathbf{q}||)^4(4||\mathbf{q}|| + 1) & \text{if } 0 \leq ||\mathbf{q}|| \leq 1, \\ 0 & \text{if } ||\mathbf{q}|| > 1. \end{cases} \tag{6}$$

The solutions of $p(\mathbf{q})$ on each patch of the cover—i.e. $p_j(\mathbf{q}) = p(\mathbf{q} \in \Omega_j)$—can be "glued together" by means of the partition of unity:

$$p(\mathbf{q}) = \sum_{j \in \Xi(\mathbf{q})} w_j(\mathbf{q}) p_j(\mathbf{q}). \tag{7}$$

$p_j(\mathbf{q})$ can be expressed in terms of the nodal values on patch j, and by linearity,

$$\mathcal{L}p(\mathbf{q}) = \sum_{j\in\Xi(\mathbf{q})}\sum_{k\in\Omega_j}\mathcal{L}\Big(w_j(\mathbf{q})\psi_k(\mathbf{q})\Big)p_k.$$

Specifically, partial derivatives can be computed according to the formula

$$\frac{\partial^{|\alpha|}}{\partial q^\alpha}p(\mathbf{q}) = \sum_{j\in\Xi(\mathbf{q})}\sum_{k\in\Omega_j}\frac{\partial^{|\alpha|}}{\partial q^\alpha}[w_j(\mathbf{q})\psi_k(\mathbf{q})]p_k = \sum_{j\in\Xi(\mathbf{q})}\sum_{k\in\Omega_j}\left[\sum_{\beta\le\alpha}\binom{\alpha}{\beta}\frac{\partial^{|\alpha-\beta|}w_j}{\partial q^{\alpha-\beta}}\frac{\partial^{|\beta|}\psi_k}{\partial q^\beta}\right]p_k,$$

where $\partial^{|\alpha|}/\partial^\alpha$ is the usual multi-index notation [4]. For instance, the angular derivative at a node with $\xi = \xi'$ and $\varphi = \varphi'$ is

$$\partial_\varphi p(\xi',\varphi') = \sum_{j\in\Xi(\mathbf{q}')}\sum_{k\in\Omega_j}[\frac{\partial w_j}{\partial q_y}(\mathbf{q}')\psi_k(\mathbf{q}') + w_j(\mathbf{q}')\frac{\partial\psi_k}{\partial q_y}(\mathbf{q}')]p_k \Rightarrow \partial_\varphi p = \partial_\varphi p(\mathbf{p}).$$

Above, we stress the fact that partial derivatives of an RBF interpolant can be expressed as a linear combination of its nodal values, with the coefficients depending only on the discretization (i.e. pointset, cover, partition of unity and RBFs), but independent of the function being differentiated. Therefore, one can explicitly compute the matrices for the nodal vectors of the required derivatives at start, and reuse them as needed. For instance, in the previous example, calling $\big((\partial_\varphi p)_1,\ldots,(\partial_\varphi p)_N\big)^T =: \partial_\varphi\mathbf{p} : [\partial_\varphi]\mathbf{p}$, with

$$[\partial_\varphi]_{mn} = \begin{cases}\sum\limits_{j\in\Xi(\mathbf{q}_n)}\left[\frac{\partial w_j}{\partial\varphi}(\mathbf{q}_m)\psi_n(\mathbf{q}_m) + w_j(\mathbf{q}_m)\frac{\partial\psi_n}{\partial\varphi}(\mathbf{q}_m)\right] & \text{if } \Xi(\mathbf{q}_m)\cap\Xi(\mathbf{q}_m)\neq\emptyset,\\ 0 & \text{otherwise,}\end{cases} \tag{8}$$

Note that matrices such as $[\partial_\varphi]$ are very sparse because only entries with indices associated to nodes in overlapping patches are nonzero. Analogously, let us define the RBF interpolants of $\dot\xi(\xi,\varphi)$ and $\dot\varphi(\xi,\varphi)$ as

$$\dot\xi(\mathbf{q}) - \sum_{j\in\Xi(\mathbf{q})}\sum_{k\in\Omega_j}\Big(w_j(\mathbf{q})\psi_k(\mathbf{q})\Big)\dot\xi_k, \qquad \dot\varphi(\mathbf{q}) = \sum_{j\in\Xi(\mathbf{q})}\sum_{k\in\Omega_j}\Big(w_j(\mathbf{q})\psi_k(\mathbf{q})\Big)\dot\varphi_k.$$

Since there are three PDEs in (1), the RBF collocation system has $3N$ equations with $3N$ unknowns $(\dot{\boldsymbol\xi},\dot{\boldsymbol\varphi},\mathbf{p})$. Let us define the *system nodal vector* as

$$\mathbf{X} = (\dot\xi_1,\ldots,\dot\xi_N,\dot\varphi_1,\ldots,\dot\varphi_N,p_1,\ldots,p_N)^T. \tag{9}$$

We shall now tackle the collocation of the complete Navier-Stokes equations. Let us start with the BCs, which are linear. We shall assume, without loss of generality, that there are $N_B < N$ nodes discretizing $\partial\Omega$ and that they are ordered first: $\mathbf{q}_i \in \partial\Omega$ iff $i \le N_B$. Moreover, the BC nodes are in turn ordered as follows: first, the set *FAR* of far-field nodes (with $\xi = 1$); then, the set *AXIS* of nodes on the x–axis (with either $\varphi = 0$ or $\varphi = \pi$); and finally, the set CYL of nodes on the cylinder (with

$\xi = 0$). For an $N \times N$ matrix $[\mathcal{L}]$ such as $[\partial_\varphi]$ in (8), let $[\mathcal{L}]_{SET \times N}$ represent the block with all the N columns and the $\#SET$ rows in the set SET. Then, collocation of the BCs in Fig. 2 yields the following contiguous matrix block B of size $\#BCs \times 3N$, where $\#BCs = 3(\#FAR + \#AXIS) + 2\#CYL$ and I is the $N \times N$ identity matrix:

$$\begin{pmatrix} I_{FAR\times N} & 0 & 0 \\ 0 & I_{FAR\times N} & 0 \\ 0 & 0 & I_{FAR\times N} \\ [\partial_\varphi]_{AXIS\times N} & 0 & 0 \\ 0 & [\partial_\varphi]_{AXIS\times N} & 0 \\ 0 & 0 & [\partial_\varphi]_{AXIS\times N} \\ I_{CYL\times N} & 0 & 0 \\ 0 & I_{CYL\times N} & 0 \end{pmatrix} \mathbf{X} =: B\mathbf{X} = \begin{pmatrix} \left[\cos\varphi_1, \ldots, \cos\varphi_{\#(FAR)}\right]^T \\ \left[-\sin\varphi_1, \ldots, -\sin\varphi_{\#(FAR)}\right]^T \\ 0 \\ \vdots \\ 0 \end{pmatrix}. \tag{10}$$

The collocations of the nonlinear operators $\mathcal{W}_1$, $\mathcal{W}_2$, and $\mathcal{W}_3$ on the interior nodes follow. They are $3N - \#BCs$ nonlinear algebraic equations in $\mathbf{X}$. (For instance, a product like $\dot{\xi}\partial_\xi\dot{\xi}$ depends quadratically on $\mathbf{X}$.) We write $W(\mathbf{q}', \mathbf{X})$ to denote the collocation of a nonlinear operator $\mathcal{W}$ acting on the RBF *interpolants* at node $\mathbf{q}'$. In sum, the discretized system of collocation equations reads (note that $\mathcal{W}_3$ is also enforced on the nodes in CYL)

$$\begin{pmatrix} B\mathbf{X} \\ W_1(\mathbf{q}_{N_B+1}, \mathbf{X}) \\ \vdots \\ W_1(\mathbf{q}_N, \mathbf{X}) \\ W_2(\mathbf{q}_{N_B+1}, \mathbf{X}) \\ \vdots \\ W_2(\mathbf{q}_N, \mathbf{X}) \\ W_3(\mathbf{q}_{N_B-\#CYL+1}, \mathbf{X}) \\ \vdots \\ W_3(\mathbf{q}_N, \mathbf{X}) \end{pmatrix} = \begin{pmatrix} \mathbf{g} \\ 0 \\ \vdots \\ 0 \end{pmatrix}, \quad \text{or} \quad \begin{pmatrix} B\mathbf{X} \\ \mathbf{W}_{123}(\mathbf{X}) \end{pmatrix} = \begin{pmatrix} \cos\varphi_1 \\ \vdots \\ -\sin\varphi_{\#FAR} \\ 0 \\ \vdots \\ 0 \end{pmatrix}. \tag{11}$$

Elimination of the BCs The block B in (10) and (11) contains only linear equations because the BCs are linear. It is advantageous to eliminate them before solving the nonlinear equations, also shrinking the size of the system to be solved. An optimally stable way of doing so is described in [5], which involves the QR decomposition of B:

$$B^T \Pi = [Q_1 Q_2] \begin{bmatrix} R \\ 0 \end{bmatrix},$$

where Π is a permutation matrix, R is upper triangular, and $Q_1 \in \mathbb{R}^{3N \times \#BCs}$ and $Q_2 \in \mathbb{R}^{3N \times (3N - \#BCs)}$ are made up of orthogonal columns. Then, the solution vector $\mathbf{X}$ can be expressed in terms of a fixed vector and a smaller vector $\mathbf{Y}$ (which remains to be found) as

$$\mathbf{X} = Q_1 R^{-T} \Pi^T \mathbf{g} + Q_2 \mathbf{Y}. \tag{12}$$

After solving for $\mathbf{Y}$, the nodal values of pressure and velocity (i.e. the vector $\mathbf{X}$ in (9)) can be found according to (12), and with them the pressure and velocity anywhere in the infinite domain can be reconstructed by virtue of (6) and (7). The nonlinear system of equations to be solved is thus

$$\mathbf{W}_{123}\left(Q_1 R^{-T} \Pi^T \mathbf{g} + Q_2 \mathbf{Y}\right) = 0. \tag{13}$$

($\#PDEs := 3N - \#BCs$ nonlinear equations in $\#PDEs$ unknowns $\mathbf{Y}$).

4 Analytic Jacobian of the RBF-PUM System

In order to solve (13), we apply the trust region algorithm, which transforms a rootfinding problem (the root being the vector solution of the system) into a minimization problem for the sum-of-squares residual in $\mathbb{R}^{\#PDEs}$. The method in the context of RBF approximations is discussed in detail in [5]. Because the residual landscape is highly nonconvex, it is critical both for convergence and for speed that the analytic Jacobian (i.e. the matrix J such that $J_{ij} = \frac{\partial (W_{123})_i}{\partial Y_j}$) be available. It is given by (see [5])

$$J = \sum_{k=1} \begin{pmatrix} diag\left[\frac{\partial W_1}{\partial\left(\mathcal{L}_k \dot{\xi}\right)}\right]\left[\mathcal{L}_k\right] & diag\left[\frac{\partial W_1}{\partial\left(\mathcal{L}_k \dot{\varphi}\right)}\right]\left[\mathcal{L}_k\right] & diag\left[\frac{\partial W_1}{\partial\left(\mathcal{L}_k p\right)}\right]\left[\mathcal{L}_k\right] \\ diag\left[\frac{\partial W_2}{\partial\left(\mathcal{L}_k \dot{\xi}\right)}\right]\left[\mathcal{L}_k\right] & diag\left[\frac{\partial W_2}{\partial\left(\mathcal{L}_k \dot{\varphi}\right)}\right]\left[\mathcal{L}_k\right] & diag\left[\frac{\partial W_2}{\partial\left(\mathcal{L}_k p\right)}\right]\left[\mathcal{L}_k\right] \\ diag\left[\frac{\partial W_3}{\partial\left(\mathcal{L}_k \dot{\xi}\right)}\right]\left[\mathcal{L}_k\right] & diag\left[\frac{\partial W_3}{\partial\left(\mathcal{L}_k \dot{\varphi}\right)}\right]\left[\mathcal{L}_k\right] & diag\left[\frac{\partial W_3}{\partial\left(\mathcal{L}_k p\right)}\right]\left[\mathcal{L}_k\right] \end{pmatrix} Q_2. \tag{14}$$

The diagonal matrices $diag[F]$ in (14) have diagonal entries $F(\mathbf{q}_{\#BCs+1}), \ldots,$ $F(\mathbf{q}_N)$ (i.e. are collocated of the interior nodes of the pointset, where the PDEs apply). Formally, the sum includes all the derivatives up to second order, but note that many of them are zero. As an example, we list the nonzero Fréchet derivatives of $\mathcal{W}_3$ (check (2) for the rest):

$$\frac{\partial W_3}{\partial(\partial_\xi \dot{\xi})} = 1 - \xi, \qquad \frac{\partial W_3}{\partial(\partial_\varphi \dot{\varphi})} = -1, \qquad \frac{\partial W_3}{\partial(\dot{\xi})} = -1, \qquad \frac{\partial W_3}{\partial(\xi)} = -\partial_\xi \dot{\xi}. \tag{15}$$

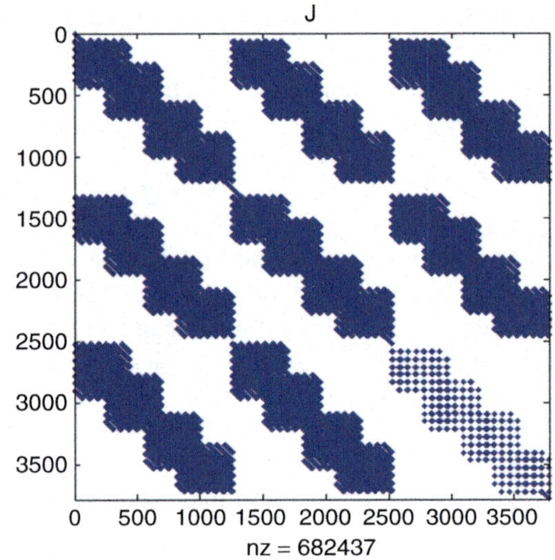

Fig. 4 Sparsity pattern of the Jacobian for the discretization in Fig. 3

After collecting all of the surviving Fréchet derivatives, it turns out that the nodal matrices involved are I (the $N \times N$ identity), $[\partial^2_{\xi\xi}]$, $[\partial^2_{\varphi\varphi}]$, $[\partial_\xi]$ and $[\partial_\varphi]$. The sparsity pattern of J for the illustrative RBF-PUM discretization shown in Fig. 3 is sketched in Fig. 4.

The analytical Jacobian does not ensure convergence, and for highly nonlinear PDEs the Hessian is required [5]. In this paper, however, we shall not consider it. The *dogleg* method for solving the trust-region subproblem is chosen due to its better conditioning and because it is already implemented as an option in Matlab's `fsolve`. Since global convergence is missing, we use the flow solved at a smaller *Re* as an initial guess for the iterations. The attainable *Re* is ultimately limited by the quality of the interpolant, the condition number of J, and by the resolution of the wake region provided by the meshless discretization.

5 Preliminary Results up to $Re = 60$

We illustrate the numerical method discussed in this paper with an example. It is preliminary because the discretization is small enough that the flow problem can be solved on a laptop, and because no effort has yet been made to optimize the location of the collocation points, which is one of the most interesting features of meshless formulations, and a well known strategy to improve performance. Thus, there are $N = 1292$ nodes forming a grid in $[0, 1] \times [0, \pi]$, corresponding to the locations in physical space shown on the left side of Fig. 3 (the physical nodes for $\xi = 1$ lie at the infinity and are obviously not shown).

Figure 4 shows the sparsity pattern of the RBF-PUM Jacobian. We report the wake structure for growing *Re* between $Re = 0.1$ and $Re = 60$. As reported in [2], recirculation starts at about $Re = 40$—see Fig. 5. After $Re = 60$, the discretization

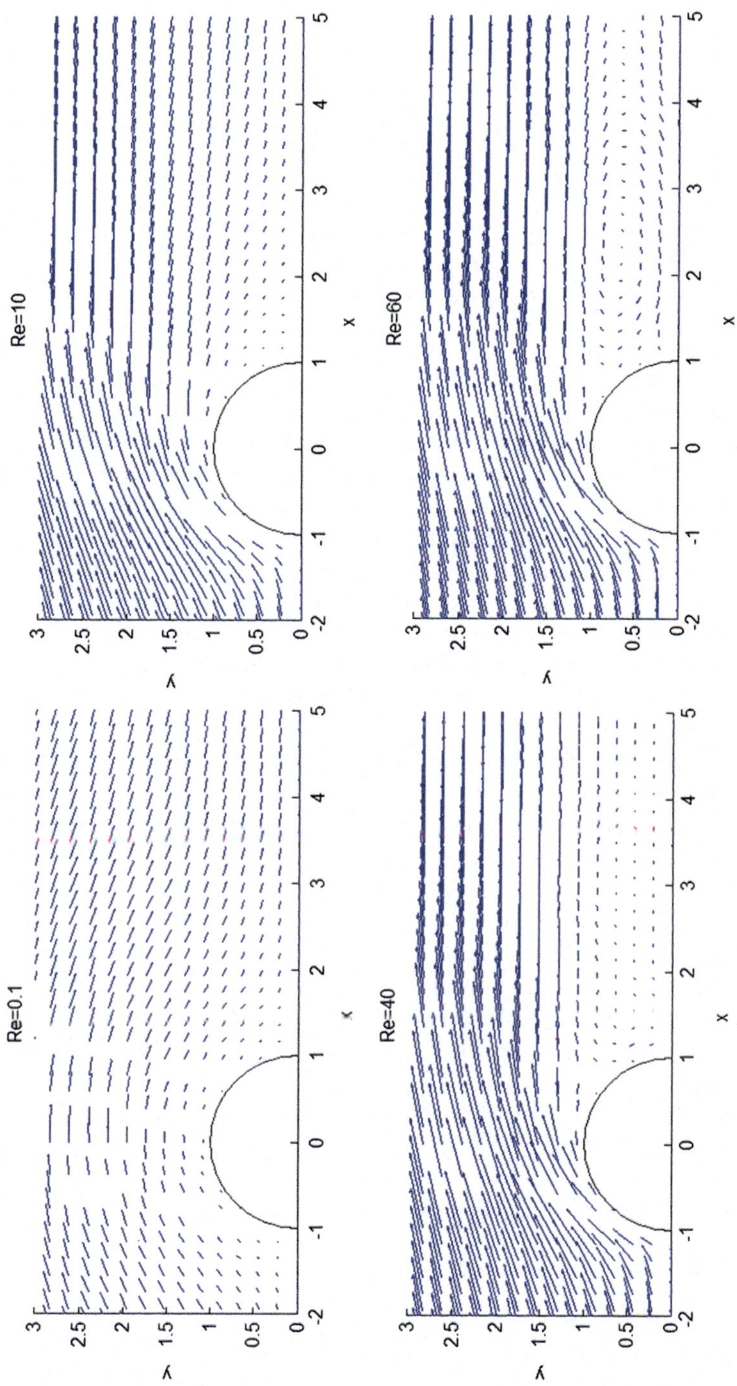

Fig. 5 Velocity field close to cylinder for increasing $Re = \{0.1, 10, 40, 60\}$

of the wake is inadequate and trust-region algorithm ceases to converge to a root, which is revealed by the fact that iterations stall at a value $\mathbf{X}_\infty$ for which $J(\mathbf{X}_\infty)$ is numerically singular (see [5] for details). In order to proceed further, a denser pointset with nodes more concentrated in the wake region is necessary, which in turn calls for a more powerful computer than a laptop; it is thus left as future work. It is, however, remarkable that $Re = 60$ can be attained with fewer than 1300 collocation nodes.

6 Conclusions and Future Work

RBF-PUM is a promising spectral, meshless method that combines the flexibility and simplicity of RBF collocation, with the possibility of tackling much larger problems than before (in terms of the discretization size), thanks to the resulting sparse structure. RBF-PUM is currently being investigated along several directions. Tailored preconditioners [6] and parallel implementations [7] have also been already proposed. In this paper, we have tested RBF-PUM on a benchmark problem in fluid dynamics. This is a challenging test due to the far-field BCs, as well as the numerical instability and nonlinearity compounding fast with growing *Re*. We have presented some numerical results from a preliminary, straightforward laptop implementation. By refining the discretization to the limit of the computational resources, we were able to solve the flow up to $Re \approx 60$, when the first eddies appear—qualitatively matching the expected flow pattern.

In order to proceed further, we plan to implement the method presented here on a parallel computer. We expect that by refining the discretization—adaptively if necessary—it will be possible to resolve the finer eddies appearing at higher *Re*. We also plan to enhance the convergence of the trust-region algorithm for the nonlinear system by incorporating Hessian information—a strategy which was deemed critical for highly nonlinear problems in [5]. With adequate computational resources, the related problem—albeit three-dimensional—of viscous flow past a sphere could be tackled with a very similar approach. Finally, by calculating and comparing the numerical drag coefficient with the experimental values over a wider range of *Re*, it will be possible to assess the convergence rate of the method for this problem.

Acknowledgements F. Bernal acknowledges support from FCT grant SFRH/BPD/79986/2011 and INESC-ID. A. Heryudono is partially supported by NSF Grant DMS 1552238.

References

1. G.G. Stokes, On the effect of the internal friction of fluids on the motion of pendulums. Trans. Camb. Philos. Soc. **9**, 8–106 (1851)
2. B. Fornberg, Steady viscous flow past a circular cylinder up to Reynolds number 600. J. Comput. Phys. **61**, 297–320 (1985)

3. F. Mandujano, R. Peralta-Fabi, On the viscous steady flow around a circular cylinder. Revista Mexicana de Física **51**(1), 87–99 (2005)
4. A. Safdari-Vaighani, A. Heryudono, E. Larsson, A radial basis function partition of unity collocation method for convection-diffusion equations arising in financial applications. J. Sci. Comput. **64**, 341–367 (2015)
5. F. Bernal, Trust-region methods for nonlinear elliptic equations with radial basis functions. Comput. Math. Appl. **72**(7), 1743–1763 (2016)
6. A. Heryudono, E. Larsson, A. Ramage, L. Von Sydow, Preconditioning for radial basis function partition of unity methods. J. Sci. Comput. **67**(3), 1089–1109 (2016)
7. I. Tominec, Parallel localized radial basis function methods for the shallow water equations on the sphere. MSc thesis, Technische Universitaet Muenchen (2017)

On Multiple Modes of Propagation of High-Order Finite Element Methods for the Acoustic Wave Equation

S.P. Oliveira

Abstract Earlier analyses of numerical dispersion of high-order finite element methods (HO-FEM) for acoustic and elastic wave propagation pointed out the presence of multiple modes of propagation. The number of modes increases with the polynomial degree of the finite element space, and since they were regarded as numerical artifacts, the use of HO-FEM was discouraged on wave propagation problems. Later on, alternative techniques showed that numerical dispersion decreases with the polynomial degree, and were supported by the success of spectral element methods on seismic wave propagation. This work concerns the interpretation of multiple propagation modes, which are solutions of an eigenvalue problem arising from the HO-FEM discretization of the wave equation as approximations to an eigenvalue problem associated with the continuous wave equation. By considering a continuous version of the standard periodic plane wave whose amplitude depends on the element grid, there are multiple combinations of the amplitude coefficients that yield exact solutions to the acoustic wave equation. Hence, modes regarded as non-physical can be associated with feasible propagation modes. Under this point of view, one can separately analyze each propagation mode or focus on the acoustical (constant amplitude) mode.

1 Introduction

Independent efforts from several fields of study such as acoustics, electromagnetism, mechanics, and seismic wave propagation have contributed to build a solid background in the numerical simulation of wave propagation. One of the concerns that most of these fields share is about how many grid points per wavelength are sufficient to guarantee that the wave will travel at the correct speed. To retrieve this information means to study the numerical dispersion of the solution scheme.

S.P. Oliveira (✉)
Departamento de Matemática, Universidade Federal do Paraná, Curitiba-PR 81531-980, Brazil
e-mail: saulopo@ufpr.br

M.L. Bittencourt et al. (eds.), *Spectral and High Order Methods for Partial Differential Equations ICOSAHOM 2016*, Lecture Notes in Computational Science and Engineering 119, DOI 10.1007/978-3-319-65870-4_36

Finite element methods (FEM) have long been popular in time-harmonic acoustics [22] and have also been successful on computational seismology in a high order version known as the spectral element method [13, 21]. The standard methodology of assessing numerical dispersion of FEM is to plug a discrete plane wave into the finite element stencil assuming an infinite, periodic mesh [4, 14]. When finite elements have interior nodes, one can separate them into sets which share the same degrees of freedom and are located at the same cyclically repeating location in the mesh pattern [11]. The numerical dispersion relation is then expressed by an eigenvalue problem. In the particular case of 1D quadratic elements, the eigenvalues have been referred as acoustical and optical branches, in analogy with the theory of wave propagation into crystal structures [4, 5]. However, one should exercise caution in attaching any physical significance to the terms "acoustical" and "optical" in the finite-element context [1].

Except for the quadratic case [1, 4, 6, 8, 10, 12], multiple modes of propagation implied by the standard analysis of high-order FEM are not fully understood. The classical interpretation is that only one eigenvalue is physically meaningful (in the case of the acoustic wave equation), while the others are regarded as computational modes [6, 14]. Since the dimension of the eigenvalue problem (and thus the number of "spurious" modes) increases with polynomial degree, FEM wave simulation should deteriorate if a high order is employed.

Alternative analyses have been proposed later on. Thompson and Pinsky [23] uses static condensation of the internal degrees of freedom, so that the eigenvalue system involves only element end nodes (see also [18]). Departing from a decomposition of the finite/spectral element space, Ainsworth and Wajid [2, 3] also formulate the discrete dispersion relation without internal degrees of freedom. Mulder [16] applies the discrete Fourier transform (DFT) sampled in the mesh nodes to the spatial operator and matches its eigenpairs with the transformed plane waves and their (normalized) wavenumbers. Under this setting, spurious modes provide reasonable approximations of particular eigenvectors of the exact operator. The eigenvalues of the spatial operator must be properly ordered to assure eigenpair matching, and it this ordering remains an open problem.

A similar identification problem is finding the acoustical branch, i.e., the eigenvalue mode that approximates the dispersion relation of the continuous wave equation. Cohen et al [7] identify the acoustical mode by a Taylor series expansion. Abboud and Pinsky [1] writes the amplitude-variable discrete plane wave as a linear combination of discrete plane waves and classify the modes with the dominating coefficient of the combination. Seriani and Oliveira [19] identify acoustical modes by a Rayleigh quotient approximation of the constant-amplitude mode. A similar analysis was done for the elastic wave equation [17, 20].

The above mentioned references provide clear evidence that the spectral element method is able to handle numerical dispersion, but do not sufficiently address the contribution and/or the interpretation of all propagation modes. Moura et al. [15] have recently presented a detailed description of the contribution of these modes and pointed out that the secondary modes not only have a smaller amplitude than the primary (acoustical) mode, but also improve the numerical approximation.

The present work revisits Abboud and Pinsky's approach of recasting the discrete plane wave as a combination of constant-amplitude modes. However, a linear combination of *continuous* plane waves is employed, as suggested by Durran [10]. Each continuous plane wave is a solution of the acoustic wave equation (given that their parameters satisfy the exact dispersion relation), thus we have multiple solutions that can be matched with the multiple modes of the discrete problem.

The paper is organized as follows: Section 2 reviews the classical dispersion analysis of high-order FEM with emphasis on 1D quadratic elements, for which analytical expressions for the eigenvalues and eigenvectors are available. Section 3 provides the linear combination of continuous plane waves associated with the discrete solutions from quadratic and N-th degree elements. Discrete and continuous modes are numerically compared.

2 Classical Dispersion Analysis of Quadratic Elements

By plugging a plane wave $u(x,t) = \exp(-\mathrm{i}\,(\omega t - \kappa x))$ into the one-dimensional acoustic wave equation with constant velocity c,

$$\frac{\partial^2 u}{\partial t^2}(x,t) = c^2 \frac{\partial^2 u}{\partial x^2}(x,t), \tag{1}$$

we find the dispersion relation $\omega = \pm c\kappa$.

Let us consider the finite-element spatial discretization of (1),

$$M\frac{\partial^2 \mathbf{u}}{\partial t^2}(t) + c^2 K\mathbf{u}(t) = \mathbf{0}, \tag{2}$$

with piecewise quadratic shape functions. In the classical analysis of numerical dispersion, we consider an infinite mesh with nodes $x_k = kh/2$, where h is the element length, and take the approximate solution $u_j(t) \approx u(x_j,t)$ as a discrete plane wave with periodic amplitude, i.e.,

$$u_j(t) := A_j e^{-\mathrm{i}\,(\omega_h t - \kappa x_j)}, \quad A_j = \begin{cases} A_0, & j \text{ is even}, \\ A_1, & j \text{ is odd}. \end{cases} \tag{3}$$

As pointed out in [1, 5], $u_j(t)$ can also be written as the combination of two discrete plane waves traveling with different velocities:

$$u_j(t) = a_0 e^{-\mathrm{i}\,(\omega_h t - \kappa x_j)} + a_1 e^{-\mathrm{i}\,(\omega_h t - (\kappa - 2\pi/h)x_j)}, \tag{4}$$

$$a_0 = \frac{A_0 + A_1}{2}, \quad a_1 = \frac{A_0 - A_1}{2}. \tag{5}$$

By substituting (3) into (2), we find the eigenvalue problem $C\mathbf{v} = \chi\mathbf{v}$, where $\mathbf{v} = [A_0, A_1]^T$, $\chi = (h\omega_h/c)^2$,

$$C = \frac{4}{3-\cos\left(\frac{4\pi}{G}\right)} \begin{bmatrix} 16+4\cos\left(\frac{4\pi}{G}\right) & -20\cos\left(\frac{2\pi}{G}\right) \\ \cos\left(\frac{2\pi}{G}\right)\left(3\cos\left(\frac{2\pi}{G}\right)^2 - 13\right) & 10 \end{bmatrix}, \tag{6}$$

and $G = \lambda/\Delta x = 4\pi/(\kappa h)$ is the number of grid points per wavelength, noting that the spacing between nodes is $\Delta x = h/2$. The solutions to $C\mathbf{v} = \chi\mathbf{v}$ are

$$\chi^{\pm} = 4\frac{13+2\cos\left(\frac{4\pi}{G}\right) \pm \sqrt{124 - 11\cos\left(\frac{4\pi}{G}\right)^2 + 112\cos\left(\frac{4\pi}{G}\right)}}{3-\cos\left(\frac{4\pi}{G}\right)} \tag{7}$$

$$\mathbf{v}^{\pm} = \begin{bmatrix} 3+2\cos\left(\frac{4\pi}{G}\right) \pm \sqrt{124 - 11\cos\left(\frac{4\pi}{G}\right)^2 + 112\cos\left(\frac{4\pi}{G}\right)} \\ 13\cos\left(\frac{2\pi}{G}\right) - 3\cos\left(\frac{2\pi}{G}\right)^3 \end{bmatrix}. \tag{8}$$

Figure 1 shows the normalized angular frequencies $\omega_h^{\pm}\Delta x/(2\pi c)$, as well as the amplitude ratios $A_1^{\pm}/A_0^{\pm}$. The solution closer to the continuous dispersion relation $\omega = \pm c\kappa$ is known as the acoustical branch, while the other is known as the optical branch [4].

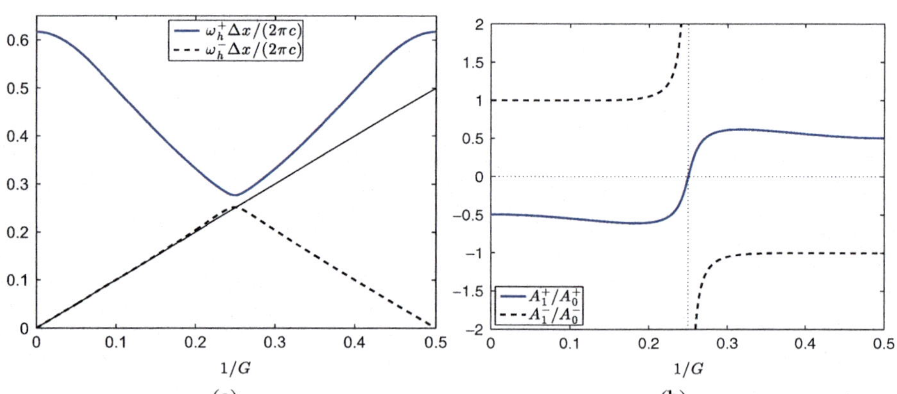

Fig. 1 Propagation modes of the quadratic finite element method: (**a**) Normalized angular frequency (the *thin, solid line* corresponds to $\omega = c\kappa$); (**b**) amplitude ratio

3 Continuous Modes Related to Variable Amplitudes

Durran [10] argues that the discrete plane wave with non-constant amplitude can be seen as grid values of a function in the form $g(x)\exp(-\mathrm{i}\,(\omega_h t - \kappa x))$, where $g(x)$ accounts for the extra spatial dependence implied by the amplitude degrees of freedom. Indeed, it follows from (4) that $u_j(t) = \tilde{u}(x_j, t)$, where

$$\begin{aligned}\tilde{u}(x,t) &= a_0 e^{-\mathrm{i}\,(\omega_h t - \kappa x)} + a_1 e^{-\mathrm{i}\,(\omega_h t - (\kappa - 2\pi/h)x)} = g(x) e^{-\mathrm{i}\,(\omega_h t - \kappa x)},\\ g(x) &= a_0 + a_1 e^{\mathrm{i}\,(2\pi/h)x}.\end{aligned} \tag{9}$$

Can the function $\tilde{u}(x,t)$ in (9) be a solution to the wave equation? This question may be recast as the following problem: find α_0, α_1, and ω such that

$$u(x,t) = \alpha_0 e^{-\mathrm{i}\,(\omega t - \kappa x)} + \alpha_1 e^{-\mathrm{i}\,(\omega t - (\kappa - 2\pi/h)x)} \tag{10}$$

satisfies (1). By substituting (10) into (1), we find the following solutions (leaving out the trivial case $\alpha_0 = \alpha_1 = 0$):

$$\alpha_0 = 1,\ \alpha_1 = 0,\ \omega = \pm c\kappa; \tag{11}$$

$$\alpha_0 = 0,\ \alpha_1 = 1,\ \omega = \pm c(\kappa - 2\pi/h). \tag{12}$$

Note that $u_j(t)$ in (3) approximates $u(x_j, t)$ in (10) if

$$\omega_h \approx \omega, \quad A_0 \approx \frac{\alpha_0 + \alpha_1}{2}, \quad A_1 \approx \frac{\alpha_0 - \alpha_1}{2}. \tag{13}$$

Each solution (7)–(8) of the eigenvalue problem $C\mathbf{v} = \chi\mathbf{v}$ approximates the exact solution $u(x,t)$ defined by one of the solutions (11)–(12) of the corresponding continuous problem (Fig. 2). Under this interpretation, neither eigenvalue mode is spurious.

The same argument applies to finite elements of degree N. In general, grid nodes are $x_j = x_{Nk+l} = kh + \zeta_l h$, where ζ_l is the l th collocation point $(0 \le l \le N)$. Let us consider equally spaced collocation points $\zeta_l = l/N$, so that $x_{Nk+l} = (Nk + l)h/N$. At an infinite, periodic grid, system (2) has N distinct equations. Substituting into these equations a discrete solution in the form

$$u_{Nk+l}(t) = A_{Nk+l} e^{-\mathrm{i}\,(\omega_h t - \kappa x_{Nk+l})}, \quad A_{Nk+l} = A_l\ (0 \le l < N), \tag{14}$$

we find an $N \times N$ eigenvalue system $C\mathbf{v} = \chi\mathbf{v}$ where, as in the quadratic case, $\chi = (h\omega_h/c)^2$ and $\mathbf{v} = [A_0, \ldots, A_{N-1}]^T$. As in (9), $u_j(t)$ can be seen as grid values

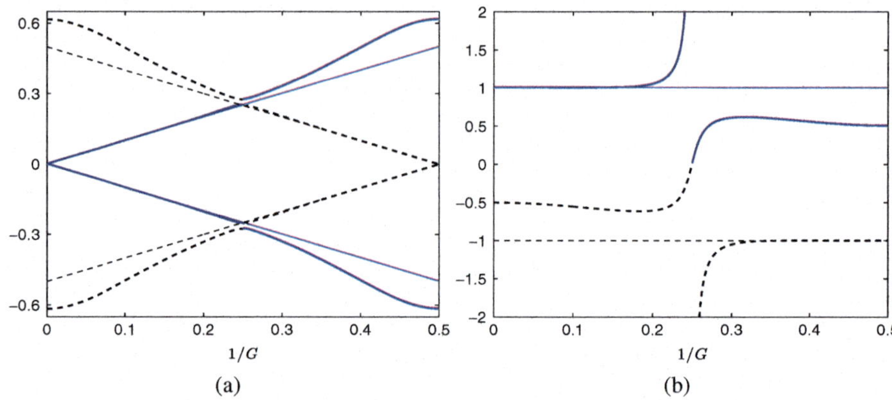

Fig. 2 Propagation modes of the quadratic finite element method: (**a**) Normalized angular frequency $\omega_h \Delta x/(2\pi c)$ with $\omega_h \approx \pm c\kappa$ (*solid*) and $\omega_h \approx \pm c(\kappa - 2\pi/h)$ (*dashed*); (**b**) amplitude ratio A_1/A_0 with A_0, A_1 approximated as in (13) and $\{\alpha_0, \alpha_1\} = \{1, 0\}$ (*solid*) and $\{\alpha_0, \alpha_1\} = \{1, 0\}$ (*dashed*). *Thinner solid* and *dashed lines* correspond to exact solutions

of the continuous function

$$\tilde{u}(x,t) = \sum_{l=0}^{N-1} a_l e^{-\mathrm{i}\left(\omega_h t - (\kappa - \frac{2\pi}{h} l)x\right)}. \tag{15}$$

It follows from (14) and (15) that coefficients a_l and A_l are related as follows:

$$A_l = \sum_{j=0}^{N-1} a_j e^{-\mathrm{i}\frac{2\pi}{N} jl}, \quad \text{i.e.,} \quad [A_0, \ldots, A_{N-1}] = N\,\mathrm{DFT}[a_0, \ldots, a_{N-1}], \tag{16}$$

where DFT denotes the discrete Fourier transform. The function corresponding to $u(x,t)$ in (10) with general N is given as

$$u(x,t) = \sum_{l=0}^{N-1} \alpha_l e^{-\mathrm{i}\left(\omega t - (\kappa - \frac{2\pi}{h} l)x\right)}. \tag{17}$$

There exist N possible coefficients $\{[\alpha_0, \ldots, \alpha_{N-1}], \omega\}$ such that function $u(x,t)$ in (17) is a solution of the acoustic wave equation (1), namely

$$\alpha_l^{(j)} = \delta_{lj} \quad (0 \le l \le N-1), \quad \omega^{(j)} = \pm c\left(\kappa - \frac{2\pi}{h} j\right), \quad 0 \le j \le N-1. \tag{18}$$

Remark 1 One can find alternative sets of solutions (with identical grid values) by replacing $\kappa - 2\pi l/h$ with $\kappa - 2\pi(l + qN)/h$ in (17), where q is an arbitrary integer.

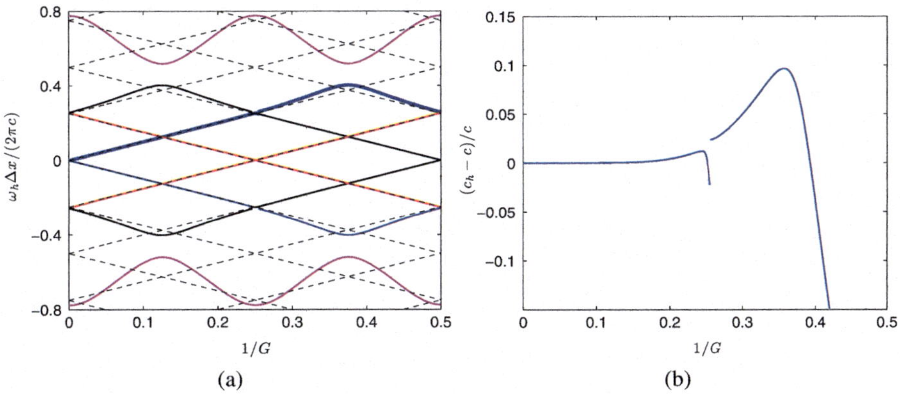

Fig. 3 Numerical dispersion the fourth-degree finite element method: (**a**) Normalized angular frequency $\omega_h \Delta x/(2\pi c)$, $\Delta x = h/4$. *Dashed lines* are $\omega = \pm(\kappa - 2\pi l/h)$, $l \in \mathbb{Z}$, and the *thicker solid line* corresponds to the acoustical mode; (**b**) Relative phase velocity error for the acoustical mode

We match each discrete solution in the form (14) with one of the continuous solutions defined by (18). Note that $u_{Nk+l}(t) \approx u(x_{Nk+l}, t)$ if $\omega_h \approx \omega$ and

$$[a_0, \ldots, a_{N-1}] \approx [\alpha_0, \ldots, \alpha_{N-1}], \quad [a_0, \ldots, a_{N-1}] = \frac{1}{N}\text{IDFT}[A_0, \ldots, A_{N-1}]. \tag{19}$$

Dashed and solid lines in Fig. 3a illustrate the matching between exact and approximate angular frequencies when the polynomial degree is $N = 4$ (additional exact solutions pointed out in Remark 1 were also considered). Unlike the quadratic case, not all numerical frequencies are equivalent to each other. Nevertheless, each mode can be seen as the approximation of a feasible solution. The relative phase velocity error of the acoustical mode $\omega_h \approx c\kappa$ is shown in Fig. 3b.

To locate the acoustical mode, one can find, for each κ, the numerical angular frequencies $\omega_h^{(1)}, \ldots, \omega_h^{(N)}$ from the eigenvalue system $C\mathbf{v} = \chi\mathbf{v}$ and then find the index that minimizes $\{|\omega_h^{(i)} - \omega_{aco}|,\ 1 \leq i \leq N\}$, where $\omega_{aco} = c\kappa$.

This procedure becomes more efficient if, rather than solving $C\mathbf{v} = \chi\mathbf{v}$ for all χ, we solve for the eigenvalue that is closer to $\chi_{aco} = (h\omega_{aco}/c)^2$. For this purpose, we use the inverse iteration algorithm [24] to compute the eigenvalue χ^* of $C - \chi_{aco}I$ with smallest magnitude and find ω_h^* such that $\chi^* + \chi_{aco} = (h\omega_h^*/c)^2$.

Remark 2 Is it not necessary to explicitly build the eigenvalue system $C\mathbf{v} = \chi\mathbf{v}$, which involves matrix inversion. For computational purposes it is more convenient to build a generalized eigenvalue system $\tilde{K}\mathbf{v} = \chi\tilde{M}\mathbf{v}$ (see, e.g., [9] for details). Accordingly, we may find the acoustical mode by applying the inverse iteration algorithm to the generalized eigenvalue problem $(\tilde{K} - \chi_{aco}\tilde{M})\mathbf{v} = \chi\tilde{M}\mathbf{v}$.

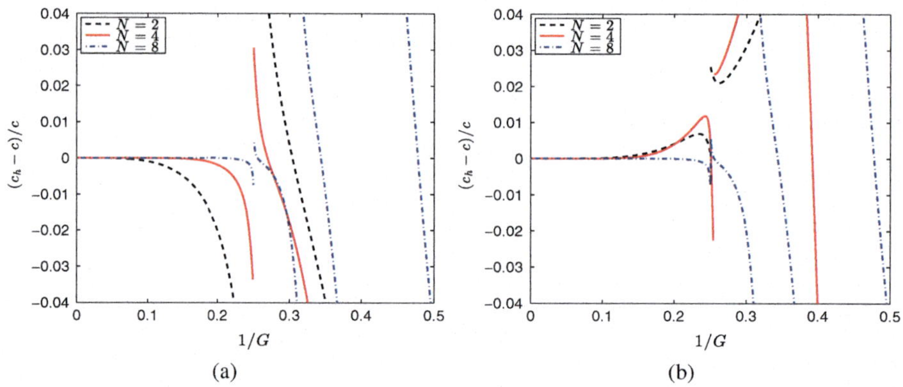

Fig. 4 Relative phase velocity error of finite elements of degree $N = 2, 4, 8$: (**a**) Gauss-Lobatto-Legendre (GLL) points; (**b**) Equally-spaced points

Remark 3 When the collocation points ζ_l are not evenly spaced, as usual on spectral element methods, the amplitude coefficients $A_0, \ldots, A_{N-1}$ are no longer related to the exact amplitudes $\alpha_0, \ldots, \alpha_{N-1}$ through the discrete Fourier transform. On the other hand, the numerical angular frequencies are handled exactly as in the case $\zeta_l = l/N$. Figure 4 shows the relative phase velocity errors for degrees $N = 2, 4, 8$ for Gauss-Lobatto-Legendre collocation points, considering that mass and stiffness matrices are computed with the quadrature defined by these points. Phase errors for consistent elements with equally-spaced collocation points are presented as well. Note that, when $G \geq 5$ (five grid points per wavelength or more), numerical dispersion consistently decreases with the polynomial degree (see also [3]).

4 Conclusions

Numerical dispersion of high-order finite elements has been widely studied, and several researchers agree that these methods are reliable despite the multiple modes of propagation resulting from variable-amplitude discrete plane waves.

In this work, all discrete modes of propagation have been interpreted as approximations to feasible solutions to the acoustic wave equation. Under this interpretation, secondary modes are not merely numerical artifacts, but numerical approximations on their own.

Such an interpretation relies on the fact that solutions in the form (14) are not grid values of a plane wave, but of a linear combination of plane waves, as it is long known in the literature. The novelty herein is to view the problem of determining the weights in the linear combination as a continuous version of the discrete eigenvalue system. In other words, multiple discrete modes approximate multiple continuous modes.

It is interesting to note that the amplitudes of the numerical plane wave and the weights of the linear combination are related by the discrete Fourier transform (DFT), when equally-spaced collocation points are employed. This establishes a connection between analyses based on linear combination of plane waves [1, 10] with those carried out in DFT space [16, 17, 19].

Acknowledgements The author is supported by CNPq under grant 306083/2014-0 and is a collaborator of INCT-GP.

References

1. N. Abboud, P. Pinsky, Finite element dispersion analysis for the three-dimensional second-order scalar wave equation. Int. J. Numer. Methods Eng. **35**(6), 1183–1218 (1992)
2. M. Ainsworth, Discrete dispersion relation for hp-version finite element approximation at high wave number. SIAM J. Numer. Anal. **42**(2), 553–575 (2004)
3. M. Ainsworth, H.A. Wajid, Dispersive and dissipative behavior of the spectral element method. SIAM J. Numer. Anal. **47**(5), 3910–3937 (2009)
4. T. Belytschko, R. Mullen, On dispersive properties of finite element solutions, in *Modern Problems in Elastic Wave Propagation*, ed. by J. Miklowitz, J. Achenbach (Wiley, New York, NY, 1978), pp. 67–82
5. L. Brillouin, *Wave Propagation in Periodic Structures. Electric Filters and Crystal Lattices*, 2nd edn. (Dover publications, New York, NY, 1953)
6. B. Cathers, B. O'Connor, The group velocity of some numerical schemes. Int. J. Numer. Methods Fluids **5**(3), 201–224 (1985)
7. G. Cohen, P. Joly, N. Tordjman, Eléments finis d'ordre élevé avec condensation de masse pour l'équation des ondes en dimension 1. Rapport de recherche RR-2323 (1994)
8. M. Cullen, The use of quadratic finite element methods and irregular grids in the solution of hyperbolic problems. J. Comput. Phys. **45**(2), 221–245 (1982)
9. J. De Basabe, M. Sen, Grid dispersion and stability criteria of some common finite-element methods for acoustic and elastic wave equations. Geophysics **72**(6), T81–T95 (2007)
10. D. Durran, Wave propagation in quadratic-finite-element approximations to hyperbolic equations. J. Comput. Phys. **159**(2), 448–455 (2000)
11. G. Gabard, R. Astley, M. Ben Tahar, Stability and accuracy of finite element methods for flow acoustics: I. General theory and application to one-dimensional propagation. Int. J. Numer. Methods Eng. **63**(7), 947–973 (2005)
12. P.M. Gresho, R.L. Lee, Comments on 'the group velocity of some numerical schemes'. Int. J. Numer. Methods Fluids **7**(12), 1357–1362 (1987)
13. D. Komatitsch, J. Tromp, Introduction to the spectral-element method for 3-D seismic wave propagation. Geophys. J. Int. **139**(3), 806–822 (1999)
14. K. Marfurt, Appendix - analysis of higher order finite-element methods, in *Numerical Modeling of Seismic Wave Propagation*, ed. by K. Kelly, K. Marfurt. Geophysics Reprint Series, vol. 13 (Society of Exploration Geophysicists, Tulsa, OK, 1990), pp. 516–520
15. R.C. Moura, S.J. Sherwin, J. Peiró, Linear dispersion–diffusion analysis and its application to under-resolved turbulence simulations using discontinuous Galerkin spectral/hp methods. J. Comput. Phys. **298**, 695–710 (2015)
16. W. Mulder, Spurious modes in finite-element discretizations of the wave equation may not be all that bad. Appl. Numer. Math. **30**(4), 425–445 (1999)
17. S.P. Oliveira, G. Seriani, DFT modal analysis of spectral element methods for the 2D elastic wave equation. J. Comput. Appl. Math. **234**(6), 1717–1724 (2009)

18. W. Scott, Errors due to spatial discretization and numerical precision in the finite-element method. IEEE Trans. Antennas Propagat. **42**(11), 1565–1570 (1994)
19. G. Seriani, S.P. Oliveira, Dispersion analysis of spectral element methods for acoustic wave propagation. J. Comput. Acoust. **16**(4), 531–561 (2008)
20. G. Seriani, S.P. Oliveira, Dispersion analysis of spectral element methods for elastic wave propagation. Wave Motion **45**(6), 729–744 (2008)
21. G. Seriani, E. Priolo, Spectral element method for acoustic wave simulation in heterogeneous media. Finite Elem. Anal. Des. **16**(3–4), 337–348 (1994)
22. L. Thompson, A review of finite-element methods for time-harmonic acoustics. J. Acoust. Soc. Am. **119**(3), 1315–1330 (2005)
23. L. Thompson, P. Pinsky, Complex wavenumber Fourier analysis of the p-version finite element method. Comput. Mech. **13**, 255–275 (1994)
24. L.N. Trefethen, D. Bau, *Numerical Linear Algebra* (SIAM, Philadelphia, 1997)

Viscous Stabilizations for High Order Approximations of Saint-Venant and Boussinesq Flows

Richard Pasquetti

Abstract Two viscous stabilization methods, namely the spectral vanishing viscosity (SVV) technique and the entropy viscosity method (EVM), are applied to flows of interest in geophysics. First, following a study restricted to one space dimension, the spectral element approximation of the shallow water equations is stabilized using the EVM. Our recent advances are here carefully described. Second, the SVV technique is used for the large-eddy simulation of the spatial and temporal development of the turbulent wake of a sphere in a stratified fluid. We conclude with a parallel between these two stabilization techniques.

1 Introduction

Simulations of flows that can develop stiff gradients generally suffer from numerical instabilities if nothing is done to stabilize the computation. Here we address shallow water flows and the turbulent wake of a sphere in a thermally stratified fluid. Shallow water flows are approximately governed by the Saint-Venant equations, i.e. by a non linear hyperbolic system that can yield shocks depending on the initial conditions. Moreover, one generally expects the numerical scheme to be well balanced and that it can support the presence of dry zones. A large literature is devoted to this questions, see e.g. [1, 24] and references herein. Our simulation of the stratified turbulent wake of a sphere relies on the incompressible Navier-Stokes equations coupled, within the Boussinesq approximation, to an advection-diffusion equation for the temperature. No shocks are in this case expected, but because the smallest scales of the flow cannot be captured by the mesh, here also a stabilization is required. Such a problem is generally addressed using the large-eddy simulation (LES) methodology, see e.g. [20]. In both cases, in the frame of high order methods,

R. Pasquetti (✉)
Université Côte d'Azur, CNRS, Inria, LJAD, France

Lab. J.A. Dieudonné (CASTOR project), Parc Valrose, 06108 Nice Cedex 2, France
e-mail: richard.pasquetti@unice.fr

M.L. Bittencourt et al. (eds.), *Spectral and High Order Methods for Partial Differential Equations ICOSAHOM 2016*, Lecture Notes in Computational Science and Engineering 119, DOI 10.1007/978-3-319-65870-4_37

typically spectrally accurate methods, standard approaches are generally not useful, because implying an unacceptable loss of accuracy.

This paper describes two stabilization techniques that allow, at least formally, to preserve the accuracy of high order methods by introducing relevant viscous terms: The entropy viscosity method (EVM) and the spectral vanishing viscosity (SVV) technique. In Sect. 2 we develop a spectral element method (SEM) for the Saint-Venant system and we stabilize it using the EVM. In Sect. 3 we use the SVV technique to carry out a Fourier-Chebyshev large-eddy simulation of the turbulent stratified wake of a sphere, on the basis of stabilized Boussinesq equations. To conclude, we provide in Sect. 4 a parallel between these two stabilization techniques.

2 Entropy Viscosity Stabilized Approximation of the Saint-Venant System

This part follows a previous paper [19] where the one-dimensional (1D) Saint-Venant system was considered and where we especially focused on problems involving dry-wet transitions. Here this work is extended to the 2D case. Moreover, the treatment of dry zones is improved and the well balanced feature of the scheme is focused on. Comparisons are done with an analytical solution involving dry-wet transitions and the result of a problem combining shocks with dry zones is presented.

The Saint-Venant system results from an approximation of the incompressible Euler equations which assumes that the pressure is hydrostatic and that the perturbations of the free surface are small compared to the water height. Then, from the mass and momentum conservation laws and with $\boldsymbol{\Omega} \subset \mathbb{R}^2$ for the computational domain, one obtains equations that describe the evolution of the height $h : \boldsymbol{\Omega} \to \mathbb{R}^+$ and of the horizontal velocity $\boldsymbol{u} : \boldsymbol{\Omega} \to \mathbb{R}^2$: For $(\boldsymbol{x}, t) \in \boldsymbol{\Omega} \times \mathbb{R}^+$:

$$\partial_t h + \nabla \cdot (h\boldsymbol{u}) = 0 \tag{1}$$

$$\partial_t (h\boldsymbol{u}) + \nabla \cdot (h\boldsymbol{uu} + gh^2\mathbb{I}/2) + gh\nabla z = 0 \tag{2}$$

with $\mathbb{I}$, identity tensor, $\boldsymbol{uu} \equiv \boldsymbol{u} \otimes \boldsymbol{u}$, g, gravity acceleration, and where $z(\boldsymbol{x})$ describes the topography, assumed such that $\nabla z \ll 1$. Let us recall the following properties:

- The system is nonlinear and hyperbolic, which means that discontinuities may develop;
- Assuming that the inlet flow-rate equals the outlet flow-rate, the total mass is preserved;
- The height h is non-negative;
- Rest solutions are stable;

- There exists a convex entropy (actually the energy E) such that

$$\partial_t E + \nabla \cdot ((E + gh^2/2)\boldsymbol{u}) \leq 0, \quad E = h\boldsymbol{u}^2/2 + gh^2/2 + ghz. \tag{3}$$

Set $\boldsymbol{q} = h\boldsymbol{u}$ and let $h_N(t)$ (resp. $\boldsymbol{q}_N(t)$) to be the piecewise polynomial continuous approximation of degree N of $h(t)$ (resp. $\boldsymbol{q}(t)$). The proposed stabilized SEM relies on the Galerkin approximation of the Saint-Venant system completed with mass and momentum viscous terms. For any $w_N, \boldsymbol{w}_N$ (scalar and vector valued functions, respectively) spanning the same approximation spaces, in semi-discrete form:

$$(\partial_t h_N + \nabla \cdot \boldsymbol{q}_N, w_N)_N = -(\nu_h \nabla h_N, \nabla w_N)_N \tag{4}$$

$$(\partial_t \boldsymbol{q}_N + \nabla \cdot I_N(\boldsymbol{q}_N \boldsymbol{q}_N / h_N) + g h_N \nabla (h_N + z_N), \boldsymbol{w}_N)_N = -(\nu_q \nabla \boldsymbol{q}_N, \nabla \boldsymbol{w}_N)_N \tag{5}$$

where $\nu_h \propto \nu_q = \nu$, with ν : entropy viscosity (in the rest of the paper we simply use $\nu_h = \nu_q$). The usual SEM approach is used here: I_N is the piecewise polynomial interpolation operator, based for each element on the tensorial product of Gauss-Lobatto-Legendre (GLL) points, and $(.,.)_N$ stands for the SEM approximation of the $L^2(\boldsymbol{\Omega})$ inner product, using for each element the GLL quadrature formula which is exact for polynomials of degree less than $2N - 1$ in each variable. The following remarks may be expressed:

- Mass conservation is ensured by the present SEM approximation: Set $w_N = 1$ in Eq. (4); If $\int_\Gamma \boldsymbol{q}_N \cdot d\boldsymbol{\Gamma} = 0$, where Γ is the boundary of $\boldsymbol{\Omega}$, then the equalities

$$\int_{\boldsymbol{\Omega}} (\partial_t h_N + \nabla \cdot \boldsymbol{q}_N)\, d\boldsymbol{\Omega} = \int_{\boldsymbol{\Omega}} \partial_t h_N \, d\boldsymbol{\Omega} + 0 = d_t \int_{\boldsymbol{\Omega}} h_N \, d\boldsymbol{\Omega} = 0$$

  still hold after the SEM discretization. Note however that it is assumed that the Jacobian determinants of the mappings from the reference element $(-1, 1)^2$ to the mesh elements are piecewise polynomials of degree less than N.
- On the contrary, the expected conservation of energy for smooth problems is approximate; This results from the presence of nonlinear terms in (3).
- A stabilization term appears in the mass equation (4). This is required when a high order approximation is involved, i.e. when the scheme numerical diffusion becomes negligible.
- In the momentum equation (5) we do not use the viscous term $\nabla \cdot (h\nu \nabla \boldsymbol{u})$, which turned out to be less efficient. Indeed, for stabilization purposes the physically relevant stabilization may not be the best suited, see e.g. [8] for the Euler equations.
- Thanks to using $\nabla \cdot I_N(g h_N^2 \mathbb{I}/2) \approx g h_N \nabla h_N$ (while h_N^2 is generally piecewise polynomial of degree greater than N), and thus grouping in (5) the pressure and topography terms, a well balanced scheme is obtained by construction: If $\boldsymbol{q}_N \equiv 0$ and $h_N \neq 0$, then $h_N + z_N = Constant$.

- Another difficulty comes from the positivity of h_N. This point is addressed at the end of the present Section.

It remains to define the entropy viscosity ν. To this end we make use of an entropy that does not depend on z but on ∇z, which is of interest, at the discrete level, to get free of the choice of the coordinate system. Taking into account the mass conservation equation (into the entropy equation) one obtains:

$$\partial_t \tilde{E} + \nabla \cdot ((\tilde{E} + gh^2/2)\boldsymbol{u}) + gh\boldsymbol{u} \cdot \nabla z \leq 0\,, \quad \tilde{E} = h\boldsymbol{u}^2/2 + gh^2/2\,. \tag{6}$$

At each time-step, we then compute the entropy viscosity $\nu(\boldsymbol{x})$ at the GLL grid points, using the following three steps procedure:

- Assuming all variables given at time t_n, compute the entropy residual, using a backward difference formula, e.g. the BDF2 scheme, to approximate $\partial_t \tilde{E}_N$

  $$r_E = \partial_t \tilde{E}_N + \nabla \cdot I_N((\tilde{E}_N + gh_N^2/2)\boldsymbol{q}_N/h_N) + g\boldsymbol{q}_N \cdot \nabla z_N$$

  where $\tilde{E}_N = \boldsymbol{q}_N^2/(2h_N) + gh_N^2/2$. Then set up a viscosity ν_E such that:

  $$\nu_E = \beta |r_E| \delta x^2 / \Delta E\,,$$

  where ΔE is a reference entropy, β a user defined control parameter and δx the local GLL grid-size, defined such that δx^2 equals the surface of the quadrilateral cell (of the dual GLL mesh) surrounding the GLL point, and using symmetry assumptions for the points at the edges and vertices of the element.
- Define a viscosity upper bound based on the wave speeds : $\lambda_{\pm} = u \pm \sqrt{gh}$:

  $$\nu_{max} = \alpha \max_{\boldsymbol{\Omega}}(|\boldsymbol{q}_N/h_N| + \sqrt{gh_N})\delta x$$

  where α is a $O(1)$ user defined parameter (recall that for the advection equation $\alpha = 1/2$ is well suited).
- Compute the entropy viscosity:

  $$\nu = \min(\nu_{max}, \nu_E)$$

  and smooth: (1) locally (in each element), e.g. in 1D: $(\nu_{i-1} + 2\nu_i + \nu_{i+1})/4 \to \nu_i$; (2) globally, by projection onto the space of the C^0 piecewise polynomials of degree N. Note that operation (2) is cheap because the SEM mass matrix is diagonal.

The positivity of h is difficult to enforce as soon as $N > 1$, so that for problems involving dry-wet transitions the present EVM methodology must be completed. The algorithm that we propose is the following: In dry zones, i.e. for any element

Q_{dry} such that at one GLL point $\min h_N < h_{thresh}$, where h_{thresh} is a user defined threshold value (typically a thousandth of the reference height):

- Modify the entropy viscosity technique, by using in Q_{dry} the upper bound first order viscosity:

$$\nu = \nu_{max} \quad \text{in} \quad Q_{dry}$$

- In the momentum equation assume that:

$$h_N g \nabla(h_N + z_N) \equiv 0 \quad \text{in} \quad Q_{dry}$$

- Moreover, notice that the upper bound viscosity $\nu_{\max}$ is not local but global, and that the entropy scaling ΔE used in the definition of ν_E is time independent. This has improved the robustness of the general approach described in [9].

The numerical results presented hereafter have been obtained using the standard SEM for the discretization in space (GLL nodes for interpolations and quadratures) and the usual forth order Runge-Kutta (RK4) scheme for the discretization in time.

To outline the efficiency of the present EVM for stabilization of Saint-Venant flows involving dry-wet transitions, we first consider a problem for which an analytical solution is available: The planar fluid surface oscillations in a paraboloid [23]. The topography is defined by $z(\boldsymbol{x}) = D_0 \boldsymbol{x}^2/L^2$ and the exact solution writes, with (x_1, x_2) for the Cartesian components of $\boldsymbol{x}$:

$$h = \max(0, 2\eta \frac{D_0}{L}(\frac{x_1}{L}\cos(\omega t) - \frac{x_2}{L}\sin(\omega t) - \frac{\eta}{2L}) + D_0 - z)$$

$$\boldsymbol{u} = -\eta\omega(\sin(\omega t), \cos(\omega t))$$

where η determines the amplitude of the motion and with $\omega = \sqrt{2gD_0}/L$. Hereafter we use $L = 1\,m$, $D_0 = 0.1\,m$ and $\eta = 0.5\,m$. The dry-wet transition being at the intersection of the paraboloid and of the planar surface of the fluid, in the (x_1, x_2) plane the rotating motion of a circle is obtained. Moreover, since $\boldsymbol{u}$ does not depend on $\boldsymbol{x}$, all fluid particles have similar circular trajectories.

The computational domain is a circle centered at the origin and of diameter 4. The discretization parameters are the following, number of elements: 2352, polynomial approximation degree: $N = 5$, resulting number of grid-points: 59,081, time-step: 2.96×10^{-4} , EVM control parameters: $\alpha = 1, \beta/\Delta E = 10$. Moreover, $h_{thresh} = \max_{\Omega} h_0/500$, with $h_0 \equiv h(t = 0)$, and at the initial time one sets $h_N = I_N(\max(h_0, h_{thresh}/2))$. Since here the exact solution is only C^0 continuous, the computation cannot be spectrally accurate and so the convergence rate would be disappointing. This is why, in order to outline the interest of using a high order method like the present stabilized SEM, we prefer to compare the EVM results to results obtained with a first order viscosity. One looks in Fig. 1 at the error $|h_N - h|(\boldsymbol{x})$ at time $t \approx 6\pi/\omega$, i.e. after three loops of the fluid surface inside the paraboloid.

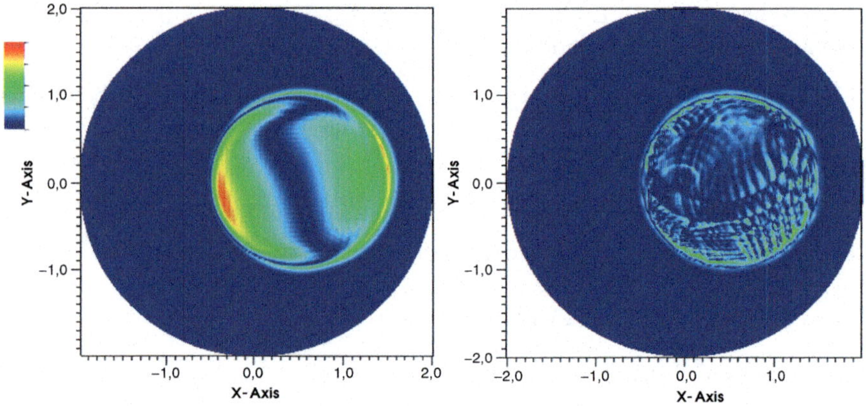

Fig. 1 Planar oscillations in a paraboloid: error on the height and at the final time for the first order viscosity (*left*) and EVM (*right*) solutions (color-scale bar: 0–0.01, from *blue* to *red*)

Figure 1 (left) gives this error field when using the first order viscosity, that is with $\nu_h = \nu_q = \nu_{\max}$, whereas Fig. 1 (right) shows the result obtained when using the EVM. Clearly, the naive approach that would consist of simply using a first order viscosity for stabilization yields a result by far worse than the one obtained with the EVM. Animations of these error fields clearly confirm such a conclusion. Another test case provided in [23] has also been investigated, namely the axisymmetric oscillations in a paraboloid, and the conclusion is quite similar.

To conclude this Section we give the results of a simulation that shows axisymmetric oscillations in a paraboloid and that involves both dry-wet transitions and shocks. The initial condition is the following:

$$h = \max(1 - \boldsymbol{x}^2, 0)\,, \quad \boldsymbol{u} = (0,0)\,,$$

and at the boundary an impermeability condition together with an homogeneous Neumann condition for h are imposed. The geometry and the space discretization are those used in the first example. Calculations have been made till time $t = 5$ with time-step 10^{-4}, and the EVM control parameters are: $\alpha = 1, \beta/\Delta E = 20$. As previously, $h_{thresh} = \max_{\boldsymbol{\Omega}} h_0/500$ and at the initial time $h_N = I_N(\max(h_0, h_{thresh}/2))$. Such a flow is alternatively expanding and then retracting towards the paraboloid axis. Figure 2 shows the flow at three different times, during the first retraction-expansion phase: At $t \approx 1.4$ the velocity field is inwards, at $t \approx 1.65$ it is close to reversal and at $t \approx 1.9$ it is outwards. The height h_N (at left) and the entropy viscosity ν (at right) are visualized. As desired, the entropy viscosity saturates in dry zones and also focuses at the shock.

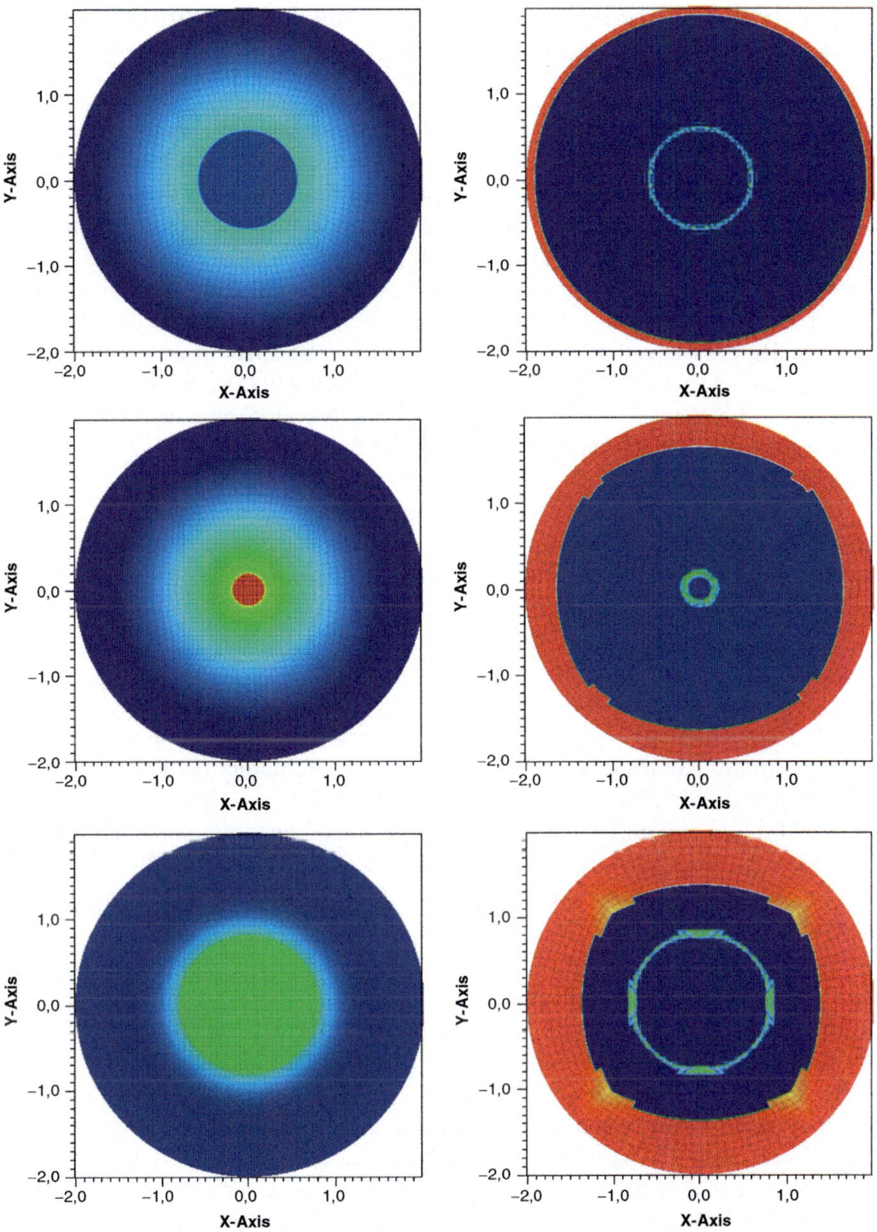

Fig. 2 Axisymmetric oscillations with shocks in a paraboloid: height (*left*, in [0, 1]) and entropy viscosity (*right*, in [0, 0.025]), at $t \approx 1.4$ (*top*), $t \approx 1.65$ (*middle*) and $t \approx 1.9$ (*bottom*)

3 Spectral Vanishing Viscosity for Large-Eddy Simulation of the Stratified Wake of a Sphere

To demonstrate the interest of using the SVV technique for the computation of turbulent flows, we consider the turbulent wake of a sphere in a thermally stratified fluid. Here we just focus on the main characteristics of the SVV technique and illustrate its capabilities for this particular geophysical flow. Details concerning the physical study may be found in [16].

The SVV technique was initially developed to solve with spectral methods (Fourier/Legendre expansions) hyperbolic problems (non-linear, scalar, 1D, typically the Burgers equation), while (1) preserving the spectral accuracy and (2) providing a stable scheme [13, 22]. Later on, say in the 2000s, the SVV technique turned out to be of interest for stabilization of the Navier-Stokes equations and so for the large-eddy simulation (LES) of turbulent flows, see e.g. [10–12, 15, 17, 25] and references herein.

The basic idea of the SVV stabilization technique is to add some artificial viscosity on the highest frequencies, i.e., to complete the conservation law of some given quantity u, assumed to be scalar for the sake of simplicity, with the SVV term:

$$V_N \equiv \epsilon_N \nabla \cdot (Q_N(\nabla u_N))$$

where N is again the polynomial degree, u_N the numerical approximation of u, ϵ_N a $O(1/N)$ coefficient and Q_N the so-called spectral viscosity operator, defined to select the highest frequencies: If Q_N is omitted in the definition of V_N, then one recovers the regular diffusion operator.

In spectral space (Fourier, Chebyshev, Legendre, or any other hierarchical basis) the operator Q_N is defined by a set of coefficients $\widehat{Q}_k$, $0 \le k \le N$. Thus, in the 1D periodic case and if Fourier expansions are concerned (trigonometric polynomials are involved in this case):

$$(\widehat{V}_N)_k = -\epsilon_N \widehat{Q}_k k^2 (\widehat{u}_N)_k\,, \quad (\widehat{V}_N)_{-k} = \overline{(\widehat{V}_N)_k}$$

with $(\widehat{.})_k$ for the k-Fourier component, $\overline{(.)}$ for complex conjugate, and where the coefficients $\widehat{Q}_k$ are such that $\widehat{Q}_k = 0$ for $k \le m_N$, with m_N a threshold value, and $0 < \widehat{Q}_k \le 1$, $\widehat{Q}_k$ being monotonically increasing, for $k > m_N$. In the seminal paper [22], $\widehat{Q}_k$ was simply chosen as a step function, i.e. with $\widehat{Q}_k = 1$ if $k > m_N$, but it quickly appeared of interest to rather use a smooth approximation. In Fig. 3 two SVV kernels are shown, using the smooth approximation proposed in [13] and with $m_N = N/2$ and $m_N = N/4$.

One may remark that if defining differently the $\widehat{Q}_k$ coefficients, one recovers alternative approaches. Thus, the hyperviscosity stabilization, i.e. such that $V_N \propto -\Delta^2 u_N$, so that $(\widehat{V}_N)_k \propto k^4 (\widehat{u}_N)_k$, can be recasted in the SVV frame by stating that $\widehat{Q}_k = (k/N)^2$. In this case, the interpolating curve $\widehat{Q}_k(k)$ is simply a parabola, see Fig. 3. Also, if choosing $\epsilon_N \propto N^{-4/3}$ and with the curve $\widehat{Q}_k(k)$ shown in Fig. 3,

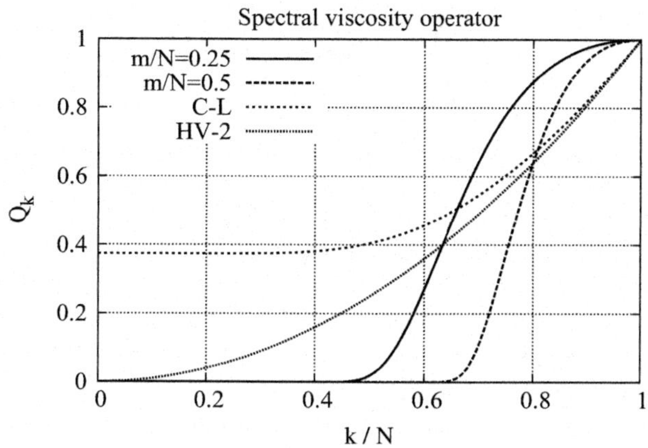

Fig. 3 $\widehat{Q}_k$ for two SVV kernels, and if using an hyperviscosity (HV) or the Chollet-Lesieur (C-L) subgrid scale model

one recovers the Chollet-Lesieur spectral viscosity [3], developed for LES in the early 80s and based on the Eddy Damped Quasi-Normal Markovian (EDQNM) theory. This latter approach strongly differs from SVV, since $\widehat{Q}_k \neq 0$ even for $k = 0$, but both approaches also differ from SVV because the associated kernels no-longer constitute an approximation of a step function. As a result, it is no longer a Laplacean which is acting in the high frequency range. Indeed, if the SVV kernel is simply a step function, then the stabilization term also writes $V_N = \epsilon_N \Delta u_N^H$, where u_N^H is the high frequency part of u_N. In this perspective, looking at the SVV kernels recently proposed in [14] is also of interest.

Results obtained with a multi-domain spectral Chebyshev-Fourier solver with SVV stabilization are presented hereafter, see [16] for details. The turbulent wake is generated by a sphere moving horizontally and at constant velocity in a stably stratified fluid, with constant temperature gradient. The flow is assumed governed by the Boussinesq equations and the control parameters are $Pr = 7, Re = 10000, F = 25$, for the Prandtl, Reynolds and internal Froude number, respectively. The study is carried out in two steps: First we make a space development study, that is the Galilean frame is associated to the moving sphere; Second, we make a time development study, with a Galilean frame at rest. Such an approach is required to compute the far wake without needing a very elongated computational domain. With respect to anterior works, see e.g. [4, 5], the originality of the present study is to make use of the result of the space development study for setting up the initial condition of the temporal development study, thus avoiding the use of a synthetic initial condition.

For the space development study, the computational domain is $(-4.5, 30.5) \times (-4, 4) \times (-4, 4)$. At the initial time, the fluid is stably stratified: $T_0 = y$, for the dimensionless initial temperature field. The sphere, of unit diameter and centered at $(0, 0)$, is modeled by using a volume penalization technique [18]. For the boundary conditions one has: Dirichlet conditions at the inlet, advection at the mean velocity at the outlet, free-slip/adiabaticity conditions on the upper and lower boundaries. The mesh makes use of 12.4×10^6 grid-points. For the velocity components, the

SVV parameters are: $m_N = N/2$, $\epsilon_N = 1/N$, where N is here associated to each of the three axis, whereas for the temperature, $m_N = \sqrt{N}$, $\epsilon_N = 1/N$. The temperature field obtained at the end of the space development study is shown in Fig. 4.

For the temporal development study the computational domain is $(-18, 18) \times (-4, 4) \times (-12, 12)$. Note it has been enlarged, since we expect, from the confinement effect due to the stratification, a strong expansion of the wake in the horizontal plane. The initial conditions are set up from the spatial development study, by extraction of the fields obtained at the final time in $(6.5, 24.5)\times(-4, 4)\times(-4, 4)$, see [16] for details. The boundary conditions are: periodicity in streamwise direction, free-slip and adiabaticity elsewhere. The mesh makes use of 27.7×10^6 grid-points. The SVV parameters have not been changed. The flow, at the beginning and at the end of the temporal study, is visualized in Fig. 5.

We conclude with some more quantitative results: In Fig. 6-left one has the evolution of the wake amplitude, both in the vertical and horizontal planes,

Fig. 4 Temperature field in the median vertical plane at the end of the space development study (*color scale*: −4 to 4, from *blue to red*) [16]

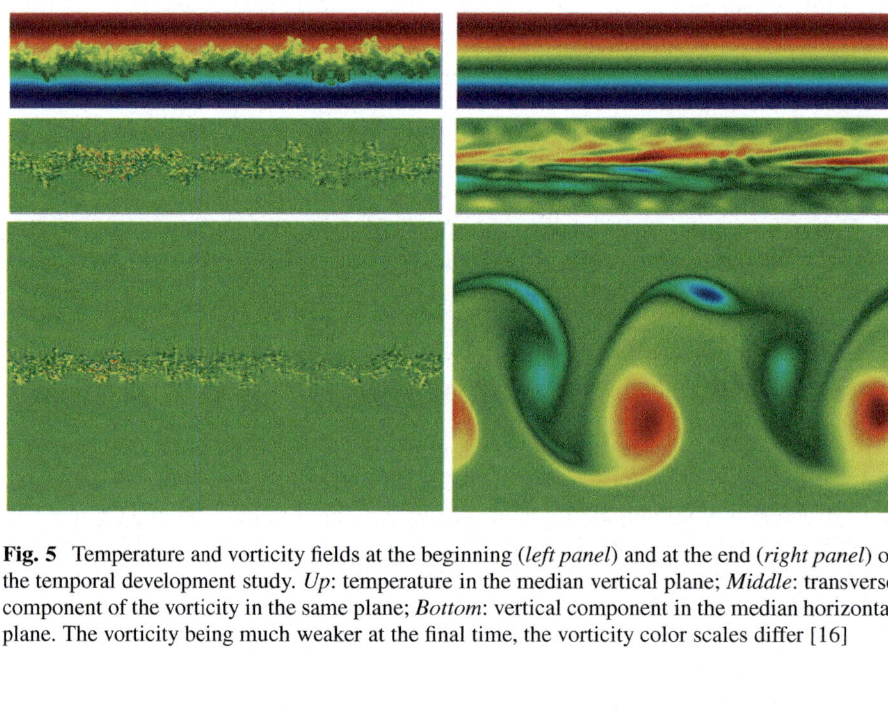

Fig. 5 Temperature and vorticity fields at the beginning (*left panel*) and at the end (*right panel*) of the temporal development study. *Up*: temperature in the median vertical plane; *Middle*: transverse component of the vorticity in the same plane; *Bottom*: vertical component in the median horizontal plane. The vorticity being much weaker at the final time, the vorticity color scales differ [16]

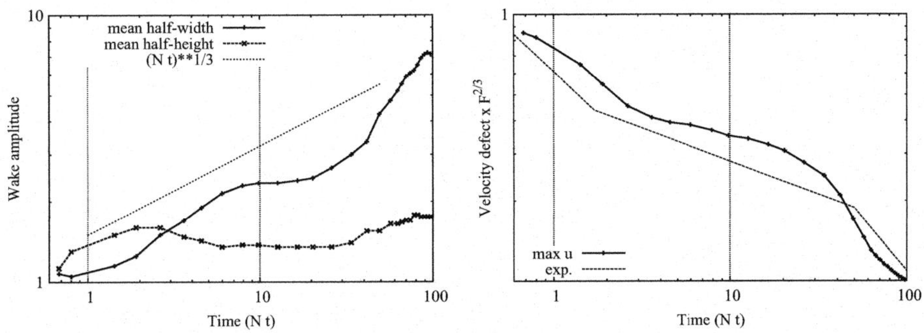

Fig. 6 Wake amplitude (*left*) and velocity deficit (*right*) vs time (N, buoyancy frequency) [16]

	Nt_I	Nt_{II}	3D rate	NEQ rate	Q2D rate
[21]	1.7 ± 0.3	50 ± 15	$-2/3$	-0.25 ± 0.4	-0.76 ± 0.12
SVV-LES	2.4	30	$\sim -2/3$	-0.2	~ -0.76

Fig. 7 Critical times Nt_I and Nt_{II} that correspond to the 3D-NEQ and NEQ-Q2D transitions, respectively, and rates of variation of the velocity defect in each of the three phases

which clearly points out the confinement effect of the stratification (here Nt is a dimensionless time, with N for the Brunt-Väisälä buoyancy angular frequency of the fluid at rest); Fig. 6-right shows the evolution of the velocity defect. This latter curve is in reasonable agreement with the "universal curve" of [21], where three phases in the development of sphere stratified wakes are described: First the 3D phase, then the non equilibrium (NEQ) phase and finally the quasi two 2D (Q2D) phase. Figure 7 compares the experimental results to the present numerical ones, in terms of characteristic quantities of the velocity defect evolution.

4 Concluding Parallel Between the EVM and SVV Stabilizations

Two viscous stabilizations, namely the entropy viscosity method (EVM) and the spectral vanishing viscosity (SVV) technique, have been successfully implemented in high order approximations of geophysical flows: With EVM shallow water flows involving dry-wet transitions have been addressed and with SVV the turbulent wake of a sphere in a thermally stratified fluid has been investigated. We conclude with a parallel between these two approaches:

- Both SVV and EVM are viscosity methods first developed for hyperbolic problems, see [7, 22].
- EVM is non-linear while SVV is linear. SVV is thus not costly, since its implementation is done in preprocessing step. Moreover, because of this linear feature it is very robust and so well adapted to the LES of turbulent flows.

- Both EVM and SVV preserve the accuracy of the numerical approximation. This is of course essential when high order methods are concerned.
- SVV is not Total Variation Diminishing (TVD) and EVM is not fully TVD, since this depends on the values of the EVM control parameters (α and β). For SVV, a post processing stage for removing spurious oscillations has been suggested [13].
- A theory exists for SVV [22], whereas no complete theory is available for EVM. Some theoretical results, restricted to some specific time schemes and space approximations, are however available [2].
- EVM may be used with various numerical methods, since based on a physical argument, including the standard finite element method (FEM), finite volume methods etc.. SVV is restricted to spectral type methods, e.g. high order FEMs like the SEM.
- SVV has proved to be of interest for LES (SVV-LES). Preliminary numerical experiments are now available for EVM [6], but additional tests and comparisons remain needed to check if EVM is not too diffusive and robust enough when turbulent flows are concerned.

Acknowledgements Part of this work was made at the Dpt of Mathematics of National Taiwan University in the frame of the Inria project AMOSS.

References

1. C. Berthon, F. Marche, A positive preserving high order VFRoe scheme for shallow water equations: a class of relaxation schemes. SIAM J. Sci. Comput. **30**, 2587–2612 (2008)
2. A. Bonito, J.L. Guermond, B. Popov, Stability analysis of explicit entropy viscosity methods for non-linear scalar conservation equations. Math. Comput. **83**, 1039–1062 (2014)
3. J.P. Chollet, M. Lesieur, Parametrisation of small scales of three-dimensional isotropic turbulence utilizing spectral closures. J. Atmos. Sci. **38**, 2747–2757 (1981)
4. P.J. Diamessis, J.A. Domaradzki, J.S. Hesthaven, A spectral multidomain penalty method model for the simulation of high Reynolds number localized incompressible stratified turbulence. J. Comput. Phys. **202**, 298–322 (2005)
5. D.G. Dommermuth, J.W. Rottman, G.E. Innis, E.V. Novikov, Numerical simulation of the wake of a towed sphere in a weakly stratified fluid. J. Fluid Mech. **473**, 83–101 (2002)
6. J.-L. Guermond, A. Larios, T. Thompson, Validation of an entropy-viscosity model for large eddy simulation. Direct and Large Eddy-Eddy Simulation IX, ECOFTAC Series **20**, 43–48 (2015)
7. J.-L. Guermond, R. Pasquetti, Entropy-based nonlinear viscosity for Fourier approximations of conservation laws. C.R. Acad. Sci. Paris Ser. I **346**, 801–806 (2008)
8. J.L. Guermond, B. Popov, Viscous regularization of the Euler equations and entropy principles. SIAM J. Appl. Math. **74**(2), 284–305 (2014)
9. J.L. Guermond, R. Pasquetti, B. Popov, Entropy viscosity method for non-linear conservation laws. J. Comput. Phys. **230**(11), 4248–4267 (2011)
10. G.S. Karamanos, G.E. Karniadakis, A spectral vanishing viscosity method for large-eddy simulation. J. Comput. Phys. **163**, 22–50 (2000)
11. R.M. Kirby, S.J. Sherwin, Stabilisation of spectral / *hp* element methods through spectral vanishing viscosity: application to fluid mechanics. Comput. Methods Appl. Mech. Eng. **195**, 3128–3144 (2006)

12. K. Koal, J. Stiller, H.M. Blackburn, Adapting the spectral vanishing viscosity method for large-eddy simulations in cylindrical configurations. J. Comput. Phys. **231**, 3389–3405 (2012)
13. Y. Maday, S.M.O. Kaber, E. Tadmor, Legendre pseudo-spectral viscosity method for nonlinear conservation laws. SIAM J. Numer. Anal. **30**, 321–342 (1993)
14. R.C. Moura, S.J. Sherwin, J. Peiró, Eigensolution analysis of spectral/hp continuous Galerkin approximations to advection-diffusion problems: insights into spectral vanishing viscosity. J. Comput. Phys. **307**, 401–422 (2016)
15. R. Pasquetti, Spectral vanishing viscosity method for large-eddy simulation of turbulent flows. J. Sci. Comput. **27**, 365–375 (2006)
16. R. Pasquetti, Temporal/spatial simulation of the stratified far wake of a sphere. Comput. Fluids **40**, 179–187 (2010)
17. R. Pasquetti, E. Séverac, E. Serre, P. Bontoux, M. Schäfer, From stratified wakes to rotor-stator flows by an SVV-LES method, Theor. Comput. Fluid Dyn. **22**, 261–273 (2008)
18. R. Pasquetti, R. Bwemba, L. Cousin, A pseudo-penalization method for high Reynolds number unsteady flows. Appl. Numer. Math. **58**(7), 946–954 (2008)
19. R. Pasquetti, J.L. Guermond, B. Popov, Stabilized spectral element approximation of the Saint-Venant system using the entropy viscosity technique, in *Lecture Notes in computational Science and Engineering: Spectral and High Order Methods for Partial Differential Equations - ICOSAHOM 2014*, vol. 106 (Springer, Berlin, 2015), pp. 397–404
20. P. Sagaut, *Large Eddy Simulation for Incompressible Flows* (Springer, Berlin, Heidelberg, 2006)
21. G.R. Spedding, F.K. Browand, A.M. Fincham, The evolution of initially turbulent bluff-body wakes at high internal Froude number. J. Fluid Mech. **337**, 283–301 (1997)
22. E. Tadmor, Convergence of spectral methods for nonlinear conservation laws. SIAM J. Numer. Anal. **26**, 30–44 (1989)
23. W.C. Thacker, Some exact solutions to the nonlinear shallow-water wave equations, J. Fluid Mech. **107**, 499–508 (1981)
24. Y. Xing, X. Zhang, Positivity-preserving well-balanced discontinuous Galerkin methods for the shallow water equations on unstructured triangular meshes. J. Sci. Comput. **57**, 19–41 (2013)
25. C.J. Xu, R. Pasquetti, Stabilized spectral element computations of high Reynolds number incompressible flows. J. Comput. Phys. **196**, 680–704 (2004)

An Adaptive Variable Order Quadrature Strategy

Paul Houston and Thomas P. Wihler

Abstract We propose a new adaptive numerical quadrature procedure which includes both local subdivision of the integration domain, as well as local variation of the number of quadrature points employed on each subinterval. In this way we aim to account for local smoothness properties of the integrand as effectively as possible, and thereby achieve highly accurate results in a very efficient manner. Indeed, this idea originates from so-called *hp*-version finite element methods which are known to deliver high-order convergence rates, even for nonsmooth functions.

1 Introduction

Numerical integration methods have witnessed a tremendous development over the last few decades; see, e.g., [2, 3, 14]. In particular, adaptive quadrature rules have nowadays become an integral part of many scientific computing codes. Here, one of the first yet very successful approaches is the application of adaptive Simpson integration or the more accurate Gauss-Kronrod procedures (see, e.g., [7]). The key points in the design of these methods are, first of all, to keep the number of function evaluations low, and, secondly, to divide the domain of integration in such a way that the features of the integrand function are appropriately and effectively accounted for.

The aim of the current article is to propose a complementary adaptive quadrature approach that is quite different from previous numerical integration schemes. In fact, our work is based on exploiting ideas from *hp*-type adaptive finite element methods (FEM); cf. [4, 6, 11, 12, 20]. These schemes accommodate and combine both traditional low-order adaptive FEM and high-order (so-called spectral) methods within a single unified framework. Specifically, their goal is to generate discrete

P. Houston (✉)
School of Mathematical Sciences, University of Nottingham, University Park, Nottingham NG7 2RD, UK
e-mail: Paul.Houston@nottingham.ac.uk

T.P. Wihler
Mathematisches Institut, Universität Bern, Sidlerstrasse 5, CH-3012 Bern, Switzerland
e-mail: wihler@math.unibe.ch

M.L. Bittencourt et al. (eds.), *Spectral and High Order Methods for Partial Differential Equations ICOSAHOM 2016*, Lecture Notes in Computational Science and Engineering 119, DOI 10.1007/978-3-319-65870-4_38

approximation spaces which allow for both adaptively refined subdomains, as well as locally varying approximation orders. In this way, the *hp*-FEM methodology is able to resolve features of an underlying unknown analytical solution in a highly efficient manner. In fact, this approach has proved to be enormously successful in the context of numerically approximating solutions of differential equations, and has been shown to exhibit high-order algebraic or exponential convergence rates even in the presence of local singularities; cf. [8, 16, 18].

With this in mind, we adopt the *hp*-adaptive FEM strategy for the purpose of introducing a variable order adaptive quadrature framework. More precisely, we propose a procedure whereby the integration domain will be subdivided adaptively in combination with a local tuning of the number of quadrature points employed on each subinterval. To drive this refinement process, we employ a smoothness estimation technique from [6, 21] (see also [11] for a related strategy), which was originally introduced in the context of *hp*-adaptive FEMs. Specifically, the smoothness test makes it possible to gain local information concerning the regularity of the integrand function, and thereby, to suitably subdivide the integration domain and select an appropriate number of quadrature points for each subinterval. By means of a series of numerical experiments we demonstrate that the proposed adaptive quadrature strategy is capable of generating highly accurate approximations at a very low computational cost. The main ideas on this new approach together with a view on practical aspects will be discussed in the subsequent section.

2 An *hp*-Type Quadrature Approach

Typical quadrature rules for the approximation of an integral $I := \int_{-1}^{1} f(x)\,\mathrm{d}x$ of a continuous function $f : [-1,1] \to \mathbb{R}$, take the form $I \approx \widehat{Q}_p(f) := \sum_{k=1}^{p} w_{p,k} f(\widehat{x}_{p,k})$, where $p \geq 1$ is a (typically prescribed) integer number, and $\{\widehat{x}_{p,k}\}_{k=1}^{p} \subset [-1,1]$ and $\{w_{p,k}\}_{k=1}^{p} \subset (0,2]$ are appropriate quadrature points and weights, respectively. When dealing with a variable number p of quadrature points and weights, we can consider one-parameter families of quadrature rules (such as, for example, Gauss-type quadrature methods); here, for each $p \in \mathbb{N}$, with $p \geq p_{\min}$, where $p_{\min}$ is a minimal number of points, there are (possibly non-hierarchical) families of quadrature points $\widehat{\boldsymbol{x}}_p = \{\widehat{x}_{p,k}\}_{k=1}^{p}$, and weights $\boldsymbol{w}_p = \{w_{p,k}\}_{k=1}^{p}$.

On an arbitrary bounded interval $[a,b]$, $a < b$, a corresponding integration formula can be obtained, for instance, by means of a simple affine scaling $\phi_{[a,b]} : [-1,1] \to [a,b]$, $\widehat{x} \mapsto x = \phi_{[a,b]}(\widehat{x}) = \frac{1}{2}h\widehat{x} + \frac{1}{2}(a+b)$, with $h = b - a > 0$. Indeed, in this case $\int_a^b f(x)\,\mathrm{d}x \approx Q_{[a,b],p}(f) := h/2 \sum_{k=1}^{p} w_{p,k}(f \circ \phi_{[a,b]})(\widehat{x}_{p,k})$, where $f : [a,b] \to \mathbb{R}$ is again continuous. As before, for any specific family of quadrature rules, the corresponding quadrature point families $\boldsymbol{x}_p$ are obtained in a straightforward way by letting $\boldsymbol{x}_p = \phi_{[a,b]}(\widehat{\boldsymbol{x}}_p)$ (noting that $\phi_{[a,b]}$ is extended componentwise to vectors).

Furthermore, the above construction allows us to define composite quadrature rules, whereby the integral of f is approximated on a collection of $n \geq 1$ disjoint (open) subintervals $\{K_i\}_{i=1}^n$ of $[a,b]$ with $[a,b] = \bigcup_{i=1}^n \overline{K}_i$, i.e., $I \approx \sum_{i=1}^n Q_{K_i,p}(f|_{K_i})$.

2.1 The Basic Idea: hp-Adaptivity

Adaptive quadrature rules usually generate a sequence of repeatedly bisected and possibly non-uniform subintervals $\{K_i\}_{i=1}^n$, $n \geq 1$, of the integration domain $[a,b]$ (i.e., each subinterval K_i may have a different length h_i), with a prescribed and uniform number p of quadrature points on each subinterval. With the aim of providing highly accurate approximations with as little computational effort as possible, the novelty of the approach presented in this article is to design an adaptive quadrature procedure, which, in addition to subdividing the original interval $[a,b]$ into appropriate subintervals, is able to adjust the number of quadrature points p_i *individually* within each subinterval K_i in an effective way. We note that this idea originates from approximation theory [5, 15] (see also [8]), and has been applied with huge success in the context of FEMs for the numerical approximation of differential equations. Indeed, under certain conditions, the judicious combination of subinterval refinements (h-refinement) and selection of local approximation orders (p-refinement), which results in the class of so-called hp-FEMs, is able to achieve high-order algebraic or exponential rates of convergence, even for solutions with local singularities; see, e.g. [18]. In an effort to automate the combined h- and p-refinement process, a number of hp-adaptive FEM approaches have been proposed in the literature; see, e.g., [13] and the references cited therein. In the current article, we pursue the smoothness estimation approach developed in [6, 21] (cf. also [11]), and translate the idea into the context of adaptive variable order numerical quadrature.

Given a subinterval K_i with p_i quadrature points, we are given a current approximation $Q_{K_i,p_i}(f|_{K_i})$ of the subintegral $\int_{K_i} f(x)\,\mathrm{d}x \approx Q_{K_i,p_i}(f|_{K_i})$. Then, with the aim of improving the approximate value $Q_{K_i,p_i}(f|_{K_i})$, in the sense of an hp-adaptive FEM methodology in one-dimension, we propose two possible refinements of K_i:

(i) h-refinement: The subinterval K_i of length h_i is bisected into two subintervals K_i^1 and K_i^2 of equal size $h_i/2$, and the number p_i of quadrature points is either inherited to both subintervals or, in order to allow for derefinement with respect to the number of local quadrature points, reduced to $p_i - 1$ points. In the latter case, we obtain the potentially improved approximation:

$$Q^{\mathrm{h}}_{K_i}(f) = Q_{K_i^1,\max(1,p_i-1)}(f) + Q_{K_i^2,\max(1,p_i-1)}(f). \tag{1}$$

(ii) p-refinement: The subinterval K_i is retained, and the number p_i of quadrature points p_i is increased by 1, i.e., $p_i \leftarrow p_i + 1$. This yields an approximation

$$Q^{\mathrm{p}}_{K_i}(f) = Q_{K_i,p_i+1}(f). \tag{2}$$

In case that $p_i = p_{\max}$, where $p_{\max}$ is a prescribed maximal number of quadrature points on each subinterval, we define

$$Q^{\mathrm{p}}_{K_i}(f) = Q_{K_i^1,p_i}(f) + Q_{K_i^2,p_i}(f), \tag{3}$$

where K_i^1 and K_i^2 result from subdividing K_i as in (i).

In order to determine which of the above refinements is more appropriate for a given subinterval K_i, we apply a smoothness estimation idea outlined below. Once a decision between h- and p-refinement for K_i has been made, the procedure is repeated iteratively for any subintervals K_i for which $Q_{K_i,p_i}(f|_{K_i})$ and its refined value (resulting from the chosen refinement) differ by at least a prescribed tolerance $\mathtt{tol} > 0$.

2.2 Smoothness Estimation

The basic idea presented in the articles [6, 11, 21] is to estimate the regularity of a function to be approximated locally. Then, following along the lines of the *hp*-approximation approach, if the function is found to be smooth, according to the underlying regularity estimation test, then a p-refinement is performed, otherwise an h-refinement is employed. In [6], the following smoothness indicator, for a (weakly) differentiable function f on an interval K_j, has been introduced (cf. [6, Eq. (3)]):

$$\mathscr{F}_{K_j}[f] := \begin{cases} \dfrac{\|f\|_{L^\infty(K_j)}}{h_j^{-1/2}\|f\|_{L^2(K_j)} + \frac{1}{\sqrt{2}}h_j^{1/2}\|f'\|_{L^2(K_j)}} & \text{if } f|_{K_j} \not\equiv 0, \\ 1 & \text{if } f|_{K_j} \equiv 0. \end{cases} \tag{F}$$

The motivation behind this definition is the well-known continuous Sobolev embedding $W^{1,2}(K_j) \hookrightarrow L^\infty(K_j)$, which implies that $\mathscr{F}_{K_j}[f] \le 1$ in (F); see [6, Proposition 1]. We classify f as being smooth on K_j if $\mathscr{F}_{K_j}[f] \ge \tau$, for a prescribed smoothness testing parameter $0 < \tau < 1$, and nonsmooth otherwise.

To begin, we first consider the special case when f is a polynomial of degree $p_j \ge 1$. Then, the derivative $f^{(p_j-1)}$ of order $p_j - 1$ of f is a linear polynomial, and the evaluation of the smoothness indicator $\mathscr{F}_{K_j}\left[f^{(p_j-1)}\right]$ from (F) is simple to obtain. In fact, let us write $f|_{K_j}$ in terms of a (finite) Legendre series, that is,

$$f|_{K_j} = \sum_{l=0}^{p_j} a_l(\widehat{L}_l \circ \phi_{K_j}^{-1}), \tag{4}$$

for coefficients $a_0, \ldots, a_{p_j} \in \mathbb{R}$. Here, $\widehat{L}_l$, $l \geq 0$, are the Legendre polynomials on $[-1, 1]$ (scaled such that $\widehat{L}_l(1) = 1$ for all $l \geq 0$), and ϕ_{K_j} is the affine scaling of $[-1, 1]$ to K_j. For f as in (4) it can be shown that

$$\mathscr{F}_{K_j}\left[f^{(p_j-1)}\right] = \frac{1+\xi_{p_j}}{\sqrt{1+\frac{1}{3}\xi_{p_j}^2+\sqrt{2}\xi_{p_j}}}, \tag{5}$$

where $\xi_{p_j} = (2p_j - 1)\left|{}^{a_{p_j}}/_{a_{p_j-1}}\right|$ (provided that $a_{p_j-1} \neq 0$); see [6, Proposition 3]. In particular, this implies that, cf. [6, §2.2],

$$\frac{1}{2} \approx \frac{\sqrt{3}}{\sqrt{6}+1} \leq \mathscr{F}_{K_j}\left[f^{(p_j-1)}\right] \leq 1. \tag{6}$$

In the context of the numerical integration rule, the above methodology can be adopted as follows: suppose we are given $p_j \geq 2$ quadrature points and weights, $\{\widehat{x}_{p_j,k}\}_{k=1}^{p_j}$ and $\{w_{p_j,k}\}_{k=1}^{p_j}$, respectively. Then,

$$\int_{K_j} f(x)\,\mathrm{d}x \approx Q_{K_j,p_j}(f|_{K_j}) = \frac{h_j}{2}\sum_{k=1}^{p_j} w_{p_j,k}(f \circ \phi_{K_j})(\widehat{x}_{p_j,k}). \tag{7}$$

We denote the uniquely defined interpolating polynomial of f of degree $p_j - 1$ at the given quadrature points by $\Pi_{K_j,p_j-1}f = \sum_{l=0}^{p_j-1} b_l(\widehat{L}_l \circ \phi_{K_j}^{-1})$. Due to orthogonality of the Legendre polynomials, we note that

$$b_l = \frac{2l+1}{h_j}\int_{K_j} \Pi_{K_j,p_j-1}f(x)(\widehat{L}_l \circ \phi_{K_j}^{-1})(x)\,\mathrm{d}x, \qquad l = 0, \ldots, p_j - 1.$$

We further assume that the quadrature rule under consideration is exact for all polynomials of degree up to $2p_j - 2$. Thereby,

$$\begin{aligned} b_l &= \frac{2l+1}{2}\sum_{k=1}^{p_j} w_{p_j,k}(\Pi_{K_j,p_j-1}f) \circ \phi_{K_j}(\widehat{x}_{p_j,k})\widehat{L}_l(\widehat{x}_{p_j,k}) \\ &= \frac{2l+1}{2}\sum_{k=1}^{p_j} w_{p_j,k}(f \circ \phi_{K_j})(\widehat{x}_{p_j,k})\widehat{L}_l(\widehat{x}_{p_j,k}). \end{aligned}$$

Consequently, we infer that

$$\xi_{K_j,p_j-1} := (2p_j-3)\left|\frac{b_{p_j-1}}{b_{p_j-2}}\right| = (2p_j-1)\frac{\sum_{k=1}^{p_j} w_{p_j,k}(f\circ\phi_{K_j})(\widehat{x}_{p_j,k})\widehat{L}_{p_j-1}(\widehat{x}_{p_j,k})}{\sum_{k=1}^{p_j} w_{p_j,k}(f\circ\phi_{K_j})(\widehat{x}_{p_j,k})\widehat{L}_{p_j-2}(\widehat{x}_{p_j,k})}, \tag{8}$$

and thus, in view of (5), we use the quantity

$$\mathsf{F}_{K_j,p_j}(f) := \frac{1+\xi_{K_j,p_j-1}}{\sqrt{1+\frac{1}{3}\xi_{K_j,p_j-1}^2+\sqrt{2}\xi_{K_j,p_j-1}}} \in \left(\frac{\sqrt{3}}{\sqrt{6}+1},1\right), \tag{9}$$

cf. (6), to estimate the smoothness of $f|_{K_j}$. We note that the computation of ξ_{K_j,p_j-1} does not require any additional function evaluations of f since the values $(f\circ\phi_{K_j})(\widehat{x}_{p_j,k})$, $k=1,\dots,p_j$, have already been determined in the application of (7).

2.3 Adaptive Variable Order Procedure

Based on the above derivations, we propose an *hp*-type adaptive quadrature method. To this end, we start by choosing a tolerance $\texttt{tol}>0$, a smoothness parameter $\tau\in\left(\sqrt{3}/(\sqrt{6}+1),1\right)$, and a maximal number $p_{\max}\geq 2$ of possible quadrature points on each subinterval. Furthermore, we define the interval $K_1=[a,b]$, and a small number p_1, $2\leq p_1\leq p_{\max}$, of quadrature points on K_1. Moreover, we initialise the set of subintervals subs, the order vector p containing the number of quadrature points on each subinterval, and the unknown value Q of the integral as follows: $\texttt{subs}=\{K_1\}$, $\texttt{p}=\{p_1\}$, $\texttt{Q}=0$. Then, the basic adaptive procedure is given by:

1: **while** $\texttt{subs}\neq\emptyset$ **do**
2: $\quad[\texttt{Q1},\texttt{subs},\texttt{p}]=\texttt{hprefine}(f,\texttt{subs},\texttt{p},p_{\max},\tau)$;
3: $\quad\texttt{Q}=\texttt{Q}+\texttt{Q1}$;
4: **end while**
5: Output Q.

Here, hprefine is a function, whose purpose is to identify those subintervals in subs, which need to be refined further for a sufficiently accurate approximation of the unknown integral. In addition, it outputs a set of subintervals (again denoted by subs), as well as an associated order vector (again denoted by p) which result from applying the most appropriate refinement, i.e., either *h*- or *p*-refinement as outlined in (i) and (ii) in Sect. 2.1 above, for each subinterval. Furthermore, hprefine returns the sum Q1 of all quadrature values corresponding to subintervals in the input set subs for which no further refinement is deemed necessary. The essential steps are summarised in Algorithm 1; here, $p_{\min}$ denotes the minimal number of quadrature points to be employed on any given subinterval.

Algorithm 1 Function [Q, subsnew, pnew]=hprefine(f, subs, p, $p_{\min}$, $p_{\max}$, τ)

Define subsnew = subs, and pnew = p. Set Q = 0.
for each subinterval $K_j \in$ subs **do**
  Evaluate the smoothness indicator $\mathsf{F}_{K_j,p_j}(f)$ from (9).
  if $\mathsf{F}_{K_j,p_j}(f) < \tau$ **then**
    Apply h-refinement to K_j, i.e., bisect K_j into two subintervals of equal size and reduce the number of quadrature points to $\max(p_j - 1, p_{\min})$ on both of them;
    Compute an improved approximation, denoted by $\widetilde{Q}_{K_j}$, of $Q_{K_j,p_j}(f|_{K_j})$ using (1) on K_j.
  else if $\mathsf{F}_{K_j,p_j}(f) \geq \tau$ and $p_j + 1 \leq p_{\max}$ **then**
    Apply p-refinement to K_j, i.e., increase the number of quadrature points to $p_j + 1$ on K_j;
    Compute an improved approximation, denoted by $\widetilde{Q}_{K_j}$, of $Q_{K_j,p_j}(f|_{K_j})$ using (2) on K_j.
  else if $\mathsf{F}_{K_j,p_j}(f) \geq \tau$ and $p_j + 1 > p_{\max}$ **then**
    Bisect K_j into two subintervals of equal size and retain the number of quadrature points p_j on both of them;
    Compute an improved approximation, denoted by $\widetilde{Q}_{K_j}$, of $Q_{K_j,p_j}(f|_{K_j})$ using (3) on K_j.
  end if
  if $|\widetilde{Q}_{K_j} - Q_{K_j,p_j}(f|_{K_j})|$ is sufficiently small **then**
    Update Q = Q + $\widetilde{Q}_{K_j}$;
    Eliminate K_j from subsnew and the corresponding entry p_j from pnew.
  else
    Replace K_j and p_j in subsnew and pnew, respectively, by the corresponding h- or p-refined subintervals as determined above.
  end if
end for

2.4 *Practical Aspects*

In this section we discuss the practical issues involved in the implementation of the procedure described in Sect. 2.3 within a given computing environment.

2.4.1 Gauss-Quadrature Rules

In principle, the adaptive procedure presented in Sect. 2.3 allows for any variable order family of quadrature rules to be exploited. For simplicity, in our numerical experiments presented in Sect. 2.5 below, we propose the use of (families of) Gauss-type quadrature schemes. We emphasise, however, that more traditional schemes, including, for example, (fixed-order) Gauss-Kronrod or Clenshaw-Curtis rules, which are naturally hierarchical, may be employed as well (where the degree of exactness $2p - 2$ is desirable with regards to an accurate computation of the smoothness estimation, cf. Sect. 2.2). Incidentally, our numerical results indicate that, although non-hierarchical rules do not support the repeated use of all previously computed function evaluations, their potentially superior degree of accuracy, compared to their embedded counterparts, can be exploited very favourably within the hp–setting. Indeed, it is a well-known feature of hp-methods that they are particularly effective on a variable high-order level, cf., e.g., [17, 18].

In the current article we employ Gauss-Legendre quadrature points and weights (with at least $p_{\min} = 2$ points and weights); these quantities can be precomputed up to any given order $p_{\max}$ (in practice $p_{\max} = 15$ is usually more than sufficient) or even be generated on the spot in an efficient way (see, e.g., [1]) if an upper bound $p_{\max}$ cannot be fixed. In addition, we note that the Gauss-Legendre rule based on p points has a degree of exactness of $2p - 1$, i.e., the smoothness indicators derived in Sect. 2.2 can be computed by means of the formula given in (8). For a given maximum number $p_{\max}$, we store the points and weights of the Gauss-Legendre rules (on the reference interval $[-1, 1]$) with up to $p_{\max}$ points in two $p_{\max} \times (p_{\max} - 1)$-matrices $\boldsymbol{X}$ and $\boldsymbol{W}$, respectively; here, for parameters $p = 2, \ldots, p_{\max}$, the p-th columns of $\boldsymbol{X}$ and $\boldsymbol{W}$ are built from the points and weights of the corresponding p-point Gauss-Legendre quadrature rule, respectively (and complementing the remaining entries in all but the last column by zeros):

$$\boldsymbol{X} = \begin{pmatrix} \widehat{x}_{2,1} & \widehat{x}_{3,1} & \cdots & \widehat{x}_{p_{\max},1} \\ \widehat{x}_{2,2} & \vdots & & \\ & \widehat{x}_{3,3} & & \vdots \\ & \mathbf{0} & \ddots & \\ & & & \widehat{x}_{p_{\max},p_{\max}} \end{pmatrix}, \quad \boldsymbol{W} = \begin{pmatrix} w_{2,1} & w_{3,1} & \cdots & w_{p_{\max},1} \\ w_{2,2} & \vdots & & \\ & w_{3,3} & & \vdots \\ & \mathbf{0} & \ddots & \\ & & & w_{p_{\max},p_{\max}} \end{pmatrix}. \tag{10}$$

We note that, for other quadrature rules, the number of rows in the above matrices may be different.

2.4.2 Vectorised Quadrature

Following the ideas of [19] we use a vectorised quadrature implementation. This means that, instead of computing the integrals on the subintervals `subs` in Algorithm 1 one at a time, they are all computed at once. This can be accomplished by using fast vector- and matrix-operations, and by carrying out all necessary function evaluations in a single operation by computing the function to be integrated for a vector of input values. Specifically, we write the composite rule $I \approx \sum_{K_i \in \texttt{subs}} Q_{K_i,p_i}(f|_{K_i}) = \sum_{K_i \in \texttt{subs}} {}^{h_i}/_2 \sum_{k=1}^{p_i} w_{p_i,k}(f \circ \phi_{K_i})(\widehat{x}_{p_i,k})$ as a dot product of a weight vector $\boldsymbol{w}$ and a function vector $f(\boldsymbol{x})$; here, the former vector contains all (scaled) weights $\{\frac{1}{2}h_i w_{p_i,k}\}_{i,k}$, and the latter vector represents the evaluation of the integrand function f on the vector $\boldsymbol{x}$ of all corresponding quadrature points $\{\phi_{K_i}(\widehat{x}_{p_i,k})\}_{i,k}$ appearing in the sum above. Evidently, these vectors can be built efficiently by extracting (and affinely mapping and scaling) the corresponding rows from the matrices $\boldsymbol{X}$ and $\boldsymbol{W}$ in (10). We emphasise that applying vectorised quadrature crucially improves the performance of the overall adaptive procedure (provided that such a technology is available in a given computing environment).

2.4.3 Smoothness Estimators

As already noted, computing the smoothness indicators from (8) does not need any additional function evaluations of the function f; they only require the values of the Legendre polynomials $\widehat{L}_{p-1}$ and $\widehat{L}_{p-2}$ at the points $\{\widehat{x}_{p,k}\}_{k=1}^{p}$, for $p = 2, \ldots, p_{\max}$. These quantities are again precomputable, and can be stored in two matrices

$$\boldsymbol{L}_1 = \begin{pmatrix} L_1(\widehat{x}_{2,1}) & L_2(\widehat{x}_{3,1}) & \cdots & L_{p_{\max}-1}(\widehat{x}_{p_{\max},1}) \\ L_1(\widehat{x}_{2,2}) & \vdots & & \\ & L_2(\widehat{x}_{3,3}) & & \vdots \\ & \boldsymbol{0} & \ddots & \\ & & & L_{p_{\max}-1}(\widehat{x}_{p_{\max},p_{\max}}) \end{pmatrix}, \tag{11}$$

and

$$\boldsymbol{L}_2 = \begin{pmatrix} L_0(\widehat{x}_{2,1}) & L_1(\widehat{x}_{3,1}) & \cdots & L_{p_{\max}-2}(\widehat{x}_{p_{\max},1}) \\ L_0(\widehat{x}_{2,2}) & \vdots & & \\ & L_1(\widehat{x}_{3,3}) & & \vdots \\ & \boldsymbol{0} & \ddots & \\ & & & L_{p_{\max}-2}(\widehat{x}_{p_{\max},p_{\max}}) \end{pmatrix}. \tag{12}$$

Then, the sums in (8) are vectorised similarly as described above. In particular, the computation of the smoothness estimators can be undertaken with an almost negligible computational cost.

2.4.4 Stopping Criterion

In order to implement the stopping-type criterion in line 14 of Algorithm 1, we exploit an idea that was proposed in the context of adaptive Simpson quadrature in [7]. More precisely, given a possibly rough approximation $\texttt{iguess} \approx \int_a^b f(x)\,\mathrm{d}x$ of the unknown integral I (e.g., obtained from a Monte-Carlo calculation such that both the approximation and the exact value are of the same magnitude; cf. [7]), and a tolerance $\texttt{tol} > 0$, we redefine $\texttt{iguess} = \texttt{iguess} * \texttt{tol}/\texttt{eps}$; here, `eps` represents the smallest (positive) machine number in a given computing environment. Then, using the comparison operator $==$, we accept the difference $|\widetilde{Q}_{K_j} - Q_{K_j,p_j}(f|_{K_j})|$ to be sufficiently small with respect to the given tolerance `tol` if the logical call $\texttt{iguess}+|\widetilde{Q}_{K_j}-Q_{K_j,p_j}(f|_{K_j})| == \texttt{iguess}$; yields a `true` value. In this way `tol` represents a reasonable approximation of the relative error.

2.5 Numerical Examples

In order to test our approach, we consider a number of benchmark problems on the interval [0, 1]. Specifically, the following functions will be studied:

$$
\begin{aligned}
f_1(x) &= \exp(x),\\
f_2(x) &= |x - 1/3|^{1/2},\\
f_3(x) &= \operatorname{sech}(10(x - 1/5))^2 + \operatorname{sech}(100(x - 2/5))^4\\
&\quad + \operatorname{sech}(1000(x - 3/5))^6 + \operatorname{sech}(1000(x - 4/5))^8,\\
f_4(x) &= \cos(1000x^2),\\
f_5(x) &= \begin{cases} 0 & \text{if } x < 1/3,\\ 1 & \text{if } x \geq 1/3. \end{cases}
\end{aligned}
$$

Whilst the first function, f_1, is analytic, f_2 is smooth except at $1/3$ (see Fig. 1). Furthermore, f_3 was proposed in [9] in the context of the CHEBFUN package [10]; this is a smooth function that exhibits several very thin spikes (see Fig. 2). Moreover, f_4 is highly oscillating towards the right end point 1, and f_5 is an example of a discontinuous function.

We perform our computations in MATLAB, and set the tolerance to $\texttt{tol} = 0.3 \times 10^{-15}$ (which is close to machine precision in MATLAB), the smoothness estimation parameter is prescribed as $\tau = 0.6$, $p_{\min} = 2$, and $p_{\max} = 15$. Within this setting, our adaptive procedure generates results that are accurate to machine precision, for all of the considered examples. In Table 1, for each of the functions $f_1, \ldots, f_5$ above, we present the number of function evaluations (counting an evaluation of the given function f_i for a vector-valued argument $\boldsymbol{x} = (x_1, \ldots, x_n)$, i.e., $f_i(\boldsymbol{x}) = (f_i(x_1), \ldots, f_i(x_n))$, as n) for the proposed *hp*-adaptive quadrature procedure, as well

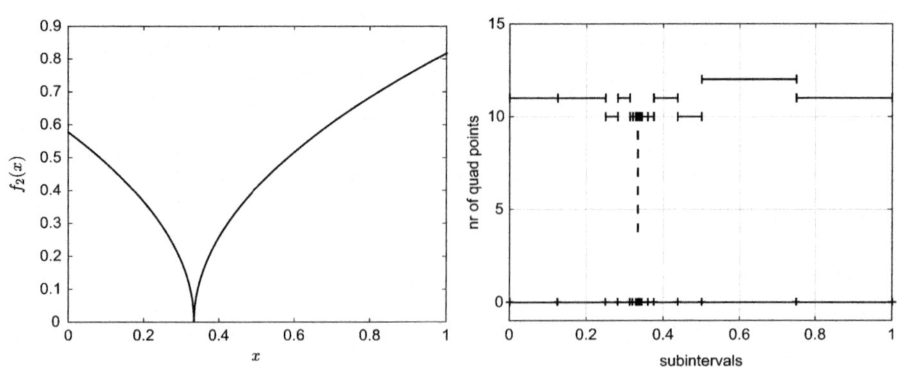

Fig. 1 Function f_2: Graph (*left*) and *hp*-mesh (*right*)

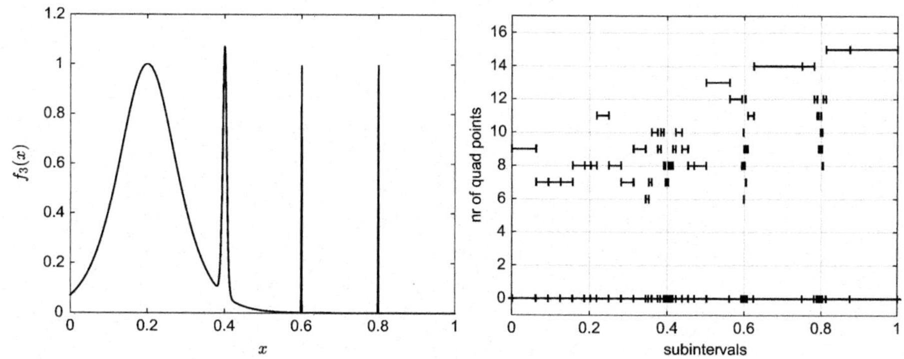

Fig. 2 Function f_3: Graph (*left*) and *hp*-mesh (*right*)

Table 1 Number of function evaluations for *hp*-type adaptive quadrature and adaptive Simpson quadrature

	hp-adaptive quadrature	Adaptive Simpson quadrature
f_1	52	4096
f_2	1718	25,488
f_3	2427	72,528
f_4	21,005	1,213,680
f_5	1493	784

as the corresponding number for a classical adaptive Simpson method from [7] (which is based on employing the two end points, as well as the midpoint on each subinterval, and reuses the former two points without recomputing), with the same tolerance value $\texttt{tol} = 0.3 \times 10^{-15}$. Except for the last function, f_5, where a low-order quadrature rule is more effective, the remarkable efficiency of the proposed *hp*-type quadrature becomes clearly visible.

In order to illustrate how the *hp*-adaptive procedure performs, we depict the final *hp*-mesh for f_2 and f_3 in Figs. 1 and 2, respectively. Here, along the horizontal axis we present the subintervals obtained as a result of the adaptive process, and on the vertical axis the number of quadrature points introduced on each subinterval is displayed. In both examples, we see that smooth regions in the underlying integrand are resolved by employing larger subintervals featuring a higher number of quadrature points, whereas close to singularities, the number of quadrature points is kept low on very small integration subdomains. It is noteworthy that this behaviour is well-known from *hp*-FEMs for differential equations, where high-order algebraic or even exponential convergence rates can be obtained by applying this type of *hp*-refinement procedure; see [18] for details.

3 Conclusions

In this article we proposed a new adaptive quadrature strategy, which features both local subdivision of the integration domain, as well as local variation of the number of quadrature points employed on each subinterval. Our approach is inspired by the *hp*-adaptive FEM methodology based on *hp*-adaptive smoothness testing. In combination with a vectorised quadrature implementation, the proposed adaptive quadrature algorithm is able to deliver highly accurate results in a very efficient manner. Since our approach is closely related to the *hp*-FEM technique, it can be extended to multiple dimensions, including, in particular, the application of anisotropic refinements of the underlying domain of integration, together with the exploitation of different numbers of quadrature points in each coordinate direction on each subinterval (based, for example, on anisotropic Sobolev embeddings as outlined in [6, §3.1]).

Acknowledgements Thomas P. Wihler acknowledges the financial support by the Swiss National Science Foundation (SNF).

References

1. I. Bogaert, Iteration-free computation of Gauss-Legendre quadrature nodes and weights. SIAM J. Sci. Comput. **36**(3), A1008–A1026 (2014). MR 3209728
2. G. Dahlquist, Å. Björck, *Numerical Methods in Scientific Computing. Volume I*. Society for Industrial and Applied Mathematics (SIAM, Philadelphia, PA, 2008)
3. P.J. Davis, P. Rabinowitz, *Methods of Numerical Integration* (Dover Publications, Inc., Mineola, NY, 2007), Corrected reprint of the 2nd edn. (1984)
4. L. Demkowicz, *Computing with hp-Adaptive Finite Elements. Volume 1*. Chapman & Hall/CRC Applied Mathematics and Nonlinear Science Series (Chapman & Hall/CRC, Boca Raton, FL, 2007). One and two dimensional elliptic and Maxwell problems
5. R. DeVore, K. Scherer, *Variable Knot, Variable Degree Spline Approximation to x^β*. Quantitative Approximation (Proceedings of International Symposium, Bonn, 1979) (Academic, New York, London, 1980), pp. 121–131
6. T. Fankhauser, T.P. Wihler, M. Wirz, The *hp*-adaptive FEM based on continuous Sobolev embeddings: isotropic refinements. Comput. Math. Appl. **67**(4), 854–868 (2014)
7. W. Gander, W. Gautschi, Adaptive quadrature—revisited. BIT **40**(1), 84–101 (2000)
8. W. Gui, I. Babuška, The h, p and $h-p$ versions of the finite element method in one-dimension, Parts I–III. Numer. Math. **49**(6), 577–683 (1986)
9. N. Hale, Spike integral (2010). http://www.chebfun.org/examples/quad/SpikeIntegral.html
10. N. Hale, L.N. Trefethen, Chebfun and numerical quadrature. Sci. China Math. **55**(9), 1749–1760 (2012)
11. P. Houston, E. Süli, A note on the design of *hp*–adaptive finite element methods for elliptic partial differential equations. Comput. Methods Appl. Mech. Eng. **194**(2–5), 229–243 (2005)
12. J.M. Melenk, B.I. Wohlmuth. On residual-based a posteriori error estimation in *hp*-FEM. Adv. Comp. Math. **15**, 311–331 (2001)
13. W.F. Mitchell, M.A. McClain, A comparison of *hp*-adaptive strategies for elliptic partial differential equations. ACM Trans. Math. Softw. **41**, 2:1–2:39 (2014)

14. W.H. Press, S.A. Teukolsky, W.T. Vetterling, B.P. Flannery, *Numerical Recipes*. The Art of Scientific Computing, 3rd edn. (Cambridge University Press, Cambridge, 2007)
15. K. Scherer, On optimal global error bounds obtained by scaled local error estimates. Numer. Math. **36**(2), 151–176 (1980/1981)
16. D. Schötzau, C. Schwab, T.P. Wihler, *hp*-DGFEM for second-order mixed elliptic problems in polyhedra. Math. Comp. **85**(299), 1051–1083, (2016).
17. C. Schwab, Variable order composite quadrature of singular and nearly singular integrals. Computing **53**(2), 173–194 (1994) MR 1300776 (96a:65035)
18. C. Schwab, *p- and hp-FEM – Theory and Application to Solid and Fluid Mechanics* (Oxford University Press, Oxford, 1998)
19. L.F. Shampine, Vectorized adaptive quadrature in Matlab. J. Comput. Appl. Math. **211**(2), 131–140 (2008)
20. P. Solin, K. Segeth, I. Dolezel, *Higher-Order Finite Element Methods*. Studies in Advanced Mathematics (Chapman & Hall/CRC, Boca Raton, London, 2004)
21. T.P. Wihler, An *hp*-adaptive strategy based on continuous Sobolev embeddings. J. Comput. Appl. Math. **235**, 2731–2739 (2011)

Recent Results on the Improved WENO-Z+ Scheme

Rafael Brandão de Rezende Borges

Abstract The WENO-Z scheme is known to achieve less dissipative results than the classical WENO scheme, especially in problems involving both shocks and smooth structures. In Acker et al. (J Comput Phys 313:726–753, 2016), the cause of the improved results of WENO-Z was shown to be its comparatively higher weights on less-smooth substencils. This knowledge was exploited to develop the fifth-order WENO-Z+ scheme, which generalizes WENO-Z by including an extra term for increasing the weights of less-smooth substencils even further. The new scheme WENO-Z+ was shown to achieve even better results than WENO-Z, while keeping the same numerical robustness. In this study, the third- and seventh-order versions of the WENO-Z+ scheme are presented and discussed. The preliminary numerical results make evident that the approach used by WENO-Z+ is also sound for orders other than 5.

1 Introduction

Weighted essentially nonoscillatory (WENO) schemes are a popular class of numerical schemes for solving hyperbolic conservation laws and, more generally, PDE whose (weak) solutions may develop discontinuities. These schemes are able to capture shocks in a sharp, essentially nonoscillatory way, and can be designed to achieve arbitrarily high orders in smooth solutions. It has been shown that, in problems containing both discontinuities and fine, complex structures in the smooth regions (such as the interaction between a shock wave and a turbulent flow), it is usually more computationally efficient to use a high-order scheme (such as WENO) than first- or second-order ones [11, 12].

WENO schemes, like the essentially nonoscillatory (ENO) schemes that preceded them [7], achieve their nonoscillatory results by avoiding interpolations across domains that contain discontinuities of the solution. The rth-order ENO scheme

R.B. de Rezende Borges (✉)
Departamento de Análise Matemática, Universidade do Estado do Rio de Janeiro, Rio de Janeiro / RJ, Brazil
e-mail: rafael.borges@ime.uerj.br

M.L. Bittencourt et al. (eds.), *Spectral and High Order Methods for Partial Differential Equations ICOSAHOM 2016*, Lecture Notes in Computational Science and Engineering 119, DOI 10.1007/978-3-319-65870-4_39

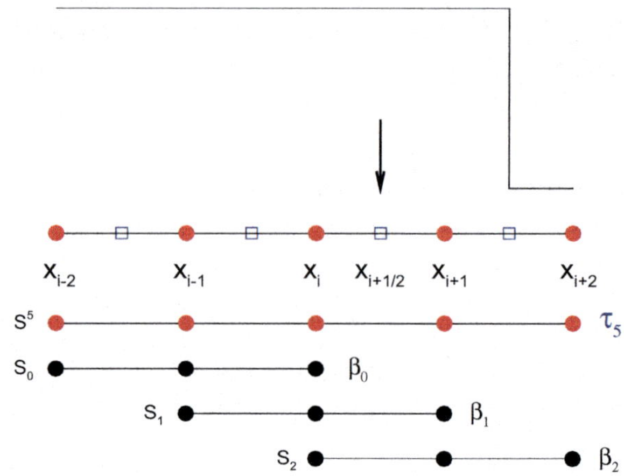

Fig. 1 The global stencil S^5 and its substencils S_0, S_1, and S_2, used in the third-order ENO / fifth-order WENO approximation procedures

does so by restricting the interpolation of the solution to the r-points substencil (out of a global stencil with $R = 2r - 1$ points) where the solution is smoothest (in some sense), see Fig. 1 for an illustration of the $r = 3$ case. WENO, on the other hand, maximizes the interpolation region by assigning a nonlinear weight ω_k to each r-points substencils S_k. The final interpolation is the weighted sum of the interpolations at each substencil S_k. The weights are designed in such a way that $\omega_k \approx 0$ if the substencil S_k contains a discontinuity, therefore avoiding oscillations; and, if the solution is smooth, the scheme uses all R points in the interpolation, achieving Rth order as a result (almost double the order of the ENO scheme with same R-points global stencil).

Currently, there are two main families of WENO schemes, each based on a different weight formula: the classical WENO-JS scheme [2, 9, 10] and its modifications, such as the WENO-M scheme [8] and many others; and the WENO-Z scheme [3–5] and schemes based on the WENO-Z formula, such as the ESWENO scheme [15], the WENO-NS scheme [6], the WENO-MZ scheme [16], and several others. It has been shown that, at least for fifth-order, the WENO-Z scheme has a higher resolution and is computationally more cost-effective than WENO-JS [3, 16]. This is particularly more pronounced near smooth extrema—see Figs. 2 and 3 for examples. In the literature, two main arguments have been used to explain this: first, WENO-Z has better accuracy than WENO-JS near critical points; second, WENO-Z assigns a higher weight to less-smooth substencils than WENO-JS does. This issue was recently investigated in [1], where the second reason was found to be the most relevant one. This is due the fact that the solution does a sharper transition in substencils containing critical points than otherwise. As such, a substencil containing a smooth extrema is inappropriately detected as less smooth than the other substencils. By assigning a larger weight to the less-smooth substencils, WENO-Z is able to achieve better results near critical points than WENO-JS.

Using this knowledge, a new WENO scheme—called WENO-Z+—was proposed in [1]. This scheme adds an extra term to the WENO-Z formula with the

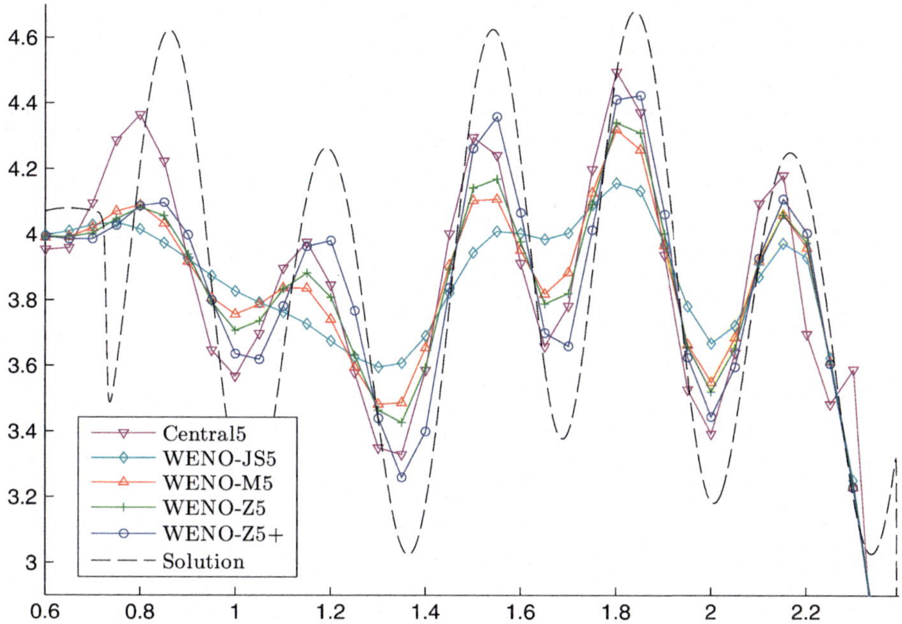

Fig. 2 Numerical solution of the shock-entropy wave problem of Shu–Osher at $t = 1.8$; density. A coarse grid with $N = 201$ points was used. Zoom in the relevant region

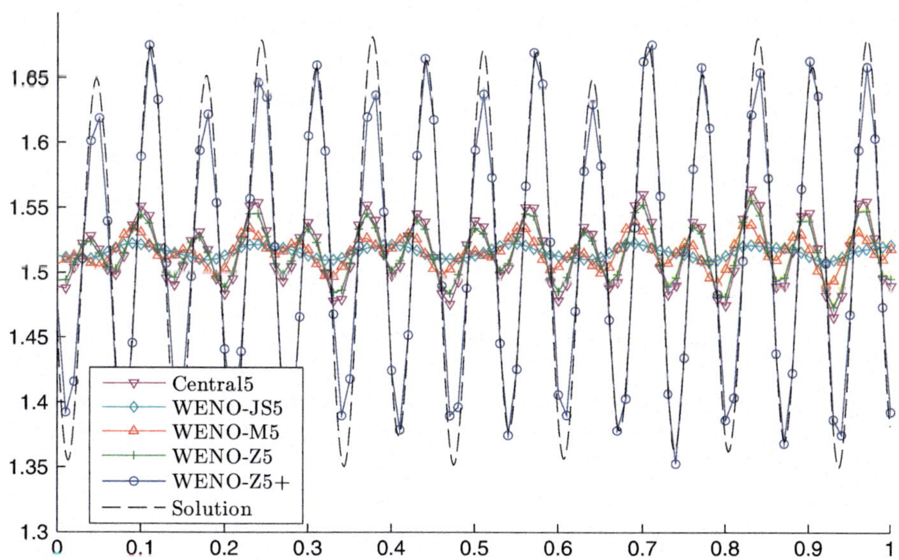

Fig. 3 Numerical solution of the shock-entropy wave problem of Titarev–Toro at $t = 5$; density. A coarse grid with $N = 1001$ points was used. Zoom in the relevant region

sole purpose of increasing the weights of less-smooth substencils even further. The results in [1], which are restricted to the fifth-order version of WENO-Z+, show that the new scheme has an even higher resolution than WENO-Z, see Figs. 2 and 3.

In the present notes, the third- and seventh-order WENO-Z+ schemes are introduced and analyzed. The general background on WENO schemes is briefly discussed in Sect. 2. Preliminary results for the third- and seventh-order WENO-Z+ schemes are shown in Sect. 3. Concluding remarks are given in Sect. 4.

2 WENO Schemes

This section briefly describes the basic concepts related to WENO schemes. Since the only difference between any of the WENO schemes presented here lies in their weight formulas, the discussion will be focused on these and not on the other building blocks of WENO schemes, such as time integration, flux splitting, numerical fluxes, characteristics decomposition, etc. More complete descriptions of WENO schemes, also covering the mentioned topics, can be found in, e.g., [1, 3, 5, 8, 9, 13, 14].

2.1 WENO Approximation

As an illustration for the WENO approximation procedure, consider the fifth-order case. Suppose we want to compute the interpolation of a given smooth by parts function $f(x)$ at the point $x_{i+\frac{1}{2}}$ using the uniformly-spaced stencil $S^5 = \{x_{i-2}, x_{i-1}, x_i, x_{i+1}, x_{i+2}\}$ with spacing Δx. This stencil is divided in three three-point substencils, $S_0 = \{x_{i-2}, x_{i-1}, x_i\}$, $S_1 = \{x_{i-1}, x_i, x_{i+1}\}$, and $S_2 = \{x_i, x_{i+1}, x_{i+2}\}$, as seen in Fig. 1. Define $\hat{f}^k(x)$ as the second-degree interpolating polynomial at the substencil S_k, $k = 0, 1, 2$, so that $\hat{f}^k(x_{i+\frac{1}{2}})$ is a third-order approximation to $f(x_{i+\frac{1}{2}})$. That is,

$$\hat{f}^k(x_{i+\frac{1}{2}}) = f(x_{i+\frac{1}{2}}) + O(\Delta x^3) = f(x_{i+\frac{1}{2}}) + A_{k,3}\Delta x^3 + A_{k,4}\Delta x^4 + O(\Delta x^5)$$

if $f(x)$ is smooth in S_k, where $A_{k,3}$ and $A_{k,4}$ are coefficients of the series expansion of $\hat{f}^k$. Using simple linear algebra, it is possible to find coefficients γ_k such that the combination of $\hat{f}^k(x_{i+\frac{1}{2}})$,

$$\gamma_0\hat{f}^0(x_{i+\frac{1}{2}}) + \gamma_1\hat{f}^1(x_{i+\frac{1}{2}}) + \gamma_2\hat{f}^2(x_{i+\frac{1}{2}}) = f(x_{i+\frac{1}{2}}) + O(\Delta x^5),$$

is fifth-order accurate if $f(x)$ is smooth in the whole stencil S^5. These coefficients γ_k are called *ideal weights*. Notice that the sum of all γ_k must be necessarily 1.

The fifth-order WENO approximation is defined as

$$\hat{f}(x_{i+\frac{1}{2}}) = \omega_0 \hat{f}^0(x_{i+\frac{1}{2}}) + \omega_1 \hat{f}^1(x_{i+\frac{1}{2}}) + \omega_2 \hat{f}^2(x_{i+\frac{1}{2}}),$$

where the weights ω_k vary with the smoothness of $f(x)$ inside S_k. Ideally, if a substencil S_d contains a discontinuity (in short, we also say that the substencil S_d is discontinuous), but there is another substencil S_c where $f(x)$ is smooth (we also say that S_c is smooth, or continuous), it is desirable that $\omega_d \approx 0$ in order to avoid computing the approximation in a domain where $f(x)$ is discontinuous. This is called the *ENO property*. However, if $f(x)$ is smooth in the whole stencil S^5, we require that $\omega_k \approx \gamma_k$ for all k, so that the WENO approximation is fifth-order accurate.

For the Rth-order WENO approximation (where R is an odd number greater than one), a global R-points stencil is divided into r substencils with r points each, where $r = (R+1)/2$. The general case is completely analogous to the fifth-order case.

2.1.1 Using the WENO Approximation for Solving PDE

In the context of numerical methods for solving PDE, the WENO approximation is generally used for computing the spatial derivative of a function (e.g., the derivative of the flux of hyperbolic conservation laws). Suppose we want to approximate the derivative of a given a function $f(x)$ at the point x_i. First, we implicitly define a special function $h(x)$ whose finite difference is exactly $f'(x_i)$:

$$f(x) = \frac{1}{\Delta x}\int_{x-\frac{\Delta x}{2}}^{x+\frac{\Delta x}{2}} h(\xi)\,d\xi \quad \therefore \quad f'(x_i) = \frac{h(x_{i+\frac{1}{2}}) - h(x_{i-\frac{1}{2}})}{\Delta x}.$$

Next, WENO is used to approximate $h(x_{i\pm\frac{1}{2}})$ through the point values of $f(x)$, resulting in the Rth-order approximations $\hat{f}(x_{i-\frac{1}{2}})$ and $\hat{f}(x_{i+\frac{1}{2}})$. It is shown [5] that

$$\frac{\hat{f}(x_{i+\frac{1}{2}}) - \hat{f}(x_{i-\frac{1}{2}})}{\Delta x} = f'(x_i) + O(\Delta x^R)$$

if $f(x)$ is smooth.

Remark 1 In what follows, the standard asymptotic symbols $O(\cdot)$, $\Omega(\cdot)$ and $\Theta(\cdot)$ will be used with their proper meanings:

- $g(\Delta x) = O(\Delta x^n)$ denotes an upper bound to $g(\Delta x)$, that is, $|g(\Delta x)| \leq C\Delta x^n$ for some $C > 0$ as $\Delta x \to 0$.
- $g(\Delta x) = \Omega(\Delta x^n)$ denotes a lower bound to $g(\Delta x)$, that is, $|g(\Delta x)| \geq C\Delta x^n$ for some $C > 0$ as $\Delta x \to 0$.
- $g(\Delta x) = \Theta(\Delta x^n)$ denotes the exact order of $g(\Delta x)$, that is, $g(\Delta x) = O(\Delta x^n)$ and $g(\Delta x) = \Omega(\Delta x^n)$ as $\Delta x \to 0$.

2.2 The WENO-JS and Related Schemes

The WENO-JS [9] and related schemes use the basic weight formula

$$\alpha_k^{\text{JS}} = \frac{\gamma_k}{(\beta_k + \varepsilon)^p}, \qquad \omega_k^{\text{JS}} = \frac{\alpha_k^{\text{JS}}}{\sum_{j=0}^{r-1} \alpha_j^{\text{JS}}}, \qquad k = 0, \ldots, r-1.$$

Here, β_k is a (local) smoothness indicator, that is,

$$\beta_k = \begin{cases} O(\Delta x^q), & \text{for some } q > 0, \text{ if } f(x) \text{ is smooth in } S_k \text{ (typically, } q = 2), \\ \Theta(1), & \text{if } f(x) \text{ is discontinuous in } S_k. \end{cases}$$

The power parameter $p \geq 1$ is used to enhance the relative ratio between the smoothness indicators β_k; the larger the p, the smaller the weights of less-smooth substencils and, as a consequence, the more dissipative is the scheme. Typically, $p = 2$ for the fifth-order WENO-JS. The sensitivity parameter $\varepsilon > 0$ is used to avoid divisions by zero in the weights formulation, but may also interfere with the smoothness detection if it is too large.

The Mapped WENO weights (WENO-M) [8] are obtained by applying the mapping function

$$g_k(\omega) = \frac{(\gamma_k + \gamma_k^2 - 3\gamma_k\omega + \omega^2)\omega}{\gamma_k^2 - 2\gamma_k\omega + \omega}, \qquad \omega \in [0, 1],$$

in the WENO-JS weights. This mapping function both improves the accuracy of WENO-JS near critical points and increases the weights of less-smooth substencils, making WENO-M considerably less dissipative than WENO-JS. The mapping function is computationally expensive, however.

2.3 The WENO-Z and Related Schemes

The WENO-Z [3] and related schemes use a different basic weight formula

$$\alpha_k^{\text{Z}} = \gamma_k \left[1 + \left(\frac{\tau}{\beta_k + \varepsilon} \right)^p \right], \qquad \omega_k^{\text{Z}} = \frac{\alpha_k^{\text{Z}}}{\sum_{j=0}^{r-1} \alpha_j^{\text{Z}}}, \qquad k = 0, \ldots, r-1. \quad (1)$$

The terms β_k, ε, and p are totally analogous to their WENO-JS counterparts. The novelty here is the inclusion of a global smoothness indicator τ in the formula, which measures the smoothness in the whole R-points stencil and has a higher order than β_k, that is,

$$\tau = \begin{cases} O(\Delta x^{q_2}) < \Theta(\beta_k) & \text{for some } q_2 > 0, \text{ if } f(x) \text{ is smooth in } S^R, \\ \Theta(1), & \text{if } f(x) \text{ is discontinuous somewhere inside } S^R. \end{cases}$$

WENO-Z behaves better near critical points than WENO-JS [5]. It also assigns a larger weight to less-smooth substencils. To see this, consider S_d a substencil where the solution is "less smooth" than in substencil S_c, meaning that $\beta_d > \beta_c$. If the same parameters ε and p and the same smoothness indicator β_k are used in both schemes, we have

$$\frac{\omega_d^Z}{\omega_c^Z} = \frac{\alpha_d^Z}{\alpha_c^Z} = \frac{\gamma_d}{\gamma_c}\frac{(\beta_c+\varepsilon)^p}{(\beta_d+\varepsilon)^p}\frac{(\beta_d+\varepsilon)^p+\tau^p}{(\beta_c+\varepsilon)^p+\tau^p} > \frac{\gamma_d}{\gamma_c}\frac{(\beta_c+\varepsilon)^p}{(\beta_d+\varepsilon)^p} = \frac{\alpha_d^{JS}}{\alpha_c^{JS}} = \frac{\omega_d^{JS}}{\omega_c^{JS}}.$$

The other WENO-Z-type schemes mentioned in the introduction [6, 15] use the same weight formula as Eq. (1), but with different smoothness indicators than the original.

2.4 *The WENO-Z+ Scheme*

The WENO-Z+ weights [1] are defined by

$$\alpha_k^{ZP} = \gamma_k\left[1 + \left(\frac{\tau+\varepsilon}{\beta_k+\varepsilon}\right)^p + \lambda\left(\frac{\beta_k+\varepsilon}{\tau+\varepsilon}\right)\right],$$

$$\omega_k^{ZP} = \frac{\alpha_k^{ZP}}{\sum_{j=0}^{r-1}\alpha_j^{ZP}}, \qquad k = 0, \ldots, r-1.$$

This is similar to the WENO-Z weight formula, with the addition of the term $\lambda(\frac{\beta_k+\varepsilon}{\tau+\varepsilon})$, which has the following properties:

- It is directly proportional to β_k; therefore, the less smooth the function is in S_k, the larger it gets.
- It is divided by the global smoothness indicator τ; as such, this term is small when there is an actual discontinuity somewhere inside the global stencil.

The parameter λ is used for controlling the size of this term. In [1], $\lambda = \Delta x^{2/3}$ was chosen for order 5 based on empiric evidence. Values much larger than that led to instabilities, and values much smaller made WENO-Z+ have almost no gain in resolution compared to WENO-Z.

Figures 2 and 3 respectively show the numerical solution of the shock-entropy wave tests of Shu–Osher and Titarev–Toro. It can be seen that WENO-Z+ has noticeably more resolution near critical points than the other WENO schemes, and even than the linear fifth-order upstream central scheme (Central5) in the Titarev–Toro test.

3 Numerical Results

Preliminary results for third- and seventh-order WENO-Z+ schemes are shown in this section. For third-order, $p = 1$ for all schemes and $\lambda = 16$ for WENO-Z+. For seventh-order, $p = 3$ for all schemes and $\lambda = 25\Delta x$ for WENO-Z+. In all tests, $\varepsilon = 10^{-40}$ and CFL = 0.5. The SSP-ERK(3,3) method was used for time integration, and is was used characteristics decomposition with Lax-Friedrichs flux splitting done in each characteristic variable separately.

The tests shown in this section are standard linear and Euler 1D tests generally used for testing WENO schemes. A complete explanation of the numerical tests can be found in the literature (e.g., [1]). For brevity, they will not be described here.

3.1 *Order 3*

For order 3, WENO-Z+ has shown a noticeably improved resolution compared to the other WENO schemes. Its resolution is very similar to the third-order linear upstream central scheme (Central3), as seen in Figs. 4 and 5, even surpassing it in the Gaussian-square-triangle-ellipse linear test (Fig. 6). Contrary to Central3, however, WENO-Z+ converges in the interacting blast waves problem (Fig. 7) and do not present relevant spurious oscillations near discontinuities (Figs. 6 and 8).

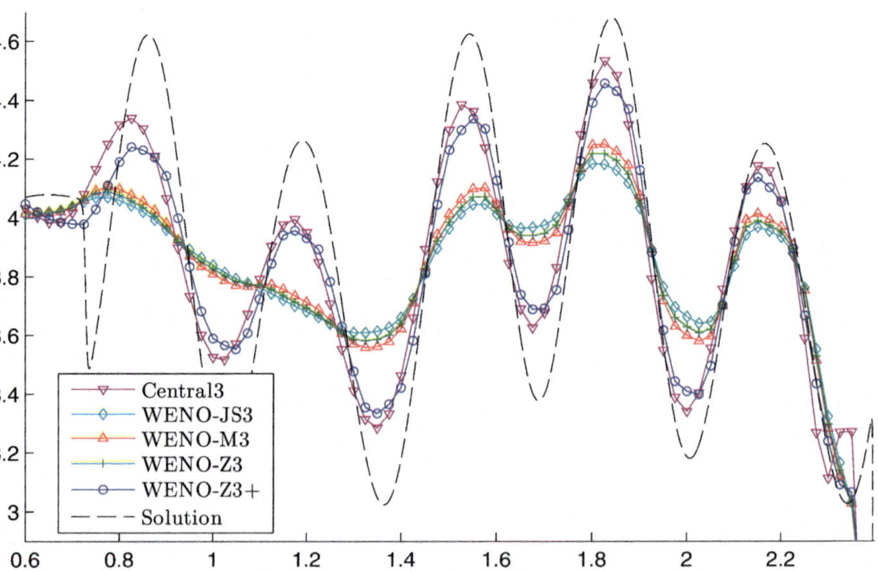

Fig. 4 Numerical solution of the shock-entropy wave problem of Shu–Osher with $N = 401$ points at $t = 1.8$; density. Zoom in the relevant region

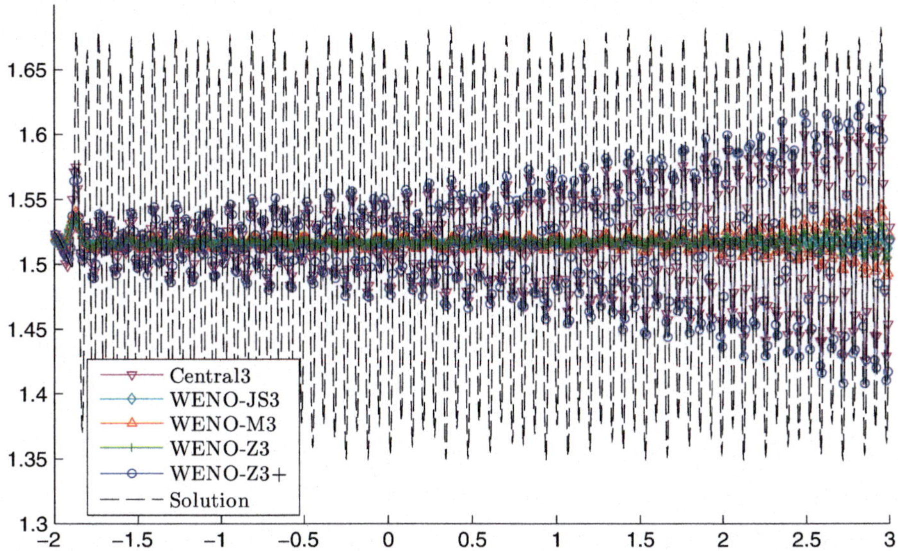

Fig. 5 Numerical solution of the shock-entropy wave problem of Titarev–Toro with $N = 2001$ points at $t = 5$; density. Zoom in the relevant region

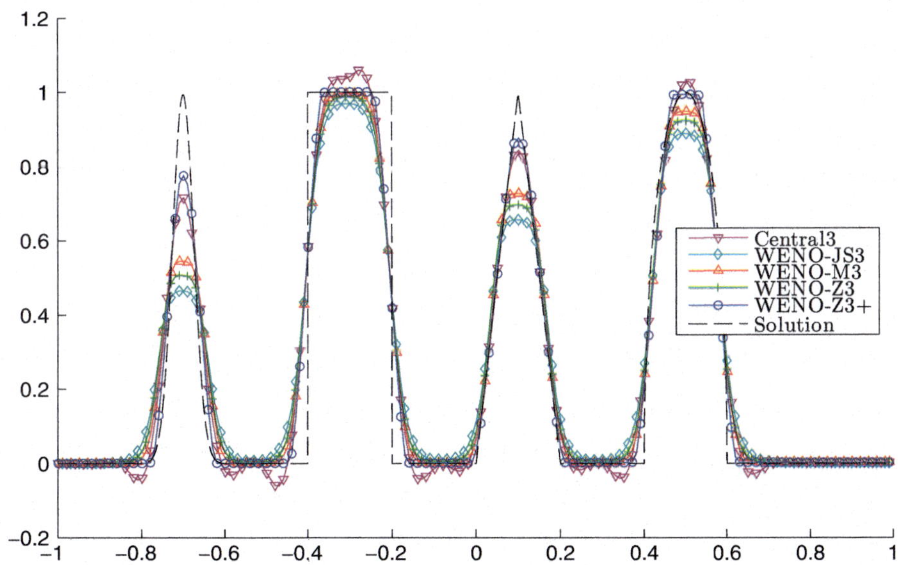

Fig. 6 Numerical solution of the Gaussian-square-triangle-ellipse linear test with $N = 200$ points at $t = 2$

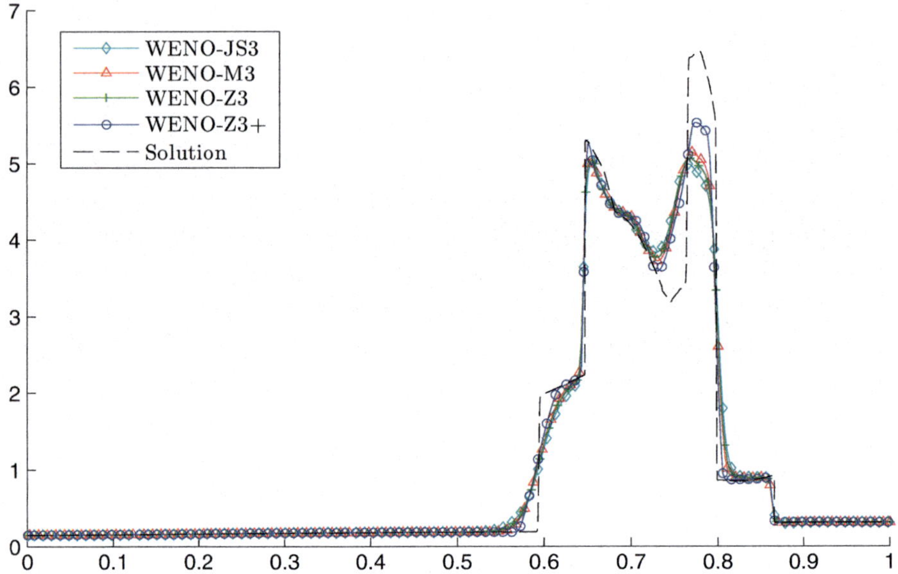

Fig. 7 Numerical solution of the interacting blast waves problem with $N = 401$ points at $t = 0.13$; density

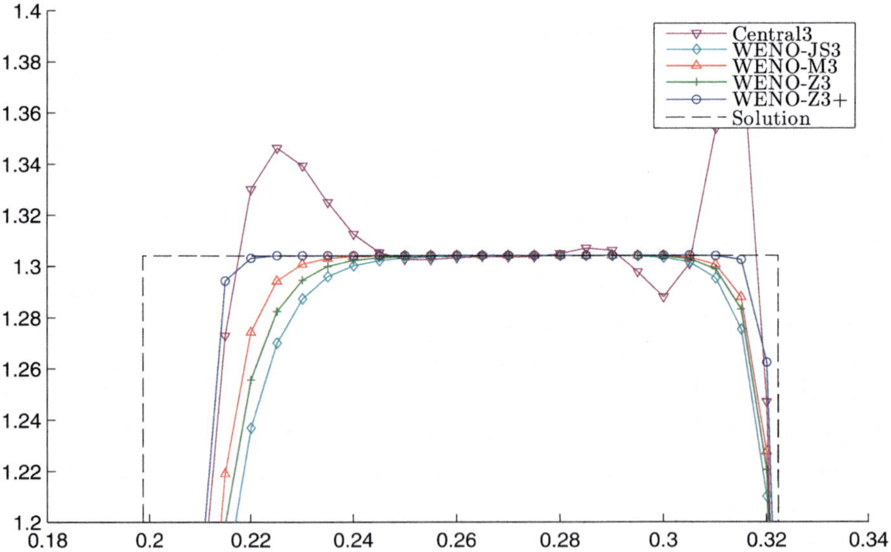

Fig. 8 Numerical solution of the Riemann problem of Lax with $N = 401$ points at $t = 0.13$; density. Zoom in the relevant region

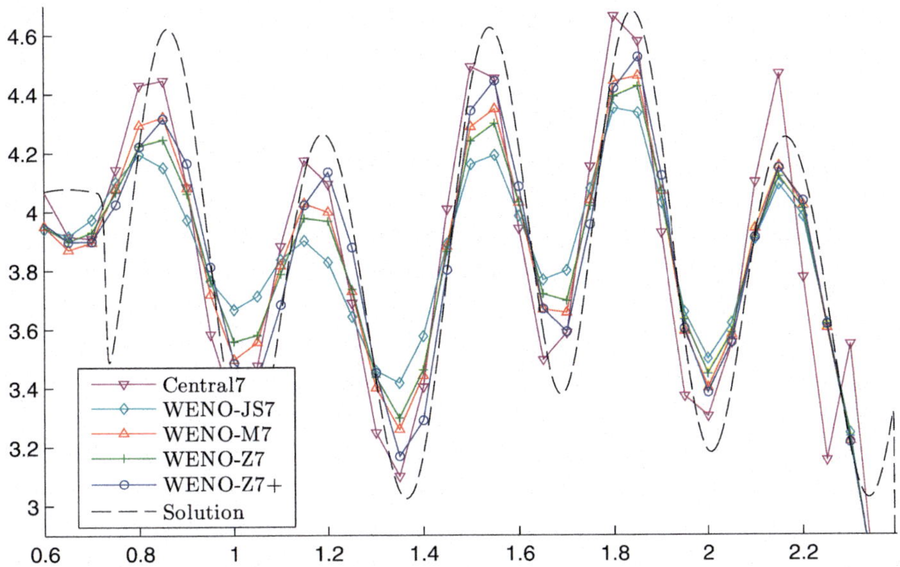

Fig. 9 Numerical solution of the shock-entropy wave problem of Shu–Osher with $N = 201$ points at $t = 1.8$; density. Zoom in the relevant region

3.2 *Order 7*

For order 7, the results of all WENO schemes are more close. The exceptions are the two shock-entropy wave interaction tests (Figs. 9 and 10), which have numerous smooth extrema. In these tests, WENO-Z+ performs better than the other WENO, even surpassing Central7 in the Titarev–Toro test. As in the orders 3 and 5, WENO-Z+ was stable and essentially oscillation-free near discontinuities with this choice of λ, as Fig. 11 illustrates.

4 Conclusions

The WENO-Z+ scheme has shown significantly better results than the other WENO schemes tested here for third and seventh orders, showing that the approach is also sound for orders other than five. While this is very promising, the scheme deserves more investigation. The ninth- and higher orders still need to be implemented. Also, 2D and 3D tests need to be run. We plan to do both things in the near future.

Regarding the parameter λ, the choices shown here were guesses corroborated by empirical tests, and they may be far from optimal. We plan to use a more systematic approach to find the optimal λ. Another issue involving λ is that it appears to

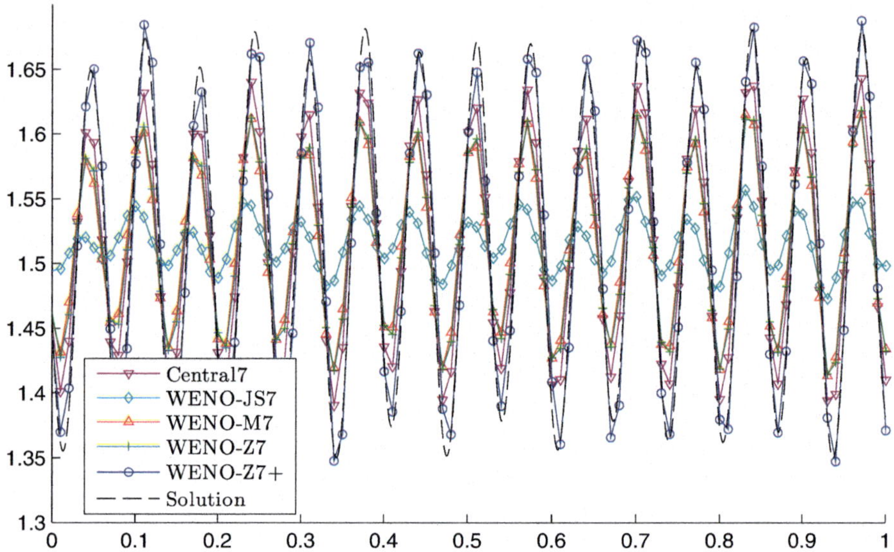

Fig. 10 Numerical solution of the shock-entropy wave problem of Titarev–Toro with $N = 1001$ points at $t = 5$; density. Zoom in the relevant region

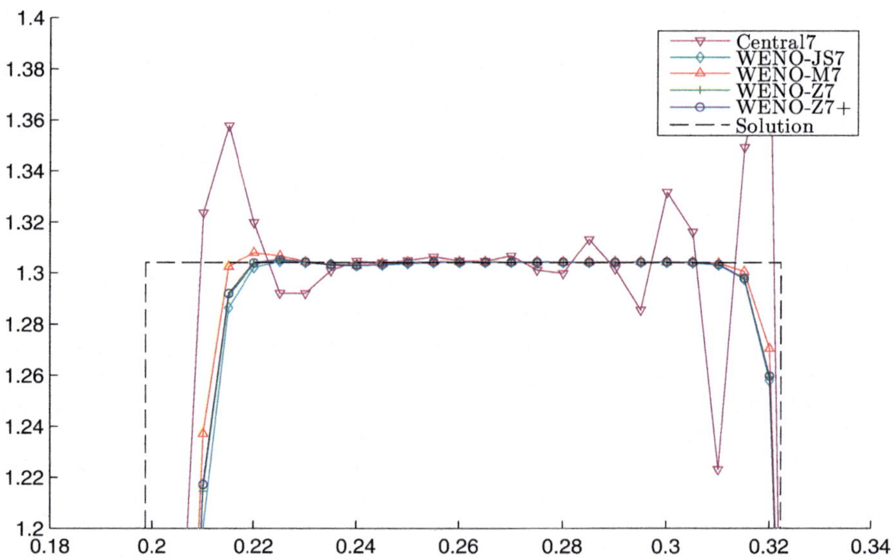

Fig. 11 Numerical solution of the Riemann problem of Lax with $N = 201$ points at $t = 0.13$; density. Zoom in the relevant region

depend on the grid size Δx, at least for fifth and seventh orders. It would be better to find a way of removing this dependence, so as to emulate the self-similarity of conservation laws.

Acknowledgements The author would like to thank Profs. Wai-Sun Don and Guus Jacobs for organizing together the minisymposium of High-order Methods at ICOSAHOM 2016; Prof. Chi-Wang Shu and the Division of Applied Mathematics of Brown University, where part of this research was done; and the organizers of ICOSAHOM 2016, especially Prof. Marco Bittencourt, for all the support.

References

1. F. Acker, R.B. de R. Borges, B. Costa, An improved WENO-Z scheme. J. Comput. Phys. **313**, 726–753 (2016)
2. B.S. Balsara, C.-W. Shu, Monotonicity preserving weighted essentially non-oscillatory schemes with increasingly high order of accuracy. J. Comput. Phys. **160**(2), 405–452 (2000)
3. R. Borges, M. Carmona, B. Costa, W.S. Don, An improved weighted essentially non-oscillatory scheme for hyperbolic conservation laws. J. Comput. Phys. **227**(6), 3191–3211 (2008)
4. M. Castro, B. Costa, W.S. Don, High order weighted essentially non-oscillatory WENO-Z schemes for hyperbolic conservation laws. J. Comput. Phys. **230**(5), 1766–1792 (2011)
5. W.-S. Don, R. Borges, Accuracy of the weighted essentially non-oscillatory conservative finite difference schemes. J. Comput. Phys. **250**, 347–372 (2013)
6. Y. Ha, C.H. Kim, Y.J. Lee, J. Yoon, An improved weighted essentially non-oscillatory scheme with a new smoothness indicator. J. Comput. Phys. **232**(1), 68–86 (2013)
7. A. Harten, B. Engquist, S. Osher, S.R. Chakravarthy, Uniformly high order accurate essentially non-oscillatory schemes, III. J. Comput. Phys. **71**(2), 231–303 (1987)
8. A.K. Henrick, T.D. Aslam, J.M. Powers, Mapped weighted essentially non-oscillatory schemes: achieving optimal order near critical points. J. Comput. Phys. **207**(2), 542–567 (2005)
9. G.-S. Jiang, C.-W. Shu, Efficient implementation of weighted ENO schemes. J. Comput. Phys. **126**(1), 202–228 (1996)
10. X.-D. Liu, S. Osher, T. Chan, Weighted essentially non-oscillatory schemes. J. Comput. Phys. **115**(1), 200–212 (1994)
11. S. Pirozzoli, On the spectral properties of shock-capturing schemes. J. Comput. Phys. **219**(2), 489–497 (2006)
12. J. Shi, Y.-T. Zhang, C.-W. Shu, Resolution of high order WENO schemes for complicated flow structures. J. Comput. Phys. **186**(2), 690–696 (2003)
13. C.-W. Shu, Essentially non-oscillatory and weighted essentially non-oscillatory schemes for hyperbolic conservation laws. NASA/CR-97-206253 ICASE Report 97–65, (1997)
14. C.-W. Shu, High order weighted essentially nonoscillatory schemes for convection dominated problems. SIAM Rev. **51**(1), 82–126 (1997)
15. N.K. Yamaleev, M.H. Carpenter, A systematic methodology for constructing high-order energy stable WENO schemes. J. Comput. Phys. **228**(11), 4248–4272 (2009)
16. S. Zhao, N. Lardjane, I. Fedioun, Comparison of improved finite-difference WENO schemes for the implicit large eddy simulation of turbulent non-reacting and reacting high-speed shear flows. Comput. Fluids **95**, 74–87 (2014)

Compact High Order Complete Flux Schemes

J.H.M. ten Thije Boonkkamp and M.J.H. Anthonissen

Abstract In this paper we outline the complete flux scheme for an advection-diffusion-reaction model problem. The scheme is based on the integral representation of the flux, which we derive from a local boundary value problem for the *entire* equation, including the source term. Consequently, the flux consists of a homogeneous part, corresponding to the advection-diffusion operator, and an inhomogeneous part, taking into account the effect of the source term. We apply (weighted) Gauss quadrature rules to derive the standard complete flux scheme, as well as a compact high order variant. We demonstrate the performance of both schemes.

1 Introduction

Conservation laws are ubiquitous in continuum physics, they occur in disciplines like fluid dynamics, combustion theory, plasma physics, semiconductor theory etc. These conservation laws are often of advection-diffusion-reaction type, describing the interplay between different processes such as advection or drift, diffusion or conduction and (chemical) reaction or recombination/generation.

In this paper we address (high order) space discretisation methods for these equations. We consider the model problem

$$\frac{\mathrm{d}}{\mathrm{d}x}\Big(u\varphi - \varepsilon\frac{\mathrm{d}\varphi}{\mathrm{d}x}\Big) = s, \tag{1}$$

where u is the advection velocity, $\varepsilon \geq \varepsilon_{\min} > 0$ a diffusion/conduction coefficient and s a source term. The unknown φ could be for example the mass fraction of a

J.H.M. ten Thije Boonkkamp (✉) • M.J.H. Anthonissen
Department of Mathematics and Computer Science, Eindhoven University of Technology, PO Box 513, 5600 MB Eindhoven, The Netherlands
e-mail: j.h.m.tenthijeboonkkamp@tue.nl; m.j.h.anthonissen@tue.nl

M.L. Bittencourt et al. (eds.), *Spectral and High Order Methods for Partial Differential Equations ICOSAHOM 2016*, Lecture Notes in Computational Science and Engineering 119, DOI 10.1007/978-3-319-65870-4_40

species in a reacting flow. Associated with (1) we introduce the flux f defined by

$$f = u\varphi - \varepsilon \frac{\mathrm{d}\varphi}{\mathrm{d}x}, \tag{2}$$

thus the conservation law can be concisely written as $\mathrm{d}f/\mathrm{d}x = s$.

For space discretisation we apply the finite volume method (FVM), thus we cover the domain with a finite set of control volumes (cells) I_j of size $h = \Delta x$. We adopt the vertex-centred approach [6], i.e., we choose $I_j = [x_{j-1/2}, x_{j+1/2}]$ where $x_{j\pm 1/2} = \frac{1}{2}\big(x_j + x_{j\pm 1}\big)$ and x_j are the grid points where φ has to be approximated. Integrating (1) over the control volume I_j, we obtain the integral conservation law

$$f(x_{j+1/2}) - f(x_{j-1/2}) = \int_{x_{j-1/2}}^{x_{j+1/2}} s(x)\,\mathrm{d}x. \tag{3}$$

To derive the discrete conservation law, we have to approximate the flux $f(x_{j+1/2})$ by a numerical flux $F_{j+1/2}$ and we have to approximate the integral in the right hand side. Thus, a generic form of the discrete conservation law reads

$$F_{j+1/2} - F_{j-1/2} = \mathrm{Q}[s; x_{j-1/2}, x_{j+1/2}], \tag{4}$$

where $F_{j+1/2}$ is the numerical flux at the cell interface $x = x_{j+1/2}$ and where $\mathrm{Q}\big[s; x_{j-1/2}, x_{j+1/2}\big]$ denotes a (high order) quadrature rule approximation for the integral in the right hand side of (3). A possible choice for the numerical flux is the standard complete flux scheme, which can be written in the form

$$F_{j+1/2} = \alpha_{j+1/2}\varphi_j - \beta_{j+1/2}\varphi_{j+1} + h\big(\gamma_{j+1/2}s_j + \delta_{j+1/2}s_{j+1}\big), \tag{5}$$

for some coefficients $\alpha_{j+1/2}$ etc., and where $\varphi_j \approx \varphi(x_j)$ denotes the numerical solution at grid point x_j and $s_j = s(x_j)$. The standard complete flux approximation results in a compact three-point scheme and is uniformly second order accurate [4]. The purpose of this paper is to derive a compact, high order variant of the complete flux scheme. The numerical flux may only depend on the two neighbouring grid point values of φ and s, and necessarily some values of s at intermediate points. This way we avoid cumbersome (W)ENO reconstruction of interface values for φ. Combined with a high order quadrature rule for s, this gives rise to a compact high order scheme. Consequently, the resulting algebraic system is straightforward to solve and the numerical solution much more accurate than the standard complete flux numerical solution.

We have organised our paper as follows. In Sect. 2 we present the integral representation of the flux, from which we derive the standard complete flux scheme. We combine the standard scheme with the midpoint rule for the source s. Next, in Sect. 3 we present a high order variant of the complete flux scheme. For the corresponding quadrature rule for s we choose the two-point Gauss-Legendre quadrature rule. We demonstrate the performance of both schemes in Sect. 4 and we end with a summary and conclusions in Sect. 5.

2 Standard Complete Flux Scheme

In this section we outline the standard complete flux scheme for Eq. (1), which is based on the integral representation of the flux; for a detailed derivation see [4].

The integral representation of the flux $f(x_{j+1/2})$ at the cell interface $x = x_{j+1/2}$ is based on the following model boundary value problem (BVP) for φ:

$$\frac{\mathrm{d}f}{\mathrm{d}x} = \frac{\mathrm{d}}{\mathrm{d}x}\Big(u\varphi - \varepsilon\frac{\mathrm{d}\varphi}{\mathrm{d}x}\Big) = s, \quad x_j < x < x_{j+1}, \tag{6a}$$

$$\varphi(x_j) = \varphi_j, \quad \varphi(x_{j+1}) = \varphi_{j+1}. \tag{6b}$$

We like to emphasize that $f(x_{j+1/2})$ corresponds to the solution of the *entire* equation, implying that $f(x_{j+1/2})$ not only depends on u and ε, but also on the source term s. We define the following variables:

$$a = \frac{u}{\varepsilon}, \quad A(x) = \int_{x_{j+1/2}}^{x} a(\xi)\,\mathrm{d}\xi, \quad S(x) = \int_{x_{j+1/2}}^{x} s(\xi)\,\mathrm{d}\xi. \tag{7}$$

Integrating Eq. (6a) from $x_{j+1/2}$ to $x \in [x_j, x_{j+1}]$ we obtain the relation

$$f(x) - f(x_{j+1/2}) = S(x). \tag{8}$$

Next, using the definition of A in (7), we rewrite the expression for the flux in its integrating factor formulation, i.e.,

$$f = -\varepsilon\frac{\mathrm{d}}{\mathrm{d}x}\Big(\varphi\,\mathrm{e}^{-A}\Big)\,\mathrm{e}^{A}. \tag{9}$$

Finally, substituting (9) in (8), integrating the resulting equation from x_j to x_{j+1} and applying the boundary conditions (6b), we obtain the following expressions for the flux

$$f(x_{j+1/2}) = f^{\mathrm{h}}(x_{j+1/2}) + f^{\mathrm{i}}(x_{j+1/2}), \tag{10a}$$

$$f^{\mathrm{h}}(x_{j+1/2}) = \Big(\mathrm{e}^{-A(x_j)}\varphi_j - \mathrm{e}^{-A(x_{j+1})}\varphi_{j+1}\Big)\Big/\int_{x_j}^{x_{j+1}} \varepsilon^{-1}\mathrm{e}^{-A}\,\mathrm{d}x, \tag{10b}$$

$$f^{\mathrm{i}}(x_{j+1/2}) = -\int_{x_j}^{x_{j+1}} \varepsilon^{-1}\mathrm{e}^{-A}S\,\mathrm{d}x\Big/\int_{x_j}^{x_{j+1}} \varepsilon^{-1}\mathrm{e}^{-A}\,\mathrm{d}x, \tag{10c}$$

where $f^{\mathrm{h}}(x_{j+1/2})$ and $f^{\mathrm{i}}(x_{j+1/2})$ are the homogeneous and inhomogeneous part of the flux, corresponding to the advection-diffusion operator and the source term, respectively.

For the inhomogeneous flux, we can derive an alternative expression. Indeed, substituting the expression for S in (7) in (10c) and changing the order of integration we obtain the relation

$$f^{\mathrm{i}}(x_{j+1/2}) = h \int_0^1 G(\sigma) s(x_j + h\sigma)\,\mathrm{d}\sigma, \tag{11}$$

where $\sigma = (x - x_j)/h$ is the normalised coordinate on $[x_j, x_{j+1}]$ and where the function G is defined by

$$G(\sigma) = \begin{cases} h \int_0^\sigma \varepsilon^{-1}\mathrm{e}^{-A}\,\mathrm{d}\eta / \int_{x_j}^{x_{j+1}} \varepsilon^{-1}\mathrm{e}^{-A}\,\mathrm{d}x & \text{for} \quad 0 \le \sigma \le \frac{1}{2}, \\ -h \int_\sigma^1 \varepsilon^{-1}\mathrm{e}^{-A}\,\mathrm{d}\eta / \int_{x_j}^{x_{j+1}} \varepsilon^{-1}\mathrm{e}^{-A}\,\mathrm{d}x & \text{for} \quad \frac{1}{2} < \sigma \le 1. \end{cases} \tag{12}$$

Note that G relates the *flux* to the source term, and therefore we refer to it as the Green's function for the flux, similar to the Green's function which relates the *solution* of (6) to the source. Summarizing, the flux is completely determined by the expressions (10a), (10b), (11) and (12).

Next, let us consider the special case of constant u and ε, the source term s is assumed to be an arbitrary function of x. We introduce the (grid) Péclet number $P = uh/\varepsilon$. In this case, the expression for the homogeneous flux reduces to

$$f^{\mathrm{h}}(x_{j+1/2}) = \frac{\varepsilon}{h}\big(B(-P)\varphi_j - B(P)\varphi_{j+1}\big). \tag{13}$$

In (13) we have used the Bernoulli function $B(z) := z/\big(\mathrm{e}^z - 1\big)$. We can evaluate all integrals involved in the expressions for G and find

$$G(\sigma; P) = \begin{cases} \frac{1-\mathrm{e}^{-P\sigma}}{1-\mathrm{e}^{-P}} & \text{for} \quad 0 \le \sigma \le \frac{1}{2}, \\ -\frac{1-\mathrm{e}^{P(1-\sigma)}}{1-\mathrm{e}^{P}} & \text{for} \quad \frac{1}{2} < \sigma \le 1; \end{cases} \tag{14}$$

see Fig. 1. Note that G explicitly depends on P as a parameter. Moreover, G is discontinuous at $\sigma = \frac{1}{2}$ and satisfies the symmetry condition $G(\sigma; P) = -G(1 - \sigma; -P)$. The flux is in this case completely determined by the expressions (10a), (13), (11) and (14).

To derive expressions for the numerical flux $F_{j+1/2}$, we have to apply quadrature rules to all integrals involved. For the general case of variable u and ε expressions for the standard complete flux scheme have been derived in [4], whereas a higher order complete flux scheme based on the 2-point Gauss-Legendre quadrature rule is presented in [1].

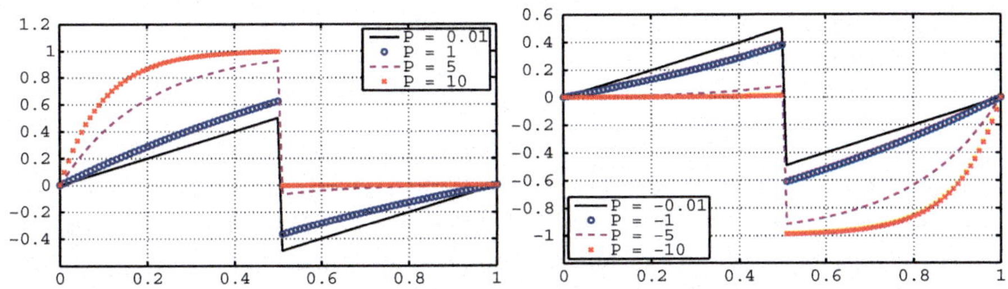

Fig. 1 Green's function for the flux for $P > 0$ (*left*) and $P < 0$ (*right*)

In the remainder of this paper we restrict ourselves to constant u and ε. It is our purpose to derive a new high order flux approximation based on *weighted* Gauss quadrature rules. As weight function we will use the function $G(\sigma; P)$ given in (14).

We start with the standard complete flux scheme. For the homogeneous numerical flux $F^{\mathrm{h}}_{j+1/2}$ we simply take the homogeneous part of the flux, i.e., $F^{\mathrm{h}}_{j+1/2} = f^{\mathrm{h}}(x_{j+1/2})$, which is exact; see (13). This approximation corresponds to the well-known exponentially fitted scheme; see for example [3] and the many references therein. To evaluate the expression (11) for the inhomogeneous flux, we need to approximate the source term on $[x_j, x_{j+1}]$. An obvious choice is the piecewise constant representation, corresponding to the midpoint rule in (4), given by

$$s(x_j + h\sigma) = \begin{cases} s_j & \text{if} \quad 0 \le \sigma \le \frac{1}{2}, \\ s_{j+1} & \text{if} \quad \frac{1}{2} < \sigma \le 1. \end{cases} \tag{15}$$

Inserting this expression in (11) and evaluating the resulting integrals, we obtain

$$F^{\mathrm{i}}_{j+1/2} = h\big(C_2(-P)s_j - C_2(P)s_{j+1}\big), \tag{16}$$

where $C_2(z) := \big(\mathrm{e}^{z/2} - 1 - z/2\big)/\big(z(\mathrm{e}^z - 1)\big)$. The total numerical flux $F_{j+1/2}$ is obviously given by $F_{j+1/2} = F^{\mathrm{h}}_{j+1/2} + F^{\mathrm{i}}_{j+1/2}$ and is referred to as the complete flux scheme (CFS).

Substituting the numerical flux in the discrete conservation law (4) and applying the midpoint rule $\mathrm{M}\big[s; x_{j-1/2}, x_{j+1/2}\big] = hs_j$, we obtain

$$F^{\mathrm{h}}_{j+1/2} - F^{\mathrm{h}}_{j-1/2} = h\big(C_2^- s_{j-1} + \big(1 - C_2^- - C_2^+\big)s_j + C_2^+ s_{j+1}\big), \tag{17}$$

where we introduced the short hand notation $C_2^{\pm} = C_2(\pm P)$. The left hand side of this equation is the discretised advection-diffusion operator, which can be written as a weighted average of the central difference and upwind discretisations, whereas the right hand side contains a weighted average of the source term values.

3 High Order Complete Flux Scheme: The Constant Coefficient Case

In this section we derive a high order approximation for the numerical flux. Consequently, we also need a high order quadrature rule $\mathrm{Q}[s; x_{j-1/2}, x_{j+1/2}]$ in (4).

Note that the homogeneous numerical flux $F^{\mathrm{h}}_{j+1/2}$ is exact for constant u and ε, thus we only have to consider the inhomogeneous numerical flux $F^{\mathrm{i}}_{j+1/2}$. Since $G(\sigma; P)$ is discontinuous at $\sigma = \frac{1}{2}$, corresponding to the interface position $x = x_{j+1/2}$, we have to split the integral in (10c) in two parts as follows

$$f^{\mathrm{i}}(x_{j+1/2}) = h\,(I_1 + I_2), \tag{18a}$$

$$I_1 = \int_0^{1/2} G(\sigma; P)\tilde{s}(\sigma)\,\mathrm{d}\sigma, \quad I_2 = \int_{1/2}^{1} G(\sigma; P)\tilde{s}(\sigma)\,\mathrm{d}\sigma, \tag{18b}$$

where $\tilde{s}(\sigma) = s(x_j + h\sigma)$. We propose the weighted Gauss (WG) quadrature rule

$$I_1 \approx \mathrm{WG}\big[\tilde{s}; 0, \tfrac{1}{2}\big] = w_1 G(\sigma_1; P)\tilde{s}(\sigma_1), \quad I_2 \approx \mathrm{WG}\big[\tilde{s}; \tfrac{1}{2}, 1\big] = w_2 G(\sigma_2; P)\tilde{s}(\sigma_2), \tag{19}$$

with weights $w_1, w_2 > 0$ and nodes $\sigma_1 \in \big(0, \frac{1}{2}\big)$ and $\sigma_2 \in \big(\frac{1}{2}, 1\big)$. We require that $I_1 = \mathrm{WG}\big[\tilde{s}; 0, \frac{1}{2}\big]$ and $I_2 = \mathrm{WG}\big[\tilde{s}; \frac{1}{2}, 1\big]$ for $\tilde{s}(\sigma) = 1$ and $\tilde{s}(\sigma) = \sigma$. For the first integral this gives rise to the equations

$$C_2(-P) = w_1 G(\sigma_1; P), \quad \tfrac{1}{2}C_2(-P) - C_3(P) = w_1 G(\sigma_1; P)\sigma_1, \tag{20}$$

where $C_3(z) := \big(C_2(z) - \frac{1}{8}B(z)\big)/z$. From the equations in (20) we find the quadrature rule

$$I_1 \approx \mathrm{WG}\big[\tilde{s}; 0, \tfrac{1}{2}\big] = \omega_1 \tilde{s}(\sigma_1), \quad \omega_1 = C_2(-P), \quad \sigma_1 = \tfrac{1}{2} - \frac{C_3(-P)}{C_2(-P)}. \tag{21a}$$

In a similar fashion we find

$$I_2 \approx \mathrm{WG}\big[\tilde{s}; \tfrac{1}{2}, 1\big] = \omega_2 \tilde{s}(\sigma_2), \quad \omega_2 = -C_2(P), \quad \sigma_2 = \tfrac{1}{2} + \frac{C_3(P)}{C_2(P)}. \tag{21b}$$

Alternatively, using the symmetry property of G, we can show that

$$I_2 = -\int_0^{1/2} G(\sigma; -P)\tilde{s}(1-\sigma)\,\mathrm{d}\sigma.$$

If we now apply the quadrature rule (21a) to this integral and replace P by $-P$, we recover (21b). The modified weights $\omega_1 = w_1 G(\sigma_1; P)$, $\omega_2 = w_2 G(\sigma_2; P)$ and

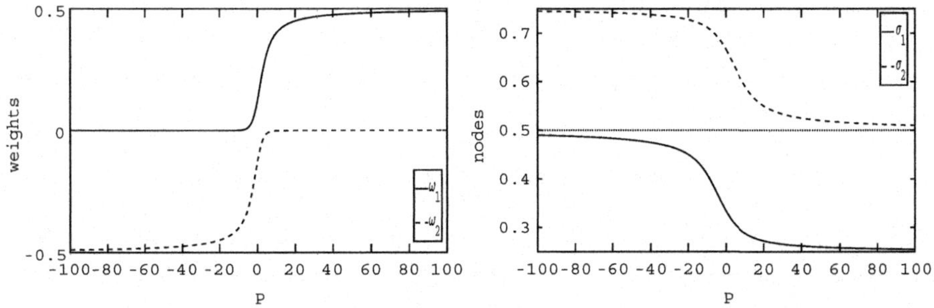

Fig. 2 The weights ω_k (*left*) and nodes σ_k (*right*) ($k = 1, 2$)

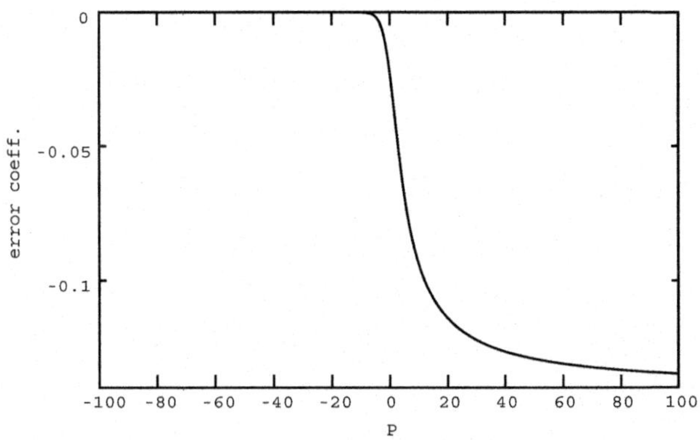

Fig. 3 Coefficient in the error term in (23) as a function of P

the corresponding nodes σ_1, σ_2 as functions of P are shown in Fig. 2. Note that $0 < \sigma_1 < \frac{1}{2}$ and $\frac{1}{2} < \sigma_2 < 1$, as anticipated.

To investigate the error of the quadrature rules (21), we have to substitute $\tilde{s}(\sigma) = \sigma^2$, since this is the lowest order monomial for which the quadrature rules are no longer exact. We restrict ourselves to (21a), thus we have

$$I_1 = \omega_1 \tilde{s}(\sigma_1) + E_1, \tag{22}$$

where the error E_1 is of the form $E_1 = C\tilde{s}''(\eta)$ for some $\eta \in (0, \frac{1}{2})$, with the prime ($'$) denoting differentiation with respect to σ. Substituting $\tilde{s}(\sigma) = \sigma^2$ we obtain $E_1 = 2C_4(-P) - C_3^2(-P)/C_2(-P)$, where $C_4(z) = (C_3(z) - \frac{1}{48}B(z))/z$. Therefore, for arbitrary $\tilde{s}(\sigma)$, we have the error term

$$E_1 = h^2\Big[C_4(-P) - \frac{C_3^2(-P)}{2C_2(-P)}\Big]\frac{\mathrm{d}^2 s}{\mathrm{d}x^2}(\xi), \quad \xi \in (x_j, x_{j+1/2}), \tag{23}$$

implying the approximation is second order accurate. The error coefficient in brackets as a function of P is shown in Fig. 3. From this figure, it is obvious that the error is negligible for $P < 0$, and small for $P > 0$. A similar result holds for the quadrature rule (21b).

Applying the quadrature rules in (21), we find the following expression for the inhomogeneous numerical flux

$$F^{\mathrm{i}}_{j+1/2} = h\big(C_2(-P)s(x_j + h\sigma_1) - C_2(P)s(x_j + h\sigma_2)\big), \tag{24}$$

which is (at least) third order accurate in view of the error term in (23). Note that this approximation is similar to (16), except for the nodes where s has to be evaluated. It is instructive to consider some limiting cases. First, for $P = 0$, i.e. no advection, the expression in (24) reduces to

$$F^{\mathrm{i}}_{j+1/2} = \tfrac{1}{8}h\big(s(x_j + \tfrac{1}{3}h) - s(x_j + \tfrac{2}{3}h)\big), \tag{25a}$$

corresponding to the piecewise linear limit function $G(\sigma) = \sigma$ for $0 \leq \sigma \leq \frac{1}{2}$ and $G(\sigma) = \sigma - 1$ for $\frac{1}{2} < \sigma \leq 1$. Alternatively, for $P \to +\infty$, i.e. $u > 0$ and no diffusion, we obtain

$$F^{\mathrm{i}}_{j+1/2} = \tfrac{1}{2}hs(x_j + \tfrac{1}{4}h), \tag{25b}$$

which is the midpoint approximation of the integral in (11) for the piecewise constant limit function $G(\sigma) = 1$ for $0 < \sigma < \frac{1}{2}$ and $G(\sigma) = 0$ for $\frac{1}{2} < \sigma < 1$. A similar expression holds when $P \to -\infty$.

To complete the discretisation, we apply the two-point Gauss-Legendre quadrature rule $\mathrm{GL2}\big[s; x_{j-1/2}, x_{j+1/2}\big]$ to the integral of s in (3) to obtain

$$\begin{aligned} F^{\mathrm{h}}_{j+1/2} - F^{\mathrm{h}}_{j-1/2} = h\big(&C_2^- s(x_{j-1} + h\sigma_1) - C_2^+ s(x_{j-1} + h\sigma_2) - \\ &C_2^- s(x_j + h\sigma_1) + C_2^+ s(x_j + h\sigma_2)\big) + \mathrm{GL2}\big[s; x_{j-1/2}, x_{j+1/2}\big]; \end{aligned} \tag{26}$$

cf. (17).

4 Numerical Example

In this section we apply the standard and high order CF schemes to a model problem to assess their (order of) accuracy.

Consider Eq. (1) defined for $0 < x < 1$. Boundary conditions and source term are chosen, such that the exact solution is given by

$$\varphi(x) = \big(1 - \tfrac{1}{5}\sin(\pi\omega)\big)\frac{\mathrm{e}^{u(x-1)/\varepsilon} - \mathrm{e}^{-u/\varepsilon}}{1 - \mathrm{e}^{-u/\varepsilon}} + \tfrac{1}{5}\sin(\pi\omega x). \tag{27}$$

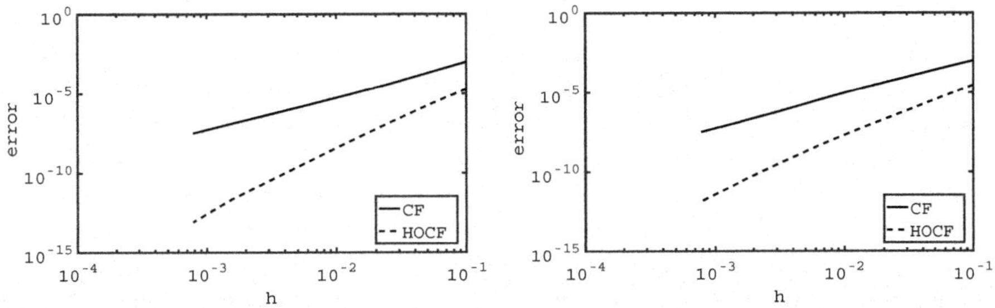

Fig. 4 1-norm of the error for $\varepsilon = 10^{-2}$ (*left*) and $\varepsilon = 10^{-3}$ (*right*)

We take the following parameter values: $\omega = 1$, $u = 1$, $\varepsilon = 10^{-2}$ or $\varepsilon = 10^{-3}$. In both cases the solution has a boundary layer at the outflow. To determine the accuracy of a numerical solution, we compute $e_h = h||\boldsymbol{\varphi} - \boldsymbol{\varphi}^*||_1$, with $\boldsymbol{\varphi}$ the numerical solution vector and $\boldsymbol{\varphi}^*$ the exact solution restricted to the grid, as a function of the grid size h, see Fig. 4. From this figure we conclude that the standard CF-scheme is second order convergent, uniformly in the Péclet number, whereas the high order CF-scheme exhibits fourth order convergence for $\varepsilon = 10^{-2}$ and roughly third order for $\varepsilon = 10^{-3}$. In both cases the high order scheme has a significant smaller error e_h than the standard scheme.

5 Concluding Remarks and Discussion

We have derived the integral representation of the flux for a model advection-diffusion-reaction equation. Applying quadrature rules to this representation, we could derive two flux approximation schemes, i.e., the standard complete flux scheme and a high order variant. The first scheme is second order accurate and the latter even fourth order, uniformly in the Péclet number. Moreover, both schemes only have a three-point coupling, albeit at the cost of a few source term evaluations at intermediate points. The compact stencil makes the discrete schemes easy to solve. A drawback is that quadrature rules for the inhomogeneous flux involving more than two weights and nodes are hard to derive.

Modifications to more complicated problems is not straightforward. This paper is a first attempt in designing high order complete flux schemes, and more research is certainly needed. A few possible modifications are the following. First, for nonlinear conservation laws the weighted Gauss quadrature rule is not feasible, and we first have to formulate a linearized BVP, analogous to (6), to derive a high order flux approximation scheme. However, this linearization is tricky and should use (the structure of) the solution of the corresponding *nonlinear* BVP. In [2] we have used this idea to derive a nonlinear (low order) flux approximation scheme for the Burgers' equation. Second, also for two-dimensional equations the scheme doesn't hold. A possible remedy is to formulate the conservation law in local flow adapted coordinates. This way we have to compute an advection-diffusion flux component aligned with the flow, for which we can use the high order scheme, and a diffusion flux component perpendicular to the flow, for which we can use a compact scheme. In [5] we have carried out this procedure for the standard complete flux scheme. Finally, extension to time-dependent problems is probably the most troublesome. We need a high order approximation for integral of the time derivative and we have to include the time derivative in the inhomogeneous flux; see [4] for details. In both cases we introduce the time derivative at intermediate points, which need to be eliminated. Moreover, we need a high order time integration method.

References

1. M.J.H. Anthonissen, J.H.M. ten Thije Boonkkamp, A compact high order finite volume scheme for advection-diffusion-reaction equations, in *Numerical Analysis and Applied Mathematics: International Conference on Numerical Analysis and Applied Mathematics 2009, AIP Conference Proceedings*, vol. 1168 (2009), pp. 410–414
2. N. Kumar, J.H.M. ten Thije Boonkkamp, B. Koren, A. Linke, A nonlinear flux approximation scheme for the viscous Burgers' equation, in *Finite Volumes for Complex Applications VIII - Hyperbolic, Elliptic and Parabolic Problems*, ed. by C. Cancès et al. Springer Proceedings in Mathematics & Statistics, vol. 200 (2017), pp. 457–465
3. K.W. Morton, *Numerical Solution of Convection-Diffusion Problems*. Applied Mathematics and Mathematical Computation, vol. 12 (Chapman & Hall, London, 1996)
4. J.H.M. ten Thije Boonkkamp, M.J.H. Anthonissen, The finite volume-complete flux scheme for advection-diffusion-reaction equations. J. Sci. Comput. **46**, 47–70 (2011)
5. J.H.M. ten Thije Boonkkamp, M.J.H. Anthonissen, R.J. Kwant, A two-dimensional complete flux scheme in local flow adapted coordinates, in *Finite Volumes for Complex Applications VIII - Hyperbolic, Elliptic and Parabolic Problems*, ed. by C. Cancès et al. Springer Proceedings in Mathematics & Statistics, vol. 200 (2017), pp. 437–445
6. P. Wesseling, *Principles of Computational Fluid Dynamics* (Springer, Berlin, 2001)

Pointwise Force Equilibrium Preserving Spectral Element Method for Structural Problems

K. Olesen, B. Gervang, J.N. Reddy, and M. Gerritsma

Abstract In structural mechanics the geometry is a crucial factor in the derivation of the governing force equilibrium equations, which describe the balance of forces in a discrete setting. In conventional discretization techniques the quantities are approximated through nodal expansions, which lead to global force equilibrium, but not local. This paper shows that by considering the geometry of the problem the equilibrium of forces can be satisfied globally as well as locally.

1 Introduction

The finite element method (FEM) is a key tool to solve structural problems in industry. The nodal FEM formulation approximates the displacement field based on Lagrange polynomials, which base their expansions on discrete nodal values, see [15]. The constitutive equations link the stress components to the strain components meaning that the discrete stress field is expanded based on the first derivative of the Lagrange polynomials for the displacement field. The discrete displacement field is typically only C^0 continuous across element boundaries which implies that the discrete stress field is discontinuous. The force equilibrium equations contain the derivatives of the stress components and the discretization of these therefore involve the second derivative of the Lagrange polynomials. The discrete force equilibrium equations are thereby satisfied globally but not locally.

K. Olesen (✉) • B. Gervang
Department of Engineering, Aarhus University, Inge Lehmanns Gade 10, 8000 Aarhus, Denmark
e-mail: keol@eng.au.dk; bge@ase.au.dk

J.N. Reddy
Department of Mechanical Engineering, Texas A & M University, College Station, TX 77843-3123, USA
e-mail: jnreddy@tamu.edu

M. Gerritsma
Faculty of Aerospace Engineering, Delft University of Technology, Kluyverweg 2, 2629 HS Delft, The Netherlands
e-mail: m.i.gerritsma@tudelft.nl

M.L. Bittencourt et al. (eds.), *Spectral and High Order Methods for Partial Differential Equations ICOSAHOM 2016*, Lecture Notes in Computational Science and Engineering 119, DOI 10.1007/978-3-319-65870-4_41

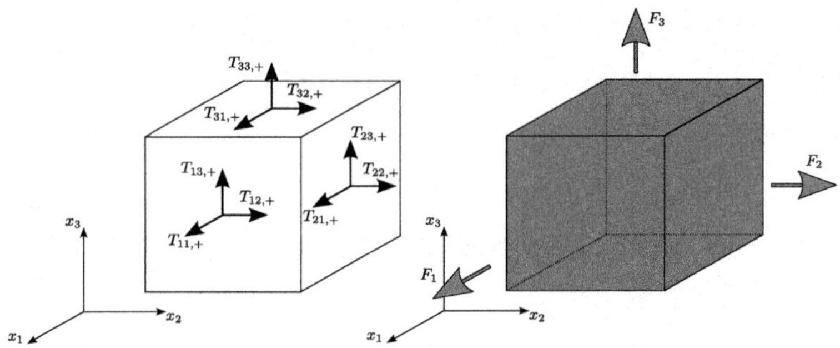

Fig. 1 The forces on a cube in $\mathbb{R}^3$. The *left picture* illustrates the components of the surface forces, T_{ij}, on the surfaces. The *right picture* shows the body force components, F_j, acting on the volume of the element

Using the ideas of Tonti, [18, 19], where integral values are treated as the Degrees of Freedom (DOFs), we will show in this paper that the discrete force equilibrium equations can be satisfied pointwise throughout the domain. Local considerations of the force equilibrium were considered by Fraeijs De Veubeke in [5–7] in the 1960s through dual analysis, where two simultaneous analyses are performed, a kinematic and a static or dynamic admissible model, which are coupled through complementary energy principles, [6, §2.2]. The static admissible model considers the tractions on the element boundaries and connections to the stresses are established. Since the stress field is coupled to the displacement field the compatibility equations are in general not satisfied, and spurious kinematic modes may be present. Such spurious modes can be eliminated by applying, for instance, a stress potential or by direct approximations of the stress field in the elements. These elements were revisited in the 1990s by Almeida and Freitas and a family of hybrid finite element schemes were developed, [12, 13]. Recently, Almeida has extended this work in [17], where the deformation of the elements is taken into account through Piola-Kirchhoff projections. In this paper we will use forces – *not stresses* – as our DOFs. The surface forces will be assigned to surfaces in the mesh, while the body forces will be assigned to the volumes in the mesh, see Fig. 1. For such particular choice of DOFs, force equilibrium reduces to a simple algebraic equation. By selecting basis functions which interpolate quantities over surfaces and volumes, discrete force equilibrium is satisfied pointwise. See [8, 10, 11] for the derivation and use of these basis functions.

2 Equilibrium of Forces

Most textbooks on continuum mechanics derive equilibrium of forces as for instance in [16]. Considering the cube in $\mathbb{R}^3$ depicted in Fig. 1 under the action of surface forces and a body force all decomposed into components, then the equilibrium of forces is written as[1]

$$T_{1j,+} - T_{1j,-} + T_{2j,+} - T_{2j,-} + T_{3j,+} - T_{3j,-} + F_j = 0 \, , \tag{1}$$

for $j = 1, 2, 3$. Here $T_{ij,+}$ and $T_{ij,-}$ denotes the surface force component with an outward unit vector in the positive and negative basis direction, respectively. In (1) the surface force components are given by

$$T_{ij} = \int_{(\partial\Omega)_{i,\pm}} \sigma_{ij} n_i \, dS \, , \tag{2}$$

where $\partial\Omega$ is the boundary of the volume in Fig. 1 with $\partial\Omega = \sum_{i=1}^{3} (\partial\Omega)_{i,+} + (\partial\Omega)_{i,-}$ and σ_{ij} are the Cauchy stress components. The body force components are given by

$$F_j = \int_{\Omega} f_j \, \mathrm{d}\Omega \, , \tag{3}$$

where Ω is any volume and f_j is the body force density component.

Inserting (2) and (3) in (1) produces

$$\int_{\partial\Omega} \sigma_{ij} \, n_i \, \mathrm{d}S + \int_{\Omega} f_j \, \mathrm{d}\Omega = 0 \, .$$

Applying Gauss' divergence theorem on the surface integral and recognizing that the volume is arbitrary leads to the differential equation

$$\frac{\partial}{\partial x_i} \sigma_{ij} + f_j = 0 \, . \tag{4}$$

The relation (4) is deprived of all geometrical relations, i.e. from this PDE it is no longer apparent that the σ_{ij} are connected to surfaces and f_j to volumes. Furthermore, notice that (1) is just a sum of finite values and is exact in a discrete space. Consider a traditional FEM grid then the nodes will span small volumes. Numbering these volumes and their bounding surfaces as in the example shown in Fig. 2 then (1) and hence (4) can be represented as

$$\mathscr{D} \boldsymbol{\Delta}_T = -\boldsymbol{F} \, . \tag{5}$$

[1]In continuum mechanics one considers the limit for the volume $\mathscr{V}$ going to zero and introduce the stresses – force per unit area – and body force density – force per unit volume or per unit mass – in this limiting case. On a finite mesh, volumes will not be zero nor do they tend to zero and therefore the physical variables 'surface force' and 'body force' are more appropriate to work with than the mathematically defined fields commonly found in books on continuum mechanics See also [18] for similar ideas.

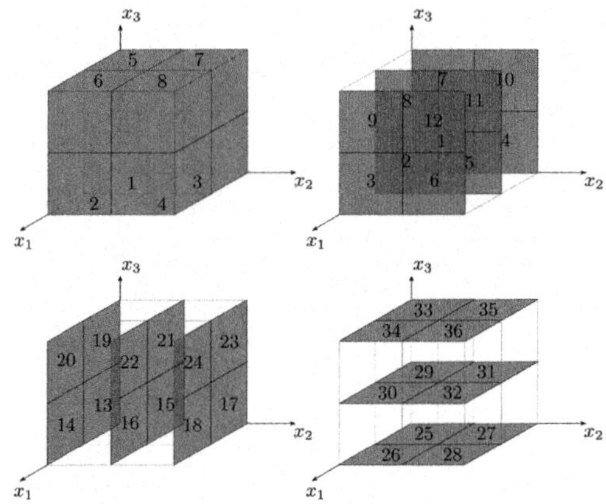

$$\boldsymbol{E}_{(3,2)} = \left[\begin{array}{cccccccccccccccccccccccccccccccccccc}
-1 & 1 & 0 & 0 & 0 & 0 & 0 & 0 & 0 & 0 & 0 & 0 & -1 & 0 & 1 & 0 & 0 & 0 & 0 & 0 & 0 & 0 & 0 & 0 & -1 & 0 & 0 & 0 & 1 & 0 & 0 & 0 & 0 & 0 & 0 & 0 \\
0 & -1 & 1 & 0 & 0 & 0 & 0 & 0 & 0 & 0 & 0 & 0 & 0 & -1 & 0 & 1 & 0 & 0 & 0 & 0 & 0 & 0 & 0 & 0 & 0 & -1 & 0 & 0 & 0 & 1 & 0 & 0 & 0 & 0 & 0 & 0 \\
0 & 0 & 0 & -1 & 1 & 0 & 0 & 0 & 0 & 0 & 0 & 0 & 0 & 0 & -1 & 0 & 1 & 0 & 0 & 0 & 0 & 0 & 0 & 0 & 0 & 0 & -1 & 0 & 0 & 0 & 1 & 0 & 0 & 0 & 0 & 0 \\
0 & 0 & 0 & 0 & -1 & 1 & 0 & 0 & 0 & 0 & 0 & 0 & 0 & 0 & 0 & -1 & 0 & 1 & 0 & 0 & 0 & 0 & 0 & 0 & 0 & 0 & 0 & -1 & 0 & 0 & 0 & 1 & 0 & 0 & 0 & 0 \\
0 & 0 & 0 & 0 & 0 & 0 & -1 & 1 & 0 & 0 & 0 & 0 & 0 & 0 & 0 & 0 & 0 & 0 & -1 & 0 & 1 & 0 & 0 & 0 & 0 & 0 & 0 & 0 & -1 & 0 & 0 & 0 & 1 & 0 & 0 & 0 \\
0 & 0 & 0 & 0 & 0 & 0 & 0 & -1 & 1 & 0 & 0 & 0 & 0 & 0 & 0 & 0 & 0 & 0 & 0 & -1 & 0 & 1 & 0 & 0 & 0 & 0 & 0 & 0 & 0 & -1 & 0 & 0 & 0 & 1 & 0 & 0 \\
0 & 0 & 0 & 0 & 0 & 0 & 0 & 0 & 0 & -1 & 1 & 0 & 0 & 0 & 0 & 0 & 0 & 0 & 0 & 0 & -1 & 0 & 1 & 0 & 0 & 0 & 0 & 0 & 0 & 0 & -1 & 0 & 0 & 0 & 1 & 0 \\
0 & 0 & 0 & 0 & 0 & 0 & 0 & 0 & 0 & 0 & -1 & 1 & 0 & 0 & 0 & 0 & 0 & 0 & 0 & 0 & 0 & -1 & 0 & 1 & 0 & 0 & 0 & 0 & 0 & 0 & 0 & -1 & 0 & 0 & 0 & 1
\end{array}\right]$$

Fig. 2 Example of the numbering of the elements and their bounding surfaces. The numbering is increasing first in the x_1- then in x_2- and lastly in the x_3-direction. The illustrations show the numbering of: *top left*: The elements, *top right*: Bounding surfaces normal to x_1, *lower left*: Bounding surfaces normal to x_2, and *lower right*: Bounding surfaces normal to x_3

Here $\mathscr{D}$ is the divergence operator given by

$$\mathscr{D} = \begin{bmatrix} \boldsymbol{E}_{(3,2)} & 0 & 0 \\ 0 & \boldsymbol{E}_{(3,2)} & 0 \\ 0 & 0 & \boldsymbol{E}_{(3,2)} \end{bmatrix} ,$$

where $E_{(3,2)}$ is an incidence matrix, see [1, 4, 10, 11]. The incidence matrix for the example in Fig. 2 is given below the illustrations. Notice that $\boldsymbol{E}_{(3,2)}$ and hence $\mathscr{D}$ only consist of the numbers 0, -1 and 1. In (5) $\boldsymbol{\Delta}_T$ is a column vector containing all the discrete surface force components in the following order

$$\boldsymbol{\Delta}_T = \left\{ \boldsymbol{T}_{11}\ \boldsymbol{T}_{21}\ \boldsymbol{T}_{31}\ \boldsymbol{T}_{12}\ \boldsymbol{T}_{22}\ \boldsymbol{T}_{32}\ \boldsymbol{T}_{13}\ \boldsymbol{T}_{23}\ \boldsymbol{T}_{33} \right\}^T ,$$

with $\boldsymbol{T}_{ij}$ as row vectors containing the individual components. $\boldsymbol{F}$ in (5) is a column vector containing all body force components in the following order

$$\boldsymbol{F} = \left\{ \boldsymbol{F}_1\ \boldsymbol{F}_2\ \boldsymbol{F}_3 \right\}^T ,$$

where $\boldsymbol{F}_j$ are row vectors with the individual components.

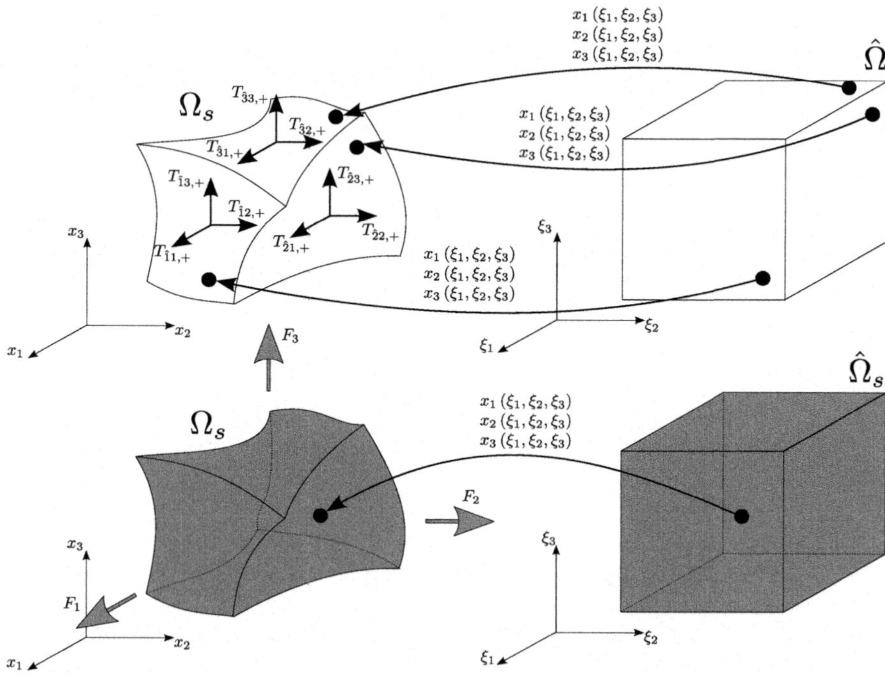

Fig. 3 The forces on a deformed element, Ω_s, in $\mathbb{R}^3$, and the mapping from a reference element, $\hat{\Omega}_s$

The relation in (5) is valid on any domain as it is a relation between integral values. The deformed domain, Ω_s, shown in Fig. 3, which represent a finite element, is geometrically described through a mapping from the reference element, $\hat{\Omega}_s$. Let $\sigma_{\hat{\imath}j}$ represent the stress components, which geometrically are associated to $\hat{\Omega}_s$, but have directions with respect to the physical basis, x_j, then the force components are given by

$$T_{\hat{\imath}j} = \int_{(\partial\hat{\Omega}_s)_{\hat{\imath},\pm}} \sigma_{\hat{\imath}j} n_{\hat{\imath}} \, d\hat{S} \,. \tag{6}$$

Here $\partial\hat{\Omega}_s$ is the boundary of $\hat{\Omega}_s$ and $(\partial\hat{\Omega}_s)_{\hat{\imath},\pm}$ is the part of boundary having an outward unit vector in the ξ_i-direction, where ξ_i are the coordinates in the reference domain (parent element), see Fig. 3. Note that $\sigma_{\hat{\imath}j}$ is similar to the first Piola-Kirchhoff stress, [16], but the Piola-Kirchhoff stress links to the undeformed configuration, while we link to a reference element. Likewise the body force components are given by

$$F_j = \int_{\hat{\Omega}_s} \hat{f}_j \, \mathrm{d}\hat{\Omega} \,,$$

where $\hat{f}_j$ is the body force density in $\hat{\Omega}_s$ having directions in the physical basis directions, x_j. The surface forces acting on the boundaries and the body force acting on the volume are invariant of the representing frame and the equilibrium of forces remains unchanged and is given by

$$T_{\hat{1}j,+} - T_{\hat{1}j,-} + T_{\hat{2}j,+} - T_{\hat{2}j,-} + T_{\hat{3}j,+} - T_{\hat{3}j,-} + F_j = 0 \,.$$

The equilibrium of forces on the finite element mesh in (5) is now given by

$$\mathscr{D}\boldsymbol{\Delta}_{\hat{T}} = -\boldsymbol{F} \,, \tag{7}$$

where

$$\boldsymbol{\Delta}_{\hat{T}} = \left\{ \boldsymbol{T}_{\hat{1}1} \; \boldsymbol{T}_{\hat{2}1} \; \boldsymbol{T}_{\hat{3}1} \; \boldsymbol{T}_{\hat{1}2} \; \boldsymbol{T}_{\hat{2}2} \; \boldsymbol{T}_{\hat{3}2} \; \boldsymbol{T}_{\hat{1}3} \; \boldsymbol{T}_{\hat{2}3} \; \boldsymbol{T}_{\hat{3}3} \right\}^T \,,$$

with $\boldsymbol{T}_{\hat{i}j}$ as row vectors containing all the surface force components in the x_j-direction acting on the surface with the outward unit vector in the ξ_i-direction. The divergence operator is the same and is purely determined by the connectivity between volumes and surfaces in the mesh.

If the surface force components, $T_{\hat{i}j,+}$, and the body force components, F_j, are discrete values in the system then the equilibrium of forces are satisfied exactly by (7).

3 The Constitutive Equations

As seen in the previous section the equilibrium of forces can be represented exactly in a discrete setting by considering the surface force components and body force components on each element boundary and volume, respectively.

The quantities in the constitutive equations are, however, not connected to a common geometrical object since the relations typically are between stress components and strain components, i.e surface force densities and relative deformation along a line. It is therefore not possible to set up a simple sum of integral relations as in the equilibrium of forces. Instead the differential versions of the relations are used, and is often denoted by

$$\boldsymbol{C}\boldsymbol{\sigma} = \boldsymbol{\varepsilon} \,, \tag{8}$$

where $\boldsymbol{\sigma}$ and $\boldsymbol{\varepsilon}$ are column vectors containing the stress and strain components, respectively, and $\boldsymbol{C}$ is a matrix containing the compliance components. In the FEM $\boldsymbol{\sigma}$ and $\boldsymbol{\varepsilon}$ only contain six components each exploiting the symmetry of the stress and strain tensors. The DOFs in this paper are force components and not stress components. Given the surface forces we use the basis functions, to be introduced in Sect. 4, to approximate the stress components. Therefore all the stress and strain components are listed in $\boldsymbol{\sigma}$ and $\boldsymbol{\varepsilon}$, and $\boldsymbol{C}$ is a 9×9 matrix.

The constitutive relation in (8) is defined as a relation between the Cauchy stress components and the strain components, which for small displacements are given by

$$\varepsilon_{ij} = \frac{1}{2}\left(\frac{\partial u_i}{\partial x_j} + \frac{\partial u_j}{\partial x_i}\right) . \tag{9}$$

However, in (6) stress components similar to the first Piola-Kirchhoff stress components are used to calculate the surface force components. The relation between the first Piola-Kirchhoff stress tensor, $\hat{\boldsymbol{\sigma}}$, and the Cauchy stress tensor, $\boldsymbol{\sigma}$ is according to [16] given by

$$\boldsymbol{\sigma} = \frac{1}{J}\begin{bmatrix} \mathscr{F} & 0 & 0 \\ 0 & \mathscr{F} & 0 \\ 0 & 0 & \mathscr{F} \end{bmatrix} \hat{\boldsymbol{\sigma}} = \frac{1}{J}\overrightarrow{\mathscr{F}}\hat{\boldsymbol{\sigma}} , \tag{10}$$

with

$$\mathscr{F} = \begin{bmatrix} \frac{\partial x_1}{\partial \xi_1} & \frac{\partial x_1}{\partial \xi_2} & \frac{\partial x_1}{\partial \xi_3} \\ \frac{\partial x_2}{\partial \xi_1} & \frac{\partial x_2}{\partial \xi_2} & \frac{\partial x_2}{\partial \xi_3} \\ \frac{\partial x_3}{\partial \xi_1} & \frac{\partial x_3}{\partial \xi_2} & \frac{\partial x_3}{\partial \xi_3} \end{bmatrix} ,$$

being the deformation gradient, $J = \det(\mathscr{F})$ and

$$\boldsymbol{\sigma} = \left\{\sigma_{11}\ \sigma_{21}\ \sigma_{31}\ \sigma_{21}\ \sigma_{22}\ \sigma_{23}\ \sigma_{31}\ \sigma_{32}\ \sigma_{33}\right\}^T ,$$

$$\hat{\boldsymbol{\sigma}} = \left\{\sigma_{\hat{1}1}\ \sigma_{\hat{2}1}\ \sigma_{\hat{3}1}\ \sigma_{\hat{2}1}\ \sigma_{\hat{2}2}\ \sigma_{\hat{2}3}\ \sigma_{\hat{3}1}\ \sigma_{\hat{3}2}\ \sigma_{\hat{3}3}\right\}^T .$$

The components of the deformation gradient in (9) are transformed by

$$\frac{\partial u_i}{\partial x_j} = \frac{\partial u_i}{\partial \xi_k}\frac{\partial \xi_k}{\partial x_j} ,$$

which in engineering notation is written as

$$\boldsymbol{D}\boldsymbol{u} = \left(\overrightarrow{\mathscr{F}}^{-1}\right)^T \boldsymbol{D}_0\boldsymbol{u} , \tag{11}$$

where

$$\boldsymbol{u} = \left\{u_1 \; u_2 \; u_3\right\}^T ,$$

$$\boldsymbol{D}_0 = \begin{bmatrix} \frac{\partial}{\partial \xi_1} & \frac{\partial}{\partial \xi_2} & \frac{\partial}{\partial \xi_3} & 0 & 0 & 0 & 0 & 0 & 0 \\ 0 & 0 & 0 & \frac{\partial}{\partial \xi_1} & \frac{\partial}{\partial \xi_2} & \frac{\partial}{\partial \xi_3} & 0 & 0 & 0 \\ 0 & 0 & 0 & 0 & 0 & 0 & \frac{\partial}{\partial \xi_1} & \frac{\partial}{\partial \xi_2} & \frac{\partial}{\partial \xi_3} \end{bmatrix}^T ,$$

and $\vec{\mathscr{F}}$ in (10). This is a common approach in FEMs, see for instance [3]. The strain is given by

$$\boldsymbol{\varepsilon} = \begin{bmatrix} 1\,0\,0\,0\,0\,0\,0\,0\,0 \\ 0\,1\,0\,1\,0\,0\,0\,0\,0 \\ 0\,0\,1\,0\,0\,0\,1\,0\,0 \\ 0\,1\,0\,1\,0\,0\,0\,0\,0 \\ 0\,0\,0\,0\,1\,0\,0\,0\,0 \\ 0\,0\,0\,0\,0\,1\,0\,1\,0 \\ 0\,0\,1\,0\,0\,0\,1\,0\,0 \\ 0\,0\,0\,0\,0\,1\,0\,1\,0 \\ 0\,0\,0\,0\,0\,0\,0\,0\,1 \end{bmatrix} \boldsymbol{D}\boldsymbol{u} = \boldsymbol{M}_{\varepsilon}\boldsymbol{D}\boldsymbol{u} , \tag{12}$$

with

$$\boldsymbol{\varepsilon} = \left\{\varepsilon_{11} \; 2\varepsilon_{21} \; 2\varepsilon_{31} \; 2\varepsilon_{12} \; \varepsilon_{22} \; 2\varepsilon_{32} \; 2\varepsilon_{13} \; 2\varepsilon_{23} \; \varepsilon_{33}\right\}^T . \tag{13}$$

Note that row two and four as well as row six and eight in $\boldsymbol{M}_{\varepsilon}$ are the same, and in e.g. [3] these are left out. However, in the present paper all stress components are considered and hence also all strain components. By doing so, symmetry of the stress tensor is weakly enforced by symmetry of the strain tensor.

To formulate a discretized version of the constitutive equations, a variational statement will be used. This can also be thought of as multiplying with an arbitrary stress field

$$\boldsymbol{\varsigma} = \left\{\varsigma_{11} \; \varsigma_{21} \; \varsigma_{31} \; \varsigma_{21} \; \varsigma_{22} \; \varsigma_{23} \; \varsigma_{31} \; \varsigma_{32} \; \varsigma_{33}\right\}^T ,$$

which produces a functional, and this is integrated over the domain, i.e.

$$\int_{\Omega} \boldsymbol{\varsigma}^T \boldsymbol{C}\boldsymbol{\sigma} \, \mathrm{d}\Omega = \int_{\Omega} \boldsymbol{\varsigma}^T \boldsymbol{\varepsilon} \, \mathrm{d}\Omega .$$

Inserting (10) and (12) gives

$$\int_{\Omega} \frac{1}{J^2} \hat{\varsigma}^T \vec{\mathscr{F}}^T \boldsymbol{C} \vec{\mathscr{F}} \hat{\boldsymbol{\sigma}} \, \mathrm{d}\Omega = \int_{\Omega} \frac{1}{J} \hat{\varsigma}^T \vec{\mathscr{F}}^T \boldsymbol{M}_\varepsilon \boldsymbol{D} \boldsymbol{u} \, \mathrm{d}\Omega \, .$$

Applying (11) and integrating with respect to $\hat{\Omega}$ yields

$$\int_{\hat{\Omega}} \frac{1}{J} \hat{\varsigma}^T \vec{\mathscr{F}}^T \boldsymbol{C} \vec{\mathscr{F}} \hat{\boldsymbol{\sigma}} \, d\hat{\Omega} = \int_{\hat{\Omega}} \hat{\varsigma}^T \vec{\mathscr{F}}^T \boldsymbol{M}_\varepsilon \left(\vec{\mathscr{F}}^{-1} \right)^T \boldsymbol{D}_0 \boldsymbol{u} \, d\hat{\Omega} \, ,$$

where the strain part can be rewritten as

$$\int_{\hat{\Omega}} \hat{\varsigma}^T \vec{\mathscr{F}}^T \boldsymbol{M}_\varepsilon \left(\vec{\mathscr{F}}^{-1} \right)^T \boldsymbol{D}_0 \boldsymbol{u} \, \mathrm{d}\hat{\Omega} =$$

$$\int_{\hat{\Omega}} \hat{\varsigma}^T \boldsymbol{D}_0 \boldsymbol{u} \, \mathrm{d}\hat{\Omega} + \int_{\hat{\Omega}} \hat{\varsigma}^T \vec{\mathscr{F}}^T (\boldsymbol{M}_\varepsilon - \boldsymbol{I}) \left(\vec{\mathscr{F}}^{-1} \right)^T \boldsymbol{D}_0 \boldsymbol{u} \, \mathrm{d}\hat{\Omega} \, ,$$

with $\boldsymbol{I}$ being a 9×9 identity matrix. Performing integration by parts on the first term yields

$$\int_{\hat{\Omega}} \hat{\varsigma}^T \boldsymbol{D}_0 \boldsymbol{u} \, \mathrm{d}\hat{\Omega} = - \int_{\hat{\Omega}} \left(\boldsymbol{D}_0 \hat{\varsigma} \right)^T \boldsymbol{u} \, \mathrm{d}\hat{\Omega} + \int_{\partial\hat{\Omega}} \hat{\varsigma}^T \boldsymbol{u} \, \mathrm{d}\hat{S} \, ,$$

and gathering all terms gives the weak formulation

$$\int_{\hat{\Omega}} \frac{1}{J} \hat{\varsigma}^T \vec{\mathscr{F}}^T \boldsymbol{C} \vec{\mathscr{F}} \hat{\boldsymbol{\sigma}} \, d\hat{\Omega} = - \int_{\hat{\Omega}} \left(\boldsymbol{D}_0 \hat{\varsigma} \right)^T \boldsymbol{u} \, \mathrm{d}\hat{\Omega} + \int_{(\partial\hat{\Omega})_u} \hat{\varsigma}^T \boldsymbol{u} \, \mathrm{d}\hat{S} + \int_{(\partial\hat{\Omega})_T} \hat{\varsigma}^T \boldsymbol{u} \, \mathrm{d}\hat{S}$$
$$+ \int_{\hat{\Omega}} \hat{\varsigma}^T \vec{\mathscr{F}}^T (\boldsymbol{M}_\varepsilon - \boldsymbol{I}) \left(\vec{\mathscr{F}}^{-1} \right)^T \boldsymbol{D}_0 \boldsymbol{u} \, \mathrm{d}\hat{\Omega} \, ,$$

where $(\partial\hat{\Omega})_u$ is the part of the boundary, where the displacements are prescribed, and $(\partial\hat{\Omega})_T$ is the part of the boundary, where the tractions are given.

Let $\mathscr{V} \subset \left[H^1(\Omega) \right]^3$ and $\mathscr{T} \subset \left[H_0(\mathrm{div})(\Omega) \right]^3$ be finite dimensional subspaces, where $\left[H^1(\Omega) \right]^3$ denotes the Sobolov space of vector functions with square-integrable gradients, and $\left[H_0(\mathrm{div})(\Omega) \right]^3$ denotes the Sobolov space of vector functions with square-integrable divergence with vanishing trace along $(\partial\hat{\Omega})_T$. The variational statement reads: Find $\left(\boldsymbol{u}^h, \hat{\boldsymbol{\sigma}}^h\right) \in \mathscr{V} \times \mathscr{T}$ such that $\forall\ \hat{\boldsymbol{\varsigma}}^h \in \mathscr{T}$:

$$\int_{\hat{\Omega}} \frac{1}{J}\left(\hat{\boldsymbol{\varsigma}}^h\right)^T \vec{\mathscr{F}}^T \boldsymbol{C} \vec{\mathscr{F}} \hat{\boldsymbol{\sigma}}^h \, d\hat{\Omega} + \int_{\hat{\Omega}} \left(\boldsymbol{D}_0 \hat{\boldsymbol{\varsigma}}^h\right)^T \boldsymbol{u}^h \, \mathrm{d}\hat{\Omega}$$
$$- \int_{\hat{\Omega}} \left(\hat{\boldsymbol{\varsigma}}^h\right)^T \vec{\mathscr{F}}^T \left(\boldsymbol{M}_\varepsilon - \boldsymbol{I}\right) \left(\vec{\mathscr{F}}^{-1}\right)^T \boldsymbol{D}_0 \boldsymbol{u}^h \, \mathrm{d}\hat{\Omega} = \int_{(\partial\hat{\Omega})_u} \left(\hat{\boldsymbol{\varsigma}}^h\right)^T \boldsymbol{u}_{bc} \, \mathrm{d}\left(\partial\hat{\Omega}\right) , \tag{14}$$

where $\boldsymbol{u}_{bc}$ is the known displacements on the $(\partial\hat{\Omega})_u$ boundary.

4 Expansion Polynomials

This section will describe the expansions of the stress and displacement fields. In Sect. 2 the DOFs of the force equilibrium equations are the surface force components and the body force components of the individual elements. Assuming that the body force components are known, e.g. a gravity load, then the unknown values are the surface force components. In Sect. 3 the constitutive relation contains the discrete stress field, so this stress field should be connected to the discrete surface force components. This can be accomplished through the use of *edge polynomials* derived in [8]. Edge polynomials are defined as

$$e_i(\xi) = -\sum_{k=0}^{i-1} \frac{dh_k(\xi)}{d\xi} , \quad i = 1, \ldots, N ,$$

where

$$h_i(\xi) = \frac{\prod_{j=0, j\neq i}^{N} (\xi - \xi_j)}{\prod_{j=0, j\neq i}^{N} (\xi_i - \xi_j)} ,$$

are the Lagrange polynomials. Just as the Lagrange polynomials have the property

$$h_i(\xi_k) = \begin{cases} 1 & \text{if } i = k \\ 0 & \text{if } i \neq k \end{cases} ,$$

the edge polynomials have the property

$$\int_{\xi_{k-1}}^{\xi_k} e_i(\xi) = \begin{cases} 1 & \text{if } i = k \\ 0 & \text{if } i \neq k \end{cases} . \tag{15}$$

The stress field is now expanded as

$$
\begin{aligned}
\sigma^h_{\hat{1}m}(\xi_1,\xi_2,\xi_3) &= \sum_{i=0}^{N}\sum_{j=1}^{N}\sum_{k=1}^{N} \left(T_{\hat{1}m}\right)_{i,j,k} h_i(\xi_1)e_j(\xi_2)e_k(\xi_3)\,, \\
\sigma^h_{\hat{2}m}(\xi_1,\xi_2,\xi_3) &= \sum_{i=1}^{N}\sum_{j=0}^{N}\sum_{k=1}^{N} \left(T_{\hat{2}m}\right)_{i,j,k} e_i(\xi_1)h_j(\xi_2)e_k(\xi_3)\,, \qquad (16)\\
\sigma^h_{\hat{3}m}(\xi_1,\xi_2,\xi_3) &= \sum_{i=1}^{N}\sum_{j=1}^{N}\sum_{k=0}^{N} \left(T_{\hat{3}m}\right)_{i,j,k} e_i(\xi_1)e_j(\xi_2)h_k(\xi_3)\,,
\end{aligned}
$$

where $(T_{\hat{n}m})_{i,j,k}$ is the discrete surface force components of the individual elements. Note that for instance

$$
\left(T_{\hat{1}m}\right)_{i,j,k} = \int_{(\xi_2)_{j-1}}^{(\xi_2)_j} \int_{(\xi_3)_{k-1}}^{(\xi_3)_k} \sigma_{\hat{1}m} \mathrm{d}\xi_2 \mathrm{d}\xi_3\,,
$$

where we have used (15) twice. This is the same as (6), however, in (6) the stress components are used to obtain the surface force components, but in (16) the surface force components are used to reconstruct the stress field.

The stress field is expanded on the mesh, which is spanned by the Gauss Lobatto Legendre (GLL) points, ξ_i, $i = 0,\ldots,N$. The displacement field is expanded using Lagrange polynomials with discrete points located in the Gauss Legendre (GL) points, $\tilde{\xi}_i$, $i = 0,\ldots,N-1$, see [2]. The GL points then satisfy $\xi_i < \tilde{\xi}_i < \xi_{i+1}$ for $i = 0,\ldots,N-1$, which means that there exists one discrete displacement component per force equilibrium equation. Let $\phi^h(\xi) = \sum_{i=0}^{N} \phi_i h_i(\xi)$ be a function in $\mathbb{R}$ expanded by Lagrange polynomials then the derivative is calculated by

$$
\frac{d\phi^h}{d\xi} = \sum_{i=1}^{N} (\phi_i - \phi_{i-1}) e_i(\xi)\,.
$$

By applying this in each direction $\boldsymbol{D}_0\boldsymbol{u}^h$ in (14) is calculated, and the integrals are evaluated using appropriate Gaussian quadratures. A global equation system is assembled using (7) and (14). For more details refer to [14].

5 Results

A numerical test is performed on a 2D domain with the plane stress case of Hooke's generalized law as the constitutive relation. This is given by

$$C = \frac{1}{E}\begin{bmatrix} 1 & 0 & 0 & -\nu \\ 0 & 2(1+\nu) & 0 & 0 \\ 0 & 0 & 2(1+\nu) & 0 \\ -\nu & 0 & 0 & 1 \end{bmatrix} ,$$

where $E = 1$ is Young's modulus, $\nu = 0.3$ is Poisson's ratio and the number 2 in the second and third row is from (13). By choosing the displacement field

$$u_1(x_1, x_2) = \sin(2\pi x_1)\cos(2\pi x_2) \quad \text{and} \quad u_2(x_1, x_2) = \cos(2\pi x_1)\sin(2\pi x_2) ,$$

the stress components are calculated from (8) and (9), while the body force density components are calculated from (4) from which the body force components are calculated. These values are used as input to the numerical test, and the error of the displacement field and the residual of the force equilibrium equations are calculated. The calculation is performed on the domain $\Omega \in [-1, 1]^2$ with the mapping

$$\begin{aligned} x_1(\xi_1, \xi_2) =& \xi_1 + c\sin(\pi\xi_1)\sin(\pi\xi_2) , \\ x_2(\xi_1, \xi_2) =& \xi_2 + c\sin(\pi\xi_1)\sin(\pi\xi_2) , \end{aligned}$$

which are plotted in Fig. 4 for $c = \{0, 0.15, 0.3\}$. The results are shown in Fig. 5, where the pointwise error of the displacement field and the pointwise residual of force equilibrium equations are evaluated in 100×100 points in each element. A problem with a smooth solution with polynomial degree P is expected to have a convergence rate of $\mathcal{O}(h_{el}^{P+1})$ according to [9], where h_{el} is the element size. This is observed for the displacement field as this has polynomials degree of $P = N$. A more interesting observation is that the residual of the discrete force equilibrium

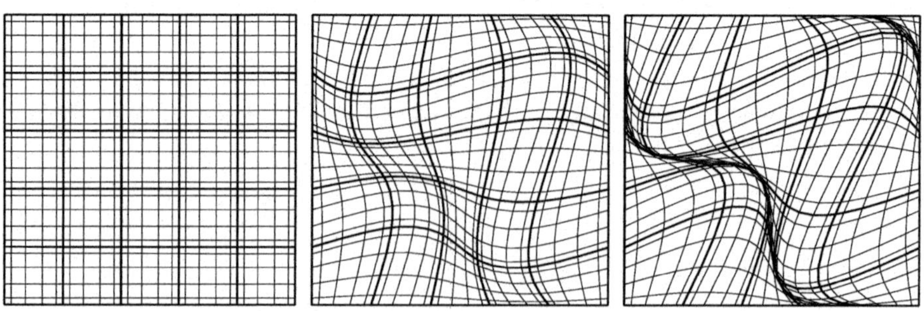

Fig. 4 Deformed grids for 5×5 elements with $N = 5$ and $c = \{0, 0.15, 0.3\}$

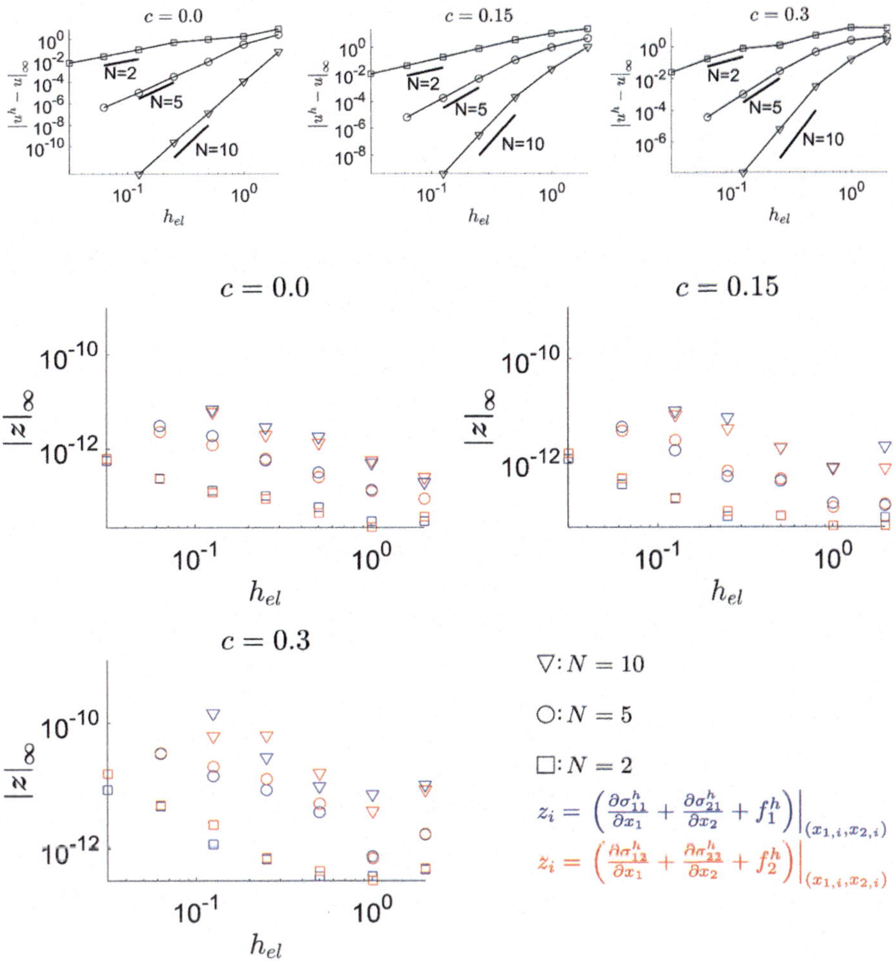

Fig. 5 *Top*: Convergence for u_i with respect to the undeformed element size h_{el} for $c = \{0, 0.15, 0.3\}$ from *left to right*. *Open square*: $N = 2$, *open circle*: $N = 5$, *down triangle*: $N = 10$. *Bottom*: The residual of the discrete force equilibrium equations with respect to the undeformed element size h_{el} for $c = \{0, 0.15, 0.3\}$

equations are around 10^{-12} to 10^{-10}, which indicates that we have pointwise force equilibrium. For details on how the force equilibrium equations are evaluated refer to [14].

6 Conclusion

In this paper we have shown that through geometrical considerations the discrete force equilibrium equations are described by the sum of surface force components on the individual elements and thereby no approximation is involved. This means that force equilibrium is satisfied to machine precision, and this should be the case independently of the constitutive relation chosen. This property could be interesting to investigate in future work.

References

1. A. Bossavit, Discretization of electromagnetic problems, in *Handbook of Numerical Analysis*, vol. 13 (North-Holland, Amsterdam, 2005), pp. 105–197
2. C. Canuto, M. Hussaini, A. Quarteroni, T. Zang, *Spectral Methods, Fundamentals in Single Domains* (Springer, Berlin, 2006)
3. R.D. Cook, D.S. Malkus, M.E. Plesha, R.J. Witt, *Concepts and Applications of Finite Element Analysis*, 4th edn. (John Wiley and sons, New York, 2001)
4. M. Desbrun, A.N. Hirani, M. Leok, J.E. Marsden, Discrete Exterior Calculus. Arxiv preprint (2005)
5. B. Fraeijs De Veubeke, Upper and Lower Bounds in Matrix Structural Analysis. AGARDograf **72**, 165–201 (1964)
6. B. Fraeijs De Veubeke, Displacements and equilibrium models in the finite elements method, in *Stress Analysis*, ed. by O.C. Zienkiewicz, G.S. Holister, Chap. 9 (Wiley, London, 1965)
7. B. Fraeijs De Veubeke, Diffusive equilibrium models, in *B.M. Fraeijs de Veubeke Memorial Volume of Selected Papers* (Sijthoff & Noordhoff, Philadelphia, 1980)
8. M. Gerritsma, Edge functions for spectral element methods, in *Spectral and High Order Methods for Partial Differential Equations*, ed. by J.S. Hesthaven, E.M. Rønquist. Lecture Notes in Computational Science and Engineering, vol. 76 (Springer, Heidelberg, 2011), pp. 199–207
9. G.E. Karniadakis, S.J. Spencer, *Spectral/hp Element Methods for Computational Fluid Dynamics*, 2nd edn. (Oxford Science Publications, Oxford, 2005)
10. J. Kreeft, A. Palha, M. Gerritsma, Mimetic Framework on Curvilinear Quadrilaterals of Arbitrary Order. Arxiv preprint (2011)
11. J. Kreeft, M. Gerritsma, Mixed mimetic spectral element method for stokes flow: a pointwise divergence-free solution. J. Comput. Phys. **240**, 284–309 (2013)
12. J.P. Moitinho De Almeida, J.A. Teixeira De Freitas, Alternative approach to the formulation of hybrid equilibrium finite elements. Comput. Struct. **40**, 1043–1047 (1991)
13. J.P. Moitinho De Almeida, J.A. Teixeira De Freitas, A set of hybrid equilibrium finite element models for the analysis of three-dimensional solids. Int. J. Numer. Methods Eng. **39**, 2789–2802 (1996)
14. K. Olesen, B. Gervang, J.N. Reddy, M. Gerritsma, Exact Force Equilibrium in Linear Elasticity. ArXiv:1605.05444 [math.NA] (2016)
15. J.N. Reddy, *An Introduction to the Finite Element Method*, 3rd edn. (McGraw-Hill, New York, 2006)
16. J.N. Reddy, *An Introduction to Continuum Mechanics*, 2nd edn. (Cambridge University Press, Cambridge, 2013)
17. H.A.F.A. Santos, J.P. Moitinho De Almeida, A family of Piola Kirchhoff hybrid stress finite elements for two-dimensional linear elasticity. Finite Elem. Anal. Des. **85**, 33–49 (2014)
18. E. Tonti, Why starting from differential equations for computational physics? J. Comput. Phys. **257**, 1260–1290 (2014)
19. E. Tonti, F. Zarantonello, Algebraic formulation of elastostatics: the cell method. Comput. Model. Eng. Sci. **39**(3), 201–236 (2009)

A Staggered Discontinuous Galerkin Method for a Class of Nonlinear Elliptic Equations

Eric T. Chung, Ming Fai Lam, and Chi Yeung Lam

Abstract In this paper, we present a staggered discontinuous Galerkin (SDG) method for a class of nonlinear elliptic equations in two dimensions. The SDG methods have some distinctive advantages, including local and global conservations, and optimal convergence. So the SDG methods have been successfully applied to a wide range of problems including Maxwell equations, acoustic wave equation, elastodynamics and incompressible Navier-Stokes equations. Among many advantages of the SDG methods, one can apply a local post-processing technique to the solution, and obtain superconvergence. We will analyze the stability of the method and derive a priori error estimates. We solve the resulting nonlinear system using the Newton's method, and the numerical results confirm the theoretical rates of convergence and superconvergence.

1 Introduction

Our aim of this paper is to extend the staggered discontinuous Galerkin (SDG) method to a class of nonlinear elliptic problems arising in, for example, hyperpolarization effects in electrostatic analysis [14], nonlinear magnetic field problems [13], subsonic flow problems [12], and heat conduction.

A detailed introduction to the SDG method is given by Chung and Engquist [4, 5]. This class of methods has been successfully applied to a wide range of problems including the Maxwell equation [6, 7], acoustic wave equation [5], elastic equations [9, 15], and incompressible Navier-Stokes equations [3]. In these applications, the approximate solutions obtain some nice properties such as energy conservation, low dispersion error and mass conservation. Recently, a connection between the SDG method and the hybridizable discontinuous Galerkin (HDG) method is obtained [8, 10]. From this perspective, the SDG method acquires some new properties, such as postprocessing and superconvergence properties, from the HDG method [11]. We

E.T. Chung (✉) • M.F. Lam • C.Y. Lam
Department of Mathematics, The Chinese University of Hong Kong, Hong Kong SAR, People's Republic of China
e-mail: tschung@math.cuhk.edu.hk; mflam@math.cuhk.edu.hk; cylam@math.cuhk.edu.hk

M.L. Bittencourt et al. (eds.), *Spectral and High Order Methods for Partial Differential Equations ICOSAHOM 2016*, Lecture Notes in Computational Science and Engineering 119, DOI 10.1007/978-3-319-65870-4_42

remark that numerical methods based on staggered meshes are important in many applications, see [17, 18].

To begin with, we let $\Omega \subset \mathbf{R}^2$ be a bounded and simply connected domain with polygonal boundary Γ. Also, we let the coefficient $\varrho : \mathbf{R}^2 \to \mathbf{R}$ be a L^∞ function satisfying certain conditions (will be specified). Then, for a given $f \in L^2(\Omega)$ we seek $u \in H_0^1(\Omega)$ such that

$$- \operatorname{div} \left(\varrho(\nabla u(x))\nabla u(x)\right) = f(x) \text{ in } \Omega, \text{ and } u(x) = 0 \quad \text{on } \Gamma, \tag{1}$$

where div is the usual divergence operator.

This paper is organized as follows. In Sect. 2, we will construct the SDG method. In Sect. 3, we will discuss the implementation of the scheme. In Sect. 4, we will prove stability estimates and an a priori error estimate of our scheme. Finally, in Sect. 5, we will numerically show the rate of convergence of our method. Throughout this paper, we use C to denote a generic positive constant, which is independent of the mesh size.

2 The SDG Formulation

We introduce new variables, the gradient $\mathbf{G} := \nabla u$ and the flux $\mathbf{U} := \rho(\mathbf{G})\mathbf{G}$. Then the problem (1) can be recasted as the following problem in Ω: Find $(\mathbf{U}, \mathbf{G}, u)$ such that,

$$\mathbf{G} = \nabla u, \quad \mathbf{U} = \rho(\mathbf{G})\mathbf{G}, \quad -\operatorname{div}\mathbf{U} = f \quad \text{in } \Omega,$$
$$u = 0 \quad \text{on } \Gamma.$$

Next we describe the staggered mesh. Following [4, 5],we first define the triangulation. Assume Ω is triangulated by a family of triangles with no hanging nodes, namely, the initial triangulation $\mathscr{T}_u$. The triangles in $\mathscr{T}_u$ are called the *first-type macro element*. We denote the set of all edges and all interior edges of $\mathscr{T}_u$ by $\mathscr{F}_u$ and $\mathscr{F}_u^0$, respectively. Then we choose an interior point υ in each first-type macro element. We denote the first-type macro element corresponding to υ by $\mathscr{S}(\upsilon)$. By connecting each of these interior points to the three vertices of the triangle, we subdivide each triangle into three subtriangles. We denote the triangulation containing all these subtriangles by $\mathscr{T}$ and assume it is shape-regular. We denote the set of all new edges in this subdivision process by $\mathscr{F}_p$. Also, we denote the set of all edges and the set of all interior edges by $\mathscr{F} := \mathscr{F}_u \cup \mathscr{F}_p$ and $\mathscr{F}_0 := \mathscr{F}_u^0 \cup \mathscr{F}_p$, respectively. For each interior edge $e_u \in \mathscr{F}_u^0$, there are two triangles $\tau_1, \tau_2 \in \mathscr{T}$ such that $e_u = \tau_1 \cap \tau_2$. We denote the union $\tau_1 \cup \tau_2$ by $\mathscr{R}(e_u)$. Also, for each boundary edge e_b, we denote the only triangle having e_b as an edge by $\mathscr{R}(e_b)$. These elements $\mathscr{R}(e_u)$ and $\mathscr{R}(e_b)$ are called the *second-type macro element*. In Fig. 1, we illustrate two first-type macro elements and a second-type macro element, which is shaded in grey, obtained from the subdividing process on two neighboring initial triangles.

Fig. 1 An illustration of the triangulation $\mathcal{T}$, where the solid edges belong to $\mathcal{F}_u$ and the dashed edges belong to $\mathcal{F}_p$

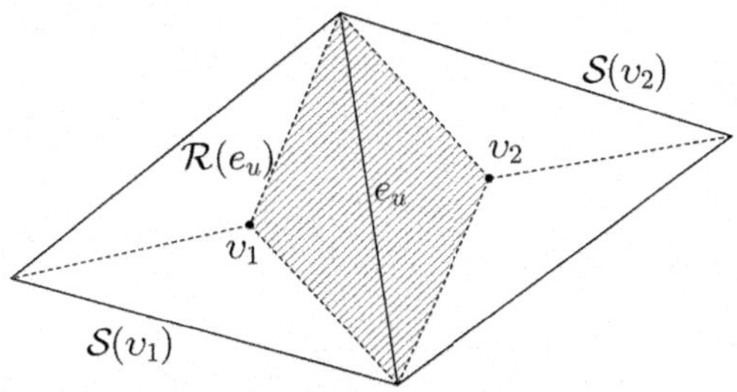

For a boundary edge e_b, we define $\mathbf{n}_e$ to be the unit normal vector pointing outside Ω. Otherwise, $\mathbf{n}_e$ is one of the two possible unit normal vectors of $e \in \mathcal{F}_0$. When it is clear which edge is being considered, we will simply use $\mathbf{n}$ instead of $\mathbf{n}_e$.

Next, we describe the finite element spaces we use in our formulation. Let $k \geq 0$ be a non-negative integer. For each triangle $\tau \in \mathcal{T}$, we denote the space of polynomials on τ with degree at most k by $P^k(\tau)$. Then we define the *locally $H^1(\Omega)$-conforming finite element* as

$$\mathcal{U}^h := \{v : v|_\tau \in P^k(\tau), \forall \tau \in \mathcal{T}; v \text{ is continuous across } e_u \in \mathcal{F}_u^0;\ v|_{\partial\Omega} = 0\},$$

and the *locally $H(\text{div};\Omega)$-conforming finite element space* as

$$\mathcal{W}^h := \{\mathbf{V} : \mathbf{V}|_\tau \in P^k(\tau)^2, \forall \tau \in \mathcal{T}; \text{the normal component } \mathbf{V}\cdot\mathbf{n}_e \text{ across } e_p \in \mathcal{F}_p \text{ is continuous}\}.$$

We consider the discrete problem in the following formulation: find $(\mathbf{U}_h, \mathbf{G}_h, u_h) \in \mathcal{W}^h \times \mathcal{W}^h \times \mathcal{U}^h$ such that,

$$\begin{aligned}
\int_{\mathcal{S}(v)} \mathbf{G}_h \cdot \mathbf{V}_h \, dx + \int_{\mathcal{S}(v)} u_h \, \text{div}_h \mathbf{V}_h \, dx - \int_{\partial\mathcal{S}(v)} \mathbf{G}_h \, (\mathbf{V}_h \cdot \mathbf{n}) \, d\sigma &= 0, \\
\int_{\mathcal{S}(v)} \mathbf{U}_h \cdot \mathbf{W}_h \, dx - \int_{\mathcal{S}(v)} \rho(\mathbf{G}_h)\mathbf{G}_h \cdot \mathbf{W}_h \, dx &= 0, \qquad (2) \\
\int_{\mathcal{R}(e)} \mathbf{U}_h \cdot \nabla_h v_h \, dx - \int_{\partial\mathcal{R}(e)} (\mathbf{U}_h \cdot \mathbf{n}) v_h \, d\sigma &= \int_{\mathcal{R}(e)} f v_h \, dx,
\end{aligned}$$

for any first-type element $\mathcal{S}(v)$ and second-type element $\mathcal{R}(e)$, any test functions $(\mathbf{V}_h, \mathbf{W}_h, v_h) \in \mathcal{W}^h \times \mathcal{W}^h \times \mathcal{U}^h$. In the above formulation, ∇_h and div_h are the elementwise gradient and divergence operators, respectively. Besides, $\mathbf{n}$ denotes outward normals on $\mathcal{S}(v)$ or $\mathcal{R}(e)$ depending on the context.

We emphasize here that $\mathbf{U}_h$ is defined via the second equation of (2), which is a new feature of our SDG method and needs special treatment (will be discussed in the next section) due to its nonlinear nature.

We define the jump operator $[\cdot]$ as follows. For $e_p \in \mathscr{F}_p$, if $\tau_1, \tau_2 \in \mathscr{T}$ such that $e_p = \tau_1 \cap \tau_2$ and $\mathbf{n}_e$ is pointing from τ_1 to τ_2, then

$$[v] := v|_{\tau_1} - v|_{\tau_2}.$$

For $e_u \in \mathscr{F}_u^0$, if $\tau_1, \tau_2 \in \mathscr{T}$ such that $e_u = \tau_1 \cap \tau_2$ and $\mathbf{n}_e$ is pointing from τ_1 to τ_2, then

$$[\mathbf{V} \cdot \mathbf{n}_e] := \mathbf{V}|_{\tau_1} \cdot \mathbf{n}_e - \mathbf{V}|_{\tau_2} \cdot \mathbf{n}_e.$$

We also introduce two bilinear forms,

$$b_h(\mathbf{V}_h, v_h) := \int_\Omega \mathbf{V}_h \cdot \nabla_h v_h \, dx - \sum_{e_p \in \mathscr{F}_p} \int_{e_p} \mathbf{V}_h \cdot \mathbf{n}[v_h] \, d\sigma,$$

$$b_h^*(v_h, \mathbf{V}_h) := -\int_\Omega v_h \nabla_h \cdot \mathbf{V}_h \, dx + \sum_{e_u \in \mathscr{F}_u^0} \int_{e_u} v_h [\mathbf{V}_h \cdot \mathbf{n}] \, d\sigma,$$

for $v_h \in \mathscr{U}^h, \mathbf{V}_h \in \mathscr{W}^h$. According to Lemma 2.4 of [5], we have

$$b_h(\mathbf{V}_h, v_h) = b_h^*(v_h, \mathbf{V}_h), \quad \forall (v_h, \mathbf{V}_h) \in \mathscr{U}^h \times \mathscr{W}^h, \tag{3}$$

which means that the bilinear forms b_h and b_h^* are adjoint to each other.

Summing the equations in (2) on $\mathscr{S}(\upsilon)$ and $\mathscr{R}(e)$, respectively, we can recast (2) into: find $(\mathbf{U}_h, \mathbf{G}_h, u_h) \in \mathscr{W}^h \times \mathscr{W}^h \times \mathscr{U}^h$ such that,

$$\int_\Omega \mathbf{G}_h \cdot \mathbf{V}_h \, dx - b_h^*(u_h, \mathbf{V}_h) = 0, \tag{4a}$$

$$\int_\Omega \mathbf{U}_h \cdot \mathbf{W}_h \, dx - \int_\Omega \rho(\mathbf{G}_h)\mathbf{G}_h \cdot \mathbf{W}_h \, dx = 0, \tag{4b}$$

$$b_h(\mathbf{U}_h, v_h) = \int_\Omega f v_h \, dx, \tag{4c}$$

for any $(\mathbf{V}_h, \mathbf{W}_h, v_h) \in \mathscr{W}^h \times \mathscr{W}^h \times \mathscr{U}^h$. This completes the definition of our SDG method.

3 Implementation

In this section we will discuss the implementation detail of our SDG method. First of all we fix a basis $\{\phi_i\}_{i=1}^{N_u}$ for $\mathscr{U}^h$ and $\{\psi_i\}_{i=1}^{N_w}$ for $\mathscr{W}^h$, and write $u_h = \sum_i (\widehat{u}_h)_i \phi_i$, $\mathbf{G}_h = \sum_i (\widehat{\mathbf{G}}_h)_i \psi_i$ and $\mathbf{U}_h = \sum_i (\widehat{\mathbf{U}}_h)_i \psi_i$, where $\widehat{u}_h$, $\widehat{\mathbf{G}}_h$ and $\widehat{\mathbf{U}}_h$ are $N_u \times 1$, $N_w \times 1$ and $N_w \times 1$ vectors, respectively. Next, we define the mass matrix M_h and the matrix B_h by $(M_h)_{ij} := \int_\Omega \psi_j \cdot \psi_i \, dx$, and$(B_h)_{ij} := b_h(\psi_j, \phi_i)$, respectively. Since $b_h(\psi_j, \phi_i) = b_h^*(\phi_i, \psi_j)$, so the matrix B_h^T can represent the bilinear form b_h^*. Then we rewrite

(4a)–(4c) as the following system:

$$M_h\widehat{\mathbf{G}}_h - B_h^T\widehat{u}_h = 0, \tag{5a}$$

$$M_h\widehat{\mathbf{U}}_h = F(\widehat{\mathbf{G}}_h), \tag{5b}$$

$$B_h\widehat{\mathbf{U}}_h = f_h, \tag{5c}$$

where f_h is a $N_u \times 1$ vector given by $(f_h)_i := \int f v_i \, dx$, and $F(\widehat{\mathbf{G}}_h)$ is a $N_w \times 1$ vector given by $F(\widehat{\mathbf{G}}_h)_i := \big(\rho(\mathbf{G}_h)\mathbf{G}_h, \psi_i\big)_{L^2(\Omega)}$. Eliminating $\widehat{\mathbf{U}}_h$ from (5a)–(5c), we obtain

$$\begin{aligned} M_h\widehat{\mathbf{G}}_h - B_h^T\widehat{u}_h &= 0, \\ B_hM_h^{-1}F(\widehat{\mathbf{G}}_h) &= f_h. \end{aligned} \tag{6}$$

Here F is not a linear function in general. Hence we use Newton's method to solve this system. Write $\widehat{\mathbf{x}}_h := (\widehat{\mathbf{G}}_h, \widehat{u}_h)^T$ and $H(\widehat{\mathbf{x}}_h) := \Big(M_h\widehat{\mathbf{G}}_h - B_h^T\widehat{u}_h, B_hM_h^{-1} F(\widehat{\mathbf{G}}_h) - f_h\Big)^T$. The Jacobian matrix of H is given by

$$J(\widehat{\mathbf{x}}_h) := \begin{pmatrix} M_h & -B_h^T \\ B_hM_h^{-1}F'(\widehat{\mathbf{G}}_h) & 0 \end{pmatrix},$$

where $F'(\widehat{\mathbf{G}}_h)$ is the derivative with respect to $\widehat{\mathbf{G}}_h$, and is given by

$$F'(\widehat{\mathbf{G}}_h)_{ij} = \big(\rho(\mathbf{G}_h)\psi_j, \psi_i\big) + \big((\nabla\rho(\mathbf{G}_h)\cdot\psi_j)\mathbf{G}_h, \psi_i\big).$$

Given an initial guess $\widehat{\mathbf{x}}_h^0$, we repeatedly update $\widehat{\mathbf{x}}_h^n$ by

$$\widehat{\mathbf{x}}_h^{n+1} = \widehat{\mathbf{x}}_h^n - [J(\widehat{\mathbf{x}}_h^n)]^{-1}H(\widehat{\mathbf{x}}_h^n),$$

until the successive error $\left\| u_h^{n+1} - u_h^n \right\|_{L^2(\Omega)}$ is less than a given tolerance δ.

4 Stability and Convergence of the SDG Method

We begin with some results from the SDG method studied in [5]. We define the discrete L^2-norm $\|\cdot\|_X$ and the discrete H^1-norm $\|\cdot\|_Z$ for any $v \in \mathscr{U}^h$ by

$$\|v\|_X^2 := \int_\Omega v^2 \, dx + \sum_{e_u\in\mathscr{F}_u^0} h_{e_u}\int_{e_u} v^2 \, d\sigma \quad \text{and}$$

$$\|v\|_Z^2 := \int_\Omega |\nabla_h v|^2 \, dx + \sum_{e_p\in\mathscr{F}_p} h_{e_p}^{-1}\int_{e_p} [v]^2 \, d\sigma,$$

respectively. We also define the discrete L^2-norm $\|\cdot\|_{X'}$ and the discrete H^1-norm $\|\cdot\|_{Z'}$ for any $\mathbf{V} \in \mathscr{W}^h$ by

$$\|\mathbf{V}\|_{X'}^2 = \int_\Omega |\mathbf{V}|^2 \, dx + \sum_{e_p \in \mathscr{F}_p} h_{e_p} \int_{e_p} (\mathbf{V} \cdot \mathbf{n})^2 \, d\sigma,$$

$$\|\mathbf{V}\|_{Z'}^2 = \int_\Omega (\nabla \cdot \mathbf{V})^2 \, dx + \sum_{e_u \in \mathscr{F}_u^0} h_{e_u}^{-1} \int_{e_u} [\mathbf{V} \cdot \mathbf{n}]^2 \, d\sigma.$$

Then we recall the following inf-sup conditions for the bilinear forms b_h and b_h^*.

Lemma 1 *There is a positive constant C independent of the mesh size h such that*

$$\inf_{\mathbf{V} \in \mathscr{W}^h} \sup_{v \in \mathscr{U}^h} \frac{b_h^*(v, \mathbf{V})}{\|v\|_X \|\mathbf{V}\|_{Z'}} \geq C,$$

$$\inf_{v \in \mathscr{U}^h} \sup_{\mathbf{V} \in \mathscr{W}^h} \frac{b_h(\mathbf{V}, v)}{\|v\|_Z \|\mathbf{V}\|_{X'}} \geq C.$$

Besides the inf-sup conditions of b_h and b_h^*, we can observe that from the definition of $\|\cdot\|_{X'}$, it is clear that for any $\mathbf{V} \in \mathscr{W}^h$,

$$\|\mathbf{V}\|_{L^2(\Omega)^2} \leq \|\mathbf{V}\|_{X'}. \tag{7}$$

Moreover, the proof in [1] shows the following discrete Poincaré–Friedrichs inequality for piecewise H^1 functions.

Lemma 2 *For any piecewise H^1 function v, there is a positive constant C independent of the mesh size h such that*

$$\|v\|_{L^2(\Omega)} \leq C\|v\|_Z.$$

Next, we impose some restrictions on the coefficient ρ. We assume ρ is bounded below by a positive number ρ_0. Moreover, we follow Bustinza and Gatica [2] to require $\rho(\mathbf{W})\mathbf{W}$ to be *strongly monotone*. In order words, there is a positive constant C independent of $\mathbf{V}, \mathbf{W} \in L^2(\Omega)^2$ such that

$$\int_\Omega [\rho(\mathbf{W})\mathbf{W} - \rho(\mathbf{V})\mathbf{V}] \cdot (\mathbf{W} - \mathbf{V}) \, dx \geq C \|\mathbf{W} - \mathbf{V}\|_{L^2(\Omega)^2}^2. \tag{8}$$

We also require $\rho(\mathbf{W})\mathbf{W}$ to be *Lipschitz continuous*. In order words, there is a positive constant C independent of $\mathbf{V}, \mathbf{W} \in L^2(\Omega)^2$ such that

$$\|\rho(\mathbf{W})\mathbf{W} - \rho(\mathbf{V})\mathbf{V}\|_{L^2(\Omega)^2}^2 \leq C \|\mathbf{W} - \mathbf{V}\|_{L^2(\Omega)^2}^2. \tag{9}$$

These two additional assumptions on ρ are essential to ensure the unique solvability of (6) (see [16]).

We will also consider the interpolants $\mathcal{I} : H^1(\Omega) \to \mathcal{U}^h$ and $\mathcal{J} : H(\text{div}; \Omega) \to \mathcal{W}^h$ discussed in [5], which are characterized by

$$\begin{aligned} b_h^*(\mathcal{I}u - u, \mathbf{V}) &= 0, \qquad \forall u \in H^1(\Omega), \mathbf{V} \in \mathcal{W}^h, \\ b_h(\mathcal{J}\mathbf{U} - \mathbf{U}, v) &= 0, \qquad \forall \mathbf{U} \in H(\text{div}; \Omega), v \in \mathcal{U}^h. \end{aligned} \tag{10}$$

It is shown that for any $v \in H^{k+1}(\Omega)$ and $\mathbf{V} \in H^{k+1}(\Omega)^2$, we have

$$\|v - \mathcal{I}v\|_{L^2(\Omega)} \leq Ch^{k+1} \|v\|_{H^{k+1}(\Omega)}, \tag{11}$$

$$\|\mathbf{V} - \mathcal{J}\mathbf{V}\|_{L^2(\Omega)^2} \leq Ch^{k+1} \|\mathbf{V}\|_{H^{k+1}(\Omega)^2}. \tag{12}$$

Theorem 1 *Let* $(u, \mathbf{G}, \mathbf{U}) \in H^{k+1}(\Omega) \times H^{k+1}(\Omega)^2 \times H^{k+1}(\Omega)^2$ *be the solution of the original problem and* $(u_h, \mathbf{G}_h, \mathbf{U}_h)$ *be the solution of the SDG scheme* (4a)–(4c). *Then we have the stability estimate*

$$\|u_h\|_{L^2(\Omega)} + \|\mathbf{U}_h\|_{L^2(\Omega)^2} + \|\mathbf{G}_h\|_{L^2(\Omega)^2} \leq C\|f\|_{L^2(\Omega)}, \tag{13}$$

and the convergence estimate

$$\begin{aligned} &\|u - u_h\|_{L^2(\Omega)} + \|\mathbf{U} - \mathbf{U}_h\|_{L^2(\Omega)^2} + \|\mathbf{G} - \mathbf{G}_h\|_{L^2(\Omega)^2} \\ &\qquad \leq Ch^{k+1} \left(\|u\|_{H^{k+1}(\Omega)} + \|\mathbf{G}\|_{H^{k+1}(\Omega)^2} \right). \end{aligned} \tag{14}$$

Proof We start by showing the stability estimate. Taking $\mathbf{W}_h = \mathbf{G}_h$, $\mathbf{V}_h = \mathbf{U}_h$, $v_h = u_h$ in (4a)–(4c), summing the three equations and applying (3), we have

$$\int_\Omega \rho(\mathbf{G}_h)\mathbf{G}_h \cdot \mathbf{G}_h dx = \int_\Omega f u_h \, dx.$$

Applying the Cauchy-Schwarz inequality and (8),

$$\|\mathbf{G}_h\|^2_{L^2(\Omega)^2} \leq C^{-1} \|f\|_{L^2(\Omega)} \|u_h\|_{L^2(\Omega)}. \tag{15}$$

Note, by Lemma 1 and Lemma 2,

$$\|u_h\|_{L^2(\Omega)} \leq C \sup_{\mathbf{V} \in \mathcal{W}^h} \frac{b_h(\mathbf{V}, u_h)}{\|\mathbf{V}\|_{X'}}. \tag{16}$$

Besides, using Eqs. (3) and (4a), we have for any $\mathbf{V} \in \mathcal{W}^h$

$$b_h(\mathbf{V}, u_h) = \int_\Omega \mathbf{G}_h \cdot \mathbf{V} \, dx \leq \|\mathbf{G}_h\|_{L^2(\Omega)^2} \, \|\mathbf{V}\|_{L^2(\Omega)^2}. \tag{17}$$

With the help of (7), (16) and (17) imply

$$\|u_h\|_{L^2(\Omega)} \leq C\|\mathbf{G}_h\|_{L^2(\Omega)^2}.$$

Combining this with (15),

$$\|\mathbf{G}_h\|_{L^2(\Omega)^2} \leq C\|f\|_{L^2(\Omega)},$$

and the stability estimate (13) follows from the Lipschitz continuity (9).

Next, we show the convergence of $\mathbf{G}_h$. Note that (4a) and (4c) still holds if we replace $\mathbf{G}_h$ by $\mathbf{G}$, $\mathbf{U}_h$ by $\mathbf{U}$ and u_h by u. Therefore,

$$\begin{aligned} \int_\Omega (\mathbf{G}-\mathbf{G}_h)\cdot\mathbf{V}\,dx - b_h^*(u-u_h,\mathbf{V}) &= 0, \\ b_h(\mathbf{U}-\mathbf{U}_h, v) &= 0, \end{aligned} \tag{18}$$

for any $(\mathbf{V}, v) \in \mathscr{W}^h \times \mathscr{U}^h$. Using the properties of $\mathscr{I}$ and $\mathscr{J}$,

$$\begin{aligned} \int_\Omega (\mathbf{G}-\mathbf{G}_h)\cdot\mathbf{V}\,dx - b_h^*(\mathscr{I}u-u_h,\mathbf{V}) &= 0, \\ b_h(\mathscr{J}\mathbf{U}-\mathbf{U}_h, v) &= 0, \end{aligned}$$

for any $(\mathbf{V}, v) \in \mathscr{W}^h \times \mathscr{U}^h$. In particular for $v = \mathscr{I}u - u_h$ and $\mathbf{V} = \mathscr{J}\mathbf{U} - \mathbf{U}_h$, adding these two equations gives

$$\int_\Omega (\mathbf{G}-\mathbf{G}_h)\cdot(\mathscr{J}\mathbf{U}-\mathbf{U}_h)\,dx = 0. \tag{19}$$

On the other hand, from the strong monotonicity (8) and using (19),

$$\begin{aligned} \|\mathscr{J}\mathbf{G}-\mathbf{G}_h\|^2_{L^2(\Omega)^2} &\leq C\int_\Omega (\mathscr{J}\mathbf{G}-\mathbf{G}_h)\cdot(\mathscr{J}\mathbf{U}-\mathbf{U}_h)\,dx \\ &\leq C\int_\Omega (\mathscr{J}\mathbf{G}-\mathbf{G})\cdot(\mathscr{J}\mathbf{U}-\mathbf{U}_h)\,dx \\ &\leq C\,\|\mathscr{J}\mathbf{G}-\mathbf{G}\|_{L^2(\Omega)^2}\,\|\mathscr{J}\mathbf{U}-\mathbf{U}_h\|_{L^2(\Omega)^2}. \end{aligned}$$

Applying the Lipschitz continuity (9),

$$\|\mathscr{J}\mathbf{G}-\mathbf{G}_h\|_{L^2(\Omega)^2} \leq C\,\|\mathscr{J}\mathbf{G}-\mathbf{G}\|_{L^2(\Omega)^2}.$$

Hence, with the help of equation (12), we have

$$\begin{aligned}\|\mathbf{G}-\mathbf{G}_h\|_{L^2(\Omega)^2} &\le \left\|\mathbf{G}-\mathscr{J}\mathbf{G}\right\|_{L^2(\Omega)^2}+\left\|\mathscr{J}\mathbf{G}-\mathbf{G}_h\right\|_{L^2(\Omega)^2}\\ &\le Ch^{k+1}\,\|\mathbf{G}\|_{H^{k+1}(\Omega)^2}\,.\end{aligned}$$

Then we show the convergence of u_h. Using equation (11),

$$\begin{aligned}\|u-u_h\|_{L^2(\Omega)} &\le \|u-\mathscr{I}u\|_{L^2(\Omega)}+\|\mathscr{I}u-u_h\|_{L^2(\Omega)}\\ &\le Ch^{k+1}\|u\|_{H^{k+1}(\Omega)}+\|\mathscr{I}u-u_h\|_{L^2(\Omega)}.\end{aligned}$$

Using the inf-sup condition in Lemma 1, equation (3), (10), (18) and (7),

$$\begin{aligned}\|\mathscr{I}u-u_h\|_{L^2(\Omega)} &\le C\sup_{\mathbf{V}\in\mathscr{W}^h}\frac{b_h(\mathbf{V},\mathscr{I}u-u_h)}{\|\mathbf{V}\|_{X'}}=C\sup_{\mathbf{V}\in\mathscr{W}^h}\frac{b_h^*(\mathscr{I}u-u_h,\mathbf{V})}{\|\mathbf{V}\|_{X'}}\\ &=C\sup_{\mathbf{V}\in\mathscr{W}^h}\frac{b_h^*(u-u_h,\mathbf{V})}{\|\mathbf{V}\|_{X'}}=C\sup_{\mathbf{V}\in\mathscr{W}^h}\frac{\int_\Omega(\mathbf{G}-\mathbf{G}_h)\cdot\mathbf{V}\,dx}{\|\mathbf{V}\|_{X'}}\\ &\le C\,\|\mathbf{G}-\mathbf{G}_h\|_{L^2(\Omega)^2}\,,\end{aligned}$$

which shows the convergence of u_h. The convergence of $\mathbf{U}_h$ follows from the Lipschitz continuity (9). This completes the proof of (14).

5 Numerical Examples

In this section, we present some numerical examples and verify the convergence rate of our SDG method. Moreover, we will obtain a postprocessed solution u_h^* which converges with higher order than u_h. We define the postprocessed solution u_h^* as follows. For each $\tau\in\mathscr{T}$, we take $u_h^*\in P^{k+1}(\tau)$ determined by

$$\int_\tau \nabla u_h^*\cdot\nabla w\,dx=\int_\tau \mathbf{G}_h\cdot\nabla w\,dx,\quad \forall w\in P^{k+1}(\tau)^0$$

$$\int_\tau u_h^*\,dx=\int_\tau u_h\,dx,$$

where $P^{k+1}(\tau)^0:=\left\{w\in P^{k+1}(\tau):\int_\tau w\,dx=0\right\}$. See [11].

For all of our numerical examples, we consider square domain $\Omega=[0,1]^2$. We divide this domain into $N\times N$ squares and divide each square into two triangles. We use this as our initial triangulation $\mathscr{T}_u$ and further subdivide each triangle taking the interior points as the centroids of the triangles following the discussion in Sect. 2.

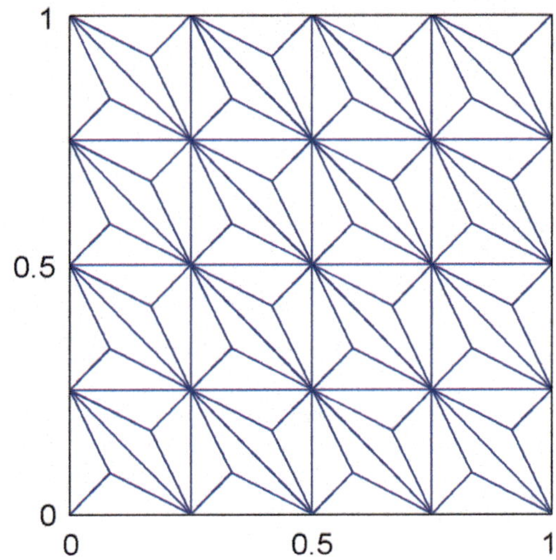

Fig. 2 Triangulation on $\Omega = [0, 1]^2$ with mesh size 1/4

We take the mesh size $h := 1/N$. We illustrate the mesh with $h = 1/4$ in Fig. 2. We consider the following solutions of equation (1).

$$u_1(x, y) = \sin(\pi x)\sin(\pi y),$$

$$u_2(x, y) = 10xy^2(1 - x)(1 - y) - \frac{e^{x-1}\sin(\pi x)\sin(\pi y)}{2}.$$

All these solutions have zero value on the boundary of Ω. We also consider the following six nonlinear coefficients to test the order of convergence.

$$\rho_1(\nabla u) := 2 + \frac{1}{1 + |\nabla u|} \qquad \rho_2(\nabla u) := 1 + \exp(-|\nabla u|)$$

$$\rho_3(\nabla u) := 1 + \exp(-|\nabla u|^2) \qquad \rho_4(\nabla u) := \frac{1}{\sqrt{1 + |\nabla u|}}$$

$$\rho_5(\nabla u) := |\nabla u| \qquad \rho_6(\nabla u) := |\nabla u|^2$$

For each u_j and ρ_ℓ, we choose f in (1) and solve for the approximate solution in the spaces of piecewise linear polynomial (i.e. $k = 1$) , using Newton's iteration with initial condition being the solution of (1) with $\rho \equiv 1$. We terminate the Newton's iteration when the successive error is less than $\delta = 10^{-10}$. Let $u_{j,h}$ be the approximate solution obtained from this Newton's iteration, and $u^*_{j,h}$ be the solution obtained from applying the above postprocessing procedure to $u_{j,h}$. Under different choices of nonlinear coefficients ρ_ℓ, we compute the L^2 error for $u_{j,h}$ and $u^*_{j,h}$, given by $\|u_j - u_{j,h}\|_{L^2(\Omega)}$ and $\|u_j - u^*_{j,h}\|_{L^2(\Omega)}$, respectively. The results are listed in Tables 1 and 2. From these results, we see clearly that the scheme gives optimal rate of convergence for the numerical solution and superconvergence for the postprocessed solution.

Table 1 The L^2 error for $u_{1,h}$ and $u^*_{1,h}$ under different choices of coefficients

Coefficient	Mesh size	$\|\|u_1 - u_{1,h}\|\|_{L^2(\Omega)}$	Order	$\|\|u_1 - u^*_{1,h}\|\|_{L^2(\Omega)}$	Order	Number of iterations
ρ_1	1/4	3.54e-2	–	2.86e-3	–	4
	1/8	9.24e-3	1.94	3.71e-4	2.95	4
	1/16	2.34e-3	1.98	4.70e-5	2.98	4
	1/32	5.86e-4	2.00	5.91e-6	2.99	4
	1/64	1.46e-4	2.00	7.40e-7	3.00	4
ρ_2	1/4	3.50e-2	–	3.00e-3	–	4
	1/8	9.23e-3	1.92	3.95e-4	2.82	4
	1/16	2.34e-3	1.98	5.07e-5	2.96	4
	1/32	5.86e-4	2.00	6.45e-6	2.98	4
	1/64	1.46e-4	2.00	8.13e-7	2.99	4
ρ_3	1/4	3.78e-2	–	4.31e-3	–	5
	1/8	9.41e-3	2.01	5.46e-4	2.98	5
	1/16	2.34e-3	2.01	5.81e-5	3.23	5
	1/32	5.86e-4	2.00	7.67e-6	2.92	5
	1/64	1.46e-4	2.00	9.84e-7	2.96	5
ρ_4	1/4	3.50e-2	–	3.13e-3	–	4
	1/8	9.21e-3	1.93	4.12e-4	2.93	5
	1/16	2.34e-3	1.98	5.30e-5	2.96	5
	1/32	5.86e-4	2.00	6.74e-6	2.98	5
	1/64	1.46e-4	2.00	8.49e-7	2.99	5
ρ_5	1/4	3.60e-2	–	3.32e-3	–	6
	1/8	9.29e-3	1.95	5.42e-4	2.62	6
	1/16	2.34e-3	1.99	8.34e-5	2.70	7
	1/32	5.86e-4	2.00	1.20e-5	2.79	8
	1/64	1.47e-4	2.00	1.67e-6	2.85	8
ρ_6	1/4	3.56e-2	–	5.98e-3	–	10
	1/8	9.29e-3	1.94	1.50e-3	2.00	10
	1/16	2.34e-3	1.99	2.28e-4	2.72	14
	1/32	5.86e-4	2.00	3.18e-5	2.84	20
	1/64	1.47e-4	2.00	4.29e-6	2.89	27

Table 2 The L^2 error for $u_{2,h}$ and $u^*_{2,h}$ under different choices of coefficients

Coefficient	Mesh size	$\|\|u_2 - u_{2,h}\|\|_{L^2(\Omega)}$	Order	$\|\|u_2 - u^*_{2,h}\|\|_{L^2(\Omega)}$	Order	Number of iterations
ρ_1	1/4	1.46e-2	–	1.78e-3	–	5
	1/8	3.91e-3	1.90	2.40e-4	2.88	5
	1/16	9.92e-4	1.98	3.11e-5	2.95	5
	1/32	2.49e-4	1.99	3.94e-6	2.98	5
	1/64	6.24e-5	2.00	5.00e-7	2.98	5
ρ_2	1/4	1.45e-2	–	1.72e-3	–	5
	1/8	3.90e-3	1.90	2.32e-4	2.89	5
	1/16	9.91e-4	1.98	3.04e-5	2.93	5
	1/32	2.45e-4	1.99	3.82e-6	2.99	5
	1/64	6.24e-5	2.00	4.94e-7	2.95	5
ρ_3	1/4	1.40e-2	–	1.90e-3	–	6
	1/8	3.94e-3	1.83	2.58e-4	2.88	6
	1/16	9.94e-4	1.99	3.22e-5	3.00	6
	1/32	2.49e-4	1.99	4.19e-6	2.94	6
	1/64	6.24e-5	2.00	5.33e-7	2.97	6
ρ_4	1/4	1.45e-2	–	1.71e-3	–	4
	1/8	3.90e-3	1.89	2.31e-4	2.89	4
	1/16	9.91e-4	1.98	3.03e-5	2.93	4
	1/32	2.49e-4	1.99	3.79e-6	3.00	4
	1/64	6.24e-5	2.00	4.89e-7	2.95	4
ρ_5	1/4	1.49e-2	–	3.97e-3	–	7
	1/8	3.94e-3	1.92	6.05e-4	2.71	8
	1/16	9.99e-4	1.98	9.24e-5	2.71	8
	1/32	2.50e-4	2.00	1.32e-5	2.81	10
	1/64	6.25e-5	2.00	1.79e-6	2.88	10
ρ_6	1/4	1.54e-2	–	6.54e-3	–	13
	1/8	3.91e-3	1.98	1.17e-3	2.49	15
	1/16	9.94e-4	1.98	1.96e-4	2.58	18
	1/32	2.50e-4	1.99	2.92e-5	2.74	21
	1/64	6.24e-5	2.00	3.91e-6	2.90	23

Acknowledgements The research of Eric Chung is partially supported by Hong Kong RGC General Research Fund (Project: 14301314).

References

1. S.C. Brenner, Poincaré–Friedrichs inequalities for piecewise H^1 functions. SIAM J. Numer. Anal. **41**(1), 306–324 (2003)
2. R. Bustinza, G.N. Gatica, A local discontinuous Galerkin method for nonlinear diffusion problems with mixed boundary conditions. SIAM J. Sci. Comput. **26**(1), 152–177 (2004)

3. S.W. Cheung, E. Chung, H.H. Kim, Y. Qian, Staggered discontinuous Galerkin methods for the incompressible Navier–Stokes equations. J. Comput. Phys. **302**, 251–266 (2015)
4. E.T. Chung, B. Engquist, Optimal discontinuous Galerkin methods for wave propagation. SIAM J. Numer. Anal. **44**(5), 2131–2158 (2006)
5. E.T. Chung, B. Engquist, Optimal discontinuous Galerkin methods for the acoustic wave equation in higher dimensions. SIAM J. Numer. Anal. **47**(5), 3820–3848 (2009)
6. E.T. Chung, C.S. Lee, A staggered discontinuous Galerkin method for the curl–curl operator. IMA J. Numer. Anal. **32**, 1241–1265 (2012)
7. E.T. Chung, P. Ciarlet, T.F. Yu, Convergence and superconvergence of staggered discontinuous Galerkin methods for the three-dimensional Maxwell's equations on Cartesian grids. J. Comput. Phys. **235**, 14–31 (2013)
8. E. Chung, B. Cockburn, G. Fu, The staggered DG method is the limit of a hybridizable DG method. SIAM J. Numer. Anal. **52**(2), 915–932 (2014)
9. E.T. Chung, C.Y. Lam, J. Qian, A staggered discontinuous Galerkin method for the simulation of seismic waves with surface topography. Geophysics **80**(4), T119–T135 (2015)
10. E. Chung, B. Cockburn, G. Fu, The staggered DG method is the limit of a hybridizable DG method. Part II: the Stokes flow. J. Sci. Comput. **66**(2), 870–887 (2016)
11. B. Cockburn, J. Guzmán, H. Wang, Superconvergent discontinuous Galerkin methods for second-order elliptic problems. Math. Comput. **78**(265), 1–24 (2009)
12. M. Feistauer, On the finite element approximation of a cascade flow problem. Numer. Math. **50**(6), 655–684 (1986)
13. B. Heise, Analysis of a fully discrete finite element method for a nonlinear magnetic field problem. SIAM J. Numer. Anal. **31**(3), 745–759 (1994)
14. L. Hu, G.-W. Wei, Nonlinear poisson equation for heterogeneous media. Biophys. J. **103**(4), 758–766 (2012)
15. J.J. Lee, H.H. Kim, Analysis of a staggered discontinuous Galerkin method for linear elasticity. J. Sci. Comput. **66**(2), 625–649 (2016)
16. J. Nečas, *Introduction to the Theory of Nonlinear Elliptic Equations*, vol. 52 (Teubner, Leipzig, 1983)
17. M. Tavelli, M. Dumbser. A staggered semi-implicit discontinuous Galerkin method for the two dimensional incompressible Navier-Stokes equations. Appl. Math. Comput. **248**, 70–92 (2014)
18. J. Virieux, P-SV wave propagation in heterogeneous media: velocity-stress finite-difference method. Geophysics **51**(4), 889–901 (1986)

Optimized High Order Explicit Runge-Kutta-Nyström Schemes

Marc Duruflé and Mamadou N'diaye

Abstract Runge-Kutta-Nyström (RKN) schemes have been developed to solve a non-linear ordinary differential equation (ODE) of the type $y'' = f(t, y)$. In Chawla and Sharma (Computing, 26:247–256, 1981), the stability condition (the Courant-Friedrichs-Lewy or CFL) associated with these schemes have been studied for order 3, 4 and 5. In this paper, we extend this study for higher orders and we propose a new algorithm to compute numerically the CFL. By using this algorithm, we compute optimal coefficients for RKN schemes of orders 6, 7, 8 and 10 which maximize the CFL. Herein, the obtained schemes are used to solve non-linear Maxwell's equations in 1-D.

1 Introduction

We consider the following ordinary differential equation (ODE)

$$\begin{cases} y''(t) = f(t, y(t)), & \forall t > 0, \\ y(0) = y_0, \\ y'(0) = y'_0. \end{cases}$$

The unknown y is vectorial, its size is equal to the number of the degrees of freedom of the system. The functional f is known and describes the dynamics of the system. This kind of ODE appears naturally in mechanical systems when the damping terms

M. Duruflé (✉)
Magique-3D, Inria Centre de Recherche Bordeaux Sud-Ouest, 200 avenue de la vieille Tour, 33 405 Talence, France
e-mail: marc.durufle@inria.fr

M. N'diaye
Laboratoire de Mathématiques et leurs Applications, University of Pau and Pays de L'Adour, avenue de l'université, 64012 Pau, France

Magique-3D, Inria Centre de Recherche Bordeaux Sud-Ouest, 200 avenue de la vieille Tour, 33 405 Talence, France
e-mail: mamadou.ndiaye@univ-pau.fr

M.L. Bittencourt et al. (eds.), *Spectral and High Order Methods for Partial Differential Equations ICOSAHOM 2016*, Lecture Notes in Computational Science and Engineering 119, DOI 10.1007/978-3-319-65870-4_43

are neglected, and also in non-linear wave equation. In order to solve this ODE, high-order Runge-Kutta-Nyström (RKN) schemes have been proposed (see [5]). They are attractive because they are explicit, one-step methods and can be applied to a non-linear operator f. A RKN scheme computes a discrete sequence y_n and y'_n, which are approximations of y and y' at time $t_n = n\Delta t$. The time step Δt is assumed to be constant in this paper. A step of the RKN scheme is performed as follows:

$$\begin{cases} k_i &= f\left(t_n + c_i\,\Delta t, \quad y_n + c_i\,\Delta t\, y'_n + \Delta t^2 \sum_j \bar{a}_{i,j}\, k_j\right), \\ y_{n+1} &= y_n + \Delta t\, y'_n + \Delta t^2 \sum_j \bar{b}_j\, k_j, \\ y'_{n+1} &= y'_n + \Delta t \sum_j b_j\, k_j, \end{cases}$$

k_i are intermediate vectors used to compute y_{n+1} and y'_{n+1}. The coefficients $\bar{a}_{i,j}, c_i, b_i, \bar{b}_i$ must satisfy the so-called order conditions such that the scheme is of order r (see [5] for a detailed description of order conditions). When it is not mentioned, the subscripts i and j vary between 0 and $s-1$ where s is the number of stages of the scheme. In this paper, only explicit schemes will be studied, the matrix $\bar{A}$ (associated with coefficients $\bar{a}_{i,j}$) is lower triangular, that is to say:

$$\bar{a}_{i,j} = 0, \ \text{if}\, j \geq i.$$

The coefficients $\bar{a}_{i,j}, c_i, b_i, \bar{b}_i$ can be obtained from the coefficients of a classical Runge-Kutta scheme of order r. But this procedure leads to RKN schemes that are less efficient (see [5]). In this paper, we are concerned to find optimal coefficients $\bar{a}_{i,j}, c_i, b_i, \bar{b}_i$ that maximize the CFL number subject to the order conditions. Such an optimization has been done for RKN schemes of order 3, 4 and 5 in [1] and [2]. In this work, we achieved to find the best coefficients for order 6, 7, 8 and 10. We propose a new algorithm to compute numerically the CFL number (stability condition) with respect to the coefficients $\bar{a}_{i,j}, c_i, b_i, \bar{b}_i$.

The remainder of this paper is organized as follows: First, we recall the stability condition as initially proposed in [1]. Next, we describe the numerical algorithm we used to compute the CFL number. Then, we propose the optimal coefficients obtained for the different schemes. Finally, we present some numerical results to show the practical interest of these schemes.

2 Stability Condition

The stability analysis is conducted for a linear functional f, which is then replaced by a matrix A:

$$f(t, y) \;=\; Ay.$$

By replacing A by its symbol $\hat{A}$ (which will be equal to an eigenvalue of A), a step of RKN scheme can be written as:

$$\begin{bmatrix} y_{n+1} \\ w_{n+1} \end{bmatrix} = D(\Delta t^2 \hat{A}) \begin{bmatrix} y_n \\ w_n \end{bmatrix}$$

where $D(\Delta t^2 \hat{A})$ is a 2×2 matrix depending on coefficients $\bar{a}_{i,j}, b_i, c_i, \bar{b}_i$. Let us note:

$$z = \Delta t^2 \hat{A}.$$

The vector w_n is equal to:

$$w_n = \frac{y_n'}{\Delta t \hat{A}}.$$

The RKN scheme is equal to:

$$\begin{cases} \Delta t^2 k_i = z\, y_n + c_i z^2 w_n + z \sum_j \bar{a}_{i,j} \Delta t^2 k_j, \\ y_{n+1} = y_n + z\, w_n + \sum_i \bar{b}_i \Delta t^2 k_i, \\ w_{n+1} = w_n + \frac{1}{z} \sum_i b_i \Delta t^2 k_i. \end{cases}$$

From these relations, it can be remarked that the entries of the 2×2 matrix $D(z)$ are polynomials in z. The amplification factor $G(z)$ is defined as:

$$G(z) = \text{Spectral radius of } D(z).$$

The stability condition is computed numerically by searching the first z such that

$$G(z) > 1.$$

The square root of this first z is defined as the CFL number:

$$\text{CFL number} = \min_{z \le 0} \{ \sqrt{-z} \text{ such that } G(z) > 1 \}.$$

3 Numerical Method to Compute the CFL

The eigenvalues of the 2×2 matrix $D(z)$ are directly computed as:

$$\lambda(D(z)) = \frac{\text{trace}(D(z)) \pm \sqrt{\text{trace}(D(z))^2 - 4 \det(D(z))}}{2}.$$

Algorithm 1 Algorithm used to compute the CFL number of RKN schemes

if $G(z_0) > 1 + \varepsilon$ **then**
  return 0
end if
$z = z_0$
while $G(z) <= 1 + \varepsilon$ **do**
  Adapt Δz_k such that any intersection of roots is not missed
  if $G(z) > \max(G(z - \Delta z_k), G(z + \Delta z_{k-1}))$ **then**
    Compute the local maximum z_m in the interval $[z - \Delta z_k, z + \Delta z_{k-1}]$
    if $G(z_m) > 1$ **then**
      $z = z_m$
      Terminate the main while loop
    end if
  end if
  $z = z - \Delta z_k$
end while
Compute z such that $G(z) = 1 + \varepsilon$ in the interval $[z - \Delta z, z]$ by bisection method
Return z

The amplification factor $G(z)$ is the maximal modulus of these two eigenvalues. From the computation of this amplification factor, the method used to compute the CFL is detailed in Algorithm 1. The computation of local maxima z_m and of the final z such that $G(z) = 1 + \varepsilon$ is performed by using a bisection method. The first float z_0 is chosen small (we have chosen $z_0 = -10^{-5}$), this first verification is needed because it happens that the amplification factor is decreasing at the origin, ie:

$$G'(0) < 0.$$

Hence for very small negative values of z, this amplification factor will be greater than one, leading to an unstable scheme. In this case, the time scheme will be unconditionally unstable.

The step Δz_k is chosen in an interval (we have selected $10^{-5} \leq \Delta z_k \leq 1$) such that the intersection of the two complex conjugate eigenvalues is not missed. This intersection occurs when the two eigenvalues get close to the real axis or when they already lie in the real axis. In Fig. 1, the amplification factor is displayed for a 6th order RKN method. In this case, the CFL is equal to 2.858 because of the presence of a local maxima above 1. It has been observed that usually the first local maximum (if present) occurs around $\sqrt{-z} \approx \pi$, the second maximum would occur around 2π, etc. In Fig. 2, we have displayed the trajectory of the two eigenvalues of $D(z)$ for $\sqrt{-z} \in [0, 4]$. These two eigenvalues start from the point $(1, 0)$ they describe an approximate circle to reach a point close to $(-1, 0)$. Then they move away from each other in the real axis, one reaches the local maximum, and the two eigenvalues get back until reaching another intersection in the real axis. Finally, they are describing a kind of hyperbole in the complex plane. With a variable Δz_k, we are able to compute the CFL with a reasonable number of evaluations of the amplification factor. Finally, ε is taken equal to $2 \cdot 10^{-13}$ for a double precision computation.

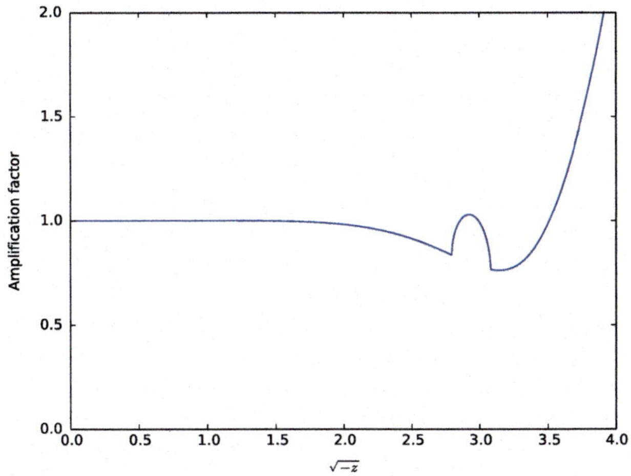

Fig. 1 Amplification factor $G(z)$ versus $\sqrt{-z}$ for a 6-th order RKN scheme, with the two free parameters equal to 0.0816464646464646 and 0.968757575757576

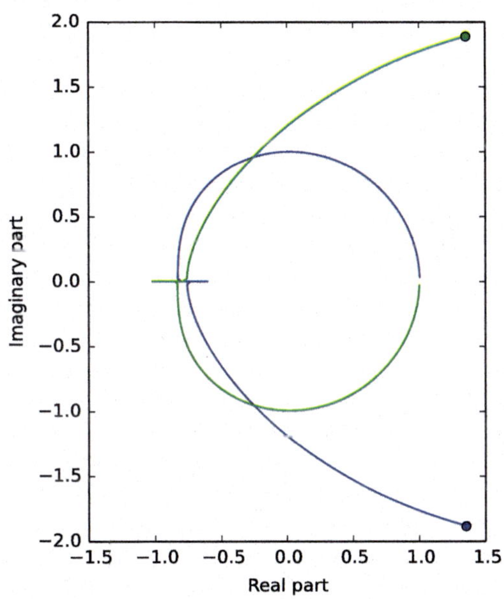

Fig. 2 Trajectory of the two eigenvalues of $D(z)$ for $\sqrt{-z} \in [0, 4]$ for a 6-th order RKN scheme, with the two free parameters equal to 0.0816464646464646 and 0.968757575757576

4 Optimization with a Minimal Number of Stages

In this section, coefficients of RKN schemes are optimized to maximize the CFL number. We consider here only schemes with a minimal number of stages (s).

4.1 Order 2 (s = 1)

For example, to obtain a second-order scheme, it is sufficient to satisfy

$$\sum_i b_i = 1, \quad \sum_i b_i c_i = \frac{1}{2}, \quad \sum_i \bar{b}_i = \frac{1}{2}.$$

Therefore, a one-stage scheme can be obtained:

$$\bar{A} = (0), \quad c = \begin{pmatrix} \frac{1}{2} \end{pmatrix}, \quad b = (1), \quad \bar{b} = \begin{pmatrix} \frac{1}{2} \end{pmatrix},$$

this scheme can be written as:

$$\begin{cases} k_0 = f\left(t^n + \frac{\Delta t}{2}, \, y_n + \frac{\Delta t}{2} y'_n\right), \\ y_{n+1} = y_n + \Delta t \, y'_n + \frac{\Delta t^2}{2} k_0, \\ y'_{n+1} = y'_n + \Delta t \, k_0. \end{cases}$$

This scheme requires only one evaluation of f (i.e. a matrix-vector product if f is linear) at each time step, which is equivalent to the cost of the classical second-order scheme (recalled below). When f is linear (replaced by a matrix A), the stability condition of this RKN scheme is:

$$\Delta t \leq \frac{2}{\sqrt{||A||_2}}.$$

This is exactly the same CFL as the classical second-order scheme:

$$\frac{y_{n+1} - 2y_n - y_{n-1}}{\Delta t^2} = f(t_n, y_n).$$

Therefore, the second-order Runge-Kutta-Nyström (RKN) scheme is optimal.

4.2 Orders 3, 4 and 5

For orders 3, 4, 5, we have found the same optimal coefficients for RKN schemes as in [1]. These coefficients are recalled below.

Order 3 ($s = 2$): A third-order RKN scheme with 2 stages is given as:

$$c_0 = \alpha, \quad c_1 = \frac{2-3\alpha}{3-6\alpha}, \qquad b_0 = \frac{\frac{c_1}{2} - \frac{1}{3}}{c_0(c_1 - c_0)}, \quad b_1 = 1 - b_0,$$

$$\bar{b}_0 = \frac{\frac{c_1}{2} - \frac{1}{6}}{c_1 - c_0}, \quad \bar{b}_1 = \frac{1}{2} - \bar{b}_0, \qquad \bar{a}_{1,0} = \frac{1}{6b_1},$$

α is a free parameter, a maximal CFL of 2.498 is obtained for $\alpha = \frac{3-\sqrt{3}}{6}$.

Order 4 ($s = 3$): A fourth-order RKN scheme with 3 stages is given as:

$$c_0 = \alpha, \quad c_1 = \frac{1}{2}, \quad c_2 = 1 - \alpha,$$

$$b_0 = \frac{1}{6(1-2\alpha)^2}, \quad b_1 = 1 - 2b_0, \quad b_2 = b_0,$$

$$\bar{b}_0 = b_0(1 - c_0), \quad \bar{b}_1 = b_1(1 - c_1), \quad \bar{b}_2 = b_2(1 - c_2),$$

$$\bar{a}_{1,0} = \frac{(1-4\alpha)(1-2\alpha)}{8(6\alpha(\alpha-1)+1)}, \quad \bar{a}_{2,0} = 2\alpha(1-2\alpha), \quad \bar{a}_{2,1} = \frac{(1-2\alpha)(1-4\alpha)}{2},$$

α is a free parameter a maximal CFL of 3.939 is obtained for

$$\alpha = \frac{1}{4\left(1 + \cos(\frac{\pi}{9})\right)}.$$

Order 5 ($s = 4$): A family of RKN schemes of order 5 with two parameters is given in [3]. A maximal CFL of 2.908 is obtained for

$$\alpha = \frac{4}{11 + \sqrt{16\sqrt{10} - 39}},$$

$$\beta = \frac{165\alpha^2 - 195\alpha + 50 - \sqrt{5\left(45\alpha^4 + 90\alpha^3 - 105\alpha^2 + 36\alpha - 4\right)}}{225\alpha^2 - 240\alpha + 60}.$$

The c_i are given as

$$c_0 = 0, \quad c_1 = \alpha, \quad c_3 = \beta, \quad c_2 = \frac{12 - 15(\alpha+\beta) + 20\alpha\beta}{15 - 20(\alpha+\beta) + 30\alpha\beta}.$$

From order 6 to 10, the optimal coefficients for RKN schemes are new. They have been computed numerically, only, in the following subsections.

4.3 *Order 6 (s = 5)*

A family with one parameter is given in [3]. Using Algorithm 1, we have obtained a maximal CFL of 3.089 for

$$\alpha \approx 0.22918326$$

The c_i are given as

$$c_0 = 0, \quad c_1 = \alpha, \quad c_2 = \frac{1}{2}, \quad c_3 = 1 - \alpha, \quad c_4 = 1.$$

Another family with two parameters can also be constructed. The maximal CFL is also equal to 3.089 for this family.

4.4 *Order 7 (s = 7)*

A family of RKN schemes of order 7 with four free parameters is given in [3]. After optimization, we have obtained a maximal CFL of 7.0875 with the following parameters:

$$\alpha_0 = 0.110451398065702, \quad \alpha_1 = 0.173816271367107$$
$$\alpha_2 = 0.459433163929695, \quad \alpha_3 = 0.652002232653235$$

The coefficients c_i are given by

$$c_0 = 0, \; c_1 = \alpha_0, \; c_2 = \alpha_1, \; c_3 = \alpha_2, \; c_4 = \alpha_3, \; c_5 = \frac{-\frac{1}{7} + \frac{\sigma_1^c}{6} - \frac{\sigma_2^c}{5} + \frac{\sigma_3^c}{4} - \frac{\sigma_4^c}{3}}{-\frac{1}{6} + \frac{\sigma_1^c}{5} - \frac{\sigma_2^c}{4} + \frac{\sigma_3^c}{3} - \frac{\sigma_4^c}{2}}, \; c_6 = 1.$$

4.5 *Order 8 (s = 8)*

A family of RKN schemes of order 8 with four free parameters is given in [3]. A maximal CFL of 7.8525 is obtained with the following parameters

$$\alpha_0 = 0.135294127286225, \quad \alpha_1 = 0.24015308384744$$

$$\alpha_2 = 0.453046953126355, \quad \alpha_3 = 0.695039606659698$$

The coefficients c_i are given by

$$c_0 = 0, \quad c_1 = \frac{\alpha_0}{2}, \quad c_2 = \alpha_0, \quad c_3 = \alpha_1, \quad c_4 = \alpha_2, \quad c_5 = \alpha_3$$

$$c_6 = \frac{-\frac{1}{8} + \frac{\sigma_1^c}{7} - \frac{\sigma_2^c}{6} + \frac{\sigma_3^c}{5} - \frac{\sigma_4^c}{4} + \frac{\sigma_5^c}{3}}{-\frac{1}{7} + \frac{\sigma_1^c}{6} - \frac{\sigma_2^c}{5} + \frac{\sigma_3^c}{4} - \frac{\sigma_4^c}{3} + \frac{\sigma_5^c}{2}}, \quad c_7 = 1.$$

4.6 *Order 10 (s = 11)*

In [4], the author presents a family of RKN schemes of order 10 with four free parameters (b_0, b_2, b_3, r_5). r_5 is an additional free parameter that we have recognized during the construction of the family, it is defined as

$$r_5 = \sum_{i=0}^{s-1} b_i c_i^3 \sum_{j=0}^{i-1} \bar{a}_{ij} c_j^5 .$$

Following the work in [4] we denote the Gauss-Lobatto nodes $\gamma_1, \gamma_2, \gamma_3, \gamma_4$:

$$\begin{cases} \gamma_1 = \dfrac{1}{2}\left(1 - \sqrt{\dfrac{7+2\sqrt{7}}{21}}\right), & \gamma_4 = 1 - \gamma_1, \\ \gamma_2 = \dfrac{1}{2}\left(1 - \sqrt{\dfrac{7-2\sqrt{7}}{21}}\right), & \gamma_3 = 1 - \gamma_2. \end{cases}$$

Among the 24 permutations choice possible for (c_3, c_4, c_5, c_6), the CFL is maximal for the following permutation

$$(c_3, c_4, c_5, c_6) = (\gamma_4, \gamma_3, \gamma_1, \gamma_2) .$$

The other c_i are given by

$$c_0 = 0, \quad c_2 = \frac{c_4\,(3c_4 - 5c_3)}{5c_4 - 10c_3}, \quad c_1 = \frac{c_2}{2}, \quad c_7 = c_3, \quad c_8 = c_2, \quad c_9 = 0, \quad c_{10} = 1.$$

For this permutation, we have obtained a maximal CFL of 4.7527 with the following parameter

$$r_5 = 0.0021632268153138$$

The CFL is maximal for this permutation only, it is strictly lower for other permutations. For other parameters, we can choose the values proposed by Hairer in [4]:

$$b_0 = 0, \quad b_2 = -0.1, \quad b_3 = 0,$$

since the CFL does not depend on these three parameters.

5 Efficiency and Numerical Results

5.1 *Efficiency*

Let s be the number of stages of the RKN scheme. The efficiency is given as:

$$\text{Efficiency} = \frac{\text{CFL number}}{2s}.$$

An optimal scheme is a scheme such that the efficiency is maximal. Since s is constant, the efficiency is maximal for a maximal CFL number. In Table 1, we have written the efficiency of the different RKN schemes.

We observe that the orders 7 and 8 are attractive since they have a correct efficiency (about 50%). It is important to have optimal schemes when the ODE to be solved is stiff, i.e. when the obtained accuracy is satisfactory when the maximal time step is chosen. However, for low orders (such as 2, 3), the accuracy is usually poor such that the time step must be chosen much smaller than the maximal time step to obtain a good solution.

Table 1 Efficiency of optimized Runge-Kutta-Nyström schemes of different orders

Order	2	3	4	5	6	7	8	10
Efficiency(%)	100	62.5	65.7	36.4	30.9	50.6	49.1	21.6

5.2 Numerical Results

Using the higher order-finite element code Montjoie, we applied the RKN schemes to solve the non-linear Maxwell's equations in 1-D, namely:

$$\begin{cases} \dfrac{\varepsilon_\infty}{c^2}\dfrac{\partial^2 E}{\partial t^2} + \dfrac{1}{c^2}\dfrac{\partial^2}{\partial t^2}\left(\sum_k P_k\right) - \dfrac{\partial^2 E}{\partial z^2} + \dfrac{\rho}{c^2}\dfrac{\partial^2}{\partial t^2}\left(|E|^2 E\right) = 0 \\ \dfrac{1}{\omega_k^2}\dfrac{\partial^2 P_k}{\partial t^2} + P_k = \nu_k E \\ E(z, t=0) = \dfrac{\partial E}{\partial t}(z, t=0) = 0 \\ E(z=0, t) = \text{Given impulsion} \end{cases}$$

Here the electric field is searched as a complex field:

$$E = E_x + iE_y,$$

where E_x and E_y are x and y-components of the electric field. P_k is the polarization, $\varepsilon_\infty, c, \rho, \nu_k, \omega_k$ are physical constants. We take the constants corresponding to silica:

$$\varepsilon_\infty = 1,\ c = 299792458,\ \nu_0 = 0.6961663,\ \nu_1 = 0.4079426,\ \nu_2 = 0.8974794$$

$$\omega_0 = \frac{2\pi c}{0.0684043\cdot 10^{-6}},\ \omega_1 = \frac{2\pi c}{0.1162414\cdot 10^{-6}},\ \omega_2 = \frac{2\pi c}{9.896161\cdot 10^{-6}},\ \gamma = 10^{-33}.$$

The impulsion is centered at $\lambda_0 = 1.053\mu m$ with a Gaussian envelope and a circular polarization:

$$\text{Given impulsion} = E_0\, e^{-\frac{1}{2}\left(\frac{t - T_{\max}}{\tau}\right)^2} e^{i\omega_L t}$$

where

$$\omega_L = \frac{2\pi c}{1.053\cdot 10^{-6}},\ T_{\max} = 6\cdot 10^{-14},\ \tau = \frac{2}{2\sqrt{2\log 2}}\cdot 10^{-14},\ E_0 = 10^9.$$

The computational domain is the 1-D interval $\Omega = [0, 1.5\cdot 10^{-4}]$, a Neumann boundary condition is set on the right extremity. 1-D finite elements are used to discretize these equations:

$$E \in V_h = \left\{u \in H^1(\Omega) \text{ such that } u|_{[z_i, z_i+1]} \in \mathbb{Q}_{10}\right\}$$

where $(z_i)_{0\le i\le 250}$ are a regular subdivision of the computational domain Ω. The mesh contains 250 elements (i.e. 2501 degrees of freedom), the numerical error

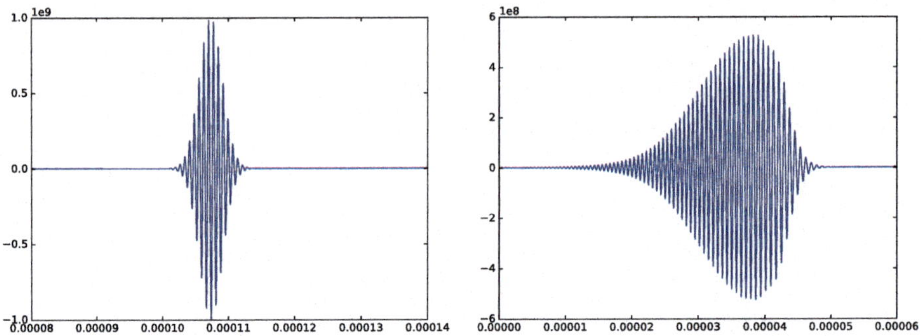

Fig. 3 Electric field E_x for $t = 10^{-12}$ and $t = 5 \cdot 10^{-11}$

due to the space discretization is around 10^{-6} (the domain contains more than 200 wavelengths). After space discretization, the system can be written in the form

$$y'' = f(t, y)$$

by using the displacement as unknown

$$D = \varepsilon_\infty E + \left(\sum_k P_k \right) + \rho |E|^2 E.$$

The electric field E is recovered from D by solving the non-linear equation written above for each degree of freedom. This equation is solved with a Newton's method, two or three iterations are sufficient to get machine precision accuracy. Gauss-Lobatto points are used both for interpolation (for the discretization of V_h) and quadrature, leading to a diagonal mass matrix. As a result the computation of $f(t, y)$ is explicit, it does not involve any solution of a linear system. The electric field is propagated from $t = 0$ until $t = 5 \cdot 10^{-11}$, in Fig. 3, the solution is plotted at two different times. The solution at the final time $t = 5 \cdot 10^{-11}$ is compared with a reference solution computed with a small time step (with tenth order RKN scheme). We try to reach an error of 0.01% for each scheme in order to compare the efficiency. In Table 2, the computational time needed to obtain this accuracy is given for each optimized RKN scheme. The simulations are performed in parallel on 20 cores on an Intel-Xeon (2 Dodeca-core Haswell E5-2680, 2.5 Ghz). From order 5, we are using the maximal time step allowed (because of the restrictive CFL), that's why the error is below 0.01% for these orders. For orders 2, 3 and 4, the time step required to obtain an error of 0.01% is much smaller than the maximal time step, that's why they are less efficient. In this case, we observe that low order schemes (2, 3, 4) are limited by the accuracy whereas high-order schemes are limited by the CFL. We see that RKN schemes of order 7 or 8 are the most efficient for this problem while order 10 is not very efficient because of its small CFL.

Table 2 Computational time needed to reach an accuracy of 0.01 % for different orders of RKN schemes

Order	2	3	4	5	6	7	8	10
Time(s)	14 164	730	144	60.9	70.8	43.2	44.8	103
Error	$1.0 \cdot 10^{-4}$	$1.0 \cdot 10^{-4}$	$1.0 \cdot 10^{-4}$	$1.5 \cdot 10^{-6}$	$7.3 \cdot 10^{-7}$	$6.5 \cdot 10^{-7}$	$1.1 \cdot 10^{-7}$	$2.0 \cdot 10^{-10}$
Δt	$9.1 \cdot 10^{-20}$	$2.9 \cdot 10^{-18}$	$2.0 \cdot 10^{-17}$	$6.2 \cdot 10^{-17}$	$6.6 \cdot 10^{-17}$	$1.5 \cdot 10^{-16}$	$1.7 \cdot 10^{-16}$	$1.0 \cdot 10^{-16}$

6 Conclusion

In this work we have proposed an algorithm to compute the CFL number of a RKN scheme. By using this algorithm, we have computed optimal coefficients for RKN schemes of order 6, 7, 8 and 10 that maximize the CFL number. The numerical results we presented show the practical interest of these schemes, in paticular order 7 and 8. In fact, we have observed that lower order schemes are limited by the accuracy while the scheme of order 10 is less efficient due to its small CFL. We think that the efficiency can be further increased by adding more stages [6, 7].

References

1. M. Chawla, S. Sharma, Families of fifth-order Nyström methods for $y'' = f(x, y)$ and intervals of periodicity. Computing **26**, 247–256 (1981)
2. M. Chawla, S. Sharma, Interval of periodicity and absolute stability of explicit Nyström methods for $y'' = f(x, y)$. BIT **21**, 455–464 (1981)
3. E. Hairer, Méthodes de Nyström pour l'équation différentielle $y'' = f(x, y)$. Numer. Math. **27**, 283–300 (1977)
4. E. Hairer, A one-step method of order 10 for $y''(t) = f(t, y)$. IMA J. Numer. Anal. **2**, 83–94 (1982)
5. E. Hairer, S.P. Norsett, G. Wanner, *Solving Ordinary Differential Equations I - Nonstiff Problems* (Springer, Berlin, 2008)
6. P. Joly, J.-C. Gilbert, Higher order time stepping for second order hyperbolic problems and optimal CFL conditions. Comput. Methods Appl. Sci. **16**, 67–93 (2008)
7. J. Neigemann, R. Diehl, K. Bush, Efficient low-storage Runge-Kutta schemes with optimized stability regions. J. Comput. Phys. **231**, 364–372 (2012)

Artificial Viscosity Discontinuous Galerkin Spectral Element Method for the Baer-Nunziato Equations

C. Redondo, F. Fraysse, G. Rubio, and E. Valero

Abstract This paper is devoted to the numerical discretization of the hyperbolic two-phase flow model of Baer and Nunziato. Special attention is paid to the discretization of interface flux functions in the framework of Discontinuous Galerkin approach, where care has to be taken to efficiently approximate the non-conservative products inherent to the model equations. A discretization scheme is proposed in a Discontinuous Galerkin framework following the criterion of Abgrall. A stabilization technique based on artificial viscosity is applied to the high-order Discontinuous Galerkin method and tested on a bench of discontinuous test cases.

1 Introduction

In this work a high order discretization method for the hyperbolic two-phase flow model of Baer and Nunziato [3] is introduced. The model is composed of seven equations in one-dimension: continuity, momentum and energy balance for each phase and a convection equation for the volume fraction. It does not make any assumption on mechanical, thermal or chemical equilibrium, thus, two pressures, velocities and temperatures are present. The main challenge of this set of equations is that it cannot be cast in conservative (or divergence) form because of the presence of non-conservative products. As a consequence, classical Rankine-Hugoniot conditions cannot be used to define the jumps across the contact discontinuities and shocks. This

C. Redondo (✉) • G. Rubio • E. Valero
School of Aeronautics (ETSIAE Universidad Politécnica de Madrid), Madrid, Spain
e-mail: carlos.redondo@upm.es; g.rubio@upm.es; eusebio.valero@upm.es

F. Fraysse
RS2N, Saint Zacharie, France
e-mail: francois.fraysse@rs2n.eu

M.L. Bittencourt et al. (eds.), *Spectral and High Order Methods for Partial Differential Equations ICOSAHOM 2016*, Lecture Notes in Computational Science and Engineering 119, DOI 10.1007/978-3-319-65870-4_44

issue has remained challenging for a long time but recently some authors published different methods in order to treat these additional terms [24, 29, 30, 32].

On the one hand, most works in the literature use a finite volume (FV) methodology, often limited to second-order of accuracy, to discretize the Baer-Nunziato equations, see [11] for a review. Attempts to reconstruct Finite Volume methodology to higher order usually suffer from a lack of compactness, which is a bottleneck for massive parallel implementation. On the other hand, discontinuous Galerkin (DG) methods take the advantages of FV approach (conservation, interface jumps, compactness) but naturally allow the solution to be represented by a high-order polynomial. DG methods, firstly introduced in [27], have emerged in recent years as an efficient and flexible method to solve convection dominated problems [8]. A nodal variant of the DG technique that uses a quad/hexa mesh topology and tensor product expansions for the polynomial spaces is known as Discontinuous Galerkin Spectral Element Method (DGSEM), as detailed in Kopriva [22]. The DGSEM has been successfully used in a wide range of applications, in particular to model one phase compressible flows [5, 20, 21, 26]. Recently some DGSEM formulations able to solve the Baer-Nunziato equations in the presence of discontinuities have been introduced by the authors [11]. One of the important aspects of this development is the special treatment to avoid oscillations in the vicinity of shocks and contact discontinuities. The method builds on work by Persson et al. [25] and uses a simple artificial viscosity technique [4, 16, 17, 34] to stabilize the solution. In this work the most successful of the formulations proposed in [11] is introduced and analyzed in detail in a one-dimensional framework.

The paper is organized as follows: in Sect. 2 the discretization of the Baer-Nunziato equations is presented. The DG method is detailed as well as the upwind fluxes, the treatment of the non-conservative products and the stabilization method. In Sect. 3, the developed numerical scheme is tested using a bench of one-dimensional test cases.

2 Discretization of the Two-Phase Two-Pressure Model of Baer and Nunziato

The one-dimensional set of Baer-Nunziato equations reads:

$$\frac{\partial \mathbf{U}}{\partial t} + \frac{\partial \mathbf{F}(\mathbf{U})}{\partial x} + \mathbf{H}(\mathbf{U})\frac{\partial \alpha_l}{\partial x} = 0, \tag{1}$$

where

$$\mathbf{U} = \begin{bmatrix} \alpha_l \\ \alpha_l\rho_l \\ \alpha_l\rho_l u_l \\ \alpha_l\rho_l E_l \\ \alpha_g\rho_g \\ \alpha_g\rho_g u_g \\ \alpha_g\rho_g E_g \end{bmatrix} \quad \mathbf{F}(\mathbf{U}) = \begin{bmatrix} 0 \\ \alpha_l\rho_l u_l \\ \alpha_l\rho_l u_l^2 + p_l \\ \alpha_l\rho_l u_l H_l \\ \alpha_g\rho_g u_g \\ \alpha_g\rho_g u_g^2 + p_g \\ \alpha_g\rho_g u_g H_g \end{bmatrix} \quad \mathbf{H}(\mathbf{U}) = \begin{bmatrix} u_{int} \\ 0 \\ -p_{int} \\ -p_{int}u_{int} \\ 0 \\ p_{int} \\ p_{int}u_{int} \end{bmatrix}$$

This system of equations is closed with an equation of state for each phase $m = g, l$ (gas and liquid) relating the internal energy $e_m = E_m - 0.5u_m^2$ to the density ρ_m and pressure p_m, the saturation condition $\alpha_l + \alpha_g = 1$ (liquid and gas volume fractions) and finally appropriate interfacial pressure and velocity. In this work, stiffened gas equation is considered for the liquid phase, $p_m = \rho_m e_m(\gamma_m - 1) + \pi_m$, and perfect gas law for the gas phase ($\pi_g = 0$). The interfacial quantities are set according to the choice of Baer and Nunziato: $u_{int} = u_l, p_{int} = p_g$.

2.1 Discontinuous Galerkin Spectral Element Method

Discontinuous Galerkin methods and, in particular the nodal variant DGSEM, were originally developed to solve conservation laws. Unfortunately, it is not possible to cast the Baer-Nunziato equations in conservative form due to the presence of non-conservative products of the form $\mathbf{H}(\mathbf{U})\frac{\partial \alpha_l}{\partial x}$. The difficulty of integrating this term over a control volume arises in the presence of a discontinuity in the volume fraction. As a result, some modifications from the original DGSEM are required. It should be noticed that the scope of this work is limited to one-dimensional approximations.

Let us rewrite the Baer-Nunziato equations,

$$\mathbf{U}_t + \mathbf{F}_x + \mathbf{H}(\alpha_l)_x = 0, \quad x \in \Omega, \tag{2}$$

where $\mathbf{U}$ is the solution and $\mathbf{U}_t$ denotes its temporal derivative. The flux function is $\mathbf{F}$, while $\mathbf{F}_x$ denotes its spatial derivative and the non conservative flux is denoted by $\mathbf{H}(\alpha_l)_x$ (notice that $\alpha_l = \mathbf{U}(1)$). In the following and to simplify the notation α_l will be shortened to α.

Discontinuous Galerkin methods tessellate the physical domain Ω into non overlapping subdomains Ω_k. The residual is forced to be orthogonal to the approximation space locally within each element,

$$\int_{\Omega_k} (\mathbf{U}_t + \mathbf{F}_x + \mathbf{H}\alpha_x)\, \psi dx = 0, \tag{3}$$

where ψ is an arbitrary locally smooth function. The physical domain Ω_k, of size Δx_k, is mapped into the computational domain, which in 1D is $[-1, 1]$,

$$\frac{\Delta x_k}{2}\int_{-1}^{1} \mathbf{U}_t \psi d\xi + \int_{-1}^{1} \mathbf{F}_\xi \psi d\xi + \int_{-1}^{1} \mathbf{H}\alpha_\xi \psi d\xi = 0. \tag{4}$$

The solution and the fluxes are approximated by polynomials of degree N. A characteristic of the DGSEM is that it approximates both the solution and the fluxes with the same polynomial degree, e.g.,

$$\mathbf{U}(\xi, t) \approx \mathbf{U}^N(\xi, t) = \sum_{i=0}^{N} \mathbf{U}^N(\xi_i, t)\ell_i(\xi). \tag{5}$$

This approximation results in a computationally efficient method with higher aliasing error. The approximation is nodal, therefore ℓ_i are Lagrange polynomials while ξ_i are chosen to be Legendre-Gauss nodes. As a result, $\mathbf{U}^N(\xi_i, t)$ is the solution at the Legendre-Gauss nodes. The nodal values of the fluxes $\mathbf{F}^N(\xi_i, t)$ and $\mathbf{H}^N(\xi_i, t)$ are computed evaluating the solution at the nodes. As the method is Galerkin, the test function can also be written as a polynomial $\psi = \sum_{i=0}^{N} \psi_i \ell_i(\xi)$. Now, substituting the polynomial expressions in Eq. (4) and taking into account that the coefficients ψ_i are linearly independent we get,

$$\begin{aligned} &\frac{\Delta x_k}{2}\int_{-1}^{1} \mathbf{U}_t^N \ell_j(\xi) d\xi + \int_{-1}^{1} \mathbf{F}_\xi^N \ell_j(\xi) d\xi \\ &+ \int_{-1}^{1} \mathbf{H}^N \alpha_\xi^N \ell_j(\xi) d\xi \; = 0, \quad j = 0, 1, \ldots, N. \end{aligned} \tag{6}$$

Equation (6) is integrated by parts to separate volume from surface contributions,

$$\begin{aligned} &\frac{\Delta x_k}{2}\int_{-1}^{1} \mathbf{U}_t^N \ell_j(\xi) d\xi + \left(\mathbf{F}^N + \mathbf{H}^N \alpha^N\right) \ell(\xi)\Big|_{-1}^{1} - \int_{-1}^{1} \mathbf{F}^N \ell_j'(\xi) d\xi \\ &- \int_{-1}^{1} \mathbf{H}_\xi^N \alpha^N \ell_j(\xi) d\xi - \int_{-1}^{1} \mathbf{H}^N \alpha^N \ell_j'(\xi) d\xi = 0, \quad j = 0, 1, \ldots, N. \end{aligned} \tag{7}$$

It should be noticed that the computation of the volume fraction derivative, α_ξ, inside the control volume is not required after the integration by parts. In order to obtain a completely discrete equation, the integrals are approximated using Gaussian quadrature. In the DSGEM the interpolation nodes are used as quadrature nodes,

$$\frac{\Delta x_k}{2}\dot{\mathbf{U}}_j^N w_j + \left(\mathbf{F}^N + \mathbf{H}^N\alpha^N\right)\ell_j(\xi)\Big|_{-1}^{1} - \sum_{i=0}^{N}\mathbf{F}_i^N\ell_j'(\xi_i)w_i$$
$$-\sum_{i=0}^{N}\mathbf{H}_i^N\alpha_j^N\ell_i'(\xi_j)w_j - \sum_{i=0}^{N}\mathbf{H}_i^N\alpha_i^N\ell_j'(\xi_i)w_i = 0, \quad j = 0,1,\ldots,N. \tag{8}$$

Finally, the elements are coupled through the definition of a numerical interface flux,

$$\frac{\Delta x_k}{2}\dot{\mathbf{U}}_j^N w_j + \left(\mathbf{F}^{N*} + \mathbf{H}^N\alpha^{N*}\right)\ell_j(\xi)\Big|_{-1}^{1} - \sum_{i=0}^{N}\mathbf{F}_i^N\ell_j'(\xi_i)w_i$$
$$-\sum_{i=0}^{N}\mathbf{H}_i^N\alpha_j^N\ell_i'(\xi_j)w_j - \sum_{i=0}^{N}\mathbf{H}_i^N\alpha_i^N\ell_j'(\xi_i)w_i = 0, \quad j = 0,1,\ldots,N. \tag{9}$$

The numerical flux $\mathbf{F}^{N*}$ and α^{N*} is a function of the element and its immediate neighbor (or a physical boundary). To calculate its value, a Riemann problem should be solved [33]. The Riemann solver calculates a value for the fluxes, taking into account the values at each side of the discontinuity and the directions of transfer of information in the equation. More information about the numerical flux computation will be given in the next section. Having obtained a suitable discrete expression for each elemental contribution, it suffices to sum over all elements in the mesh and apply the boundary conditions weakly to finalize the DGSEM method, see details in Kopriva [22].

2.2 *Interface Flux Approximation. Criterion of Abgrall*

In this section we present the approximate Riemann solver employed to compute the intercell fluxes $\mathbf{F}^{N*}$ and α^{N*} of the 7-equation Baer-Nunziato two-phase flow model.

For the conservative flux $\mathbf{F}^{N*}$, the Rusanov flux is chosen. The Rusanov flux [23, 28] only uses one wave speed S_{max} which is the maximum absolute eigenvalue of left and right states of the Jacobian matrix. The main advantage of the Rusanov flux is its simplicity and low dependence on the eigenstructure of the flux Jacobian. Thus, it is particularly easy to implement when the flux Jacobian is difficult to formulate, for example when a complex equation of state is used. Its main disadvantage is its high diffusion of discontinuities, in particular the contact discontinuities. This effect is diminished if a high order approximation is used.

The volume fraction interface flux approximation α^{N*} will be computed following the so-called Abgrall criterion. The criterion of Abgrall [1] states that a two-phase flow uniform in velocity and pressure should remain uniform in these variables with time evolution. In order to satisfy the criterion of Abgrall for the Rusanov flux, some choices are to be made for the flux $\mathbf{F}^{N*}$ and the liquid volume fraction at the interface α^{N*}. The classical flux holds the conservative part of the system and is here augmented to hold a contribution from the liquid volume fraction equation, denoted by $F^*_{1_b}$. The scheme is chosen such that:

$$\begin{cases} F^*_{1_b} &= -\dfrac{S_{max}}{2}(\alpha_R - \alpha_L) \\ \alpha^{N*} &= \dfrac{\alpha_R + \alpha_L}{2} \end{cases} \quad \text{Rusanov flux with Abgrall criterion} \tag{10}$$

It should be noticed that an interpolation from the interior Gauss points to the interface points ± 1 is required to obtain left and right states (e.g. α_L,α_R) for the intercell flux computation.

2.3 Stabilization Using an Artificial Viscosity Method

The upwind scheme presented earlier may yield high oscillations in the vicinity of discontinuities due to Gibbs phenomena [14]. The objective here is to search for a method that detects the occurrence of the Gibbs phenomena and attenuates it. Several methods can be found in the literature to stabilize the solution in the presence of discontinuities. Classical limiters work well in order to avoid the local creation of extrema, however they severely degrade the accuracy, often reduced to one in the entire cell. Artificial viscosity methods, firstly introduced in the scope of finite differences in the fifties by von Neumann and Richtmyer [34], add a controlled amount of viscosity to the governing equations in the vicinity of strong gradients, such as shock waves or contact discontinuities. In this way the discontinuity may be resolved in the space of interpolating polynomials. Other variants of artificial viscosity methods exist as well. A particularly important one is the method of Spectrally Vanishing Viscosity (SVV) [18, 31], which is similar in spirit, but the smoothing is limited to the high frequency components of the solution.

In this work, we construct an stabilization method for the multiphase flow based on the single phase work of Persson et al. [25], where the mitigation is attained through an artificial viscosity technique. The new set of equations, with the artificial viscosity term included, reads:

$$\frac{\partial \mathbf{U}}{\partial t} + \frac{\partial \mathbf{F}}{\partial x} + \mathbf{H}\frac{\partial \alpha_l}{\partial x} = \frac{\partial}{\partial x}\left(\epsilon \frac{\partial \mathbf{U}}{\partial x}\right) \tag{11}$$

Then the discontinuity should spread over a layer of thickness ϵ. The definition of the parameter ϵ, that controls the amount of viscosity introduced, as well as the definition of a sensor to capture the regions where the stabilizing viscosity should be added, are key aspects in the development of the artificial viscosity method.

The discontinuity sensor is built following [25]. Spectral methods represent the solution of the problem as a sum of basis functions multiplied by some coefficients. In particular the DGSEM uses the Legendre orthogonal polynomials as a basis and therefore a one dimensional solution of order N, can be represented in each element as a sum of local modes,

$$\mathbf{U}^N(x) = \sum_{i=0}^{N} \tilde{\mathbf{U}}_i^N L_i(x), \tag{12}$$

where $\tilde{\mathbf{U}}_i^N$ is the projection of the solution onto the Legendre orthogonal polynomial $L_i(x)$. It should be noticed that the DGSEM is a nodal method, and therefore the coefficients, $\mathbf{U}_i^N$, obtained in Eq. (9) are the values of the solution at the collocation nodes and not the modal coefficients $\tilde{\mathbf{U}}_i^N$. However, they can be computed from the nodal values as:

$$\mathcal{V}\tilde{\mathbf{U}}^N = \mathbf{U}^N, \tag{13}$$

where matrix $\mathcal{V}$ is a generalized Vandermonde matrix [15]. A particularity of spectral methods is that for smooth solutions the coefficients $\tilde{\mathbf{U}}_i$ decay very quickly (exponential convergence), while the convergence rate is poor (algebraic convergence) for non smooth solutions [6, 13].

A truncated expansion of order $N-1$ of the solution, $\mathbf{U}^N(x)$, is also constructed as:

$$\hat{\mathbf{U}}^{N-1}(x) = \sum_{i=0}^{N-1} \tilde{\mathbf{U}}_i^N L_i(x). \tag{14}$$

The difference between the truncated expansion of the solution, $\hat{\mathbf{U}}^{N-1}$, and the solution itself, $\mathbf{U}^N$, is small for smooth solutions and big for discontinuous solutions due to spectral convergence. In order to measure the difference between the two functions the following indicator is computed within each element:

$$s = \log_{10} \max \left(\frac{(\mathbf{U}^N - \hat{\mathbf{U}}^{N-1}, \mathbf{U}^N - \hat{\mathbf{U}}^{N-1})}{(\mathbf{U}^N, \mathbf{U}^N)} \right), \tag{15}$$

where $(u, v) = \int_{-1}^{1} uv dx$ represents the usual L^2 inner product and can be approximated using Legendre-Gauss quadrature. It should be noticed that the maximum value among all the equations is taken, which is justified as the objective of the indicator is to capture discontinuities in any of the equations.

Finally, ϵ, the amount of viscosity imposed in each element, is computed as:

$$\epsilon = \begin{cases} 0 & \text{if } s < s_0 - \kappa \\ \frac{\epsilon_0}{2}\left(1 + \sin\frac{\pi(s - s_0)}{2\kappa}\right) & \text{if } s_0 - \kappa \leq s \leq s_0 + \kappa \\ \epsilon_0 & \text{if } s > s_0 + \kappa \end{cases} \tag{16}$$

In this work, the values of $\epsilon_0 = \frac{h}{(N+1)}$ (being h the size of the elements), $s_0 = \log_{10}\frac{1}{(N+1)^4}$ and $\kappa = 5$ were chosen empirically and demonstrated very satisfactory results. As it is explained in [25], the selection of these values for the parameters introduces viscosity only when the solution is not continuous and the profiles of the discontinuities are sharp but smooth. The effectiveness of these parameters to correctly capture the discontinuities and stabilize the solution in the multiphase framework will be shown in Sect. 3.

The artificial viscosity method produces an *a posteriori* stabilization, i.e. the solution is not stabilized until the oscillation is generated. In general there is no problem with this, however if the amplitude of the oscillation is too high it can transiently produce unphysical values of the variables, e.g. negative densities or volume fractions outside the interval [0, 1]. This is inadmissible as the computation of some quantities, e.g. the speed of sound, would result in invalid operations. The development of a robust artificial viscosity method requires the introduction of relaxation iterations. If any of the aforementioned variables acquire an unphysical value as a result of an oscillation, a relaxation iteration is performed instead of the regular iteration. In a regular iteration, the time derivative of the solution is computed with Eq. (11), while in a relaxation iteration only the diffusive terms,

$$\frac{\partial \mathbf{U}}{\partial t} = \frac{\partial}{\partial x}\left(\epsilon \frac{\partial \mathbf{U}}{\partial x}\right), \tag{17}$$

are computed. It should be noticed that relaxation iterations do not produce an advance in physical time but only a filtering of the solution.

A comment should be made about the major drawback of the artificial viscosity approach: the reduction in the stable time step for explicit time stepping schemes [7, 19]. The scaling of the explicit time step is given by:

$$\Delta t \sim \left(S_{max}N^2/h + ||\epsilon||_{L^\infty}N^4/h^2\right)^{-1}, \tag{18}$$

where S_{max} is the absolute value of the largest characteristic velocity, ϵ is the magnitude of the viscosity, h is the size of the element and N is the approximation's polynomial degree [15]. Therefore if the maximum value of the artificial viscosity is used (see Eq. (16)), the time step is given by:

$$\Delta t \sim \left(N^2/h(S_{max} + N)\right)^{-1}, \tag{19}$$

therefore S_{max} is increased by N because of the artificial viscosity method. To overcome this limitation, several approaches are available. The cost of explicit time stepping methods can be reduced, for example, by using local time stepping [36] or adaptive time stepping [9] techniques. A different approach is to circumvent the time stability limit by using implicit methods [35]. Finally, a similar effect to artificial viscosity can be obtained by filtering the solution, thus not affecting the time stability limit [12, 15].

The derivation of the DGSEM performed in Sect. 2.1 does not include second order derivatives. However, the stabilization using artificial viscosity requires them. Several methods are available in the literature to perform the discretization of elliptic problems, see [2] for a thorough review. In this work the Symmetric Interior Penalty Discontinuous Galerkin [10] has been chosen to discretize the second order derivatives.

3 Numerical Experiments

In this section, our aim is to test the developed method. Several shock tube problems are used to test the capturing properties of the scheme in the presence of discontinuities. We consider seven test problems which are classical benchmark, see for instance [32]. The initial data consists of two constant states separated by a discontinuity located at $x = x_0$, all the parameters are listed in Table 1. Transmissive boundary conditions are imposed at $x = 0$ and $x = 1$.

In test 1 the liquid phase wave pattern consists of a left rarefaction, a right shock wave and a right traveling liquid contact, while the gas phase consists of a left rarefaction, a contact and a right shock wave. The equations of state for both phases are assumed ideal, with $\gamma_g = \gamma_l = 1.4$. Test 2 is more demanding than test 1 as it includes large variations of initial data and non-ideal equation of state. In test 3 the solution, for both phases, consists of a right shock wave, a right traveling contact discontinuity and a left sonic rarefaction wave. The correct resolution of the sonic point is very important in assessing the entropy satisfaction property of the numerical scheme. In test 4 both solid and gas phases consist of a two symmetric rarefaction waves and a trivial stationary contact wave. The region between the rarefaction waves is close to vacuum, therefore this test case is useful to assess the pressure positivity in different numerical methods. Test 5 was designed to assess the ability of numerical methods to resolve the stationary isolated contact waves. The exact solution allows the existence of the stationary contact waves in the solid and gaseous phases when the volume fraction and solid pressure gradients are present across the solid contact. The solution of this test problem contains isolated contacts in both solid and gas phases.

A comparison is made between a first order discretization (which corresponds to a classical Finite Volume approach) and a sixth order discretization, both with 100 elements. A solution obtained with a first order full non linear Riemann solver [30] on a mesh consisting of 2000 elements is shown for comparison. Results are

Table 1 One-dimensional shock tubes. EOS parameters, initial discontinuity position and initial data for liquid and gas phases

(a) EOS parameters and initial discontinuity position

	Test 1	Test 2	Test 3	Test 4	Test 5
γ_l	1.4	3.0	1.4	1.4	3.0
γ_g	1.4	1.35	1.4	1.4	1.4
π_l	0.0	3400.0	0.0	0.0	10.0
π_g	0.0	0.0	0.0	0.0	0.0
x_0	0.5	0.5	0.5	0.5	0.5

(b) Liquid phase

Test	α_{L_l}	ρ_{L_l}	u_{L_l}	p_{L_l}	α_{R_l}	ρ_{R_l}	u_{R_l}	p_{R_l}
1	0.8	1.0	0.0	1.0	0.3	1.0	0.0	1.0
2	0.2	1900.0	0.0	10.0	0.9	1950.0	0.0	1000.0
3	0.8	1.0	0.75	1.0	0.3	0.125	0.0	0.1
4	0.8	1.0	−2.0	0.4	0.5	1.0	2.0	0.4
5	0.6	1.4	0.0	2.0	0.3	1.0	0.0	3.0

(c) Gas phase

Test	α_{L_g}	ρ_{L_g}	u_{L_g}	p_{L_g}	α_{R_g}	ρ_{R_g}	u_{R_g}	p_{R_g}
1	0.2	0.2	0.0	0.3	0.7	1.0	0.0	1.0
2	0.8	2.0	0.0	3.0	0.1	1.0	0.0	1.0
3	0.2	1.0	0.75	1.0	0.7	0.125	0.0	0.1
4	0.2	1.0	−2.0	0.4	0.5	1.0	2.0	0.4
5	0.4	1.4	0.0	1.0	0.7	1.0	0.0	1.0

shown in Fig. 1 where the mixture density is displayed ($\rho_m = \alpha_l \rho_l + \alpha_g \rho_g$). Final time has been set to $t = 0.15$ for tests 1 to 5. When the spatial discretization uses a first order representation of the solution, the Rusanov flux, although robust, does not give satisfactory results in the sense that it dissipates too much discontinuities. On the contrary, when the polynomial degree $N = 5$ is used, the solution is almost indistinguishable from the reference solution. It is remarkable how the artificial viscosity approach, detailed in Sect. 2.3, achieves to impose a very controlled amount of viscosity, keeping very sharp fronts and almost no oscillations.

4 Conclusions

In this work a discontinuous Galerkin discretization of the Baer-Nunziato equations that takes the DGSEM as a basis was introduced. The condition of Abgrall was used to extend the Rusanov flux to high order and to treat the non-conservative products. A stabilization technique based on local artificial viscosity was adapted

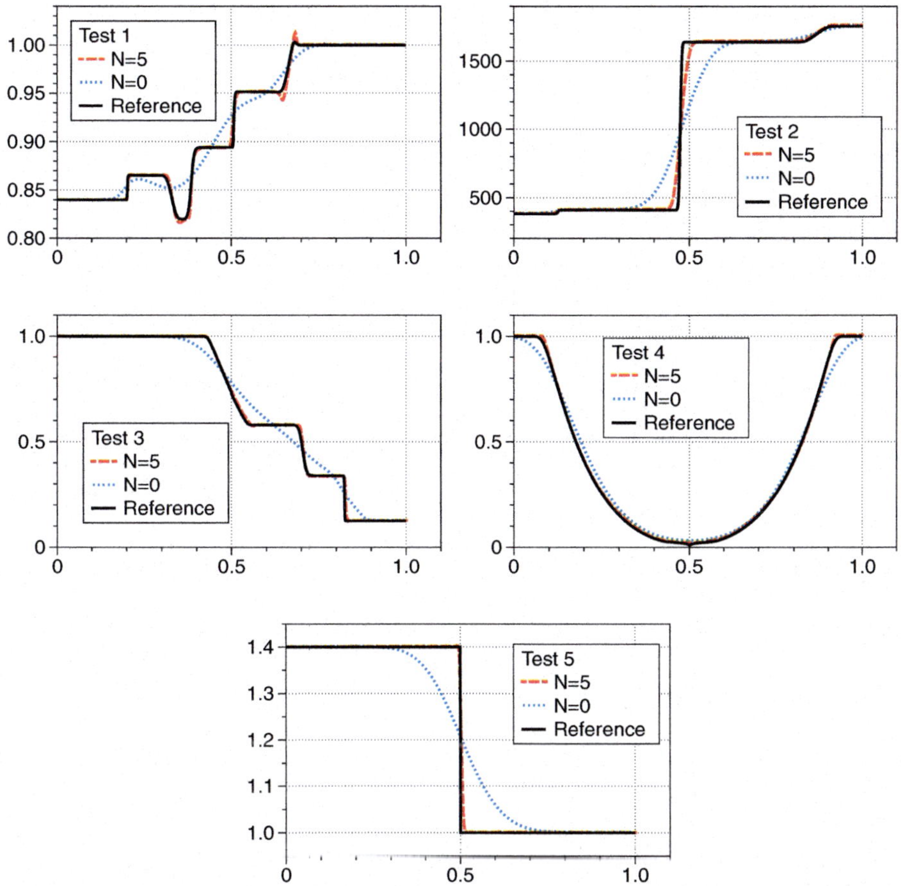

Fig. 1 Shock tube problems. Comparison between first order Rusanov with 100 elements, sixth order Rusanov with 100 elements and full non linear Riemann solver with 2000 elements. Density mixture

to the Baer-Nunziato equations to deal with the inherent oscillations caused by high-order discretizations in the vicinity of discontinuities. This approach allowed to smooth the discontinuities in a very thin region and thus resolve them in the space of polynomials. The numerical experiments showed that the proposed discretization allows very high-order solutions in the presence of discontinuities. It was also shown that the accuracy of these solutions is comparable to the ones obtained with a full non linear Riemann solver with more than three times the number of degrees of freedom of the high-order counterpart.

Acknowledgements This work has been partially supported by REPSOL under the research grant P130120150 monitored by Dr. Angel Rivero. This work has been partially supported by Ministerio de Economía y Competitividad (Spain) under the research grant TRA2015-67679-C2-2-R. The authors would like to thank the anonymous reviewers for their comments and suggestions which greatly improved this work.

References

1. R. Abgrall, How to prevent pressure oscillations in multicomponent flow calculations: a quasi conservative approach. J. Comput. Phys. **125**, 150–160 (1996)
2. D.N. Arnold, F. Brezzi, B. Cockburn, L.D. Marini, Unified analysis of discontinuous Galerkin methods for elliptic problems. SIAM J. Numer. Anal. **39**(5), 1749–1779 (2001)
3. M. Baer, J. Nunziato, A two-phase mixture theory for the deflagration to detonation transition (DDT) in reactive granular materials. Int. J. Multiphase Flow **12**, 861–889 (1986)
4. B. Baldwin, R. MacCormack, Interaction of strong shock wave with turbulent boundary layer, in *Proceedings of the Fourth International Conference on Numerical Methods in Fluid Dynamics* (Springer, Berlin, 1975), pp. 51–56
5. K. Black, Spectral element approximation of convection–diffusion type problems. Appl. Numer. Math. **33**(1–4), 373–379 (2000)
6. C. Canuto, M.Y. Hussaini, A. Quarteroni, T.A. Zang, Spectral methods, in *Fundamentals in Single Domains* (Springer, Berlin, 2006)
7. A. Chaudhuri, G. Jacobs, W. Don, H. Abbassi, F. Mashayek, Explicit discontinuous spectral element method with entropy generation based artificial viscosity for shocked viscous flows. J. Comput. Phys. **332**, 99–117 (2017)
8. B. Cockburn, G.E. Karniadakis, C.W. Shu, *The Development of Discontinuous Galerkin Methods* (Springer, Berlin, 2000)
9. J.R. Dormand, P.J. Prince, A family of embedded Runge-Kutta formulae. J. Comput. Appl. Math. **6**(1), 19–26 (1980)
10. J. Douglas, T. Dupont, Interior penalty procedures for elliptic and parabolic Galerkin methods, in *Computing Methods in Applied Sciences* (Springer, Berlin, 1976), pp. 207–216
11. F. Fraysse, C. Redondo, G. Rubio, E. Valero, Upwind methods for the Baer–Nunziato equations and higher-order reconstruction using artificial viscosity. J. Comput. Phys. **326**, 805–827 (2016)
12. D. Gottlieb, J.S. Hesthaven, Spectral methods for hyperbolic problems. J. Comput. Appl. Math. **128**(1), 83–131 (2001)
13. D. Gottlieb, S.A. Orszag, *Numerical Analysis of Spectral Methods: Theory and Applications* (SIAM, Philadelphia, 1977)
14. D. Gottlieb, C.W. Shu, On the Gibbs phenomenon and its resolution. SIAM Rev. **39**(4), 644–668 (1997)
15. J.S. Hesthaven, T. Warburton, *Nodal Discontinuous Galerkin Methods: Algorithms, Analysis, and Applications* (Springer Science and Business Media, New York, 2008)
16. T.J. Hughes, L. Franca, M. Mallet, A new finite element formulation for computational fluid dynamics: I. Symmetric forms of the compressible Euler and Navier-Stokes equations and the second law of thermodynamics. Comput. Methods Appl. Mech. Eng. **54**(2), 223–234 (1986)
17. A. Jameson, W. Schmidt, E. Turkel, Numerical solution of the Euler equations by finite volume methods using Runge Kutta time stepping schemes, in *14th fluid and Plasma Dynamics Conference* (1981), p. 1259
18. R.M. Kirby, S.J. Sherwin, Stabilisation of spectral/hp element methods through spectral vanishing viscosity: application to fluid mechanics modelling. Comput. Methods Appl. Mech. Eng. **195**(23), 3128–3144 (2006)

19. A. Klöckner, T. Warburton, J.S. Hesthaven, Viscous shock capturing in a time-explicit discontinuous Galerkin method. Math. Model. Nat. Phenom. **6**(3), 57–83 (2011)
20. M. Kompenhans, G. Rubio, E. Ferrer, E. Valero, Adaptation strategies for high order discontinuous Galerkin methods based on Tau-estimation. J. Comput. Phys. **306**, 216–236 (2016)
21. M. Kompenhans, G. Rubio, E. Ferrer, E. Valero, Comparisons of p-adaptation strategies based on truncation-and discretisation-errors for high order discontinuous Galerkin methods. Comput. Fluids **139**, 36–46 (2016)
22. D.A. Kopriva, *Implementing Spectral Methods for Partial Differential Equations: Algorithms for Scientists and Engineers* (Springer, Berlin, 2009)
23. P.D. Lax, Weak solutions of nonlinear hyperbolic equations and their numerical computation. Commun. Pure Appl. Math. **7**(1), 159–193 (1954)
24. C. Parès, Numerical methods for nonconservative hyperbolic systems: a theoretical framework. SIAM J. Numer. Anal. **44**, 300–321 (2006)
25. P.O. Persson, J. Peraire, Sub-cell shock capturing for discontinuous Galerkin methods, in *Proceedings of the 44th AIAA Aerospace Sciences Meeting and Exhibit*, AIAA-2006-112 (2006)
26. P. Rasetarinera, M.Y. Hussaini, An efficient implicit discontinuous spectral Galerkin method. J. Comput. Phys. **172**(2), 718–738 (2001)
27. W.H. Reed, T.R. Hill, Triangular mesh methods for the neutron transport equation. Los Alamos Report LA-UR-73-479 (1973)
28. V.V. Rusanov, The calculation of the interaction of non-stationary shock waves and obstacles. USSR Comput. Math. Math. Phys. **1**(2), 304–320 (1961)
29. R. Saurel, R. Abgrall, A multiphase Godunov method for compressible multifluid and multiphase flows. J. Comput. Phys. **150**(2), 425–467 (1999)
30. D. Schwendeman, C. Wahle, A. Kapila, The Riemann problem and a high-resolution Godunov method for a model of compressible two-phase flow. J. Comput. Phys. **212**(2), 490–526 (2006)
31. E. Tadmor, Convergence of spectral methods for nonlinear conservation laws. SIAM J. Numer. Anal. **26**(1), 30–44 (1989)
32. S. Tokareva, E. Toro, HLLC-type Riemann solver for the Baer–Nunziato equations of compressible two-phase flow. J. Comput. Phys. **229**(10), 3573–3604 (2010)
33. E.F. Toro, *Riemann Solvers and Numerical Methods for Fluid Dynamics. A Practical Introduction* (Springer, Berlin, 2009)
34. J. Von Neumann, R.D. Richtmyer, A method for the numerical calculation of hydrodynamic shocks. J. Appl. Phys. **21**(3), 232–237 (1950)
35. Z. Wang, High-order methods for the Euler and Navier–Stokes equations on unstructured grids. Prog. Aerosp. Sci. **43**(1), 1–41 (2007)
36. A.R. Winters, D.A. Kopriva, High-order local time stepping on moving DG spectral element meshes. J. Sci. Comput. **58**(1), 176–202 (2014)

Summation-by-Parts and Correction Procedure via Reconstruction

Hendrik Ranocha, Philipp Öffner, and Thomas Sonar

Abstract The correction procedure via reconstruction (CPR, also known as flux reconstruction), is a framework of high order methods for conservation laws, unifying some discontinuous Galerkin, spectral difference and spectral volume methods. These methods are embedded in the framework of summation-by-parts (SBP) operators with simultaneous approximation terms (SATs), recovering the linearly stable methods of Vincent et al. (J Comput Phys 230(22): 8134–8154, 2011; J Sci Comput 47(1):50–72, 2011; Comput Methods Appl Mech Eng 296:248–272, 2015). The introduction of new correction terms enables stability for Burgers' equation using nodal bases not including boundary nodes, i.e. Gauss nodes. Extended notions of SBP operators and split-forms are used to obtain stability.

1 Introduction

The correction procedure via reconstruction (CPR), also known as flux reconstruction (FR), has been introduced by Huynh [9] as a framework of high order methods for conservation laws, unifying some discontinuous Galerkin (DG), spectral difference, and spectral volume methods with appropriate choice of parameters. It is a polynomial collocation framework and resembles strong form DG methods. There are several results about linear stability in a semidiscrete setting [10, 23–25] and the framework has been implemented in the high performance industry targeting open source framework PyFR [27]. However, there are far less results about nonlinear stability [11].

Summation-by-parts operators originate in the framework of finite difference (FD) operators, as described inter alia in the review articles [4, 13, 20], and have been used to obtain rigorous results about linear stability in bounded domains, mimicking analytical proofs in the continuous PDE setting using integration by parts to obtain energy (L^2) stability. This framework has recently been applied to nodal DG methods [6] and general nodal bases with appropriate quadrature strength [3].

H. Ranocha (✉) • P. Öffner • T. Sonar
Institute Computational Mathematics, TU Braunschweig, Braunschweig, Germany
e-mail: h.ranocha@tu-bs.de; p.oeffner@tu-bs.de; t.sonar@tu-bs.de

M.L. Bittencourt et al. (eds.), *Spectral and High Order Methods for Partial Differential Equations ICOSAHOM 2016*, Lecture Notes in Computational Science and Engineering 119, DOI 10.1007/978-3-319-65870-4_45

Here, the concept of SBP operators is applied to CPR methods, resulting in a reformulation enabling stability results in a semidiscrete setting with norms adapted to the discrete inner product by quadrature. The family of energy stable CPR methods [24, 25] can be recovered and nonlinear stability for Burgers' equation can be obtained by a skew-symmetric form [18]. Additionally, the concept of SBP operators is generalised to nodal bases not including boundary nodes and modal bases [19]. Finally, artificial dissipation and modal filtering are formulated in this framework, enhancing stability of these methods, yielding fully discrete stability for Euler's method [7, 17].

2 Correction Procedure via Reconstruction

The correction procedure via reconstruction (CPR) is a polynomial collocation method. For a scalar conservation law

$$\partial_t u + \partial_x f(u) = 0 \tag{1}$$

in one space dimension, equipped with periodic boundary conditions or compactly supported initial data, the method can be formulated as follows.

The domain Ω is partitioned into non-overlapping intervals Ω_i, with $\bigcup_i \overline{\Omega_i} = \Omega$. On each element Ω_i, the solution u^i is approximated by a polynomial of degree $\leq p \in \mathbb{N}_0$ in a nodal basis, i.e. the values $\underline{u}_j^i$ at certain points $x_0^i, \dots, x_p^i \in \Omega_i$ are used. In this collocation framework, the flux $f(u^i)$ is computed pointwise at the nodes x_j^i and interpreted as a polynomial of degree $\leq p$, too, yielding $\underline{f}_j^i = f(\underline{u}_j^i) = f(u^i(x_j^i))$.

The divergence of the flux can thus be calculated as the discrete derivative $\underline{\underline{D}}\,\underline{u}$ of a polynomial, where $\underline{\underline{D}}$ is the discrete derivative matrix and $\underline{u} = (\underline{u}_0, \dots, \underline{u}_p)^t$ the representation in the nodal basis. However, to incorporate the coupling with the neighbouring elements, the flux $\underline{f} = (\underline{f}_0, \dots, \underline{f}_p)^t$ is interpolated to the left and right boundary, yielding the values f_L, f_R. At each boundary, the two adjacent cells give boundary values u_-, u_+ of the solutions by interpolation, and a common numerical flux $f^{\text{num}}(u_-, u_+)$ is computed. To enforce this corrected flux at the boundaries, left and right correction functions g_L, g_R are used, approximating zero in each element g_L vanishes at the right boundary and has the value 1 at the left boundary, and g_R is obtained as reflection of g_L at the cell centre. These correction functions are polynomials of degree $\leq p+1$, i.e. of one degree higher than the numerical solution u. Finally, the semidiscrete approximation of (1) in each element is obtained as

$$\partial_t \underline{u} + \underline{\underline{D}}\,\underline{f} + (f_L^{\text{num}} - f_L)\underline{g}'_L + (f_R^{\text{num}} - f_R)\underline{g}'_R = 0, \tag{2}$$

where $g'_{L/R}$ is the derivative of the correction function $g_{L/R}$ and $\underline{g}'_{L/R}$ the corresponding representation in the nodal basis. In order to simplify the implementation, each cell is mapped to a standard / reference element $[-1, 1] \subset \mathbb{R}$ and the computation is performed there.

3 Generalised Summation-by-Parts Operators

A general analytical setting of summation-by-parts (SBP) operators is given by the following ingredients, as described in [19].

Functions on the volume (interval) Ω are approximated by vectors in a finite-dimensional (real) Hilbert space X_V with basis $\mathscr{B}_V$. The mass matrix $\underline{\underline{M}}$ is symmetric positive definite and induces the scalar product on X_V, approximating the L^2 scalar product

$$\underline{u}^T \underline{\underline{M}}\, \underline{v} = \langle \underline{u}, \underline{v} \rangle_M \approx \int_\Omega u\, v = \langle u, v \rangle_{L^2} . \tag{3}$$

The derivative is represented in the basis $\mathscr{B}_V$ by the matrix $\underline{\underline{D}}$.

Additionally, there is a finite-dimensional (real) Hilbert space X_B with basis $\mathscr{B}_B$, representing functions on boundary $\partial\Omega$. In one space dimension, this Hilbert space is two-dimensional and a basis is given by the values at both boundary nodes $\{-1, 1\}$ in the standard element. On this Hilbert space, there is a bilinear form, approximating integration with respect to the outer normal as in the divergence theorem, i.e. in one space dimension

$$\underline{u}_B^T \underline{\underline{B}}\, \underline{f}_B = B(u_B, f_B) = u_B f_B \Big|_{-1}^{1}, \qquad \underline{\underline{B}} = \operatorname{diag}(-1, 1) . \tag{4}$$

These Hilbert spaces are coupled via a restriction operator $\underline{\underline{R}}$, performing interpolation of functions on the volume to the boundary. In this setting, the SBP property reads

$$\underline{\underline{M}}\, \underline{\underline{D}} + \underline{\underline{D}}^T \underline{\underline{M}} = \underline{\underline{R}}^T \underline{\underline{B}}\, \underline{\underline{R}}, \tag{5}$$

mimicking integration by parts on a discrete level

$$\int_\Omega u\, (\partial_x v) + \int_\Omega (\partial_x u)\, v = u\, v \Big|_{\partial\Omega} . \tag{6}$$

For polynomials u, v of degree $\le p$, the product uv is in general a polynomial of degree $\le 2p$. Thus, the discrete linear operator describing multiplication with a function u on the volume is denoted by $\underline{\underline{u}}^+$ and maps in general to a bigger Hilbert space $X_v^+ \supset X_V$. Therefore, some projection has to be used to get a multiplication operator $\underline{\underline{u}}$ mapping X_V to X_V. For a nodal basis, this projection is given by pointwise evaluation, i.e. collocation, whereas for a modal Legendre basis, this projection will be the exact L^2 projection. Thus, the operator $\underline{\underline{u}}$ on X_V performing multiplication with u in the collocation approach for a nodal basis is given by $\underline{\underline{u}} = \operatorname{diag}\left(\underline{u}_0, \dots, \underline{u}_p\right)$.

This framework can also be extended to multiple dimensions, not relying on, but including tensor product formulations [15], similarly to the numerical setting of [8].

4 Correction Procedure via Reconstruction Using Summation-by-Parts Operators

The semidiscretisation (2) can be reformulated as

$$\partial_t \underline{u} + \underline{\underline{D}}\,\underline{f} + \underline{\underline{C}} \left(\underline{f}^{\mathrm{num}} - \underline{\underline{R}}\,\underline{f} \right) = 0, \tag{7}$$

where the correction matrix $\underline{\underline{C}} = \left(\underline{g}_L', \underline{g}_R' \right)$ contains the derivatives of the correction functions as columns, and $\underline{f}^{\mathrm{num}} = \left(f_L^{\mathrm{num}}, f_R^{\mathrm{num}} \right)^T$, $\underline{\underline{R}}\,\underline{f} = \left(f_L, f_R \right)^T$. Then, due to the SBP property (5), one gets

Lemma 1 (Lemma 1 in [18]) *If* $\underline{1}^T \underline{\underline{M}}\,\underline{\underline{C}} = \underline{1}^T \underline{\underline{R}}^T \underline{\underline{B}}$, *then the semidiscretisation* (7) *is conservative across elements.*

Proof Denoting the constant function $x \mapsto 1$ by $\underline{1}$, in each element

$$\begin{aligned} \tfrac{\mathrm{d}}{\mathrm{d}t} \int u =& \underline{1}^T \underline{\underline{M}}\, \partial_t \underline{u} = -\underline{1}^T \underline{\underline{M}}\,\underline{\underline{D}}\,\underline{f} - \underline{1}^T \underline{\underline{R}}^T \underline{\underline{C}} \left(\underline{f}^{\mathrm{num}} - \underline{\underline{R}}\,\underline{f} \right) \\ =& -\underline{1}^T \underline{\underline{R}}^T \underline{\underline{B}}\,\underline{\underline{R}}\,\underline{f} + \underline{1}^T \underline{\underline{D}}^T \underline{\underline{M}}\,\underline{f} - \underline{1}^T \underline{\underline{R}}^T \underline{\underline{B}} \left(\underline{f}^{\mathrm{num}} - \underline{\underline{R}}\,\underline{f} \right) = -\underline{1}^T \underline{\underline{R}}^T \underline{\underline{B}}\,\underline{f}^{\mathrm{num}}, \end{aligned} \tag{8}$$

where the SBP property (5), the assumption, and exact differentiation of constant function $\underline{\underline{D}}\,\underline{1} = 0$ have been used. Thus, summing the contributions of all elements and using periodic boundary conditions, the terms sum up to zero and the method is conservative.

5 Linear Stability

Since linear problems with constant coefficients

$$\partial_t u + \partial_x u = 0 \tag{9}$$

are usually investigated regarding stability in L^2, it is natural to look for semidiscrete stability in the discrete norm induced by the mass matrix $\underline{\underline{M}}$. Using the SBP property (5), proofs can be transferred from the continuous level of PDE analysis to the semidiscrete level.

As proposed by Jameson [10] in the context of a spectral difference method, stability can also be obtained in some other norm, since all norms in finite dimensional spaces are equivalent. Using this idea, Vincent et al. discovered a family of linearly stable CPR methods [24, 25]. Their results have been transferred to the new formulation of CPR methods in the SBP framework

Lemma 2 (Lemma 2 of [18], see also Theorem 1 of [25]) *If the semidiscretisation*

$$\partial_t \underline{u} + \underline{\underline{D}}\,\underline{u} + \underline{\underline{C}}\left(\underline{f}^{\mathrm{num}} - \underline{\underline{R}}\,\underline{u}\right) = 0 \tag{10}$$

of (9) *is used with* $\underline{\underline{C}} = \left(\underline{\underline{M}} + \underline{\underline{K}}\right)^{-1} \underline{\underline{R}}^T \underline{\underline{B}}$, *where* $\underline{\underline{M}} + \underline{\underline{K}}$ *is positive definite and* $\underline{\underline{M}}\,\underline{\underline{K}}$ *is antisymmetric, then the SBP CPR method is linearly stably in the discrete norm* $\|\cdot\|_{M+K}$ *induced by* $\underline{\underline{M}} + \underline{\underline{K}}$, *if an adequate numerical flux* f^{num} *is chosen.*

In [18], the authors provided a translation of the one-parameter family of Vincent et al. [24] to a corresponding one-parameter family in the new setting using discrete norms. Additionally, the multi-parameter family of Vincent et al. [25] can be translated similarly.

The discrete norm $\|\cdot\|_{M+K}$ can be interpreted as some kind of Sobolev norm. However, this equivalence of norms has to be used very carefully. Looking at convergence results or stability under mesh refinement, the dimension and therefore the constants involved in the equivalence of norms may blow up. Additionally, in the spirit of SBP operators, stability results of numerical methods should mimic corresponding well-posedness results in the continuous setting. For a linear scalar conservation law with constant coefficients (9) and periodic boundary conditions, the initial data is simply transported as it is. Therefore, both the L^2 norm and Sobolev norms of u remain constant. However, this is not valid for nonlinear conservation laws anymore. Thus, it is recommended to use the discrete norms approximating the continuous norms used to obtain well-posedness of the PDE.

Numerical calculations of the linear advection equation (9) using $N = 4$ uniform elements with polynomials of degree $\leq p = 7$ to evolve the initial condition $u_0(x) = \exp(-20x^2)$ in the domain $[-1, 1]$ have been conducted using the classical fourth order Runge-Kutta method with $5,000$ time steps in the interval $[0, 20]$. As numerical flux, a central flux $f^{\mathrm{num}}(u_-, u_+) = \frac{u_- + u_+}{2}$ has been chosen, yielding $\frac{\mathrm{d}}{\mathrm{d}t}\|u\|^2 = 0$ in the semidiscrete setting on a periodic domain.

The numerical solutions at $t = 20$ using Gauss and Lobatto nodes with associated canonical correction matrices $\underline{\underline{C}} = \underline{\underline{M}}^{-1}\underline{\underline{R}}^T\underline{\underline{B}}$, corresponding to the parameters $c = c_0 = 0$ (Gauss) and $c = c_{Hu}$ (Lobatto) in [24], are plotted in Fig. 1.

Very interesting regarding the stability result above is the plot of the energy $\|u\|^2$ of the solution in Fig. 2, computed via Gauss and Lobatto quadrature using the same number of nodes $p = 7$. The energy computed via Gauss nodes remains constant for the solution computed with the canonical correction matrix corresponding to Gauss nodes, whereas the energy computed via Lobatto nodes remains constant (due to equivalence of norms) but oscillatory, and vice versa. This can be explained by

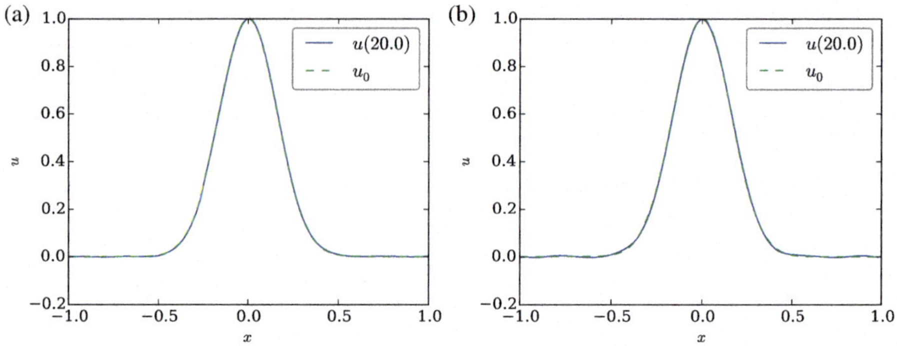

Fig. 1 The numerical solutions of constant velocity linear advection at $t = 20$. (**a**) $c = c_0$, Gauss nodes. (**b**) $c = c_{Hu}$, Lobatto nodes

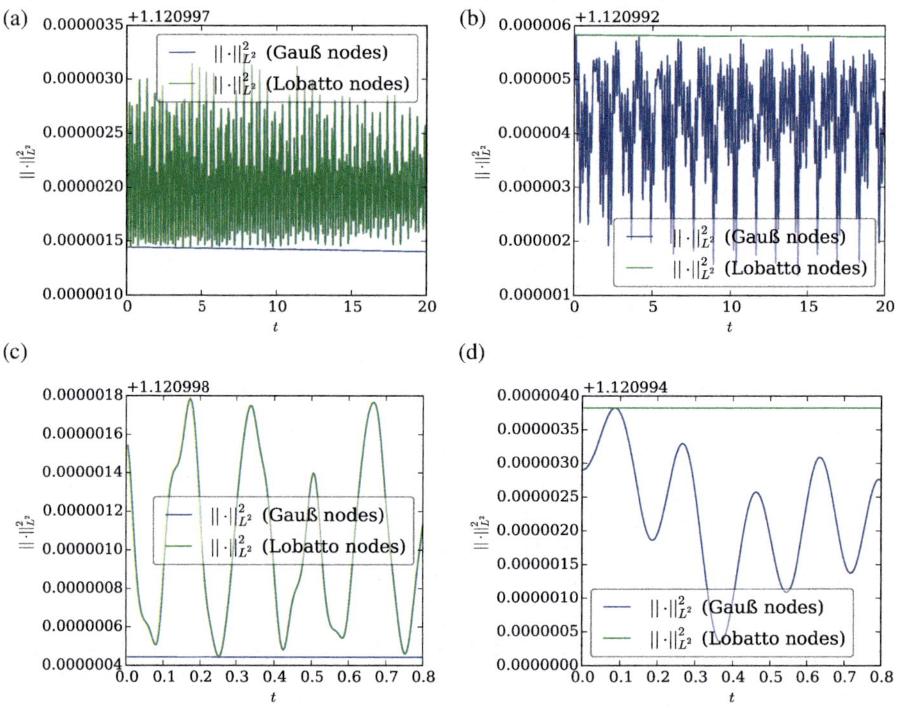

Fig. 2 Energies of the numerical solutions of constant velocity linear advection. The Fig. (**c**) and (**d**) provide zoomed in views of the Figs. (**a**) and (**b**), respectively. (**a**) $c = c_0$, Gauss nodes. (**b**) $c = c_{Hu}$, Lobatto nodes. (**c**) $c = c_0$, Gauss nodes. (**d**) $c = c_{Hu}$, Lobatto nodes

Lemma 2, since the strong stability statement $\frac{\mathrm{d}}{\mathrm{d}t}\|u\|^2_{M+K} \le 0$ for the solution using periodic boundary conditions does only hold for the norm induced by $\underline{\underline{M}} + \underline{\underline{K}}$. For $c = c_0$, this corresponds to the correct L^2 norm computed via Gauss quadrature, whereas $c = c_{Hu}$ corresponds to Lobatto quadrature.

6 Nonlinear Stability for Burgers' Equation

Skew-symmetric forms have been known for a long time to yield (entropy) stability results for conservation laws [14, 21]. Additionally, Fisher et al. [5] provided the justification to use these skew-symmetric forms in combination with diagonal norm SBP operators with respect to the Lax-Wendroff theorem.

The nonlinear flux of Burgers' equation

$$\partial_t u + \partial_x \frac{u^2}{2} = 0 \tag{11}$$

does not allow the simple cancellation of terms used to prove L^2 stability for the advection equation with constant coefficients (9). However, using the split form

$$\partial_t u + \frac{1}{3}\partial_x u^2 + \frac{1}{3} u \partial_x u = 0, \tag{12}$$

integration by parts can be used to obtain L^2 stability $\frac{\mathrm{d}}{\mathrm{d}t}\|u\|^2 \leq 0$ for appropriately chosen boundary conditions. Similarly, using SBP operators on a nodal basis including boundary nodes, the splitting

$$\partial_t \underline{u} + \frac{1}{3}\underline{\underline{D}}\,\underline{u}^2 + \frac{1}{3}\underline{\underline{u}}\,\underline{\underline{D}}\,\underline{u} + \underline{\underline{M}}^{-1}\underline{\underline{R}}^T\underline{\underline{B}}\left(\underline{f}^{\mathrm{num}} - \frac{1}{2}\underline{\underline{R}}\,\underline{u}^2\right) \tag{13}$$

yields a conservative (across elements) and stable (in the discrete norm $\|\cdot\|_M$) method, if an adequate numerical flux f^{num} and appropriate boundary conditions are chosen, see inter alia [5, 6, 18].

This kind of split form has been seen often as some kind of correction of the product rule $\partial_x(uv) = (\partial_x u)v + u(\partial_x v)$, that is invalid for weak solutions and in the discrete setting. However, it should be emphasised that it is *multiplication* which is invalid in the discrete setting, not only the product rule. Using this idea, the authors [19] extended the split form also to boundary terms, and to a more general correction of the volume terms, resulting in

Theorem 1 (Theorem 2 of [19]) *For a general SBP basis, the semidiscretisation*

$$\partial_t \underline{u} + \frac{1}{3}\underline{\underline{D}}\,\underline{u}^2 + \frac{1}{3}\underline{\underline{u}}^*\underline{\underline{D}}\,\underline{u} + \underline{\underline{M}}^{-1}\underline{\underline{R}}^T\underline{\underline{B}}\left(\underline{f}^{\mathrm{num}} - \frac{1}{3}\underline{\underline{R}}\,\underline{\underline{u}}\,\underline{u} - \frac{1}{6}\left(\underline{\underline{R}}\,\underline{u}\right)^2\right) = 0 \tag{14}$$

is conservative. Moreover, it is stable in the discrete norm $\|\cdot\|_M$ induced by $\underline{\underline{M}}$, if an appropriate numerical flux fulfilling the entropy stability condition of Tadmor [22]

$$(u_+ - u_-) f^{\mathrm{num}}(u_-, u_+) - \frac{1}{6}\left(u_+^3 - u_-^3\right) \leq 0 \tag{15}$$

is chosen, e.g. an entropy conservative flux, a local Lax-Friedrichs flux, or Osher's flux.

Here, the $\underline{\underline{M}}$-adjoint $\underline{\underline{u}}^*$ of $\underline{\underline{u}}$ is given by $\underline{\underline{u}}^* = \underline{\underline{M}}^{-1}\underline{\underline{u}}^T\underline{\underline{M}}$, and the correction for the volume term can be justified by the fact, that the multiplication operator $\underline{\underline{u}}$, representing multiplication with a real function u, should be self-adjoint, at least if an appropriate domain is chosen, which is trivial in the finite dimensional case.

These reformulations allow many nodal basis without diagonal norm, e.g. Chebyshev bases, as well as modal bases. For a diagonal norm nodal basis, the multiplication operators are self-adjoint, since $\underline{\underline{M}}, \underline{\underline{u}}$ are diagonal and commute. Additionally, the multiplication operators in a modal Legendre basis using exact L^2 projection for the multiplication are self-adjoint, since the Legendre polynomials are orthogonal, i.e. for three polynomials u, v, w of degree $\leq p$,

$$\int \operatorname{proj}(uv)w = \int (uv)w = \int v(uw) = \int v \operatorname{proj}(uw). \tag{16}$$

As another argument not to rely on the equivalence of norms in finite dimensional spaces and the use of another correction matrix than $\underline{\underline{C}} = \underline{\underline{M}}^{-1}\underline{\underline{R}}^T\underline{\underline{B}}$ as described in Sect. 5, the authors have not been able to get nonlinear stability results for Burgers' equation using norms different from $\|\cdot\|_M$.

7 Enhancing Stability

Artificial viscosity has long been known as a means to enhance the stability of numerical methods [26], inspired by corresponding results in the continuous setting of PDE analysis.

Inspired by the framework of spectral viscosity, the conservation law (1) is enhanced by a right hand side

$$\partial_t u(t,x) + \partial_x f\left(u(t,x)\right) = (-1)^{s+1}\epsilon(\partial_x a(x)\partial_x)^s u(t,x), \tag{17}$$

where s is the order, ϵ the strength, a a suitable function, and $(\partial_x a(x)\partial_x)^s$ the s-th power of the linear operator mapping $u \mapsto \partial_x (a\partial_x u)$.

Similarly to results in the context of FD methods [12], it is very important how the discretisation of the right hand side in (17) is performed. If a basis not including boundary values is used, conservation across elements and stability might be lost, if a naive discretisation is used.

Lemma 3 (Lemma 1 of [17]) *Augmenting a conservative and stable SBP semidiscretisation of the scalar conservation law* (1) *with the right hand side*

$$-\epsilon\left(\underline{\underline{M}}^{-1}\underline{\underline{D}}^T\underline{\underline{M}}\,\underline{\underline{a}}\,\underline{\underline{D}}\right)^s \underline{u} \tag{18}$$

where $a \geq 0$ is a polynomial on the element Ω fulfilling $a\big|_{\partial\Omega} = 0$, results in a stable and conservative semidiscretisation, if a nodal basis with diagonal mass matrix or a modal basis with exact L^2 norm and projection for multiplication is used.

The eigenvalues of the discrete operator on the right hand side (18) depend on the quadrature strength of the associated mass matrix $\underline{\underline{M}}$ and mimic the ones of the continuous operator in the Sturm-Liouville problem

$$\partial_x \left((1 - x^2) \partial_x \phi_n(x) \right) = -n(n+1)\phi_n(x), \tag{19}$$

where $a(x) = 1 - x^2$ has been chosen in the reference element $[-1, 1]$, and ϕ_n is the n-th Legendre polynomial [17].

In a fully discrete setting, using a forward Euler method, the new value after one time step Δt is $\underline{u}^+ = \underline{u} + \Delta t\, \partial_t \underline{u}$. Thus, the fully discrete stability estimate becomes

$$\|\underline{u}^+\|_M^2 = \|\underline{u}\|_M^2 + 2\Delta t \left\langle \underline{u}, \partial_t \underline{u} \right\rangle + (\Delta t)^2 \|\partial_t \underline{u}\|_M^2, \tag{20}$$

where the second term on the right hand side has been estimated for a stable semidiscretisation, but the last term is non negative and might destroy the desired stability. Thus, the authors of [17] proposed a simple estimate of the artificial dissipation strength ϵ, needed to dissipate this undesired influence of the time discretisation, see Lemma 3 in [17].

Similarly, the same authors used a classical operator splitting approach to convert the artificial dissipation to modal filtering [7]. A similar estimate for fully discrete stability using Euler's method has been obtained, see Lemmas 2 and 3 of [7].

Additionally, they compared several possibilities of modal filtering:

1. Use modal filtering in an operator splitting approach, i.e. compute $\underline{u}^+ = \underline{\underline{F}}\, \underline{\tilde{u}}^+$, where $\underline{\tilde{u}}^+$ has been obtained by the Euler method applied to a stable and conservative semidiscretisation.
2. Filter the time derivative $\partial_t \underline{u}$ in Euler's method.
3. Filter the solution $\underline{u}$ used to compute the time derivative in Euler's method.

The first possibility is similar to the artificial dissipation approach as described above, conservative, stable, and recommended. The second approach corresponds to the use of another norm used to prove stability in the CPR framework, similarly to results in [1, 2]. However, as described in Sect. 5, this equivalence of norms in finite dimensional spaces may not be adequate for all problems. Finally, the third possibility can be seen as a combination of the other two, followed by an application of an inverse filter $\underline{\underline{F}}^{-1}$ (if existing), and is thus not recommended.

8 Summary and Outlook

The formulation of CPR schemes in the framework of SBP operators and SATs has been presented following [18]. The linearly stable schemes of [24, 25] can be recovered in this formulation and nonlinear stability for Burgers' equation can be obtained using split-forms of the equation. Due to the introduction of new correction terms, this nonlinear stability can be extended to generalised SBP bases, both nodal bases not including boundary nodes and modal bases [19]. Moreover, the stability of the schemes can be enhanced by artificial dissipation and modal filtering, if the correct formulation is chosen for the general SBP bases [7, 17].

An extension of these results to systems of hyperbolic equations is possible in some cases [16]. However, these schemes become increasingly complicated. Therefore, it may be questionable, whether improvements in accuracy and stability can justify the complexity of schemes not using Lobatto nodes as basis.

References

1. Y. Allaneau, A. Jameson, Connections between the filtered discontinuous Galerkin method and the flux reconstruction approach to high order discretizations. Comput. Methods Appl. Mech. Eng. **200**(49), 3628–3636 (2011)
2. D. De Grazia, G. Mengaldo, D. Moxey, P.E. Vincent, S.J. Sherwin, Connections between the discontinuous Galerkin method and high-order flux reconstruction schemes. Int. J. Numer. Methods Fluids **75**(12), 860–877 (2014)
3. D.C.D.R. Fernández, P.D. Boom, D.W. Zingg, A generalized framework for nodal first derivative summation-by-parts operators. J. Comput. Phys. **266**, 214–239 (2014)
4. D.C.D.R. Fernández, J.E. Hicken, D.W. Zingg, Review of summation-by-parts operators with simultaneous approximation terms for the numerical solution of partial differential equations. Comput. Fluids **95**, 171–196 (2014)
5. T.C. Fisher, M.H. Carpenter, J. Nordström, N.K. Yamaleev, C. Swanson, Discretely conservative finite-difference formulations for nonlinear conservation laws in split form: theory and boundary conditions. J. Comput. Phys. **234**, 353–375 (2013)
6. G.J. Gassner, A skew-symmetric discontinuous Galerkin spectral element discretization and its relation to SBP-SAT finite difference methods. SIAM J. Sci. Comput. **35**(3), A1233–A1253 (2013)
7. J. Glaubitz, H. Ranocha, P. Öffner, T. Sonar, Enhancing stability of correction procedure via reconstruction using summation-by-parts operators II: modal filtering (2016). arXiv: 1606.01056 [math.NA] (Submitted)
8. J.E. Hicken, D.C. Del Rey Fernández, D.W. Zingg, Multidimensional summation-by-parts operators: general theory and application to simplex elements. SIAM J. Sci. Comput. **38**(4), A1935–A1958 (2016)
9. H. Huynh, A flux reconstruction approach to high-order schemes including discontinuous Galerkin methods. AIAA Paper **4079**, 2007 (2007)
10. A. Jameson, A proof of the stability of the spectral difference method for all orders of accuracy. J. Sci. Comput. **45**(1–3), 348–358 (2010)
11. A. Jameson, P.E. Vincent, P. Castonguay, On the non-linear stability of flux reconstruction schemes. J. Sci. Comput. **50**(2), 434–445 (2012)

12. K. Mattsson, M. Svärd, J. Nordström, Stable and accurate artificial dissipation. J. Sci. Comput. **21**(1), 57–79 (2004)
13. J. Nordström, P. Eliasson, New developments for increased performance of the SBP-SAT finite difference technique, in *IDIHOM: Industrialization of High-Order Methods-A Top-Down Approach* (Springer, Berlin, 2015), pp. 467–488
14. P. Olsson, J. Oliger, Energy and maximum norm estimates for nonlinear conservation laws. Technical Report NASA-CR-195091, NASA, Research Institute for Advanced Computer Science; Moffett Field, CA, United States (1994)
15. H. Ranocha, SBP operators for CPR methods. Master's thesis, TU Braunschweig (2016)
16. H. Ranocha, Shallow water equations: split-form, entropy stable, well-balanced, and positivity preserving numerical methods. GEM–Int. J. Geomath. (2016). doi:10.1007/s13137-016-0089-9. arXiv: 1609.08029 [math.NA]
17. H. Ranocha, J. Glaubitz, P. Öffner, T. Sonar, Enhancing stability of correction procedure via reconstruction using summation-by-parts operators I: artificial dissipation. (2016). arXiv: 1606.00995 [math.NA] (Submitted)
18. H. Ranocha, P. Öffner, T. Sonar, Summation-by-parts operators for correction procedure via reconstruction. J. Comput. Phys. **311**, 299–328 (2016). doi:10.1016/j.jcp.2016.02.009. arXiv: 1511.02052 [math.NA]
19. H. Ranocha, P. Öffner, T. Sonar, Extended skew-symmetric form for summation-by-parts operators and varying Jacobians. J. Comput. Phys. **342**, 13–28 (2017). doi:10.1016/j.jcp.2017.04.044. arXiv: 1511.08408 [math.NA]
20. M. Svärd, J. Nordström, Review of summation-by-parts schemes for initial-boundary-value problems. J. Comput. Phys. **268**, 17–38 (2014)
21. E. Tadmor, Skew-selfadjoint form for systems of conservation laws. J. Math. Anal. Appl. **103**(2), 428–442 (1984)
22. E. Tadmor, The numerical viscosity of entropy stable schemes for systems of conservation laws. I. Math. Comput. **49**(179), 91–103 (1987)
23. P.E. Vincent, P. Castonguay, A. Jameson, Insights from von Neumann analysis of high-order flux reconstruction schemes. J. Comput. Phys. **230**(22), 8134–8154 (2011)
24. P.E. Vincent, P. Castonguay, A. Jameson, A new class of high-order energy stable flux reconstruction schemes. J. Sci. Comput. **47**(1), 50–72 (2011)
25. P.E. Vincent, A.M. Farrington, F.D. Witherden, A. Jameson, An extended range of stable-symmetric-conservative flux reconstruction correction functions. Comput. Methods Appl. Mech. Eng. **296**, 248–272 (2015)
26. J. von Neumann, R.D. Richtmyer, A method for the numerical calculation of hydrodynamic shocks. J. Appl. Phys. **21**(3), 232–237 (1950)
27. F.D. Witherden, A.M. Farrington, P.E. Vincent, PyFR: an open source framework for solving advection-diffusion type problems on streaming architectures using the flux reconstruction approach. Comput. Phys. Commun. **185**(11), 3028–3040 (2014)

Three-Dimensional Flow Stability Analysis Based on the Matrix-Forming Approach Made Affordable

Daniel Rodríguez and Elmer M. Gennaro

Abstract Theoretical developments for hydrodynamic instability analysis are often based on eigenvalue problems, the size of which depends on the dimensionality of the reference state (or base flow) and the number of coupled equations governing the fluid motion. The straightforward numerical approach consisting on spatial discretization of the linear operators, and numerical solution of the resulting matrix eigenvalue problem, can be applied today without restrictions to one-dimensional base flows. The most efficient implementations for one-dimensional problems feature spectral collocation discretizations which produce dense matrices. However, this combination of theoretical approach and numerics becomes computationally prohibitive when two-dimensional and three-dimensional flows are considered. This paper proposes a new methodology based on an optimized combination of high-order finite differences and sparse algebra, that leads to a substantial reduction of the computational cost. As a result, three-dimensional eigenvalue problems can be solved in a local workstation, while other related theoretical methods based on the WKB expansion, like global-oscillator instability or the Parabolized Stability Equations, can be extended to three-dimensional base flows and solved using a personal computer.

1 Introduction

Matrix-forming approaches for the solution of the multidimensional eigenvalue problems (EVPs) appearing in the study of the instability of complex flows have

D. Rodríguez (✉)
Graduate program in Mechanical Engineering, Department of Mechanical Engineering, Universidade Federal Fluminense, Niterói, RJ 24210-240, Brazil

Pontifical Catholic University of Rio de Janeiro, Rio de Janeiro, RJ 22451-900, Brazil
e-mail: danielrodriguez@id.uff.br, danielrodalv84@gmail.com

E.M. Gennaro
São Paulo State University (UNESP), Campus of São João da Boa Vista, São João da Boa Vista, São Paulo, SP 13874-149, Brazil
e-mail: elmer.gennaro@sjbv.unesp.br

M.L. Bittencourt et al. (eds.), *Spectral and High Order Methods for Partial Differential Equations ICOSAHOM 2016*, Lecture Notes in Computational Science and Engineering 119, DOI 10.1007/978-3-319-65870-4_46

traditionally been based on spectral collocation discretizations, on account of their convergence properties and the computer memory limitations. The discretization of 2D (biglobal) EVP [17] using spectral methods and dense storage may easily require of O(Terabytes) for the discretization of problems with moderate resolutions. A limit in this direction was reached in [12], in which parallelization on 2048 cores of a Blue Gene/P enabled the computation of eigenmodes dependent on two spatial directions and periodic on the third. The combination of spectral methods with sparse storage for this class of problems does not result in significant cost reductions, but combinations of spectral and high-order finite differences increase the sparsity of the discretized operators so that remarkable improvements in the numerical efficiency are obtained: 2D EVPs that required using a supercomputer 6 years ago can be solved today on a personal computer [5]. In the present contribution we go a step further, optimizing the combination of the sparse storage and the discretization scheme in the formation of the matrices in order to study flow stability problems in which the velocity field is fully 3D. This includes EVPs in which the eigenmodes are fully dependent on the three spatial directions (triglobal EVP [17]), and extensions of weakly non-parallel (WKB) approach and the parabolized stability equations to flows with a strong dependence on the cross-stream directions and a mild dependence on the streamwise one.

2 Theory

Hydrodynamic instability studies the behavior of small disturbances introduced or superimposed upon a well-defined reference state, referred to as base flow. Let the flow field be characterized by the vector $\mathbf{q} = (\mathbf{v}, p, T, \ldots)^T$, containing the three velocity components $\mathbf{v}$, the pressure p, temperature T, and any other field variable required to define the flow state. In the most general case, $\mathbf{q}$ is a function of the three spatial coordinates ($\mathbf{x} = (x, y, z)^T$) and time t. The flow is then decomposed into a time-invariant or time-averaged solution of the governing equations, the aforementioned base flow $\bar{\mathbf{q}}$ plus time-dependent fluctuations or perturbations $\mathbf{q}'$:

$$\mathbf{q}(\mathbf{x}, t) = \bar{\mathbf{q}}(\mathbf{x}) + \mathbf{q}'(\mathbf{x}, t). \tag{1}$$

Temporal stability analysis considers the evolution in time of perturbations introduced at a given initial instant. In this context, upon substitution of the flow decomposition into the Navier-Stokes equations and subtraction of the terms involving exclusively the base flow—as they verify the governing equations themselves—we arrive at the system of equations governing the fluctuations. These equations can be written in matrix form as:

$$\mathscr{R}\partial\mathbf{q}'/\partial t = \mathscr{L}\mathbf{q}' + \mathbf{F}(\mathbf{q}', \mathbf{q}'), \tag{2}$$

where $\mathscr{R}, \mathscr{L}$ are linear operators depending on the base flow and its spatial derivatives, physical parameters defining the flow at hand like Reynolds number or Mach number, and comprising first and second order spatial derivatives. Finally, $\mathbf{F}(\mathbf{q}', \mathbf{q}')$ comprises the quadratic nonlinearities between the perturbations. Provided that the instability behavior of the flow is determined by the evolution of infinitesimally small disturbances, the nonlinear term can be neglected leading to a linear, homogeneous problem.

While considering the long-time evolution of the disturbances, it is natural to assume an exponential dependence of them with time,

$$\mathbf{q}'(\mathbf{x}, t) = \hat{\mathbf{q}}(\mathbf{x}) \exp(-\mathrm{i}\omega t) + c.c., \tag{3}$$

where $\omega = \omega_r + \mathrm{i}\omega_i$. The real part, ω_r, is a circular frequency oscillation while the imaginary part ω_i corresponds to a temporal growth rate. $c.c.$ denotes the complex conjugate, that has to be added so that $\mathbf{q}'$ is a real quantity. Substituting the modal form in the linearized Navier-Stokes equations, we arrive at a generalized matrix eigenvalue problem (EVP) of the form

$$-\mathrm{i}\omega\mathscr{R}\hat{\mathbf{q}} = \mathscr{L}\hat{\mathbf{q}}. \tag{4}$$

The solution of the EVP recovers a set of eigenmodes, formed by eigenvalues ω and their corresponding eigenfunctions $\hat{\mathbf{q}}$. If all eigenmodes present have $\omega_i < 0$, then any linearly small perturbation decays for long times, and the base flow is said to be linearly stable. On the contrary, if at least one eigenmode has $\omega_i > 0$, a random initial condition will be amplified, recovering for long times the exponentially-growing modal behavior; in this case, the base flow is linearly unstable, and disturbances will grow until nonlinear interactions set in.

Due to the nonmodal nature of the linearized Navier-Stokes equations, the modal scenario does not suffice to predict the evolution of the disturbances for finite times: the eigenfunctions are in general non-orthogonal, and linear combinations of even damped eigenmodes can result in transient disturbance amplifications of several orders of magnitude. Though linearly stable, some flows can sustain transient linear amplifications large enough to reach nonlinear amplitudes, producing a sub-critical transition to a different state. A theoretical approach for the determination of the maximum amplification and the so-called optimal initial condition for transient growth is based on the computation of a large number of eigenmodes by *first* solving the EVP (4) [15].

In the most general case in which the base flow is dependent on the three spatial directions, the linear operators describing the EVP (4) contains coefficients with the same dimensionality. The linear operators $\mathscr{R}$ and $\mathscr{L}$ represent fully-coupled three-dimensional partial differential equations, and an appropriate spatial discretization and numerical solution are required. Considering a structured spatial discretization with $N_x \times N_y \times N_z$ discretization points and a system of N_e equations, the EVP has $N_x \times N_y \times N_z \times N_e$ degrees of freedom. This numbers being prohibitively large even for moderate resolutions, simplifications of the problem are sought for that allow for the reduction of the problem dimension. If the base flow is homogeneous along the z-spatial direction, the operators $\mathscr{R}$ and $\mathscr{L}$ are consequently homogeneous on the same direction and a Fourier transformation is allowed for. A three-dimensional perturbation takes then the form

$$\mathbf{q}'(\mathbf{x},t) = \hat{\mathbf{q}}(x,y)\exp[\mathrm{i}(\beta z - \omega t)] + c.c., \tag{5}$$

where β is a wavenumber in the homogeneous z–direction. The resulting EVP is then two-dimensional, with both linear operators and eigenfunctions discretized in a two-dimensional domain. The leading dimension of the problem is then reduced to $N_x \times N_y \times N_e$.

A further degree of simplification is introduced if the base flow is *parallel* or *nearly-parallel*, as is the case for boundary layers developing on flat plates or laminar mixing layers at high Reynolds numbers. In this case, the base flow can be assumed to be one-dimensional, depending on the transversal direction y alone. A second Fourier transform is introduced then for the streamwise direction, and modal perturbations take the form

$$\mathbf{q}'(\mathbf{x},t) = \hat{\mathbf{q}}(y)\exp[\mathrm{i}(\alpha x + \beta z - \omega t)] + c.c., \tag{6}$$

where α is the streamwise wavenumber. In this case, the EVP is one-dimensional and the leading dimension of the problem is $N_y \times N_e$.

These problems are sometimes referred to in the literature as triglobal (for $\bar{\mathbf{q}}$ depending on three dimensions), biglobal ($\bar{\mathbf{q}}$ depending on two dimensions) and *local* for base flows depending only upon the cross-stream direction[17]. However this definition of *local* as 1D problems is not precise, as cross-stream perturbations can result in 2D problems if the flow is inhomogeneous along the two cross-stream directions in a quasi-parallel approximation.

3 Numerical Methods for the Matrix-Forming Approach

One-dimensional EVPs for hydrodynamic instability were the first to be addressed, on account of their relative simplicity. First solutions were obtained by means of analytical expansions for few canonical base flows. With the introduction of computers, shooting methods were applied, that allow for the determination of eigenvalues and eigenfunctions of any kind of one-dimensional flow. It was not until 1971 that Orszag [9] presented the first *Matrix-forming* solution of the one-dimensional EVP. The matrix-forming approach consists on spatially discretizing the linear operators describing the EVP, and computing its eigenvalues and eigenfunctions using numerical linear algebra techniques. While its extension to two- and three-dimensional EVPs is straightforward, the resulting matrices are, even for moderate resolutions, remarkably large. First solutions for two-dimensional eigenmode problems appeared in the 1980s [10, 18], but it was not until more a decade later that they became a common tool for fluid-dynamicists [17]. The extension to three-dimensional problems of the matrix-forming approach by straightforward modification of the same numerical techniques is impractical: the computational requirements widely exceed the computational power of today's computers.

3.1 Previous Experiences

Most of the matrix-forming solutions of one-dimensional EVPs even to date, and of the first two-dimensional problems, considered a spatial discretization using spectral collocation methods, specially Chebyshev-Gauss-Lobatto points. Spectral collocation provides the higher accuracy for a given number of discretization points, thus reducing the memory requirements. The resulting matrices were stored as dense, and the eigensolutions were computed using variants of the QR algorithm [9]. First experiences with two-dimensional EVPs considered the same combination of spectral collocation and direct method for the eigenspectrum computation. While the memory requirements increased substantially, it was the computational time associated with the QR/QZ algorithms that became the limiting factor. Subspace iteration methods like the Arnoldi algorithm [2] were then introduced, drastically reducing the CPU time to, at leading order, the cost of performing a LU factorization of the discretized matrix $\mathscr{L}$ matrix.

The combination of dense matrices and shift-and-invert Arnoldi algorithm was found to be very limited for the solution of two-dimensional EVPs, and impractical for three-dimensional ones. Table 1 shows memory and FLOPs estimations corresponding to few representative problems. It was concluded that, for large two-dimensional and three-dimensional problems to be solved, a different numerical methodology was required.

Rodríguez and Theofilis [12] presented a massively parallel solution of the stability EVP, implementing the spectral collocation discretization, dense matrices and Arnoldi algorithm, but with distributed-memory storage and operation using MPI communications and the ScaLAPACK linear algebra package. Consequently, the maximum problem dimension becomes a function of the number of processors available, and the CPU-time was scaled accordingly. This implementation was shown to scale correctly using up to 4096 cores in a Blue Gene/P (JUGENE,

Table 1 Memory and FLOP estimations for the solution of representative stability EVPs ($N_e = 4$), using dense matrix storage and Arnoldi algorithm

Coupled dimensions	Spatial resolution	Memory	Floating-point operations (FLOPs)
1	250	15 MB	10^9
2	250×250	1 TB	10^{16}
3	$240 \times 50 \times 25$	22 TB	10^{18}

Forschunzentrum Juelich, Germany), and enabled the solution of EVP problems with matrices as large as 1 TB.

A different direction was pursued by Gennaro et al. [5], in which serial sparse solutions were studied. Considering two-dimensional stability problems, it was found that a discretization considering pseudo-spectral methods (Chebyshev polynomials) results into matrices too dense for the sparse treatment to be efficient. However, considering high-order finite differences or combinations of Chebyshev polynomials on one direction and high-order finite differences on the other, order-of-magnitude reductions were achieved both in memory and CPU-time.

3.2 The Method Proposed

Based on our previous experiences [5, 12], a new code was developed for the solution of one-, two- and three-dimensional EVPs, using combinations of high-order finite differences and pseudo-spectral discretizations, sparse storage and operation, and an in-house implementation of the shift-and-invert Arnoldi algorithm [2, 6]. The multifrontal sparse linear algebra MUMPS [1] is used for the LU factorization of the sparse matrices and for performing the required substitutions. Matrix-lines reordering is previously applied using the library METIS.

Variable-stencil finite differences are implemented and used as default discretization method. The stencil varies from a maximum of seven or nine points in the inner points to forward or backward differences with four or five points at boundaries. This discretization has the benefits of producing very sparse and banded matrix-blocks, optimizing the sparse algebra efficiency, while presenting a convergence much faster than low order discretization methods. Our earlier experiences have shown that sixth order finite differences deliver the optimum balance between the sparse efficiency (better for the lowest order) and resolution power (better for higher order). A Fourier collocation discretization is also implemented, to be used for periodic directions only. Coordinate transformations are introduced to concentrate the computational mesh in the spatial regions were field gradients are stronger and at solid domain boundaries.

The treatment of the domain boundaries deserves special attention. One known problem of spectral methods is the appearance of point-wise oscillations of the solution when the resolution is insufficient. This is especially problematic in flow stability problems considering outflow sections, as very thin artificial boundary layers appear [13]. On the other hand, the outflow region seldom affects the *converged* structure of the eigenfunctions, and devoting a large number of discretization points to prevent the point-wise oscillation is not practical. Using a relative low-order discretization at open boundaries prevents the numerical oscillations from contaminating the rest of the computational domain.

4 Three-Dimensional EVPs

The proposed methodology enabled the solution of three-dimensional EVPs using a local workstation machine, featuring 256 GB of shared memory and 16 dual core processors. Validations were done by performing comparisons with results delivered by two-dimensional EVPs. A spanwise-homogeneous base flow, corresponding to a two-dimensional laminar separation bubble on a flat plate boundary layer, was considered. It is known [14] that this flow is modally stable for two-dimensional disturbances and becomes unstable for spanwise-periodic perturbations, with a preferred periodicity wavenumber $\beta_c \approx 0.16$. Figure 1a compares the eigenspectra corresponding to two-dimensional problems for $\beta = 0$, and β_c, with the one computed for the same base flow considering a three-dimensional EVP with spanwise domain length of $L_z = 2\pi/\beta_c$. Accurate convergence of the eigenvalues is attained for a small number of collocation points ($N_z = 15$) on the spanwise

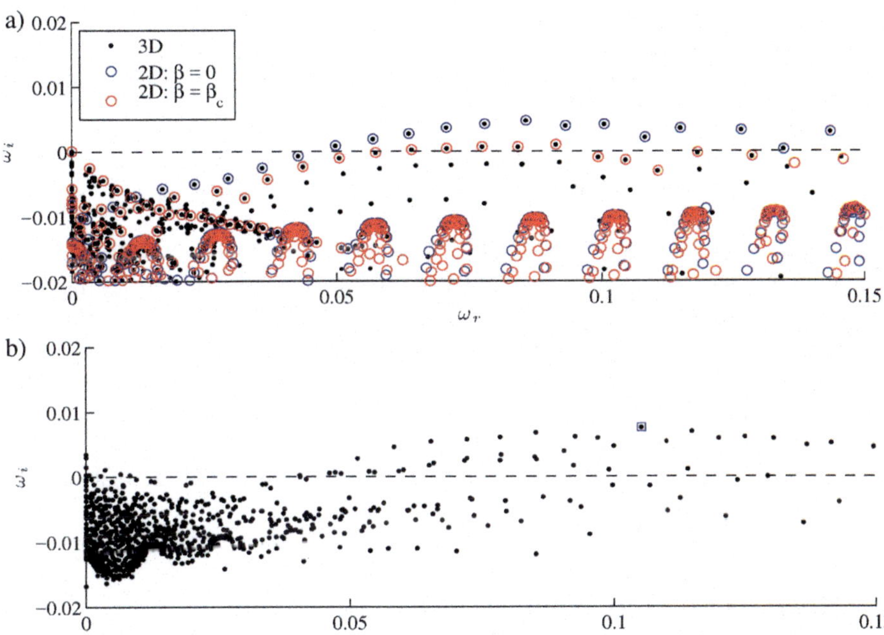

Fig. 1 (**a**) Stability eigenspectra for a two-dimensional laminar separation bubble base flow. *Blue and Red circles*: solution of the 2D EVP for plane ($\beta = 0$) and spanwise-periodic ($\beta = \beta_c$) perturbations, respectively. *Black dots*: solution of the 3D EVP with spanwise domain $z \in [0, 2\pi/\beta_c]$. (**b**) Stability eigenspectra for a three-dimensional laminar separation bubble resulting from the amplification of the primary instability. *Blue square* denotes the most unstable eigenmode, corresponding to a secondary instability. *Dashed horizontal line* separates unstable and stable regions

Table 2 Convergence history of the leading eigenmode in Fig. 1b

Spatial resolution	Memory (GB)	CPU-time (mins)	ω	Relative difference
$131 \times 41 \times 15$	44	123	0.08529 + i 0.005572	0.188
$151 \times 51 \times 17$	81	179	0.09178 + i 0.006532	0.126
$181 \times 51 \times 19$	106	878	0.09918 + i 0.007221	0.055
$221 \times 51 \times 25$	224	1426	0.10500 + i 0.007518	–

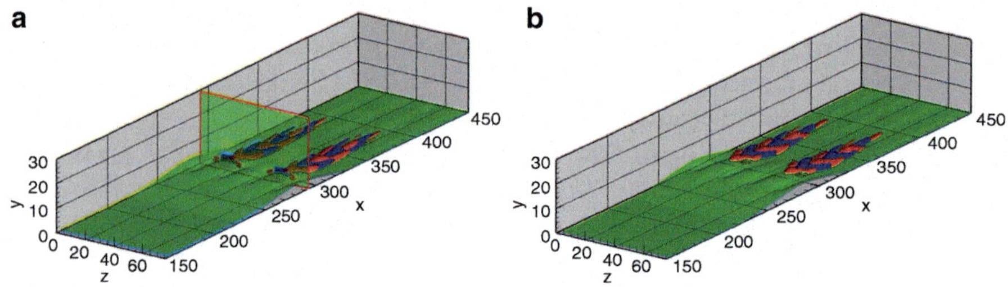

Fig. 2 Streamwise perturbations corresponding to a secondary instability of a three-dimensional separation bubble. (**a**) Obtained by WKB method on cross-planes, (**b**) Obtained as a three-dimensional eigenmode (*blue square* in Fig. 1b). The slice shown in the *left panel* corresponds to the projection on the real space of the wavemaker plane, X_s

direction. The maximum resolution used in the computations was $N_x \times N_y \times N_z = 221 \times 51 \times 25$.

Rodríguez and Gennaro (2015) [11] showed that the primary instability of separation bubbles leads to a supercritical pitchfork bifurcation resulting into fully three-dimensional steady flows. The present three-dimensional EVP solver is pertinent for the analysis of modal secondary instabilities associated with this flow. Figure 1b shows the eigenspectrum corresponding to such a fully three-dimensional separated flow. The leading eigenmode found corresponds to $\omega = 0.105 + \mathbf{i}0.0075$. Table 2 shows the convergence of the leading eigenmode with respect to spatial resolution and the respective computational cost. Figure 2b depicts the corresponding eigenfunctions.

5 Other 3D Stability Analysis Made Affordable

The remarkable improvement in the computational efficiency attained with the proposed combination of variable-stencil high-order finite differences and sparse storage and operation allows, in addition to the solution of three-dimensional EVPs on a workstation, the solution of two-dimensional problems with an associated expenditure of few seconds and O(Megabytes) per problem. This makes possible the consideration of two additional approaches for the instability analysis of three-dimensional base flows, that can deliver useful results on today's personal computers.

Many flows exist in which the variation of the mean flow properties along the dominant streamwise direction take place along distances long compared with those of the disturbances. Consequently, we can investigate the *local* instability properties at each X-section (capital letter is used to denote a large scale, $X = \varepsilon x$ with $\varepsilon \ll 1$), and then stitch them together using the multiple-scales or WKB approximation. This methodology can be applied to base flows with strong variations in one or the two cross-stream directions. Two particular methods of this class are relevant here.

5.1 Global Oscillator Based on WKB Based on Cross-Stream Planes

This approach studies the existence of self-excited global oscillations, synchronized along the X direction, based on the existence of local regions of absolute instability [8]. From the solution of the EVP for varying X cross-stream sections ($\bar{\mathbf{q}}(y, z; X)$), the dispersion relation $\mathscr{D}(\alpha, \omega; X) = 0$ is obtained. First, the absolute frequency $\omega_0(X)$ is obtained at each X as the saddle point, where $\partial\omega/\partial\alpha|_{\omega_0} = 0$. Then, the *wavemaker* location X_s is determined from the saddle-point condition $\partial\omega_0/\partial X = 0$. The global oscillation frequency is $\omega_g = \omega_0(X_s)$. At leading order, the spatial structure of the oscillator is obtained as:

$$\mathbf{q}'(x, y, z, t) \sim A(X)\hat{\mathbf{q}}^{\pm}(X, y, z) \exp\left(\int_{x'} \mathrm{i}\alpha^{\pm}(x')dx' - \mathrm{i}\omega_g t\right) + c.c. \tag{7}$$

The computation of a three-dimensional global oscillator requires the solution of a large number of *local* two-dimensional, cross-sectional EVPs. Spatial resolutions comparable to those used in the three-dimensional EVP for the secondary instability of the separation bubble ($N_y \times N_z = 51 \times 16$) deliver converged results, requiring of $O(400\,\mathrm{MB})$ and 4 s for the solution of each EVP serially on a laptop computer. Consequently, this approach can be used as an inexpensive alternative to the solution of three-dimensional EVPs for the calculation of the secondary instability of the separation bubble: Figure 2a shows the global oscillator obtained for the same base flow, corresponding to a global frequency $\omega_g = 0.1039 + \mathbf{i}0.0078$, agreeing with the three-dimensional eigenmode ($\omega = 0.105 + \mathbf{i}0.0075$, Fig. 2b).

5.2 3D Parabolized Stability Equations

The second method based on the multiple-scales approach considered here studies the spatial evolution of time-periodic convective instabilities, introduced at a given X-section, as they propagate downstream. The classic Parabolized Stability Equations (PSE) [7] recast the Navier-Stokes equations in disturbance form as a parabolic marching problem along the slow, streamwise coordinate X. The solution is computed at each cross-section by an iterative solution procedure that involves the solution of a number of one-dimensional linear problems, described by operators akin to the one-dimensional stability EVP. The classic PSE approach is extended here to three-dimensional base flows that depend strongly on two spatial directions and only mildly on the streamwise one: disturbances of the form

$$\mathbf{q}'(x, y, z, t) = \hat{\mathbf{q}}(X, y, z) \exp\left[\mathrm{i}(\int_{x'} \alpha(X')dx' - \omega t) \right] + c.c. \tag{8}$$

are considered, where $\hat{\mathbf{q}}(X, y, z)$ is the shape function and $\alpha(X')$ is a streamwise wavenumber which depends on the slow variable X. Introduction of this decomposition into the Navier Stokes equations in disturbance form, one obtains the matrix problem

$$\mathscr{R}\frac{\partial \hat{\mathbf{q}}}{\partial X} = \mathscr{L}\hat{\mathbf{q}} + \mathbf{F}(\hat{\mathbf{q}}, \hat{\mathbf{q}}). \tag{9}$$

PSE can take into account non-linear interactions between the different frequency Fourier modes, through the coupling term **F**. The marching algorithm in PSE requires of a normalization condition to isolate the slow variations of the shape function $\hat{\mathbf{q}}$ from the fast-scale oscillations and spatial growth. Here, the following normalization condition [7] is used

$$\int_y \int_z \hat{\mathbf{q}}^* \frac{\partial \hat{\mathbf{q}}}{\partial X}\, dydz, \tag{10}$$

which provides a condition for the iterative calculation of α. The superscript $*$ denotes complex conjugation. This approach, being an straight-forward extension of the classic PSE, was not successfully implemented until Broadhurst and Sherwin (2008) [3] due to its computational cost. The numerical methodology presented herein enables the routinary use of the 3D PSE equations for stability analyses.

The amplification of externally generated Tollmien-Schlichting waves by the same three-dimensional steady bubble considered in the previous examples is considered here to illustrate the 3D PSE methodology. The cross-plane resolution $N_y \times N_z = 51 \times 16$ also delivered converged spatial amplification curves. The iterative marching procedure requires a relatively large number of matrix inversions, each one requiring of about 3.6 s. The complete calculation for a single frequency requires about 20 min on a laptop computer.

Figure 3a shows the spatial amplification curves in terms of the N-factor ($N = \log(A(x)/A(x_0))$) for a range of dimensionless frequencies from $\omega = 0.05$ up to 0.15. Maximum amplitude is attained for $\omega = 0.12$, and the corresponding disturbance streamwise velocity field is shown in Fig. 3b.

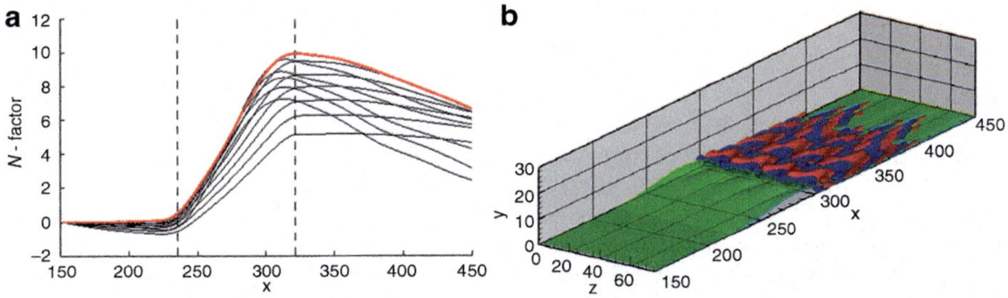

Fig. 3 Amplification and distortion of incoming plane Tollmien-Schlichting waves by the steady three-dimensional laminar separation bubble: N-factor curves for frequencies $\omega = .05(0.01)0.15$. *Red curve* corresponds to $\omega = 0.12$ (a). The *vertical dashed lines* show the spanwise-averaged separation and reattachment lines; streamwise velocity perturbation field for $\omega = 0.12$ (a)

6 Concluding Remarks

Matrix-forming approaches for the study of linear hydrodynamic instability are the most straightforward application of the theoretical developments, but unfortunately are also invariably affected by the "curse of dimensionality" that renders their computational cost prohibitive when the reference state or base flow is two- or three-dimensional. The classic solution approach, devised for the one-dimensional Orr-Sommerfeld equation, considered a pseudo-spectral discretization and resulted into dense matrices. However, for two- and three-dimensional problems the matrices naturally present some degree of sparsity. Combining spectral discretizations with sparse algebra improves only slightly the algorithm efficiency. A new code was developed that combines high-order finite differences with sparse storage and operation, resulting in substantial reductions both in memory and CPU-time. This new approach enables the solution of fully three-dimensional EVPs in a local size workstation with up to 256 GB and in times of few hours, thus being competitive with respect to *Matrix-free* techniques [4, 16], which deliver stability results based on a time-stepping code. Whether one methodology should be preferred over the other remains an open question, and a fair comparison would depend on the kind of problem and results (e.g. number of eigenmodes) under consideration.

In addition to 2D and 3D EVPs, other pre-existing approaches (Global oscillator based on local regions of absolute instability and Parabolized Stability Equations), based on the solution of a large number of cross-stream one-dimensional problems, can be extended to three-dimensional base flows using the present methodology; as a result, the instability analysis of fully three-dimensional base flows can be achieved on a personal computer.

Acknowledgements The work of D.R. was funded by the Brazilian Science without borders/CAPES "Attraction of Young Talents" Fellowship, grant 88881.064930/2014-01. This work is also supported by the São Paulo State Research Foundation (FAPESP) grants 2014/24782-0 and 2017/01586-0 and Brazilian National Counsel of Technological and Scientific Development (CNPq) grants 423846/2016-7, 405144/2016-4 and 305512/2016-1.

References

1. P.R. Amestoy, I.S. Duff, J.Y. L'xcellent, J. Koster, A fully asynchronous multifrontal solver using distributed dynamic scheduling. SIAM J. Matrix Anal. Appl. **23**(1), 15–41 (2001)
2. W.E. Arnoldi, The principle of minimized iterations in the solution of the matrix eigenvalue problem. Q. Appl. Math. **9**, 17–29 (1951)
3. M. Broadhurst, S. Sherwin, The parabolised stability equations for 3D-flows: implementation and numerical stability. Appl. Numer. Math. **58**(7), 1017–1029 (2008)
4. S. Chiba, Global stability analysis of incompressible viscous flow. J. Jpn. Soc. Comput. Fluid Dyn. **7**, 20–48 (1998)
5. E.M. Gennaro, D. Rodríguez, M.A.F. de Medeiros, V. Theofilis, Sparse techniques in global flow instability with application to compressible leading-edge flow. AIAA J. **51**(9), 2295–2303 (2013)
6. G.H. Golub, C.F. Van Loan, *Matrix Computations* (John Hopkins University Press, Baltimore, 1989)
7. T. Herbert, Parabolized stability equations. Annu. Rev. Fluid Mech. **29**, 245–283 (1997)
8. P. Huerre, P. Monkewitz, Local and global instabilities in spatially developing flows. Annu. Rev. Fluid Mech. **22**, 473–537 (1990)
9. S.A. Orszag, Accurate solution of the Orr-Sommerfeld stability equation. J. Fluid Mech. **5**(4), 689–703 (1971)
10. R.T. Pierrehumbert, S.E. Widnall, The two- and three-dimensional instabilities of spatially periodic shear layer. J. Fluid Mech. **114**, 59–82 (1982)
11. D. Rodríguez, E.M. Gennaro, On the secondary instability of forced and unforced laminar separation bubbles. Procedia IUTAM **14**, 78–87 (2015)
12. D. Rodríguez, V. Theofilis, Massively parallel numerical solution of the biglobal linear instability eigenvalue problem using dense linear algebra. AIAA J. **47**(10), 2449–2459 (2009)
13. D. Rodríguez, A. Tumin, V. Theofilis, Towards the foundation of a global mode concept, in *6th AIAA Theoretical Fluid Mechanics Conference, 27–30 June, Honolulu, USA* (AIAA, 2011), pp. 2011–3603.
14. D. Rodríguez, E.M. Gennaro, M.P. Juniper, The two classes of primary modal instability in laminar separation bubbles. J. Fluid Mech. **734**, R4 (2013)
15. P.J. Schmid, D.S. Henningson, *Stability and Transition in Shear Flows* (Springer, New York, 2001)
16. S. Sherwin, H. Blackburn, Three-dimensional instabilities and transition of steady and pulsatile axisymmetric. J. Fluid Mech. **533**, 297–327 (2005)
17. V. Theofilis, Global linear stability. Annu. Rev. Fluid Mech. **43**, 319–352 (2011)
18. A. Zebib, Instabilities of viscous flow past a circular cylinder. J. Eng. Math. **21**, 155–165 (1987)

Fast Spectral Methods for Temporally-Distributed Fractional PDEs

Mehdi Samiee, Ehsan Kharazmi, and Mohsen Zayernouri

Abstract Temporally-distributed fractional partial differential equations appear as rigorous mathematical models that solve the probability density function of non-Markovian processes coding multi-physics diffusion-to-wave and multi-rate ultra slow-to-super diffusion dynamics (Chechkin et al, Phys Rev E 66(4):046129, 2002). We develop a Petrov-Galerkin spectral method for high dimensional temporally-distributed fractional partial differential equations with two-sided derivatives in a *space-time* hypercube. We employ Jacobi poly-fractonomials given in (Zayernouri and Karniadakis, J Comput Phys 252:495–517, 2013) and Legendre polynomials as the temporal and spatial basis/test functions, respectively. Moreover, we formulate a fast linear solver for the corresponding Lyapunov system. Furthermore, we perform the corresponding discrete stability and error analysis of the numerical scheme. Finally, we carry out several numerical test cases to examine the efficiency and accuracy of the method.

1 Introduction

Anomalous transport, which manifests in power-law distribution, non-local behavior and memory effects, have been studied in many applications such as turbulence [10, 23], fluid flows in porous media [3, 4], and bioengineering [18, 24, 25]. Fractional PDEs appear as rigorous models that naturally incorporate such non-local features. Moreover, distributed FPDEs provide a powerful modeling tool to describe complex multi-physics multi-rate processes. For instance, temporally-distributed fractional diffusion equations rigorously solve the probability density function of the multi-fractal random processes, subordinated to Wiener process (retarding sub-diffusion) [6].

M. Samiee • E. Kharazmi • M. Zayernouri (✉)
Department of Computational Mathematics, Science, and Engineering, Michigan State University, 428 S Shaw Ln, East Lansing, MI 48824, USA

Department of Mechanical Engineering, Michigan State University, 428 S Shaw Ln, East Lansing, MI 48824, USA
e-mail: zayern@msu.edu

M.L. Bittencourt et al. (eds.), *Spectral and High Order Methods for Partial Differential Equations ICOSAHOM 2016*, Lecture Notes in Computational Science and Engineering 119, DOI 10.1007/978-3-319-65870-4_47

Discretization of FPDEs is challenging due to the history dependence and non-local feature of such fractional models. In addition to finite difference and finite volume methods [5, 7, 12, 17, 31], other numerical methods such as finite element [13, 21] and spectral/spectral element methods [8, 9, 14, 20, 28–30, 32–34] have been extensively studied. In addition to the aforementioned challenges in solving FPDEs, the distributed order FPDEs usually add at least one extra order of magnitude to the complexity of numerical methods; see [2, 19, 27]. In addition to the computational challenges, the analysis of distributed order operators requires a proper mathematical framework (characterizing the underlying function spaces, norms, etc.) that allows incorporating real data for predictive simulations.

In this study, we introduce distributed Sobolev space with the equivalent associated norms. We construct Petrov-Galerkin spectral methods with a unified fast solver for a class of temporally-distributed FPDEs with constant coefficients subject to Dirichlet boundary/initial conditions. We develop the fast linear solver based on the eigensolutions of the corresponding temporal/spatial mass and stiffness matrices. We carry out the discrete stability and error analysis of the PG method for the two-dimensional case. Eventually, we illustrate the spectral convergence and the efficiency of the method by performing several numerical simulations.

This study is organized as follows: in Sect. 2, we introduce the preliminaries on fractional calculus and define the distributed fractional Sobolev spaces. We define the problem and the corresponding variational form in Sect. 3. In Sect. 4, we construct the PG methods and formulate the fast solver. The discrete stability and error analysis are discussed in Sect. 5. In Sect. 6, we provide some numerical tests. We end the paper with a summary and conclusion.

2 Preliminaries

Following [22, 32], we denote the left- and right-sided Reimann-Liouville fractional derivatives by ${}^{RL}_{a}\mathcal{D}^{\nu}_{x}f(x)$ and ${}^{RL}_{x}\mathcal{D}^{\nu}_{b}g(x)$, respectively, in which $g(x) \in C^n[a,b]$. We recall from [1] that ${}^{RL}_{a}\mathcal{D}^{\nu}_{x}g(x) = {}^{C}_{a}\mathcal{D}^{\nu}_{x}g(x) = {}_{a}\mathcal{D}^{\nu}_{x}g(x)$, $\nu \in (0,1)$, when homogeneous Dirichlet initial and boundary conditions are enforced. Following [14], we analytically obtain the fractional derivatives of the Jacobi poly-fractonomials [29], which are later used in developing the numerical scheme, as

$$ {}^{RL}_{-1}\mathcal{D}^{\sigma}_{\xi}\left\{(1+\xi)^{\mu}P^{-\mu,\mu}_{n-1}(\xi)\right\} = \frac{\Gamma(n+\mu)}{\Gamma(n+\mu-\sigma)}(1+\xi)^{\mu-\sigma}P^{-\mu+\sigma,\mu-\sigma}_{n-1}(\xi) \tag{1} $$

$$ {}^{RL}_{\xi}\mathcal{D}^{\sigma}_{1}\left\{(1-\xi)^{\mu}P^{\mu,-\mu}_{n-1}(\xi)\right\} = \frac{\Gamma(n+\mu)}{\Gamma(n+\mu-\sigma)}(1-\xi)^{\mu-\sigma}P^{\mu-\sigma,-\mu+\sigma}_{n-1}(\xi) \tag{2} $$

in which $\mu, \sigma > 0$ and $P_{n-1}^{\mu,-\mu}(\xi)$ is the standard Jacobi polynomial of order $n-1$. Similarly, the μ-th order fractional derivatives of the Legendre polynomials are given as

$$
{}_{-1}\mathcal{D}_x^{\mu} P_n(x) = \frac{\Gamma(n+1)}{\Gamma(n-\mu+1)} P_n^{\mu,-\mu}(x)\,(1+x)^{-\mu},
$$
$$
{}_{x}\mathcal{D}_1^{\mu} P_n(x) = \frac{\Gamma(n+1)}{\Gamma(n-\mu+1)} P_n^{-\mu,\mu}(x)\,(1-x)^{-\mu},
$$

in which $P_n(x)$ represents the Legendre polynomial of order n.

2.1 Distributed Fractional Sobolev Spaces

According to [16], the usual Sobolev space associated with the real index ν_1 on bounded interval $\Lambda_1 = (a_1, b_1)$, is denoted by $H^{\nu_1}(\Lambda_1)$. Due to Lemma 2.6 in [16], $\|\cdot\|_{H^{\nu_1}(\Lambda_1)} \equiv \|\cdot\|_{{}^cH^{\nu_1}(\Lambda_1)}$, where $\|\cdot\|_{{}^cH^{\nu_1}(\Lambda_1)} = \Big(\|{}_{x_1}\mathcal{D}_{b_1}^{\nu_1}(\cdot)\|_{L^2(\Lambda_1)}^2 + \|{}_{a_1}\mathcal{D}_{x_1}^{\nu_1}(\cdot)\|_{L^2(\Lambda_1)}^2 + \|\cdot\|_{L^2(\Lambda_1)}^2\Big)^{\frac{1}{2}}$. Let $\Lambda_i = (a_i, b_i) \times \Lambda_{i-1}$ for $i = 2, \cdots, d$, and $\mathcal{X}_1 = H_0^{\nu_1}(\Lambda_1)$, which is associated with the norm $\|\cdot\|_{{}^cH^{\nu_1}(\Lambda_1)}$. Therefore, $\mathcal{X}_d$ is constructed such that $\mathcal{X}_d = H_0^{\nu_d}\Big((a_d, b_d); L^2(\Lambda_{d-1})\Big) \cap L^2(I; \mathcal{X}_{d-1})$, associated with norm $\|\cdot\|_{\mathcal{X}_d} = \Big\{\|\cdot\|_{L^2(\Lambda_d)}^2 + \sum_{i=1}^d \Big(\|{}_{x_i}\mathcal{D}_{b_i}^{\nu_i}(\cdot)\|_{L^2(\Lambda_d)}^2 + \|{}_{a_i}\mathcal{D}_{x_i}^{\nu_i}(\cdot)\|_{L^2(\Lambda_d)}^2\Big)\Big\}^{\frac{1}{2}}$, where

$$
\mathcal{X}_{d-1} = H_0^{\nu_{d-1}}\Big((a_{d-1}, b_{d-1}); L^2(\Lambda_{d-2})\Big) \cap L^2(I; \mathcal{X}_{d-2}),
$$
$$
\vdots
$$
$$
\mathcal{X}_2 = H_0^{\nu_2}\Big((a_2, b_2); L^2(\Lambda_1)\Big) \cap L^2(I; \mathcal{X}_1). \tag{3}
$$

Following [14], we denote by $H^{\varphi}(\mathbb{R})$ the *distributed* fractional Sobolev space on $\mathbb{R}$, which is endowed with the following norm $\|\cdot\|_{H^{\varphi}(\mathbb{R})} = \Big(\int_{\alpha_1}^{\alpha_2} \varphi(\alpha)\, \|(1+|\omega|^2)^{\frac{\alpha}{2}} \mathcal{F}(\cdot)(\omega)\|_{L^2(\mathbb{R})}^2\, d\alpha\Big)^{\frac{1}{2}}$, where $\varphi \in L^1([\alpha_1, \alpha_2])$, $0 \le \alpha_1 < \alpha_2$. Subsequently, we denote by $H^{\varphi}(I)$ the *distributed* fractional Sobolev space on the finite closed interval $I = (0, T)$, which is defined as $H^{\varphi}(I) = \{v \in L^2(I) |\ \exists \tilde{v} \in H^{\varphi}(\mathbb{R})\ s.t.\ \tilde{v}|_I = v\}$, with the equivalent norms $\|\cdot\|_{{}^lH^{\varphi}(I)}$ and $\|\cdot\|_{{}^rH^{\varphi}(I)}$ in [14],

where

$$\|\cdot\|_{{}^lH^{\varphi}(I)} = \left(\|\cdot\|^2_{L^2(I)} + \int_{\alpha_1}^{\alpha_2} \varphi(\alpha) \ \| \ {}^{RL}_{0}\mathcal{D}^{\alpha}_{t}(\cdot)\|^2_{L^2(I)} \ d\alpha \right)^{\frac{1}{2}},$$

$$\|\cdot\|_{{}^rH^{\varphi}(I)} = \left(\|\cdot\|^2_{L^2(I)} + \int_{\alpha_1}^{\alpha_2} \varphi(\alpha) \ \| \ {}^{RL}_{t}\mathcal{D}^{\alpha}_{T}(\cdot)\|^2_{L^2(I)} \ d\alpha \right)^{\frac{1}{2}}. \tag{4}$$

Let $\Omega = I \times \Lambda_d$. We define

$${}^l_0H^{\varphi}\Big(I; L^2(\Lambda_d)\Big) := \Big\{u \mid \|u(t,\cdot)\|_{L^2(\Lambda_d)} \in H^{\varphi}(I), u|_{t=0} = u|_{x=a_i} = u|_{x=b_i} = 0, \ i = 1, \cdots, d\Big\},$$

which is equipped with the norm

$$\|u\|_{{}^lH^{\tau}(I;L^2(\Lambda_d))} = \Big\| \ \|u(t,\cdot)\|_{L^2(\Lambda_d)} \Big\|_{{}^lH^{\varphi}(I)} = \left(\|u\|^2_{L^2(\Omega)} + \int_{\alpha_1}^{\alpha_2} \varphi(\alpha) \ \| \ {}^{RL}_{0}\mathcal{D}^{\alpha}_{t}(u)\|^2_{L^2(\Omega)} \ d\alpha \right)^{\frac{1}{2}}.$$

Similarly,

$${}^r_0H^{\varphi}\Big(I; L^2(\Lambda_d)\Big) := \Big\{v \mid \|v(t,\cdot)\|_{L^2(\Lambda_d)} \in H^{\varphi}(I), v|_{t=T} = v|_{x=a_i} = v|_{x=b_i} = 0, \ i = 1, \cdots, d\Big\},$$

which is equipped with the norm

$$\begin{aligned} \|v\|_{{}^rH^{\varphi}(I;L^2(\Lambda_d))} &= \Big\| \ \|v(t,\cdot)\|_{L^2(\Lambda_d)} \Big\|_{{}^rH^{\varphi}(I)} \\ &= \left(\|v\|^2_{L^2(\Omega)} + \int_{\alpha_1}^{\alpha_2} \varphi(\alpha) \ \| \ {}^{RL}_{t}\mathcal{D}^{\alpha}_{T}(v)\|^2_{L^2(\Omega)} \ d\alpha \right)^{\frac{1}{2}}. \end{aligned}$$

We define the solution space $\mathcal{B}^{\varphi,\nu_1,\cdots,\nu_d}(\Omega) := {}^l_0H^{\tau}\Big(I; L^2(\Lambda_d)\Big) \cap L^2(I; \mathcal{X}_d)$, endowed with the norm $\|u\|_{\mathcal{B}^{\varphi,\nu_1,\cdots,\nu_d}} = \Big\{ \|u\|^2_{{}^lH^{\varphi}(I;L^2(\Lambda_d))} + \|u\|^2_{L^2(I;\mathcal{X}_d)} \Big\}^{\frac{1}{2}}$. Therefore,

$$\begin{aligned} &\|u\|_{\mathcal{B}^{\varphi,\nu_1,\cdots,\nu_d}} \\ &= \Big\{ \|u\|^2_{L^2(\Omega)} + \int_{\alpha_1}^{\alpha_2} \varphi(\alpha) \| \ {}^{RL}_{0}\mathcal{D}^{\alpha}_{t}(u)\|^2_{L^2(\Omega)} \ d\alpha + \sum_{i=1}^{d} \Big(\| \ {}_{x_i}\mathcal{D}^{\nu_i}_{b_i}(u)\|^2_{L^2(\Omega)} + \| \ {}_{a_i}\mathcal{D}^{\nu_i}_{x_i}(u)\|^2_{L^2(\Omega)} \Big) \Big\}^{\frac{1}{2}}. \end{aligned} \tag{5}$$

Likewise, we define the test space $\mathfrak{B}^{\varphi,\nu_1,\cdots,\nu_d}(\Omega) := {}^rH^{\varphi}\Big(I; L^2(\Lambda_d)\Big) \cap L^2(I; \mathcal{X}_d)$, endowed with the norm $\|v\|_{\mathfrak{B}^{\tau,\nu_1,\cdots,\nu_d}} = \Big\{ \|v\|^2_{{}^rH^{\tau}(I;L^2(\Lambda_d))} + \|v\|^2_{L^2(I;\mathcal{X}_d)} \Big\}^{\frac{1}{2}}$. Therefore,

$$\|v\|_{\mathfrak{B}^{\varphi,\nu_1,\cdots,\nu_d}}$$

$$= \Big\{\|v\|^2_{L^2(\Omega)} + \int_{\alpha_1}^{\alpha_2} \varphi(\alpha) \| {}^{RL}_{t}\mathcal{D}^{\alpha}_{T}(v)\|^2_{L^2(\Omega)}\, d\alpha + \sum_{i=1}^{d} \big(\| {}_{x_i}\mathcal{D}^{\nu_i}_{b_i}(v)\|^2_{L^2(\Omega)} + \| {}_{a_i}\mathcal{D}^{\nu_i}_{x_i}(v)\|^2_{L^2(\Omega)}\big)\Big\}^{\frac{1}{2}}. \tag{6}$$

We note that in general, φ can be defined in any possible subset of the interval $[\alpha_1, \alpha_2]$ and thus arbitrarily confines the domain of integration, where the theoretical framework of the problem remains invariant while requiring the solution to have less regularity. The following lemma is useful in construction of the proposed numerical scheme.

Lemma 2.1 ([15]) *For all $0 < \alpha \leq 1$, if $u \in H^1([a,b])$ such that $u(a) = 0$, and $w \in H^{\alpha/2}([a,b])$, then $({}_a\mathcal{D}_s{}^{\alpha} u, w)_\Omega = ({}_a\mathcal{D}_s{}^{\alpha/2} u\,,\, {}_s\mathcal{D}_b{}^{\alpha/2} w)_\Omega$, where $(\cdot,\cdot)_\Omega$ represents the standard inner product in $\Omega = [a,b]$.*

3 Problem Definition

Let $\alpha \mapsto \varphi(\alpha)$ be a continuous mapping in $[\alpha_1, \alpha_2]$. Then, we define the distributed order fractional derivative as

$${}^{D}\mathcal{D}_{\varphi} u(t,x) = \int_{\alpha_1}^{\alpha_2} \varphi(\alpha)\, {}^{*}_{a}\mathcal{D}^{\alpha}_{t} u(t,x)\, d\alpha, \quad t > a, \tag{7}$$

where ${}^{*}_{a}\mathcal{D}^{\alpha}_{t}$ denotes the Riemann-Liouville fractional derivative of order α. Next, Let $u \in \mathcal{B}^{\varphi,\nu_1,\cdots,\nu_d}(\Omega)$ for some positive integer d and $\Omega = [0,T] \times [a_1, b_1] \times [a_2, b_2] \times \cdots \times [a_d, b_d]$, where

$${}^{D}\mathcal{D}_{\varphi} u + \sum_{j=1}^{d} \big[c_{l_j}\, {}_{a_j}\mathcal{D}^{2\mu_j}_{x_j} u + c_{r_j}\, {}_{x_j}\mathcal{D}^{2\mu_j}_{b_j} u\big] - \sum_{j=1}^{d} \big[\kappa_{l_j}\, {}_{a_j}\mathcal{D}^{2\nu_j}_{x_j} u + \kappa_{r_j}\, {}_{x_j}\mathcal{D}^{2\nu_j}_{b_j} u\big] + \gamma\, u = f, \tag{8}$$

in which all the coefficients γ, c_{l_j}, c_{r_j}, κ_{l_j}, and κ_{r_j} are constant, $2\mu_j \in (0,1)$, $2\nu_j \in (1,2)$ for $j = 1, 2, \cdots, d$, and $0 < \alpha_1 < \alpha_2 \leq 1$. Problem (8) is subject to the Dirichlet initial and boundary conditions, i.e. $u|_{t=0} = 0$ and $u|_{x_j=a_j} = u|_{x_j=b_j} = 0$ for $j = 1, 2, \cdots, d$. According to (5), the norm associated with $\mathcal{B}^{\varphi,\nu_1,\cdots,\nu_d}(\Omega)$ can be reduced to

$$\|u\|_{\mathcal{B}^{\varphi,\nu_1,\cdots,\nu_d}(\Omega)}$$

$$= \Big\{\int_{\alpha_1}^{\alpha_2} \varphi(\alpha) \underbrace{\| {}_0\mathcal{D}^{\alpha}_{t}(u)\|^2_{L^2(\Omega)}}_{U_I^{\varphi}}\, d\alpha + \sum_{j=1}^{d} \Big[\underbrace{\| {}_{a_j}\mathcal{D}^{\nu_j}_{x_j}(u)\|^2_{L^2(\Omega)}}_{U_{II}^{j}} + \underbrace{\| {}_{x_j}\mathcal{D}^{\nu_j}_{b_j}(u)\|^2_{L^2(\Omega)}}_{U_{III}^{j}} \Big]\Big\}^{1/2},$$

and similarly, the norm, associated with $\mathfrak{B}^{\varphi,\nu_1,\cdots,\nu_d}(\Omega)$, in (6) is equivalent to

$$\|v\|_{\mathfrak{B}^{\varphi,\nu_1,\cdots,\nu_d}(\Omega)}$$
$$= \Big\{ \int_{\alpha_1}^{\alpha_2} \underbrace{\varphi(\alpha)\, \|{}_t\mathcal{D}_T^{\alpha}(v)\|^2_{L^2(\Omega)}\, d\alpha}_{V_I^{\varphi}} + \sum_{j=1}^{d} \Big[\underbrace{\|{}_{x_j}\mathcal{D}_{b_j}^{\nu_j}(v)\|^2_{L^2(\Omega)}}_{V_{II}^j} + \underbrace{\|{}_{a_j}\mathcal{D}_{x_j}^{\nu_j}(v)\|^2_{L^2(\Omega)}}_{V_{III}^j} \Big] \Big\}^{1/2}.$$

In order to obtain the variational form of problem, we multiply (8) by a proper test function v and integrate over the computational domain. The corresponding continuous bilinear form $a : \mathcal{B}^{\varphi,\nu_1,\cdots,\nu_d}(\Omega) \times \mathfrak{B}^{\varphi,\nu_1,\cdots,\nu_d}(\Omega) \to \mathbb{R}$ takes the form

$$\begin{aligned} a_{\varphi}(u, v) &= \int_{\alpha_1}^{\alpha_2} \varphi(\alpha)\, ({}_0\mathcal{D}_t^{\alpha/2}\, u, {}_t\mathcal{D}_T^{\alpha/2}\, v)_{\Omega}\, d\alpha \\ &+ \sum_{j=1}^{d} \big[c_{l_j} ({}_{a_j}\mathcal{D}_{x_j}^{\mu_j}\, u, {}_{x_j}\mathcal{D}_{b_j}^{\mu_j}\, v)_{\Omega} + c_{r_j} ({}_{x_j}\mathcal{D}_{a_j}^{\mu_j}\, u, {}_{a_j}\mathcal{D}_{x_j}^{\mu_j}\, v)_{\Omega} \big] \\ &- \sum_{j=1}^{d} \big[\kappa_{l_j} ({}_{a_j}\mathcal{D}_{x_j}^{\nu_j}\, u, {}_{x_j}\mathcal{D}_{b_j}^{\nu_j}\, v)_{\Omega} + \kappa_{r_j} ({}_{x_j}\mathcal{D}_{b_j}^{\nu_j}\, u, {}_{a_j}\mathcal{D}_{x_j}^{\nu_j}\, v)_{\Omega} \big] + \gamma(u, v)_{\Omega}, \end{aligned} \tag{9}$$

where $(\cdot,\cdot)_{\Omega}$ represents the usual L^2-product. Thus, the problem reads as: find $u \in \mathcal{B}^{\varphi,\nu_1,\cdots,\nu_d}(\Omega)$ such that

$$a_{\varphi}(u, v) = (f, v)_{\Omega}, \quad \forall v \in \mathfrak{B}^{\varphi,\nu_1,\cdots,\nu_d}(\Omega). \tag{10}$$

Next, we choose proper finite-dimensional subspaces of $U_N \subset \mathcal{B}^{\varphi,\nu_1,\cdots,\nu_d}(\Omega)$ and $V_N \subset \mathfrak{B}^{\varphi,\nu_1,\cdots,\nu_d}(\Omega)$; thus, the discrete problem reads as: find $u_N \in U_N$ such that

$$a_{\varphi}(u_N, v_N) = (f, v_N)_{\Omega}, \quad \forall v_N \in V_N. \tag{11}$$

4 Petrov Galerkin Mathematical Formulation

We construct a Petrov-Galerkin spectral method for the discrete problem $u_N \in U_N$, satisfying the weak form (11). We first define the proper finite-dimensional basis/test spaces and then implement the numerical scheme.

4.1 Space of Basis (U_N) and Test (V_N) Functions

We employ the Legendre polynomials as the spatial basis, given in the standard domain $\xi \in [-1, 1]$ as $\phi_m(\xi) = \sigma_m\big(P_{m+1}(\xi) - P_{m-1}(\xi)\big)$, $m = 1, 2, \cdots$. We also employ the poly-fractonomial of first kind [29, 33] as the temporal basis function, given in the standard domain $\eta \in [-1,1]$ as $\psi_n^\tau(\eta) = \sigma_n(1+\eta)^\tau P_{n-1}^{-\tau,\tau}(\eta)$, $n = 1,2,\cdots$. The coefficients σ_m are defined as $\sigma_m = 2+(-1)^m$. Therefore, we construct the trial space as

$$U_N = span\Big\{\Big(\psi_n^\tau \circ \eta\Big)(t)\prod_{j=1}^{d}\Big(\phi_{m_j} \circ \xi_j\Big)(x_j) : n = 1,2,\cdots,\mathcal{N},\ m_j = 1,2,\cdots,\mathcal{M}_j\Big\},$$

where $\eta(t) = 2t/T - 1$ and $\xi_j(s) = 2\frac{s-a_j}{b_j-a_j} - 1$. The temporal and spatial basis functions naturally satisfy the initial and boundary conditions, respectively. Moreover, we define the temporal and spatial test functions in the standard domain as $\Psi_r^\tau(\eta) = \widetilde{\sigma}_r(1-\eta)^\tau P_{r-1}^{\tau,-\tau}(\eta)$, $r = 1,2,\cdots$ (poly-fractonomial of second kind) and $\Phi_k^\mu(\xi) = \widetilde{\sigma}_k\big(P_{k+1}(\xi) - P_{k-1}(\xi)\big)$, $k = 1,2,\cdots$, respectively. The coefficients $\widetilde{\sigma}_k$ are defined as $\widetilde{\sigma}_k = 2\,(-1)^k + 1$. Hence, we construct the corresponding test space as

$$V_N = span\Big\{\Big(\Psi_r^\tau \circ \eta\Big)(t)\prod_{j=1}^{d}\Big(\Phi_{k_j} \circ \xi_j\Big)(x_j) : r = 1,2,\cdots,\mathcal{N},\ k_j = 1,2,\cdots,\mathcal{M}_j\Big\}.$$

4.2 Implementation of PG Spectral Method

We represent the solution of (11) as a linear combination of elements of the solution space U_N. Therefore,

$$u_N(x,t) = \sum_{n=1}^{\mathcal{N}}\sum_{m_1=1}^{\mathcal{M}_1}\cdots\sum_{m_d=1}^{\mathcal{M}_d} \hat{u}_{n,m_1,\cdots,m_d}\Big[\psi_n^\tau(t)\prod_{j=1}^{d}\phi_{m_j}(x_j)\Big] \tag{12}$$

in Ω. By substituting the expansion (12) into (11) and choosing $v_N = \Psi_r^\tau(t)\prod_{j=1}^{d}\Phi_{k_j}(x_j)$, $r = 1,2,\ldots,\mathcal{N}$, $k_j = 1,2,\ldots,\mathcal{M}_j$, we obtain the following Lyapunov system

$$\Big(S_\tau^\varphi \otimes M_1 \otimes M_2 \cdots \otimes M_d + \sum_{j=1}^{d}[M_\tau \otimes M_1 \otimes \cdots \otimes M_{j-1} \otimes S_j^{Tot} \otimes M_{j+1} \cdots \otimes M_d]$$
$$+ \gamma M_\tau \otimes M_1 \otimes M_2 \cdots \otimes M_d\Big)\mathcal{U} = F, \tag{13}$$

in which $\otimes$ represents the Kronecker product, F denotes the multi-dimensional load matrix whose entries are given as

$$F_{r,k_1,\cdots,k_d} = \int_\Omega f(t,x_1,\cdots,x_d)\Big(\Psi_r^\tau \circ \eta\Big)(t)\prod_{j=1}^{d}\Big(\Phi_{k_j} \circ \xi_j\Big)(x_j)\,d\Omega, \tag{14}$$

and $S_j^{Tot} = c_{l_j} \times S_{\mu_j,l} + c_{r_j} \times S_{\mu_j,r} - \kappa_{l_j} \times S_{\nu_j} - \kappa_{r_j} \times S_{\nu_j,r}$. The matrices S_τ^φ and M_τ denote the temporal stiffness and mass matrices, respectively; S_{ν_j}, S_{μ_j} and M_j denote the spatial stiffness and mass matrices, respectively. The entries of spatial mass matrix M_j are computed analytically, while we employ proper quadrature rules to accurately compute the entries of spatial stiffness S_{ν_j}, S_{μ_j} and temporal mass matrices M_τ. We note that due to the choices of basis/test functions, the obtained mass and stiffness matrices are symmetric. Moreover, we accurately compute the entries of temporal stiffness matrix, S_τ^φ, using theorem (3.1) in [14].

4.3 Unified Fast FPDE Solver

We develop a unified fast solver in terms of the generalized eigensolutions in order to formulate a closed-form solution to the Lyapunov system (13).

Theorem 4.1 *Let $\{\vec{e}^{\,j}_{m_j}, \lambda^j_{m_j}\}_{m_j=1}^{\mathcal{M}_j}$ be the set of general eigen-solutions of the spatial stiffness matrix S_j^{Tot} with respect to the mass matrix M_j. Moreover, let $\{\vec{e}_n^{\,\tau}, \lambda_n^\tau\}_{n=1}^{\mathcal{N}}$ be the set of general eigen-solutions of the temporal mass matrix M_τ with respect to the stiffness matrix S_τ^φ. Then the matrix of unknown coefficients $\mathcal{U}$ is explicitly obtained as*

$$\mathcal{U} = \sum_{n=1}^{\mathcal{N}} \sum_{m_1=1}^{\mathcal{M}_1} \cdots \sum_{m_d=1}^{\mathcal{M}_d} \kappa_{n,m_1,\cdots,m_d}\, \vec{e}_n^{\,\tau} \otimes \vec{e}^{\,1}_{m_1} \otimes \cdots \otimes \vec{e}^{\,d}_{m_d}, \tag{15}$$

where $\kappa_{n,m_1,\cdots,m_d}$ is given by

$$\kappa_{n,m_1,\cdots,m_d} = \frac{(\vec{e}_n^{\,\tau}\, \vec{e}^{\,1}_{m_1} \cdots \vec{e}^{\,d}_{m_d})F}{\left[(\vec{e}_n^{\,\tau T} S_\tau^\varphi \vec{e}_n^{\,\tau}) \prod_{j=1}^d (\vec{e}^{\,jT}_{m_j} M_j \vec{e}^{\,j}_{m_j})\right] \Lambda_{n,m_1,\cdots,m_d}}, \tag{16}$$

in which the numerator represents the standard multi-dimensional inner product, and $\Lambda_{n,m_1,\cdots,m_d}$ is obtained in terms of the eigenvalues of all mass matrices as $\Lambda_{n,m_1,\cdots,m_d} = \left[(1 + \gamma\, \lambda_n^\tau) + \lambda_n^\tau \sum_{j=1}^d (\lambda^j_{m_j})\right]$.

Proof Consider the following generalised eigenvalue problems as

$$S_j^{Tot} \vec{e}^{\,j}_{m_j} = \lambda^j_{m_j} M_j \vec{e}^{\,j}_{m_j}, \quad m_j = 1, 2, \cdots, \mathcal{M}_j, \quad j = 1, 2, \cdots, d, \tag{17}$$

$$M_\tau \vec{e}_n^{\,\tau} = \lambda_n^\tau S_\tau^\varphi \vec{e}_n^{\,\tau}, \quad n = 1, 2, \cdots, \mathcal{N}. \tag{18}$$

Having the spatial and temporal eigenvectors determined in Eqs. (18) and (17), we can represent the unknown coefficient matrix $\mathcal{U}$ in (12) in terms of the aforementioned eigenvectors as $\mathcal{U} = \sum_{n=1}^{\mathcal{N}} \sum_{m_1=1}^{\mathcal{M}_1} \cdots \sum_{m_d=1}^{\mathcal{M}_d} \kappa_{n,m_1,\cdots,m_d}\, \vec{e}_n^{\,\tau} \otimes \vec{e}^{\,1}_{m_1} \otimes \cdots \otimes \vec{e}^{\,d}_{m_d}$, where $\kappa_{n,m_1,\cdots,m_d}$ is obtained as follows. Following [26], we substitute $\mathcal{U}$ in the corresponding Lyapunov equation and then, take the inner product of both sides of equation by $\vec{e}_q^{\,\tau}\vec{e}_{p_1}^{\,1}\cdots\vec{e}_{p_d}^{\,d}$. Therefore, by rearranging the terms, we obtain

$$\kappa_{n,m_1,\cdots,m_d}=\frac{(\vec{e}_n^{\,\tau}\vec{e}_{m_1}^{\,1}\cdots\vec{e}_{m_d}^{\,d})F}{\Big[(\vec{e}_n^{\,\tau^T}S_\tau^\varphi\vec{e}_n^{\,\tau})\prod_{j=1}^d(\vec{e}_{m_j}^{\,j^T}M_j\vec{e}_{m_j}^{\,j})\Big]\times\Big[(1+\gamma\,\lambda_n^\tau)+\lambda_n^\tau\sum_{j=1}^d(\lambda_{m_j}^j)\Big]}.$$

Since the spatial Mass M_j and temporal stiffness matrices S_τ^φ are diagonal, we have $(\vec{e}_q^{\,\tau^T}S_\tau^\varphi\vec{e}_n^{\,\tau})=0$ if $q\neq n$, and also $(\vec{e}_{p_j}^{\,j^T}M_j\vec{e}_{m_j}^{\,j})=0$ if $p_j\neq m_j$, which completes the proof. □

5 Stability Analysis

The following theorem provides the discrete stability analysis of the scheme for (1+1)-dimensional temporally-distributed fractional diffusion problem. Such a stability analysis can be extended to the problem of (1+d)-dimensional with both-sided derivatives, which we will be carried out in our future work.

Theorem 5.1 *The Petrov-Gelerkin spectral method for (1+1)-D temporally-distributed and space-fractional diffusion problem $a_\varphi(u,v)=l(v)$ is stable, i.e., the discrete* inf-sup *condition*

$$\inf_{\substack{u_N\in U_N\\u_N\neq 0}}\ \sup_{\substack{v_N\in V_N\\v_N\neq 0}}\frac{|a(u_N,v_N)|}{\|v_N\|_{\mathfrak{B}^{\varphi,\nu_1,\cdots,\nu_d}(\Omega)}\|u_N\|_{\mathcal{B}^{\varphi,\nu_1,\cdots,\nu_d}(\Omega)}}\geq\beta>0, \tag{19}$$

holds with $\beta>0$ and independent of N.

Proof Let $\psi_n^\tau(\eta)=(1+\eta)^\tau P_n^{-\tau,\tau}(\eta)$, $\Psi_n^\tau(\eta)=(1-\eta)^\tau P_n^{\tau,-\tau}(\eta)$, and $u_N=\sum_{n=1}^N\sum_{m=0}^{M+1}\bar{u}_{n,m}\,\psi_n^\tau(t)P_m(x)$, where $u_N\in U_N$. Hence,

$$U_I^\varphi=\int_{-1}^{+1}\int_0^T\sum_{n=1}^N\sum_{m=0}^{M+1}\sum_{k=1}^N\sum_{r=0}^{M+1}\bar{u}_{k,r}\bar{u}_{n,m}\,{}_0\mathcal{D}_t^{\alpha/2}\psi_n^\tau(t)\,{}_0\mathcal{D}_t^{\alpha/2}\psi_k^\tau(t)P_m(x)P_r(x)dtdx$$

$$=\sum_{n=1}^N\sum_{m=0}^{M+1}\sum_{k=1}^N\sum_{r=0}^{M+1}\bar{u}_{k,r}\bar{u}_{n,m}\underbrace{\int_{-1}^{+1}P_m(x)P_r(x)dx}_{C_m^{0,0}\delta_{m,r}}\,(\frac{T}{2})^{1-2\tau_1}\Gamma_{n-1}^{\tau_1,-\tau_1}\Gamma_{k-1}^{\tau_1,-\tau_1}\times$$

$$\int_{-1}^{+1}(1+\eta)^{\tau_1}P_{n-1}^{-\tau_1,\tau_1}(\eta)(1+\eta)^{\tau_1}P_{k-1}^{-\tau_1,\tau_1}(\eta)d\eta, \tag{20}$$

where $\Gamma_{n-1}^{\tau_1,-\tau_1} = \Gamma_{n-1}^{-\tau_1,\tau_1} = \frac{\Gamma(n+\tau_1)}{\Gamma(n)}$ and $\tau_1 = \tau - \frac{\alpha}{2}$. Take $P_n^{-\tau_1,\tau_1}(\eta) = \sum_{q=0}^{n} a_q^{\tau_1,n} P_q^{0,2\tau_1}(\eta)$ then,

$$U_I^{\varphi} = \sum_{n=1}^{N}\sum_{k=1}^{N}\sum_{m=0}^{M+1}\sum_{q_3=1}^{n}\sum_{q_4=1}^{r} \bar{u}_{k,m}\bar{u}_{n,m}\, C_m^{0,0}\,(\frac{T}{2})^{1-2\tau_1}\,\Gamma_{n-1}^{\tau_1,-\tau_1}\,\Gamma_{k-1}^{\tau_1,-\tau_1}\, a_{q_3}^{\tau_1,n} a_{q_4}^{\tau_1,r}$$

$$\times \underbrace{\int_{-1}^{+1}(1+\eta)^{2\tau_1} P_{q_3}^{0,2\tau_1}(\eta) P_{q_4}^{0,2\tau_1}(\eta)d\eta}_{C_{q_3}^{0,2\tau_1}\delta_{q_3,q_4}}$$

$$= \sum_{m=0}^{M+1}\sum_{q_3=1}^{N} {}^{(1)}\check{u}_{q_3,m}^{2}\, C_{q_3}^{0,2\tau_1}\, C_m^{0,0}(\frac{T}{2})^{1-2\tau_1} = \sum_{m=0}^{M+1}\sum_{n=1}^{N} {}^{(1)}\check{u}_{n,m}^{2}(\frac{T}{2})^{1-2\tau_1}\, C_n^{0,2\tau_1}\, C_m^{0,0},$$

in which ${}^{(1)}\check{u}_{n,m} = \sum_{q=0}^{M+1-q} \bar{u}_{q,m}\, a_n^{\tau_1,q}\, \Gamma_{q-1}^{\tau_1,-\tau_1}$. Besides,

$$U_{II}^{1} = \int_{-1}^{+1}\int_{0}^{T}\sum_{n=1}^{N}\sum_{m=0}^{M+1}\sum_{k=1}^{N}\sum_{r=0}^{M+1} \bar{u}_{k,r}\bar{u}_{n,m}\psi_n^{\tau}(t)\psi_k^{\tau}(t)\ {}_{-1}\mathcal{D}_x^{\nu}P_m(x)\ {}_{-1}\mathcal{D}_x^{\nu}P_r(x)dtdx$$

$$= \sum_{n=1}^{N}\sum_{m=0}^{M+1}\sum_{k=1}^{N}\sum_{r=0}^{M+1} \bar{u}_{k,r}\bar{u}_{n,m}(\frac{T}{2})\int_{-1}^{+1}(1+\eta)^{2\tau}P_{n-1}^{-\tau,\tau}(\eta)P_{k-1}^{-\tau,\tau}(\eta)d\eta\times$$

$$\int_{-1}^{+1}(1+x)^{-2\nu}\Gamma_m^{\nu}\Gamma_r^{\nu}P_m^{\nu,-\nu}(x)P_r^{\nu,-\nu}(x)dx, \tag{21}$$

where $\Gamma_m^{\nu} = \frac{m+1}{m-\nu+1}$. By substituting $P_i^{\nu,-\nu}(x) = \sum_{q=0}^{i} b_q^{2\nu,i} P_q^{-2\nu,0}(x)$ and $P_n^{-\tau,\tau}(\eta) = \sum_{q=0}^{n} a_q^{\tau,n} P_q^{0,2\tau}(\eta)$ into (21) and reorganizing, we obtain

$$U_{II}^{1} = \sum_{n=1}^{N}\sum_{k=1}^{N}\sum_{m=0}^{M+1}\sum_{q_3=1}^{n}\sum_{q_4=1}^{k} {}^{(2)}\check{u}_{n,m}\,{}^{(2)}\check{u}_{k,m} C_m^{-2\nu,0}(\frac{T}{2}) a_{q_3}^{\tau,n} a_{q_4}^{\tau,k}$$

$$\times \underbrace{\int_{-1}^{+1}(1+\eta)^{2\tau} P_{q_3}^{0,2\tau}(\eta) P_{q_4}^{0,2\tau}(\eta)d\eta}_{C_{q_3}^{0,2\tau}\delta_{q_3,q_4}}$$

$$= \sum_{m=0}^{M+1}\sum_{q_3=1}^{N} {}^{(L)}\check{u}_{q_3,m}^{2}\, C_{q_3}^{0,2\tau}\, C_m^{-2\nu,0}(\frac{T}{2}) = \sum_{m=0}^{M+1}\sum_{n=1}^{N} {}^{(L)}\check{u}_{n,m}^{2}(\frac{T}{2}) C_n^{0,2\tau}\, C_m^{-2\nu,0}, \tag{22}$$

where ${}^{(2)}\check{u}_{n,m} = \sum_{q=0}^{M+1-q} \bar{u}_q\, b_m^{2\nu,q}\, \Gamma_q^{\nu}$ and ${}^{(L)}\check{u}_{n,m} = \sum_{q=1}^{N-n} {}^{(2)}\check{u}_{q,m} a_n^{\tau,q}$.

Let $v_N = \sum_{k=1}^{N} \sum_{n=0}^{M+1} \bar{u}_{k,r}(-1)^{k+r} \Psi_k^\tau(t) P_r(x)$. Following the same steps as in U_I^φ, for the norm of the test function we have

$$
\begin{aligned}
V_I^\varphi &= \int_{-1}^{+1} \int_0^T \sum_{n=1}^{N} \sum_{m=0}^{M+1} \sum_{k=1}^{N} \sum_{r=0}^{M+1} \bar{u}_{k,r} \bar{u}_{n,m} \, (-1)^{n+k} \, {}_t\mathcal{D}_T^{\alpha/2} \Psi_n^\tau(t) \, {}_t\mathcal{D}_T^{\alpha/2} \Psi_k^\tau(t) P_m(x) \\
&\quad \times P_r(x) (-1)^{r+m} \, dt \, dx \\
&= \sum_{m=0}^{M+1} \sum_{n=1}^{N} {}^{(1)}\check{v}_{n,m}^2 \left(\frac{T}{2}\right)^{1-2\tau_1} C_n^{0,2\tau_1} C_m^{0,0},
\end{aligned} \tag{23}
$$

in which we employ $P_n^{\tau_1,-\tau_1}(\eta) = \sum_{q=0}^{n} a_q^{-\tau_1,n} P_q^{2\tau_1,0}(\eta)$ and ${}^{(1)}\check{v}_{n,m} = \sum_{q=0}^{M+1-q} \bar{u}_{n,q} a_n^{-\tau_1,q} \Gamma_q^{\tau_1,-\tau_1}$. Besides,

$$
\begin{aligned}
V_{II}^1 &= \int_{-1}^{+1} \int_0^T \sum_{n=1}^{N} \sum_{m=0}^{M+1} \sum_{k=1}^{N} \sum_{r=0}^{M+1} \bar{u}_{k,r} \bar{u}_{n,m} \, (-1)^{n+k} \Psi_n^\tau(t) \, \Psi_k^\tau(t) \, (-1)^{r+m} \, {}_{-1}\mathcal{D}_x^\nu \\
&\quad \times P_m(x) \, {}_{-1}\mathcal{D}_x^\nu P_r(x) \, dt \, dx \\
&= \sum_{n=1}^{N} \sum_{m=0}^{M+1} \sum_{k=1}^{N} \sum_{r=0}^{M+1} \bar{u}_{k,r} \bar{u}_{n,m} \, \left(\frac{T}{2}\right) (-1)^{n+k} \int_{-1}^{+1} (1-\eta)^{2\tau} P_{n-1}^{\tau,-\tau}(\eta) P_{k-1}^{\tau,-\tau}(\eta) d\eta \times \\
&\quad (-1)^{m+r} \int_{-1}^{+1} (1+x)^{-2\nu} \, \Gamma_m^\nu \Gamma_r^\nu \, P_m^{\nu,-\nu}(x) \, P_r^{\nu,-\nu}(x) dx, \\
&= \sum_{n=1}^{N} \sum_{k=1}^{N} \sum_{m=0}^{M+1} \sum_{q_3=1}^{n} \sum_{q_4=1}^{r} {}^{(2)}\check{v}_{n,m} \, {}^{(2)}\check{v}_{k,m} C_m^{-2\nu,0} \left(\frac{T}{2}\right) a_{q_3}^{-\tau,n} a_{q_4}^{-\tau,r} (-1)^{n+k} \\
&\quad \times \underbrace{\int_{-1}^{+1} (1-\eta)^{2\tau} P_{q_3}^{2\tau,0}(\eta) P_{q_4}^{2\tau,0}(\eta) d\eta}_{C_{q_3}^{2\tau,0} \delta_{q_3,q_4} = C_{q_3}^{0,2\tau} \delta_{q_3,q_4}} \\
&= \sum_{m=0}^{M+1} \sum_{q_3=1}^{N} {}^{(L)}\check{v}_{q_3,m}^2 C_{q_3}^{0,2\tau} C_m^{-2\nu,0} \left(\frac{T}{2}\right) = \sum_{n=1}^{N} \sum_{m=0}^{M+1} {}^{(L)}\check{v}_{n,m}^2 \left(\frac{T}{2}\right) C_n^{0,2\tau} C_m^{-2\nu,0},
\end{aligned} \tag{24}
$$

where ${}^{(2)}\check{v}_{n,m} = \sum_{q=0}^{M+1-m} (-1)^q \, \bar{u}_{n,q} \, b_m^{2\nu,q} \, \Gamma_q^\nu$, ${}^{(L)}\check{v}_{n,m} = \sum_{i=1}^{N-n} {}^{(2)}\check{v}_{i,m} a_n^{-\tau,i} (-1)^i$, and $P_n^{\tau,-\tau}(\eta) = \sum_{q=0}^{n} a_q^{-\tau,n} P_q^{2\tau,0}(\eta)$. Let $A_I^\varphi = ({}_0\mathcal{D}_t^{\alpha/2} u_N, \, {}_t\mathcal{D}_T^{\alpha/2} v_N)_\Omega$ and $A_{II} = \kappa_l({}_{-1}\mathcal{D}_x^\nu u_N, \, {}_x\mathcal{D}_1^\nu u_N)_\Omega$. By employing $P_{n-1}^{\tau_1,-\tau_1}(x) = \sum_{q=0}^{n-1} a_q^{2\tau_1,n} P_q^{\tau_1,\tau_1}(x)$ and

$P_{k-1}^{-\tau_1,\tau_1}(x) = \sum_{q=0}^{k-1}(-1)^{q+k} a_q^{2\tau_1,k} P_q^{\tau_1,\tau_1}(x)$, we obtain

$$A_I^{\varphi} = \int_0^T \int_{-1}^{+1} \sum_{n=1}^{N}\sum_{k=1}^{N}\sum_{m=0}^{M+1}\sum_{r=0}^{M+1} \bar{u}_{n,m}\,\bar{u}_{k,r}\,(-1)^k\,{}_0\mathcal{D}_t^{\alpha/2}\psi_n^{\tau}(t)\,{}_t\mathcal{D}_T^{\alpha/2}\Psi_k^{\tau}(t)(-1)^r$$

$$\times P_m(x)\,P_r(x)\,dxdt$$

$$= \sum_{n=1}^{N}\sum_{k=1}^{N}\sum_{m=0}^{M+1}\sum_{r=0}^{M+1} \bar{u}_{n,m}\,\bar{u}_{k,r}\,(-1)^r \underbrace{\int_{-1}^{+1} P_m(x)\,P_r(x)dx}_{C_m^{0,0}\delta_{m,r}} (-1)^k(\frac{T}{2})^{1-2\tau_1}\,\Gamma_{n-1}^{\tau_1,-\tau_1}\,\Gamma_{k-1}^{\tau_1,-\tau_1}\times$$

$$\int_{-1}^{1}(1-\eta^2)^{\tau_1} P_{n-1}^{-\tau_1,\tau_1}(\eta)P_{k-1}^{\tau_1,-\tau_1}(\eta)d\eta = \sum_{n=1}^{N}\sum_{m=0}^{M+1} {}^{(3)}\check{u}_{n,m}^2\,(-1)^{m+k}(\frac{T}{2})^{1-2\tau_1}\,C_m^{0,0}\,C_n^{\tau_1,\tau_1}, \tag{25}$$

where ${}^{(3)}\check{u}_{n,m} = \sum_{q=1}^{N} a_n^{2\tau_1,q}\Gamma_{q-1}^{\tau_1,-\tau_1}\,\bar{u}_{q,m}$. Moreover, based on $P_{n-1}^{\tau,-\tau}(\eta) = \sum_{q=0}^{n-1} a_q^{2\tau,n} P_q^{\tau,\tau}(\eta)$, $P_{k-1}^{-\tau,\tau}(\eta) = \sum_{q=0}^{k-1}(-1)^{q+k} a_q^{2\tau,k} P_q^{\tau,\tau}(\eta)$, $P_i^{\nu,-\nu}(x) = \sum_{q=0}^{i} b_q^{2\nu,i} P_q^{-2\nu,0}(x)$, and $P_i^{-\nu,\nu}(x) = \sum_{q=0}^{i}(-1)^{i+q} b_q^{2\nu,i} P_q^{-2\nu,0}(x)$, we get

$$A_{II} = \int_0^T \int_{-1}^{+1} \sum_{n=1}^{N}\sum_{k=1}^{N}\sum_{m=0}^{M+1}\sum_{r=0}^{M+1} \bar{u}_{n,m}\,\bar{u}_{k,r}\,\psi_n^{\tau}(t)\Psi_k^{\tau}(t)(-1)^{r+k}\,{}_{-1}\mathcal{D}_x^{\nu}P_m(x)\,{}_x\mathcal{D}_1^{\nu}P_r(x)\,dxdt$$

$$= \sum_{n=1}^{N}\sum_{k=1}^{N}\sum_{m=0}^{M+1}\sum_{q_3=1}^{n}\sum_{q_4=1}^{r} {}^{(1)}\tilde{u}_{n,m}\,{}^{(1)}\tilde{u}_{k,m}(-1)^m C_m^{-2\nu,0}(\frac{T}{2})a_{q_3}^{2\tau,n}a_{q_4}^{2\tau,k}(-1)^{q_4}$$

$$\times \underbrace{\int_{-1}^{+1}(1+\eta)^{2\tau}P_{q_3}^{0,2\tau}(\eta)P_{q_4}^{0,2\tau}(\eta)d\eta,}_{C_{q_3}^{0,2\tau}\delta_{q_3,q_4}}$$

which can be simplified to $A_{II} = \sum_{m=0}^{M+1}\sum_{n=1}^{N} {}^{(L)}\tilde{u}_{n,m}^2(-1)^{n+m}(\frac{T}{2})C_n^{0,2\tau}C_m^{-2\nu,0}$, where ${}^{(L)}\tilde{u}_{n,m} = \sum_{q3=1}^{N-n} {}^{(1)}\tilde{u}_{q_3,m}a_n^{\tau,q_3}$ and ${}^{(1)}\tilde{u}_{n,m} = \sum_{i=0}^{M+1}\bar{u}_{n,i}b^{2\nu,q}$. On the other hand, we have $|a(u_N, v_N)| \geq \bar{c}\big[\int_{\alpha_1}^{\alpha_2}\varphi(\alpha)|A_I^{\varphi}| + \kappa_l|A_{II}|\big]$. To compare $|a(u_N, v_N)|$ with $\|u_N\|_{\mathcal{B}^{\varphi,\nu_1,\cdots,\nu_d}(\Omega)}\|v_N\|_{\mathfrak{B}^{\varphi,\nu_1,\cdots,\nu_d}(\Omega)}$,

$$|A_I^{\varphi}| = |\sum_{n=1}^{N}\sum_{m=0}^{M+1}(-1)^{m+k}\underbrace{\frac{{}^{(3)}\check{u}_{n,m}^2\,(\Gamma_{n-1}^{\tau_1,-\tau_1})^2\,C_m^{0,0}\,C_n^{\tau_1,\tau_1}}{{}^{(1)}\check{u}_{n,m}^2 C_n^{0,2\tau_1}C_m^{0,0}}}_{{}^{(1)}\tilde{\beta}_{n,m}}(\frac{T}{2})^{1-2\tau_1}\,{}^{(1)}\check{u}_{n,m}^2 C_n^{0,2\tau_1}C_m^{0,0}|$$

$$\geq \alpha_1\,{}^{(1)}\tilde{\beta}\,U_I^{\varphi}$$

and

$$|A_{II}| = |\sum_{m=0}^{M+1}\sum_{n=1}^{N}(-1)^{n+m}\underbrace{\frac{{}^{(L)}\tilde{u}^2_{n,m}(\frac{T}{2})C_n^{\tau,\tau}C_m^{-\nu,-\nu}}{{}^{(L)}\check{u}^2_{n,m}(\frac{T}{2})C_n^{0,2\tau}C_m^{-2\nu,0}}}_{{}^{(2)}\tilde{\beta}_{n,m}}\,{}^{(L)}\check{u}^2_{n,m}(\frac{T}{2})C_n^{0,2\tau}C_m^{-2\nu,0}| \geq \alpha_2\ {}^{(2)}\tilde{\beta}\,U_{II},$$

where ${}^{(2)}\tilde{\beta} = \min\{{}^{(2)}\tilde{\beta}_{n,m}\}$. Besides, we can have

$${}^{(\alpha)}V_I = \sum_{m=0}^{M+1}\frac{{}^{(1)}\check{v}^2_m}{{}^{(1)}\check{u}^2_m}\,{}^{(1)}\check{u}^2_m\,C_m^{-2\nu,0}(\frac{T}{2})C_n^{0,2\tau_1} = \sum_{m=0}^{M+1}{}^{(1)}\check{\beta}_m\ {}^{(1)}\check{u}^2_m\,C_m^{-2\nu,0}(\frac{T}{2})C_n^{0,2\tau_1} \leq {}^{(1)}\check{\beta}\,U_I^{\varphi},$$

$$V_{II} = \sum_{m=0}^{M+1}\frac{{}^{(R)}\check{v}^2_{n,m}}{{}^{(R)}\check{u}^2_{n,m}}\,{}^{(R)}\check{u}^2_{n,m}(\frac{T}{2})C_n^{0,2\tau}C_m^{-2\nu,0} = \sum_{m=0}^{M+1}{}^{(2)}\check{\beta}_{n,m}\ {}^{(R)}\check{u}^2_{n,m}\,C_m^{-2\nu,0}(\frac{T}{2})C_n^{0,2\tau} \leq {}^{(2)}\check{\beta}\,U_{II},$$

where ${}^{(1)}\check{\beta} = max\{{}^{(1)}\check{\beta}_m\}$ and ${}^{(2)}\check{\beta} = max\{{}^{(2)}\check{\beta}_{n,m}\}$. This results in

$$\|v_N\|^2_{\mathfrak{B}^{\varphi,\nu_1,\cdots,\nu_d}(\Omega)} \leq \underbrace{max\{{}^{(2)}\check{\beta},\ {}^{(1)}\check{\beta}\}}_{\tilde{\beta}^2}\|u_N\|^2_{\mathcal{B}^{\varphi,\nu_1,\cdots,\nu_d}(\Omega)}.$$

$u \in U$, A_I^{φ}, and A_{II} has finite values, therefore

$$\begin{aligned}
|a(u_N, v_N)| &\geq \alpha\left[|A_I^{\varphi}| + \kappa_l|A_{II}||\right] \geq \alpha\big[\alpha_1\ {}^{(1)}\tilde{\beta}\,U_I^{\varphi} + \alpha_2\ {}^{(2)}\tilde{\beta}\kappa_l\,U_{II}\big] \\
&\geq \underbrace{\alpha\, min\{\alpha_1\ {}^{(1)}\tilde{\beta},\ \alpha_2\ {}^{(2)}\tilde{\beta}\kappa_l\}}_{\tilde{\alpha}}\ \|u_N\|^2_{\mathcal{B}^{\varphi,\nu_1,\cdots,\nu_d}(\Omega)} \\
&\geq \tilde{\alpha}\,\tilde{\beta}\|u_N\|_{\mathcal{B}^{\varphi,\nu_1,\cdots,\nu_d}(\Omega)}\ \|v_N\|_{\mathfrak{B}^{\varphi,\nu_1,\cdots,\nu_d}(\Omega)},
\end{aligned} \tag{26}$$

which shows that discrete *inf-sup* condition holds for the time-dependent fractional diffusion problem. □

6 Error Analysis

Kharazmi et al. [14] performed the error analysis of the distributed order differential equations, where they employed Jacobi polyfractonomials of first kind as the basis function. Following similar steps, we can show that the projection error in time and space takes the same form. Let $\mathcal{D}^{(r)}u = \frac{\partial^r u}{\partial t^{r_0}\,\partial x_1{}^{r_1}\cdots\partial x_d{}^{r_d}}$, where $r = \sum_{i=0}^{d} r_i$. Thus, if $\mathcal{D}^{(r)}u \in U$ for some integer $r \geq 1$, that is, $\int_{\alpha_1}^{\alpha_2}\varphi(\alpha)\|\,{}_0\mathcal{D}_t^{\frac{\alpha}{2}}(\mathcal{D}^{(r)}u)\|_{L^2}d\alpha < \infty$, and

u_N denotes the projection of the exact solution u, then

$$\|u-u_N\|_U \le \beta M^{-r}\Big\{\|\mathcal{D}^{(r)}u\|^2_{L^2(\Omega)} + \int_{\alpha_1}^{\alpha_2} \varphi(\alpha)\ \|\ {}^{RL}_{0}\mathcal{D}^{\alpha}_{t}\,(\mathcal{D}^{(r)}u)\|^2_{L^2(\Omega)}\ d\alpha$$

$$+\sum_{i=1}^{d}\big(\|\ {}_{x_i}\mathcal{D}^{\nu_i}_{b_i}\,(\mathcal{D}^{(r)}u)\|^2_{L^2(\Omega)} + \|\ {}_{a_i}\mathcal{D}^{\nu_i}_{x_i}\,(\mathcal{D}^{(r)}u)\|^2_{L^2(\Omega)}\big)\Big\}^{\frac{1}{2}} \tag{27}$$

Since the *inf-sup* condition holds in theorem 5.1, by the Banach-Nečas-Babuška theorem in [11], the error in the numerical scheme is less than or equal to a constant times the projection error.

7 Numerical Simulations

We provide numerical examples of the spectral scheme we have proposed. We consider the exact solution of the form $u^{ext} = u_t \prod_{j=1}^{d} u_{\xi_j}$ with finite regularity, where $u_t = t^{p_1+\tau}$, $t \in [0,T]$, and $u_{\xi_j} = (1+\xi_j)^{p_2+\beta}(1-\xi_j)^{p_3+\beta}$, $\xi_j \in [-1,1]$. We obtain the force function by substituting u^{ext} into (8), where the advection and diffusion coefficients are considered to be unity in all dimensions.

Figure 1 shows the convergence of error via spatial and temporal p-refinement for (1+2)-D problem. In the left sub-figure, $u^{ext} = t^{3+1/2}\prod_{j=1}^{2}(1+\xi_j)^{4+1/2}(1-\xi_j)^{4+1/2}$, for which we choose $N = 4$ to control the error in time and perform p-refinement in space for different values of fractional orders $\{2\mu, 2\nu\} = \{0.5, 1.1\}$ and $\{2\mu, 2\nu\} = \{0.5, 1.9\}$. The results show the expected spectral convergence. In the right sub-figure, we perform p-refinement in time for $u^{ext} = t^{3+\tau}\prod_{j=1}^{2}(1+\xi_j)^4(1-\xi_j)^4$, where $\tau = 0.1, 0.9$ and we choose $\mathcal{M}_1 = \mathcal{M}_2 = 8$ to control the error in space. The choice of ploy-fractonomials as the temporal basis enable the scheme to accurately capture the singularity in time. The obtained results show the convergence of error to machine precision with $N = 4$. Moreover, in Table 1, we

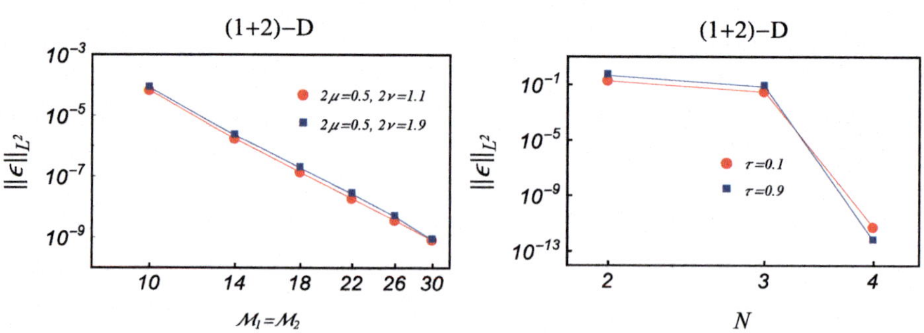

Fig. 1 PG spectral method, temporal and spatial p-refinement for (1+2)-D problem

Table 1 PG spectral method, CPU time (in min) and L^2-norm error for multi-dimensional problems

	(1+1)-D		(1+2)-D		(1+3)-D	
$N = \mathcal{M}_1 = \mathcal{M}_2 = \mathcal{M}_3$	L^2-norm error	CPU time	L^2-norm error	CPU time	L^2-norm error	CPU time
2	6.2067×10^{-1}	0.6	5.9428×10^{-1}	1	5.1307×10^{-1}	1.7
6	2.7852×10^{-2}	1	2.9233×10^{-2}	1.5	2.6720×10^{-2}	4
10	6.7506×10^{-5}	3.13	7.089×10^{-5}	4.5	6.4714×10^{-5}	27.9
14	1.7541×10^{-6}	20.3	1.8463×10^{-6}	27.5	1.6876×10^{-6}	149

show the CPU time (which includes the construction of the linear system and load vector) as well as the computed L^2-norm error for the problems of (1+1)- to (1+3)-dimensions, where $p_1 = 3,\ \tau = 0.5,\ p_2 = p_3 = 4,\ \beta = 0.5,\ 2\mu = 0.5,\ 2\nu = 1.5$.

8 Summary

We developed a Petrov-Galerkin spectral method for high dimensional temporally-distributed fractional partial differential equations with two-sided derivatives in a space-time hypercube. We employed Jacobi poly-fractonomials and Legendre polynomials as the temporal and spatial basis/test functions, respectively. To solve the corresponding Lyapunov linear system, we further formulated a fast linear solver and performed the corresponding discrete stability and error analysis. At last, we carried out several numerical simulations to examine the performance of the method.

Acknowledgements This work was supported by the AFOSR Young Investigator Program (YIP) award on: "Data-Infused Fractional PDE Modeling and Simulation of Anomalous Transport" (FA9550-17-1-0150).

References

1. R. Askey, J. Fitch, Integral representations for jacobi polynomials and some applications. J. Math. Anal. Appl. **26**, 411–437 (1969)
2. T.M. Atanackovic, S. Pilipovic, D. Zorica, Time distributed-order diffusion-wave equation. I. Volterra-type equation. Proc. R. Soc. A Math. Phys. Eng. Sci. **465**(2106), 1869–1891 (2009)
3. D.A. Benson, R. Schumer, M.M. Meerschaert, S.W. Wheatcraft, Fractional dispersion, Lévy motion, and the MADE tracer tests, in *Dispersion in Heterogeneous Geological Formations* (Springer, Netherlands, 2001), pp. 211–240
4. D.A. Benson, S.W. Wheatcraft, M.M. Meerschaert, Application of a fractional advection-dispersion equation. Water Resour. Res. **36**(6), 1403–1412 (2000)
5. J. Cao, C. Li, Y. Chen, Compact difference method for solving the fractional reaction–subdiffusion equation with Neumann boundary value condition. Int. J. Comput. Math. **92**(1), 167–180 (2015)

6. A. Chechkin, R. Gorenflo, I. Sokolov, Retarding subdiffusion and accelerating superdiffusion governed by distributed-order fractional diffusion equations. Phys. Rev. E **66**(4), 046129 (2002)
7. M. Chen, W. Deng, A second-order numerical method for two-dimensional two-sided space fractional convection diffusion equation. Appl. Math. Model. **38**(13), 3244–3259 (2014)
8. F. Chen, Q. Xu, J.S. Hesthaven, A multi-domain spectral method for time-fractional differential equations. J. Comput. Phys. **293**, 157–172 (2015)
9. S. Chen, J. Shen, L.L. Wang, Generalized Jacobi functions and their applications to fractional differential equations. Mathematics of Math. Comput. **85**(300), 1603–1638 (2016)
10. D. del Castillo-Negrete, B. Carreras, V. Lynch, Fractional diffusion in plasma turbulence, Phys. Plasmas (1994-present) **11**(8), 3854–3864 (2004)
11. A. Ern, J. Guermond, *Theory and Practice of Finite Elements*, vol. 159 (Springer Science & Business Media, New York, 2013)
12. H. Hejazi, T. Moroney, F. Liu, A finite volume method for solving the two-sided time-space fractional advection-dispersion equation. Open Phys. **11**(10), 1275–1283 (2013)
13. B. Jin, R. Lazarov, J. Pasciak, Z. Zhou, Error analysis of semidiscrete finite element methods for inhomogeneous time-fractional diffusion. IMA J. Numer. Anal. **35**(2), 561–582 (2014)
14. E. Kharazmi, M. Zayernouri, G.E. Karniadakis, Petrov-Galerkin and spectral collocation methods for distributed order differential equations. SIAM J. Sci. Comput. **39**(3), A1003–A1037 (2017)
15. X. Li, C. Xu, A space-time spectral method for the time fractional diffusion equation. SIAM J. Numer. Anal. **47**(3), 2108–2131 (2009)
16. X. Li, C. Xu, Existence and uniqueness of the weak solution of the space-time fractional diffusion equation and a spectral method approximation. Commun. Comput. Phys. **8**(5), 1016 (2010)
17. C. Lubich, Discretized fractional calculus. SIAM J. Math. Anal. **17**(3), 704–719 (1986)
18. R.L. Magin, *Fractional Calculus in Bioengineering* (Begell House Redding, West Redding, 2006)
19. F. Mainardi, G. Pagnini, R. Gorenflo, Some aspects of fractional diffusion equations of single and distributed order. Appl. Math. Comput. **187**(1), 295–305 (2007)
20. Z. Mao, J. Shen, Efficient spectral–Galerkin methods for fractional partial differential equations with variable coefficients. J. Comput. Phys. **307**, 243–261 (2016)
21. W. McLean, K. Mustapha, Convergence analysis of a discontinuous Galerkin method for a sub-diffusion equation. Numer. Algorithms **52**(1), 69–88 (2009)
22. M.M. Meerschaert, A. Sikorskii, *Stochastic Models for Fractional Calculus*, vol. 43 (Walter de Gruyter, Berlin, 2012)
23. M.M. Meerschaert, F. Sabzikar, M.S. Phanikumar, A. Zeleke, Tempered fractional time series model for turbulence in geophysical flows. J. Stat. Mech: Theory Exp. **2014**(9), P09023 (2014)
24. M. Naghibolhosseini, Estimation of outer-middle ear transmission using DPOAEs and fractional-order modeling of human middle ear, Ph.D. thesis, City University of New York, NY, 2015
25. P. Perdikaris, G.E. Karniadakis, Fractional-order viscoelasticity in one-dimensional blood flow models. Ann. Biomed. Eng. **42**(5) 1012–1023 (2014)
26. M. Samiee, M. Zayernouri, M.M. Meerschaert, A unified spectral method for FPDEs with two-sided derivatives; part I: a fast solver. J. Comput. Phys. (In Press)
27. T. Srokowski, Lévy flights in nonhomogeneous media: distributed-order fractional equation approach. Phys. Rev. E **78**(3), 031135 (2008)
28. M. Zayernouri, W. Cao, Z. Zhang, G.E. Karniadakis, Spectral and discontinuous spectral element methods for fractional delay equations. SIAM J. Sci. Comput. **36**(6), B904–B929 (2014)
29. M. Zayernouri, G.E. Karniadakis, Fractional Sturm–Liouville eigen-problems: theory and numerical approximation. J. Comput. Phys. **252**, 495–517 (2013)
30. M. Zayernouri, G.E. Karniadakis, Fractional spectral collocation methods for linear and nonlinear variable order FPDEs. J. Comput. Phys. **293**, 312–338 (2015)

31. M. Zayernouri, A. Matzavinos, Fractional Adams–Bashforth/Moulton methods: an application to the fractional Keller–Segel chemotaxis system, J. Comput. Phys. **317**, 1–14 (2016)
32. M. Zayernouri, M. Ainsworth, G.E. Karniadakis, A unified Petrov–Galerkin spectral method for fractional PDEs. Comput. Methods Appl. Mech. Eng. **283**, 1545–1569 (2015)
33. M. Zayernouri, M. Ainsworth, G.E. Karniadakis, Tempered fractional Sturm–Liouville eigenproblems. SIAM J. Sci. Comput. **37**(4), A1777–A1800 (2015)
34. L. Zhao, W. Deng, J.S. Hesthaven, Spectral methods for tempered fractional differential equations. arXiv:1603.06511 (arXiv preprint)

Supercritical-Order Mimetic Operators on Higher-Dimensional Staggered Grids

Eduardo J. Sanchez, Guillermo F. Miranda, Jose M. Cela, and Jose E. Castillo

Abstract We focus on the construction of 2- and 3D mimetic gradient, divergence, curl, and Laplacian operators. We base this work on the method by Castillo and Grone, which constructs mimetic gradient and divergence operators via a discrete instance of Gauss' divergence theorem. This method can not construct tenth-order gradient nor eighth-order divergence operators (nor higher) because the computed weights discretizing the corresponding weighted inner products are not all positive for these cases. Thus, we define the tenth order and the eighth order thresholds as critical orders of accuracy for the gradient and divergence operators, respectively. In previous works, we introduced the Castillo–Blomgren–Sanchez algorithm. This algorithm constructs supercritical-order mimetic operators. The contribution of this work is the extension to higher dimensions of the operators constructed by this algorithm. This includes detailing the mathematics of this extension. We also detail the construction of a mimetic curl operator via a linear combination of the divergence of auxiliary Gaussian fluxes. This avoids any interpolation from classic discretization approaches based on Stokes' theorem. We validate our operators by solving higher-dimensional elliptic problems.

E.J. Sanchez (✉)
Computer Applications for Science and Engineering, Barcelona Supercomputing Center (BSC), Carrer de Jordi Girona 29, Barcelona 08034, Cataluña, Spain

Computational Science Research Center, San Diego State University, 5500 Campanile Drive, San Diego, CA 92182-1245, USA
e-mail: eduardo.sanchez@bsc.es

G.F. Miranda • J.E. Castillo
Computational Science Research Center, San Diego State University, 5500 Campanile Drive, San Diego, CA 92182-1245, USA

J.M. Cela
Computer Applications for Science and Engineering, Barcelona Supercomputing Center (BSC), Carrer de Jordi Girona 29, Barcelona 08034, Cataluña, Spain

M.L. Bittencourt et al. (eds.), *Spectral and High Order Methods for Partial Differential Equations ICOSAHOM 2016*, Lecture Notes in Computational Science and Engineering 119, DOI 10.1007/978-3-319-65870-4_48

1 Introduction

Differential equations derived from conservation laws model many physical phenomena. Thus, we seek numerical solutions to these equations that preserve conservation and symmetry properties. In particular, **mimetic finite difference operators** satisfy a discrete instance of Gauss' divergence theorem [1]. Thus, they provide discrete models that preserve conservation and symmetry properties [1, 2].

Research on methods to construct mimetic operators is extensive, and we do not intend to present an exhaustive review. In previous works [3] we have studied both the Castillo–Grone method and the Castillo–Runyan method. We showed that these methods are not able to construct tenth-order gradient nor eighth-order divergence operators (nor higher).

This is because the computed weights discretizing the corresponding weighted inner products are not all positive. Therefore, we define these thresholds as **critical orders of accuracy** for each operator, respectively.

To overcome the aforementioned restrictions and to achieve supercritical-order accuracy, in [3, 4] we restated the problem of constructing supercritical-order mimetic operators as a constrained linear optimization problem. This produced a new algorithm, hereby referred to as the **Castillo–Blomgren–Sanchez algorithm**.

The contribution of this work is the extension to higher dimensions of the operators constructed by the Castillo–Blomgren–Sanchez algorithm. This includes detailing the mathematics of this extension. This also includes detailing the mathematics behind the construction of a mimetic curl operator via a linear combination of the divergence of auxiliary Gaussian fluxes, thus avoiding the need for interpolation from classic discretization approaches based on Stokes' theorem. Finally, we present preliminary examples to validate the correct functioning of the produced operators.

2 Supercritical-Accuracy Operators in One Dimension

In this work, mimetic operators are written using matrix notation. We will denote a k-th order (k even and positive) mimetic gradient, divergence, curl, and Laplacian operator evaluated on 2D and 3D domains as $\breve{\mathbf{G}}^k_{\{xy,xyz\}}$, $\breve{\mathbf{D}}^k_{\{xy,xyz\}}$, $\breve{\mathbf{C}}^k_{\{xy,xyz\}}$, and $\breve{\mathbf{L}}^k_{\{xy,xyz\}}$, respectively. Also, let n_i denote the number of cells discretizing in the i direction, $i \in \{x, y, z\}$.

The Castillo–Blomgren–Sanchez algorithm uses a constrained linear optimization method to compute positive weights, thus enforcing this constraint, in order to construct supercritical-order mimetic operators. Given the linearity of this problem, the Castillo–Blomgren–Sanchez algorithm uses the Simplex method [5] to solve the optimization problem that is constructed from a modified form of the system of equations used to compute the quadrature coefficient matrices $\mathbf{P}$ and $\mathbf{Q}$ needed to evaluate the following instance of Gauss' divergence theorem [1, 4]: $\Delta x\langle \breve{\mathbf{G}}\tilde{f}, \tilde{\mathbf{v}}\rangle_{\mathbf{P}} + \Delta x\langle \tilde{f}, \breve{\mathbf{D}}\tilde{\mathbf{v}}\rangle_{\mathbf{Q}} = \langle \tilde{f}, \breve{\mathbf{B}}\tilde{\mathbf{v}}\rangle$, where the boundary operator $\breve{\mathbf{B}}$ accounts for effect of the discretization near and at the boundaries arising from the selection of the discrete step-size, Δx.

The Castillo–Blomgren–Sanchez algorithm computes the quadrature coefficients as the solution of the following constrained optimization problem [4]:

$$\text{Find } \check{\tilde{\mathbf{q}}} \text{ such that (minimize) } \mathbf{r}_i^T \check{\tilde{\mathbf{q}}} = \min_{\tilde{\mathbf{q}} \in \mathbb{R}^k} \mathbf{r}_i^T \tilde{\mathbf{q}} \tag{1}$$

$$\text{subject to } \boldsymbol{\Phi}_i \check{\tilde{\mathbf{q}}} \geq \boldsymbol{\Lambda}_i, \text{ with } \check{\tilde{\mathbf{q}}} > \mathbf{0}, \tag{2}$$

where $\mathbf{r}_i$, $\check{\tilde{\mathbf{q}}} \in \mathbb{R}^{k \times 1}$, $\boldsymbol{\Phi}_i \in \mathbb{R}^{k \times k}$, $\boldsymbol{\Lambda}_i \in \mathbb{R}^{k \times 1}$, and $i \in [1, k+1]$. This optimization problem introduces a surplus quantity ϵ. In this context, ϵ is a parameter, or mimetic threshold, controlling the effect of $\mathbf{P}$ and $\mathbf{Q}$ on the conservative feature of the operators, while preserving a uniform order of numerical accuracy.

Note that the accuracy requirements for mimetic gradient and divergence operators are [6]:

$$(\breve{\mathbf{D}}_x^k x^j)_{i+1/2} - j((i+1/2)h)^{j-1} = 0, 0 \leq i \leq N-1, \tag{3}$$

$$(\breve{\mathbf{G}}_x^k x^j)_i - j(ih)^{j-1} = 0, 0 \leq i \leq N, \tag{4}$$

for $0 \leq j \leq k$ and $N \in \mathbb{Z}^+$. 1D mimetic operators built via the Castillo–Blomgren–Sanchez algorithm do achieve the required order of accuracy, as shown in Tables 1 and 2. In these tables, the relative error of the differences (3) and (4) is produced, for each order of accuracy, for different values of the ϵ parameter.

Table 1 Results measuring the accuracy of the constructed divergence operators

k	Relative error		
	$\epsilon = 1 \times 10^{-3}$	$\epsilon = 1 \times 10^{-6}$	$\epsilon = 1 \times 10^{-9}$
2	5.10756e−14	5.10756e−14	5.10756e−14
4	7.55433e−14	7.55433e−14	7.55433e−14
6	6.23919e−14	6.23919e−14	6.23919e−14
8	1.55481e−13	1.57592e−13	1.52383e−13
10	2.79119e−12	2.80687e−10	4.43036e−07
12	7.27814e−11	7.2776e−11	7.27843e−11
14	8.11899e−10	8.11943e−10	8.11913e−10

Table 2 Results measuring the accuracy of the constructed gradient operators

k	Relative error		
	$\epsilon = 1 \times 10^{-3}$	$\epsilon = 1 \times 10^{-6}$	$\epsilon = 1 \times 10^{-9}$
2	5.8315e−14	5.8315e−14	5.8315e−14
4	7.85105e−14	7.85105e−14	7.85105e−14
6	6.27153e−14	6.27153e−14	6.27153e−14
8	7.00361e−13	7.00361e−13	7.00361e−13
10	5.67544e−12	5.67483e−12	5.67361e−12
12	8.11966e−11	1.08737e−09	4.31647e−07
14	1.55235e−09	7.54992e−09	1.98015e−05

3 Supercritical-Accuracy Operators in Higher Dimensions

Assuming a 2D domain discretized via a uniform staggered grid, and assuming the existence of supercritical-accuracy 1D operators $\breve{\mathbf{G}}_i^k \in \mathbb{R}^{(n_i+1)\times(n_i+2)}$ and $\breve{\mathbf{D}}_x^k \in \mathbb{R}^{(n_i+2)\times(n_i+1)}$ have been built [3], then:

$$\breve{\mathbf{G}}_{xy}^k = \left[(\hat{\mathbf{I}}_y^\top \otimes \breve{\mathbf{G}}_x^k)\,(\breve{\mathbf{G}}_y^k \otimes \hat{\mathbf{I}}_x^\top) \right]^\top, \; \breve{\mathbf{D}}_{xy}^k = \left[(\hat{\mathbf{I}}_y \otimes \breve{\mathbf{D}}_x^k)\,(\breve{\mathbf{D}}_y^k \otimes \hat{\mathbf{I}}_x) \right], \tag{5}$$

where $\hat{\mathbf{I}}_i \in \mathbb{R}^{(n_i+2)\times(n_i+2)}$ is a zero-padded identity matrix, built according to the discretization in the i direction. Simple algebra will show that:

$$\breve{\mathbf{G}}_{xy}^k \in \mathbb{R}^{[(n_y+2)(n_x+1)+(n_y+1)(n_x+2)]\times(n_y+2)(n_x+2)}, \tag{6}$$

$$\breve{\mathbf{D}}_{xy}^k \in \mathbb{R}^{(n_y+2)(n_x+2)\times[(n_y+2)(n_x+1)+(n_y+1)(n_x+2)]}. \tag{7}$$

Assuming a 3D domain discretized via a uniform staggered grid, then:

$$\breve{\mathbf{G}}_{xyz}^k = \left[(\hat{\mathbf{I}}_z^\top \otimes \hat{\mathbf{I}}_y^\top \otimes \breve{\mathbf{G}}_x^k)\,(\hat{\mathbf{I}}_z^\top \otimes \breve{\mathbf{G}}_y^k \otimes \hat{\mathbf{I}}_x^\top)\,(\breve{\mathbf{G}}_z^k \otimes \hat{\mathbf{I}}_y^\top \otimes \hat{\mathbf{I}}_x^\top) \right]^\top, \tag{8}$$

$$\breve{\mathbf{D}}_{xyz}^k = \left[(\hat{\mathbf{I}}_z \otimes \hat{\mathbf{I}}_y \otimes \breve{\mathbf{D}}_x^k)\,(\hat{\mathbf{I}}_z \otimes \breve{\mathbf{D}}_y^k \otimes \hat{\mathbf{I}}_l)\,(\breve{\mathbf{D}}_z^k \otimes \hat{\mathbf{I}}_y \otimes \hat{\mathbf{I}}_x) \right]. \tag{9}$$

Simple algebra will also show that:

$$\breve{\mathbf{G}}_{xy}^k \in \mathbb{R}^{[n_y n_z(n_x+1)+n_x n_z(n_y+1)+n_x n_y(n_z+1)]\times[(n_x+2)+(n_y+2)+(n_z+2)]}, \tag{10}$$

$$\breve{\mathbf{D}}_{xy}^k \in \mathbb{R}^{[(n_x+2)+(n_y+2)+(n_z+2)]\times[n_y n_z(n_x+1)+n_x n_z(n_y+1)+n_x n_y(n_z+1)]}. \tag{11}$$

The domain of the mimetic gradient operator is the set of points on the staggered grid onto which the scalar fields are bound, and its range is the set of points on the staggered grid onto which the vector fields are bound. Conversely, the domain of the mimetic divergence operator is the set of points on the staggered grid onto which the vector fields are bound, and its range is the set of points on the staggered grid onto which the scalar fields are bound.

For both 2D and 3D domains, and following the works [1, 3], the mimetic Laplacian is built as the matrix product:

$$\breve{\mathbf{L}}_{\{xy,xyz\}}^k = \breve{\mathbf{D}}_{\{xy,xyz\}}^k \breve{\mathbf{G}}_{\{xy,xyz\}}^k. \tag{12}$$

From (6) and (7) we note that $\breve{\mathbf{L}}_{xy}^k \in \mathbb{R}^{(n_x+2)(n_y+2)\times(n_x+2)(n_y+2)}$. Analogously, from (10) and (11) we note that $\breve{\mathbf{L}}_{xyz}^k \in \mathbb{R}^{(n_x+2)(n_y+2)(n_z+2)\times(n_x+2)(n_y+2)(n_z+2)}$. Note that by leveraging the supercritical-accuracy of the gradient and divergence operators, we do not need any additional projection mechanisms, such as the use of the Hodge start operator, for example.

3.1 Poisson's Equation on a 2D Staggered Grid

We solve:

$$\nabla^2 u(x, y) = F(x, y), \tag{13}$$

for $(x, y) \in (0, 1)^2$, with

$$F(x, y) = xy \exp\left(-\frac{1}{2}x^2 - \frac{1}{2}y^2\right)(x^2 + y^2 - 6). \tag{14}$$

Boundary conditions for a domain $\Omega = [0 : \Delta x : x_m] \times [0 : \Delta y : y_n]$:

$$u(x, 0) = u(0, y) = 0, \tag{15}$$

$$\nabla u(x, y_n) = -y_n \exp\left(-\frac{1}{2}x^2 - \frac{1}{2}y_n^2\right)(x^2 - 1), \tag{16}$$

$$\nabla u(x_m, y) = -x_m \exp\left(-\frac{1}{2}x_m^2 - \frac{1}{2}y^2\right)(y^2 - 1), \tag{17}$$

where (for our example) $x_m = y_n = 1$. Convergence of relative error between computed and analytic solutions are showed in Fig. 1. This shows that the attained

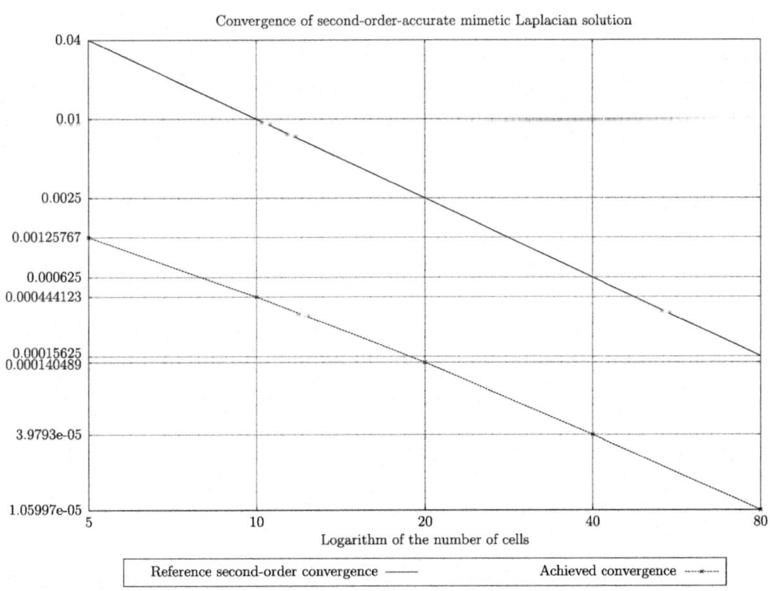

Fig. 1 Convergence plot for second-order accuracy exhibited by the mimetic discretization of equation (13). We show a reference line for ideal second-order accuracy, as well as the convergence of the relative error between the analytic and the computed solutions to the problem

accuracy for the mimetic Laplacian is indeed second order, which is consistent with the results in Tables 1 and 2. These results were computed using the Mimetic Methods Toolkit (MTK)—a C++11 library for mimetic numerical methods [7, 8].

4 Supercritical-Accuracy Curl Operator

Let $\mathbf{v}(\mathbf{x}) = p(\mathbf{x})\mathbf{i} + q(\mathbf{x})\mathbf{j} + r(\mathbf{x})\mathbf{k}$ be some smooth 3D vector field. Classically, the curl operator is defined as follows: Let S be a surface with a normal $\mathbf{n}$, whose boundary C is a closed path. If $A(S)$ denotes the area of S, then, by Stokes' theorem (Fig. 2):

$$\operatorname{curl} \mathbf{v}(\mathbf{x}) \cdot \mathbf{n} = \lim_{A(S) \longmapsto 0} \frac{1}{A(S)} \oint_C \mathbf{v} \cdot dC. \tag{18}$$

If S is a 2D rectangle (for example) then C is a set of four rectilinear edges, and the evaluation of the circulation of $\mathbf{v}$ along C needs an estimation for the tangential components of $\mathbf{v}$. This implies the introduction of dual spaces in the context of the general Stokes' theorem on manifolds. Furthermore, common discretizations for the curl operator, such as the one proposed in [9], are based on (18). However, this proposed discretization requires the interpolation of the argument vector field.

In this work, we first propose an alternative definition for the curl operator, so that the introduction of dual spaces becomes unnecessary. We then use this alternative definition to construct a mimetic curl operator in two and three dimensions that profits from the uniform supercritical-order accuracy exhibited by mimetic divergence operators, without requiring any interpolation.

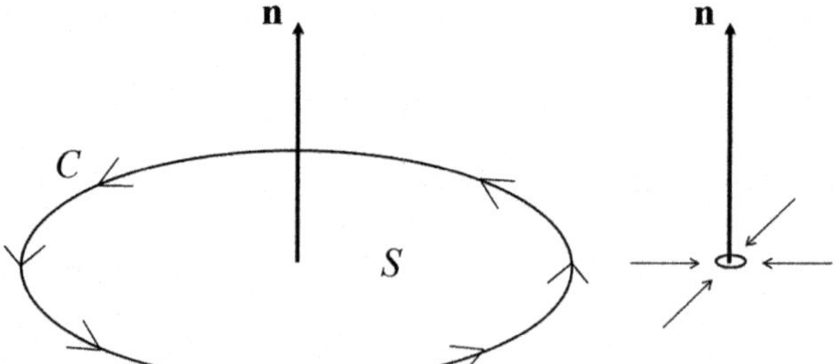

Fig. 2 A small rotating disk S, bounded by C, with an orienting normal $\mathbf{n}$. A limiting process then takes place by collapsing $A(S)$ to 0, thus allowing for a definition for the curl operator via "Stokian circulations"

4.1 Redefining the Curl Operator Via "Gaussian Fluxes"

In order to consider an alternative definition for the curl operator, we first consider the more practical statement:

$$\nabla \times \mathbf{v}(\mathbf{x}) = \left(\frac{\partial r}{\partial y} - \frac{\partial q}{\partial z}\right)\mathbf{i} + \left(\frac{\partial p}{\partial z} - \frac{\partial r}{\partial x}\right)\mathbf{j} + \left(\frac{\partial q}{\partial x} - \frac{\partial p}{\partial y}\right)\mathbf{k}. \tag{19}$$

Mathematically, a closed circuit such as C is a 1D object, and can be thought of as modeling a 3D thin wire having a cross-section with an infinitely small diameter, thus collapsing its three dimensions to only one dimension. But there is an alternative way for collapsing a 3D object down to a 2D object having a 1D boundary C.

Instead of a 2D plane surface S with oriented normal $\mathbf{n}$, think of a thin 3D cylindrical plate, with cylindrical axis along $\mathbf{n}$, and having S as its uniform cross-section (Fig. 3). If this 3D cylindrical plate becomes infinitely thin, then it becomes a 2D surface with 1D boundary C, but now C can be regarded as an object which is a limiting form for a 2D band or cylindrical mantle M, through which some vector field $\mathbf{v}$ can flow, and some flux can be computed. This mantle M, together with two surfaces parallel to S: S_u above S and S_d below S, with S_u and S_d having the same area $A(S)$ and being very close to one another, constitute the total surface boundary of a 3D thin plate P.

Naturally, if we consider some 3D vector field which is normal to $\mathbf{n}$, and therefore, also parallel to S_d and S_u, then its Gaussian flux through the boundary of P would reduce to the flux through the mantle M. Since M is a 2D band, with a width w equal to the distance between the parallel surfaces S_d and S_u, it follows that

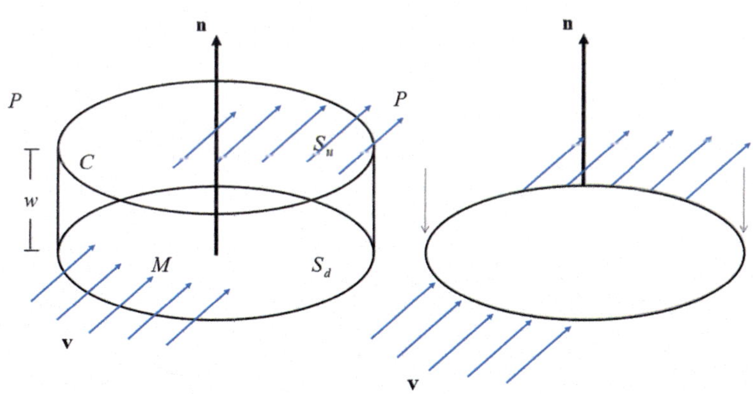

Fig. 3 A limiting process for an infinitesimally thin disk S with boundary C and orienting normal $\mathbf{n}$ created upon collapsing surfaces S_u and S_d, aligned through a mantle M of width w, which is then considered to tend to 0. Also, a Gaussian-like flux $\mathbf{v}$ flows through the infinitely thin disk S

when w tends to zero, S_d and S_u collapse to S, so the 2D band M collapses to the 1D closed circuit C.

This visualization of geometric dimensional collapse will allow us, in the next subsection, to numerically estimate the scalar components of a 3D curl vector field from some adequate 2D fluxes, rather than from 1D circulations. This will be possible by means of some auxiliary 2D vector fields. In turn, these 2D fluxes will be related to supercritical-order 2D mimetic divergence operators.

4.2 Auxiliary 2D Vector Fields

The basic definitions are described in [1, 4]. Furthermore, in [1, 4], the type of 2D staggering needed in order to compute the 2D curl is worked out in detail, but the combination of simultaneous discretizations in the x, y and z directions needed in the 3D case is only hinted at graphically (see Fig. 4.10 in [1]). In this work, we present a more detailed explanation and we make explicit their relation to supercritical-order mimetic operators.

Considering (19), we define the following auxiliary vector fields: $\mathbf{v}^*_{xy} = \mathbf{i}q - \mathbf{j}p = \mathbf{i}P^*_{xy} + \mathbf{j}Q^*_{xy}$, $\mathbf{v}^*_{yz} = \mathbf{j}r - \mathbf{k}q = \mathbf{j}Q^*_{yz} + \mathbf{k}R^*_{yz}$, and $\mathbf{v}^*_{zx} = \mathbf{k}p - \mathbf{i}r = \mathbf{k}R^*_{zx} + \mathbf{i}P^*_{zx}$. It follows immediately that:

$$\nabla \times \mathbf{v}(\mathbf{x}) = (\nabla \cdot \mathbf{v}^*_{yz}(\mathbf{x}))\mathbf{i} + (\nabla \cdot \mathbf{v}^*_{zx}(\mathbf{x}))\mathbf{j} + (\nabla \cdot \mathbf{v}^*_{xy}(\mathbf{x}))\mathbf{k}. \tag{20}$$

Therefore, the 3D vector expression for curl$\mathbf{v}(\mathbf{x})$ depends upon three scalar 2D divergences evaluated at $\mathbf{x}$. These 2D divergences, simultaneously needed for the 3D curl of $\mathbf{v}$, all arise from 2D fluxes of vector fields $\mathbf{v}^*_{xy}$, $\mathbf{v}^*_{yz}$, and $\mathbf{v}^*_{zx}$. These vector fields lie in planes orthogonal to the coordinate axis, passing through $\mathbf{x}$. This fact yields the following definitions for 2D vector fields: $(\nabla \times \mathbf{v}(\mathbf{x})) \cdot \mathbf{i} = \nabla \cdot \mathbf{v}^*_{yz}(\mathbf{x})$, $(\nabla \times \mathbf{v}(\mathbf{x})) \cdot \mathbf{j} = \nabla \cdot \mathbf{v}^*_{zx}(\mathbf{x})$, and $(\nabla \times \mathbf{v}(\mathbf{x})) \cdot \mathbf{k} = \nabla \cdot \mathbf{v}^*_{xy}(\mathbf{x})$. Therefore, the mimetic counterparts of the 2D and 3D curl operators are:

$$\breve{\mathbf{C}}^k_{xyz}\tilde{\mathbf{v}}(\mathbf{x}) = \breve{\mathbf{D}}^k_{yz}\tilde{\mathbf{v}}^*_{yz}(\mathbf{x})\mathbf{i} + \breve{\mathbf{D}}^k_{zx}\tilde{\mathbf{v}}^*_{zx}(\mathbf{x})\mathbf{j} + \breve{\mathbf{D}}^k_{zy}\tilde{\mathbf{v}}^*_{zy}(\mathbf{x})\mathbf{k}, \tag{21}$$

$$\breve{\mathbf{C}}^k_{zy}\tilde{\mathbf{v}}(\mathbf{x}), = \breve{\mathbf{D}}^k_{xy}\tilde{\mathbf{v}}^*_{xy}(\mathbf{x}). \tag{22}$$

4.3 Compatibility with Stokes' Theorem

In this subsection, we go back to the relation between the Stokes-based and the Gaussian-based definitions for curl $\mathbf{v}$, while considering the component along z.

From Stokes' theorem:

$$\iint_S (\nabla \times \mathbf{v}(\mathbf{x})) \cdot \mathbf{k}dxdy = \oint_C (p(x, y, z)dx + q(x, y, z)dy). \tag{23}$$

When $\mathbf{i}dx + \mathbf{j}dy$ is a tangent vector of length ds along a counterclockwise oriented circuit C in the $x-y$ plane, then $\mathbf{i}dy - \mathbf{j}dx$ is a normal field $\mathbf{n}ds$, outwardly directed to C. Therefore, the previous Stokes' formula can be now read "Gauss-like" as follows:

$$\iint_S (\nabla \cdot \mathbf{v}^*_{xy}(x,y,z))dxdy = \oint_C (P^*_{xy}(x,y,z)dy - Q^*_{xy}(x,y,z)dx) \tag{24}$$

$$= \oint_C < \mathbf{i}P^*_{xy} + \mathbf{j}Q^*_{xy}, \mathbf{i}dy - \mathbf{j}dx > \tag{25}$$

$$= \oint_C < \mathbf{v}^*_{xy}(x,y,z), \mathbf{n}(x,y,z) > ds. \tag{26}$$

Since these expressions also equal the mean value of $< \mathbf{k}, \mathrm{curl}\mathbf{v}(x,y,z) >$ times the area of the surface surrounded by C, then we obtain the following mean value:

$$\oint < \mathbf{v}^*_{xy}(x,y,z), \mathbf{n}(x,y,z) > ds. \tag{27}$$

The mean value in (27) equals the outward flux of $\mathbf{v}^*_{xy}$ through C, divided by the above surface area. Therefore, we conclude that this approach preserves the original behavior inherent to the functioning of Stokes' theorem. Furthermore, this approach is simple, in the sense that, in its foundation, it is just a change of variables.

4.4 A Simplified Model of a Vortex

Let $\mathbf{v}(x,y,z) = -\mathbf{i}y + \mathbf{j}x$. This vector field is plotted in Fig. 4a. Since $\mathbf{v} = \mathbf{k} \times (\mathbf{i}x + \mathbf{j}y)$, we have $p(x,y,z) = -y$, $q(x,y,z) = x$, and $r(x,y,z) = 0$. In this case, we know that

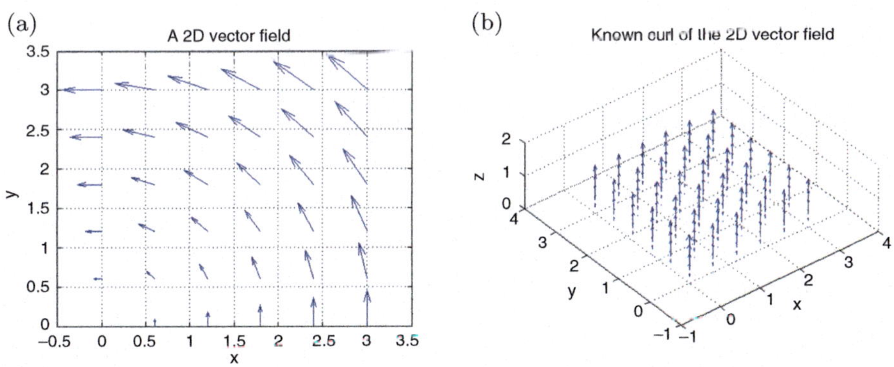

Fig. 4 Reference solution. (**a**) Vector field: $\mathbf{v}(\mathbf{x}) = -y\mathbf{i} + x\mathbf{j}$. (**b**) Known curl: $\nabla \times \mathbf{v} = 2\mathbf{k}$

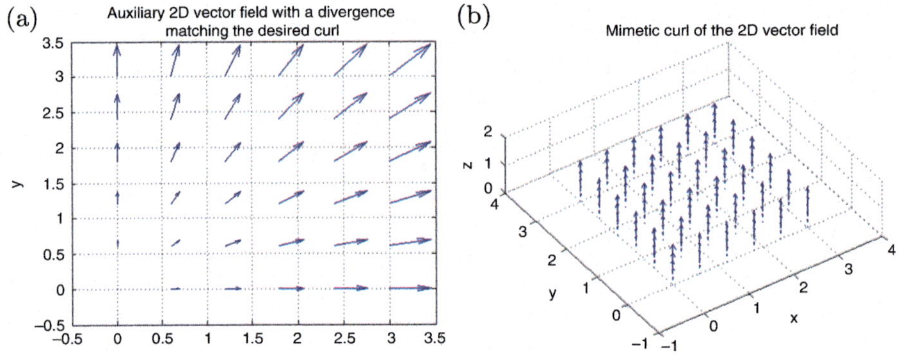

Fig. 5 Computed mimetic curl via the proposed Gaussian approach. (**a**) Auxiliary vector field: $\mathbf{v}^*_{xy}$. (**b**) $\breve{\mathbf{D}}^k_{xy}\tilde{\mathbf{v}}^*_{xy} = \breve{\mathbf{C}}^k_{xy}\tilde{\mathbf{v}}$

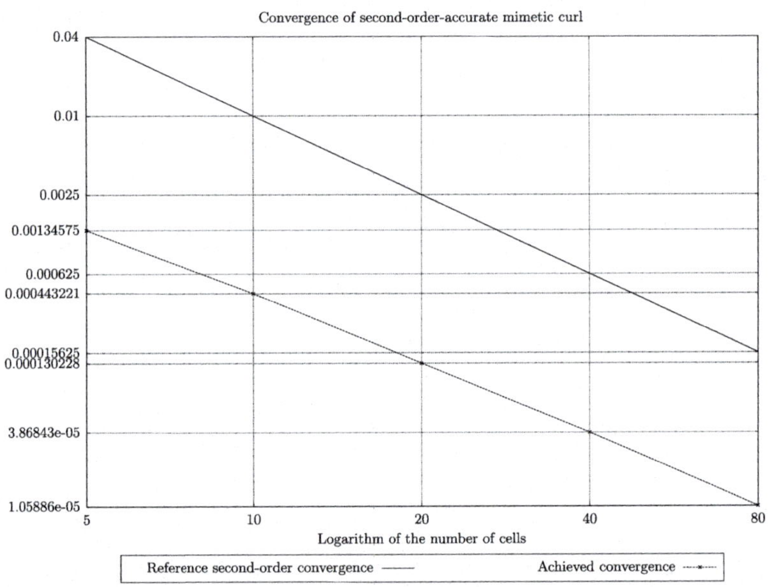

Fig. 6 Convergence plot for second-order accuracy exhibited by the mimetic discretization of the curl operator. We show a reference line for ideal second-order accuracy, as well as the convergence of the relative error between the analytic and the computed solutions to the problem

$\mathbf{v} \times \mathbf{k} = \mathbf{i} \times x + \mathbf{j}y = \mathbf{v}^*_{xy}(x, y, z)$. Thus, $\mathrm{div}\mathbf{v}^*_{xy} = 1 + 1 = 2$, which is grid-wise constant. The known curl is rendered in Fig. 4b. The suggested vector field, was then discretized, on a staggered grid, and the auxiliary vector field was also discretized. Figure 5a plots the auxiliary field. Figure 5b shows the computed mimetic curl, through the Gaussian approach. This plot has to be compared with that of Fig. 4b. Figure 6 shows the second-order convergence achieved via successive mesh grid refinement.

5 Concluding Remarks

We have detailed the construction of mimetic operators with supercritical accuracy in higher dimensions, based on their 1D counterparts resulting from the Castillo–Blomgren–Sanchez algorithm. Results of the 1D operators show that the 1D operators yield the required accuracy. Kronecker products with these 1D supercritical-order operators are used as the main tool to formulate the construction of the higher-dimensional supercritical-order operators.

Two test cases include an elliptic problem and the direct computation of a curl vector field. The results were successful, although further testing is required in order to truly render the benefit of supercritical accuracy in higher-dimensional contexts. We intend to expand this article with additional numerical experiments to fully study the impact of higher-orders of numerical accuracy; however, in this work, the theory behind this extension has been detailed and supported with preliminary results.

Immediate future work will focus on adapting the operators, constructed via the Castillo–Blomgren–Sanchez algorithm to non-uniform rectangular grids. This will prompt the study of the application of these operators in frequency-domain problems, for which very high orders of numerical accuracy are desired.

The authors would like to extend their gratitude to Dr. Josep de la Puente, and Dr. Otilio Rojas, for many fruitful discussions about this topic.

References

1. J.E. Castillo, G.F. Miranda, *Mimetic Discretization Methods*, 1st edn. (CRC Press, Boca Raton, 2013)
2. J.E. Castillo, J.M. Hyman, M.J. Shashkov, S. Steinberg, The sensitivity and accuracy of fourth order finite difference schemes on nonuniform grids in one dimension. Comput. Math. Appl. **30**(8), 41–55 (1995)
3. E. Sanchez, C. Paolini, P. Blomgren, J. Castillo, Algorithms for higher-order mimetic operators, in *Spectral and High Order Methods for Partial Differential Equations ICOSAHOM 2014: Selected papers from the ICOSAHOM conference, June 23–27, 2014, Salt Lake City, Utah, USA* (Springer International Publishing, Cham, 2015), pp. 425–434
4. E.J. Sanchez, Mimetic finite differences and parallel computing to simulate carbon dioxide subsurface mass transport, PhD thesis, Claremont Graduate University and San Diego State University, 2015
5. G.B. Dantzig, Linear programming and extensions, Technical report, August 1963
6. J.E. Castillo, J.M. Hyman, M. Shashkov, S. Steinberg, Fourth- and sixth-order conservative finite difference approximations of the divergence and gradient. Appl. Numer. Math. **37**, 171–187 (2001)
7. E.J. Sanchez, C.P. Paolini, J.E. Castillo, The mimetic methods toolkit: an object-oriented api for mimetic finite differences. J. Comput. Appl. Math. **270**, 308–322 (2014). Fourth International Conference on Finite Element Methods in Engineering and Sciences (FEMTEC 2013)
8. E.J. Sanchez, Mimetic Methods Toolkit (MTK) (2012), http://www.csrc.sdsu.edu/mtk/,
9. J.B. Runyan, A novel higher order finite difference time domain method based on the Castillo-Grone mimetic curl operator with application concerning the time-dependent Maxwell equations, Master's thesis, San Diego State University, San Diego, CA, 2011

An *hp* Finite Element Method for Fourth Order Singularly Perturbed Problems

Christos Xenophontos, Philippos Constantinou, and Charalambia Varnava

Abstract We present an *hp* Finite Element Method (FEM) for the approximation to the solution of singularly perturbed fourth order problems in one-dimension. In (Panaseti et al, Appl Numer Math 104:81–97, 2016) it was shown that the *hp* version of the FEM, on the so-called *Spectral Boundary Layer Mesh* (Melenk et al, IMA J Numer Anal 33(2):609–628, 2013) yields robust exponential convergence when the error is measured in the energy norm. This result is sharpened by showing that the same method gives robust exponential convergence in a stronger, more *balanced* norm. As a corollary, we also get exponential convergence in the maximum norm. A numerical example illustrating the theory is also presented.

1 Introduction

The numerical solution of singularly perturbed problems has been studied extensively over the last few decades (see, e.g., the books [8, 15] and the references therein). A main difficulty in these problems is the presence of *boundary layers* in the solution and the numerical method should be tailored for their effective approximation, in order to be considered *robust* (meaning it converges independently of the singular perturbation parameter). The solution of such problems is usually decomposed into a smooth part and a boundary layer part and the numerical method aims at approximating both components equally well. Classical methods are able to approximate smooth (enough) solutions, but they fail in approximating the boundary layer. In the context of the Finite Element Method (FEM) and Finite Difference Methods (FDM), the robust approximation of boundary layers requires the use of layer adapted, parameter-dependent meshes (cf. [1] and [19] for FDM and [5, 17] for *hp*-FEM, both methods applied to second order problems). In the aforementioned references, the error estimates obtained are given in the

C. Xenophontos (✉) • P. Constantinou • C. Varnava
Department of Mathematics and Statistics, University of Cyprus, Nicosia, Cyprus
e-mail: xenophontos@ucy.ac.cy; constantinouphilippos@gmail.com; harisv_8@hotmail.com

M.L. Bittencourt et al. (eds.), *Spectral and High Order Methods for Partial Differential Equations ICOSAHOM 2016*, Lecture Notes in Computational Science and Engineering 119, DOI 10.1007/978-3-319-65870-4_49

(natural) energy norm associated with the boundary value problem. However, the energy norm for some problems is rather weak in that it does not "see" the boundary layers. By that we mean that the energy norm of the layer vanishes as the singular perturbation parameter $\varepsilon \to 0$, whereas the energy norm of the smooth part of the solution does not. Recently, for second order problems, the convergence of h-FEM methods has been analyzed in norms stronger than the energy norm [4, 12, 13]. The analysis was performed in an ε-weighted H^1-norm which is *balanced* in the sense that both the smooth part and the layer part are bounded away from zero uniformly in ε. The weight in this ε-weighted H^1-norm is $\varepsilon^{1/2}$ as opposed to ε as in the energy norm. Robust convergence in this balanced norm is shown in [4, 12, 13], if Shishkin meshes are used. In [6] the same ideas were used for an *hp*–FEM on the *Spectral Boundary Layer Mesh* (see Definition 1 ahead) and robust exponential convergence in this balanced norm was established.

When one considers fourth order singularly perturbed problems, the available results are not as plentiful (notable exceptions are the works [9, 11, 14, 18, 20]). Recently, it was shown in [10] that the *hp* version of the FEM on the *Spectral Boundary Layer Mesh* applied to fourth order singularly perturbed problems yields robust exponential convergence when the error is measured in the energy norm. Following [6], the analysis of [10] was extended in [3] where it was shown that the *hp* version of the FEM on the *Spectral Boundary Layer Mesh* applied to fourth order singularly perturbed problems, yields robust exponential convergence in this balanced norm as well.

The rest of the article is organized as follows: In Sect. 2 we describe the model problem, its variational formulation and its discretization by an *hp*–FEM on the *Spectral Boundary Layer Mesh.* In Sect. 3 we present the analysis of the method in the balanced norm, obtaining in addition robust exponential convergence in the maximum norm. Finally, Sect. 4 contains the results of a numerical experiment to illustrate the theory.

We will utilize the usual Sobolev space notation $H^k(I)$ to denote the space of functions on the interval I with $0, 1, 2, \ldots, k$ generalized derivatives in $L^2(I)$, equipped with the norm $\|\cdot\|_{k,I}$ and seminorm $|\cdot|_{k,I}$. We will also use the space $H_0^k(I) = \left\{ u \in H^k(I) : u^{(i)}\big|_{\partial I} = 0, i = 0, \ldots, k-1 \right\}$, where ∂I denotes the boundary of I. The norm of the space $L^\infty(I)$ of essentially bounded functions is denoted by $\|\cdot\|_{\infty,I}$. The letters C, c will be used to denote generic positive constants, independent of any discretization or singular perturbation parameters and possibly having different values in each occurrence.

2 The Model Problem and Its Discretization

2.1 Variational Formulation and Regularity

We consider the following model problem: Find u such that

$$\mathscr{L}_\varepsilon u(x) := \varepsilon^2 u^{(4)}(x) - \big(a(x)u'(x)\big)' + b(x)u(x) = f(x), \quad x \in I = (0,1), \tag{1}$$

along with the boundary conditions

$$u(0) = u'(0) = u'(1) = u(1) = 0. \tag{2}$$

The parameter $0 < \varepsilon \leq 1$ is given, as are the functions $a, b > 0$ and f, which are assumed to be analytic on $\bar{I} = [0,1]$. In particular, we assume that there exist constants $C_f, \gamma_f, C_a, \gamma_a, C_b, \gamma_b > 0$, independent of ε, such that

$$\left\| f^{(n)} \right\|_{\infty,I} \leq C_f \gamma_f^n n!, \left\| a^{(n)} \right\|_{\infty,I} \leq C_a \gamma_a^n n!, \left\| b^{(n)} \right\|_{\infty,I} \leq C_b \gamma_b^n n! \quad \forall\, n = 0,1,2,\ldots. \tag{3}$$

The variational formulation of (1)–(2) reads: Find $u \in H_0^2(I)$ such that

$$\mathscr{B}_\varepsilon(u,v) = \mathscr{F}(v) \quad \forall\, v \in H_0^2(I), \tag{4}$$

where, with $\langle \cdot, \cdot \rangle_I$ the usual $L^2(I)$ inner product,

$$\mathscr{B}_\varepsilon(u,v) = \varepsilon^2 \langle u'', v'' \rangle_I + \langle au', v' \rangle_I + \langle bu, v \rangle_I \;, \; \mathscr{F}(v) = \langle f, v \rangle_I. \tag{5}$$

It follows that the bilinear form $\mathscr{B}_\varepsilon(\cdot,\cdot)$ given by (5) is coercive with respect to the *energy norm*

$$\|u\|_{E,I}^2 := \mathscr{B}_\varepsilon(u,u),$$

i.e., $\mathscr{B}_\varepsilon(u,u) \geq \|u\|_{E,I}^2 \quad \forall\, u \in H_0^2(I)$. The solution u is analytic in $\bar{I}$ and its derivative features boundary layers at the endpoints. The following result from [10] describes the regularity of u (see also [2]).

Theorem 1 *Assume (3) and let $u \in H_0^2(I)$ be the solution of (5). Then*

(i) there exist constants $C, K > 0$ depending only on the data such that

$$\left\| u^{(n)} \right\|_{\infty,I} \leq CK^n \max\{n^n, \varepsilon^{1-n}\} \quad \forall\, n = 0,1,2,\ldots$$

(ii) *u can be decomposed as* $u = w + u_{BL} + r$, *where for some constants* C_w, γ_w, C_{BL}, γ_{BL}, C_r, γ_r, $\beta > 0$ *independent of* ε, *there holds*

$$\left\| w^{(n)} \right\|_{\infty,I} \leq C_w \gamma_w^n n^n \quad \forall\, n = 0, 1, 2, \ldots$$

$$\left| u_{BL}^{(n)}(x) \right| \leq C_{BL} \gamma_{BL}^n \varepsilon^{1-n} e^{-\beta dist(x,\partial I)/\varepsilon} \quad \forall\, n = 1, 2, \ldots$$

$$\|r\|_{\infty,\partial I} + \|r'\|_{\infty,\partial I} + \|r\|_{E,I} \leq C_r e^{-\gamma_r/\varepsilon}.$$

2.2 *Discretization by a* C^1 *hp-FEM*

The discrete version of (4) reads: find $u_{FEM} \in S \subset H_0^2(I)$ such that

$$\mathscr{B}_\varepsilon (u_{FEM}, v) = \mathscr{F}(v) \quad \forall\, v \in S \subset H_0^2(I),$$

with the finite dimensional subspace S defined as follows: partition the interval $I = (0, 1)$ using an arbitrary mesh $\Delta = \{0 = x_0 < x_1 < \ldots < x_{\mathscr{M}} = 1\}$ and set $I_j = (x_{j-1}, x_j)$, $h_j = x_j - x_{j-1}, j = 1, \ldots, \mathscr{M}$. Also, define the reference (or standard) element $I_{ST} = (-1, 1)$, and note that it can be mapped onto the jth element I_j by the linear mapping $x = Q_j(t) = \frac{1}{2}(1-t)x_{j-1} + \frac{1}{2}(1+t)x_j$. With $\Pi_p(I_{ST})$ the space of polynomials of degree less than or equal to $p \geq 3$ on I_{ST} (and with $\circ$ denoting composition of functions), we define the finite dimensional subspaces

$$S^p(\Delta) = \left\{ v \in H^2(I) \cap H_0^1(I) : v \circ Q_j^{-1} \in \Pi_{p_j}(I_{ST}), j = 1, \ldots, \mathscr{M} \right\},$$
$$S_0^p(\Delta) = S^p(\Delta) \cap H_0^2(I),$$

and take $S = S_0^p(\Delta)$. For simplicity, we assume constant polynomial degree p for all elements, i.e., $p_j = p, j = 1, \ldots, \mathscr{M}$. The above discretization was introduced in [10], where appropriate hierarchical basis functions were constructed for the space $S_0^p(\Delta)$ and whose approximation properties were studied (see Lemma 7 in [10] for details).

The following definition describes the mesh that will be used for the resolution of the layers (cf. [7]).

Definition 1 (Spectral Boundary Layer Mesh) For $\kappa > 0, p \in \mathbf{N}$ and $0 < \varepsilon \leq 1$, define the Spectral Boundary Layer mesh $\Delta_{BL}(\kappa, p)$ as

$$\Delta_{BL}(\kappa, p) = \begin{cases} \{0, \kappa p\varepsilon, 1 - \kappa p\varepsilon, 1\} & \textit{if } \kappa p\varepsilon < 1/4 \\ \{0, 1\} & \textit{if } \kappa p\varepsilon \geq 1/4 \end{cases}$$

The spaces $S(\kappa, p)$ and $S_0(\kappa, p)$ of piecewise polynomials of degree $p \geq 3$ are given by

$$S(\kappa, p) := S^p(\Delta_{BL}(\kappa, p)),$$
$$S_0(\kappa, p) := S_0^p(\Delta_{BL}(\kappa, p)) = S(\kappa, p) \cap H_0^2(I).$$

In [10] the following was established.

Theorem 2 *Assume (3) holds. Let u be the solution to (4) and $u_{FEM} \in S_0(\kappa, p)$ its finite element approximation based on the Spectral Boundary Layer Mesh. Then, there exists positive constants C, σ, independent of ε, u and p such that*

$$\|u_{FEM} - u\|_{E,I} \approx \|u_{FEM} - u\|_{1,I} + \varepsilon \left\|(u_{FEM} - u)''\right\|_{0,I} \leq Ce^{-\sigma p}.$$

Our next goal is to obtain the following estimate:

$$\|u_{FEM} - u\|_{1,I} + \varepsilon^{1/2} \left\|(u_{FEM} - u)''\right\|_{0,I} \leq Ce^{-\sigma p}, \tag{6}$$

for some constants $C, \sigma > 0$ independent of ε. Comparing (6) with the result of Theorem 2, we see that

$$\|u_{BL}\|_{1,I} + \varepsilon \|u_{BL}''\|_{0,I} = O(\varepsilon^{1/2}),$$

while

$$\|u_{BL}\|_{1,I} + \varepsilon^{1/2} \|u_{BL}''\|_{0,I} = O(1).$$

This shows that the layer contribution goes to 0 as $\varepsilon \to 0$, if the energy norm is used (hence the phrase 'the energy norm does not see the layer'). The 'balanced norm' estimate (6) sharpens the result of Theorem 2.

In oder the achieve our goal, we need a more refined approximation result. In [3] the following was established.

Lemma 1 *Assume that (3) holds and let u be the solution to (4). Then there are constants C, κ_0, $\beta > 0$ (depending only on the data) such that for every $\kappa \in (0, \kappa_0]$ there exists an approximation $\mathscr{I}_p u \in S_0(\kappa, p)$ that satisfies*

$$\|u - \mathscr{I}_p u\|_{\infty,I} + \|\left(u - \mathscr{I}_p u\right)'\|_{\infty,I} \leq Ce^{-\beta\kappa p}, \tag{7}$$

$$\|u - \mathscr{I}_p u\|_{1,I} + \sqrt{\kappa p \varepsilon} \|\left(u - \mathscr{I}_p u\right)''\|_{0,I} \leq Ce^{-\beta\kappa p}. \tag{8}$$

3 Error Estimates in the Balanced Norm

We begin by defining the bilinear form $\mathscr{B}_0 : H_0^1(I) \times H_0^1(I) \to \mathbf{R}$ by

$$\mathscr{B}_0(u, v) = \langle au', v' \rangle_I + \langle bu, v \rangle_I, \tag{9}$$

corresponding to the reduced/limit problem. We also introduce the operator $\mathscr{P}_0$: $H_0^1(I) \to S_0(\kappa, p)$ by the orthogonality condition

$$\mathscr{B}_0(u - \mathscr{P}_0 u, v) = 0 \ \forall\ v \in S_0(\kappa, p). \tag{10}$$

Then, by Galerkin orthogonality, satisfied by $u - u_{FEM}$ with respect to the bilinear form $\mathscr{B}_\varepsilon$ and by (10), we have

$$\begin{aligned}
\|u_{FEM} - \mathscr{P}_0 u\|_{E,I}^2 &= \mathscr{B}_\varepsilon(u_{FEM} - \mathscr{P}_0 u, u_{FEM} - \mathscr{P}_0 u) = \mathscr{B}_\varepsilon(u - \mathscr{P}_0 u, u_{FEM} - \mathscr{P}_0 u) \\
&= \varepsilon^2 \langle (u - \mathscr{P}_0 u)'', (u_{FEM} - \mathscr{P}_0 u)'' \rangle_I \\
&\leq \varepsilon^2 \| (u - \mathscr{P}_0 u)'' \|_{0,I} \| (u_{FEM} - \mathscr{P}_0 u)'' \|_{0,I}, \\
&\leq \varepsilon \| (u - \mathscr{P}_0 u)'' \|_{0,I} \|u_{FEM} - \mathscr{P}_0 u\|_{E,I},
\end{aligned}$$

hence

$$\varepsilon \left\| (u_{FEM} - \mathscr{P}_0 u)'' \right\|_{0,I} \leq \|u_{FEM} - \mathscr{P}_0 u\|_{E,I} \leq \varepsilon \left\| (u - \mathscr{P}_0 u)'' \right\|_{0,I}.$$

The triangle inequality will then allow us to infer exponential convergence in the balanced norm, provided we can show that

$$\left\| (u - \mathscr{P}_0 u)'' \right\|_{0,I} \leq C\varepsilon^{-1/2} e^{-\sigma p},$$

for some positive constants C, σ independent of ε, u and p.

We begin by noting the following stability estimate

$$\|\mathscr{P}_0 z\|_{1,I} \leq \|z\|_{1,I} \ \forall z \in H_0^1(I), \tag{11}$$

which follows by taking $v = \mathscr{P}_0 u$ in (10). Next, we assume that $\kappa p \varepsilon < 1/4$ and define the layer region

$$I_\varepsilon := [0, \kappa p \varepsilon] \cup [1 - \kappa p \varepsilon, 1].$$

Now, with $z \in H_0^1(I)$, we note that $\mathscr{P}_0 z \in S_0(\kappa, p)$ satisfies

$$|\mathscr{P}_0 z(\kappa p\varepsilon)| \le C \int_0^{\kappa p\varepsilon} |(\mathscr{P}_0 z)'(x)|dx \le C\kappa p\varepsilon \|(\mathscr{P}_0 z)'\|_{\infty,I_\varepsilon} \le C\kappa p\varepsilon \|(\mathscr{P}_0 z)'\|_{\infty,I}.$$

Using an inverse estimate [16, Theorem 3.92]

$$\|\pi\|_{\infty,I} \le Cp\, \|\pi\|_{0,I} \;\; \forall\, \pi \in \Pi_p(I), \tag{12}$$

we get

$$|\mathscr{P}_0 z(\kappa p\varepsilon)| \le C\kappa p\varepsilon p \|\mathscr{P}_0 z\|_{1,I}, \tag{13}$$

and similarly for $|\mathscr{P}_0 z(1 - \kappa p\varepsilon)|$. The following lemma describes a decomposition of $\mathscr{P}_0 z$ (see [3] for details) and it is needed for the proof of Lemma 3.

Lemma 2 *There exists a constant $c > 0$ such that under the assumption*

$$p\sqrt{\kappa p\varepsilon} \le c, \tag{14}$$

the following is true: For each $z \in H_0^1(I)$, the decomposition of $\mathscr{P}_0 z = z_1 + z_\varepsilon$ into the components $z_1, z_\varepsilon \in \Pi_p(I) \cap H_0^1(I)$ satisfies

$$\|z_1'\|_{0,I} \le C\|z'\|_{0,I} \tag{15}$$

$$\|z_\varepsilon'\|_{0,I} \le C\|z'\|_{0,I_\varepsilon} + p\sqrt{\kappa p\varepsilon}\|z'\|_{0,I}. \tag{16}$$

We are now in the position to prove the following

Lemma 3 *Assume that (14) holds and that κ is sufficiently small (depending only on the data). Then*

$$\|(u - \mathscr{P}_0 u)''\|_{0,I} \le C\varepsilon^{-1/2} e^{-\sigma p}, \tag{17}$$

where the constants $C, \sigma > 0$ depend on κ but are independent of ε and p.

Proof By Lemma 1, we can find an approximation $\mathscr{I}_p u \in S(\kappa, p)$ with $(u - \mathscr{I}_p u)^{(k)}(0) = (u - \mathscr{I}_p u)^{(k)}(1) = 0, k = 0, 1$ such that

$$\|u - \mathscr{I}_p u\|_{1,I} + \sqrt{\kappa p\varepsilon}\|(u - \mathscr{I}_p u)''\|_{0,I} \le Ce^{-\sigma p}. \tag{18}$$

Since $\mathscr{P}_0$ is a projection on $S_0(\kappa, p)$, we can write $u - \mathscr{P}_0 u = u - \mathscr{I}_p u - \mathscr{P}_0(u - \mathscr{I}_p u)$ and

$$\| (u - \mathscr{P}_0 u)'' \|_{0,I} \leq \| \left(u - \mathscr{I}_p u\right)'' \|_{0,I} + \| \left(\mathscr{P}_0(u - \mathscr{I}_p u)\right)'' \|_{0,I}.$$

The first term is already treated in (18). For the second term $\mathscr{P}_0(u - \mathscr{I}_p u) \in S_0(\kappa, p)$, we utilize the decomposition $\mathscr{P}_0(u - \mathscr{I}_p u) = z_1 + z_\varepsilon$ and use Lemma 2, to get

$$\|z_1''\|_{0,I} \leq Cp^2 \|z_1'\|_{0,I} \leq Cp^2 \|(u - \mathscr{I}_p u)'\|_{0,I} \leq Ce^{-\sigma p}$$

$$\|z_\varepsilon''\|_{0,I} \leq C\frac{p^2}{\kappa p \varepsilon} \|z_\varepsilon'\|_{0,I} \leq C\frac{p^2}{\kappa p \varepsilon} \left\{ \|(u - \mathscr{I}_p u)'\|_{0,I_\varepsilon} + p\sqrt{\kappa p \varepsilon} \|(u - \mathscr{I}_p u)'\|_{0,I} \right\}.$$

Since $(u - \mathscr{I}_p u)'(0) = (u - \mathscr{I}_p u)'(1) = 0$ we use $z'(x) = \int_0^x z''(t)\, dt$ and obtain

$$\| \left(u - \mathscr{I}_p u\right)' \|_{0,I_\varepsilon} \leq C\kappa p \varepsilon \|(u - \mathscr{I}_p u)''\|_{0,I_\varepsilon}.$$

Hence,

$$\|z_\varepsilon''\|_{0,I} \leq Cp^2 \left\{ \|(u - \mathscr{I}_p u)''\|_{0,I_\varepsilon} + p(\kappa p \varepsilon)^{-1/2} \| \left(u - \mathscr{I}_p u\right)' \|_{0,I} \right\} \leq C\varepsilon^{-1/2} e^{-\sigma p}.$$

Combining the above gives the result.

We are now in a position present our main result.

Theorem 3 *There is a $\kappa_0 > 0$ depending only on the data a, b and f such that for every $\kappa \in (0, \kappa_0]$, the hp-FEM space $S_0(\kappa, p)$ on the Spectral Boundary Layer mesh leads to an approximation $u_{FEM} \in S_0(\kappa, p)$ to the solution u of (4), such that*

$$\|u - u_{FEM}\|_{E,I} + \sqrt{\varepsilon} \|(u - u_{FEM})''\|_{0,I} \leq Ce^{-\sigma p},$$

where the constants $C, \sigma > 0$ depend on the choice of κ but are independent of ε and p.

Proof Since the energy norm bound $\|u - u_{FEM}\|_{E,I} \leq Ce^{-\sigma p}$ was shown in [10], we focus on the control of $\sqrt{\varepsilon} \|(u - u_{FEM})''\|_{0,I}$. If (14) holds then Lemma 3 yields the result. If not, i.e. $p\sqrt{\kappa p \varepsilon} \geq c$ for the constant c appearing in (14), we have

$$\sqrt{\varepsilon} \|(u - u_{FEM})''\|_{0,I} \leq \varepsilon^{-1/2} \|u - u_{FEM}\|_{E,I} \leq c^{-1} p^{3/2} \kappa^{1/2} \|u - u_{FEM}\|_{E,I} \leq Ce^{-\sigma p}.$$

As a corollary, we get exponential convergence in the maximum norm for u and u'.

Corollary 1 *Let u be the solution of (4) and let $u_{FEM} \in S_0(\kappa, p)$ be its finite element approximation on the Spectral Boundary Layer mesh. Then*

$$\left\| u^{(k)} - u_{FEM}^{(k)} \right\|_{\infty,I} \leq C p^{1/2} e^{-\sigma p}, \ k \in \{0, 1\}, \tag{19}$$

where the constants $C, \sigma > 0$ are independent of ε and p.

Proof Let $k \in \{0, 1\}$. Then

$$\left| u^{(k)}(x) - u_{FEM}^{(k)}(x) \right| = \left| \int_0^x \left(u^{(k)}(t) - u_{FEM}^{(k)}(t) \right)' dt \right|.$$

Assume first that $x \in (0, \kappa p \varepsilon]$. Then by the Cauchy-Schwarz inequality

$$\left| u^{(k)} - u_{FEM}^{(k)}(x) \right| \leq \sqrt{\kappa p \varepsilon} \left\| \left(u^{(k)} - u_{FEM}^{(k)} \right)' \right\|_{0,I} \leq C \left\{ \begin{array}{l} \sqrt{\kappa p \varepsilon} e^{-\sigma p} \ , \ k = 0 \\ \sqrt{\kappa p} e^{-\sigma p} \ , \ k = 1 \end{array} \right. .$$

The same technique works if $x \in [1 - \kappa p \varepsilon, 1]$. To complete the proof it remains to consider $x \in [\kappa p \varepsilon, 1 - \kappa p \varepsilon] =: J$. We note that

$$\left\| (u - u_{FEM})^{(k)} \right\|_{\infty,J} \leq \left\| (u - \mathscr{I}_p u)^{(k)} \right\|_{\infty,J} + \left\| (\mathscr{I}_p u - u_{FEM})^{(k)} \right\|_{\infty,J},$$

with $\mathscr{I}_p$ the operator defined in Lemma 1. By (18) and Sobolev's embeding theorem, the first term on the right hand side above satisfies

$$\left\| (u - \mathscr{I}_p u)^{(k)} \right\|_{\infty,J} \leq C e^{-\sigma_1 p}, \ C, \sigma_1 \in \mathbf{R}^+.$$

For the second term, the inverse estimate (12) gives

$$\left\| (\mathscr{I}_p u - u_{FEM})^{(k)} \right\|_{\infty,J} \leq C p^2 \left\| (\mathscr{I}_p u - u_{FEM})^{(k)} \right\|_{0,J} \leq C \left\| u - u_{FEM} \right\|_{E,I} \leq C e^{-\sigma p}.$$

Combining the above completes the proof.

4 Numerical Experiments

In this section we illustrate our theoretical findings by considering one example; we refer to [3] for additional numerical computations. We use the *Spectral Boundary Layer Mesh* with $\kappa = 1$, i.e. $\Delta = \{0, p\varepsilon, 1 - p\varepsilon, 1\}$ and polynomials of degree $p \geq 3$

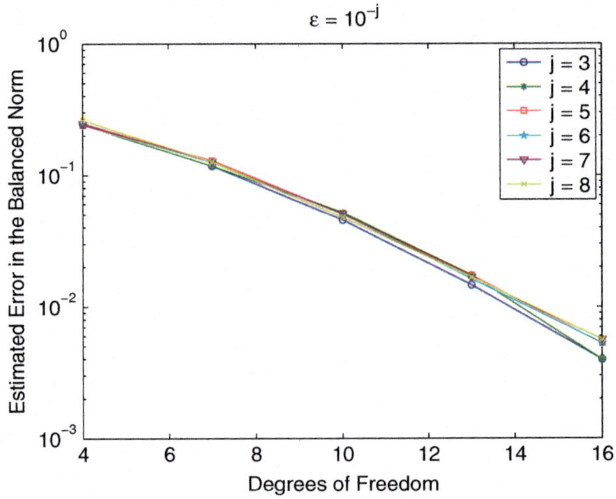

Fig. 1 Balanced norm convergence

which we increase to improve accuracy. The number of degrees of freedom, i.e. the dimension of the finite dimensional subspace, is then given by $DOF = 3p - 5$. We will measure the error $(u - u_{FEM})$ in the balanced norm as well as the error $(u - u_{FEM})^{(k)}, k = 0, 1$ in the maximum norm. We choose $\varepsilon = 10^{-j}, j = 3, \ldots, 8$ for the computations, in order to show the robustness of the proposed method.

The data is selected as $a(x) = b(x) = 1, f(x) = (x + 1/2)^{-1}$. We note that this choice for f allows us to see how the (lack of) smoothness of the right hand side may affect the computational results. Since no exact solution is available, we use a reference solution for the computations, obtained with polynomials of degree $2p$. We show, in Fig. 1, the estimated error in the balanced norm, versus the number of degrees of freedom, in a semi-log scale. We see that the method converges robustly at an exponential rate, as predicted by Theorem 3. In the case of a fixed order h–version FEM, the optimal convergence rate would be affected by the (lack of) smoothness of the data, something that does not occur for the hp–version considered here.

In Figs. 2 and 3 we show the error in u and in u', respectively, using the estimated maximum norm. The results are in accordance with our theoretical findings as stated in Corollary 1.

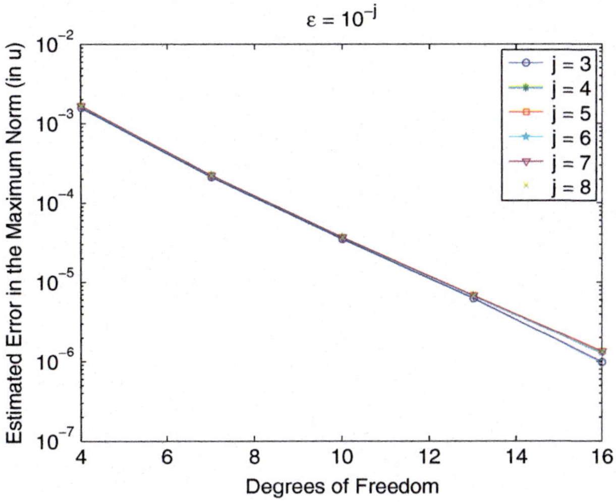

Fig. 2 Maximum norm convergence in u

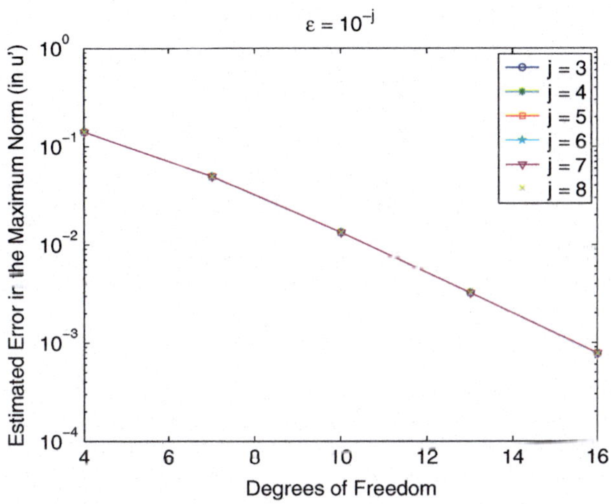

Fig. 3 Maximum norm convergence in u'

References

1. N.S. Bakhvalov, Towards optimization of methods for solving boundary value problems in the presence of boundary layers, (in Russian). Zh. Vychisl. Mat. Mat. Fiz. **9**, 841–859 (1969)
2. P. Constantinou, The *hp* Finite Element Method for fourth order singularly perturbed problems, Doctoral Dissertation, Department of Mathematics and Statistics, University of Cyprus, in preparation, 2016

3. P. Constantinou, C. Varnava, C. Xenophontos, An *hp* finite element method for fourth order singularly perturbed problems. Numer. Algorithms **73**, 567–590 (2016). doi:10.1007/s11075-016-0108-9
4. R. Lin, M. Stynes, A balanced finite element method for singularly perturbed reaction-diffusion problems. SIAM J. Numer. Anal. **50**(5), 2729–2743 (2012)
5. J.M. Melenk, On the robust exponential convergence of *hp* finite element methods for problems with boundary layers. IMA J. Numer. Anal. **17**, 577–601 (1997)
6. J.M. Melenk, C. Xenophontos, Robust exponential convergence of *hp*-FEM in balanced norms for singularly perturbed reaction-diffusion equations. Calcolo **53**, 105–132 (2016)
7. M.J. Melenk, C. Xenophontos, L. Oberbroeckling, Robust exponential convergence of *hp*-FEM for singularly perturbed systems of reaction-diffusion equations with multiple scales. IMA J. Numer. Anal. **33**(2), 609–628 (2013)
8. J.J. Miller, E. O'Riordan, G.I. Shishkin, *Fitted Numerical Methods For Singular Perturbation Problems* (World Scientific, Singapore, 1996)
9. Z.-X. Pan, The difference and asymptotic methods for a fourth order equation with a small parameter, in *BAIL IV Proceedings* (Book Press, Dublin, 1986), pp. 392–397
10. P. Panaseti, A. Zouvani, N. Madden, C. Xenophontos, A C^1–conforming *hp* finite element method for fourth order singularly perturbed boundary value problems. Appl. Numer. Math. **104**, 81–97 (2016)
11. H.-G. Roos, A uniformly convergent discretization method for a singularly perturbed boundary value problem of the fourth order. Rev. Res. Fac. Sci. Math. Ser. Univ. Novi Sad **19**, 51–64 (1989)
12. H.G. Roos, S. Franz, Error estimation in a balanced norm for a convection-diffusion problems with two different boundary layers. Calcolo **51**, 423–440 (2014)
13. H.G. Roos, M. Schopf, Convergence and stability in balanced norms of finite element methods on Shishkin meshes for reaction-diffusion problems. ZAMM **95**, 551–565 (2015)
14. H.-G. Roos, M. Stynes, A uniformly convergent discretization method for a fourth order singular perturbation problem. Bonn. Math. Schr. **228**, 30–40 (1991)
15. H.G. Roos, M. Stynes, L. Tobiska, in *Robust Numerical Methods for Singularly Perturbed Differential Equations*. Springer Series in Computational Mathematics, vol. 24 (Springer, Berlin, 2008)
16. C. Schwab, *p/hp Finite Element Methods* (Oxford University Press, Oxford, 1998)
17. C. Schwab, M. Suri, The *p* and *hp* versions of the finite element method for problems with boundary layers. Math. Comput. **65**, 1403–1429 (1996)
18. G.I. Shishkin, A difference scheme for an ordinary differential equation of the fourth order with a small parameter at the highest derivative. Differ. Equ. (in Russian) **21**, 1743–1742 (1985)
19. G.I. Shishkin, Grid approximation of singularly perturbed boundary value problems with a regular boundary layer. Sov. J. Numer. Anal. Math. Model. **4**, 397–417 (1989)
20. G. Sun, M. Stynes, Finite-element methods for singularly perturbed high order elliptic two point boundary value problems I: reaction-diffusion-type problems. IMA J. Numer. Anal. **15**, 117–139 (1995)

Editorial Policy

1. Volumes in the following three categories will be published in LNCSE:

i) Research monographs
ii) Tutorials
iii) Conference proceedings

Those considering a book which might be suitable for the series are strongly advised to contact the publisher or the series editors at an early stage.

2. Categories i) and ii). Tutorials are lecture notes typically arising via summer schools or similar events, which are used to teach graduate students. These categories will be emphasized by Lecture Notes in Computational Science and Engineering. **Submissions by interdisciplinary teams of authors are encouraged.** The goal is to report new developments – quickly, informally, and in a way that will make them accessible to non-specialists. In the evaluation of submissions timeliness of the work is an important criterion. Texts should be well-rounded, well-written and reasonably self-contained. In most cases the work will contain results of others as well as those of the author(s). In each case the author(s) should provide sufficient motivation, examples, and applications. In this respect, Ph.D. theses will usually be deemed unsuitable for the Lecture Notes series. Proposals for volumes in these categories should be submitted either to one of the series editors or to Springer-Verlag, Heidelberg, and will be refereed. A provisional judgement on the acceptability of a project can be based on partial information about the work: a detailed outline describing the contents of each chapter, the estimated length, a bibliography, and one or two sample chapters – or a first draft. A final decision whether to accept will rest on an evaluation of the completed work which should include

- at least 100 pages of text;
- a table of contents;
- an informative introduction perhaps with some historical remarks which should be accessible to readers unfamiliar with the topic treated;
- a subject index.

3. Category iii). Conference proceedings will be considered for publication provided that they are both of exceptional interest and devoted to a single topic. One (or more) expert participants will act as the scientific editor(s) of the volume. They select the papers which are suitable for inclusion and have them individually refereed as for a journal. Papers not closely related to the central topic are to be excluded. Organizers should contact the Editor for CSE at Springer at the planning stage, see *Addresses* below.

In exceptional cases some other multi-author-volumes may be considered in this category.

4. Only works in English will be considered. For evaluation purposes, manuscripts may be submitted in print or electronic form, in the latter case, preferably as pdf- or zipped ps-files. Authors are requested to use the LaTeX style files available from Springer at http://www.springer.com/gp/authors-editors/book-authors-editors/manuscript-preparation/5636 (Click on LaTeX Template → monographs or contributed books).

For categories ii) and iii) we strongly recommend that all contributions in a volume be written in the same LaTeX version, preferably LaTeX2e. Electronic material can be included if appropriate. Please contact the publisher.

Careful preparation of the manuscripts will help keep production time short besides ensuring satisfactory appearance of the finished book in print and online.

5. The following terms and conditions hold. Categories i), ii) and iii):

Authors receive 50 free copies of their book. No royalty is paid.
Volume editors receive a total of 50 free copies of their volume to be shared with authors, but no royalties.

Authors and volume editors are entitled to a discount of 33.3 % on the price of Springer books purchased for their personal use, if ordering directly from Springer.

6. Springer secures the copyright for each volume.

Addresses:

Timothy J. Barth
NASA Ames Research Center
NAS Division
Moffett Field, CA 94035, USA
barth@nas.nasa.gov

Michael Griebel
Institut für Numerische Simulation
der Universität Bonn
Wegelerstr. 6
53115 Bonn, Germany
griebel@ins.uni-bonn.de

David E. Keyes
Mathematical and Computer Sciences
and Engineering
King Abdullah University of Science
and Technology
P.O. Box 55455
Jeddah 21534, Saudi Arabia
david.keyes@kaust.edu.sa

and

Department of Applied Physics
and Applied Mathematics
Columbia University
500 W. 120 th Street
New York, NY 10027, USA
kd2112@columbia.edu

Risto M. Nieminen
Department of Applied Physics
Aalto University School of Science
and Technology
00076 Aalto, Finland
risto.nieminen@aalto.fi

Dirk Roose
Department of Computer Science
Katholieke Universiteit Leuven
Celestijnenlaan 200A
3001 Leuven-Heverlee, Belgium
dirk.roose@cs.kuleuven.be

Tamar Schlick
Department of Chemistry
and Courant Institute
of Mathematical Sciences
New York University
251 Mercer Street
New York, NY 10012, USA
schlick@nyu.edu

Editor for Computational Science
and Engineering at Springer:
Martin Peters
Springer-Verlag
Mathematics Editorial IV
Tiergartenstrasse 17
69121 Heidelberg, Germany
martin.peters@springer.com

Lecture Notes in Computational Science and Engineering

1. D. Funaro, *Spectral Elements for Transport-Dominated Equations.*
2. H.P. Langtangen, *Computational Partial Differential Equations.* Numerical Methods and Diffpack Programming.
3. W. Hackbusch, G. Wittum (eds.), *Multigrid Methods V.*
4. P. Deuflhard, J. Hermans, B. Leimkuhler, A.E. Mark, S. Reich, R.D. Skeel (eds.), *Computational Molecular Dynamics: Challenges, Methods, Ideas.*
5. D. Kröner, M. Ohlberger, C. Rohde (eds.), *An Introduction to Recent Developments in Theory and Numerics for Conservation Laws.*
6. S. Turek, *Efficient Solvers for Incompressible Flow Problems.* An Algorithmic and Computational Approach.
7. R. von Schwerin, ***M**ulti **B**ody **S**ystem **SIM**ulation.* Numerical Methods, Algorithms, and Software.
8. H.-J. Bungartz, F. Durst, C. Zenger (eds.), *High Performance Scientific and Engineering Computing.*
9. T.J. Barth, H. Deconinck (eds.), *High-Order Methods for Computational Physics.*
10. H.P. Langtangen, A.M. Bruaset, E. Quak (eds.), *Advances in Software Tools for Scientific Computing.*
11. B. Cockburn, G.E. Karniadakis, C.-W. Shu (eds.), *Discontinuous Galerkin Methods.* Theory, Computation and Applications.
12. U. van Rienen, *Numerical Methods in Computational Electrodynamics.* Linear Systems in Practical Applications.
13. B. Engquist, L. Johnsson, M. Hammill, F. Short (eds.), *Simulation and Visualization on the Grid.*
14. E. Dick, K. Riemslagh, J. Vierendeels (eds.), *Multigrid Methods VI.*
15. A. Frommer, T. Lippert, B. Medeke, K. Schilling (eds.), *Numerical Challenges in Lattice Quantum Chromodynamics.*
16. J. Lang, *Adaptive Multilevel Solution of Nonlinear Parabolic PDE Systems.* Theory, Algorithm, and Applications.
17. B.I. Wohlmuth, *Discretization Methods and Iterative Solvers Based on Domain Decomposition.*
18. U. van Rienen, M. Günther, D. Hecht (eds.), *Scientific Computing in Electrical Engineering.*
19. I. Babuška, P.G. Ciarlet, T. Miyoshi (eds.), *Mathematical Modeling and Numerical Simulation in Continuum Mechanics.*
20. T.J. Barth, T. Chan, R. Haimes (eds.), *Multiscale and Multiresolution Methods.* Theory and Applications.
21. M. Breuer, F. Durst, C. Zenger (eds.), *High Performance Scientific and Engineering Computing.*
22. K. Urban, *Wavelets in Numerical Simulation.* Problem Adapted Construction and Applications.
23. L.F. Pavarino, A. Toselli (eds.), *Recent Developments in Domain Decomposition Methods.*

24. T. Schlick, H.H. Gan (eds.), *Computational Methods for Macromolecules: Challenges and Applications.*

25. T.J. Barth, H. Deconinck (eds.), *Error Estimation and Adaptive Discretization Methods in Computational Fluid Dynamics.*

26. M. Griebel, M.A. Schweitzer (eds.), *Meshfree Methods for Partial Differential Equations.*

27. S. Müller, *Adaptive Multiscale Schemes for Conservation Laws.*

28. C. Carstensen, S. Funken, W. Hackbusch, R.H.W. Hoppe, P. Monk (eds.), *Computational Electromagnetics.*

29. M.A. Schweitzer, *A Parallel Multilevel Partition of Unity Method for Elliptic Partial Differential Equations.*

30. T. Biegler, O. Ghattas, M. Heinkenschloss, B. van Bloemen Waanders (eds.), *Large-Scale PDE-Constrained Optimization.*

31. M. Ainsworth, P. Davies, D. Duncan, P. Martin, B. Rynne (eds.), *Topics in Computational Wave Propagation.* Direct and Inverse Problems.

32. H. Emmerich, B. Nestler, M. Schreckenberg (eds.), *Interface and Transport Dynamics.* Computational Modelling.

33. H.P. Langtangen, A. Tveito (eds.), *Advanced Topics in Computational Partial Differential Equations.* Numerical Methods and Diffpack Programming.

34. V. John, *Large Eddy Simulation of Turbulent Incompressible Flows.* Analytical and Numerical Results for a Class of LES Models.

35. E. Bänsch (ed.), *Challenges in Scientific Computing - CISC 2002.*

36. B.N. Khoromskij, G. Wittum, *Numerical Solution of Elliptic Differential Equations by Reduction to the Interface.*

37. A. Iske, *Multiresolution Methods in Scattered Data Modelling.*

38. S.-I. Niculescu, K. Gu (eds.), *Advances in Time-Delay Systems.*

39. S. Attinger, P. Koumoutsakos (eds.), *Multiscale Modelling and Simulation.*

40. R. Kornhuber, R. Hoppe, J. Périaux, O. Pironneau, O. Wildlund, J. Xu (eds.), *Domain Decomposition Methods in Science and Engineering.*

41. T. Plewa, T. Linde, V.G. Weirs (eds.), *Adaptive Mesh Refinement – Theory and Applications.*

42. A. Schmidt, K.G. Siebert, *Design of Adaptive Finite Element Software.* The Finite Element Toolbox ALBERTA.

43. M. Griebel, M.A. Schweitzer (eds.), *Meshfree Methods for Partial Differential Equations II.*

44. B. Engquist, P. Lötstedt, O. Runborg (eds.), *Multiscale Methods in Science and Engineering.*

45. P. Benner, V. Mehrmann, D.C. Sorensen (eds.), *Dimension Reduction of Large-Scale Systems.*

46. D. Kressner, *Numerical Methods for General and Structured Eigenvalue Problems.*

47. A. Boriçi, A. Frommer, B. Joó, A. Kennedy, B. Pendleton (eds.), *QCD and Numerical Analysis III.*

48. F. Graziani (ed.), *Computational Methods in Transport.*

49. B. Leimkuhler, C. Chipot, R. Elber, A. Laaksonen, A. Mark, T. Schlick, C. Schütte, R. Skeel (eds.), *New Algorithms for Macromolecular Simulation.*

50. M. Bücker, G. Corliss, P. Hovland, U. Naumann, B. Norris (eds.), *Automatic Differentiation: Applications, Theory, and Implementations.*

51. A.M. Bruaset, A. Tveito (eds.), *Numerical Solution of Partial Differential Equations on Parallel Computers.*

52. K.H. Hoffmann, A. Meyer (eds.), *Parallel Algorithms and Cluster Computing.*

53. H.-J. Bungartz, M. Schäfer (eds.), *Fluid-Structure Interaction.*

54. J. Behrens, *Adaptive Atmospheric Modeling.*

55. O. Widlund, D. Keyes (eds.), *Domain Decomposition Methods in Science and Engineering XVI.*

56. S. Kassinos, C. Langer, G. Iaccarino, P. Moin (eds.), *Complex Effects in Large Eddy Simulations.*

57. M. Griebel, M.A Schweitzer (eds.), *Meshfree Methods for Partial Differential Equations III.*

58. A.N. Gorban, B. Kégl, D.C. Wunsch, A. Zinovyev (eds.), *Principal Manifolds for Data Visualization and Dimension Reduction.*

59. H. Ammari (ed.), *Modeling and Computations in Electromagnetics: A Volume Dedicated to Jean-Claude Nédélec.*

60. U. Langer, M. Discacciati, D. Keyes, O. Widlund, W. Zulehner (eds.), *Domain Decomposition Methods in Science and Engineering XVII.*

61. T. Mathew, *Domain Decomposition Methods for the Numerical Solution of Partial Differential Equations.*

62. F. Graziani (ed.), *Computational Methods in Transport: Verification and Validation.*

63. M. Bebendorf, *Hierarchical Matrices.* A Means to Efficiently Solve Elliptic Boundary Value Problems.

64. C.H. Bischof, H.M. Bücker, P. Hovland, U. Naumann, J. Utke (eds.), *Advances in Automatic Differentiation.*

65. M. Griebel, M.A. Schweitzer (eds.), *Meshfree Methods for Partial Differential Equations IV.*

66. B. Engquist, P. Lötstedt, O. Runborg (eds.), *Multiscale Modeling and Simulation in Science.*

67. I.H. Tuncer, Ü. Gülcat, D.R. Emerson, K. Matsuno (eds.), *Parallel Computational Fluid Dynamics 2007.*

68. S. Yip, T. Diaz de la Rubia (eds.), *Scientific Modeling and Simulations.*

69. A. Hegarty, N. Kopteva, E. O'Riordan, M. Stynes (eds.), *BAIL* 2008 – *Boundary and Interior Layers.*

70. M. Bercovier, M.J. Gander, R. Kornhuber, O. Widlund (eds.), *Domain Decomposition Methods in Science and Engineering XVIII.*

71. B. Koren, C. Vuik (eds.), *Advanced Computational Methods in Science and Engineering.*

72. M. Peters (ed.), *Computational Fluid Dynamics for Sport Simulation.*

73. H.-J. Bungartz, M. Mehl, M. Schäfer (eds.), *Fluid Structure Interaction II - Modelling, Simulation, Optimization.*

74. D. Tromeur-Dervout, G. Brenner, D.R. Emerson, J. Erhel (eds.), *Parallel Computational Fluid Dynamics 2008.*

75. A.N. Gorban, D. Roose (eds.), *Coping with Complexity: Model Reduction and Data Analysis.*

76. J.S. Hesthaven, E.M. Rønquist (eds.), *Spectral and High Order Methods for Partial Differential Equations.*

77. M. Holtz, *Sparse Grid Quadrature in High Dimensions with Applications in Finance and Insurance.*

78. Y. Huang, R. Kornhuber, O.Widlund, J. Xu (eds.), *Domain Decomposition Methods in Science and Engineering XIX.*

79. M. Griebel, M.A. Schweitzer (eds.), *Meshfree Methods for Partial Differential Equations V.*

80. P.H. Lauritzen, C. Jablonowski, M.A. Taylor, R.D. Nair (eds.), *Numerical Techniques for Global Atmospheric Models.*

81. C. Clavero, J.L. Gracia, F.J. Lisbona (eds.), *BAIL 2010 – Boundary and Interior Layers, Computational and Asymptotic Methods.*

82. B. Engquist, O. Runborg, Y.R. Tsai (eds.), *Numerical Analysis and Multiscale Computations.*

83. I.G. Graham, T.Y. Hou, O. Lakkis, R. Scheichl (eds.), *Numerical Analysis of Multiscale Problems.*

84. A. Logg, K.-A. Mardal, G. Wells (eds.), *Automated Solution of Differential Equations by the Finite Element Method.*

85. J. Blowey, M. Jensen (eds.), *Frontiers in Numerical Analysis - Durham 2010.*

86. O. Kolditz, U.-J. Gorke, H. Shao, W. Wang (eds.), *Thermo-Hydro-Mechanical-Chemical Processes in Fractured Porous Media - Benchmarks and Examples.*

87. S. Forth, P. Hovland, E. Phipps, J. Utke, A. Walther (eds.), *Recent Advances in Algorithmic Differentiation.*

88. J. Garcke, M. Griebel (eds.), *Sparse Grids and Applications.*

89. M. Griebel, M.A. Schweitzer (eds.), *Meshfree Methods for Partial Differential Equations VI.*

90. C. Pechstein, *Finite and Boundary Element Tearing and Interconnecting Solvers for Multiscale Problems.*

91. R. Bank, M. Holst, O. Widlund, J. Xu (eds.), *Domain Decomposition Methods in Science and Engineering XX.*

92. H. Bijl, D. Lucor, S. Mishra, C. Schwab (eds.), *Uncertainty Quantification in Computational Fluid Dynamics.*

93. M. Bader, H.-J. Bungartz, T. Weinzierl (eds.), *Advanced Computing.*

94. M. Ehrhardt, T. Koprucki (eds.), *Advanced Mathematical Models and Numerical Techniques for Multi-Band Effective Mass Approximations.*

95. M. Azaïez, H. El Fekih, J.S. Hesthaven (eds.), *Spectral and High Order Methods for Partial Differential Equations ICOSAHOM 2012.*

96. F. Graziani, M.P. Desjarlais, R. Redmer, S.B. Trickey (eds.), *Frontiers and Challenges in Warm Dense Matter.*

97. J. Garcke, D. Pflüger (eds.), *Sparse Grids and Applications – Munich 2012.*

98. J. Erhel, M. Gander, L. Halpern, G. Pichot, T. Sassi, O. Widlund (eds.), *Domain Decomposition Methods in Science and Engineering XXI.*

99. R. Abgrall, H. Beaugendre, P.M. Congedo, C. Dobrzynski, V. Perrier, M. Ricchiuto (eds.), *High Order Nonlinear Numerical Methods for Evolutionary PDEs - HONOM 2013.*

100. M. Griebel, M.A. Schweitzer (eds.), *Meshfree Methods for Partial Differential Equations VII.*

101. R. Hoppe (ed.), *Optimization with PDE Constraints - OPTPDE 2014.*

102. S. Dahlke, W. Dahmen, M. Griebel, W. Hackbusch, K. Ritter, R. Schneider, C. Schwab, H. Yserentant (eds.), *Extraction of Quantifiable Information from Complex Systems.*

103. A. Abdulle, S. Deparis, D. Kressner, F. Nobile, M. Picasso (eds.), *Numerical Mathematics and Advanced Applications - ENUMATH 2013.*

104. T. Dickopf, M.J. Gander, L. Halpern, R. Krause, L.F. Pavarino (eds.), *Domain Decomposition Methods in Science and Engineering XXII.*

105. M. Mehl, M. Bischoff, M. Schäfer (eds.), *Recent Trends in Computational Engineering - CE2014.* Optimization, Uncertainty, Parallel Algorithms, Coupled and Complex Problems.

106. R.M. Kirby, M. Berzins, J.S. Hesthaven (eds.), *Spectral and High Order Methods for Partial Differential Equations - ICOSAHOM'14.*

107. B. Jüttler, B. Simeon (eds.), *Isogeometric Analysis and Applications 2014.*

108. P. Knobloch (ed.), *Boundary and Interior Layers, Computational and Asymptotic Methods – BAIL 2014.*

109. J. Garcke, D. Pflüger (eds.), *Sparse Grids and Applications – Stuttgart 2014.*

110. H. P. Langtangen, *Finite Difference Computing with Exponential Decay Models.*

111. A. Tveito, G.T. Lines, *Computing Characterizations of Drugs for Ion Channels and Receptors Using Markov Models.*

112. B. Karazösen, M. Manguoğlu, M. Tezer-Sezgin, S. Göktepe, Ö. Uğur (eds.), *Numerical Mathematics and Advanced Applications - ENUMATH 2015.*

113. H.-J. Bungartz, P. Neumann, W.E. Nagel (eds.), *Software for Exascale Computing - SPPEXA 2013-2015.*

114. G.R. Barrenechea, F. Brezzi, A. Cangiani, E.H. Georgoulis (eds.), *Building Bridges: Connections and Challenges in Modern Approaches to Numerical Partial Differential Equations.*

115. M. Griebel, M.A. Schweitzer (eds.), *Meshfree Methods for Partial Differential Equations VIII.*

116. C.-O. Lee, X.-C. Cai, D.E. Keyes, H.H. Kim, A. Klawonn, E.-J. Park, O.B. Widlund (eds.), *Domain Decomposition Methods in Science and Engineering XXIII.*

117. T. Sakurai, S. Zhang, T. Imamura, Y. Yusaku, K. Yoshinobu, H. Takeo (eds.), *Eigenvalue Problems: Algorithms, Software and Applications, in Petascale Computing.* EPASA 2015, Tsukuba, Japan, September 2015.

118. T. Richter (ed.), *Fluid-structure Interactions.* Models, Analysis and Finite Elements.

119. M.L. Bittencourt, N.A. Dumont, J.S. Hesthaven (eds.), *Spectral and High Order Methods for Partial Differential Equations - ICOSAHOM 2016.*

120. Z. Huang, M. Stynes, Z. Zhang (eds.), *Boundary and Interior Layers, Computational and Asymptotic Methods - BAIL 2016.*

For further information on these books please have a look at our mathematics catalogue at the following URL: `www.springer.com/series/3527`

Monographs in Computational Science and Engineering

1. J. Sundnes, G.T. Lines, X. Cai, B.F. Nielsen, K.-A. Mardal, A. Tveito, *Computing the Electrical Activity in the Heart.*

For further information on this book, please have a look at our mathematics catalogue at the following URL: www.springer.com/series/7417

Texts in Computational Science and Engineering

1. H. P. Langtangen, *Computational Partial Differential Equations.* Numerical Methods and Diffpack Programming. 2nd Edition
2. A. Quarteroni, F. Saleri, P. Gervasio, *Scientific Computing with MATLAB and Octave.* 4th Edition
3. H. P. Langtangen, *Python Scripting for Computational Science.* 3rd Edition
4. H. Gardner, G. Manduchi, *Design Patterns for e-Science.*
5. M. Griebel, S. Knapek, G. Zumbusch, *Numerical Simulation in Molecular Dynamics.*
6. H. P. Langtangen, *A Primer on Scientific Programming with Python.* 5th Edition
7. A. Tveito, H. P. Langtangen, B. F. Nielsen, X. Cai, *Elements of Scientific Computing.*
8. B. Gustafsson, *Fundamentals of Scientific Computing.*
9. M. Bader, *Space-Filling Curves.*
10. M. Larson, F. Bengzon, *The Finite Element Method: Theory, Implementation and Applications.*
11. W. Gander, M. Gander, F. Kwok, *Scientific Computing: An Introduction using Maple and MATLAB.*
12. P. Deuflhard, S. Röblitz, *A Guide to Numerical Modelling in Systems Biology.*
13. M. H. Holmes, *Introduction to Scientific Computing and Data Analysis.*
14. S. Linge, H. P. Langtangen, *Programming for Computations* - A Gentle Introduction to Numerical Simulations with MATLAB/Octave.
15. S. Linge, H. P. Langtangen, *Programming for Computations* - A Gentle Introduction to Numerical Simulations with Python.
16. H.P. Langtangen, S. Linge, *Finite Difference Computing with PDEs* - A Modern Software Approach.

For further information on these books please have a look at our mathematics catalogue at the following URL: www.springer.com/series/5151

Printed by Printforce, the Netherlands